STABLE ISOTOPES
IN ECOLOGY AND
ENVIRONMENTAL SCIENCE

T0137953

ECOLOGICAL METHODS AND CONCEPTS SERIES

This series, successor to the Methods in Ecology series edited by John Lawton and Gene Likens, presents the latest ideas and techniques across the whole field of ecology and their application, from genetic to the global, from pest management to policy development. Books may be single- or multi-authored and will address emerging new areas within the field as well as updating well-established areas of endeavour. The new Series Editor is Professor Roger Kitching of Griffiths University, Brisbane, who will welcome suggestions for works within the series. Email: r.kitching@griffin.edu.au

Ecological Methods and Concepts Series

Stable Isotopes in Ecology and Environmental Science
Second edition, 2007
Edited by Robert Michener and Kate Lajtha

Forthcoming

Litter Decomposition in Aquatic Ecosystems
Edited by Mark Gessner

Vegetation Classification and Survey
Andrew Gillison

An Introduction to Ecological and Evolutionary Modelling: Time and Space
Michael Gillman

Canopy Science: Concepts and Methods
Edited by John Pike and James Morison

Methods in Ecology Series

Insect Sampling in Forest Ecosystems
2005
Edited by Simon Leather

Molecular Methods in Ecology
2000
Edited by Allan J Baker

Population Parameters: Estimation for Ecological Models
2000
Hamish McCallum

Biogenic Trace Gases: Measuring Emissions from Soils and Water
1995
Edited by PA Matson and RC Harriss

Geographical Population Analysis: Tools for the Analysis of Biodiversity
1994
Brian A Maurer

Stable Isotopes in Ecology and Environmental Science

SECOND EDITION

EDITED BY

ROBERT MICHENER
AND KATE LAJTHA

Blackwell
Publishing

© 2007 by Blackwell Publishing Ltd

BLACKWELL PUBLISHING
350 Main Street, Malden, MA 02148-5020, USA
9600 Garsington Road, Oxford OX4 2DQ, UK
550 Swanston Street, Carlton, Victoria 3053, Australia

The right of Robert Michener and Kate Lajtha to be identified as the Authors of the Editorial Material in this Work has been asserted in accordance with the UK Copyright, Designs, and Patents Act 1988.

All rights reserved. No part of this publication may be reproduced, stored in a retrieval system, or transmitted, in any form or by any means, electronic, mechanical, photocopying, recording or otherwise, except as permitted by the UK Copyright, Designs, and Patents Act 1988, without the prior permission of the publisher.

Designations used by companies to distinguish their products are often claimed as trademarks. All brand names and product names used in this book are trade names, service marks, trademarks, or registered trademarks of their respective owners. The publisher is not associated with any product or vendor mentioned in this book.

This publication is designed to provide accurate and authoritative information in regard to the subject matter covered. It is sold on the understanding that the publisher is not engaged in rendering professional services. If professional advice or other expert assistance is required, the services of a competent professional should be sought.

First published 2007 by Blackwell Publishing Ltd

1   2007

Library of Congress Cataloging-in-Publication Data

Stable isotopes in ecology and environmental science / edited by Robert Michener and Kate Lajtha. – 2nd ed.
p.   cm.
Includes bibliographical references and index.
ISBN-13: 978-1-4051-2680-9 (pbk. : alk. paper)
ISBN-10: 1-4051-2680-9 (pbk. : alk. paper)
1. Stable isotopes in ecological research.   I. Michener, Robert H.   II. Lajtha, Kate.

QH541.15.S68L35 2007
577.072–dc22
2006100366

A catalogue record for this title is available from the British Library.

Set in 9.5 on 12 pt Meridien
by SNP Best-set Typesetter Ltd., Hong Kong

The publisher's policy is to use permanent paper from mills that operate a sustainable forestry policy, and which has been manufactured from pulp processed using acid-free and elementary chlorine-free practices. Furthermore, the publisher ensures that the text paper and cover board used have met acceptable environmental accreditation standards.

For further information on
Blackwell Publishing, visit our website:
www.blackwellpublishing.com

Contents

14 Compound-specific stable isotope analysis in ecology and paleoecology, 480

Richard P. Evershed, Ian D. Bull, Lorna T. Corr, Zoe M. Crossman, Bart E. van Dongen, Claire J. Evans, Susan Jim, Hazel R. Mottram, Anna J. Mukherjee, and Richard D. Pancost

Contributors

J. Renée Brooks — US EPA/NHEERL, Western Ecology Division, 200 SW 35th St, Corvallis, OR 97333, USA

Deanne J. Brice — Oak Ridge National Laboratory, Environmental Sciences Division, Oak Ridge, TN 37831-6038, USA

Ian D. Bull — Organic Geochemistry Unit, Biogeochemistry Research Centre, School of Chemistry, University of Bristol, Cantock's Close, Bristol BS8 1TS, UK

Lorna T. Corr — Organic Geochemistry Unit, Biogeochemistry Research Centre, School of Chemistry, University of Bristol, Cantock's Close, Bristol BS8 1TS, UK

Zoe M. Crossman — Organic Geochemistry Unit, Biogeochemistry Research Centre, School of Chemistry, University of Bristol, Cantock's Close, Bristol BS8 1TS, UK

William S. Currie — University of Michigan, School of Natural Resources & Environment, Dana Building, 440 Church Street, Ann Arbor, MI 48109-1041, USA

Emily M. Elliott — Department of Geology and Planetary Science, 4107 O'Hara Street, University of Pittsburgh, Pittsburgh, PA 15260-3332, USA

Claire J. Evans — Organic Geochemistry Unit, Biogeochemistry Research Centre, School of Chemistry, University of Bristol, Cantock's Close, Bristol BS8 1TS, UK

R. Dave Evans — Washington State University, School of Biological Sciences, P.O. Box 644236, Washington State University, Pullman, WA 99164-4236, USA

Richard P. Evershed — Organic Geochemistry Unit, Biogeochemistry Research Centre, School of Chemistry, University of Bristol, Cantock's Close, Bristol BS8 1TS, UK

Jacques C. Finlay
University of Minnesota, Department of Ecology, Evolution and Behavior, 1987 Upper Buford Circle, St Paul, MN 55108, USA

Charles T. Garten
Oak Ridge National Laboratory, Environmental Sciences Division, Oak Ridge, TN 37831-6038, USA

Paul J. Hanson
Oak Ridge National Laboratory, Environmental Sciences Division, Oak Ridge, TN 37831-6038, USA

Keith A. Hobson
Environment Canada, 11 Innovation Blvd, Saskatoon, SK S7N 3H5, Canada

Susan Jim
Organic Geochemistry Unit, Biogeochemistry Research Centre, School of Chemistry, University of Bristol, Cantock's Close, Bristol BS8 1TS, UK

Les Kaufman
Boston University, Department of Biology, Boston MA 02215, USA

Carol Kendall
US Geological Survey, 345 Middlefield Rd, MS 434, Menlo Park, CA 94025, USA

Paul Koch
University of California, Santa Cruz, Isotope Biogeochemistry and Vertebrate Paleontology, Earth Sciences, E&MS A250, Santa Cruz, CA 95064, USA

Kate Lajtha
Editor in Chief, Biogeochemistry, Director, Environmental Sciences Program, Department of Botany and Plant Pathology, Oregon State University, Corvallis, OR 97331, USA

*Bonnie B. Lu
Oak Ridge National Laboratory, Environmental Sciences Division, Oak Ridge, TN 37831-6038, USA

John D. Marshall
University of Idaho, Department of Forest Resources, Moscow, ID 83844, USA

Jeff McDonnell
Oregon State University, Department of Forest Engineering, 015 Peavy Hall, Corvallis, OR 97331-5706, USA

*Retired.

Kevin J. McGuire — USDA Forest Service, Northeastern Research, Northeastern Research Station, Center for the Environment, Plymouth State University, 208 Boyd Science Center, MSC 63, 17 High Street, Plymouth, NH 03264, USA

Robert H. Michener — IRMS Laboratory Manager, Boston University Stable Isotope Laboratory, Department of Biology, 5 Cummington St, Boston, MA 02215, USA

Joseph P. Montoya — Georgia Institute of Technology, School of Biology, 310 Ferst Drive, Atlanta GA 30332-0230, USA

Hazel R. Mottram — Organic Geochemistry Unit, Biogeochemistry Research Centre, School of Chemistry, University of Bristol, Cantock's Close, Bristol BS8 1TS, UK

Anna J. Mukherjee — Organic Geochemistry Unit, Biogeochemistry Research Centre, School of Chemistry, University of Bristol, Cantock's Close, Bristol BS8 1TS, UK

Richard D. Pancost — Organic Geochemistry Unit, Biogeochemistry Research Centre, School of Chemistry, University of Bristol, Cantock's Close, Bristol BS8 1TS, UK

Elizabeth W. Sulzman — Oregon State University, Department of Crop and Soil Science, 3017 Agricultural and Life Science Building, Oregon State University, Corvallis, OR 97331, USA

Donald E. Todd — Oak Ridge National Laboratory, Environmental Sciences Division, Oak Ridge, TN 37831-6038, USA

Bart E. van Dongen — Organic Geochemistry Unit, Biogeochemistry Research Centre, School of Chemistry, University of Bristol, Cantock's Close, Bristol BS8 1TS, UK

Cindy Lee Van Dover — Director, Duke University Marine Laboratory, Nicholas School of the, Environment and Earth Sciences, 135 Marine Lab Rd, Beaufort NC 28516, USA

Scott D. Wankel — U. S. Geological Survey, 345 Middlefield Road, MS 434, Menlo Park, CA 94025, USA

Abbreviations

ANPP	above-ground net primary production
ANCA-MS	automated nitrogen and carbon analyzer mass spectrometry
BMP	best management practice
BSR	Bottom Simulating Reflector
BSTFA	N,O-bis(trimethylsiyl)trifluoroacetamide
BTEX	benzene, toluene, ethylbenzene, and xylenes
CAM	Crassulacean acid metabolism
CDT	Canyon Diablo Troilite
CF-GC/C/IRMS	continuous-flow gas chromatography/combustion/isotope ratio mass spectrometry
CF-IRMS	continuous flow isotope ratio mass spectrometer
CP/MAS	cross polarization magic-angle-spinning
DHAP	dihydroxyacetone phosphate
DI-IRMS	dual-inlet isotope ratio mass spectrometer
DIC	dissolved inorganic carbon
DIN	dissolved inorganic nitrogen
DNRA	dissimilatory reduction of NO_3^- to ammonium
DOC	dissolved organic carbon
DON	dissolved organic nitrogen
EA-IRMS	elemental analyzer isotope ratio mass spectrometry
EBM	ecosystem-based management
EMMA	End Member Mixing Analysis
FAME	fatty-acid methyl esters
FEBS	Federation of European Biochemical Societies
GC-C-IRMS	gas chromatography combustion-IRMS
GIS	geographic information systems
GMWL	global meteoric water line
GNIP	global network of isotopes in precipitation
HPLC	high-performance liquid chromatograph
IAEA	International Atomic Energy Association
ICP-MS	inductively coupled plasma mass spectrometry
IHS	isotopic hydrograph separations
IRGA	infrared gas analyzer

IRMS	isotope ratio mass spectrometer
KE	kinetic energy
KIE	kinetic isotope effect
LAG	lines of arrested growth
LAVD	leaf area vapor deficit
LMWL	local meteoric water lines
LTER	Long Term Ecological Research
MDF	mass dependent fractionations
MIF	mass independent fractionation
MPA	Marine Protected Area
MTBE	methyl *tert*-butyl ether
NADP	National Atmospheric Deposition Program
*N*AIP	*N*-acetyl iso-propyl
NESIS	Non-Equilibrium Stable Isotope Simulator
NICCCE	Nitrogen Isotopes and Carbon Cycling in Coniferous Ecosystems
NMR	nuclear magnetic resonance
*N*TFA-IP	*N*-trifluoroacetyl iso-propyl
PAN	peroxyacetyl nitrate
PCR	polymerase chain reaction
PEP	phosphoenolpyruvate
PMIs	pentamethylicosanes
POC	particulate organic carbon
POM	particulate organic matter
PON	particulate organic nitrogen
PN	particulate nitrogen
PLFAs	phospholipids fatty acids
RCC	River Continuum Concept
SEM	standard errors of measurement
SIA	stable isotope analysis
SIMS	secondary ion mass spectrometry
SIP	stable isotope probing
SOB	sulfide-oxidizing bacteria
SOM	soil organic matter
SRB	sulfate-reducing bacteria
tBDMS	*tert*-butyldimethylsilyl
TCA	tricarboxylic acid
TDLAS	tunable diode laser absorption spectroscopy
TEF	trophic enrichment factor
TFA	trifluoroacetyl
TLE	total lipid extraction
TMS	trimethylsilylation
TOF-SIMS	time-of-flight secondary ion mass spectrometry
TRACE	Tracer Redistributions Among Compartments in Ecosystems

TTD	travel time distribution
V-PDB	Vienna Pee Dee Belemnite
V-SMOW	Vienna Standard Mean Ocean Water
WMO	World Meteorological Organization
WUE	water-use efficiency

Introduction

Since the first edition of our book, the field of stable isotopes has expanded tremendously. From its earliest uses, geochemists and paleooceanographers have developed a rigorous theoretical and empirical basis for the integration of isotopes into studies of global element cycles, past climatic conditions, hydrothermal vent systems, and tracing rock sources. Similarly, plant biologists, ecologists, and environmental chemists have developed the theoretical framework and the empirical database for the use of isotopes to study plants and animals. Natural abundance isotope signatures can be used to find patterns and mechanisms at the single organism level as well as to trace food webs, understand paleodiets, and follow whole ecosystem nutrient cycling in both terrestrial and marine systems. As a consequence, isotopic analysis has really become almost a standard tool for physiologists, ecologists, and all scientists studying element or material cycling in the environment.

As access to isotope ratio mass spectrometers has increased and prices for sample analysis have decreased, ecologists from a broad range of disciplines, not necessarily trained as isotope chemists, have increasingly added stable isotope analysis as another tool in their research. This second edition is intended as a review and assessment of the theory and practice of stable isotope analysis in a variety of ecological disciplines, with suggestions for both generalist ecologists who might be considering including such analyses to their studies, as well as for the more experienced isotope ecologist who is pioneering new uses and new directions in the field. We have taken a look at the field, and have chosen topics that are basic to ecologists, as well as new and emerging uses of stable isotope analysis in a variety of ecological subdisciplines. We have started with an excellent primer by Elizabeth Sulzman for those of you who are new to the field or who are teaching an introductory course in stable isotopes. From here, the book is divided into several broad areas beginning with terrestrial systems. John Marshall and his colleagues look at the variation of stable isotopes in plants. The next chapter by Charles Garten and colleagues looks at forest status and soil carbon dynamics, followed by Dave Evan's in-depth discussion of nitrogen isotopes in soil systems. Moving on to things both alive and fossilized, Paul Koch discusses the use of isotopes in the biology of vertebrates. Keith Hobson then discusses how migratory organisms can be traced using stable isotopes. The next three chapters involve the marine environment, beginning with Joe Montoya's discussion of nitrogen in planktonic systems. Cindy Van Dover looks at the extreme environments of chemoautotrophic-based systems. Bob Michener and Les Kaufman discuss the use of stable isotopes in marine food webs, and how isotopes can be applied to marine conservation and management. Back on land, we take a look at freshwater systems and hydrology. Jacques Finlay and Carol Kendall apply isotopes to freshwater ecosystems, looking at both temporal and spatial variability in organic matter sources. Kevin McGuire and Jeff McDonnell take a look at where water goes when it rains, using isotopes as tracers in watershed hydrology. Following this, Carol Kendall and her colleagues discuss where the nitrogen goes, looking at the impact of anthropogenic N to ecosystems, especially as it applies to watersheds. Next, given the volume of data scientists can now generate, Bill Currie explores how to use that data to model ecosystem dynamics. We finish the book with a comprehensive chapter on compound-specific uses of stable isotopes put together by Richard Evershed and his colleagues.

Methods from a historical perspective

Many ecologists using stable isotopes will, and perhaps should, choose to send their samples to outside laboratories that specialize in the analysis of stable isotopes. Although the price of instrumentation has decreased, the costs for an individual to set up this type of laboratory are still quite high; typical startup budgets can be as high as $500,000. Maintenance of the mass spectrometer and the costs of having a full time, trained laboratory manager to run the laboratory (a necessity) are also steep. One can contrast this to the analysis costs of a typical study, which could be about $1000–$10,000 (with per sample charges averaging $7–$80, depending on the sample matrix and isotope in question). For many scientists it is much more cost-effective to use an outside laboratory.

Since the first edition, many of the methods to analyze samples have been automated, allowing for larger numbers of samples to be measured more quickly and inexpensively. More of this is covered in greater detail in chapters 1 and 14, and will be briefly discussed here. However, it is also useful to take a look at how scientists used to measure stable isotopes just 20 years ago, to give one an appreciation for how far the field has come, as well as seeing how it was done in the "dark ages" of stable isotope analysis. Fortunately, we have come a long way from the chart recorder and ruler! For those of you who are brand-new to the field and are not quite sure what mass spectrometers or stable isotopes are, we encourage you to first read chapter 1, an introduction to the terminology and chemistry of stable isotopes. After that, feel free to come back and get a historical perspective on stable isotope analysis.

Methods of sample preparation vary for each isotope. The goal in stable isotope analysis is to convert a sample quantitatively to a suitable purified gas (typically CO_2, N_2, or H_2) that the mass spectrometer can analyze. Sulfur can be analyzed as SO_2 or SF_6. Usually, organic samples are first dried (either in a 60°C oven or freeze-dried) and then ground to a fine powder. The samples can then be stored indefinitely in closed containers (such as scintillation vials or plastic bags), provided they are kept dry. If the investigator is interested in carbon isotopes for samples that may contain inorganic carbonates, the samples must first be acidified (usually with 1 N HCl, although some investigators are using dilute H_3PO_4; Showers & Angle 1986), since carbonate isotopic values are quite distinct from organic values and will skew the results (Haines & Montague 1979; Fry 1988).

Carbon and nitrogen in organic matter

In the early days of stable isotope measurement, most researchers used an oxidation reaction either "off-line" (sealed tubes in a muffle furnace, referred to as a Dumas combustion) or "on-line" (sample preparation line connected directly to the mass spectrometer) to combust the organic sample to a gas. The off-line combustion involves mixing the sample (typically 5–20 mg, depending on the sample's organic content) with an oxidant, usually cupric oxide, in a vycor (quartz) tube. In this procedure the sample must be in intimate contact with the CuO, which can be done in several ways: shaking the sample vigorously with the CuO within the tube (Fogel "shake method", M. Fogel, pers. comm.), grinding both in a mortar and pestle, or using a Wig-L-bug (Crescent Dental Manufacturing, Lyons, Illinois, USA). Shaking or using a Wig-L-bug is preferred, since there is less chance of sample cross-contamination. Approximately 1 g of CuO is used, then about 0.5 g of Cu is placed on top of the sample mix within the vycor tube in order to absorb the excess oxygen and convert N_2O to N_2. Once all sample tubes are prepared, they are then sealed under vacuum and combusted at 900°C for 1–2 h in a muffle furnace, and allowed to cool overnight to room temperature. It is then possible to cryogenically separate and purify the combined gases of CO_2, N_2, and H_2O (Stump & Frazer 1983; Boutton et al. 1983; Minagawa et al. 1984; Nevins et al. 1985). With manual samples, this can be done on a vacuum line using liquid nitrogen and an ethanol/dry ice slush. An important point to note for these and all following procedures is that the combustions and collections must be quantitative and close to 100% efficiency in order to prevent any fractionation. At this point the purified gas samples can be introduced into the isotope ratio mass spectrometer.

From this time-consuming, laborious process (generally 10–15 samples per day), the isotope ratio mass spectrometer (IRMS) manufacturers developed semi-automated combustion systems using elemental (CN) analyzers coupled to cryogenic purification systems that reduced sample preparation times and cost per analysis (Fry et al. 1992) and allowed simultaneous analysis of carbon and nitrogen isotopic compositions. This type of system was appropriate for most organic tissue samples, sediment and soil samples containing sufficient organic matter, as well as materials such as collagen and some plankton samples. Note that encapsulators were also available which allowed liquid samples to be analyzed. Some of the next generation of cryogenic systems were able to analyze samples containing as little as $1\,\mu$mol N and $1\,\mu$mol C. Depending on the type of system used and the type of sample being analyzed, 1–20 mg of material is loaded into a tin boat, folded, then placed in the sample carousel. In automated systems, combustion and separation of the gases is, in principle, similar to the manual method. The sample is flash-combusted at 1600–1800°C in an oxygen stream, then the combustion gases are carried in the helium stream through a series of cryogenic traps, which are maintained at specific temperatures to collect H_2O, CO_2, and N_2. The gas of interest is then introduced into the mass spectrometer for analysis by appropriate timing of a valve that shunts the gases either to waste or to the mass spectrometer.

The next evolution of automated combustion system involved introducing the helium stream containing the combusted gases directly into the mass spectrometer, otherwise known as continuous flow analysis, which is very rapid and can analyze around 100 samples per day.

Carbonates and dissolved inorganic carbon

Inorganic carbonate samples (e.g., foraminifera for paleotemperature studies) are reacted under vacuum with 100% phosphoric acid, which results in a complete conversion of carbonate to purified CO_2 (Craig 1953). This allows for the analysis of both $\delta^{13}C$ and $\delta^{18}O$ from the same sample, provided the phosphoric acid is pure and contains no water (Coplen et al. 1983).

Dissolved inorganic carbon (DIC) in water samples is prepared by acidifying a water sample and stripping the water with CO_2-free gas under a partial vacuum (Kroopnick 1985; McCorkle & Emerson 1988), then isolating and purifying the gas. The same principle can be applied to samples of bicarbonate in blood for tracer studies (Moulton-Barrett et al. 1993).

The latest methods for both types of samples have evolved with the development of automated continuous flow systems (Revesz & Landwehr 2004). Instead of evacuating the vials completely for carbonates, a helium stream displaces any atmosphere in the vial before the acid is added. After a set reaction time, the CO_2 is transferred into a sampling loop before being introduced into the mass spectrometer through either a helium stream or a dual inlet system. A similar technique is used for DIC in water.

Ammonia and nitrate $\delta^{15}N$ in water samples

In the dark days of dissolved inorganic nitrogen (DIN) analysis, ammonium in water samples was isolated using various steam distillation techniques (Velinsky et al. 1989) or later using passive diffusion within a closed container (Brooks et al. 1989). Both procedures involve making the pH of the water sample basic, then trapping or collecting the ammonium in an acid trap. Steam distillation techniques are good for large water samples containing low levels of NH_3, can be used on salt water solutions, and take about 30 min per sample (Velinsky et al. 1989). Once the ammonia is collected in an acid trap, zeolite is used to remove NH_3 from solution. The zeolite is then dried and can be analyzed using the sealed tube Dumas combustion method (see above). As with all methods, care is needed to trap all of the NH_3 in all steps in order to avoid fractionation. Nitrate-N can be distilled using the same techniques after first reducing the nitrate in the water to ammonia with Devarda's Alloy, a chemical reagent.

These passive diffusion techniques work well for samples such as soil solutions or Kjeldahl digests, and can be done in a batch fashion. Two different procedures are used, one involving suspending an acidified filter paper (usually with H_2SO_4) above the solution, the other wrapping the filter paper in Teflon tape and floating the packet in the solution (Downs et al. 1999). The solution is made basic and, using the same principle as the distillation technique, the ammonia diffuses onto the acidic filter paper. After the diffusion is complete (anywhere from 3–5 days), the filter paper is dried and can be combusted using the automated CN-mass spectrometer system.

These are by no means the only techniques to measure NH_3 and NO_3 in water samples, and we refer the reader to the volume by Knowles & Blackburn (1993) for further details on these and other methods. The latest techniques involve using denitrifying bacteria and a gas concentration system interfaced to the mass spectrometer; interested readers should see chapter Chapter 12 and papers by Chang et al. (1999), Sigman et al. (2001) and Casciotti et al. (2002).

Oxygen in water

The measurement of ^{18}O in water samples can be accomplished using several different procedures. One of the earliest procedures used by oceanographers, applicable to larger volume water samples (such as groundwater), uses 200 µL to 1 mL of water (Socki et al. 1992; Wong et al. 1987; Taylor 1973). The water sample is first placed in a suitable vessel such as a vacutainer or serum bottle. After removing the headspace atmosphere, a measured aliquot of CO_2 of known isotopic composition is introduced into the headspace. The water is incubated at a controlled temperature for a period of time that allows the oxygen in the water to completely exchange with the oxygen in the CO_2, after which the headspace CO_2 is removed using cryogenic techniques, then analyzed on the mass spectrometer. Modifications of this technique, using automated continuous flow analysis, routinely use 100–200 μL of water with excellent reproducibility (Horita & Kendall 2004).

Small-volume samples in the range of 3–10 μL, which may be generated from samples such as small animal metabolic studies, plant water, or combusted organic matter, are more problematic, given the difficulty of balancing the amount of headspace CO_2 to water volume. One technique that was used to get around this was a chemical procedure utilizing guanidine hydrochloride to release the oxygen. For the details of the technique, see Wong et al. (1987). This was an incredibly time-consuming and labor-intensive procedure, but produced good results.

The latest technique utilizes pyrolysis, where the water sample is combusted in an oxygen-free environment and the oxygen converted to CO, which is then analyzed by the mass spectrometer (Farquhar et al. 1997). The debate rages concerning carryover between water samples, but many laboratories have developed protocols to eliminate this problem (P. Brooks pers. comm.; Ghosh & Brand 2003; Gehre et al. 2004).

Deuterium

In the past, in order to measure ^2H, or deuterium, from organic tissue, the sample was combusted using an off-line, sealed tube procedure and the resulting water collected quantitatively (Schiegl & Vogel 1970). The water was then reduced to H_2 using either a vacuum line and uranium furnace, or using zinc in a sealed vessel (Krishnamurthy & DeNiro 1982, Coleman et al. 1982). The procedure could be used for other types of water samples, such as plant water, ground water, and water obtained for metabolic studies. Many investigators used the zinc method, as it could be done in a batch fashion and avoided any problems associated with obtaining uranium for the furnace.

Modern procedures now use one of two automated techniques. The first is similar to the CO_2 equilibration procedure for water samples. Instead of flushing the water vials with a CO_2/helium mix gas, the technique uses a hydrogen/helium mix gas; the vials also contain Hokko beads (platinum on a polymer base) that are suspended out of the water (Coplen et al. 1991). The beads act to enhance the exchange of hydrogen in the water with the gaseous hydrogen in the headspace. After a fixed time, the hydrogen is extracted and introduced via continuous flow into the mass spectrometer. The second technique is a pyrolysis procedure involving chromium. Water is injected onto hot chromium, which is contained in a combustion column in an elemental analyzer. The water reacts with the chromium and is converted to hydrogen gas, which goes directly into the mass spectrometer. This procedure is very rapid and very precise (Kelly et al. 2001; Nelson & Dettman 2001). Certainly a long way from the off-line uranium furnace days!

Sulfur

The analysis of sulfur isotopes depends on the starting matrix, but in essence involves converting sulfur in the sample to SO_2 or SF_6. Sulfur hexafluoride has the advantage that fluorine has only one isotope, but the techniques involved are somewhat hazardous, therefore most laboratories use SO_2 gas. Most of the early procedures to isolate sulfur from its matrix (water, plant and animal tissue, soils, sulfides) generally involved oxidizing sulfur to sulfate in solution. The sulfate could then be precipitated as $BaSO_4$ using a 10% barium chloride solution. From here the sample was oxidized to SO_2 gas and introduced into the mass spectrometer via a dual inlet. These procedures were generally not done in the laboratory of an ecologist, due to the labor, materials, and time involved. For a more detailed description of early sulfur preparation, see Krouse & Tabatabai (1986).

Once again, continuous flow has really revolutionized sulfur analysis. It is still not easy, but it certainly has evolved from those early days. Samples are combusted in an elemental analyzer and then passed through a gas chromatography (GC) column to separate the various combustion gases. Sulfur is much stickier and will take longer to elute. One has to make quite a few modifications to the elemental analyzer, due to the amount of water produced and the possibility that the water will make the sulfur "stick" in the system (C. Cook, pers. comm.; Giesemann et al. 1994).

The reproducibility of isotope measurements will depend on the procedure and laboratory techniques of the investigator, but is typically ±0.2‰ or better for carbon, oxygen, nitrogen, and sulfur, and 0.3–2‰ for hydrogen. This methods section is but a brief introduction to the procedures involved in preparing stable isotope samples. Our aim was to give you an idea of how modern day procedures have evolved out of the early days of stable isotope analysis. For more elaboration and further details on other methods, we refer the reader to volumes by Coleman & Fry (1991), Knowles & Blackburn (1993), and de Groot (2004).

Acknowledgments

In preparation of this book, we asked many colleagues to review the chapters. They are too numerous to list, but we would like to acknowledge each and every one of them. For without their input and careful critiques, this volume would not have been possible. We would also like to thank our families, and our colleagues at Boston University and Oregon State University for their patience and understanding in the final stages of this volume.

Finally, it is with great sadness that we report the death of Elizabeth Sulzman on June 10, 2007. She was 40 years old, and was a scientist, colleague, teacher, mentor, mother, and wife. We will remember her kindness, her laughter, and great conversations about science and life. We will miss her greatly.

ROBERT MICHENER AND KATE LAJTHA

References

Boutton, T.W., Wong, W.W., Hachey, D.L., Lee, L.S., Cabrera, M.P. & Klein, P.D. (1983) Comparison of quartz and pyrex tubes for combustion of organic samples for stable carbon isotope analysis. *Analytical Chemistry*, **55**, 1832–1833.

Brooks, P.D., Stark, J.M., McInteer, B.B. & Preston, T. (1989) Diffusion method to prepare soil extracts for automated nitrogen-15 analysis. *Soil Science Society of America Journal*, **53**, 1707–1711.

Casciotti, K.L., Sigman, D.M., Hastings, M.G., Bohlke, J.K. et al. (2002) Measurement of the oxygen isotopic composition of nitrate in seawater and freshwater using the denitrifier method. *Analytical Chemistry*, **74**, 4905–4912.

Chang, C.C.Y., Langston, J., Riggs, M., Campbell, D.H. et al. (1999) A method for nitrate collection for $\delta^{15}N$ and $\delta^{18}O$ analysis from waters with low nitrate concentrations. *Canadian Journal of Fisheries and Aquatic Sciences*, **56**, 1856–1864.

Coleman, D.C. & Fry, B. (Eds) (1991) *Carbon Isotope Techniques*. Academic Press, San Diego, CA.

Coleman, M.L., Shepherd, T.J., Durham, J.J., Rouse, J.E. & Moore, G.R. (1982) Reduction of water with zinc for hydrogen isotope analysis. *Analytical Chemistry*, **54**, 995–998.

Coplen, T.B., Kendall, C. & Hopple, J. (1983) Comparison of stable isotope reference samples. *Nature*, **302**, 236–238.

Coplen, T.B., Wildman, J.D. & Chen, J. (1991) Improvements in the gaseous hydrogen-water equilibration technique for hydrogen isotope ratio analysis. *Analytical Chemistry*, **54**, 2611–2612.

Craig, H. (1953) The geochemistry of the stable carbon isotopes. *Geochimica et Cosmochimica Acta*, **3**, 53–92.

De Groot, P. (Ed.) (2004) *Handbook of Stable Isotope Analytical Techniques*, Vol. 1. Elsevier, Amsterdam, 1234 pp.

Downs, M., Michener, R., Fry, B. & Nadelhoffer, K. (1999) Routine measurement of dissolved inorganic15N in precipitation and streamwater. *Environmental Monitoring and Assessment*, **55**, 211–220.

Farquhar, G.D., Henry, B.K. & Styles, J.M. (1997) A rapid on-line technique for determination of oxygen isotope composition of nitrogen-containing organic matter and water. *Rapid Communcations in Mass Spectrometry*, **11**, 1554–1560.

Fry, B. (1988) Food web structure on Georges Bank from stable C, N, and S isotopic compositions. *Limnology and Oceanography*, **33**, 1182–1190.

Fry, B., Brandt, W., Mersch, F.J., Tholke, K. & Garritt, R. (1992) Automated analysis system for coupled $\delta^{13}C$ and $\delta^{15}N$ measurements. *Analytical Chemistry*, **64**, 288–291.

Gehre, M., Geilmann, H., Richter, J., Werner, R.W. & Brand, W. (2004) Continuous flow $^2H/^1H$ and $^{18}O/^{16}O$ analysis of water samples with dual inlet precision. *Rapid Communications in Mass Spectrometry*, **18**, 2650–2660.

Giesemann, A., Jäger, H.-J., Norman, A.L., Krouse, H.R. & Brand, W.A. (1994) On-line sulfur-isotope determination using an elemental analyzer coupled to a mass spectrometer. *Analytical Chemistry*, **66**, 2816–2819.

Ghosh, P. & Brand, W. (2003) Stable isotope ratio mass spectrometry in global climate change research. *International Journal of Mass Spectrometry*, **228**, 1–33.

Haines, E.B. & Montague, C.L. (1979) Food sources of estuarine invertebrates analyzed using $^{13}C/^{12}C$ ratios. *Ecology*, **60**, 48–56.

Horita, J. & Kendall, C. (2004) Stable isotope analysis of water and aqueous solutions by conventional dual-inlet mass spectrometry. In: *Handbook of Stable Isotope Analytical Techniques* (Ed. P.A. de Groot), pp. 1–37. Elsevier, Amsterdam.

Kelly, S.D., Heaton, K.D. & Brereton, P. (2001) Deuterium/hydrogen isotope ratio measurement of water and organic samples by continuous-flow isotope ratio mass spectrometry using chromium as the reducing agen in an elemental analyser. *Rapid Communcations in Mass Spectrometry*, **15**, 1283–1286.

Knowles, R. & Blackburn, T.H. (Eds) (1993) *Nitrogen Isotope Techniques*. Academic Press, San Diego, CA.

Krishnamurthy, R.V. & DeNiro, M.J. (1982) Sulfur interference in the determination of hydrogen concentration and stable isotopic composition in organic matter. *Analytical Chemistry*, **54**, 153–154.

Kroopnick, P.M. (1985) The distribution of ^{13}C of TCO_2 in the world oceans. *Deep Sea Research*, **32**, 57–84.

Krouse, H.R. & Tabatabai, M.A. (1986) Stable sulfur isotopes. In: *Sulfur in Agriculture* (Ed. M.A. Tabatabai), pp. 169–201. Academic Press, New York.

McCorkle, D.C. & Emerson, S.R. (1988) The relationship between pore water carbon isotopic composition and bottom water oxygen concentration. *Geochimica et Cosmochimica Acta*, **52**, 1169–1178.

Minagawa, M., Winter, D.A. & Kaplan, I.R. (1984) Comparison of Kjeldahl and combustion methods for measurement of nitrogen isotope ratios in organic matter. *Analytical Chemistry*, **56**, 1859–1861.

Moulton-Barrett, R., Triadafilopoulos, G., Michener, R. & Gologorsky, D. (1993) Serum ^{13}C-bicarbonate in the assessment of gastric *Helicobacter pylori* urease activity. *American Journal of Gastroenterology*, **88**, 369–374.

Nelson, S.T. & Dettman, D. (2001) Improving hydrogen isotope ratio measurements for on-line chromium reduction systems. *Rapid Communcations in Mass Spectrometry*, **15**, 2301–2306.

Nevins, J.L., Altabet, M.A. & McCarthy, J.J. (1985) Nitrogen isotope ratio analysis of small samples: sample preparation and calibration. *Analytical Chemistry*, **57**, 2143–2145.

Revesz, K., & Landwehr, J.M. (2004) Measurement of $\delta^{13}C$ and $\delta^{18}O$ isotopic ratios of $CaCO_3$ by Thermoquest-Finnigan GasBench II and Delta Plus XL continuous flow isotope ratio mass spectrometer with application to Devils Hole Core DH-11 calcite. In: *Isotope Hydrology and Integrated Water Resources Management*. IAEA-CSP-23, International Atomic Energy Agency, Vienna, p. 485.

Schiegl, W.E. & Vogel, J.C. (1970) Deuterium content of organic matter. *Earth and Planetary Science Letters*, **7**, 307–313.

Showers, W.J. & Angle, D.G. (1986) Stable isotopic characterization of organic carbon accumulation on the Amazon continental shelf. *Continental Shelf Research*, **6**, 227–244.

Sigman, D.M., Casciotti, K.L., Andreani, M., Barford, C., Galanter, M. & Böhlke, J.K. (2001) A bacterial method for the nitrogen isotopic analysis of nitrate in seawater and freshwater. *Analytical Chemistry*, **73**, 4145–4153.

Socki, R.A., Karlsson, H.R. & Gibson, E.K., Jr. (1992) Extraction technique for the determination of oxygen-18 in water using preevacuated glass vials. *Analytical Chemistry*, **64**, 829–831.

Stump, R.K. & Frazer, J.W. (1983) *Simultaneous Determination of Carbon, Hydrogen and Nitrogen in Organic Compounds*. Report 1973, UCID-16198, University of California, Livermore, CA.

Taylor, C.B. (1973) *Measurement of Oxygen-18 Ratios in Environmental Waters using the Epstein–Mayeda Technique. Part 1: Theory and Experimental Details of the Equilibration Technique*. Publication 556, Institution of Nuclear Science, Low Hutt, New Zealand.

Velinsky, D.J., Pennock, J.R., Sharp, J.H., Cifuentes, L.A. & Fogel, M.L. (1989) Determination of the isotopic composition of ammonium-nitrogen at the natural abundance level from estuarine waters. *Marine Chemistry*, **26**, 351–361.

Wong, W.W., Lee, L.L. & Klein, P.D. (1987) Oxygen isotope ratio measurements on carbon dioxide generated by reaction of microliter quantities of biological fluids with guanidine hydrochloride. *Analytical Chemistry*, **59**, 690–693.

CHAPTER 1

Stable isotope chemistry and measurement: a primer

ELIZABETH W. SULZMAN

Introduction

Stable isotopes have been commonly used to: (i) identify sources (e.g., pollutants to a stream), (ii) infer processes (e.g., heterotrophic nitrification), (iii) estimate rates (e.g., soil C turnover), (iv) determine proportional inputs (e.g., percent contribution of a particular prey item to a predator's diet), and (v) confirm, reject, or constrain models derived from the use of other techniques. This chapter will address some of the fundamentals behind the biological uses of stable isotopes. I will address what stable isotopes are, why some stable isotopes are better for ecological process studies than others, and how they are measured. The remaining chapters will provide examples of the ecological uses of stable isotopes.

What isotopes are, what makes them distinct

Isotopes are atoms with the same number of protons and electrons but differing numbers of neutrons (Figure 1.1). Isotopes are denoted by an atomic "formula." For example, the most common isotope of carbon is $^{12}_{6}C$, where 12 is the atomic mass, or the sum of neutrons and protons, and 6 is the atomic number (number of protons/electrons); the number of neutrons can be determined by difference. Stable isotopes are defined as those that are energetically stable and do not decay; thus, they are not radioactive. An isotope tends to be stable when the number of neutrons (N) and the number of protons (Z) are quite similar ($N/Z \leq 1.5$; Figure 1.2). There are roughly 300 stable isotopes, over 1200 radioactive isotopes, and only 21 elements that are known to have only one isotope (Hoefs 1997). Table 1.1 lists the relative abundances of the stable isotopes most common in ecological research. The isotopes in ecological research are dominated by the lighter elements both because they dominate biological compounds and because the percent increase in mass caused by the addition of a single neutron is greatest for these elements (see below). Iron and strontium are among the heaviest isotopes used in ecological studies, and currently their use is not common. Typically, an isotope ratio mass spectrometer (IRMS, see below) is configured

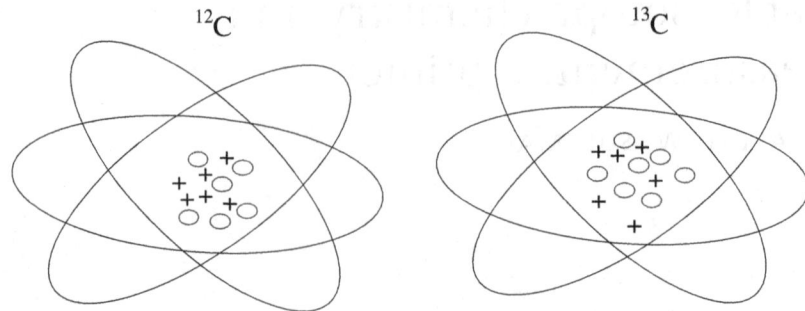

Figure 1.1 Idealized structure of the two most common stable isotopes of carbon, ^{12}C (left) and ^{13}C (right). Plus symbols represent protons, which are positively charged, open circles represent neutrons (neutral), and ellipses represent the path taken by the six negatively charged electrons (not shown) as they orbit the nucleus, balancing the charge of the protons.

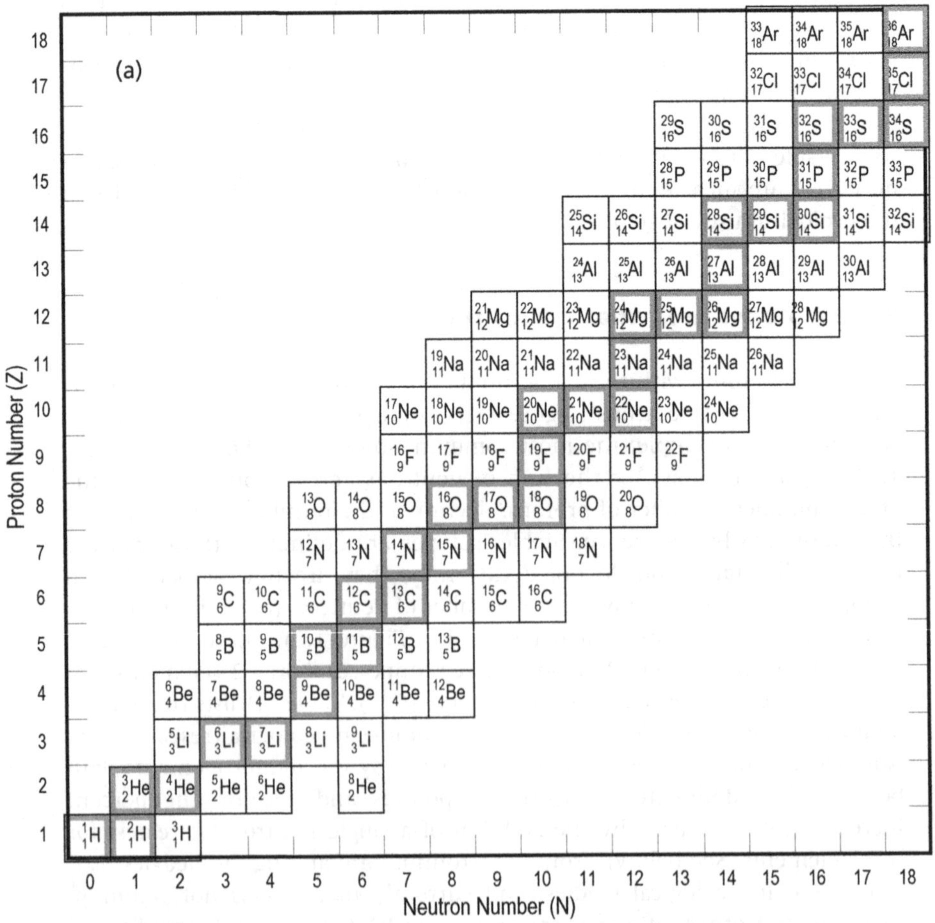

Figure 1.2 Partial periodic table of the stable and unstable isotopes: (a) the lighter isotopes; (b) some of the heavier isotopes. Shaded borders indicate stable isotopes, unshaded borders indicate short-lived unstable isotopes; long-lived unstable nuclides are indicated with a triangular border (e.g., ^{40}K). Note that the stable isotopes generally have a proton-to-neutron ratio of 1.5 or less. (Modified after Faure 1986.)

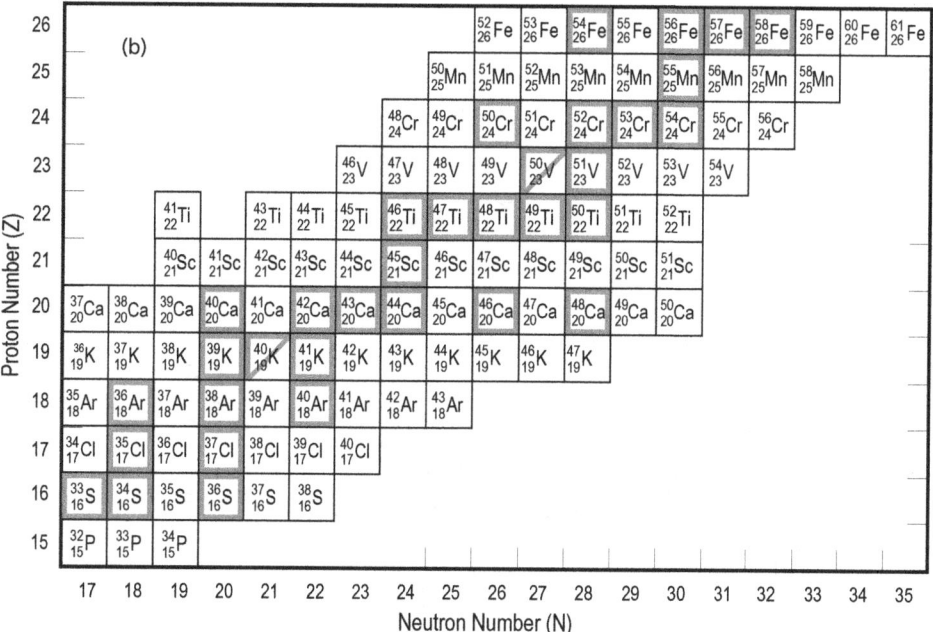

Figure 1.2 *Continued*

for analysis of either the lighter elements (e.g., H, C, O, N, and maybe S) or heavier elements, but not both.

Notation and terminology

The isotopic differences between various materials (e.g., leaves, minerals, seawater) are exceedingly small, so isotopic composition is reported relative to an internationally accepted standard and expressed in parts per thousand deviation from that standard by:

$$\delta(\text{\textperthousand}) = \left(\frac{R_{\text{sample}}}{R_{\text{standard}}} - 1\right) \times 1000 \tag{1.1}$$

where R is the ratio of heavy-to-light (typically, but not always, rare-to-abundant) isotope, R_{sample} is that ratio in the sample, and R_{standard} is that in the standard (see Table 1.1). The focus of this chapter is on natural abundance isotopes, but it should be noted that isotopes are also used as tracers. As tracers isotopes are at elevated levels, and so are often expressed in Atom%, where

$$\text{Atom\%} = 100 \times [R_{\text{sample}}/(1 + R_{\text{sample}})] \tag{1.2}$$

Table 1.1 Relative abundances of the stable isotopes most common in ecological research in order of increasing mass.

Element	Isotope	Abundance (%)	Relative mass difference (%)	International Standard	Absolute abundance of the standard ($R_{standard}$)
Hydrogen	^{1}H	99.985	100	Vienna Standard Mean Ocean Water (VSMOW)	^{2}H:^{1}H = 0.00015576
	^{2}H (also D)	0.0155			
Carbon	^{12}C	98.892	8.3	Vienna Pee Dee Belemnite (VPDB)*	^{13}C:^{12}C = 0.0112372
	^{13}C	1.108			
Nitrogen	^{14}N	99.635	7.1	Atmospheric nitrogen (air)	^{15}N:^{14}N = 0.0036765
	^{15}N	0.365			
Oxygen	^{16}O	99.759	12.5 (^{18}O:^{16}O)	VSMOW in water, generally VPDB in CO_2 or carbonate	VSMOW = 0.0020052 VPDB = 0.0020672 both for ^{18}O:^{16}O
	^{17}O	0.037			
	^{18}O	0.204			
Silicon†	^{28}Si	92.21	7.1 (^{30}Si:^{28}Si)	NBS-28	^{30}Si:^{28}Si has not been reported
	^{29}Si	4.70			
	^{30}Si	3.09			
Sulfur	^{32}S	95	6.3 (^{34}S:^{32}S)	Vienna Cañon Diablo meteorite troilite (VCDT)	^{34}S:^{32}S = 0.0450045
	^{33}S	0.75			
	^{34}S	4.21			
	^{36}S	0.014			

Iron‡	^{54}Fe	5.82	3.7	Average terrestrial	^{56}Fe:^{54}Fe = 15.7028
	^{56}Fe	91.66	(^{56}Fe:^{54}Fe)	and lunar rocks§	
	^{57}Fe	2.19			
	^{58}Fe	0.33			
Strontium¶	^{84}Sr	0.56	1.1	US Geological	^{87}Sr:^{86}Sr = 0.709249
	^{86}Sr	9.87	(^{87}Sr:^{86}Sr)	Survey *Tridacna***	
	^{87}Sr	7.04			
	^{88}Sr	82.53			

* The defined reference standard for ^{13}C has been Belemnite of the Pee Dee Formation in South Carolina, USA (PDB). Because PDB is no longer available, a new reference standard, Vienna-PDB (VPDB), has been defined by its relationship to NBS19. The reported, although not universally accepted, ratio for VPDB is 0.0111797 (http://deuterium.nist.gov/standards.html).

† δ^{30}Si values are being used to assess the terrestrial (plant and rock weathering) contribution to the global silicon cycle (Ziegler et al., pers. comm.).

‡ Use of iron isotopes in ecological studies is quite new (first results published by Bullen & McMahon, 1998). Iron fractionation by bacteria is implicated as a potential means to trace the distribution of microorganisms in modern and ancient Earth (Beard et al. 1999).

§ Other recognized standards for iron include IRMM-14, which is commercially available and certified by the Institute for Reference Materials and Measurement, European Commission Joint Research Centre, and US Geological Survey rock standard BIR-1, an Icelandic basalt.

¶ Most of the studies involving Sr have been geological; however, the similarity between Sr and Ca has allowed use of δ^{86}Sr of bones and teeth to infer the geographic region an animal or human inhabited (Beard & Johnson 2000; Kennedy et al. 2000; Perakis et al. 2006).

** Other materials, such as seawater (Capo et al. 1998), are also used.

Because the international standards are too expensive for daily use, each laboratory has one or more internal working standards, which are compared against the international standard (and often "blind tested" by several other laboratories). The final isotopic composition of a sample, then, is converted to its value via:

$$\delta_{sa/istd} = \delta_{sa/ws} + \delta_{ws/istd} + 10^{-3} (\delta_{sa/ws} \times \delta_{ws/istd}) \qquad (1.3)$$

where $\delta_{sa/istd}$ signifies the isotopic value of the sample relative to the international standard, $\delta_{sa/ws}$ refers to the isotopic value of the sample relative to the working standard, and $\delta_{ws/istd}$ refers to the isotopic value of the working standard relative to the international standard (all values in ‰; Craig 1957). In a typical measurement series reference compounds (internal working standards) are processed the same way as the sample, and so the term $\delta_{ws/istd}$ is measured in its reverse form, $\delta_{istd/ws}$. These two can be converted using:

$$\delta_{ws/istd} = -1 \times \cfrac{1}{\left(\cfrac{1}{\delta_{istd/ws}} + 10^{-3}\right)} \qquad (1.4)$$

A positive δ ("delta") indicates that the sample has more of the heavy isotope than does the standard whereas a negative delta value indicates the sample has less of the heavy isotope than the standard. Note that the units (‰) are parts per thousand, sometimes referred to as "per mil", sometimes as "permil", sometimes as "permill", sometimes as "per mille", or even as "permille"! There are several common means to compare the isotopic composition of two materials, including "heavy" vs. "light", high vs. low values (can be confusing as we are typically talking in negative numbers), more/less positive, and "enriched" vs. "depleted". The last should only be used with clarification as to which isotope the sample is enriched or depleted in, the lighter isotope, or the heavier one. The standard approach when using this terminology is to refer in terms of the heavy isotope (which is often, but not always, more rare); for example, "sample X is depleted in ^{15}N whereas sample Y is enriched in ^{15}N."

Reactivity of isotopes and their compounds

The number of electrons in an atom's outer shell controls the chemical reactions the atom undergoes; thus, the chemical behavior of two isotopes is **qualitatively** similar. However, atomic mass determines the vibrational energy of the nucleus, thus differences in mass lead to differences in both reaction rate and bond strength. The physical behavior of two isotopes is therefore **quantitatively** different, with the greatest differences between isotopes occurring among the lightest elements where the percent change in mass is the greatest (Table 1.1). For example, a deuterium atom has twice the mass of a hydrogen atom, whereas the extra neutron in ^{33}S is only a 3 percent increase in mass over ^{32}S. The behavioral differences between hydrogen and deuterium therefore will be much greater than between those of ^{33}S and ^{32}S. The reason mass differences lead to physical behavior differences is because kinetic energy (KE) is constant for a given element in fixed environmental surroundings:

$$KE = \tfrac{1}{2} mv^2 \tag{1.5}$$

where m is mass and v is velocity. Molecules in the same physical environment (namely temperature) have the same kinetic energy, so a molecule of larger mass will travel at slower velocity. Note that the above equation applies only in a vacuum. Craig (1953, p. 73) provides the more complicated version for diffusion through air. Rearranging the simple equation above, we can see that different masses of the same molecule (**isotopomers**, e.g., $H_2^{16}O$ vs. $H_2^{18}O$) react with different velocities, where L denotes the lighter isotope and H, the heavier isotope:

$$\frac{v_L}{v_H} = \left(\frac{m_H}{m_L}\right)^{1/2} \quad \text{so} \quad \frac{v_L}{v_H} = \sqrt{\frac{m_H}{m_L}} \quad \text{or} \quad \frac{v_{160}}{v_{180}} = \sqrt{\frac{20}{18}} \tag{1.6}$$

Thus, the velocity of $H_2^{16}O$ is 1.05 times faster than that of $H_2^{18}O$, regardless of temperature.

Another physical law that determines the behavior of isotopes states that the vibrational energy of a molecule is controlled by the frequency of its vibration. Because heavy atoms vibrate more slowly than lighter ones, the energy of the molecule with the heavy isotope is lower, and therefore it forms more stable, stronger bonds. These velocity and bond strength differences among isotopes and isotopomers lead to **fractionations**, or isotopic differences between the source and product compounds of a chemical transformation. For example, the vapor pressure of ^2H_2O (also known as deuterated water, or D_2O) is nearly 40 torr lower than that of ^1H_2O (Hoefs 1997) because vapor pressure is inversely proportional to intermolecular forces and the ^2H—O bonds are stronger than the ^1H—O bonds. Thus, evaporation will lead to observable fractionation, yielding relatively light (lower δ) vapor and ^2H-rich (higher δ) water left behind.

Fractionation mechanisms

Three mechanisms lead to isotopic fractionation: equilibrium (also called thermodynamic or exchange), kinetic, and nuclear spin. **Equilibrium fractionation reactions** are those in which the distribution of isotopes differs between chemical substances (reactant vs. product) or phases (e.g., vapor vs. liquid) when a reaction is in equilibrium. In these reactions the reactants and products remain in close contact in a closed, well-mixed system such that back reactions can occur and chemical equilibrium can be attained. An example of an equilibrium exchange reaction is that between carbon dioxide and water in a closed container:

$$C^{16}O_2 + H_2^{18}O \leftrightarrow C^{18}O^{16}O + H_2^{16}O \tag{1.7}$$

Here the reactants and products are the same, but have final masses that differ from their initial masses because ^{18}O forms a stronger covalent bond with carbon than does ^{16}O (Faure 1986). The extent of the difference in final and initial masses in this and other equilibrium fractionation reactions is temperature dependent, with the greatest differences occurring at the lowest temperatures. The rule of thumb for equilibrium exchange reactions is that the heavier isotope tends to accumulate where bonds are the strongest (Bigeleisen 1965), which generally means in the denser phase, the compound having the largest molecular mass (e.g., $CaCO_3$ vs. CO_2), and/or the phase with the highest oxidation state (e.g., CO_2 vs. CH_4, Craig 1953).

Kinetic isotope effects arise in irreversible or unidirectional reactions because for some reason the reverse reaction is inhibited or not occurring, such as evaporation in an open system when water vapor moves away from the liquid water pool. In kinetic reactions, both bond strength and isotope velocity are important. Kinetic fractionation reactions are normally associated with processes such as evaporation, diffusion, dissociation reactions, and enzymatic effects. Kinetic fractionations are often quite large, usually much larger than equilibrium fractionations, and result in the lighter isotope accumulating in the product (lighter goes faster). Figure 1.3 illustrates the isotopic change in reactant and product in an irreversible kinetic reaction. Note that if the reaction goes to completion the cumulative product will have the same isotopic composition as the initial substrate. Also note that the isotopic difference between substrate and instantaneous product (magnitude of the isotopic fractionation, ε, see below) is a constant, although because of the changing slope it does not appear that way.

Unlike equilibrium and kinetic fractionation mechanisms, **nuclear spin isotope effects** are not mass dependent; rather, they arise because of differences in the nuclear structure among isotopes and lead to differences in nuclear spin. It is not clear how important nuclear spin fractionation is in most circumstances, but it does allow one to couple tracers (e.g., a ^{13}C-labeled substrate) with nuclear magnetic resonance (NMR) techniques (see "Technological Advances," below).

Quantifying fractionation

The process of isotopic fractionation is quantified with fractionation factors (α), which can be defined as the ratio of two isotope ratios:

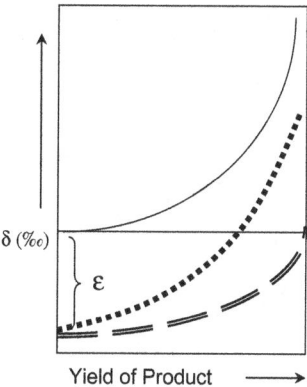

δ (‰)

ε

Yield of Product →

Figure 1.3 Relative changes in δ values of substrate, instantaneous product, and cumulative product during unidirectional kinetic fractionation processes. The solid curve (upper) represents the substrate, the thick dotted curve (middle) represents the instantaneous product, and the dashed curve (lowest δ values) represents the cumulative product. The horizontal line is drawn to highlight the fact that initial substrate and cumulative product have the same isotopic composition if the reaction goes to completion (i.e., all substrate is consumed). The fractionation factor, ε, is constant. (Modified from Kendall & Caldwell 1998.)

$$\alpha_{p-s} = R_p / R_s \qquad (1.8)$$

where R is the ratio of heavy to light isotopes in the instantaneous product (R_p) and substrate (R_s). When reactions are bidirectional (equilibrium exchange reactions) the subscripts may indicate the phase rather than which side of the reaction a substance is listed (e.g., $\alpha_{l-v} = R_l / R_v$, with l representing liquid, and v representing vapor). Values of α usually are near 1.00, except for hydrogen isotopes, where they can be as large as 4 at room temperature (Kendall & Caldwell 1998). If an α value is >1, it means that the instantaneous product is enriched in the heavier isotope relative to the substrate. For example, α_{l-v} for ^{18}O in water at 20°C is 1.0098 (Majoube 1971), indicating that liquid water is 9.8‰ more enriched in ^{18}O than is water vapor at equilibrium (i.e., water vapor is depleted in ^{18}O at equilibrium). Equilibrium fractionation factors are highly dependent on temperature, decreasing as temperature increases (Figure 1.4). The degree of fractionation is 10 times higher for hydrogen than it is for oxygen and carbon because of the larger relative mass difference between hydrogen isotopes.

Fractionation factors and the isotopic composition of two substances are related according to:

$$\alpha_{A-B} = (1000 + \delta_A)/(1000 + \delta_B) \qquad (1.9)$$

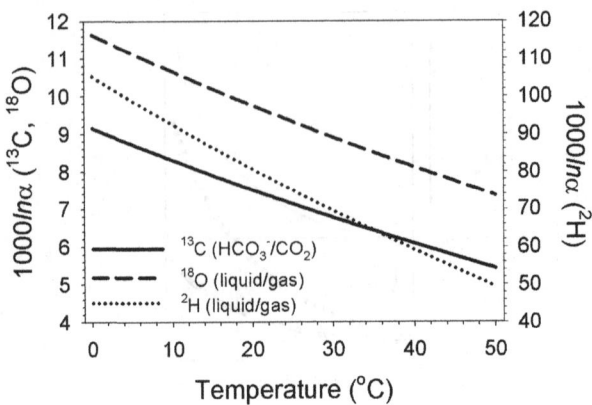

Figure 1.4 The temperature dependence of the equilibrium fractionation factor for ^{13}C exchange during the bicarbonate–carbon-dioxide reaction, ^{18}O exchange during the liquid–gas phase change for water, and ^{2}H exchange during the liquid–gas phase change for water. Hydrogen has a separate larger scale (right-hand axis) compared with oxygen and carbon.

where A and B represent different substances. As noted above, isotope effects are generally small (i.e., $\alpha \approx 1$), so authors often report the deviation of the fractionation factor from unity:

$$\varepsilon_{A-B} = (\alpha_{A-B} - 1) \times 1000 \qquad (1.10)$$

where ε is the <u>result</u> of the fractionation process, and is referred to as **enrichment** (when $\varepsilon > 0$) or **depletion** (when $\varepsilon < 0$) of the heavy isotope in compound B with respect to compound A, expressed in ‰. Another common formulation for isotopic fractionation is:

$$\varepsilon_{A-B} = 1000 \ln \alpha_{A-B} \qquad (1.11)$$

Also common in the ecological literature, but somewhat confusing, is the term "discrimination," which refers to the degree to which a process (e.g., photosynthesis) "avoids" the heavy isotope. One formulation for discrimination is:

$$\Delta_{A-B} = \delta_A - \delta_B \qquad (1.12)$$

while another is:

$$\Delta_{A-B} = (\delta_A - \delta_B)/(1 + \delta_B/1000) \qquad (1.13)$$

Those interested in plant- and canopy-level isotopic discrimination (including those in the global carbon modeling community) commonly use equation

Table 1.2 Comparison among the different formulations used to compare the isotopic composition of two materials (Modified after Hoefs 1997).

δ_A	δ_B	α_{A-B} (eqn 9), $(1000 + \delta_A)/(1000+\delta_B)$	ε_{A-B} (eqn 10), $10^3 (\alpha_{A-B} - 1)$	ε_{A-B} (eqn 11), $10^3 \ln \alpha_{A-B}$	Δ_{A-B} (eqn 12), $\delta_A - \delta_B$	Δ_{A-B} (eqn 13), $(\delta_A - \delta_B)/(1 + \delta_B/1000)$
−10	−5	0.994975	−5.02513	−5.03779	−5	−5.02513
−8	−27	1.019527	19.52724	19.33903	19	19.52724
1	5	0.996020	−3.9801	−3.98804	−4	−3.9801
1	10	0.991089	−8.91089	−8.95083	−9	−8.91089
1	20	0.981373	−18.6275	−18.8031	−19	−18.6275
10	10	1	0	0	0	0
30	10	1.019802	19.80198	19.60847	20	19.80198

1.12. It should be noted that equations 1.12 and 1.13 do not give identical mathematical solutions if the delta values (δ) of substances A and B vary by more than 10‰ (Hoefs 1997; Table 1.2). Equations 1.10 (fractionation) and 1.13 (discrimination) are mathematically equivalent. Authors are encouraged to explicitly state which formulae and symbols they used. For those desiring more background in the mathematical relations among fractionation factors, there is an excellent treatment of the subject in Faure (1991).

Properties of ecologically useful stable isotopes

All stable isotopes are not created equal. Several characteristics are common to stable isotopes useful in ecological research, including the following (modified from http://epswww.unm.edu/facstaff/zsharp/bio1.htm):

1 large mass difference between the rare and abundant isotope (see Table 1.1, all but Fe and Sr have mass differences greater than 6%, and these have only become useful recently, with improved technologies);
2 low atomic mass (leads to above point);
3 rare isotope comprises small fraction of total elemental occurrence (more important for older machines with lower sensitivity);
4 element has more than one oxidation state.

How isotopes are measured

Stable isotope ratios are typically measured via a technique called isotope ratio mass spectrometry (IRMS), which was invented by J.J. Thompson in

1910; the first definitive demonstration that isotopes exist in nature followed in 1922 (Aston 1922). A mass spectrometer is an instrument that separates charged atoms or molecules on the basis of their mass-to-charge-ratio, *m/z*. There are two basic types of IRMS, dual-inlet (DI-IRMS) and continuous flow (CF-IRMS). In general, precision is higher when samples are analyzed with a dual-inlet system. A continuous flow system allows a researcher to introduce multiple component samples (e.g., atmospheric air, soil, leaves) and obtain isotopic information for individual elements or compounds within the mixture.

Recent advances in instrumentation, particularly following the development of CF-IRMS with regard to sample automation, has greatly increased the accessibility of mass spectrometry outside the fields of geochemistry and physical chemistry. Whether the system is DI-IRMS or CF-IRMS, there are four basic components to isotope ratio mass spectrometry: inlet system, ion source, mass analyzer, and ion detector (Figure 1.5). All differences between a CF and DI system are in the inlet component. Below I outline each

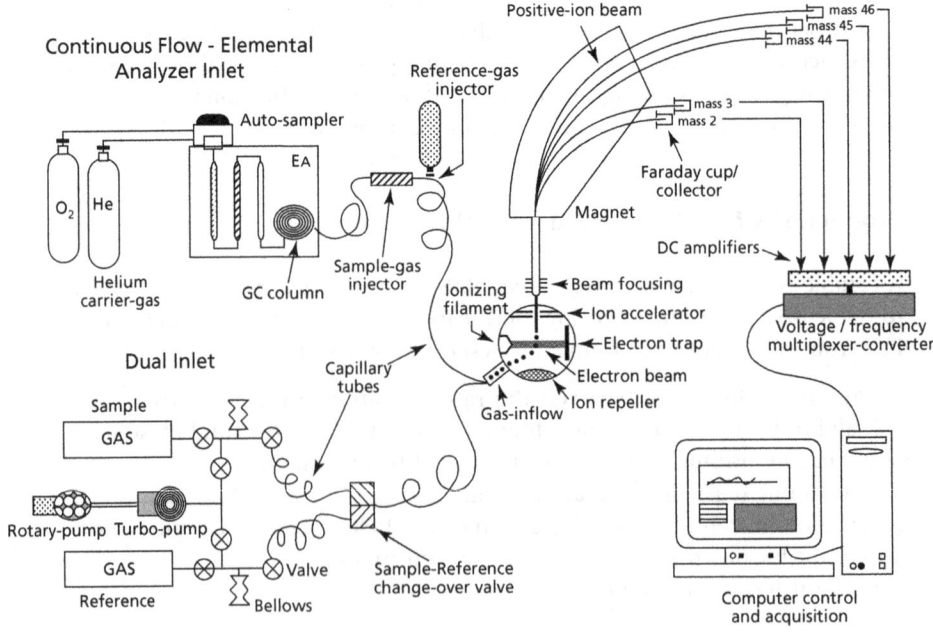

Figure 1.5 From Dawson & Brooks (2001). Diagrammatic representation of the continuous flow (top) and dual inlet (bottom) interfaces to the ion source of an isotope ratio mass spectrometer (center). Ionized gases leaving the ionization chamber are accelerated and focused (with the aid of the magnet) towards the collector array. Not to scale.

component briefly, but those with particular interest in the instrumentation are referred to Barrie & Prosser (1996) and Brand (2004) for a much more in-depth treatment including discussion of linearity, memory effects, background correction, and other topics.

Samples are introduced to the mass spectrometer via the inlet system as a gas. The most important aspect of the inlet system is the use of capillary tubes to assure there is no isotope separation (i.e., fractionation) during introduction of the gases into the mass spectrometer. If the mean free path (flight without collision with other molecules) is long, molecules will move according to **molecular flow**, and will separate, leaving the heavier isotopes behind in the reservoir while the lighter isotopes flow into the mass spectrometer. Capillary tubing induces **viscous flow**, in which there are many molecular collisions and therefore no fractionation (because the sample is well-mixed). In a dual inlet system, the sample gas flows to a variable volume reservoir (also known as the bellows) where it is condensed or expanded such that its pressure equals that of the reference gas. The sample then flows through a changeover valve, which shunts one gas (either the sample or the working standard) to "waste," and the other into the ion source. This system maintains constant flow of both gases and allows direct and repeated (usually 10 replicates of each sample) measures of the sample and standard under identical temperature and pressure conditions, yielding very high precision (in air: 0.01‰ for ^{13}C, 0.03‰ for ^{18}O, and 0.05‰ for ^{15}N). In a continuous flow system samples are carried in a helium stream through a chromatographic column. If samples are solids (e.g., soil, animal, or plant material) they are combusted prior to entry into the chromatographic column. The timing of a "heart-cut" valve is set by the operator to send the elutant of interest (e.g., CO_2) into the mass spectrometer, while other gases (e.g., N_2) are sent to waste. Each sample is typically only measured once (although some inlet systems, such as the GasBench II®, allow repeated sampling). Precision is typically an order of magnitude less for CF-IRMS than for DI-IRMS (e.g., 0.07‰ for ^{13}C, 0.1‰ for ^{18}O, and 0.2‰ for ^{15}N).

In the ion source, electrons are released under high vacuum ($\sim 10^{-8}$ torr) where a wire of tungsten, rhenium, or thoriated iridium is electrically heated. The electrons are then accelerated by electrostatic potentials to an energy between 50 and 150 eV before entering the ionization box, where they impact sample gas, forming positively charged particles. The resulting ion beam is repelled by an electrical field toward a flight tube, further accelerated to 3–10 kV and focused with two parallel plates to form a thin beam.

The ion beam then enters the mass analyzer, where a magnetic field perpendicular to the flight path bends the beams (positively charged ions repelled), with the lighter isotope beam bending more than the heavier isotope beam because the radius of curvature is proportional to the square root of the mass-to-charge ratio ($\sqrt{m/z}$). Ions of identical mass are then focused once more by passing through an aperature referred to as the α-split.

The beams produced are broad, flat-topped peaks, thereby making the system insensitive to drift. Broad peaks also ensure optimal capture in the specialized Faraday cups of the <u>ion detector</u>. Current-generation isotope ratio mass spectrometers have three or more Faraday cups, positioned to capture specific masses (e.g., 44, 45, 46) simultaneously. Faraday cups are long, narrow metal tubes that prevent ions or secondary electrons from bouncing out; their internal surfaces are coated with gold or colloidal graphite to reduce secondary electron emission, and they are paired with electron suppressor plates to repel those that are produced. The cups are connected to ground via a high ohmic resistor. The ion current flowing through the resistor creates a voltage that is used as the output from the mass spectrometer. This voltage is fed to the computer system, which converts relative signal strength to a ratio, and then a δ value in parts per thousand relative to a standard (see above).

Technological advances and current trends in the ecological use of isotopes

Until the late 1980s samples of all sorts had to be purified manually, an often complicated and/or tedious process that allowed analysis of only ca. 2–15 samples per day. The advent of CF-IRMS and numerous automated online preparation techniques (e.g., combustion, pyrolysis, cryoseparation) has led to the ability to process ca. 60–200 samples per day, depending on sample type. It should be noted, however, that samples are still purified and converted to a pure gas phase for analysis (CO_2, N_2, H_2, SO_2), but this is now done automatically in the inlet system before a sample enters the mass spectrometer. In addition to faster sample analysis, techniques today require less sample, and so allow molecular-scale studies including the use of ^{13}C-tagged phospholipids fatty acids (PLFAs) to study selective utilization of specific substrates by different functional groups (e.g., fungi) (Boschker et al. 1998; Brant et al. 2006). Smaller sample size and faster sample throughput, in addition to lower per sample cost, has also led to the use of isotopes in regional to global studies that integrate ecosystem processes via modeling (e.g., Ciais et al. 1999; Hobbie et al. 1999). Some examples are presented below of cutting edge, as well as more common techniques, including: gas chromatography combustion-IRMS (GC-C-IRMS) and some of its derivatives, pyrolysis GC-IRMS and ^{13}C-PLFA. Brief mention is also made of stable isotope probing (SIP), secondary ion mass spectrometry (SIMS), ^{13}C-NMR, and tunable diode laser absorption spectroscopy (TDLAS). Several chapters of this book focus on compound-specific techniques, so they will not be covered here beyond the examples cited under pyrolysis GC-IRMS and ^{13}C-PLFA techniques.

Gas chromatography followed by isotope ratio mass spectrometry is a powerful combination of techniques that has allowed a rapid expansion of research in the past decade. Gas chromatography combustion isotope ratio mass spectrometry was first implemented by Sano et al. in 1976, and became commercially available in 1988 (Brand 1996); it allows isotopic identification of individual compounds within a mixture. In GC-C-IRMS a complex sample is injected onto a capillary column gas chromatograph, which separates the compounds in the sample, after which they are combusted to CO_2 or N_2 and fed in a stream of helium into the ion source of the mass spectrometer. Because the heavy isotope elutes first (less interaction with the CG column), each peak is integrated over its entire width to obtain the correct isotope ratio. A working reference is injected into the GC column roughly every tenth sample in the same manner that the samples were injected to standardize the sample runs. This approach has almost supplanted the traditional use of DI-IRMS on purified samples, and has been applied to address a variety of environmental/ecological questions ranging from soil–atmosphere gas exchange (Miller et al. 1999; Menyailo et al. 2003; Pendall et al. 2003) to estimating fire severity and fire-induced loss of carbon (Schuur et al. 2003), to paleoclimate and its rate of change (e.g., Street-Perrott & Huang 1997; Kuypers et al. 1999; Schefuß et al. 2003), to food web analysis (e.g., Ben-David et al. 1997; Beaudoin et al. 1999; O'Brien et al. 2002; Fry & Ewel 2003), among many others. It should be noted that the combustion of solid samples (e.g., animal or plant tissues) is based on the same principles as GC-C-IRMS, but the samples are first combusted in an on-line elemental analyzer and the products then enter a CG column before analysis in the isotope ratio mass spectrometer instead of first entering the gas chromatograph and then being combusted. This now commonly used tool is referred to as either ANCA-MS (automated nitrogen and carbon analyzer mass spectrometry) or EA-IRMS (elemental analyzer isotope ratio mass spectrometry).

A recent extension of GC-C-IRMS is known as pyrolysis GC-IRMS (py-GC-IRMS), in which a complex substrate is thermally degraded under vacuum to form smaller volatile fragments (pyrolysate) which are directed into an isotope ratio mass spectrometer, yielding a unique fingerprint of individual compounds. This technique was first adapted for $\delta^{13}C$ in organic substances in the early 1990s (Goñi et al. 1994) and has been applied successfully to determine molecular turnover times of soil components such as proteins and lignin (Gleixner et al. 1999), and to determine the origin of dissolved organic matter in a bog (Kracht & Gleixner 2000). Because oxygen is not introduced with pyrolysis, online measurements of $\delta^{18}O$ in organic matter are now possible (e.g., Werner et al. 1996; Farquhar et al. 1997). Tobias et al. (1997) adapted py-GC-IRMS for analysis of δD in organic samples, and some laboratories are now simultaneously measuring $\delta^{18}O$ and δD in water (Gehre et al. 2004).

Another derivative of GC-C-IRMS technology useful in ecological studies is a compound-specific method referred to as ^{13}C-phospholipid fatty acid anaylsis (^{13}C-PLFA). Since the late 1990s there have be a number of investigations into substrate usage by different microbial communities using ^{13}C incorporation into fatty acids (e.g., Abraham et al. 1998; Boschker et al. 1998; Hanson et al. 1999; Butler et al. 2004; Londry et al. 2004). Briefly, the method is carried out via an extraction using a $2:1:0.8$ solution of methanol, chloroform, and phosphate buffer followed by filtration, then separation on solid phase extraction columns. Separated phospholipids are saponified and methylated to fatty-acid methyl esters (FAME), which are introduced to a gas chromatograph (GC) where compounds are separated prior to combustion and introduced into an isotope ratio mass spectrometer. The peaks are then compared with known peaks from a reference library to identify the general class of organisms that took up the ^{13}C labeled substrate. For example, preferential incorporation of ^{13}C-labeled glucose into the $18:2\omega6,9$ fatty acid would indicate the fungal population was active in metabolizing this substrate. Other known markers include $20:4\omega6$, a marker for protozoa, and $18:1\omega8c$, a marker for methanotrophs, among many others. Boschker et al. (1998) incubated marine sediments with ^{13}C-labeled acetate and found that none of the 17-carbon PLFAs characteristic of many of the gram-negative sulfate reducers were highly labeled, but several bacterial biomarkers were highly labeled, questioning the importance of the gram-negative group in marine sediments. Despite its great potential, there are currently two shortcomings of the ^{13}C-PLFA method: (i) the identification database is currently limited to ~100 fatty acids; (ii) interpretation is not straightforward because some markers have more than one meaning. For example $17:0cy$, is <u>both</u> a general stress biomarker <u>and</u> a marker for gram-negative bacteria.

Another approach to gain insight into small-scale processes is a suite of methods that involve bombardment of a sample with a primary ion beam to produce secondary ions that are subsequently introduced into a mass spectrometer. One example of such an approach is secondary ion mass spectrometry (SIMS), which is a technique primarily applied in mineralogical studies thus far, but which has potential for use in environmental studies. Secondary ion mass spectrometry uses a Cs^+ primary beam to ablate the surface of a sample. The main advantages of this technique are its high sensitivity and small sample size. However, the sputtering process produces secondary ions that interfere with the atomic ions of interest. Furthermore, the matrix in which the ion of interest resides has a strong influence on the ionization efficiency, making quantitative analysis difficult. Despite the challenges, Orphan et al. (2001) recently used a SIMS approach in combination with fluorescence *in situ* hybridization to identify a single archaeal group responsible for methane consumption in anoxic marine sediments. Their results clearly demonstrate the utility of SIMS for studies of microbial function. A modification of the method, time-of-flight secondary ion mass spectrometry (TOF-SIMS), was recently applied by Cliff et al. (2002) to detect ^{13}C and ^{15}N assimilation by individual bacterial cells, and shows promise as a tool for exploring microsite heterogeneities in soil and other media.

Yet another recent technological advance involving the use of stable isotopes in ecological studies is stable isotope probing (SIP). This technique involves incubation of soils or sediments with ^{13}C-labeled substrates (e.g., acetate, phenol, or methanol), extraction of labeled nucleic acids (DNA or rRNA) by bead beating, and separation of light and heavy fractions using density gradient centrifugation, which occurs because of differences in the buoyant density of ^{13}C- vs. ^{12}C-DNA (Meselson & Stahl 1958). The extracted DNA or rRNA is then amplified using polymerase chain reaction (PCR), and analyzed using molecular techniques. For example, PCR products can be cloned and sequenced to allow phylogenetic characterization of the organisms that took up the labeled substrate by comparison with gene libraries (e.g., GenBank). Radajewski et al. (2002) applied the stable isotope probing approach to identify the active methylotroph populations in an acid forest soil. They unexpectedly found ammonia oxidizing bacteria (*Nitrosomonas* and *Nitrosospira*) incorporated the labeled methanol, highlighting the tight coupling of carbon and nitrogen cycles in these soils. Recent advances in rRNA methods, which require less label and smaller sample size (Lueders et al. 2004), will encourage future application of the SIP technique for soil community analysis.

Solid state cross polarization magic-angle-spinning ^{13}C nuclear magnetic resonance (CP/MAS ^{13}C-NMR) is a tool potentially useful for determining both the chemical structure of organic material and the changes to that structure through the decomposition process. This non-destructive technique is different from the above techniques in that this method does not quantify isotope ratios. It works because nuclei with an odd number of neutrons and/or protons spin, generating a magnetic dipole, which can be induced to resonate. The induced resonance, measured as a chemical shift in the absorption of photons, is a function of the chemical environment of the nuclei (e.g., C—H vs. C—C bonds). Similar ranges of chemical shift (e.g., 0–40 ppm) are characteristic of distinct carbon groups (e.g., alkyls), with signal intensity indicating the proportion of the sample comprised of that carbon group. Chen & Chiu (2003) analyzed separate particle size fractions using CP/MAS ^{13}C-NMR and reported a decrease in aromaticity with a decrease in particle size, which they attributed to high rainfall in their subalpine study area. Keeler & Maciel (2003) cautioned that a substantial fraction of the carbon content of soil samples is missed by CP/MAS ^{13}C-NMR methods, and that the treatment of organic matter to remove interfering paramagnetics (2% HF) can alter the chemical nature of the sample. Thus, it is unclear how useful ^{13}C-NMR analyses will be in analysis of soils and sediments in the next several years.

A very recent development (Bowling et al. 2003) is the use of tunable diode laser absorption spectroscopy (TDLAS) to measure $\delta^{13}C$ of atmospheric CO_2. While still in its infancy, this technique holds great promise for providing high-resolution detail on isotopic fluxes used to constrain models of the global carbon budget. Like ^{13}C-NMR techniques, TDLAS does not involve a mass spectrometer (and so does not measure isotope ratios); however, there the similarity between the two methods ends. The use of a tunable diode laser is possible because it measures absorption of infrared energy, which is proportional to molecular density following Beer's law. The diode laser is tuned to a particular narrow emission band by controlling the temperature and current applied to the laser, and the source wave number is selected to match an individual absorption line on a particular molecule of interest. The sample mol fraction is inferred by comparing sample and reference gas absorbances (note the critical need to keep the two at the same temperature and pressure). Bowling et al. (2003) modified a commercially available tunable diode laser (TDL) to measure $[^{12}CO_2]$ and $[^{13}CO_2]$ simultaneously using a jump-scan technique wherein the laser scans first one isotope and then the other, every 0.1 s. The modified laser was used under field conditions continuously for nearly 5 days, yielding an unprecedented analysis of the temporal variability of the isotopic composition of ecosystem respiration with a precision of 0.3‰ (similar to standard CF-IRMS, although typical resolution of a TDL system is 1‰). The 2-min time-step of TDL is two-orders of magnitude greater sample frequency than has been achieved by the alternate means to measure ecosystem respiration, which involves filling gas flasks and plotting the isotopic composition of CO_2 vs. the inverse of its concentration (i.e., a "Keeling" plot; see, e.g., Yakir & Wang 1996). Recent application of this technique over a 26-day period (Griffis et al. 2004) indicates considerable promise for applying this approach for continuous year-long measurements, whereas all conventional studies to date have only provided snapshots of ecosystem processes.

Acknowledgments

Thanks to the editors for inviting my contribution, to D. Myrold for helpful discussions while co-teaching a course in stable isotopes at Oregon State University, to the "Ehleringer course" (and its new equivalent, the "Högberg course") instructors for access to their lecture notes. Thanks also to Shawna McMahon and Stephanie Boyle for sharing their thoughts on the most relevant ^{13}C-PLFA and "isotope probing" literature, respectively. An earlier draft of this chapter was greatly improved by thoughtful reviews by J. Renee Brooks, Nate McDowell, and an anonymous reviewer. The Agricultural Experiment Station, Oregon State University provided funding.

References

Abraham, W.-R., Hesse, C. & Pelz, O. (1998) Ratios of carbon isotopes in microbial lipids as an indicator of substrate usage. *Applied and Environmental Microbiology*, **64**, 4202–4209.

Aston, F.W. (1922) *Isotopes*. Longmans, Green & Co., London.

Barrie, A. & Prosser, S.J. (1996) Automated analysis of light-element stable isotopes by isotope ratio mass spectrometry. In: *Mass Spectrometry of Soils* (Eds T.W. Boutton & S. Yamaski), pp. 1–46. Marcel Dekker, New York.

Beard, B.L. & Johnson, C.M. (2000) Strontium isotope composition of skeletal material can determine the birth place and geographic mobility of humans and animals. *Journal of Forensic Science*, **September**, 1049–1061.

Beard, B.L., Johnson, C.M., Cox, L., Sun, H., Nealson, K.H. & Aguilar, C. (1999) Iron isotope biosignatures. *Science*, **285**, 1889–1892.

Beaudoin, C.P., Tonn, W.M., Prepas, E.E. & Wassenaar, L.I. (1999) Individual specialization and trophic adaptability of northern pike (*Esox lucius*): an isotope and dietary analysis. *Oecologia*, **120**, 386–396.

Ben-David, M., Flynn, R.W. & Schell, D.M. (1997) Annual and seasonal changes in diets of martens: evidence from stable isotope analysis. *Oecologia*, **111**, 280–291.

Bigeleisen, J. (1965) Chemistry of isotopes. *Science*, **147**, 463–471.

Boschker, H.T.S., Nold, S.C., Wellsbury, P., et al. (1998) Direct linking of microbial populations to specific biogeochemical processes by ^{13}C-labelling of biomarkers. *Nature*, **392**, 801–805.

Bowling, D.R., Sargent, S.D., Tanner, B.D. & Ehleringer, J.R. (2003) Tunable diode laser absorption spectroscopy for stable isotope studies of ecosystem–atmosphere CO_2 exchange. *Agricultural and Forest Meteorology*, **118**, 1–19.

Brand, W.A. (1996) High precision isotope ratio monitoring techniques in mass spectrometry. *Journal of Mass Spectrometry*, **31**, 225–235.

Brand, W.A. (2004) Mass spectrometer hardware for analyzing stable isotope ratios. In: *Handbook of Stable Isotope Analytical Techniques* (Eds W.A. Brand & P. deGroot), pp. 835–856. Elsevier, Amsterdam.

Brant, J.B., Myrold, D.D. & Sulzman, E.W. (2006) Microbial community utilization of added carbon substrates in response to long-term carbon input manipulation. *Soil Biology and Biochemistry*, **38**, 2219–2232.

Bullen, T.D. & Mcmahon, P.M. (1998) Using stable Fe isotopes to assess microbially-mediated Fe^{3+} reduction in a jet-fuel contaminated aquifer. *Minerology Magazine*, **62A**, 255–256.

Butler, J.L., Bottomley, P.J., Griffith, S.M. & Myrold, D.D. (2004) Distribution and turnover of recently fixed photosynthate in ryegrass rhizospheres. *Soil Biology and Biochemistry*, **36**, 371–382.

Capo, R.C., Stewart, B.W. & Chadwick, O.A. (1998) Strontium isotopes as tracers of ecosystem processes: theory and methods. *Geoderma*, **82**, 197–225.

Chen, J.-S. & Chiu, C.-Y. (2003) Characterization of soil organic matter in different particle-size fractions in humid subalpine soils by CP/MAS ^{13}C NMR. *Geoderma*, **117**, 129–141.

Ciais, P., Friedlingstein, P., Schimel, D.S. & Tans, P.P. (1999) A global calculation of the $\delta^{13}C$ of soil respired carbon: implications for the biospheric uptake of anthropogenic CO_2. *Global Biogeochemical Cycles*, **13**, 519–530.

Cliff, J.B., Gaspar, D.J., Bottomley, P.J. & Myrold, D.D. (2002) Exploration of inorganic C and N assimilation by soil microbes with time-of-flight secondary ion mass spectrometry. *Applied and Environmental Microbiology*, **68**, 4067–4073.

Craig, H. (1953) The geochemistry of the stable carbon isotopes. *Geochimica et Cosmochimica Acta*, **3**, 53–92.

Craig, H. (1957) Isotopic standards for carbon and oxygen and correction factors for mass-spectrometric analysis of carbon dioxide. *Geochimica et Cosmochimica Acta*, **12**, 133–149.

Dawson, T.E. & Brooks, P.D. (2001) Fundamentals of stable isotope chemistry and measurement. In: *Stable Isotope Techniques in the Study of Biological Processes and Functioning of Ecosystems* (Eds M. Unkovich, J. Pate, A. McNeill & D.J. Gibbs), pp. 1–18. Kluwer Academic Publishers, Dordrecht.

Farquhar, G.D., Henry, B.K. & Styles, J.M. (1997) A rapid on-line technique for determination of oxygen isotope composition of nitrogen-containing organic matter and water. *Rapid Communications in Mass Spectrometry*, **11**, 1554–1560.

Faure, G. (1986) *Principles of Isotope Geology*. John Wiley & Sons, New York.

Faure, G. (1991) *Principles and Applications of Geochemistry*. Prentice-Hall, New Jersey.

Fry, B. & Ewel, K.C. (2003) Using stable isotopes in mangrove fisheries research: a review and outlook. *Isotopes and Environmental Health Studies*, **39**, 191–196.

Gehre, M., Geilmann, H., Richter, J., Werner, R.A. & Brand, W.A. (2004) Continuous flow $^2H/^1H$ and $^{18}O/^{16}O$ analysis of water samples with dual inlet precision. *Rapid Communications in Mass Spectrometry*, **18**, 2650–2660.

Gleixner, G., Bol, R. & Balesdent, J. (1999) Molecular insight into soil carbon turnover. *Rapid Communications in Mass Spectrometry*, **13**, 1278–1283.

Goñi, M.A. & Eglinton, T.I. (1994) Analysis of kerogens and kerogen precursors by flash pyrolysis in combination with isotope-ratio-monitoring gas chromatography-mass spectrometry (irm-GC-MS). *Journal of High Resolution Chromatography*, **17**, 476–488.

Griffis, T.J., Baker, J.M., Sargent, S.D., Tanner, B.D. & Zhang, J. (2004) Measuring field-scale isotopic CO_2 fluxes with tunable diode laser absorption spectroscopy and micrometeorological techniques. *Agricultural and Forest Meteorology*, **124**, 15–29.

Hanson, J.R., Macalady, J.L., Harris, D. & Scow, K.M. (1999) Linking toluene degradation with specific mictobial populations in soil. *Applied and Environmental Microbiology*, **65**, 5403–5408.

Hobbie, E.A., Macko, S.A. & Shugart, H.H. (1999) Interpretation of nitrogen isotope signatures using the NIFTE model. *Oecologia*, **120**, 405–415.

Hoefs, J. (1997) *Stable Isotope Geochemistry*. Springer-Verlag, New York.

Keeler, C. & Maciel, G.E. (2003) Quantitation in the solid-state ^{13}C NMR analysis of soil and organic soil fractions. *Analytical Chemistry*, **75**, 2421–2432.

Kendall, C. & Caldwell, E.A. (1998) Fundamentals of isotope geochemistry. In: *Isotope Tracers in Catchment Hydrology* (eds C. Kendall & J. J. McDonnell), pp. 51–86. Elsevier Science, Amsterdam.

Kennedy, B.P., Blum, J.D., Folt, C.L. & Nislow, K.H. (2000) Using natural strontium isotopic signatures as fish markers: methodology and application. *Canadian Journal of Fisheries and Aquatic Science*, **57**, 2280–2292.

Kracht, O. & Gleixner, G. (2000) Isotope analysis of pyrolysis products from Sphagnum peat and dissolved organic matter from bog water. *Organic Geochemistry*, **31**, 645–654.

Kuypers, M.M.M., Pancost, R.D. & Sinninghe Dansté, J.S. (1999) A large and abrupt fall in atmospheric CO_2 concentration during Cretaceous times. *Nature*, **399**, 342–345.

Londry, K.L., Jahnke, L.L. & De Marais, D.J. (2004) Stable carbon isotope ratios of lipid biomarkers of sulfate-reducing bacteria. *Applied and Environmental Microbiology*, **70**, 745–751.

Lueders, T., Manefield, M. & Friedrich, M.W. (2004) Enhanced sensitivity of DNA- and rRNA-based stable isotope probing by fractionation and quantitative analysis of isopycnic centrifugation gradients. *Environmental Microbiology*, **6**, 73–78.

Majoube, M. (1971) Fractionnement en oxygène-18 et en deutérium entre l'eau et sa vapeur. *Journal of Chemical Physics*, **197**, 1423–1436.

Menyailo, O.V., Hungate, B.A., Lehmann, J., Gebauer, G. & Zech, W. (2003) Tree species of the central amazon and soil moisture alter stable isotope composition of nitrogen and oxygen in nitrous oxide evolved from soil. *Isotopes in Environmental Health Studies*, **39**, 41–52.

Meselson, M. & Stahl, F.W. (1958) The replication of DNA in *Escherichia coli*. *Proceedings of the National Academy of Sciences*, **44**, 671–682.

Miller, J.B., Yakir, D., White, J.W.C. & Tans, P.P. (1999) Measurement of $^{18}O/^{16}O$ in the soil-atmosphere CO_2 flux. *Global Biogeochemical Cycles*, **13**, 761–774.

O'Brien, D.M., Fogel, M.L. & Boggs, C.L. (2002) Renewable and nonrenewable resources: Amino acid turnover and allocation to reproduction in Lepidoptera. *Proceedings of the National Academy of Sciences*, **99**, 4413–4418.

Orphan, V.J., House, C.H., Hinrichs, K.-U., Mckeegan, K.D. & Delong, E.F. (2001) Methane-consuming Archaea revealed by directly coupled isotopic and phylogenetic analysis. *Science*, **293**, 484–487.

Pendall, E., Del Grosso, S., King, J.Y., et al. (2003) Elevated atmospheric CO_2 effects and soil water feedbacks on soil respiration components in a Colorado grassland. *Global Biogeochemical Cycles*, **17**, doi:10.1029/2001GB001821.

Perakis, S.S., Maguire, D.A., Bullen, T.D., Cromack, K., Jr., Waring, R.H. & Boyle, J.R. (2006) Coupled nitrogen and calcium cycles in forests of the Oregon Coast Range. *Ecosystems*, **9**, 63–74.

Radajewski, S., Webster, G., Reay, D.S., et al. (2002) Identification of active methylotroph populations in an acidic forest soil by stable isotope probing. *Microbiology*, **148**, 2331–2342.

Sano, M., Yotsui, Y., Abe, H. & Sasaki, S. (1976) A new technique for the detection of metabolites labeled by the isotope ^{13}C using mass fragmentography. *Biomedicine and Mass Spectrometry*, **3**, 1–3.

Schefuß, E., Schouten, S., Jansen, J.H.F. & Sinninghe Dansté, J.S. (2003) African vegetation controlled by tropical sea surface temperatures in the mid-Pleistocene period. *Nature*, **422**, 418–421.

Schuur, E.A.G., Trumbore, S.E., Mack, M.C. & Harden, J.W. (2003) Isotopic composition of carbon dioxide from a boreal forest fire: Inferring carbon loss from measurements and modeling. *Global Biogeochemical Cycles*, **17**, 1001, doi:10.1029/2001GB001840.

Street-Perrott, F.A. & Huang, Y. (1997) Impact of lower atmospheric carbon dioxide on tropical mountain ecosystems. *Science*, **278**, 1422–1426.

Tobias, D.J., Tu, K. & Klein, M.L. (1997) Atomic-scale molecular dynamics simulations of lipid membranes. *Current Opinions in Colloid and Interface Science*, **2**, 15–26.

Werner, R.A., Kornexl, B.E., Roßmann, A. & Schmidt, H.-L. (1996) On-line determination of $\delta^{18}O$ values of organic substances. *Analytica Chimica Acta*, **319**, 159–164.

Yakir, D. & Wang, X.F. (1996) Fluxes of CO_2 and water between terrestrial vegetation and the atmosphere estimated from isotope measures. *Nature*, **380**, 515–517.

CHAPTER 2

Sources of variation in the stable isotopic composition of plants*

JOHN D. MARSHALL, J. RENÉE BROOKS,
AND KATE LAJTHA

Introduction

The use of stable isotopes of carbon, nitrogen, oxygen, and hydrogen to study physiological processes has increased exponentially in the past three decades. When Harmon Craig (1953, 1954), a geochemist and early pioneer of natural abundance stable isotopes, first measured isotopic values of plant materials, he found that plants tended to have a fairly narrow $\delta^{13}C$ range of −25 to −35‰. In these initial surveys, he was unable to find large taxonomic or environmental effects on these values. Since that time ecologists have identified clear isotopic signatures based not only on different photosynthetic pathways, but also on ecophysiological differences, such as photosynthetic water-use efficiency (WUE) and sources of water and nitrogen used. As large empirical databases have accumulated and our theoretical understanding of isotopic composition has improved, scientists have continued to discover mismatches between theoretical and observed values, as well as confounding effects from sources and factors not previously considered. In the best tradition of science, these discoveries have led to important new insights into physiological or ecological processes, as well as new uses of stable isotopes in plant ecophysiology. This chapter reviews the most common applications of stable isotope analysis in plant ecophysiology.

Carbon isotopes

Photosynthetic pathways

Plants contain less ^{13}C than the atmospheric CO_2 on which they rely for photosynthesis. They are therefore "depleted" of ^{13}C relative to the atmosphere. This depletion is caused by enzymatic and physical processes that discriminate against ^{13}C in favor of ^{12}C. Discrimination varies among plants using different photosynthetic pathways. The Calvin cycle (C3), Hatch–Slack

*Disclaimer: This document has been reviewed in accordance with U.S. Environmental Protection Agency policy and approved for publication. Mention of trade names or commercial products does not constitute endorsement or recommendation for use.

cycle (C4) and Crassulacean acid metabolism (CAM) photosynthetic pathways differ so profoundly and so consistently (O'Leary 1981, 1988) that ecologists have used isotopic signatures to distinguish them in large-scale surveys of plant species (Teeri & Stowe 1976; Sage & Monson 1999).

The C3 pathway begins with the diffusion of CO_2 from the atmosphere into the air-filled spaces within the leaf. This diffusion occurs through the still air occupying stomatal pores. Such diffusion has an apparent fractionation ($\Delta\delta$) of ~4.4‰ due to the slower motion of the heavier ^{13}C-containing CO_2 molecules. Within the leaf, the carboxylating enzyme ribulose bisphosphate carboxylase/oxygenase (rubisco) discriminates further against the ^{13}C, with a $\Delta\delta$ of about 29‰. If atmospheric diffusion were the sole limitation for CO_2 uptake, i.e., if rubisco did not discriminate against ^{13}C, then we would expect to see only the fractionation of 4.4‰. This 4.4‰ would be subtracted from the $\delta^{13}C$ value for CO_2 in the atmosphere, which is about −8‰, yielding a $\delta^{13}C$ of −12.4‰. At the opposite extreme, if enzyme activity were the sole limitation for CO_2 uptake, i.e., if diffusion did not discriminate, then only the rubisco fractionation would be observed. These conditions would yield a predicted leaf $\delta^{13}C$ value of about −37‰. In fact, $\delta^{13}C$ values for C3 plants lie between these extremes, with a median of about −27‰. Variation about this median depends on the balance between diffusive supply and enzymatic demand for CO_2. Estimates of this balance point have been exploited by ecologists as an index of water-use efficiency (see following section).

Isotopic composition of C4 plants differs substantially from that of C3 plants. The initial step in C4 photosynthesis is the same: the diffusion of CO_2 from the atmosphere into the leaf via stomata. However, C4 photosynthesis is catalyzed by a different enzyme, phosphoenolpyruvate (PEP) carboxylase, which has a different discrimination, approximately −6‰, for the fixation of CO_2 (Farquhar 1983). If this enzymatic fractionation were fully and exclusively expressed relative to atmospheric CO_2, it would yield tissue values around −2‰. Diffusion-limited uptake would be the same as that for C3 plants, namely −12.4‰. One might expect real C4 plants to lie between these extremes, analogous to C3 plants. In fact, measured $\delta^{13}C$ values for C4 plants lie below this range, clustering around −14‰.

These surprisingly negative values result from the unique physiology of C4 photosynthesis. The C4 compounds produced by PEP carboxylase are transported into the bundle sheath, which is the cylinder of vascular tissue enclosed in the center of the leaf. Inside the bundle sheath, the C4 compounds are catabolized to C3 compounds, releasing CO_2, which accumulates to high concentrations. The released CO_2 is then refixed by rubisco, the same enzyme used by C3 photosynthesis. The surprisingly negative $\delta^{13}C$ values are caused by a slow leak of enriched CO_2 from the bundle sheath. The leaking CO_2 pool is enriched in ^{13}C by the preference of rubisco for the light isotope. As the enriched CO_2 leaks out, it depletes the $\delta^{13}C$ of the CO_2 left behind

(Ehleringer & Pearcy 1983; Berry 1989). Support for this mechanism comes from evidence that C4 plants with the most leaky bundle sheaths tend to exhibit the most negative $\delta^{13}C$ (Hattersley 1982; Henderson et al. 1992; Sandquist & Ehleringer 1995). These interacting controls over C4 isotope fractionation serve as an example of the sometimes complex metabolic feedbacks influencing stable isotope composition.

Plants using the CAM photosynthetic pathway rely on the same carboxylating enzymes as C4 plants and in the same sequence. However, they segregate the activities of the enzymes between night and day. Initially, CO_2 is fixed by PEP carboxylase into C4 acids at night; the acids are stored in the vacuoles of the bulky leaves and stems (e.g., in cacti). When the sun comes up, the carbon is released from the C4 acids and refixed by rubisco. Because the enzyme sequence is similar, obligate-CAM plants discriminate against ^{13}C much as C4 plants do (Farquhar 1983), though perhaps with less leakage. Their $\delta^{13}C$ values cluster around −11‰. Some species use the CAM pathway facultatively, conducting C3 photosynthesis when conditions are favorable, switching to the CAM sequence when drought strikes. Facultative-CAM plants have $\delta^{13}C$ values intermediate between −11‰ and those of obligate-C3 plants (−27‰). These intermediate carbon isotope signatures can be used to estimate the proportion of CAM vs. C3 photosynthesis during the period when the tissue was produced (Osmond et al. 1976; Teeri & Gurevitch 1984; Kalisz & Teeri 1986; Smith et al. 1986; Mooney et al. 1989; Kluge et al. 1991).

Carbon isotopes cannot effectively distinguish CAM from C4 plants because of their similarity. However, empirical evidence suggests that CAM plants are considerably enriched in deuterium relative to source water, providing yet another means of distinguishing them from C4 plants (Figure 2.1; Ziegler et al. 1976; Sternberg & DeNiro 1983; Sternberg et al. 1984a,b,c; Sternberg 1989). Traditionally the distinction between C4 and CAM has relied on anatomical methods, specifically succulence (CAM) vs. Kranz anatomy (C4).

Plant water-use efficiency

Ecologists also use carbon isotope ratios to infer photosynthetic water-use efficiency (WUE). Traditionally WUE has been defined as the ratio of net photosynthesis to transpiration (A/E). Farquhar et al. (1982) demonstrated that $\delta^{13}C$ in C3 plant tissues provides a reliable index of water-use efficiency because both WUE and $\delta^{13}C$ are controlled by intercellular CO_2 levels. The relationship can be described as:

$$\delta^{13}C_{leaf} = \delta^{13}C_{atm} - a - (b - a)\, c_i/c_a \qquad (2.1)$$

where $\delta^{13}C_{atm}$ is −8.1‰ at this writing (http://cdiac.esd.ornl.gov/trends/co2/iso-sio/data/iso-data.html); it continues to fall as fossil-fuel sources of C are

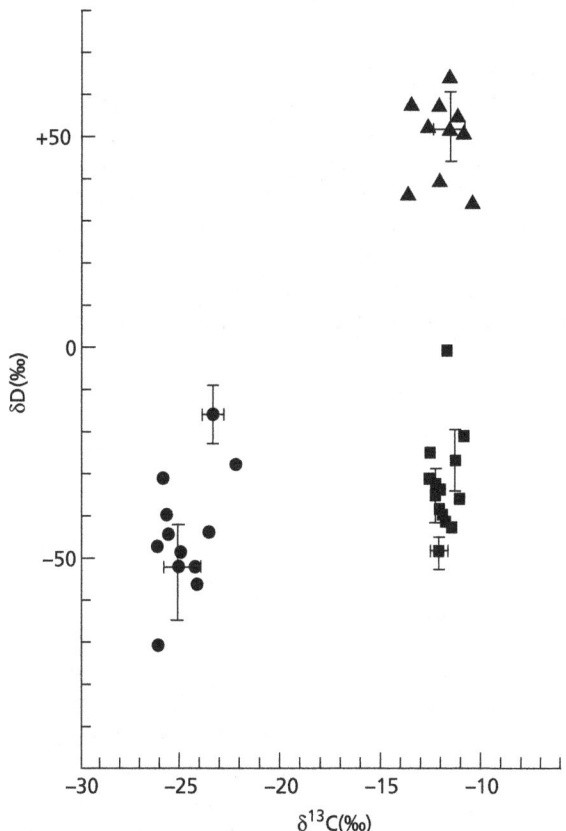

Figure 2.1 Hydrogen vs. carbon isotope ratios of plant cellulose nitrate for plants having different photosynthetic pathways. Squares represent C4, circles C3, and triangles Crassulacean acid metabolism. (From Sternberg et al. 1984a.)

added to the atmosphere. Term a is the fractionation caused by diffusion (4.4‰), b is the fractionation associated with carbon dioxide fixation (27‰), and c_i/c_a is the ratio of intercellular to ambient concentrations of CO_2. This form of the equation has been simplified to enhance utility; it is easily parameterized and adequate for many applications. We discuss its limitations and present a more complex form of the equation below.

It was noted earlier that $\delta^{13}C$ is also related to WUE. Using traditional symbols to describe photosynthetic gas exchange, we describe these relationships as:

$$A = (c_a - c_i)g/1.53 \tag{2.2}$$

$$E = g(\text{LAVD}) \tag{2.3}$$

$$\text{WUE} = A/E = (c_a - c_i)/[1.53(\text{LAVD})] \tag{2.4}$$

where A is net photosynthetic rate, E is transpiration rate, g is stomatal conductance to water vapor, 1.53 is the ratio of diffusivities of water vapor and CO_2 in air (Campbell & Norman 1998), and leaf-to-air vapor difference (LAVD) is the difference in water vapor concentration between the interior of the leaf and the surrounding atmosphere (Farquhar & Richards 1984). Because c_a is nearly constant in the global troposphere within a given year, WUE varies primarily with c_i and LAVD. If LAVD can be assumed constant among the plants being considered at a given site, then plant $\delta^{13}C$ will be linearly correlated with WUE. However, LAVD is not always constant, especially when one compares across leaves that differ in morphology and function (e.g., Goldstein et al. 1989). When LAVD cannot be assumed constant or is unknown, many ecologists instead infer A/g, termed the intrinsic water-use efficiency, from stable isotope data:

$$A/g = (c_a - c_i)/1.53 \tag{2.5}$$

where the terms have already been defined above and the result is no longer dependent on assumed LAVD.

The relationship between $\delta^{13}C$ and WUE can be understood intuitively if one considers the behavior of CO_2 molecules as they enter a photosynthesizing leaf. The CO_2 molecules diffuse down a concentration gradient into the leaf. The CO_2 is diffusing against a countervailing diffusive flux of water vapor out of the leaf due to transpiration. Partial closure of the stomata reduces stomatal conductance, which reduces both gas fluxes – but not equally. The reduction in transpiration is proportional to the reduction in stomatal conductance because the water vapor concentration gradient is unaffected by stomatal closure. The air within the leaf remains saturated, the air outside the leaf remains at ambient, and the only change is in the ability of the water molecules to diffuse down this gradient. Net photosynthetic rates also decline, but the decline is less than proportional. This smaller decline occurs because the photosynthetic consumption of CO_2 within the leaf continues at a high rate even as the diffusive transport of CO_2 into the leaf is reduced. The resulting shift in the balance between diffusive supply and biochemical demand reduces c_i, offsetting some of the loss in photosynthesis that would otherwise have resulted from stomatal closure. Thus, the decline in net photosynthesis is less than the decline in transpiration, and water-use efficiency increases. Such a reduction in c_i would also modify the isotopic composition of photosynthate. As c_i falls, the $\delta^{13}C$ of the CO_2 inside the leaf is progressively enriched and the photosynthate produced is likewise enriched. The observed correlation between $\delta^{13}C$ and WUE thus results from independent, but correlated responses of these variables to falling c_i. As C4 plants show no evidence of variation in the balance between enzymatic assimilation and diffusion through the stomata, their $\delta^{13}C$ cannot be interpreted in terms of water-use efficiency.

A particular advantage of using $\delta^{13}C$ to estimate WUE is its long integration time. Closed-system gas-exchange techniques generally measure A/E on an instantaneous (15–30 s) basis, and on a few leaves. They can be programmed to integrate over a longer time course (Field et al. 1989), but the measurement itself represents a period of seconds. Carbon isotope analysis provides an estimate of WUE integrated over the time during which carbon in the plant was fixed, often weeks to months, and can do so for large numbers of independent samples.

Several complications have arisen related to integration time. The integration time depends on the time course of carbon emplacement, which varies during a tissue's lifetime. Meinzer et al. (1992) found that the integration heavily favors the period during which the leaf expanded. Another complication is that early stages of leaf construction are sometimes fueled by "heterotrophic" photosynthate (produced elsewhere in the plant). This imported photosynthate would contaminate the "autotrophic" carbon signal, and would need to be corrected to estimate WUE (Terwilliger et al. 2001). Finally, in evergreen leaves, small amounts of carbon continue to be added in the years after the leaves are first produced, which may similarly interfere with the interpretation of the isotopic signal (Hobbie et al. 2002).

Integration time may also vary on a shorter time scale, as where diel gas-exchange patterns differ. Net photosynthesis might sample different LAVD, e.g., if a leaf were active only during cool portions of the day, when LAVD is low. This would reduce leaf temperature (and therefore LAVD) during periods of carbon gain, increasing WUE. Such a pattern has been observed in a comparison of coastal to continental genotypes of Douglas-fir (Figure 2.2; Zhang et al. 1993). Inland genotypes maintained high conductance and high transpiration in mid-afternoon, despite high LAVD. In contrast, coastal genotypes closed their stomata almost completely in mid-afternoon, reducing gas exchange almost to zero. Average LAVD, weighted by net photosynthetic rates, was therefore different, as was WUE. These differences in diurnal pattern would change WUE, but would not be reflected in $\delta^{13}C$.

These caveats related to integration time are usually minor and are often viewed as advantages of the isotopic technique. They are presented here only to ensure that isotopic data are used knowledgeably. The use of $\delta^{13}C$ analysis to estimate long-term WUE has now been routine for more than a decade (Ehleringer 1989, 1991; Ehleringer & Osmond 1989; Farquhar et al. 1989a,b).

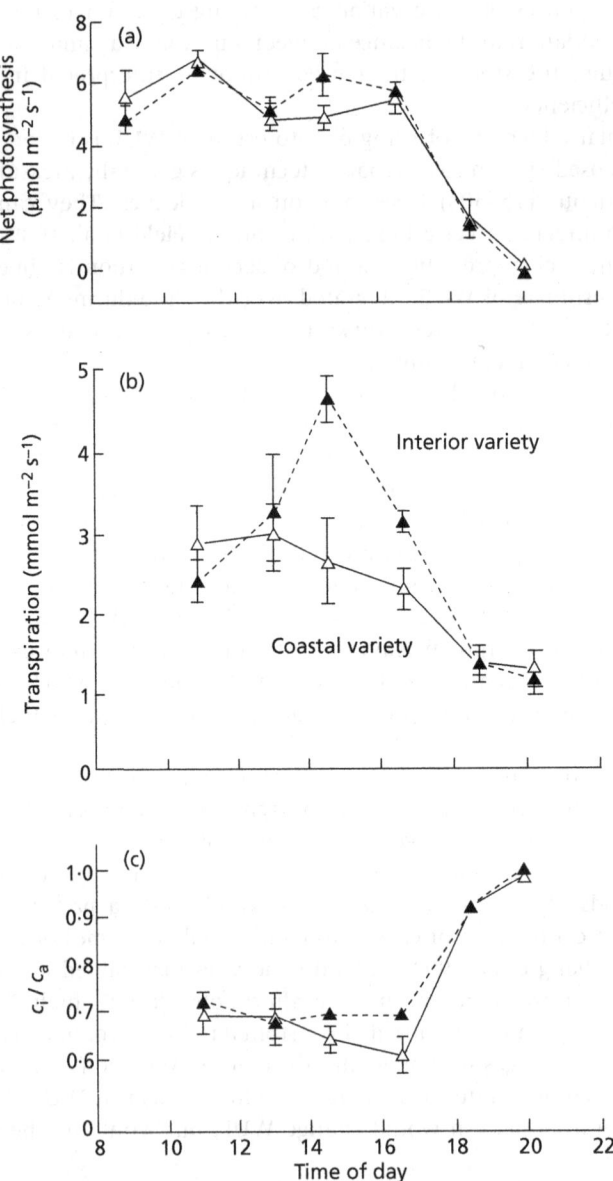

Figure 2.2 Diel patterns of (a) net photosynthesis (A), (b) transpiration (E), and (c) c_i/c_a from an experiment comparing the coastal and interior varieties of Douglas-fir in a common garden. (From Zhang et al. 1993.)

Landscape and population patterns of plant $\delta^{13}C$

Soon after the isotopic method of estimating WUE became available, ecologists began to gain important insights from surveys of carbon isotope ratios in natural ecosystems. Much of this research focused on water gradients in desert ecosystems and altitude gradients in montane ecosystems. A survey of co-occurring species found that short-lived annual or herbaceous species had significantly lower $\delta^{13}C$ values than long-lived perennial species (Smedley et al. 1991). Species active during the wetter, more favorable months discriminated against ^{13}C more than did species that persisted over dry seasons, reflecting a lower WUE in the less drought-tolerant species. Desert plants that spanned a gradient from wash (an intermittent streambed) to a drier upland slope expressed higher WUE in the drier upland habitat (Ehleringer & Cooper 1988) than in the wash. This same pattern was observed in a survey of forest tree species in the mesic Appalachian Mountains (Garten & Taylor 1992).

In global surveys of $\delta^{13}C$ over altitudinal gradients, Körner et al. (1988, 1991) found that $\delta^{13}C$ increased with altitude. Körner et al. (1991) suggested that decreased oxygen partial pressures inhibited photorespiration, decreasing c_i. Morecroft & Woodward (1990) attributed the altitudinal gradient primarily to temperature effects on gas exchange, based on extrapolations from controlled-environment studies. Similarly consistent altitude gradients were observed in the Rocky Mountainst (Marshall & Zhang 1994). Because of shifts in $\delta^{13}C$ and LAVD, WUE increased threefold over 2000 m of altitude (Figure 2.3).

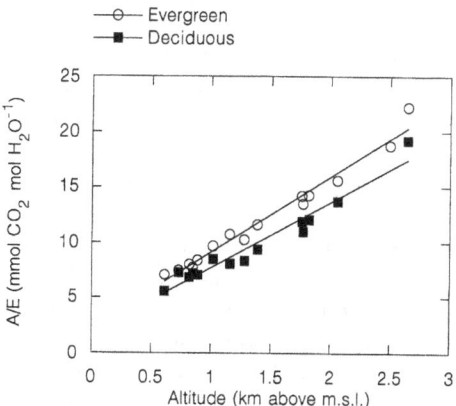

Figure 2.3 Water-use efficiency (A/E) of evergreen and deciduous species occurring along an altitude gradient in the northern Rocky Mountains. (From Marshall & Zhang 1994.)

Some of the variation observed in these surveys has a genetic basis. At the species level, there are consistent differences at a site that can be related to leaf morphology (Marshall & Zhang 1994). Deciduous species are generally more depleted in ^{13}C (have more negative $\delta^{13}C$) than evergreens. Among evergreens, scale-leaved species (e.g., members of the Cupressaceae) are most enriched (less negative $\delta^{13}C$).

Further genetic variation is observed within species. Populations within a species often vary in $\delta^{13}C$ when planted into common environments. Such population differences have been studied especially in dryland crops, rangeland grasses, and forest trees. In most cases, the isotope data have compared well with other methods of estimating WUE and the results have been remarkably consistent across environments (Farquhar & Richards 1984; Condon et al. 1987; Hubick et al. 1988; Ehleringer et al. 1990; Johnson et al. 1990; Ehdaie et al. 1991; Read et al. 1991; Zhang et al. 1993, 1994). Because these measurements are made with a single species in a common environment, variation in temperature and LAVD are minimized and the isotopic differences are readily interpreted as WUE differences. Reported heritabilities are often high (Hubick et al. 1988; Farquhar et al. 1989b; Hall et al. 1990; Ehdaie et al. 1991), suggesting that WUE is a trait with strong genetic control, and for which breeding programs could readily be designed. The application of molecular genetic techniques to map and analyze the genetic basis of variation in WUE provides opportunities for deeper insights into this trait.

These surveys sometimes found variation that was not explained by the simple model presented earlier (equation 2.1). For example, Vitousek et al. (1990) found that foliar $\delta^{13}C$ of *Metrosideros polymorpha* increased with elevation on wet lava flows but not dry lava flows on the Mauna Loa volcano. But these patterns were not correlated with c_i, as measured by leaf gas-exchange. They were, however, correlated with specific leaf weight, or leaf mass per area, a measure of leaf thickness and density. Specific leaf weight increased significantly with elevation. The authors hypothesized that this discrepancy might be attributed to high internal resistance to CO_2 diffusion within the leaves. This resistance might explain the discrepancy because gas-exchange techniques estimate c_i in the substomatal cavities, just beneath the leaf surface. In contrast, stable carbon isotope ratios are deter-mined by the CO_2 concentration in the chloroplasts (c_{ch}), deep within the leaves. In the thick *Metrosideros* leaves at the high elevation sites, c_{ch} was substantially lower than c_i and therefore WUE and $\delta^{13}C$ were decoupled. This observation has subsequently been confirmed several times, particularly in thick-leaved species (Lloyd et al. 1992; Hultine & Marshall 2000; Warren et al. 2003) . This decoupling, expressed as a varying offset from the equation relating $\delta^{13}C$ to WUE, might be expected in any species with high leaf mass per area.

A detailed model of isotope fractionation in plants

In their detailed model, Farquhar et al. (1982) divided the path of photosynthetic CO_2 transfer into four distinct parts (boundary layer, stomata, transfer from stomata to chloroplast, and chloroplast) and used the following equation to describe each:

$$\delta^{13}C_{leaf} = \delta^{13}C_{atm} - a_b(p_a - p_s)/p_a - a(p_s - p_i)/p_a$$
$$- (e_s + a_i)(p_i - p_c)/p_a - b(p_c/p_a) + (eR_d/k + f\Gamma^*)/p_a \qquad (2.6)$$

where p represents partial pressures, which are closely related to concentrations, and subscripts s, i, and c designate measurements made at the leaf surface, in the intercellular space within the leaf, and within the chloroplast, respectively. The term a_b is the fractionation caused by the mixture of diffusion and convection in the boundary layer (2.9‰); e_s and a_i are the fractionations due to dissolution into and diffusion through water, respectively; e and f are the fractionations due to dark respiration and photorespiration, respectively; R_d is the dark respiration rate, k is the carboxylation efficiency, and Γ^* is the the CO_2 compensation point in the absence of dark respiration. The simplified form of the equation presented earlier includes only the first, third, and fifth terms in this equation. The second term, $a_b(p_a - p_s)/p_a$, accounts for the discrimination as CO_2 moves across the leaf boundary layer. When the boundary layer is thin, as at high windspeeds, in rough canopies, or over narrow leaves, this term is negligible. The fourth term, $(e_s + a_i)(p_i - p_c)/p_a$, accounts for the fractionation that occurs as gas-phase CO_2 dissolves in the cell wall and then diffuses through liquid water to the chloroplast. Because this discrimination is small and relatively constant (<1%) it is often neglected as well.

The sixth term has been the focus of intense research since the model was first published. The fractionation due to respiration (e) has been addressed in a series of recent papers describing the difference between respiratory substrates and the CO_2 produced. Protoplasts extracted from plant tissue and forced to deplete a substrate *in vitro* show no fractionation (Lin & Ehleringer 1997). However, intact plants nearly always fractionate, with CO_2 produced in respiration enriched by 2–7‰ relative to their substrate (Duranceau et al. 1999; Ghashghaie et al. 2001; Tcherkez et al. 2002; Ocheltree & Marshall 2004). If total respiration consumes on the order of 50 percent of photosynthate, this release of ^{13}C-enriched CO_2 would significantly deplete the $\delta^{13}C$ of the tissue produced. In fact, plant tissue is often depleted in ^{13}C relative to the carbohydrate pools from which it is produced (Park & Epstein 1961; Ghashghaie et al. 2003; Ocheltree & Marshall 2004).

Photorespiratory fractionation (f) has been difficult to measure, but may be significant; estimates range from 0 to 9‰, with most estimates on the higher end of this range (Gillon & Griffiths 1997; Ghashghaie et al. 2003). It is straightforward to eliminate photorespiration experimentally by lowering the oxygen concentration. The difficulty is in measuring the fractionation; photorespiratory CO_2 is produced within the chloroplast; where much of the CO_2 is immediately recycled in photosynthesis rather than leaving the leaf. Likewise, the fractionated dihydroxyacetone phosphate (DHAP) is recycled into ribulose and then consumed in photosynthesis. If such recycling were complete, no fractionation would be observed (Ghashghaie et al. 2003). However, photorespiration leaves its signature in another unique pattern; the C4 of the glucose molecule would be expected to vary with photorespiratory activity (Hobbie & Werner 2004). This insight may help to constrain estimates of apparent photorespiratory discrimination in the future.

There are many opportunities for respired CO_2 to be refixed in chlorophyllous tissues other than leaves. These tissues include fruits, floral parts, twigs, and in some cases, large-diameter stems (Aschan & Pfanz 2003). Refixed C can be quite depleted because it begins at the $\delta^{13}C$ of photosynthate and is further depleted by the fractionation due to rubisco (Cernusak et al. 2001). Although refixation does not constitute a net gain of C, it can eliminate as much as 100 percent of the diffusive loss from respiring tissues. Shading experiments show that the refixed C is incorporated into biosynthetic pathways, decreasing the $\delta^{13}C$ of the tissues in which it occurs (Cernusak et al. 2001).

A remaining source of variation in gas exchange is the transfer conductance, sometimes referred to as the mesophyll conductance, which was described earlier. The existence of a diffusive resistance across the leaf was recognized in the fourth term of equation 2.6 (Farquhar et al., 1982). In the simplified form of the equation, c_i is used to predict $\delta^{13}C$, though c_i is not at the end of the gaseous diffusion path across the leaf. The transfer conductance, which describes the remainder of the diffusion path to the chloroplast, has been related to chloroplast surface area (Evans et al. 1994). In the leaves of Douglas-fir, the transfer conductance is similar in magnitude to stomatal conductance, and must therefore be accounted for. Fortunately, it is uniform throughout the canopy (Warren et al. 2003).

One last source of variation is related to the fixation of CO_2 by enzymes other than rubisco. Even C3 plants fix small amounts of CO_2 using PEP carboxylase, the enzyme we associate with the first steps in C4 and CAM photosynthesis. In C3 plants, PEP carboxylase activity is usually small in comparison with rubisco activity, but the isotopic fractionation is so different (29‰ vs. −6‰) that a small amount of PEP carboxylase activity could significantly influence $\delta^{13}C$. Brugnoli et al. (1988) speculated that variation in b (equation 2.1) might be caused by differences in the proportion of CO_2 fixed by these two enzymes.

The observed correlation between – gas-exchange measurements derived from an infrared gas analyzer (IRGA) and carbon isotope ratios was in part the basis upon which Farquhar et al. (1982) constructed their theory. The term b in equation 2.1, which describes photosynthetic discrimination, was parameterized by fitting a regression equation. If one were to construct the theory from its mechanistic basis upward, the term b would account for discrimination due to dissolution of CO_2 in water, diffusion of CO_2 through water, and rubisco; its value would be around 29–30‰ (Roeske & O'Leary 1984; Guy et al. 1987). In the simplified form of the equation presented above the empirically derived b-value is about 27‰. This difference is due to the numerous minor influences on carbon isotope ratios discussed above (Farquhar et al. 1982). The variable magnitudes of these influences would lead to variation in this parameter.

Source of carbon

This chapter has focused so far on physiological processes that modify $\delta^{13}C$. Another source of variation in plant $\delta^{13}C$ is the source of CO_2 that they use for photosynthesis, which we have designated $\delta^{13}C_{atm}$, the first term in equation 2.1. These issues were first raised in dense, closed-canopy forests, where foliar $\delta^{13}C$ often increases with canopy height (Vogel 1978; Medina & Minchin 1980; Medina et al. 1986, 1991; Sternberg et al. 1989). Such a vertical gradient in foliar $\delta^{13}C$ might be due to changes in $\delta^{13}C_{atm}$. In fact, several authors have found only limited vertical gradients in the isotopic composition of air from the canopy to the forest floor during the daytime when surface heating supports rapid air mixing. In these systems, the refixation of soil CO_2 is limited to a narrow zone near the soil surface. Respired CO_2 represented 18 percent of total CO_2 at 0.5 m above the forest floor in a tropical forest (Sternberg et al. 1989) and was substrate for 5–6 percent of total canopy photosynthesis in a boreal conifer forest (Brooks et al. 1997). This suggests that even in dense closed-canopy forests, estimates of WUE from foliar $\delta^{13}C$ do not need to be corrected for vertical changes in $\delta^{13}C_{atm}$, except perhaps for leaves produced near the soil surface.

More recently, source CO_2 has become particularly important in studies of CO_2 enrichment. Here the effects are often pronounced. Experimental enrichment of CO_2 is generally accomplished by adding fossil-fuel-derived carbon dioxide, which is depleted in ^{13}C, to ambient carbon dioxide, which is relatively enriched in ^{13}C. The CO_2 enrichment significantly leads to ^{13}C depletion relative to ambient (Ellsworth 1999). Few researchers have had the foresight and the funding to determine $\delta^{13}C$ of atmospheric CO_2 as frequently as necessary to characterize it well in these experiments. Several solutions have been developed. Ellsworth (1999) inferred $\delta^{13}C$ from CO_2 concentration. An alternate has been the deployment of C4 plants within the experimental atmosphere (Marino & McElroy 1991; Beerling 1999).

In long-term studies, variation in atmospheric $\delta^{13}C$ must be accounted for. The atmospheric concentration of CO_2 has increased by 30 percent over the past 100 years, due primarily to fossil fuel combustion (Figure 2.4). This rapid

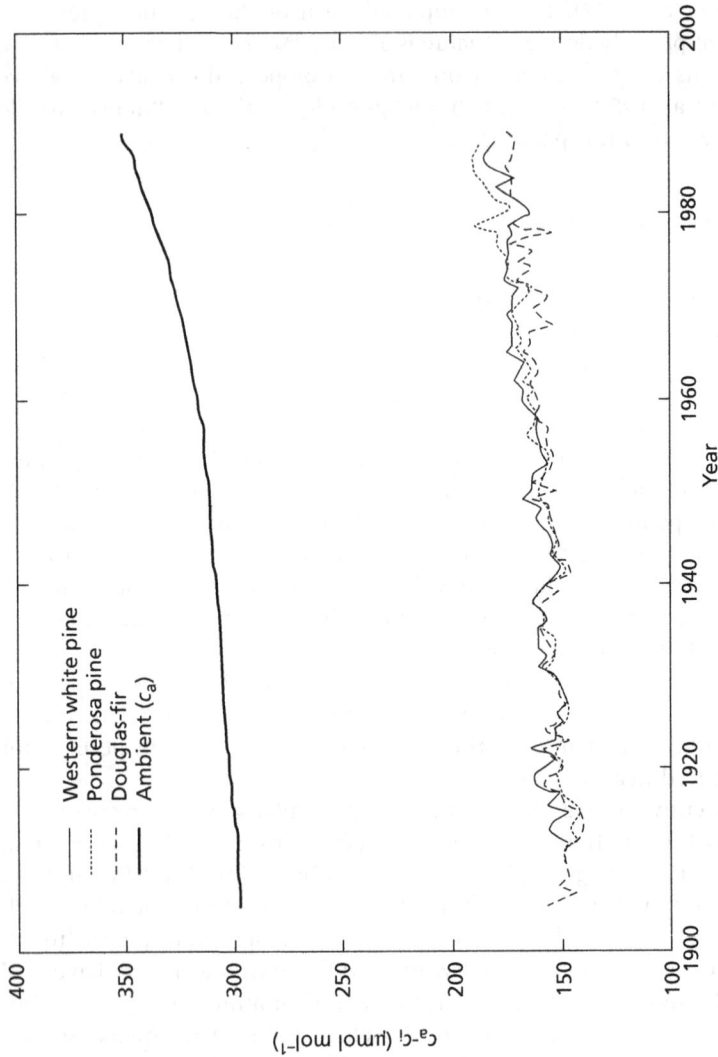

Figure 2.4 Ambient CO_2 concentration (c_a) and time series for internal leaf concentration (c_i) by species. The difference $c_a - c_i$ is proportional to intrinsic water-use efficiency. Estimates are based on $\delta^{13}C$ measured in ice cores and tree rings, respectively. (From Monserud & Marshall 2001.)

change must be accounted for in multi-decade tree-ring studies (Freyer 1979; Marshall & Monserud 1996; Feng 1998; Monserud & Marshall 2001). The atmosphere is so well mixed that global averages can often be used to infer $\delta^{13}C_{atm}$ (http://cdiac.esd.ornl.gov/trends/co2/iso-sio/data/iso-data.html).

Variation among tissues and compounds

Stable carbon isotope composition also varies among plant tissues (O'Leary 1981). Some of this variation is due to differences among the chemical components of plant tissue. Lipids can be as much as 10‰ lighter than the whole tissue (O'Leary 1981; Badeck et al. 2005). In contrast, cellulose and other carbohydrates are typically 1–2‰ heavier than whole tissue (e.g., Leavitt & Long 1986) and lignin is typically 1–2‰ lighter. In detailed work, sample variation can be reduced by extracting a single compound or class of compounds. Because cellulose is the most abundant single compound in plant tissue and its structure is homogeneous, it is frequently used where such control of variation is necessary (Park & Epstein 1961; Freyer 1979).

Isotopic composition of tree rings is often enriched by 1.5–2‰ relative to foliage (Leavitt & Long 1986). This difference is important because 2‰ would represent a shift of over 25 percent if one were to use it to infer water-use efficiency of most C3 plants. It is presently unclear which tissue better reflects the $\delta^{13}C$ of canopy photosynthate or how these differences come about. Several possible mechanisms have been proposed to explain the discrepancy, including fractionation of the phloem contents with vertical distance (Damesin & LeLarge 2003) and enrichment of cellulose due to the production of lignin (Hobbie & Werner 2004). This mechanism needs to be identified before stable carbon isotope data from tree rings can be used to generate reliable gas-exchange inferences. Once resolved, it should be possible to use stable carbon isotope data to parameterize gas-exchange algorithms in ecosystem models (Katul et al. 2000).

Nitrogen isotopes

Although the use of natural abundance ratios of N in plants is not as well established as that of C isotope ratios, there is a great deal to be learned from a comparison of $\delta^{15}N$ among plants within an ecosystem, between plants and their source of N, and among plant components. Many studies have assumed that plant $\delta^{15}N$ reflected the bulk N source, but this assumption has not been universally supported. Differences between N source and plant isotope signatures has been an effective tool in understanding plant nutrient physiology. The physiological mechanisms that influence plant N isotopic signatures have been most recently reviewed by Evans (2001), and earlier by Högberg (1997) and Handley & Raven (1992). Some key applications using $\delta^{15}N$ in plant tissues include assessing contributions of various N sources to plant N uptake in the field, including symbiotic nitrogen fixation and atmospheric deposition, the role of mycorrhizal infection, uptake of dissolved N, and the interpretation of $\delta^{15}N$ profiles in soils. Two other chapters in this book discuss specific applications of the use of $\delta^{15}N$ natural abundance measurements in ecology and environmental science (see Garten et al. and Evans); we will limit the discussion here to natural abundance measurements in plant tissues.

Sources of plant nitrogen

Although a number of pot culture studies have reported significant discrimination between plant tissues and the N in solution, there is general agreement that discrimination is only observed when plant N demand is low compared with N supply (Evans et al. 1996; Högberg et al. 1999). Although nitrate reductase (the enzyme responsible for nitrate assimilation and transformation to ammonium) and the glutamine-synthetase–glutamate-synthetase pathway (responsible for assimilation of ammonium) both discriminate against ^{15}N, this discrimination will only be observed if there is a pool of enriched inorganic N that can leak from plant roots after uptake, which is unlikely if plant N demand is high relative to N supply (Evans et al. 1996). Because N demand frequently exceeds N supply in natural systems, this suggests that plant $\delta^{15}N$ is a good approximation of $\delta^{15}N$ of the available N source(s), under most field conditions.

With this as a baseline assumption, much can be discerned from a comparison of measured ecosystem N pools and plant $\delta^{15}N$. However, we offer three precautions. First, it should be noted that $\delta^{15}N$ of either bulk soil or soil organic matter cannot be used as an indicator of source N to plants. Most N in soil is highly recalcitrant and unavailable to plants, the dissolved labile N pool is small, transient, and may have a significantly different isotopic composition than bulk soil (Bergersen et al. 1990). Secondly, plants may take up either NO_3^-, NH_4^+, or dissolved organic nitrogen (DON), with many (but not all) plants showing distinct preferences. Because nitrification has a fairly large isotope effect (α) (Shearer et al. 1974; Delwiche & Steyn 1970), NH_4^+ in soils with significant nitrification will be enriched in ^{15}N compared with NO_3^-. Thus a simple analysis of soil solution, or extractable N, will not conclusively identify the source of N to a specific plant, although Hobbie et al. (1998) used isotopic differences between NH_4^+ and NO_3^- to infer differences in plant preference for NH_4^+ vs. NO_3^-. Plants can change their preference for NO_3^- vs. NH_4^+ with environmental conditions; factors such as forest harvest (e.g., Pardo et al. 2002) that increase nitrification, or the application of fertilizer, can shift the relative uptake rates of NH_4^+ and NO_3^- (Högberg 1997). Taken together, these factors may be the reason that different studies have reported both depletions (e.g., Virginia & Delwiche 1982; Vitousek et al. 1989; Gebauer &

Schulze 1991) and enrichments in ^{15}N with plant uptake of soil N in the field.

A third precaution when comparing $\delta^{15}N$ values of plants and presumed sources of N is that foliar N cannot always be used as a proxy of whole plant N (Kolb & Evans 2002), which can be problematic for studies of large woody species. Different organic compounds can have quite different $\delta^{15}N$ values, dependent on their biosynthetic pathways (reviewed in Werner & Schmidt 2002). Due to fractionation associated with glutamate, proteins are generally ^{15}N-enriched relative to the bulk $\delta^{15}N$ of the plant cell, while secondary products such as chlorophyll, lipids, amino sugars and alkaloids are depleted in [15]N.

Nitrogen fixation

Within these sideboards on the interpretation of plant isotopic signatures, many authors have used $\delta^{15}N$ data to draw inferences regarding N sources. The degree of N fixation has been assessed in a number of different ecosystems and species, using techniques first developed by Shearer and colleagues (Shearer et al. 1983; Shearer & Kohl 1986). Their technique relied on finding local reference species that would integrate the signal from available soil N that could then be compared with the signature of the presumed fixing species. Since the atmospheric signal is defined as 0, and they assumed no fractionation, a simple two-component mixing model could determine the percent contribution by N_2 fixation This technique has been extensively reviewed, used, modified, and criticized (Peoples et al. 1991; Binkley et al. 1985; Lajtha & Schlesinger 1986; Pate et al. 1993; Bowman et al. 1996; Spriggs et al. 2003), with the general conclusion that this technique should be used with great caution, if at all. Handley et al. (1994) point out that plant available N cannot be treated as a single source at any one site, and that large variations in $\delta^{15}N$ can be found in reference plants due to variations in rooting depth, timing and preference of NH_4^+ vs. NO_3^- uptake, and possibly mycorrhizal status, making comparisons between fixing and non-fixing species difficult at best. Still, authors have used plant $\delta^{15}N$ as a qualitative indication of diazotrophy (Hobbie et al. 1998).

Atmospheric sources of nitrogen

One recent application of the $\delta^{15}N$ of plant tissues has been to assess the contribution of atmospheric sources of N to plants. The ability of plants to take up N directly via foliage has been recognized for some time through experiments using ^{15}N in various gas or liquid sources as tracers (Boyce et al. 1996; Wilson & Tiley 1998) or from leaf-chamber input–output budgets (Sparks et al. 2001, 2003). However, quantifying direct N uptake from atmospheric sources vs. soil sources in the field has proven difficult. Although published $\delta^{15}N$ of specific pollutant N sources vary widely from +24.9‰ for NH_3 in barnyard samples to -13‰ for NO_x in automobile exhaust (Heaton 1986, 1990; Moore 1977), mixing models using soil and atmospheric sources can be confounded by multiple fractionations upon uptake or metabolism (Gebauer et al. 1994).

Although it might be difficult to quantify the importance of atmospheric sources of N to plants using natural abundance $\delta^{15}N$ values, general inferences can be made. For example, Evans & Ehleringer (1994) showed that relatively depleted N in marine fog was not a significant source of N to plants in the fog zone of the Atacama Desert of Chile, as $\delta^{15}N$ values of plants in the fog zone were significantly more positive than the fog. In contrast, Tozer et al. (2005) argued that the extremely depleted isotopic signatures found in lithophytes near Rotorua, New Zealand, were due both to significant fractionation upon volatilization of marine ammonia (NH_3) into the air and fractionation upon assimilation of NH_3 into plants. Because the volatilization of NH_3 to the atmosphere is accompanied by a very large fractionation, it should be easy to trace this source of N to plants where NH_3 might be significant, such as near intensive animal husbandry, guano deposits, or grazed land. Indeed, Erskine et al. (1998) suggested that the wide variation in plant $\delta^{15}N$ on subantarctic Macquarie Island reflected locations near either highly enriched scavenger excrement or significantly depleted NH_3 volatilized from penguin guano. Similarly, Frank et al. (2004) used isotope data to show that shoots of grassland plants in Yellowstone National Park directly absorbed [15]N – depleted NH_3–N that was volatilized from urine patches. Tozer et al. (2005) hypothesized that the uptake of highly depleted NH_3 is more widespread than currently thought and may be most significant in ecosystems that are highly N-limited; this hypothesis certainly deserves to be explored with additional measurements.

Mycorrhizal status

The idea that mycorrhizal status can affect the relationship between plant $\delta^{15}N$ and plant N source has been debated by many authors. Differences between mycorrhizal plants and non-mycorrhizal plants have been observed in numerous studies (Bardin et al. 1977; Högberg 1990; Chang & Handley 2000), due either to differences in N source (such as organic N) or due to differences in isotopic fractionation during N uptake. Since ectomycorrhizal fungi can directly use small organic compounds, the mineralization/nitrification pathway can be "short-circuited" (Trudell et al. 2004) and different factors can control isotopic abundance of this fungal-derived N. Many studies have shown that $\delta^{15}N$ in ectomycorrhizal fungi is generally greater than that of plant foliage and of their substrates in soil (Gebauer & Dietrich 1993; Högberg et al. 1999; Henn & Chapela 2001; Spriggs et al. 2003), but the exact mechanism of this difference is not clear, and could relate either to discrimination within fungi or upon transfer to the host plant, or else due to differences in substrate accessibility.

Hobbie et al. (1998, 1999, 2000) described a range of sites representing different post-deglaciation ages at Glacier Bay, Alaska. They hypothesized that the low $\delta^{15}N$ values in foliage from plants with a high dependence on mycorrhizal fungi were due to a large isotopic fractionation within the fungi. The fungi produced isotopically depleted amino acids, which were subsequently passed on to plant symbionts. Thus, the observed difference between soil mineral nitrogen $\delta^{15}N$ and foliar $\delta^{15}N$ in later succession could be a consequence of greater reliance on mycorrhizae under N-limited conditions. Spriggs et al. (2003) also observed a large discrimination against ^{15}N during transfer of N from the fungus to the host plant, leaving the latter with a more negative $\delta^{15}N$ value. However; as Högberg (1990) points out, these studies have not excluded uptake of organic N as a cause of the differences. Many of these studies have found that the fungal rhizomorphs were enriched in ^{15}N, yet Högberg et al. (1999) calculated that the limited biomass of the fungus could cause only a marginal decrease in $\delta^{15}N$ of the N passing from the substrate through the fungus to the host. This conclusion was supported by the high transfer efficiency of N between the fungus and the plant. Clearly, more mechanistic studies are needed to resolve this conflict.

Hydrogen and oxygen isotopes

Hydrogen and oxygen atoms in plant tissues predominantly come from water, thus processes that affect the isotopic ratio of water will influence the hydrogen and oxygen isotope ratios in plants. The main sources of isotopic variation in plant water come from isotopic variation in precipitation inputs and enrichment of the heavy isotope (both ^{18}O and D) in water from evaporation from the soil surface, and evaporation from the leaf surface during transpiration.

Isotopic fractionation in water

As mentioned in McGuire & McDonnell (this volume, pp. 334–374), the isotopic variation in water is predominantly due to fractionation associated with phase changes from solid to liquid to vapor and *vice versa* (equilibrium fractionation), and to the diffusion processes of water vapor (kinetic fractionation). Both kinetic and equilibrium fractionation are important depending on where and how the phase changes occur.

Kinetic fractionation of hydrogen and oxygen isotopes is exactly analogous to that of carbon, and can be attributed to the faster diffusion of molecules containing the lighter isotopes. Kinetic fractionation factors for H_2O/DHO and $H_2^{16}O/H_2^{18}O$ are 1.016 and 1.032, respectively (Cappa et al. 2003); these factors describe the 2–3 percent faster diffusion of light water relative to heavy water.

When the liquid and vapor phases of water are in equilibrium, the equilibrium fractionation factor describes the isotopic difference between the two phases. In the liquid phase, the heavier isotope is bound more tightly through hydrogen bonding than the lighter isotope, thus the lighter isotope is more readily released to the vapor phase. Equilibrium fractionation factors increase with decreasing temperature. For example at 25°C, fractionation values are 1.076 and 1.0092 for H_2O/DHO and $H_2^{16}O/H_2^{18}O$, respectively (Majoube 1971); these values describe the isotope ratio of the vapor divided by that of the liquid phase. Thus, at 25°C isotope ratios of water vapor would be 76‰ and 9.2‰ more negative than those of liquid water at equilibrium for δD and $δ^{18}O$, respectively. However at 0°C fractionation values are 1.108 and 1.011 for H_2O/DHO and $H_2^{16}O/H_2^{18}O$, respectively, and the vapor would be 108‰ and 11.0‰ more negative than the liquid water.

Isotopic composition of water available for plant uptake

Isotopic ratios of hydrogen and oxygen in precipitation worldwide are closely related, lying on a single line known as the global meteoric water line (GMWL), where δD = 8 $δ^{18}O$ + 10 (Craig 1961; Gat et al. 2001) (Figure 2.5). One end of the GMWL lies near the origin where the isotopic values of precipitation are similar to mean ocean water (V-SMOW, the isotopic standard for δD and $δ^{18}O$ in water). Most of the line lies below the origin, reflecting the depletion of heavy isotopes in the vapor phase upon evaporation (see grey shaded arrow in Figure 2.5). The specific slope of the GMWL results because condensation in rain clouds occurs under equilibrium conditions, and kinetic fractionation is generally not a factor. The slope is simply the ratio of the equilibrium fractionation factors for the two atoms (δD and $δ^{18}O$), which has the value of 8.

If evaporation from the ocean occurred under equilibrium conditions (RH = 100%) the intercept of the GMWL would be zero. However, the relative humidity (RH) is generally lower, thus kinetic fractionation through diffusion plays a role in evaporation of ocean waters. Kinetic fractionation is greater for $δ^{18}O$ compared with equilibrium fractionation (1.032 vs. 1.0092 at 25°C, respectively), but the fractionation difference for δD is the opposite (1.016 vs. 1.076). The degree to which kinetic fractionation affects evaporation depends on relative humidity. This kinetic fractionation during evaporation causes the evaporation slope to be less than 8 and the slope decreases with increasing kinetic effect (lower RH). For example, in Figure 2.5, if vapor was formed completely under equilibrium conditions at 25°C, its isotopic ratios would be −76 and −9.2‰ for δD and $δ^{18}O$, respectively (black solid point in Figure 2.5). However, since relative humidity is less than 100 percent (approximately 85% over the ocean) kinetic fractionation is

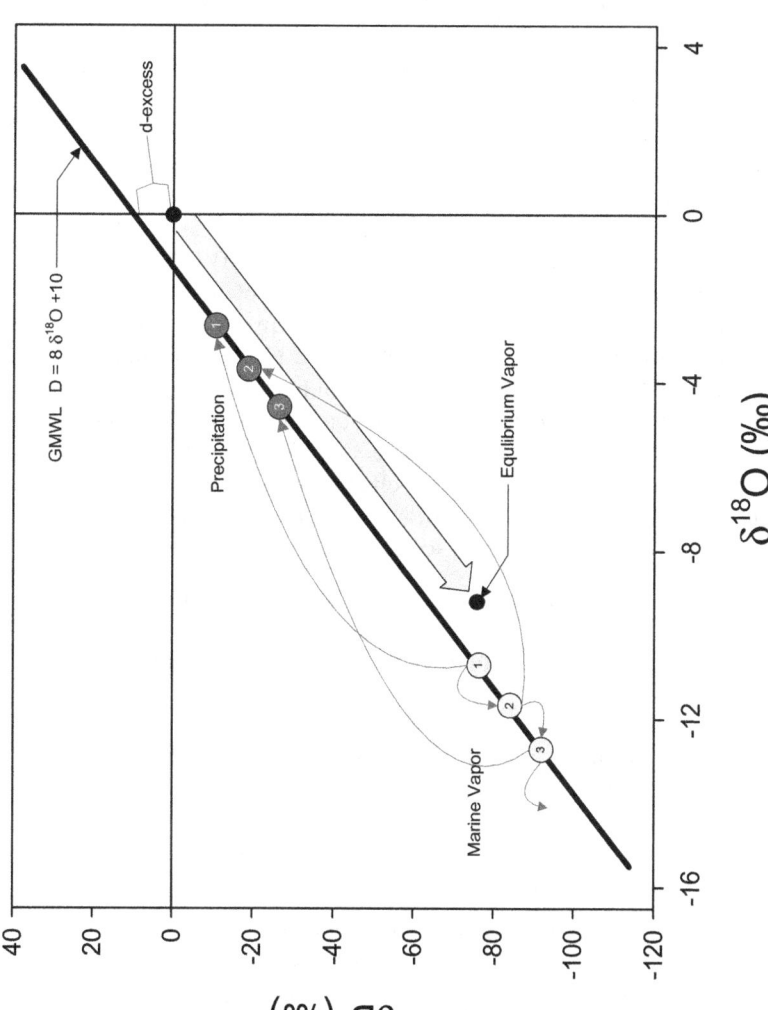

Figure 2.5 Schematic representation of the GMWL from evaporation from the ocean. The black dot at 0,0 represents ocean water, and the grey shaded arrow represents evaporation under equilibrium conditions (100% RH). The grey circles are the actual isotopic value of marine vapor since ocean evaporation occurs at RH lower than 100 percent (~85%), which leads to a d-excess of approximately 10. The numbered circles (1–3) for marine vapour (light circles) and precipitation (dark circles) illustrate the increasing depletion as precipitation forms under equilibrium conditions and falls from the cloud (rainout effect). (After Gat et al. 2001.)

involved, and the $\delta^{18}O$ value is more depleted than for the equilibrium case (see Figure 2.5 Marine Vapor 1). Note that the δD is about the same value in both cases. When precipitation is formed from this vapor (under equilibrium conditions because rain cloud RH = 100%, so the slope will again be 8), the resulting precipitation will have an excess of deuterium (d-excess in Figure 2.5) that causes the intercept of the GMWL to be around 10‰ (Clark & Fritz 1997).

If precipitation falls through relatively dry air and evaporates as it falls, kinetic fractionation again influences the respective evaporation rates of DHO and $H_2^{18}O$, drawing the water off the global meteoric water line. The deviations give rise to local meteoric water lines (LMWL) that usually have slopes less than 8. The slope of a LMWL can be used to determine the extent of evaporation after raindrops are formed (Gat et al. 2001).

The isotopic ratio of precipitation will change along the GMWL because of temperature effects on fractionation, and the depletion of precipitation with continued rainfall. Since clouds and precipitation form under equilibrium conditions, heavier isotopes accumulate in the liquid phase and will fall first in the precipitation. Consequently, rain clouds will be come isotopically lighter over time. This "rainout" depletion of rain clouds follows a Rayleigh distillation process (see McGuire & McDonnell, this volume, pp. 334–374, for more details). These temperature and rainout effects underlie pronounced geographical and seasonal variation in the isotope ratios of precipitation that can be summarized as follows (influencing variables noted in parentheses):

Latitude effect   more negative with increasing latitude (temperature and rainout)

Elevation effect   more negative with increasing altitude (temperature and rainout)

Continental effect   more negative as an air mass moves inland (rainout)

Seasonal effect   more negative in winter, less negative in summer (temperature)

Amount effect   more negative when more precipitation falls (rainout and decreased evaporation of rain due to higher relative humidity with more rain)

Within a particular ecosystem, the isotopic ratios of precipitation vary from storm to storm depending on the origin of the storm (high vs. low latitudes) and the cloud temperature during condensation. Generally, more depleted isotopic values of precipitation are found during the winter and enriched precipitation during the summer. For example, values for δD in one study of precipitation in Austria range from −180‰ in December to −5‰ in May (Libby et al. 1976). Similar data from Utah range from −220‰ for April snow to −15‰ from a summer thundershower (Dawson & Ehleringer, 1991). Precipitation also interacts with plant canopies as it falls, which can alter the isotopic signal of precipitation reaching the soil through evaporation of intercepted water (see McGuire & McDonnell, this volume, pp. 334–374). This variation in precipitation isotopic values along with precipitation amount can create unique isotopic patterns within the soil profile from which plants obtain their moisture.

Generally, deep soil water or ground water has been reported as approximately the average of mean annual precipitation (Clark & Fritz 1997). However, the actual value may vary from this mean depending on seasonal patterns of water inputs (precipitation) and water losses (evapotranspiration). For example, evapotranspiration is usually less during the winter dormant season so winter precipitation has a greater chance of percolating to deeper soil layers rather than being evaporated or transpired compared with summer precipitation, leading to more lighter isotopic values deeper within the soil profile. Summer precipitation tends to be enriched isotopically, leading to heavier isotopic signals in the upper soil during the active growing season. Evaporation can also occur from the soil surface, further enriching the surface soil water. However, evaporation from the soil surface generally occurs within the top 10 to 20 cm, and not much deeper.

Water uptake by plants

Water is not isotopically fractionated when taken up by the plant (Dawson & Ehleringer 1993), except perhaps under exceedingly unusual conditions such as salt-water uptake by salt-excluding halophytes (Lin & Sternberg 1993). As a result, water in plant tissues carries the same isotopic signal as the source water – until it reaches the sites of evaporation, generally in the leaves. Therefore, the isotope ratio of xylem water can be used as a measure of the isotopic signature of the soil-water being utilized. This measurement can be particularly useful, provided the various water sources are sufficiently distinct in their isotopic composition, e.g., surface water vs. ground water vs. fog. For two sources, a simple two-ended mixing model is used with the two sources being the end members, and the plant-water value being somewhere in between. With more isotopically distinct sources, the precision for percent water use from each source will increase (Phillips & Gregg 2001). With two isotopes for water (δD and $\delta^{18}O$), theoretically, three sources could be separated using a simple mixing model; however, as mentioned above, δD and $\delta^{18}O$ are highly correlated (GMWL), so in reality two sources can be separated using water isotopes (Phillips & Gregg 2001). Phillips & Gregg (2003) have also worked out techniques to deal with more sources than can be uniquely solved for, so it is possible to work with more than two sources of water. In addition, other information can be used to help constrain possibilities such as information on soil-water potential. For example, for a particular soil layer, if soil-water potential was lower than leaf-water potential, that soil layer could not be a source for water. One problem with the two-ended mixing model technique is that isotopic profiles in soils are a continuum rather than two discrete isotopic values (Allison et al. 1983; Thorburn & Ehleringer 1995; Meinzer et al. 1999; Plamboeck et al. 1999; Moreira et al. 2000).

In situations where sources are isotopically distinct, the natural abundance of isotopes in water has proved extremely useful for understanding patterns of plant water uptake. For example, this technique was used to demonstrate that co-existing plant life forms with varying rooting morphologies can utilize different sources of water (Ehleringer et al. 1991). Williams & Ehleringer (2000) used this technique to separate surface and deep soil water usage for three tree species over 2 years along a summer monsoonal gradient. They found that one species (*Quercus gambelii*) did not use summer precipitation where summer precipitation was not a reliable resource. However, the two conifer species (*Juniperus osteosperma* and *Pinus edulis*) did utilize surface summer rains when available along the entire gradient. In a classic study, Dawson & Ehleringer (1991) demonstrated that riparian trees do not necessarily use stream water, but instead may rely on deeper aquifer water. Stable isotopes of stem water have also been useful for documenting seasonal use of fog water in a redwood forest (Dawson 1998), and in epiphytic plants (Field & Dawson 1998). Meinzer et al. (1999) reported that smaller tropical trees relied more on deep water than did larger co-located trees. Thorburn & Ehleringer (1995) also sampled roots with depth, and found that roots in a particular soil layer did not always take up water from that layer. Dawson & Pate (1996) found that the isotopic ratio in surface lateral roots shifted seasonally, with water in lateral roots matching the isotopic value in deep sinker roots during the dry season rather than the isotopic value of water in the surrounding soil, which indicated that during the dry season, the sinker roots were providing water to the surface roots as well as the tree. In cases where natural abundance differences within the soil are not sufficient for determining source utilization, using an isotope label can also be highly effective to examine source utilization (Plamboeck et al. 1999; Moreira et al. 2000; Brooks et al. 2002, 2006), but tracer enrichment studies are beyond the scope of this book.

Isotope ratios of leaf water

As mentioned earlier, transpiration from plant leaves leads to fractionation of the xylem water. As a result, leaf water is often considerably enriched in the heavier isotopes. The problem of describing leaf-water enrichment is an elaboration of the more general problem of water evaporation; the seminal work in this area was conducted by Craig & Gordon (1965). Flanagan et al. (1991) expanded on this framework by incorporating leaf boundary layer effects, resulting in the following Craig–Gordon model to describe the isotopic composition of water in a transpiring leaf at steady state:

$$R_{wl} = \alpha^* \left[\alpha_k R_{wx} \left(\frac{e_i - e_s}{e_i} \right) + \alpha_{kb} R_{wx} \left(\frac{e_s - e_a}{e_i} \right) + R_a \left(\frac{e_a}{e_i} \right) \right]$$ (2.7)

where R is the molar ratio of the heavy to light isotope, e is partial pressure of water vapor and the subscripts "wl", "wx", "i" "s" and "a" refer to leaf water, xylem water, intercellular spaces, leaf surface and bulk air, respectively; α^* is the equilibrium fractionation factor; α_k is the kinetic fractionation factor; α_{kb} is the kinetic fractionation factor associated with diffusion through a turbulent boundary layer (α_k to the 2/3 power). This model can also be written in terms of enrichment above the source water (Farquhar et al. 1989b; Farquhar & Lloyd 1993; Farquhar & Gan 2003):

$$\Delta_e = \alpha^* + \alpha_k + (\Delta_v - \alpha_k) \frac{e_a}{e_i}$$ (2.8)

where Δ_v is the oxygen isotope composition of the atmospheric water relative to the xylem water source. If isotopic composition of the xylem and atmospheric water and leaf temperature are assumed constant, which would hold α^*, e_s and e_i constant, then R_{wl} and Δ_e would be linearly related to e_a/e_i, which is the relative humidity of the atmosphere. Because leaf water is the source for oxygen and hydrogen in leaf organic matter, leaf organic matter can contain information about relative humidity during the leaf's lifespan.

If bulk leaf water comprised a single homogeneous, equilibrated pool, one would expect the isotope ratio of leaf water to match predictions based on the equations above. However, bulk leaf water does not match this simple model (Yakir et al. 1989; Flanagan et al. 1991) for several reasons. First, it takes a matter of hours for leaves to approach isotopic steady state, the point at which the isotopic composition of the water entering the leaf is equal to that of the water diffusing out (White 1989; Cernusak et al. 2002). Second, leaves often vary in water volume on a diel basis, and third, the various pools of water within a leaf are seldom well-mixed, resulting in isotopic heterogeneity within the leaf (Yakir 1998; Barbour et al. 2000). One reason for heterogeneous leaf water is that gradients of enriched water can result from convection of unenriched xylem water towards the site of evaporation and the opposing diffusion of enriched water away from the site of evaporation (Farquhar & Lloyd 1993), which is known as the Péclet effect ($\wp$). The influence of the Péclet effect on the isotopic composition of bulk leaf water can be estimated with the following equations:

$$\Delta_L = \frac{\Delta_e(1 - e^{-\wp})}{\wp}$$ (2.9)

$$\wp = \frac{LE}{CD}$$ (2.10)

where Δ_L is the isotopic composition of bulk leaf water relative to source water, L is the effective path length, E is transpiration, C is the molar density of water, and D is the diffusivity of $H_2{}^{18}O$ (Farquhar & Lloyd 1993; Barbour et al. 2004). The Péclet effect varies with the rate of transpiration from the leaf, with enrichment being relatively lower with higher transpiration rates. This can affect the $\delta^{18}O$ of sucrose because sucrose is not synthesized at the site of evaporation within a leaf but in the chloroplast, which will have a different leaf-water signature than the site of evaporation (Barbour et al. 2000).

For a broad range of environmental conditions, the basic Craig–Gordon model, which does not account for the Péclet effect, provides a useful interpretation of the $\delta^{18}O$ and δD of leaf water and the organic matter derived from leaf water (Roden & Ehleringer 1999b). However, depending on the particular application, accounting for the Péclet effect may improve estimates of transpiration rates (Wang & Yakir 2000; Barbour & Farquhar 2003; Barbour et al. 2004). More elaborate models have recently been developed to predict bulk leaf water while accounting for the Péclet effect (Barbour & Farquhar 2003; Farquhar & Gan 2003) and for non-steady state conditions (Cernusak et al. 2002; Farquhar & Cernusak 2005). As always, more elaborate models are more difficult to parameterize. The optimal modeling approach depends on the objectives of the user.

How are leaf water enrichment data used? First, as mentioned above, the degree of enrichment reflects the relative humidity of the environment and may allow inference of transpiration rates. These isotopic signals are preserved in organic matter and so may provide long-lived records of the distant past. In addition to the climate signal, $\delta^{18}O$ in leaf organic matter might provide insights as to whether changes in $\delta^{13}C$ were caused by differences in stomatal conductance or photosynthetic capacity (Scheidegger et al. 2000). Recall that changes in $\delta^{13}C$ can be through changes in stomatal supply or photosynthetic demand. Knowing the relative humidity can help one deduce which variable was more likely responsible for variation in $\delta^{13}C$, and thus, c_i. Second, CO_2 entering the leaf exchanges oxygen with enriched leaf water; a portion of this CO_2 then diffuses back out of the leaf, affecting the $\delta^{18}O$ of atmospheric CO_2 (Farquhar et al. 1993). Oxygen evolved during photosynthesis also contains the leaf enrichment signature. This isotopic information in atmospheric gases is used by global modelers to separate terrestrial and marine productivity.

Hydrogen and oxygen isotopes in plant tissues

The primary interest in $\delta^{18}O$ and δD in plant tissue has been to obtain climate information. The above discussion elaborates on how leaf-water isotopic signatures are influenced by relative humidity. Recall that xylem water has the same isotopic composition as the soil water it is derived from. The $\delta^{18}O$ and δD of plant tissue is influenced by both these pools of water (leaf water and soil/xylem water). Several authors have noted correlations between treering cellulose isotope data and temperature (Gray & Thompson 1976; Yapp & Epstein 1982), relative humidity (Yapp & Epstein 1982; Edwards & Fritz 1986, 1988), or precipitation amount (Lawrence & White 1984; Krishnamurthy & Epstein 1985; White et al. 1994; Saurer et al. 1997b); however, these correlations differ in the relative influence of each factor. A number of investigators began to realize that two water sources were important for determining the $\delta^{18}O$ and δD of plant tissue: xylem water and atmospheric water vapor. Thus plant tissue samples could contain information on both temperature and relative humidit (Luo & Sternberg 1992; White et al. 1994; Saurer et al. 1997a).

To determine past climate signals, it is important for the tissue to maintain the isotopic signature over time. Some authors have argued that plant tissues need to be extracted to yield α-cellulose for oxygen and nitrated cellulose for hydrogen to remove all oxygen and hydrogen atoms not bound to carbon (Leavitt & Danzer 1993; Loader et al. 1997). Atoms not bound to carbon continue to exchange in the presence of water, thus losing the $\delta^{18}O$ or δD signal at the time the cellulose was formed. However, a recent study has indicated that whole wood contains isotopic data of sufficient quality to analyze environmental signals (Barbour et al. 2001), thus for some purposes extraction may not be necessary, but further research into the necessity of extraction is needed. A larger problem exists in separating the temperature and relative humidity signals contained in $\delta^{18}O$ and δD of tree rings for climate reconstruction.

In a landmark series of papers, Roden, Ehleringer and others (Roden & Ehleringer 1999a, 1999b, 2000; Roden et al. 2000) elucidated the relative influence of leaf water and source water on tree ring $\delta^{18}O$ and δD composition as follows:

$$\delta D_{cx} = f_H(\delta D_{wx} + \varepsilon_{HH}) + (1 - f_H)(\delta D_{wl} + \varepsilon_{HA})$$

$$\delta^{18}O_{cx} = f_o(\delta^{18}O_{wx} + \varepsilon_o) + (1 - f_o)(\delta^{18}O_{wl} + \varepsilon_o)$$

(2.11)

where f_H and f_o refer to the fraction of carbon-bound hydrogen and oxygen, respectively, that undergo exchange with the water source during cellulose formation. Leaf water is subscripted 'wl' and xylem water is subscripted 'wx'. These f values were estimated to be 0.36 and 0.42, respectively. Large biochemical fractionation factors are associated with the formation of sucrose and cellulose, and are different for oxygen and hydrogen isotopes. The hydrogen isotope fractionation factor associated with autotrophic carbohydrate metabolism (ε_{HA}) was determined to be between -120 and $-171‰$ (Yakir & DeNiro 1990), and for heterotrophic carbohydrate metabolism (ε_{HH}), between $+144$ and $+166‰$ (Yakir & DeNiro 1990; Luo & Sternberg 1992). The oxygen isotope ratio has just one fractionation factor (ε_o) of $+27‰$ associated with the water/carbonyl interactions (Sternberg 1989; Yakir & DeNiro 1990). In some situations, the water in which cellulose is being formed may be a mixture of xylem water and enriched leaf water (Barbour & Farquhar 2000), and thus an equation using xylem water values would not be correct. For oxygen, Barbour & Farquhar (2000) proposed the following model:

$$\Delta_{cx} = \Delta_L(1 - p_{ex}p_x) + \varepsilon_o \qquad (2.12)$$

where p_{ex} is the proportional exchange (equivalent to f_o) and p_x is the proportion of unenriched water (xylem water) at the site of cellulose formation. Sternberg et al. (2003) proposed a method for separating leaf and xylem water signals by examining only oxygens in cellulose that completely exchange with xylem water at the time of cellulose formation; however, the authors express the need for more work on this method before it will be useful in tree-ring studies.

Separating the influence of humidity (enriched leaf water) and temperature (xylem water) will be a challenge; however, significant and interesting environmental trends have been found in tree-ring chronologies of $\delta^{18}O$ or δD. Saurer et al. (2000) found that the $\delta^{18}O$ in a latewood tree-ring chronology of *Abies* in Switzerland was highly correlated with $\delta^{18}O$ in June/July precipitation. Saurer et al. (1997b) noted similar results for *Fagus* trees in Switzerland. However, in both studies the correlation between temperature and $\delta^{18}O$ in tree rings was weak. Subdividing rings is a promising approach for examining seasonal dynamics. Barbour et al. (2002) documented both site and seasonal differences related to relative humidity in $\delta^{18}O$ and $\delta^{13}C$ chronologies of *Pinus radiata* that were subdivided. White et al. (1994) also found seasonal changes in δD on subdivided rings with increasing enrichment through the season. Tree rings are not the only material that can be used to study climate. Helliker & Ehleringer (2002) demonstrated that the $\delta^{18}O$ in sections of grass blades could record shifts in relative humidity during leaf expansion. Jäggi et al. (2003) noted that short-term climate change was more strongly reflected in the $\delta^{18}O$ of needle tissue compared with either early or late wood. These studies and others indicate the potential power of $\delta^{18}O$ and δD in plant tissues to record environmental information (McCarroll & Loader 2004).

Separating evapotranspiration using stable isotopes

Another emerging area using $\delta^{18}O$ and δD in water are studies which have attempted to quantify evaporation from soil relative to transpiration from leaves (Moreira et al. 1997; Wang & Yakir 2000; Yepez et al. 2003; Williams et al. 2004). Water evaporated from leaves and from soil have different isotopic signatures. Leaf water is closer to steady state and enriched to a greater degree than soil water. By definition, water vapor diffusing from a leaf at steady state must have the same isotopic signature as the source xylem water (Yakir et al. 1993), whereas water vapor from soil is more depleted. By measuring the isotopic ratio of soil water where evaporation is taking place and using the Craig–Gordon model (equation 2.7), the isotopic composition of soil-water vapor can be estimated. The third source of water vapor is from the atmosphere. Separating evaporated and transpired water relies on both a mixing model approach and a technique known as Keeling plots (Pataki et al. 2003). The Keeling-plot approach will determine the signature of evapotranspired water, and the mixing model will determine the relative proportion of evaporated and transpired water. In the Amazon forest, transpiration was the dominant source of water vapor (Moreira et al. 1997). Similarly, Wang & Yakir (2000) found that soil evaporation was only 1.5–3.5 percent of the evapotranspiration flux from crops in a desert environment. Williams et al. (2004) observed that soil evaporation changed from 0 percent in an olive orchard prior to irrigation to 14–31 percent for the 5 days following irrigation. This approach is still in its infancy and is also complicated by non-steady state transpiration. Lai et al. (2006) found it necessary to include a non-steady state model for transpiration to tease appart diurnal patterns in the isoflux (isotope signal of water and the water flux) from a 400-year-old coniferous forest.

Conclusions

The use of natural abundance stable isotopes to elucidate physiological processes in plants is one of the most common, and one of the oldest, applications of isotope analysis in ecology. Plants display particularly strong isotopic signals because they construct their tissues from such small molecules. These molecules, e.g., CO_2, NO_3^-, NH_4^+, H_2O, are small enough that the presence of an extra neutron in a heavy atom affects the behavior of the whole molecule. These isotopic effects are often less pronounced in animals, which tend to use bigger molecules as substrates, and therefore tend to retain the signals imposed on them by the plants they consume.

Isotope analysis is now a well-established tool used to determine carbon fixation pathways of plant species, plant water-use efficiency, and source of water used; new uses are rapidly being developed. Understanding the physiological processes behind stable isotope signatures of primary producers has given researchers new tools to analyze animal paleodiets (see Evans, this volume, pp. 83–98), to trace food webs in both marine (see Montoya, this volume, pp. 176–201) and terrestrial ecosystems, and to trace carbon sources in estuarine systems (see Finlay & Kendall, this volume, pp. 283–333), among other applications.

As with any other research tool, it is critical to understand the assumptions and where they might be violated. In this chapter we have described the theoretical assumptions that underlie the interpretation of stable isotope signatures in plant ecophysiology and the types of empirical data that are commonly collected. We have described several cases where these assumptions might be violated. These caveats notwithstanding, we fully expect that the potential of these powerful new isotope techniques will continue to expand in the future. It is our intention to encourage the use of these techniques by ensuring that users are knowledgeable. As work continues to identify the detailed mechanisms underlying isotopic composition, the range of applications is sure to expand.

References

Allison, G.B., Barnes, C.J. & Hughes, M.W. (1983) The distribution of deuterium and ^{18}O in dry soil. *Experimental Journal of Hydrology*, **64**, 377–397.

Aschan, G. & Pfanz, H. (2003) Non-foliar photosynthesis: a strategy of additional carbon acquisition. *Flora*, **198**, 81–97.

Badeck, F.-W., Tcherkez, G., Nogués, S., et al. (2005) Post-photosynthetic fractionation of stable carbon isotopes between plant organs – a widespread phenomenon. *Rapid Communications in Mass Spectrometry*, **19**, 1381–1391.

Barbour, M.M. & Farquhar, G.D. (2000) Relative humidity- and ABA-induced variation in carbon and oxygen isotope ratios of cotton leaves. *Plant, Cell and Environment*, **23**, 473–485.

Barbour, M.M. & Farquhar, G.D. (2003) Do pathways of water movement and leaf anatomical dimensions allow development of gradients in $H_2^{18}O$ between veins and the sites of evaporation within leaves? *Plant, Cell and Environment*, **27**, 107–121.

Barbour, M.M., Schurr, U., Henry, B.K., Wong, S.C. & Farquhar, G.D. (2000) Variation in the oxygen isotope ratio of phloem sap sucrose from castor bean. Evidence in support of the Péclet effect. *Plant Physiology*, **123**, 671–679.

Barbour, M.M., Andrews, T.J. & Farquhar, G.D. (2001) Correlations between oxygen isotope ratios of wood constituents of *Quercus* and *Pinus* samples from around the world. *Australian Journal of Plant Physiology*, **28**, 335–348.

Barbour, M.M., Walcroft, A.S. & Farquhar, G.D. (2002) Seasonal variation in $\delta^{13}C$ and $\delta^{18}O$ of cellulose from growth rings of *Pinus radiata*. *Plant, Cell and Environment*, **25**, 1483–1499.

Barbour, M.M., Roden, J.S., Farquhar, G.D. & Ehleringer, J.R. (2004) Expressing leaf water and cellulose oxygen isotope ratios as enrichment above source water reveals evidence of a Péclet effect. *Oecologia*, **138**, 426–435.

Bardin, R., Domenach, A.-M. & Chalamet, A. (1997) Rapports isotopiques de l'azote. II. *Revue d'Ecologie et de Biologie du Sol*, **14**, 395–402.

Beerling, D.J. (1999) Long-term responses of boreal vegetation to global change: an experimental and modeling investigation. *Global Change Biology*, **5**, 55–74.

Bergersen, F.J., Peoples, M.B. & Turner, G.L. (1990) Measurement of N_2 fixation by ^{15}N natural abundance in the management of legumes: roles and precautions. In: *Nitrogen Fixation: Achievements and Objectives. Proceedings of the 8th International Congress on Nitrogen Fixation* (Eds P. M. Greshoff, L. E. Roth, G. Stacey & W. E. Newton), pp. 315–322. Chapman & Hall, New York.

Berry, J.A. (1989) Studies of mechanisms affecting the fractionation of carbon isotopes in photosynthesis. In: *Stable Isotopes in Ecological Research* (Eds P.W. Rundel, J.R. Ehleringer & K.A. Nagy), pp. 82–94. Springer-Verlag, New York.

Binkley, D., Sollins, P. & McGill, W.B. (1985) Natural abundance of nitrogen-15 as a tool for tracing alder-fixed nitrogen. *Soil Science Society of America Journal*, **49**, 444–447.

Bowman, W.D., Schardt, J.C. & Schmidt, S.K. (1996) Symbiotic N_2 fixation in alpine tundra: ecosystem input and variation in fixation rates among communities. *Oecologia*, **108**, 345–350.

Boyce, R.L., Friedland, A.J., Chamberlain, C.P. & Poulson, S.R. (1996) Direct canopy nitrogen uptake from ^{15}N-labeled wet deposition by mature red spruce. *Canadian Journal of Forest Research*, **26**, 1539–1547.

Brooks, J.R., Flanagan, L.B., Varney, G.T. & Ehleringer, J.R. (1997) Vertical gradients in photosynthetic gas exchange characteristics and refixation of respired CO_2 within boreal forest canopies. *Tree Physiology*, **17**, 1–12.

Brooks, J.R., Meinzer, F.C., Coulombe, R. & Gregg, J.W. (2002) Hydraulic redistribution of soil water during summer drought in two contrasting Pacific Northwest coniferous forests. *Tree Physiology*, **22**, 1107–1117.

Brooks, J.R., Meinzer, F.C., Warren, J.M., Domec, J.C. & Coulombe, R. (2006) Hydraulic redistribution in a Douglas-fir forest: lessons from system manipulations. *Plant, Cell and Environment*, **29**, 138–150.

Brugnoli, E., Hubick, K.T., von Caemmerer, S., Wong, S.C. & Farquhar, G.D. (1988) Correlation between the carbon isotope discrimination in leaf starch and sugars of C3 plants and the ratio of intercellular and atmospheric partial pressures of carbon dioxide. *Plant Physiology*, **88**, 1418–1424.

Campbell, G.S. & Norman, J.M. (1998) *An Introduction to Environmental Biophysics*, 2nd edn. Springer-Verlag, New York, 286 pp.

Cappa, C.D., Hendricks, M.B., DePaolo, D.J. & Cohen, R.C. (2003) Isotope fractionation of water during evaporation. *Journal of Geophysical Research*, **108**(D16), 4525, doi 10.1029/2003JD003597.

Cernusak, L.A., Marshall, J.D., Balster, N.J. & Comstock, J. (2001) Carbon isotope discrimination in photosynthetic bark. *Oecologia*, **128**, 34–45.

Cernusak, L.A., Pate, J.S. & Farquhar, G.D. (2002) Diurnal variation in the stable isotope composition of water and dry matter in fruiting *Lupinus angustifolius* under field conditions. *Plant, Cell and Environment*, **25**, 893–907.

Chang, S.X. & Handley, L.L. (2000) Site history affects soil and plant ^{15}N natural abundances in forests of northern Vancouver Island, British Columbia. *Functional Ecology*, **14**, 273.

Clark, I.D. & Fritz, P. (1997) *Environmental Isotopes in Hydrogeology*. Lewis Publishers, New York.

Condon, A.G., Richards, R.A. & Farquhar, G.D. (1987) Carbon isotope discrimination is positively correlated with grain yield and dry matter production in field-grown wheat. *Crop Science*, **2**, 996–1001.

Craig, H. (1953) The geochemistry of the stable carbon isotopes. *Geochimica et Cosmochimica Acta*, **3**, 53–92.

Craig, H. (1954) Carbon-13 in plantsand relationships between carbon-13 and carbon-14 variations in nature. *Journal of Geology*, **62**, 115–149.

Craig, H. (1961) Isotopic variations in meteoric waters. *Science*, **133**, 1702–1703.

Craig, H. & Gordon, L.I. (1965) Deuterium and oxygen-18 variation in the ocean and the marine atmosphere. In: *Proceedings of a Conference on Stable Isotopes in Oceanographic Studies and Paleotemperatures, Spoleto, Italy* (Ed. E. Tongiorgi), pp 9–130.

Damesin, C. & Lelarge, C. (2003). Carbon isotope composition of currenty-year shoots from Fagus sylvatica in relation to growth, respiration and use of reserves. *Plant Cell and Environment*, **26**, 207–219.

Dawson, T.E. (1998) Fog in the California redwood forest: Ecosystem inputs and use by plants. *Oecologia*, **117**, 476–485.

Dawson, T.E. & Ehleringer, J.R. (1991) Streamside trees that do not use streamside water. *Nature*, **350**, 335–337.

Dawson, T.E. & Ehleringer, J.R. (1993) Isotopic enrichment of water in the "woody" tissues of plants: Implications for plant water source, water uptake and other studies which use the stable isotopic composition of cellulose. *Geochimica et Cosmochimica Acta*, **57**, 3487–3492.

Dawson, T.E. & Pate, J.S. (1996) Seasonal water uptake and movement in root systems of Australian phraeatophytic plants of dimorphic root morphology: a stable isotope investigation. *Oecologia*, **107**, 13–20.

Delwiche, C.C. & Steyn, P.L. (1970) Nitrogen isotope fractionation in soils and microbial reactions. *Environmental Science and Technology*, **4**, 929–935.

Duranceau, M., Ghashghaie, J., Badeck, F., Deleens, E. & Cornic, G. (1999) $\delta^{13}C$ of CO_2 respired in the dark in relation to $\delta^{13}C$ of leaf carbohydrates in *Phaseolus vulgaris* L. under progressive drought. *Plant Cell and Environment*, **22**, 515–523.

Edwards, T.W.D. & Fritz, P. (1986) Assessing meteoric water composition and relative humidity from 18O and 2H inwood cellulose: paleoclimatic implications for southern Ontario, Canada. *Applied Geochemistry*, **1**, 715–723.

Edwards, T.W.D. & Fritz, P. (1988) Stable-isotope paleoclimate records for southern Ontario, Canada: comparison of results from marl and wood. *Canadian Journal of Earth Science*, **25**, 1397–1406.

Ehdaie, B., Hall, A.E., Farquhar, G.D., Nguyen, H.T. & Waines, J.G. (1991) Water-use efficiency and carbon isotope discrimination in wheat. *Crop Science*, **31**, 1282–1288.

Ehleringer, J.R. (1989) Carbon isotope ratios and physiological processes in arid land plants. In: *Stable Isotopes in Ecological Research* (Eds P.W. Rundel, J.R. Ehleringer & K.A. Nagy), pp. 41–54. Springer-Verlag, New York.

Ehleringer, J.R. (1991) $^{13}C/^{12}C$ fractionation and its utility in terrestrial plant studies. In: *Carbon Isotope Techniques* (Eds D.C. Coleman & B. Fry), pp. 187–200. Academic Press, New York.

Ehleringer, J.R. & Cooper, T.A. (1988) Correlations between carbon isotope ratio and micro-habitat in desert plants. *Oecologia*, **76**, 562–566.

Ehleringer, J.R. & Osmond, C.B. (1989) Stable isotopes. In: *Plant Physiological Ecology* (eds R.W. Pearcy, J.R. Ehleringer, H.A. Mooney & P.W. Rundel), pp. 281–300. Chapman & Hall, New York.

Ehleringer, J.R. & Pearcy, R.W. (1983) Variation in quantum yields for CO_2 uptake among C3 and C4 plants. *Plant Physiology*, **73**, 555–559.

Ehleringer, J.R., White, J.W., Johnson, D.A. & Brick, M. (1990) Carbon isotope discrimination, photosynthetic gas exchange, and transpiration efficiency in beans and range grasses. *Acta Oecologia*, **11**, 611–625.

Ehleringer, J.R., Phillips, S.L., Schuster, W.S.F. & Sandquist, D.R. (1991) Differential utilization of summer rains by desert plants: Implications for competition and climate change. *Oecologia*, **88**, 430–434.

Ellsworth, D.S. (1999) CO_2 enrichment in a maturing pine forest: are CO_2 exchange and water status in the canopy affected? *Plant Cell and Environment*, **22**, 461–472.

Erskine, P.D., Bergstrom, D.M., Schmidt, S., Stewart, G.R., Tweedie, C.E. & Shaw, J.D. (1998) Subantarctic Macquarie Island – a model ecosystem for studying animal-derived nitrogen sources using ^{15}N natural abundance. *Oecologia*, **117**, 187–193.

Evans, J.R., von Caemmer, S., Setchell, B.A. & Hudson, G.S. (1994) The relationship between CO_2 transfer conductance and leaf anatomy in transgenic tobacco with a reduced content of Rubisco. *Australian Journal of Plant Physiology*, **21**, 475–495.

Evans, R.D. (2001) Physiological mechanisms influencing plant nitrogen isotope composition. *Trends in Plant Science*, **6**, 121–126.

Evans, R.D. & Ehleringer, J.R. (1994) Plant $\delta^{15}N$ values along a fog gradient in the Atacama Desert, Chile. *Journal of Arid Environments*, **28**, 189–193.

Evans, R.D., Bloom, A.J., Sukrapanna, S.S. & Ehleringer, J.R. (1996) Nitrogen isotope composition of tomato (*Lycopersicon esculentum* Mill. cv. T-5) grown under ammonium or nitrate nutrition. *Plant Cell and Environment*, **19**, 1317–1323.

Farquhar, G.D. (1983) On the nature of carbon isotope discrimination in C4 species. *Australian Journal of Plant Physiology*, **10**, 205–226.

Farquhar, G.D. & Cernusak, L.A. (2005) On the isotopic composition of leaf water in the non-steady state. *Functional Plant Biology*, **32**, 293–303.

Farquhar, G.D. & Gan, K.S. (2003) On the progressive enrichment of the oxygen isotope composition of water along a leaf. *Plant, Cell and Environment*, **26**, 1579–1597.

Farquhar, G.D. & Lloyd, J. (1993) Carbon and oxygen isotope effects in the exchange of carbon dioxide between terrestrial plants and the atmosphere. In: *Stable Isotopes and Plant Carbon–Water Relations* (Eds J.R. Ehleringer, A.E. Hall & G.D. Farquhar), pp. 47–70. Academic Press, San Diego.

Farquhar, G.D. & Richards, R.A. (1984) Isotopic composition of plant carbon correlates with water–use efficiency of wheat genotypes. *Australian Journal of Plant Physiology*, **11**, 539–552.

Farquhar, G.D., O'Leary, M.H. & Berry, J.A. (1982) On the relationship between carbon isotope discrimination and the intercellular carbon dioxide concentration in leaves. *Australian Journal of Plant Physiology*, **9**, 121–137.

Farquhar, G.D., Ehleringer, J.R. & Hubick, K.T. (1989a) Carbon isotope discrimination and photosynthesis. *Annual Review of Plant Physiology and Plant Molecular Biology*, **40**, 503–537.

Farquhar, G.D., Hubick, K.T., Condon, A.G. & Richards, R.A. (1989b) Carbon isotope fractionation and plant water-use efficiency. In: *Stable Isotopes in Ecological Research* (Eds P.W. Rundel, J.R. Ehleringer & K.A. Nagy), pp. 21–40. Springer-Verlag, New York.

Farquhar, G.D., Lloyd, J., Taylor, J.A., et al. (1993) Vegetation effects on the isotope composition of oxygen in atmospheric CO_2. *Nature*, **363**, 439–443.

Feng, X. (1998) Long-term c_i/c_a response of trees in western North America to atmospheric CO_2 concentration derived from carbon isotope chronologies, *Oecologia*, **117**, 19–25.

Field, T.S. & Dawson, T.E. (1998) Water sources used by *Didymopanax pittieri* at different life stages in a tropical cloud forest. *Ecology*, **79**, 1448–1452.

Field, C.B., Ball, J.T. & Berry, J.A. (1989) Photosynthesis: principles and field techniques. In: *Plant Physiological Ecology* (Eds R.W. Pearcy, J.R. Ehleringer, H.A. Mooney & P.W. Rundel,), pp. 209–253. Chapman & Hall, New York.

Flanagan, L.B., Comstock, J.P. & Ehleringer, J.R. (1991) Comparison of modeled and observed environmental influences on the stable oxygen and hydrogen isotope composition of leaf water in *Phaseolus vulgaris* L. *Plant Physiology*, **96**, 588–596.

Frank, D.A., Evans, R.D. & Tracy, B.F. (2004) The role of ammonia volatilization in controlling the natural ^{15}N abundance of a grazed grassland. *Biogeochemistry*, **68**, 169–178.

Freyer, H.D. (1979) On the ^{13}C record in tree rings. Part I. ^{13}C variation in northern hemispheric trees during the last 150 years. *Tellus*, **31**, 124–137.

Garten, C.T., Jr. & Taylor, G.E., Jr. (1992) Foliar $\delta^{13}C$ within a temperate deciduous forest: spatial, temporal, and species sources of variation. *Oecologia*, **90**, 1–7.

Gat, J., Mook, W.G. & Meijer, H.A.J. (2001) *Environmental Isotopes in the Hydrological Cycle: Principles and Applications*. Volume II: *Atmospheric Water*. International Atomic Energy Agency, http://www.iaea.or.at/programmes/ripc/ih/volumes/volumes.htm.

Gebauer, G. & Dietrich, P. (1993) Nitrogen isotope ratios in different compartments of a mixed stand of spruce, larch and beech trees and of understorey vegetation including fungi. *Environmental and Health Studies*, **29**, 35–44.

Gebauer, G. & Schulze, E.-D. (1991) Carbon and nitrogen isotope ratios in different compartments of a healthy and a declining (*Picea abies*) forest in the Fichtelgebirge, NE Bavaria. *Oecologia*, **87**, 198–207.

Gebauer, G., Giesemann, A., Schulze, E.-D. & Jager, H.-J. (1994) Isotope ratios and concentrations of sulfur and nitrogen in needles and soils of *Picea abies* stands as influenced by atmospheric deposition of sulfur and nitrogen compounds. *Plant and Soil*, **164**, 267–281.

Ghashghaie, J., Duranceau, M., Badeck, F., Cornic, G., Adeline, M.T. & Deleens, E. (2001) $\delta^{13}C$ of CO_2 respired in the dark in relation to leaf metabolites: comparisons between *Nicotiana sylvestris* and *Helianthus annuus* under drought. *Plant Cell and Environment*, **24**, 505–515.

Ghashghaie, J., Badeck, F.-W., Lanigan, G., et al. (2003) Carbon isotope fractionation during dark respiration and photosrespiration in C3 plants. *Phytochemistry Reviews*, **2**, 145–161.

Gillon, J.S. & Griffiths, H. (1997) The influence of (photo)respiration on carbon isotope discrimination in plants. *Plant Cell and Environment*, **20**, 1217–1230.

Goldstein, G., Rada, F., Sternberg, L., et al. (1989) Gas exchange and water balance of a mistletoe species and its mangrove hosts. *Oecologia*, **78**, 176–183.

Gray, J., Thompson, P. (1976) Climatic information from $^{18}O/^{16}O$ ratios of cellulose in tree rings. *Nature*, **262**, 481–482.

Guy, R.D., Fogel, M.F., Berry, J.A. & Hoering, T.C. (1987) Isotope fractionation during oxygen production and consumption by plants. In: *Progress in Photosynthesis Research* (Ed. J. Biggins), pp. 597–600. Martinus Nijhoff, Dordrecht.

Hall, A.E., Mutters, R.G., Hubick, K.T. & Farquhar, G.D. (1990) Genotypic differences in carbon isotope discrimination by cowpea under wet and dry field conditions. *Crop Science*, **30**, 300–305.

Handley, L.L. & Raven, J.A. (1992) The use of natural abundance of nitrogen isotopes in plant physiology and ecology. *Plant, Cell and Environment*, **15**, 965–985.

Handley, L.L., Odee, D. & Scrimgeour, C.M. (1994) $\delta^{15}N$ and $\delta^{13}C$ patterns in savanna vegetation: dependence on water availability and disturbance. *Functional Ecology*, **8**, 306–314.

Hattersley, P.W. (1982) $\delta^{13}C$ values of C4 types in grasses. *Australian Journal of Plant Physiology*, **9**, 139–154.

Heaton, T.H.E. (1986) Isotopic studies of nitrogen pollution in the hydrosphere and atmosphere: a review. *Chemical Geology*, **59**, 87–102.

Heaton, T.H.E. (1990) $^{15}N/^{14}N$ ratios of NO_x from vehicle engines and coal-fired stations. *Tellus*, **42B**, 304–307.

Helliker, B.R. & Ehleringer, J.R. (2002) Grass blades as tree rings: environmentally induced changes in the oxygen isotope ratio of cellulose along the length of grass blades. *New Phytologist*, **155**, 417–424.

Henderson, S.A., von Caemmerer, S. & Farquhar, G.D. (1992) Short-term measurements of carbon isotope discrimination in several C4 species. *Australian Journal of Plant Physiology*, **19**, 263–285.

Henn, M.R. & Chapela, I.H. (2001) Ecophysiology of ^{13}C and ^{15}N isotopic fractionation in forest fungi and the roots of the saprotrophic-mycorrhizal divide. *Oecologia*, **128**, 480–487.

Hobbie, E.A. & Werner, R.A. (2004) Intramolecular, compound-specific, and bulk carbon isotope patterns in C3 and C4 plants: a review and synthesis. *New Phytologist*, **161**, 371–385.

Hobbie, E.A., Macko, S.A. & Shugart, H.H. (1998) Patterns in N dynamics and N isotopes during primary succession in Glacier Bay, Alaska. *Chemical Geology*, **152**, 3–11.

Hobbie, E.A., Macko, S.A. & Shugart, H.H. (1999) Insights into nitrogen and carbon dynamics of ectomycorrhizal and saprotrophic fungi from isotopic evidence. *Oecologia*, **118**, 353–360.

Hobbie, E.A., Macko, S.A. & Williams, M. (2000) Correlations between foliar $\delta^{15}N$ and nitrogen concentrations may indicate plant-mycorrhizal interactions. *Oecologia*, **122**, 273–283.

Hobbie, E.A., Gregg, J., Olszyk, D.M., Rygiewicz, P.T. & Tingey, D.T. (2002) Effects of climate change on labile and structural carbon in Douglas-fir needles as estimated by $\delta^{13}C$ and C_{area} measurements. *Global Change Biology*, **8**, 1072–1084.

Högberg, P. (1990) ^{15}N natural abundance as a possible marker of the ectomycorrhizal habit of trees in mixed African woodlands. *New Phytologist*, **115**, 483–486.

Högberg, P. (1997) Tansley Review No. 95: ^{15}N natural abundance in soil–plant systems. *New Phytologist*, **137**, 179–203.

Högberg, P., Högberg, M.N., Quist, M.E., Ekblad, A. & Näsholm, T. (1999) Nitrogen isotope fractionation during nitrogen uptake by ectomycorrhizal and non-mycorrhizal *Pinus sylvestris*. *New Phytologist*, **142**, 569–576.

Hubick, K.T., Shorter, R. & Farquhar, G.D. (1988) Heritability and genotype X environment interactions of carbon isotope discrimination and transpiration efficiency in peanut (*Arachis hypogaea* L.). *Australian Journal of Plant Physiology*, **15**, 799–813.

Hultine, K.R. & Marshall, J.D. (2000) Altitude trends in conifer leaf morphology and stable carbon isotope composition. *Oecologia*, **123**, 32–40.

Jäggi, M., Saurer, M., Fuhrer, J. & Siegwolf, R. (2003) Seasonality of $\delta^{18}O$ in needles and wood of *Picea abies*. *New Phytologist*, **158**, 51–59.

Johnson, D.A., Asay, K.H., Tieszen, L.L., Ehleringer, J.R. & Jefferson, P.G. (1990) Carbon isotope discrimination: potential in screening cool-season grasses for water-limited environments. *Crop Science*, **30**, 338–343.

Kalisz, S. & Teeri, J.A. (1986) Population-level variation in photosynthetic metabolism and growth in *Sedum wrightii*. *Ecology*, **67**, 20–26.

Katul, G.G., Ellsworth, D.S. & Lai, C.-T. (2000) Modelling assimilation and intercellular CO2 from measured conductance: a synthesis of approaches. *Plant Cell Environment*, **23**, 1313–1328.

Kluge, M., Brulfert, J., Ravelomanana, D., Lipp, J. & Ziegler, H. (1991) Crassulacean acid metabolism in Kalanchoë species collected in various climatic zones of Madagascar: a survey by $\delta^{13}C$ analysis. *Oecologia*, **88**, 407–414.

Kolb, K.J. & Evans, R.D. (2002) Implications of leaf nitrogen recycling on the nitrogen isotope composition of deciduous plant tissues. *New Phytologist*, **156**, 57–64.

Körner, Ch., Farquhar, G.D. & Roksandic, Z. (1988) A global survey of carbon isotope discrimination in plants from high altitude. *Oecologia*, **74**, 623–632.

Körner, Ch., Farquhar, G.D. & Wong, S.C. (1991) Carbon isotope discrimination by plants follows latitudinal and altitudinal trends. *Oecologia*, **88**, 30–40.

Krishnamurthy, R.V. & Epstein, S. (1985) Tree ring D/H ratio from Kenya, East Africa and its palaeoclimatic significance. *Nature*, **317**, 160–161.

Lai, C.-T., Ehleringer, J.R., Bond, B.J. & Paw, K.T.U. (2006) Contributions of evaporation, isotopic non-steady state transpiration and atmospheric mixing on the $\delta^{18}O$ of water vapour in Pacific Northwest coniferous forests. *Plant, Cell and Environment*, **29**, 77–94.

Lajtha, K. & Schlesinger, W.H. (1986) Plant response to variations in nitrogen availability in a desert shrubland community. *Biogeochemistry*, **2**, 29–37.

Lawrence, J.R. & White, J.W.C. (1984) Growing season precipitation from D/H ratios of Eastern White Pine. *Nature*, **311**, 558–562.

Leavitt, S.W. & Danzer, S.R. (1993) Methods for batch processing small wood samples to holocellulose for stable-carbon isotope analysis. *Analytical Chemistry*, **65**, 87–89.

Leavitt, S.W. & Long, A. (1986) Stable carbon isotope variability in tree foliage and wood. *Ecology*, **67**, 1002–1010.

Libby, L.M., Pandolfi, L.J., Payton, P.H., et al. (1976) Isotopic tree thermometers. *Nature*, **261**, 284–288.

Lin, G. & Ehleringer, J.R. (1997) Carbon isotopic fractionation does not occur during dark respiration in C3 and C4 plants. *Plant Physiology*, **114**, 391–394.

Lin, G. & Sternberg, L. (1993) Hydrogen isotope factionation by plant roots during water uptake in coastal wetland plants. In: *Stable Isotopes and Plant Carbon–Water Relations* (Eds J.R. Ehleringer, A.E. Hall & G.D. Farquhar), pp. 497–510. Academic Press, San Diego.

Lloyd, J., Syvertsen, J.P., Kriedemann, P.E. & Farquhar, G.D. (1992) Low conductances for carbon dioxide diffusion from stomata to the sites of carboxylation in leaves of woody species. *Plant Cell Environment*, **15**, 873–899.

Loader, N.J., Robertson, I., Barker, A.C., Switsur, V.R. & Waterhouse, J.S. (1997) An improved technique for the batch processing of small wholewood samples to a-cellulose. *Chemical Geology*, **136**, 313–317.

Luo, Y. & Sternberg, L. (1992) Hydrogen and oxygen isotope fractionation during heterotropic cellulose synthesis. *Journal of Experimental Botany*, **43**, 47–50.

Majoube, M. (1971) Fractionnement en oxygene-18 et en deuterium entre l'eau et sa vapeur. *Journal of Chemical Physics*, **58**, 1423–1436.

Marino, B.D. & McElroy, M.B. (1991) Isotopic composition of atmospheric CO_2 inferred from carbon in C4 plant cellulose. *Nature*, **349**, 127–131.

Marshall, J.D. & Monserud, R.A. (1996) Homeostatic gas-exchange parameters inferred from $^{13}C/^{12}C$ in tree rings of conifers during the twentieth century. *Oecologia*, **105**, 13–21.

Marshall, J.D. & Zhang, J. (1994) Carbon isotope discrimination and water-use efficiency in native plants of the north-central Rockies. *Ecology*, **75**, 1887–1895.

McCarroll, D. & Loader, N.J. (2004) Stable isotopes in tree rings. *Quaternary Science Reviews*, **23**, 771–801.

Medina, E. & Minchin, P. (1980) Stratification of $\delta^{13}C$ values of leaves in Amazonian rain forests. *Oecologia*, **45**, 377–378.

Medina, E., Montes, G., Cuevas, E. & Roksandic, Z. (1986) Profiles of CO_2 concentration and $\delta^{13}C$ values in tropical rain forests of the Upper Rio Negro Basin, Venezuela. *Journal of Tropical Ecology*, **2**, 207–217.

Medina, E., Sternberg, L. & Cuevas, E. (1991) Vertical stratification of $\delta^{13}C$ values in closed natural and plantation forests in the Luquillo mountains, Puerto Rico. *Oecologia*, **87**, 369–372.

Meinzer, F.C., Saliendra, N.Z. & Crisosto, C.H. (1992) Carbon isotope discrimination and gas exchange in *Coffea arabica* during adjustment to different soil moisture regimes. *Australian Journal of Plant Physiology*, **19**, 171–184.

Meinzer, F.C., Andrade, J.l., Goldstein, G., Holbrook, N.M., Cavelier, J. & Wright, S.J. (1999) Partitioning of soil water among canopy trees in a seasonally dry tropical forest. *Oecologia*, **121**, 293–301.

Monserud, R.A. & Marshall, J.D. (2001) Time-series analysis of $\delta^{13}C$ from tree rings. I. Time trends and autocorrelation. *Tree Physiology*, **21**, 1087–1102.

Mooney, H.A., Bullock, S.H. & Ehleringer, J.R. (1989) Carbon isotope ratios of plants of a tropical dry forest in Mexico. *Functional Ecology*, **3**, 137–142.

Moore, H. (1977) The isotopic composition of ammonia, nitrogen dioxide and nitrate in the atmosphere. *Atmospheric Environment*, **11**, 1239–1243.

Morecroft, M.D. & Woodward, F.I. (1990) Experimental investigations on the environmental determination of $\delta^{13}C$ at different altitudes. *Journal of Experimental Botany*, **41**, 1303–1308.

Moreira, M.Z., Sternberg, L., Martinelli, L.A., et al. (1997) Contribution of transpiration to forest ambient vapor based on isotopic measurements. *Global Change Biology*, **3**, 439–450.

Moreira, M.Z., Sternberg, L. & Nepstad, D.C. (2000) Vertical patterns of soil water uptake by plants in a primary forest and an abandoned pasture in the eastern Amazon: an isotopic approach. *Plant and Soil*, **222**, 95–107.

Ocheltree, T.W. & Marshall, J.D. (2004) Apparent respiratory discrimination is correlated with growth rate in the shoot apex of sunflower (*Helianthus annuus*). *Journal of Experimental Botany*, **55**, 2599–2605.

O'Leary, M.H. (1981) Carbon isotope fractionations in plants. *Phytochemistry*, **20**, 553–567.

O'Leary, M.H. (1988) Carbon isotopes in photosynthesis. *Bioscience*, **38**, 328–336.

Osmond, C.B., Bender, M.M. & Burris, R.H. (1976) Pathways of CO_2 fixation in the CAM plant Kalanchoë daigremontiana. III. Correlation with $\delta^{13}C$ value during growth and water stress. *Australian Journal of Plant Physiology*, **3**, 787–799.

Pardo, L.H., Hemond, H.F., Montoya, J.P., Fahey, T.J. & Siccama, T.G. (2002) Response of the natural abundance of ^{15}N in forest soils and foliage to high nitrate loss following clear-cutting. *Canadian Journal of Forest Research*, **32**, 1126–1136.

Park, R. & Epstein, S. (1961) Metabolic fractionation of ^{13}C and ^{12}C in plants. *Plant Physiology*, **36**, 133–138.

Pataki, D.E., Ehleringer, J.R., Flanagan, L.B. et al. (2003) The application and interpretation of Keeling plots in terrestrial carbon cycle research. *Global Biogeochemical Cycles*, **17**, 22-1–22-14.

Pate, J.S., Stewart, G.R. & Unkovich, M. (1993) ^{15}N natural abundance of plant and soil components of a Banksia woodland ecosystem in relation to nitrate utilization, life form, mycorrhizal status and N_2-fixing abilities of component species. *Plant, Cell and Environment*, **16**, 365–373.

Peoples, M.B., Bergersen, F.J., Turner, G.L., et al. (1991) Use of the natural enrichment of ^{15}N in plant available soil N for the measurement of symbiotic N_2 fixation. In: *Stable Isotopes in Plant Nutrition, Soil Fertility and Environmental Studies*, pp. 117–123. International Atomic Energy Agency and Food and Agriculture Organization, Vienna.

Phillips, D.L. & Gregg, J.W. (2001) Uncertainty in source partitioning using stable isotopes. *Oecologia*, **127**, 171–179.

Phillips, D.L. & Gregg, J.W. (2003) Source partitioning using stable isotopes: coping with too many sources. *Oecologia*, **136**, 261–269.

Plamboeck, A.H., Grip, H. & Nygren, U. (1999) A hydrological tracer study of water uptake depth in a Scots pine forest under two different water regimes. *Oecologia*, **119**, 452–460.

Read, J.J., Johnson, D.A., Asay, K.H. & Tieszen, L.L. (1991) Carbon isotope discrimination, gas exchange, and water-use efficiency in crested wheatgrass clones. *Crop Science*, **31**, 1203–1208.

Roden, J.S. & Ehleringer, J.R. (1999a) Hydrogen and oxygen isotope ratios of tree-ring cellulose for riparian trees grown long-term under hydroponically controlled environments. *Oecologia*, **121**, 467–477.

Roden, J.S. & Ehleringer, J.R. (1999b) Observations of hydrogen and oxygen isotopes in leaf water confirm the Craig–Gordon Model under wide-ranging environmental conditions. *Plant Physiology*, **120**, 1165–1173.

Roden, J.S. & Ehleringer, J.R. (2000) Hydrogen and oxygen isotope ratios of tree rings cellulose for field-grown riparian trees. *Oecologia*, **123**, 481–489.

Roden, J.S., Lin, G. & Ehleringer, J.R. (2000) A mechanistic model for interpretation of hydrogen and oxygen isotope ratios in tree-ring cellulose. *Geochimica et Cosmochimica Acta*, **64**, 21–35.

Roeske, C.A. & O'Leary, M.H. (1984) Carbon isotope effect on carboxylation of ribulose bisphosphate catalyzed by ribulose bisphosphate carboxylase from *Rhodospirillum rubrum*. *Biochemistry*, **24**, 1603–1607.

Sage, R.F., Li, M. & Monson, R.K. (1999) The taxonomic distribution of C4 photosynthesis. In: *C4 Plant Biology* (Eds R.F. Sage & R.K. Monson), pp. 551–581. Academic Press, San Diego.

Sandquist, D.R. & Ehleringer, J.R. (1995) Carbon isotope discrimination in the C4 shrub *Atriplex confertifolia* along a salinity gradient. *Great Basin Naturalist*, **55**, 135–141.

Saurer, M., Aellen, K. & Siegwolf, R. (1997a) Correlating $\delta^{13}C$ and $\delta^{18}O$ in cellulose of trees. *Plant, Cell and Environment*, **20**, 1543–1550.

Saurer, M, Borella, S. & Leuenberger, M. (1997b) $\delta^{18}O$ if tree rings of beech (*Fagus silvatica*) as a record of $d^{18}O$ of the growing season precipitation. *Tellus*, **49B**, 80–92.

Saurer, M., Cherubini, P. & Siegwolf, R. (2000) Oxygen isotopes in tree rings of *Abies alba*: The climatic significance of interdecadal variations. *Journal of Geophysical Research*, **105**, 461–470.

Scheidegger, Y., Saurer, M., Bahn, M. & Siegwolf, R. (2000) Linking stable oxygen and carbon isotopes with stomatal conductance and photosynthetic capacity: a conceptual model. *Oecologia*, **125**, 350–357.

Shearer, G. & Kohl, D.H. (1986) N_2 fixation in field settings: estimations based on natural ^{15}N abundance. *Australian Journal of Plant Physiology*, **13**, 699–756.

Shearer, G., Duffy, J., Kohl, D. & Commoner, B. (1974) A steady state model of isotopic fractionation accompanying nitrogen transformation in soils. *Soil Science Society of America Proceedings*, **38**, 315–322.

Shearer, G., Virginia, R.A., Bryan, B.A., et al. (1983) Estimates of N_2-fixation from variation in the natural abundance of ^{15}N in Sonoran Desert ecosystems. *Oecologia*, **56**, 365–373.

Smedley, M.P., Dawson, T.E., Comstock, J.P., et al. (1991) Seasonal carbon isotope discrimination in a grassland community. *Oecologia*, **85**, 314–320.

Smith, J.A.C., Griffiths, H. & Luttge, G. (1986) Comparative ecophysiology of CAM and C3 bromeliads. I. The ecology of the Bromeliaceae in Trinidad. *Plant, Cell, and Environment*, **91**, 359–376.

Sparks, J.P., Monson, R.K., Sparks, K.L. & Lerdau, M. (2001) Leaf uptake of nitrogen dioxide (NO_2) in a tropical wet forest: Implications for tropospheric chemistry. *Oecologia*, **127**, 214–221.

Sparks, J.P., Roberts, J.M. & Monson, R.K. (2003) The uptake of gaseous organic nitrogen by leaves: A significant global nitrogen transfer process. *Geophysical Research Letters*, **30**, ASC 4.

Spriggs, A.C., Stock, W.D. & Dakora, F.D. (2003) Influence of mycorrhizal associations on foliar $\delta^{15}N$ values of legume and non-legume shrubs and trees in the fynbos of South Africa: Implications for estimating N_2 fixation using the ^{15}N natural abundance method. *Plant and Soil*, **255**, 495–502.

Sternberg, L. (1989) Oxygen and hydrogen isotope measurements in plant cellulose analysis. In: *Modern Methods of Plant Analysis*, Vol. 10: *Plant Fibers* (Eds H.F. Linskens & J.F. Jackson), pp. 89–99. Springer-Verlag, Berlin.

Sternberg, L. & DeNiro, M.J. (1983) Isotopic composition of cellulose from C3, C4, and CAM plants growing in the vicinity of one another. *Science*, **220**, 947–948.

Sternberg, L., DeNiro, M.J. & Johnson, H.B. (1984a) Isotope ratios of cellulose from lants having different photosynthetic pathways. *Plant Physiology*, **74**, 557–561.

Sternberg, L., DeNiro, M.J. & Ting, I.P. (1984b) Carbon, hydrogen and oxygen isotope ratios of cellulose from plants having intermediate photosynthetic modes. *Plant Physiology*, **74**, 104–107.

Sternberg, L., DeNiro, M.J. & Keeley, J.E. (1984c) Hydrogen, oxygen and carbon isotope ratios of cellulose from submerged aquatic Crassulacean acid metabolism and non-Crassulacean acid metabolism plants. *Plant Physiology*, **76**, 68–70.

Sternberg, L., Mulkey, S.S. & Wright, S.J. (1989) Ecological interpretation of leaf carbon isotope ratios: influence of respired carbon dioxide. *Ecology*, **70**, 1317–1324.

Sternberg, L., Anderson, W.T. & Morrison, K. (2003) Separating soil and leaf water 18O isotope signals in plant stem cellulose. *Geochimica et Cosmochimica Acta*, **67**, 2561–2566.

Tcherkez, G., Nogus, S., Bleton, J., Cornic, G., Badeck, F.-W. & Gahshghaie, J. (2002) Metabolic origian of carbon isotope composition of leaf dark-respired CO_2 in *Phaseolus vulgaris* L. *Plant Physiology*, **131**, 237–244.

Teeri, J.A. & Gurevitch, J. (1984) Environmental and genetic control of Crassulacean acid metabolism in two Crassulacean species and an F_1 hybrid with differing biomass $\delta^{13}C$ values. *Plant Cell and Environment*, **7**, 589–596.

Teeri, J.A. & Stowe, L.G. (1976) Climatic patterns and the distribution of C4 grasses in North America. *Oecologia*, **23**, 1–12.

Terwilliger, V.J., Kitajima, K., Le Roux-Swarthout, D.J., Mulkey, S. & Wright, S.J. (2001) Intrinsic water-use efficiency and heterotrophic investment in tropical leaf growth of two Neotropical pioneer tree species as estimated from $\delta^{13}C$ values. *New Phytologist*, **152**, 267–281.

Thorburn, P.J. & Ehleringer, J.R. (1995) Root water uptake of field-growing plants indicated by measurements of natural-abundance deuterium. *Plant and Soil*, **177**, 225–233.

Tozer, W.C., Hackell, D., Miers, D.B. & Silvester, W.B. (2005) Extreme isotopic depletion of nitrogen in New Zealand lithophytes and epiphytes; the result of diffusive uptake of atmospheric ammonia? *Oecologia*, **144**, 628–635.

Virginia, R. & Delwiche, C.C. (1982) Natural ^{15}N abundance of presumed N_2-fixing and non-N_2-fixing plants from selected ecosystems. *Oecologia*, **54**, 317–325.

Vitousek, P.M., Shearer, G. & Kohl, D.H. (1989) Foliar ^{15}N natural abundance in Hawaiian rainforest: patterns and possible mechanisms. *Oecologia*, **78**, 383–388.

Vitousek, P.M., Field, C.B. & Matson, P.A. (1990) Variation in foliar $\delta^{13}C$ in Hawaiian *Metrosideros polymorpha*: a case of internal resistance? *Oecologia*, **84**, 362–388.

Vogel, J.C. (1978) Recycling of carbon in a forest environment. *Oecologia Plantarum*, **13**, 89–94.

Wang, X.-F. & Yakir, D. (2000) Using stable isotopes of water in evapotranspiration studies. *Hydrological Processes*, **14**, 1407–1421.

Warren, C.R., Ethier, G.J., Livingston, N.J., et al. (2003) Transfer conductance in second growth Douglas-fir (*Pseudotsuga menziesii* (Mirb.) Franco) canopies. *Plant, Cell and Environment*, **26**, 1215–1227.

Werner, R.A. & Schmidt, H.L. (2002) The *in vivo* nitrogen isotope discrimination among organic plant compounds. *Phytochemistry*, **61**, 465–484.

White, J.W.C. (1989) Stable hydrogen isotope ratios in plants: a review of current theory and some potential applications. In: *Applications of Stable Isotopes in Ecological Research* (Eds P.W. Rundel, J.R. Ehleringer & K.A. Nagy), pp. 142–160. Springer-Verlag, New York.

White, J.W.C., Lawrence, J.R. & Broecker, W.S. (1994) Modeling and interpreting D/H ratios in tree rings: A test case of white pine in the northern United States. *Geochimica et Cosmochimica Acta*, **58**, 851–862.

Williams, D.G. & Ehleringer, J.R. (2000) Intra- and interspecific variation for summer precipitation use in Pinyon–Juniper woodlands. *Ecological Monographs*, **70**, 517–537.

Williams, D.G., Cable, W.L., Hultine, K.R., et al. (2004) Evapotranspiration components determined by stable isotope, sap flow and eddy covariance techniques. *Agricultural and Forest Meteorology*, **125**, 241–258.

Wilson, E.J. & Tiley, C. (1998) Foliar uptake of wet-deposited nitrogen by Norway spruce: An experiment using ^{15}N. *Atmospheric Environment*, **32**, 513–518.

Yakir, D. (1998) Oxygen-18 of leaf water: a crossroad for plant-associated isotopic signals. In: *Stable Isotopes: Integration of Biological, Ecological and Geochemical Processes* (Ed. H. Griffiths), pp. 147–168. BIOS Scientific Publishers, Oxford.

Yakir, D. & DeNiro, M.J. (1990) Oxygen and hydrogen isotope fractionation during cellulose metabolism in Lemna gibba L. *Plant Physiology*, **93**, 325–332.

Yakir, D., DeNiro, M.J. & Rundel, P.W. (1989) Isotopic inhomogeneity of leaf water: Evidence and implications for the use of isotopic signals transduced by plants. *Geochimica et Cosmochimica Acta*, **53**, 2769–2773.

Yakir, D., Berry, J.A., Giles, L. & Osmond, C.B. (1993) The $\delta^{18}O$ of water in the metabolic compartment of transpiring leaves. In: *Stable Isotopes and Plant Carbon–Water Relations* (Eds J.R. Ehleringer, A.E. Hall & G.D. Farquhar), pp. 529–540. Academic Press, San Diego.

Yapp, C. & Epstein, S. (1982) Climatic significance of the hydrogen isotope ratio in tree cellulose. *Nature*, **297**, 636–639.

Yepez, E.A., Williams, D.G., Scott, R.L. & Lin, G. (2003) Partitioning overstory and understory evapotranspiration in a semiarid savanna woodland from the isotopic compostion of water vapor. *Agricultural and Forest Meteorology*, **119**, 53–68.

Zhang, J., Marshall, J.D. & Jaquish, B.C. (1993) Genetic differentiation in carbon isotope discrimination and gas exchange in *Pseudotsuga menziesii*: a common garden experiment. *Oecologia*, **93**, 80–87.

Zhang, J.W., Fins, L. & Marshall, J.D. (1994) Stable carbon isotope discrimination, photosynthetic gas exchange, and growth differences among western larch families. *Tree Physiology*, **14**, 531–540.

Ziegler, H., Osmond, C.B., Stickler, W. & Trimborn, D. (1976) Hydrogen isotope discrimination in higher plants: correlation with photosynthetic pathways and environment. *Planta*, **128**, 85–92.

CHAPTER 3

Natural ^{15}N- and ^{13}C-abundance as indicators of forest nitrogen status and soil carbon dynamics

CHARLES T. GARTEN, JR, PAUL J. HANSON,
DONALD E. TODD, JR, BONNIE B. LU, AND
DEANNE J. BRICE

Introduction

The purpose of this chapter is to examine how natural abundance measurements of stable N and C isotope ratios might be used as indicators of environmental processes that impact soil C storage in forest ecosystems. This is important because increasing atmospheric CO_2 concentrations have created substantial recent interest in science and technology needs for increasing global C sequestration (Lal 2004). Soil C balance, in particular, has been at the center of many science questions related to the exchange of CO_2 between the terrestrial biosphere and the atmosphere because most of the global C inventory resides in soils (Post et al. 1990; Schimel 1995). Environmental factors that produce even a small change in global soil C stocks (through changes in soil C inputs or outputs) have the potential to produce a disproportionately large change in levels of atmospheric CO_2.

Enhanced uptake of atmospheric CO_2 by terrestrial ecosystems through management of croplands, grasslands, and forests has been suggested as a low-cost and technologically achievable strategy to partially offset CO_2 emissions to the atmosphere from the continuing use of fossil fuels (Lal 2004; Post et al. 2004). However, because below-ground processes are difficult to observe and measure, we have a relatively poor understanding of (i) soil C dynamics, (ii) the degree of certainty associated with estimates of below-ground processes, and (iii) the factors that regulate soil C sequestration potential at regional and global scales (Metting et al. 2001).

Forest soil C dynamics can be simply described as the difference between soil C inputs and outputs (Six & Jastrow 2002). Litterfall and rhizodeposition (i.e., root mortality and root exudation) are the primary contributors to soil C inputs. Decomposition of organic matter inputs by heterotrophic soil microorganisms is the primary contributor to soil C loss. Quantitative

*The submitted manuscript has been authored by a contractor of the U.S. Government under contract DE-AC05-00OR22725. Accordingly, the U.S. Government retains a nonexclusive, royalty-free license to publish or reproduce the published form of this contribution, or allow others to do so, for U.S. Government purposes.

determinations of processes controlling forest soil C storage can be both costly and time-consuming. However, continuous-flow, light stable isotope ratio mass spectrometers (CF-IRMS) allow for rapid and precise measurements of C or N concentrations and isotope ratios ($^{13}C/^{12}C$, $^{15}N/^{14}N$) at natural abundance levels in both soils and plants. Our hypothesis is that such measurements may be useful for preliminary assessments of environmental factors affecting forest soil C sequestration potential.

In practice, soil C sequestration potential has two broad aspects: (i) retention and (ii) accumulation. Retention of existing soil C stocks means implementation of land management practices that alter soil processes to minimize long-term reductions in soil organic matter (SOM). Accumulation involves application of management practices or technologies that alter soil processes to maximize the likelihood of increasing long-term soil C storage. Accumulation involves altering the soil C balance such that inputs exceed C losses, while the objective of retention is to ensure that soil C losses do not exceed inputs.

The focus of this chapter is forests because forests occupy ca. 25 percent of the total land area in the USA. Approximately 83 percent of forest resources in the southern USA are classified as "natural" forest while only 17 percent are classified as planted (i.e., tree plantations or "augmented forests" with varying degrees of forest management) (Smith et al. 1997). The latter distribution mirrors a national trend and indicates that most forests in the USA are not intensively managed. Natural forest resources have an important role in strategies for soil C sequestration. Active management practices that promote soil C sequestration on a relatively small coverage of planted forests may be augmented or offset by minor changes in soil C sequestration under a large coverage of natural forests. One of the more important questions in evaluating the role of natural forest resources in C sequestration strategies is: What can natural abundance measurements of stable C and N isotope ratios tell us about forest soil C storage and dynamics?

Although there are multiple environmental factors that can impact forest soil C dynamics (such as temperature, moisture, and mineralogy), many studies indicate an important role of N availability and organic matter decomposition rates. The latter two factors are not necessarily independent; nonetheless we first examine the potential importance of N availability on processes determining soil C dynamics and the use of ^{15}N natural abundance measurements as an indicator of site N status. Second, we examine the use of ^{13}C natural abundance measurements in soil profiles as an indicator of differences in SOM dynamics. Rapid assessments of forest soil C sequestration potential require methodologies and tools that will provide easily measured, reliable indicators of both soil C dynamics and site N status. Such information may permit relative comparisons of soil C sequestration potential across different forest ecosystems, soil types, and climate regimes. Aside from issues surrounding C sequestration, measurements of stable N and C isotopes may also provide a window into biogeochemical processes, such as N cycling and SOM dynamics, which are important to rapid assessments of forest health.

Significance of ^{15}N-abundance to soil carbon sequestration

The role of nitrogen in forest soil carbon dynamics

The importance of ^{15}N measurements in plants and soils as an ecosystem indicator emerges when considering the control that N availability has over soil C dynamics. Reviews of multiple studies demonstrate N fertilization generally increases forest soil C stocks (Johnson 1992; Johnson & Curtis 2001) through increased inputs and decreased losses of SOM. In addition to a widely reported increase in forest growth following N fertilization, studies across both hardwood and coniferous forest stands show that ca. 50 percent of the variation in above-ground net primary production (ANPP) is explained by variation in annual net soil N mineralization (Reich et al. 1997). Some studies along soil N availability gradients indicate annual leaf litter production increases with annual net soil N mineralization (e.g., Nadelhoffer et al. 1983). Greater soil N availability also appears to increase forest fine root production and turnover (Aber et al. 1985; Nadelhoffer 2000). Both increased leaf litter production and greater fine root turnover can directly contribute to increased soil C inputs in N-rich forests.

There is also a growing body of evidence that N availability indirectly controls forest soil C dynamics through effects on organic matter decomposition. Regional-scale studies in forest ecosystems indicate a significant effect of soil N availability on measures of litter quality, such as C-to-N ratios (Stump & Binkley 1993). Greater soil N availability potentially reduces the C-to-N ratio of above- and below-ground litter inputs by increasing plant tissue N concentrations. Various studies (Taylor et al. 1989; Janssen 1996; Kuperman 1999; Silver & Miya 2001) indicate decomposition rates are inversely related to litter C-to-N ratios. While low litter C-to-N ratios accelerate initial stages of litter decomposition, high litter N concentrations (i.e., low litter C-to-N ratios) appear to inhibit the latter stages, resulting in a greater amount of organic matter remaining near the terminus of decomposition (Berg & Matzner 1997; Berg 2000; Berg & Meentemeyer 2002). In some forests, N-rich leaf litter inputs appear to significantly increase humus accumulation (Berg et al. 2001).

Inhibition of phenol oxidase, a lignin-degrading soil enzyme, in high N environments is one mechanism that may promote more humus formation and possibly greater soil C storage when litter inputs have low C-to-N ratios (Berg et al. 2001; Berg & Meentemeyer 2002). Even though responses are sometimes ecosystem specific, elevated levels of inorganic soil N can alter soil microbial composition (Frey et al. 2004; Gallo et al. 2004), suppress phenol oxidase activity (Carreiro et al. 2000; Saiya-Cork et al. 2002; DeForest et al. 2004; Frey et al. 2004; Gallo et al. 2004; Matocha et al. 2004; Waldrop et al. 2004a) and inhibit soil respiration (Fisk & Fahey 2001; Franklin et al. 2003; Bowden et al. 2004; Burton et al. 2004), indicating an overall lower rate of soil microbial activity and reduced organic matter decomposition in N-rich forests. Short-term soil C gains in some upland forest stands appear to be related to the suppression of phenol oxidase activity (Waldrop et al. 2004b).

In summary, greater soil C sequestration in N-rich forests results from increasing soil C inputs and/or decreasing rates of organic matter decomposition. Field studies indicate both processes, altered inputs and decomposition rates, contribute to greater soil C sequestration under N_2-fixing tree species (Resh et al. 2002). However, at least two studies (Neff et al. 2002; Swanston et al. 2004) indicate that enhanced soil N (through fertilization) promotes stabilization of organic matter in soil pools that have long turnover times. Thus, greater N availability has the potential to shift soil C partitioning in favor of more refractory SOM, reduce the overall rate of total soil C turnover, and increase forest soil C sequestration potential. Site N status is a potentially important factor controlling forest soil C dynamics and high N availability can contribute to greater soil C storage.

Use of ^{15}N-abundance for indicating site nitrogen status

Nitrogen-15 case study

Measurements of foliar ^{15}N-abundance and soil-to-plant ^{15}N enrichment factors (*EF*) have been shown to be useful indicators of forest N status. The isotopic composition of soil N may vary from one site to another, thus the relationship between the N isotope composition of plants and soil is frequently presented in terms of an isotopic "enrichment factor" (e.g., Mariotti et al. 1981). Because soil N (the substrate) is a large reservoir relative to plant foliage (the product), the enrichment factor can be approximated as the difference between ^{15}N-abundance in the substrate and the product, or $EF = \delta^{15}N_{leaf} - \delta^{15}N_{soil}$. The use of enrichment factors helps to adjust potential differences in natural foliar ^{15}N-abundance for existing differences in soil ^{15}N-abundance.

Here we present a site-specific example for the use of natural ^{15}N-abundance as an indicator of site N status based on ongoing studies at three forest stands on the U.S. Department of Energy's Oak Ridge Reservation (36°58′N; 84°16′W) near Knoxville, Tennessee. Overstory trees at the upland sites (one ridge and one slope) are predominantly oak (*Quercus* spp.) with scattered pine (*Pinus echinata* and *P. virginiana*) and mesophytic hardwoods (*Liriodendron tulipifera, Fagus grandifolia, Acer rubrum*). One site is located in a valley and dominated by mesophytic hardwoods (primarily *L. tulipifera*). The difference in elevation between ridge and valley at this location is ca. 100 m.

We made conventional measurements (Hart et al. 1994) of potential net N mineralization in surface (0–10 cm) mineral soil samples collected during the spring of 2002 and 2003 from the three forest sites. Composite soil samples from nine replicate plots at each site were used for the analysis. Results were expressed as μg N produced g^{-1} dry soil per week and were normalized for bulk soil N concentrations (μg N g^{-1} dry soil) to estimate a weekly rate of potential net soil N mineralization.

Potential net soil N mineralization in 12-week aerobic laboratory incubations, expressed as either N production (μg N g^{-1}) or production normalized for soil N concentration (i.e., percent soil N mineralized), was significantly different between study sites and years (Table 3.1). Site disparities in production of inorganic N indicated greater N availability in soils from the valley ($F_{2,48} = 78$, $P \leq 0.001$). Significant site differences in N production normalized for soil N concentrations followed a similar pattern ($F_{2,48} = 37$, $P \leq 0.001$), and indicated N availability across the three sites increased in the following order: ridge < slope < valley.

Site-to-site measurements of soil N availability were consistent in 2002 and 2003 (Table 3.1). Nitrate accounted for all of the inorganic N produced during aerobic incubations of valley soils and net soil nitrification was significantly greater in the valley soils than at the ridge or slope sites ($F_{2,48} = 224$, $P \leq 0.001$). The observed topographic patterns in soil N availability were in agreement with prior studies that indicate riparian forests on the Oak Ridge Reservation are more N-rich than upland forests (Garten 1993; Garten et al. 1994).

Table 3.1 Mean (±SE) potential net N mineralization and potential net nitrification in aerobic laboratory incubations (12 weeks) of surface (0–10 cm) mineral soil samples collected from ridge, slope, and valley forests in 2002 and 2003. Means in the same row with different alphabetic superscripts are significantly different ($P \leq 0.05$).

Variable	Units	Year	n	Ridge	Slope	Valley
Potential net soil N	μg N g^{-1}	2002	9	1.23[a] ± 2.06	9.32[a] ± 4.22	51.0[b] ± 2.46
mineralization		2003	9	4.42[a] ± 2.83	5.21[a] ± 1.00	22.7[b] ± 3.90
N production	%	2002	9	0.13[a] ± 0.18	1.34[b] ± 0.55	4.13[c] ± 0.26
normalized for		2003	9	0.43[a] ± 0.29	0.95[b] ± 0.14	1.86[c] ± 0.35
soil N						
concentration						
Potential net soil	μg N g^{-1}	2002	9	0.43[a] ± 0.30	2.12[a] ± 1.04	53.8[b] ± 2.85
nitrification		2003	9	0.59[a] ± 0.25	0.71[a] ± 0.27	24.7[b] ± 4.10

Leaf litterfall was also collected at the ridge, slope, and valley site in the autumn of 2002 and 2003. Litterfall samples were combined by plot (three sampling plots at each site), oven-dried (70°C), and subsampled. Soils (passed through a 2-mm sieve to remove gravel, live roots, and coarse debris) were ball milled and litterfall samples were ground to a powder in a sample mill prior to analysis by combustion methods for total C and N using a LECO CN-2000 (LECO Corporation, St Joseph, MI). Stable isotope ratios for C and N were measured by continuous-flow, isotope ratio mass spectrometry (Integra-CN, SerCon Ltd, Cheshire, UK).

Forest N status was evaluated using ^{15}N enrichment factors (EF) in leaf litterfall: $EF = \delta^{15}N_{litterfall} - \delta^{15}N_{soil}$. The ^{15}N-abundance in the surface mineral soil was used to calculate the enrichment factor. Prior studies (Garten 1993; Kolb & Evans 2002) have shown there is little isotopic discrimination against ^{15}N associated with foliar N reabsorption prior to leaf senescence. Consequently, analysis of leaf litterfall yields an accurate estimate of foliar ^{15}N-abundance in the forest canopy for a particular growing season.

The three forest sites were significantly different in leaf litterfall N concentrations, C-to-N ratios, ^{15}N-abundance, and enrichment factors (Table 3.2). Leaf litterfall C-to-N ratios were inversely related to soil N availability. Leaf litterfall N concentrations were significantly greater ($F_{2,77} = 156$, $P \leq 0.001$) and litterfall C-to-N ratios were significantly less ($F_{2,77} = 156$, $P \leq 0.001$) in the valley (Table 3.2). Differences between years in leaf litterfall N concentrations were also statistically significant ($F_{1,77} = 7.1$, $P \leq 0.01$), but they were consistent in both years (thus data from both 2002 and 2003 were combined for the analysis). Greater soil N availability in the valley produced higher leaf litterfall N concentrations and lower litterfall C-to-N ratios.

Table 3.2 Mean (±SE) surface soil ^{15}N abundance, leaf litterfall enrichment factors (EF), ^{15}N abundance, N concentrations, and C-to-N ratios at three forest sites on the Oak Ridge Reservation. Means in the same column with different alphabetic superscripts are significantly different.

Site	Surface soil		Leaf litterfall				
	n	$\delta^{15}N$ (‰)	n	EF	$\delta^{15}N$ (‰)	N (%)	C:N
Ridge	3	3.42 ± 0.63	15	−7.21[a] ± 0.12	−3.79[a] ± 0.12	0.75[a] ± 0.01	65.5[a] ± 1.1
Slope	3	4.11 ± 0.50	13	−8.25[b] ± 0.10	−4.14[a] ± 0.10	0.81[a] ± 0.04	60.8[a] ± 2.4
Valley	3	3.82 ± 0.42	14	−4.99[c] ± 0.20	−1.17[b] ± 0.20	1.27[b] ± 0.05	37.8[b] ± 1.6
Statistics							
F-value				124.2	119.7	64.3	79.3
Probability				0.001	0.001	0.001	0.001

There were also statistically significant differences in leaf litterfall ^{15}N-abundance and enrichment factors among the three forest sites (Table 3.2). Even though surface soil ^{15}N-abundance was similar among the three locations, site-specific surface soil $\delta^{15}N$ values were used to calculate leaf litterfall ^{15}N enrichment factors. Both the litterfall $\delta^{15}N$ value and the ^{15}N enrichment factor were less negative at the valley site (Table 3.2). More positive (or less negative) foliar $\delta^{15}N$ values and ^{15}N enrichment factors closer to zero are indicative of more N-rich environments (Garten & Van Miegroet 1994), but how do such spatial patterns relate to soil C storage?

There was a tendency toward greater mineral soil C stocks (0–30 cm) at the valley site, relative to the upland forests, but the site differences are not statistically significant (Table 3.3). However, the C stock associated with soil silt and clay (i.e., mineral-associated organic matter) was significantly greater in the valley, as was the total mineral soil N stock (Table 3.3). As discussed in the following section, greater soil N availability (in addition to other environmental factors) may contribute to the partitioning of soil C to mineral-associated organic matter and greater stabilization of mineral soil C stocks in the N-rich valley. Carbon associated with silt and/or clay has been shown to have a longer turnover time than C in more labile fractions of SOM (Balesdent 1996).

In summary, the N-rich valley had higher potential rates of net soil N mineralization, lower leaf litterfall C-to-N ratios, more positive ^{15}N enrichment factors (i.e., *EF* approached zero), and a greater partitioning of C to the soil silt and clay fraction than N-poor forests located on ridges or slopes. An observed relationship between soil N availability and foliar ^{15}N-abundance or soil-to-plant ^{15}N enrichment factors is not new (see e.g., Garten 1993; Garten & Van Miegroet 1994). Garten & Van Miegroet (1994) proposed several possible mechanisms that might underlie changes in the isotopic composition of

Table 3.3 Mean (±SE) N stocks and C stocks in whole mineral soil or the silt and clay fraction at three forest sites on the Oak Ridge Reservation. Means in the same column with different alphabetic superscripts are significantly different (NS = not statistically significant).

Site	n	$g\,N\,m^{-2}$	$g\,C\,m^{-2}$	
		Mineral soil	Mineral soil	Silt and clay
Ridge	12	$133^a \pm 7$	$2941^a \pm 127$	$2045^a \pm 90$
Slope	12	$114^a \pm 5$	$2637^a \pm 150$	$1839^a \pm 88$
Valley	12	$257^b \pm 17$	$3123^a \pm 224$	$2568^b \pm 198$
Statistics				
F-value		50.2	2.0	7.7
Probability		0.001	NS	0.01

foliar N along gradients of N availability in addition to a model that predicts how foliar $\delta^{15}N$ values are affected by (i) varying uptake of soil ammonium-N and nitrate-N, (ii) the isotopic composition of different soil N pools, and (iii) relative rates of soil N transformations. In particular, foliar $\delta^{15}N$ values appear to increase in direct association with the relative importance of net soil nitrification. In this case study, soils from the valley had the highest rates of net soil nitrification (Table 3.1) and the least negative foliar $\delta^{15}N$ values (Table 3.2). Results from three forest sites on the Oak Ridge Reservation exemplify the utility of natural ^{15}N-abundance for distinguishing N-rich and N-poor forests that share similar climates.

Evidence from other studies

Numerous studies lend support to the use of natural abundance measurements of stable N isotopes in plants or calculated ^{15}N enrichment factors as indicators of ecosystem N cycling and/or site N status (Högberg 1990; Garten 1993; Högberg & Johannisson 1993; Garten & Van Miegroet 1994; Nasholm et al. 1997; Emmett et al. 1998; Martinelli et al. 1999; Vervaet et al. 2002; Koba et al. 2003). Several possible mechanisms are potentially responsible for such an association. First, the product of nitrification is ^{15}N-depleted nitrate-N (Mariotti et al. 1981). High rates of net soil nitrification and elevated levels of nitrate-N leaching can contribute to a gradual enrichment in ^{15}N-abundance in forests with open and leaky N cycles because there is a chronic loss of ^{15}N-depleted nitrate (Högberg 1997). In an experimental test of the foregoing hypothesis, nitrate leaching from soils following forest clear-cutting was associated with more positive foliar $\delta^{15}N$ values and changes in foliar ^{15}N-abundance through time coincided with temporal changes in streamwater nitrate concentrations (Pardo et al. 2002). Second, in soils prone to high nitrification rates, denitrification is a natural process of nitrate loss characterized by a relatively large isotopic fractionation (Mariotti et al. 1981) that leaves remaining soil nitrate isotopically enriched in ^{15}N. The utilization of that nitrate by plants can also contribute to more positive foliar $\delta^{15}N$ values in N-rich settings. Last, in N-poor environments, greater reliance on N derived from mycorrhizal fungi contributes to lower foliar $\delta^{15}N$ values while in N-rich environments, less reliance on N derived from mycorrhizal fungi results in higher foliar $\delta^{15}N$ values (Hobbie et al. 2000; Hobbie & Colpaert 2003). In field studies, the relative importance of these various mechanisms (nitrate leaching, denitrification, and mycorrhizal N fractionation) can be difficult to ascertain.

The foregoing mechanisms may be site-specific and they may work independently or in combination, but the overall relationship of foliar $\delta^{15}N$ to site N status is unchanged. Despite the complexity of the N cycle and potential isotopic fractionations associated with various soil N transformations (Högberg 1997; Bedard-Haughn et al. 2003), N-rich forests (particularly sites with elevated net soil nitrification) have been repeatedly differentiated from N-poor forests by more positive foliar δ^{15}N values and more positive ^{15}N-enrichment factors (Garten 1993; Garten & Van Miegroet 1994; Pardo et al. 2002; Koba et al. 2003). For example, δ^{15}N values are generally more negative in N-poor temperate forests with closed N cycles than in relatively N-rich tropical forests with more open N cycles (Martinelli et al. 1999). The utility of natural abundance ^{15}N measurements in distinguishing site N status has also been demonstrated experimentally using gradients of both N fertilization (Högberg 1991; Johannisson & Högberg 1994) and atmospheric N deposition (Emmett et al. 1998).

Vertical changes in soil ^{13}C-abundance and soil carbon dynamics

Decomposition and vertical soil profiles of ^{13}C-abundance

Organic matter decomposition causes a decline in soil C concentration leaving progressively older C and eventually terminating at a steady-state pool of refractory soil C (Wang et al. 1996). This is one reason why rates of organic matter decomposition decline with increasing soil depth (Paul et al. 1997; Van Dam et al. 1997). Carbon-14 measurements demonstrate that both the age and turnover time of forest SOM change over the soil profile such that the oldest and most refractory soil C is generally found in the deepest soils (Wang et al. 1996; Paul et al. 1997; Van Dam et al. 1997; Gaudinski et al. 2000; Torn et al. 2002; Gaudinski & Trumbore 2003). Carbon in mineral-associated organic matter (i.e., soil silt and clay) is also usually older than bulk soil C (Gaudinski & Trumbore 2003).

There is a widely reported occurrence of increasing ^{13}C-abundance with greater soil depth or a similar change along a continuum of progressively more decomposed fractions of SOM (e.g., see Balesdent et al. 1993; Ehleringer et al. 2000; Garten et al. 2000; Powers & Schlesinger 2002; Schweizer et al. 1999). Possible causal mechanisms for greater ^{13}C-abundance with increasing soil depth have been discussed in detail by other authors (Nadelhoffer & Fry 1988; Ehleringer et al. 2000). Briefly, these mechanisms include:

1 changing isotopic ratios in atmospheric CO_2 over the past 200+ years;
2 the enrichment of soil C in ^{13}C as a result of fractionation during organic matter decomposition;
3 the mixing of new C inputs with older SOM that has a different δ^{13}C value;
4 preferential decomposition of organic matter by soil microorganisms.

Opinions vary, but there are a large number of recent studies indicating discrimination against ^{13}C, relative to ^{12}C, during decomposition of organic substrates by heterotrophic soil microorganisms (Mary et al. 1992; Schweizer et al. 1999; Henn & Chapela 2000; Santruckova et al. 2000; Fernandez & Cadisch 2003; Fernandez et al. 2003). Such studies point to a likely mechanism for increasing ^{13}C-abundance during SOM decomposition and are important to the interpretation of vertical profiles of ^{13}C-abundance in forest soils.

The slope of the regression, or the regression coefficient (b), between natural ^{13}C-abundance (Y) and log C concentration (X) has been used by some investigators to describe the rate of change in C isotope composition along a continuum of organic matter decomposition in forest soils (Garten et al. 2000; Powers & Schlesinger 2002). The regression coefficient (b), from the equation $Y = a + b \, (X)$, represents the rate of change in ^{13}C-abundance at a particular forest site or the per mil enrichment of ^{13}C as one proceeds from an initial substrate (generally the forest O-horizon) to deeper and older C in the mineral soil. Hence, b is similar to the isotopic difference between product and source, or an isotopic enrichment factor. Although others (Powers & Schlesinger 2002) have referred to the slope from such a regression as a "beta value", it is referred to here simply as a regression coefficient (which is the statistical term used to refer to the slope of a straight line between two associated variables).

Use of ^{13}C-abundance for indicating soil carbon dynamics

Carbon-13 case study

Relationships between ^{13}C-abundance in vertical soil profiles and soil C concentrations (mg C g^{-1}) may indicate the decomposition rate of SOM (Accoe et al. 2002, 2003) or its turnover time (Garten 2006). Here we present a particular case study of how regression coefficients indicate relative differences in soil C dynamics at seven forest sites on the Oak Ridge Reservation. The mean annual temperature and precipitation at this location is 14°C and 136 cm, respectively (Johnson & Van Hook 1989).

In addition to soil samples from the three sites used in the ^{15}N case study, soil sampling to a 90 cm depth using a 10-cm-diameter corer was conducted on four widely spaced ridges (WB, TV, PR, HR) in March 2001. Soil types on the Oak Ridge Reservation are diverse because of the underlying complexity of the near-surface geology (Hatcher et al. 1992). Sites WB and TV were located on Ultisols while sites PR and HR were located on Inceptisols. The highly weathered soils on the Oak Ridge Reservation are consistently acidic (pH 4.5–5.7) with a high silt and clay content (65–85%).

Three cores were taken from each of eight replicate plots at sites WB, TV, PR, and HR and comparable depth increments were pooled by plot. The soil cores were separated into the following depth increments: O_i, $O_e + O_a$, and four mineral soil depths (0–15, 15–30, 30–60, 60–90 cm). Following procedures described in the preceding ^{15}N case study, the samples were prepared and analyzed for total C concentrations and ^{13}C-abundance. Differences between regression coefficients were tested using Prism 4 (GraphPad Software, Inc., San Diego, CA).

Regressions of ^{13}C-abundance against the logarithm of C concentration were highly significant ($P \leq 0.001$) at sites WB, TV, PR, and HR and indicated an enrichment in ^{13}C-abundance along a continuum from soil C inputs (O-horizons) to more highly decomposed organic matter in deeper soils (Figure 3.1). Similar significant ($P \leq 0.05$) regressions were observed at the ridge, slope, and valley site despite a smaller number of sample types and a shallower sampling depth (Figure 3.2). For the sites examined in this case study, there was significant variation between sites in the slope of the regression of ^{13}C-abundance against the log C concentration. A comparison of regression coefficients from the seven forest sites (Figure 3.3) indicated that the differences were highly significant ($F_{6,25} = 9.4$; $P \leq 0.001$). Soil ^{13}C profiles described in this way were significantly steeper in the valley than at other forest sites.

Regression coefficients from the seven forest sites on the Oak Ridge Reservation were negatively correlated ($P \leq 0.01$) with surface mineral soil N stocks (Figure 3.4) indicating different soil C dynamics in the N-rich valley forest. Disparities in soil moisture are another likely contributor to between-site variations in regression coefficients. Other studies on the Oak Ridge Reservation demonstrate distinctly wetter conditions in valleys and more xeric conditions on ridges and slopes (Hanson et al. 1993). A greater prevalence of anaerobic microsites throughout the soil profile may also promote soil C stabilization through greater humification of organic matter at the valley site.

What might these patterns mean? One interpretation is that within the same climate regime, steeper slopes (i.e., increasingly negative regression coefficients) indicate faster processing of fresh soil C inputs, less new soil C remaining, and possibly greater stabilization of forest soil organic C. Because it is widely established that older, more stabilized, soil organic C has a longer turnover time than fresh C inputs, steeper slopes (like those measured at the valley site) may be indicative of conditions that favor rapid decomposition of fresh C inputs but overall longer soil C turnover times and greater soil C sequestration. Conditions contributing to soil C sequestration in the valley site include both greater N availability (see preceding case study) and higher soil moisture content. In other studies along co-varying gradients of increasing altitude, N availability, and precipitation, greater rates of change in ^{13}C-abundance through forest soil profiles are associated with soil C partitioning (i.e., smaller labile soil C pools) and faster whole soil C turnover times (Garten 2006).

In summary, forest soils in valleys on the Oak Ridge Reservation are both wetter and more N-rich than those on ridges and slopes and vertical patterns in soil ^{13}C-abundance indicate topographic variability in the rate of change

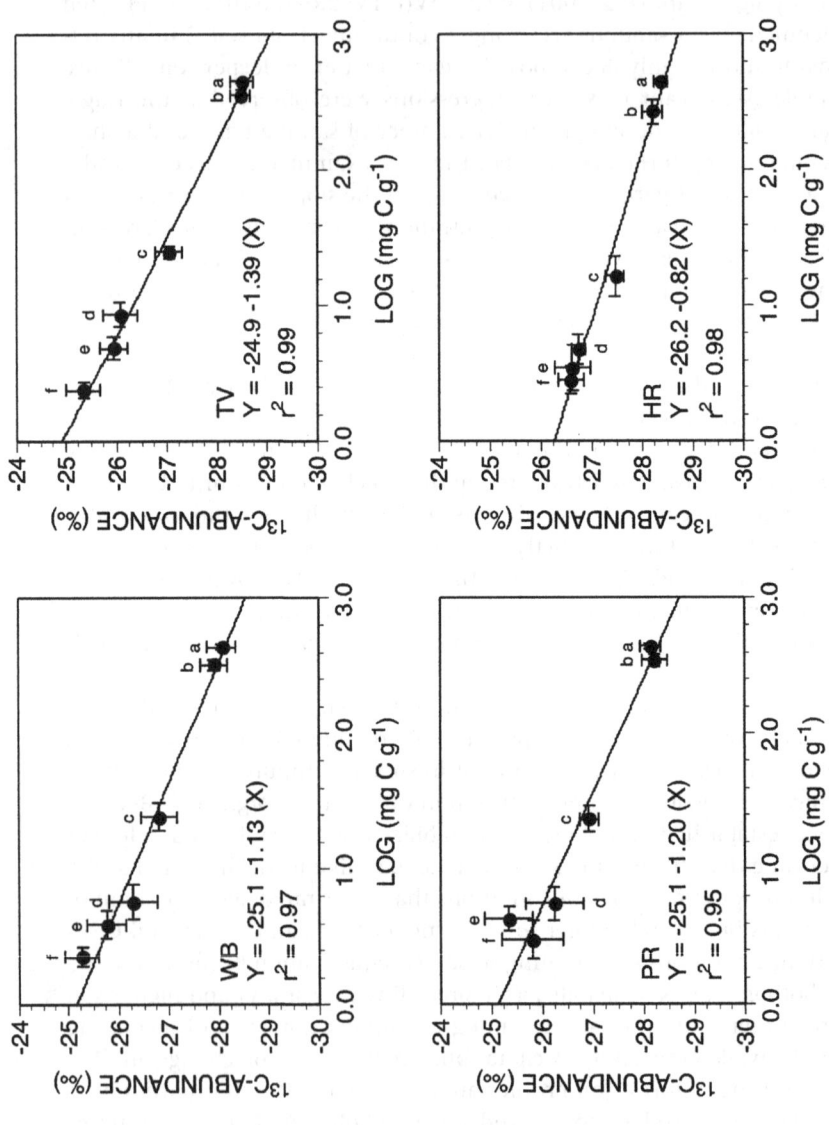

Figure 3.1 Regressions of mean ^{13}C-abundance against the logarithm of mean C concentration in four ridge-top forest soils on the Oak Ridge Reservation. Error bars about each mean are ±1 SE. Various soil parts are indicated by the following letters: (a) O_i horizon, (b) O_e + O_a horizon, (c) soil (0–15 cm), (d) soil (15–30 cm), (e) soil (30–60 cm), (f) soil (60–90 cm).

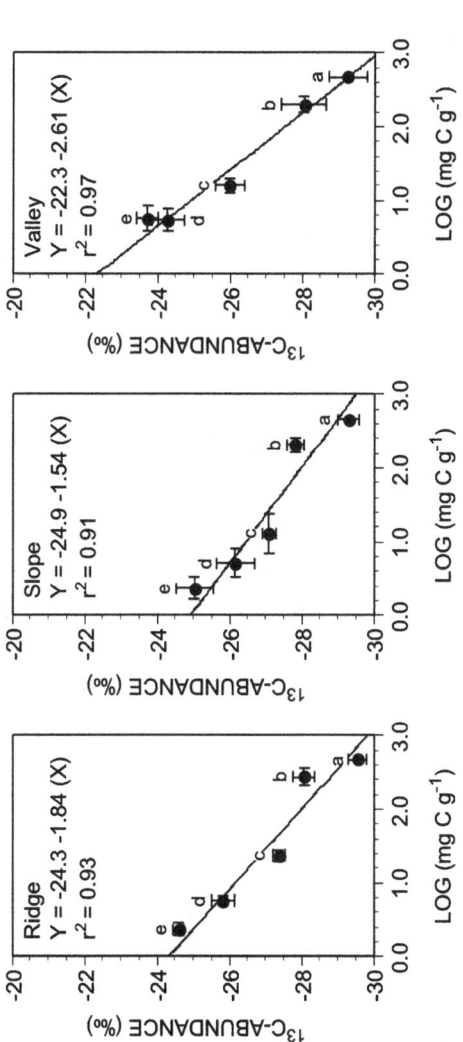

Figure 3.2 Regressions of mean ^{13}C-abundance against the logarithm of mean C concentration in a ridge, slope, and valley soil on the Oak Ridge Reservation. Error bars about each mean are ±1 SE. Various soil parts are indicated by the following letters: (a) leaf litterfall. (b) O-horizon, (c) soil (0–10 cm), (d) soils (10–20 cm), (e) soil (20–30 cm).

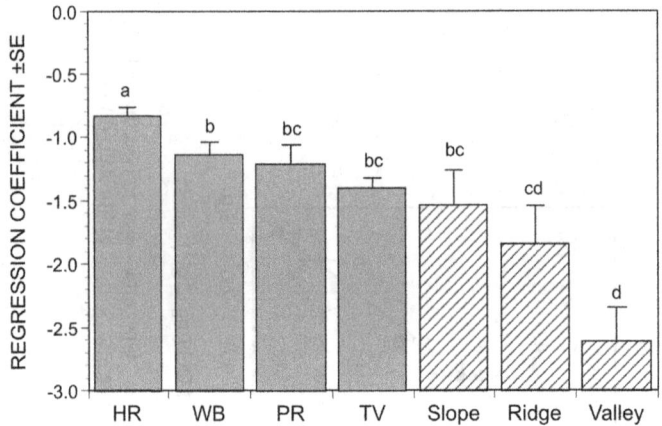

Figure 3.3 Inter-site comparisons of regression coefficients (b) from regressions of mean ^{13}C-abundance against the logarithm of mean C concentrations in seven forest soils on the Oak Ridge Reservation. Sites with the same letter above the error bar are not significantly different.

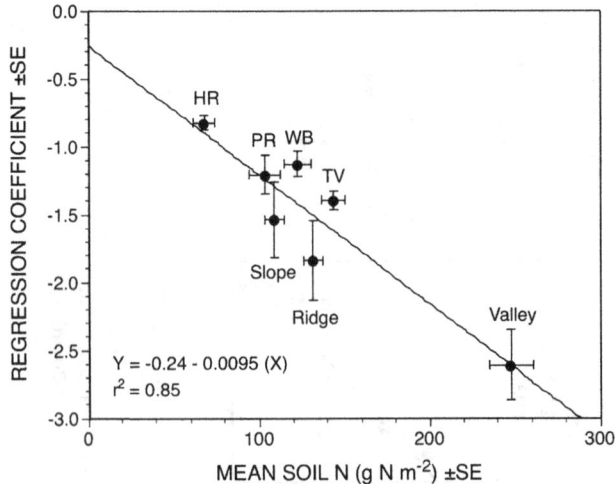

Figure 3.4 Relationship between the regression coefficient (b) and mean surface (0–30 cm) soil N stocks at various forest sites on the Oak Ridge Reservation. Error bars indicate ±1 SE.

in ^{13}C-abundance from the forest floor to deeper SOM. Confounding factors do not necessarily invalidate the regression approach as an indicator of SOM dynamics. Multiple environmental factors can affect decomposition and contribute to between-site differences in the regression coefficients even within the same climate regime. The consideration of findings from other research is also helpful to further an understanding of the observed patterns of soil ^{13}C-abundance in this second case study.

Evidence from other studies

Although soil C dynamics can be derived from differences in ^{13}C-abundance following changes in plant community composition (e.g., a switch from C3 to C4 plants), the utility of natural abundance ^{13}C measurements as an indicator of soil C dynamics, in the absence of plant community changes, is still open to interpretation. Other investigators have inferred faster forest soil C turnover at sites with more negative regression coefficients (i.e., steeper slopes) and reported statistically significant associations between regression coefficients and abiotic variables in tropical settings (Powers & Schlesinger 2002). Here, based on prior published research, we attempt to further illuminate some unanswered questions about what the variation in regression coefficients means with respect to forest soil C dynamics.

Leaf litterfall and rhizodeposition are the initial substrates in a continuum of organic matter decomposition summarized by the regression coefficient. Other research (Garten et al. 2000), in addition to studies on the Oak Ridge Reservation, indicate that forest O-horizons and fine roots are similar in their C isotope composition but differ from leaf litterfall inputs. For example, at sites HR, PR, TV, and WB in the second case study, there was $\approx$1‰ difference in leaf litterfall (δ^{13}C = −29.4‰) and the forest floor organic matter (δ^{13}C = −28.2‰) or fine roots from the forest floor and the mineral soil (δ^{13}C = −28.2‰). Foliar ^{13}C-abundance exhibits little within-year variation but significant interannual variation in forests near site WB (Garten & Taylor 1992). Nonetheless, based on measurements of leaf litterfall, the mean foliar ^{13}C-abundance on the Oak Ridge Reservation in 2004 (−29.4‰) was similar to the ^{13}C-abundance previously measured in tree leaves (−29.5‰) at xeric sites over multiple years (Garten & Taylor 1992). Measurement of forest floor ^{13}C-abundance provides some short-term integration over interannual variation in the isotopic composition of leaf litterfall. The extent of the difference in ^{13}C-abundance between the forest floor and fine roots appears to be ecosystem specific (Powers & Schlesinger 2002), but the forest O-horizon is often a zone of intense fine root activity. Carbon-13 in forest soil C inputs can also be approximated from the regression equations (see e.g., Figures 3.1 & 3.2) using input C concentrations.

Studies of soil ^{13}C-abundance under continuous C3 grasslands indicate that a progressive enrichment of ^{13}C with increasing soil depth is well fitted by the Rayleigh equation (Accoe et al. 2002, 2003). The Rayleigh equation (Mariotti et al. 1981) describes the relationship between δ^{13}C and $\ln(C_d/C_o)$, where C_o denotes the surface soil C concentration and C_d represents the C concentration in increasingly deeper soil increments. The slope of the relationship, ε, is the isotopic enrichment factor and, like the regression coefficient (b), it reflects the per mil enrichment of ^{13}C as one proceeds from an initial C substrate to deeper and older C in the mineral soil. Furthermore, the foregoing studies by Accoe et al. show that the average rate of change in soil δ^{13}C is directly related to organic matter decomposition rates in different parts of the soil profile. A greater change in soil δ^{13}C values corresponds to a faster rate of organic matter decomposition. Thus, the extent of change in ^{13}C-abundance with increasing soil depth may indicate the quality or stability of SOM under continuous C3 vegetation (Accoe et al. 2002, 2003). More recent studies indicate that applicability of the Rayleigh equation to vertical profiles of soil δ^{13}C values may depend on soil texture (Wynn et al. 2005). Enrichment factors, from the Rayleigh equation, and regression coefficients, from relationships between soil ^{13}C-abundance and log C concentration, are highly correlated when calculated for the same soil profile because both quantities are a function of soil C concentration.

Prior studies along altitudinal gradients in the southern Appalachian Mountains indicate that the slope of the regression of ^{13}C-abundance against log C concentration is related to both temperature and the C-to-N ratio of soil C inputs (Garten et al. 2000). Within a particular climate regime, steeper slopes (i.e., more negative regression coefficients) appear to be associated with higher litter quality as indicated by lower litter C-to-N ratios that may reflect differences in site N status (see figure 4 in Garten et al. 2000) and faster soil C turnover (Garten 2006). Similar patterns were found in the second case study on soil δ^{13}C profiles where more negative regression coefficients were measured in the mesic, N-rich valley (Figure 3.3) that had leaf litterfall inputs with low C-to-N ratios (Table 3.2). Different degrees of isotopic fractionation during decomposition of soil organic matter is a likely mechanism contributing to observed site-to-site variation in the measured regression coefficients.

Differences in regression coefficients have also been observed across climate regimes. Field studies indicate declining decomposition rates of labile soil C (Garten et al. 1999; Garten 2004), greater soil C storage (Garten et al. 1999; Garten & Hanson 2006), and increasing soil C turnover times (Garten 2006) with increasing altitude in the southern Appalachian Mountains. The slope of the regression of soil ^{13}C-abundance against log C concentration in cool, N-rich, high-elevation forests is less steep than regression coefficients measured in warm, N-poor, low-elevation forests (Garten et al. 2000). Along elevation gradients, flatter slopes and greater soil C storage at high elevations probably result from an overall decline in the rate of SOM decomposition (including the decomposition of labile forest soil C) which can be attributed, in part, to a 6–7°C decline in mean annual temperature with increasing altitude (Garten et al. 1999). High-elevation forests tend to have large stocks of labile soil C in the O-horizon and the vulnerability of these stocks to decomposition in a warming environment has been discussed in other publications

(Joslin & Johnson 1998; Garten et al. 1999). Steeper slopes at warmer, low-elevation forests still reflect faster decomposition of labile soil C and possibly greater stabilization of older SOM. Thus, regression coefficients may reflect a difference in soil C partitioning which is ultimately a strong control of overall forest soil C dynamics. These types of associations may not exist in soils with buried horizons, colluviums, or otherwise historically disturbed soil profiles.

Conclusions

Many natural, mature forest ecosystems experience negligible soil distur-bance aside from atmospheric deposition of anthropogenically derived pol-lutants. For example, soils appear to be the primary sink for atmospheric N deposition in many temperate forest ecosystems (e.g., Nadelhoffer et al. 1999) and chronic N additions may gradually alter both soil N availability and soil C dynamics (e.g., Waldrop et al. 2004a). The particular case studies discussed here, in addition to the brief review of results from other published research, indicate natural abundance measurements of stable N and C isotopes in forest ecosystems hold considerable promise as indicators of site N status and soil C dynamics, respectively. Additional research is warranted to:

1 verify the findings presented here regarding associations between soil ^{13}C-abundance and log C concentration, the relationship of forest soil δ^{13}C profiles to N availability, the utility of regression coefficients as indicators of soil C dynamics and/or forest soil C sequestration potential, the utility of ^{15}N-abundance as an indicator of site N status;
2 further explore possible mechanisms that are responsible for the observed patterns.

Acknowledgments

This research was sponsored by U.S. Department of Energy, Office of Science, Biological and Environmental Research (Terrestrial Carbon Processes Pro-gram) through funding to the Consortium for Research on Enhancing Carbon Sequestration in Terrestrial Ecosystems (CSiTE), to the Enriched Background Isotope Study (EBIS), and to the Forest Soil C Studies Project under contract DE-AC05-00OR22725 with Oak Ridge National Laboratory (ORNL), man-aged by UT-Battelle, LLC for the U.S. Department of Energy.

References

Aber, J.D., Melillo, J.M., Nadelhoffer, K.J., McClaugherty, C.A. & Pastor, J. (1985) Fine root turnover in forest ecosystems in relation to quantity and form of nitrogen availabil-ity: a comparison of two methods. *Oecologia*, **66**, 317–321.

Accoe, F., Boeckx, P., Van Cleemput, O., et al. (2002) Evolution of the $\delta^{13}C$ signature related to total carbon contents and carbon decomposition rate constants in a soil profile under grassland. *Rapid Communications in Mass Spectrometry*, **16**, 2184–2189.

Accoe, F., Boeckx, P., Van Cleemput, O. & Hofman, G. (2003) Relationship between soil organic C degradability and the evolution of the $\delta^{13}C$ signature in profiles under permanent grassland. *Rapid Communications in Mass Spectrometry*, **17**, 2591–2596.

Balesdent, J. (1996) The significance of organic separates to carbon dynamics and its modeling in some cultivated soils. *European Journal of Soil Science*, **47**, 485–493.

Balesdent, J., Girardin, C. & Mariotti, A. (1993) Site-related $\delta^{13}C$ of tree leaves and soil organic matter in a temperate forest. *Ecology*, **74**, 1713–1721.

Bedard-Haughn, A., Van Groenigen, J.W. & Van Kessel, C. (2003) Tracing ^{15}N through landscapes: potential uses and precautions. *Journal of Hydrology*, **272**, 175–190.

Berg, B. (2000) Litter decomposition and organic matter turnover in northern forest soils. *Forest Ecology and Management*, **133**, 13–22.

Berg, B. & Matzner, E. (1997) Effect of N deposition on decomposition of plant litter and soil organic matter in forest systems. *Environmental Reviews*, **5**, 1–25.

Berg, B. & Meentemeyer, V. (2002) Litter quality in a north European transect versus carbon storage potential. *Plant and Soil*, **242**, 83–92.

Berg, B., McClaugherty, C., De Santo, AV. & Johnson, D. (2001) Humus buildup in boreal forests: effects of litterfall and its N concentration. *Canadian Journal of Forest Research*, **31**, 988–998.

Bowden, R.D., Davidson, E., Savage, K., Arabia, C. & Steudler, P. (2004) Chronic nitrogen additions reduce total soil respiration and microbial respiration in temperate forest soils at the Harvard Forest. *Forest Ecology and Management*, **196**, 43–56.

Burton, A.J., Pregitzer, K.S., Crawford, J.N., Zogg, G.P. & Zak, D.R. (2004) Simulated chronic NO_3^- deposition reduces soil respiration in northern hardwood forests. *Global Change Biology*, **10**, 1080–1091.

Carreiro, M.M., Sinsabaugh, R.L. & Parkhurst, D.F. (2000) Microbial enzyme shifts explain litter decay responses to simulated nitrogen deposition. *Ecology*, **81**, 2359–2365.

DeForest, J.L., Zak, D.R., Pregitzer, K.S. & Burton, A.J. (2004) Atmospheric nitrate deposition, microbial community composition, and enzyme activity in northern hardwood forests. *Soil Science Society of America Journal*, **68**, 132–138.

Ehleringer, J.R., Buchmann, N. & Flanagan, L.B. (2000) Carbon isotope ratios in belowground carbon cycle processes. *Ecological Applications*, **10**, 412–422.

Emmett, B.A., Kjonaas, O.J., Gundersen, P., Koopmans, C., Tietema, A. & Sleep, D. (1998) Natural abundance of ^{15}N in forests across a nitrogen deposition gradient. *Forest Ecology and Management*, **101**, 9–18.

Fernandez, I. & Cadisch, G. (2003) Discrimination against ^{13}C during degradation of simple and complex substrates by two white rot fungi. *Rapid Communications in Mass Spectrometry*, **17**, 2614–2620.

Fernandez, I., Mahieu, N. & Cadisch, G. (2003) Carbon isotopic fractionation during decomposition of plant materials of different quality. *Global Biogeochemical Cycles*, **17**, article no. 1075.

Fisk, M.C. & Fahey, T.J. (2001) Microbial biomass and nitrogen cycling responses to fertilization and litter removal in young hardwood forests. *Biogeochemistry*, **53**, 201–223.

Franklin, O., Högberg, P., Ekblad, A. & Agren, G.I. (2003) Pine forest floor carbon accumulation in response to N and PK additions: bomb ^{14}C modeling and respiration studies. *Ecosystems*, **6**, 644–658.

Frey, S.D., Knorr, M., Parrent, J.L. & Simpson, R.T. (2004) Chronic nitrogen enrichment affects the structure and function of the soil microbial community in temperate hardwood and pine forests. *Forest Ecology and Management*, **196**, 159–171.

Gallo, M., Amonette, R., Lauber, C., Sinsabaugh, R.L. & Zak, D.R. (2004) Microbial community structure and oxidative enzyme activity in nitrogen-amended north temperate forest soils. *Microbial Ecology*, **48**, 218–229.

Garten, Jr., C.T. (1993) Variation in foliar ^{15}N abundance and the availability of soil nitrogen on Walker Branch Watershed. *Ecology*, **74**, 2098–2113.

Garten, Jr., C.T. (2004) Potential net soil N mineralization and decomposition of glycine-^{13}C in forest soils along an elevation gradient. *Soil Biology and Biochemistry*, **36**, 1491–1496.

Garten, Jr., C.T. (2006) Relationships among forest soil C isotopic composition, partitioning, and turnover times. *Canadian Journal of Forest Research*, **36**, 2157–2167.

Garten, Jr., C.T. & Hanson, P.J. (2006) Measured forest soil C stocks and estimated turnover times along an elevation gradient. *Geoderma*, **136**, 342–352.

Garten, Jr., C.T. & Taylor, G.E. (1992) Foliar δ^{13}C within a temperate deciduous forest: spatial, temporal, and species sources of variation. *Oecologia*, **90**, 1–7.

Garten, Jr., C.T. & Van Miegroet, H.M. (1994) Relationships between soil nitrogen dynamics and natural ^{15}N-abundance in plant foliage from the Great Smoky Mountains National Park. *Canadian Journal of Forest Research*, **24**, 1636–1645.

Garten, Jr., C.T., Huston, M.A. & Thoms, C.A. (1994) Topographic variation of soil nitrogen dynamics at Walker Branch Watershed, Tennessee. *Forest Science*, **40**, 497–512.

Garten, Jr., C.T., Post, III, W.M., Hanson, P.J. & Cooper L.W. (1999) Forest soil carbon inventories and dynamics along an elevation gradient in the southern Appalachian Mountains. *Biogeochemistry*, **45**, 115–145.

Garten, Jr., C.T., Cooper, L.W., Post, III, W.M. & Hanson, P.J. (2000) Climate controls on forest soil C isotope ratios in the southern Appalachian Mountains. *Ecology*, **81**, 1108–1119.

Gaudinski, J.B. & Trumbore, S.E. (2003) Soil carbon turnover. In: *North American Temperate Deciduous Forest Responses to Changing Precipitation Regimes* (eds P.J. Hanson & S.D.Wullschleger), pp. 190–209. Springer-Verlag, New York.

Gaudinski, J.B., Trumbore, S.E., Davidson, E.A. & Zheng, S.H. (2000) Soil carbon cycling in a temperate forest: radiocarbon-based estimates of residence times, sequestration rates and partitioning of fluxes. *Biogeochemistry*, **51**, 33–69.

Hanson, P.J., Wullschleger, S.D., Bohlman, S.A. & Todd, D.E. (1993) Seasonal and topographic patterns of forest floor CO_2 efflux from an upland oak forest. *Tree Physiology*, **13**, 1–15.

Hart, S.C., Stark, J.M., Davidson, E.A. & Firestone, M.K. (1994) Nitrogen mineralization, immobilization, and nitrification. In: *Methods of Soil Analysis, Part 2. Microbiological and Biochemical Processes* (Eds R.W. Weaver, S. Angle, P. Bottomley, et al.), pp. 985–1018. Soil Science Society of America, Madison, WI.

Hatcher, Jr., R.D., Lemiszki, P.J., Dreier, R.B., et al. (1992) *Status report on the geology of the Oak Ridge Reservation*. Oak Ridge National Laboratory, Oak Ridge, TN.

Henn, M.R. & Chapela, I.H. (2000) Differential C isotope discrimination by fungi during decomposition of C3- and C4-derived sucrose. *Applied and Environmental Microbiology*, **66**, 4180–4186.

Hobbie, E.A. & Colpaert, J.V. (2003) Nitrogen availability and colonization by mycorrhizal fungi correlate with nitrogen isotope patterns in plants. *The New Phytologist*, **157**, 115–126.

Hobbie, E.A., Macko, S.A. & Williams, M. (2000) Correlations between foliar δ^{15}N and nitrogen concentrations may indicate plant-mycorrhizal interactions. *Oecologia*, **122**, 273–283.

Högberg, P. (1990) Forests losing large quantities of nitrogen have elevated ^{15}N:^{14}N ratios. *Oecologia*, **84**, 229–231.

Högberg, P. (1991) Development of ^{15}N enrichment in a nitrogen-fertilized forest soil–plant system. *Soil Biology and Biochemistry*, **23**, 335–338.

Högberg, P. (1997) Tansley review No. 95 – ^{15}N natural abundance in soil–plant systems. *The New Phytologist*, **137**, 179–203.

Högberg, P. & Johannisson, C. (1993) ^{15}N abundance of forests is correlated with losses of nitrogen. *Plant and Soil*, **157**, 147–150.

Janssen, B.H. (1996) Nitrogen mineralization in relation to C:N ratio and decomposability of organic materials. *Plant and Soil*, **181**, 39–45.

Johannisson, C. & Högberg, P. (1994) ^{15}N abundance of soils and plants along an experimentally induced forest nitrogen supply gradient. *Oecologia*, **97**, 322–325.

Johnson, D.W. (1992) Effects of forest management on soil carbon storage. *Water, Air, and Soil Pollution*, **64**, 83–120.

Johnson, D.W. & Curtis, P.S. (2001) Effects of forest management on soil C and N storage: meta-analysis. *Forest Ecology and Management*, **140**, 227–238.

Johnson, D.W. & Van Hook, R.I. (eds.) (1989) *Analysis of Biogeochemical Cycling Processes in Walker Branch Watershed*. Springer-Verlag, New York.

Joslin, J.D. & Johnson, D.W. (1998) Effects of soil warming, atmospheric deposition, and elevated carbon dioxide on forest soils in the southeastern United States. In: *The Productivity and Sustainability of Southern Forest Ecosystems in a Changing Environment* (Eds R.A. Mickler & S.A. Fox), pp. 571–587. Springer-Verlag, New York.

Koba, K., Hirobe, M., Koyama, L., et al. (2003) Natural ^{15}N abundance of plants and soil N in a temperate coniferous forest. *Ecosystems*, **6**, 457–469.

Kolb, K.J. & Evans, R.D. (2002) Implications of leaf nitrogen recycling on the nitrogen isotope composition of deciduous plant tissues. *The New Phytologist*, **156**, 57–64.

Kuperman, R.G. (1999) Litter decomposition and nutrient dynamics in oak–hickory forests along an historic gradient of nitrogen and sulfur deposition. *Soil Biology and Biochemistry*, **31**, 237–244.

Lal, R. (2004) Soil carbon sequestration to mitigate climate change. *Geoderma*, **123**, 1–22.

Mariotti, A., Germon, J.C., Hubert, P., et al. (1981) Experimental determination of nitrogen kinetic isotope fractionation: some principles; illustration for the denitrification and nitrification processes. *Plant and Soil*, **62**, 413–430.

Martinelli, L.A., Piccolo, M.C., Townsend, A.R., et al. (1999) Nitrogen stable isotopic composition of leaves and soil: tropical versus temperate forests. *Biogeochemistry*, **46**, 45–65.

Mary, B., Mariotti, A. & Morel, J.L. (1992) Use of ^{13}C variations at natural abundance for studying the biodegradation of root mucilage, roots and glucose in soil. *Soil Biology and Biochemistry*, **24**, 1065–1072.

Matocha, C.J., Haszler, G.R. & Grove, J.H. (2004) Nitrogen fertilization suppresses soil phenol oxidase enzyme activity in no-tillage systems. *Soil Science*, **169**, 708–714.

Metting, F.B., Smith, J.L., Amthor, J.S. & Izaurralde, R.C. (2001) Science needs and new technology for increasing soil carbon sequestration. *Climatic Change*, **51**, 11–34.

Nadelhoffer, K.J. (2000) The potential effects of nitrogen deposition on fine-root production in forest ecosystems. *The New Phytologist*, **147**, 131–139.

Nadelhoffer, K.J. & Fry, B. (1988) Controls on natural nitrogen-15 and carbon-13 abundances in forest soil organic matter. *Soil Science Society of America Journal*, **52**, 1633–1640.

Nadelhoffer, K.J., Aber, J.D. & Melillo, J.M. (1983) Leaf-litter production and soil organic matter dynamics along a nitrogen-availability gradient in southern Wisconsin (USA). *Canadian Journal of Forest Research*, **13**, 12–21.

Nadelhoffer, K.J., Emmett, B.A., Gundersen, P., et al. (1999) Nitrogen deposition makes a minor contribution to carbon sequestration in temperate forests. *Nature*, **398**, 145–148.

Nasholm, T., Nordin, A., Edfast, A.B. & Högberg, P. (1997) Identification of coniferous forests with incipient nitrogen saturation through analysis of arginine and nitrogen-15 abundance of trees. *Journal of Environmental Quality*, **26**, 302–309.

Neff, J.C., Townsend, A.R., Gleixner, G., Lehman, S.J., Turnbull, J. & Bowman, W.D. (2002) Variable effects of nitrogen additions on the stability and turnover of soil carbon. *Nature*, **419**, 915–917.

Pardo, L.H., Hemond, H.F., Montoya, J.P., Fahey, T.J. & Siccama, T.G. (2002) Response of the natural abundance of ^{15}N in forest soils and foliage to high nitrate losses following clear-cutting. *Canadian Journal of Forest Research*, **32**, 1126–1136.

Paul, E.A., Follett, R.F., Leavitt, S.W., Halvorson, A., Peterson, G.A. & Lyon, D.J. (1997) Radiocarbon dating for determination of soil organic matter pool sizes and dynamics. *Soil Science Society of America Journal*, **61**, 1058–1067.

Post, W.M., Peng, T.H., Emanuel, W.R., King, A.W., Dale, V.H. & DeAngelis, D.L. (1990) The global carbon cycle. *American Scientist*, **78**, 310–326.

Post, W.M., Izaurralde, R.C., Jastrow, J.D., et al. (2004) Enhancement of carbon sequestration in US soils. *Bioscience*, **54**, 895–908.

Powers, J.S. & Schlesinger, W.H. (2002) Geographic and vertical patterns of stable carbon isotopes in tropical rain forests of Costa Rica. *Geoderma*, **109**, 141–160.

Reich, P.B., Grigal, D.F., Aber, J.D. & Gower, S.T. (1997) Nitrogen mineralization and productivity in 50 hardwood and conifer stands on diverse soils. *Ecology*, **78**, 335–347.

Resh, S.C., Binkley, D. & Parrotta, J.A. (2002) Greater soil carbon sequestration under nitrogen-fixing trees compared with Eucalyptus species. *Ecosystems*, **5**, 217–231.

Saiya-Cork, K.R., Sinsabaugh, R.L. & Zak, D.R. (2002) The effects of long term nitrogen deposition on extracellular enzyme activity in an *Acer saccharum* forest soil. *Soil Biology and Biochemistry*, **34**, 1309–1315.

Santruckova, H., Bird, M.I. & Lloyd, J. (2000) Microbial processes and carbon-isotope fractionation in tropical and temperate grassland soils. *Functional Ecology*, **14**, 108–114.

Schimel, D.S. (1995) Terrestrial ecosystems and the carbon cycle. *Global Change Biology*, **1**, 77–91.

Schweizer, M., Fear, J. & Cadisch, G. (1999) Isotopic (^{13}C) fractionation during plant residue decomposition and its implications for soil organic matter studies. *Rapid Communications in Mass Spectrometry*, **13**, 1284–1290.

Silver, W.L. & Miya, R.K. (2001) Global patterns in root decomposition: comparisons of climate and litter quality effects. *Oecologia*, **129**, 407–419.

Six, J. & Jastrow, J.D. (2002) Organic matter turnover. In: *Encyclopedia of Soil Science* (Ed. R. Lal), pp. 936–942. Marcel Dekker, New York.

Smith, W.B., Vissage, J.S., Darr, D.R. & Sheffield, R.M. (1997) *Forest Resources of the United States, 1997*. United States Department of Agriculture, Forest Service, North Central Research Station, St Paul, MN.

Stump, L.M. & Binkley, D. (1993) Relationships between litter quality and nitrogen availability in Rocky Mountain forests. *Canadian Journal of Forest Research*, **23**, 492–502.

Swanston, C., Homann, P.S., Caldwell, B.A., Myrold, D.D., Ganio, L. & Sollins, P. (2004) Long-term effects of elevated nitrogen on forest soil organic matter stability. *Biogeochemistry*, **70**, 227–250.

Taylor, B.R., Parkinson, D. & Parsons, W.F.J. (1989) Nitrogen and lignin content as predictors of litter decay rates: a microcosm test. *Ecology*, **70**, 97–104.

Torn, M.S., Lapenis, A.G., Timofeev, A., Fischer, M.L., Babikov, B.V., Harden, J.W. (2002) Organic carbon and carbon isotopes in modern and 100-year-old-soil archives of the Russian steppe. *Global Change Biology*, **8**, 941–953.

Van Dam, D., Veldkamp, E. & Van Breemen, N. (1997) Soil organic carbon dynamics: variability with depth in forested and deforested soils under pasture in Costa Rica. *Biogeochemistry*, **39**, 343–375.

Vervaet, H., Boeckx, P., Unamuno, V., Van Cleemput, O. & Hofman, G. (2002) Can $\delta^{15}N$ profiles in forest soils predict NO_3^- losses and net N mineralization rates? *Biology and Fertility of Soils*, **36**, 143–150.

Waldrop, M.P., Zak, D.R., Sinsabaugh, R.L., Gallo, M. & Lauber, C. (2004a) Nitrogen deposition modified soil carbon storage through changes in microbial enzymatic activity. *Ecological Applications*, **14**, 1172–1177.

Waldrop, M.P., Zak, D.R. & Sinsabaugh, R.L. (2004b) Microbial community response to nitrogen deposition in northern forest ecosystems. *Soil Biology and Biochemistry*, **36**, 1443–1451.

Wang, Y., Amundson, R. & Trumbore, S. (1996) Radiocarbon dating of soil organic matter. *Quaternary Research*, **45**, 282–288.

Wynn, J.G., Bird, M.I. & Wong, V.N.L. (2005) Rayleigh distillation and the depth profile of $^{13}C/^{12}C$ ratios of soil organic carbon from soils of disparate texture in Iron Range National Park, Far North Queensland, Australia. *Geochimica et Cosmochimica Acta*, **69**, 1961–1973.

CHAPTER 4

Soil nitrogen isotope composition

R. DAVE EVANS

Introduction

The isotope composition of a reaction product is determined by the isotope ratio of the substrate and fractionation during chemical transformations. Fractionation may not always occur so isotope ratios of some elements can be used as natural tracers, for example, the δ^2H and $\delta^{18}O$ of stem- and soil-water can be used to infer patterns of water use in plants (Dawson & Ehleringer 1991; Ehleringer et al. 1991). In other cases such as plant $\delta^{13}C$, fractionation events predominate, but mechanistic models have been developed to explain and predict product isotope ratios (Farquhar et al. 1982; Flanagan et al. 1991). Early interpretation of soil $\delta^{15}N$ was heavily influenced by enriched ^{15}N labeling methodologies that were common in agriculture and ecology to understand soil nitrogen transformations, and it was hoped that ^{15}N at natural abundance levels could be used as a natural tracer. Fractionation events can be ignored in labeling studies, and it was hoped a similar assumption could be made for ^{15}N at natural abundance. For example, early natural abundance ^{15}N studies attempted to trace fertilizer and soil nitrate into groundwater, or quantify the contribution of nitrogen fixation to total plant nitrogen (reviewed in Shearer & Kohl 1986; Högberg 1997; Robinson 2001). Scientists now realize that soil $\delta^{15}N$ is very complex, with multiple sources of nitrogen inputs, numerous internal transformations with associated fractionation events, and multiple sources of nitrogen loss that all potentially discriminate against ^{15}N (Figure 4.1). It is because of this complexity and the reliance on assumptions and methods for enrichment studies that our knowledge of the mechanisms controlling soil $\delta^{15}N$ has not progressed as rapidly as those for other elements. Nonetheless scientists have made significant progress in understanding soil $\delta^{15}N$ in the past decade and it is clear that it serves as a valuable integrator of soil processes (Robinson 2001). This chapter provides an overview of these recent advances.

Sources of variation in soil $\delta^{15}N$

The isotopic composition of nitrogen inputs and fractionation during nitrogen transformations and loss determine soil $\delta^{15}N$. It is difficult to measure *in situ*

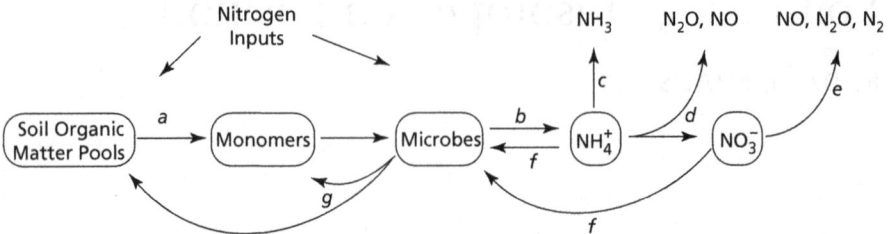

Figure 4.1 Soil nitrogen transformations based on the conceptual model of Schimel & Bennet (2004). Nitrogen inputs can enter into the organic and inorganic pools. The transformations are: *a*, depolymerization; *b*, gross mineralization; *c*, volatilization; *d*, nitrification; *e*, denitrification; *f*, microbial immobilization; *g*, death of soil microbes.

Table 4.1 Observed discriminations for transformations in the nitrogen cycle. Process refers to the transformations indicated by letters in Figure 4.1. Values are taken from reviews by Shearer & Kohl (1986), Wada & Ueda (1996), Högberg (1997) and Robinson (2001).

Transformation	Process	Discrimination (‰)
Gross mineralization	*b*	0–5
Nitrification	*d*	0–35
$NH_4^+ \leftrightarrow NH_3$ equilibrium		20–27
Volatilization	*c*	29
N_2O and NO production during nitrification	*d*	0–70
N_2O and N_2 production during denitrification	*e*	0–39
NO_3^- immobilization	*f*	13
NH_4^+ immobilization	*f*	14–20

fractionation factors for most soil processes so the observed discrimination ($\delta^{15}N_{substrate} - \delta^{15}N_{product}$) is commonly used instead. The observed discrimination exhibits considerable variation for individual transformations (Table 4.1). Shearer & Kohl (1993) and Högberg (1997) identified several reasons for this variation. First, a process may exhibit positive discrimination when substrates are plentiful, but discrimination will not be observed when a reaction is substrate limited because all substrate is converted to product. Second, there can be multiple substrates with different $\delta^{15}N$ for each product. For example, N_2O and NO are produced by both nitrification and denitrification, which results in different isotope ratios depending upon the dominant process of formation. Third, some substrates such as NH_4^+ and NO_3^- are exposed to multiple, competing reactions with different fractionation factors. Fourth, related functional groups of organisms may have slightly different discrimination values for the same process. Finally, it is difficult to predict the effects of interactions between abiotic and biotic factors, and they may vary between ecosystems or even within a site.

Isotopic composition of nitrogen inputs

Nitrogen fixation

Prior to the industrial age the primary external source of nitrogen input for soils was biological nitrogen fixation (Galloway et al. 1995). Scientists initially hoped that quantifying $\delta^{15}N$ could be used to trace the relative contribution of nitrogen fixation to plants and soils. One assumption of this approach is that fractionation does not occur during nitrogen fixation, therefore the $\delta^{15}N$ of nitrogen-fixing organisms will reflect that of the atmospheric source. This assumption is often not met because the discrimination observed during nitrogen fixation can vary from 0 to 3‰ (Shearer & Kohl 1986, 1993), so the $\delta^{15}N$ of organisms that derive the nitrogen from biological fixation varies from −3 to 0‰ (Fry 1991). Soil $\delta^{15}N$ can, however, be used to infer the dominant sources of nitrogen input in very simple soil systems. In arid ecosystems the primary source of nitrogen input is the biological soil crust dominated by cyanobacteria and lichens that are capable of nitrogen fixation. The crusts form a continuous cover in undisturbed plant communities, and spatial coverage is often higher than for vascular plants (see Evans & Johansen 1999). Surface disturbance in arid ecosystems is widespread and eliminates biological soil crusts and nitrogen fixation. Therefore identifying whether nitrogen input is dominated by a physical (atmospheric deposition) or biological (N_2-fixation) process is important to determine the potential impacts of surface disturbance on ecosystem nitrogen cycles. Evans & Ehleringer (1993) used a Rayleigh relationship to assess the relative contribution of physical and biological processes in a Pinyon–Juniper community on the Colorado Plateau. The predicted linear relationship between soil $\delta^{15}N$ and the log of soil nitrogen content was established using soil values (Figure 4.2). Values for the biological soil crust fell immediately along this relationship while values for atmospheric deposition fell well off the relationship. This indicates that the primary source of nitrogen input was biological nitrogen fixation, and land-use change may alter the balance between nitrogen input and loss by eliminating this source. This result was confirmed by Evans & Belnap (1999), who observed lower soil nitrogen contents and greater soil $\delta^{15}N$ values in disturbed versus undisturbed sites.

Atmospheric deposition

An initial goal of early studies measuring soil nitrogen isotope composition was to identify and trace sources of atmospheric input into ecosystems.

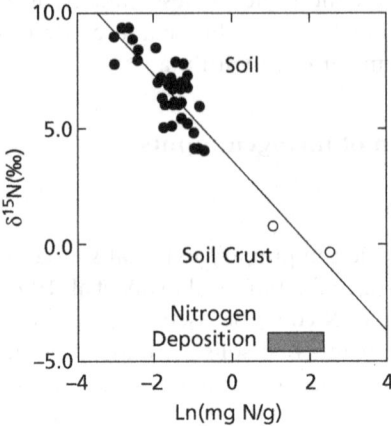

Figure 4.2 Relationship between soil $\delta^{15}N$ and ln(soil nitrogen content) in soils and in two potential sources of N in a cold desert ecosystem in southern Utah.

Heaton (1987) pointed out the early promise of this approach by measuring the $\delta^{15}N$ of oxidized and reduced nitrogen gases. The $\delta^{15}N$ ranged from an extreme of −150‰ for oxidized nitrogen gases from the stack of a nitric acid plant to 5.2‰ for emissions from a coal-burning power station. However, the $\delta^{15}N$ of atmospheric deposition can vary significantly depending upon the nitrogen form and season. Heaton (1986) reviewed $\delta^{15}N$ of NO_3^- and NH_4^+ in wet and dry deposition. In general, values for wet deposition were <0‰ while values for dry deposition were >0‰. Weighted values for wet and dry deposition estimated the $\delta^{15}N$ of total atmospheric deposition to be −3‰ for both NO_3^- and NH_4^+. A drawback to point measurements of atmospheric deposition is that significant temporal variation is common. The $\delta^{15}N$ of NO_3^- and NH_4^+ ranged from −7 to +4‰ and −9 to −2‰, respectively, in a single year, and could vary as much as 5‰ in a single storm.

The observed temporal changes in the $\delta^{15}N$ of atmospheric deposition can be caused by several factors including changes in the source or intensity of precipitation. Pichlmayer et al. (1998) reconstructed sources of NO_3^- in atmospheric deposition in European alpine ecosystems by correlating $\delta^{15}N$ in snow and ice with known storm tracks. Storm tracks were calculated twice daily so that specific strata in the snow could be correlated with individual precipitation events and their region of origin. The observed seasonal variation was ca. 5‰ and appeared to be the result of storms originating from north–south tracks versus east–west. All atmospheric values were <1‰ suggesting anthropogenic origin, because samples taken from pre-industrial ice cores in the same region had $\delta^{15}N$ values >2.6‰.

Bragazza et al. (2004, 2005) recently adopted a novel approach to integrate temporal and source changes at 16 sites across 11 European countries and identifying sources of atmospheric deposition by measuring the $\delta^{15}N$ of mosses. Rates of atmospheric deposition ranged from 1 to $20\,kg\,N\,ha^{-1}\,yr^{-1}$, and moss $\delta^{15}N$ varied from -8 to -3‰ across the gradient. The $\delta^{15}N$ values were not correlated with total deposition, annual temperature, or annual precipitation. Instead, the isotope compositions were significantly correlated with the ratio of reduced to oxidized nitrogen (NH_x/NO_x) in the deposition. Mosses located in areas with greater emissions of NH_3 from agricultural activities had lower $\delta^{15}N$ values than those more heavily influenced by NO_x emissions from industrial activity. This study is among the first to provide integrated ^{15}N values and to also identify potential agricultural and industrial sources and their impact on the $\delta^{15}N$ of nitrogen inputs.

Fertilizers

Consideration of the rates of input and $\delta^{15}N$ of fertilizer N is essential to understand current and future patterns of soil $\delta^{15}N$ in agricultural soils. Preindustrial global rates of nitrogen input, primarily from biological nitrogen fixation, are estimated at 90 to $130\,Tg\,N\,yr^{-1}$ (Galloway et al. 1995). In contrast, fertilizer application was about $80\,Tg\,N\,yr^{-1}$ in 1990 and is predicted to exceed rates of biological nitrogen fixation by 2020 (Galloway et al. 1995; Vitousek et al. 1997; Steffen et al. 2004). Vitoria et al. (2004) recently reviewed published values for fertilizer isotope composition. The isotope composition of fertilizers reflects their origins from atmospheric nitrogen (0‰) and oxygen (22.5‰). The $\delta^{15}N$ of total nitrogen ranged from -3 to $+4\text{‰}$ with the highest frequency from -1 to $+1\text{‰}$. The $\delta^{15}N$ in NO_3^- and NH_4^+ exhibited much greater variation; $\delta^{15}N$ of NO_3^- varied from -8 to $+7\text{‰}$ and values were primarily enriched in [15]N. In contrast, $\delta^{15}N$ of NH_4^+ were primarily all negative and ranged from -7 to $+3\text{‰}$. The difference between NO_3^- and NH_4^+ is thought to be due to fractionation during oxidation of NH_3 (Freyer & Aly 1974). Chilean nitrates can be a significant source of fertilizer nitrogen in some regions. The $\delta^{15}N$ of these fertilizers was similar to the pattern observed for synthetic fertilizers, but the $\delta^{18}O$ was enriched from $+40$ to $+50\text{‰}$.

Soil transformations

Mineralization and organism-available nitrogen

Fractionation during mineralization has been estimated by comparing bulk soil and NH_4^+ $\delta^{15}N$. The differences are often small and it is assumed that the observed discrimination is negligible (Table 4.1). Further research is needed on the processes and fractionation events that produce organism-available nitrogen in light of our changing understanding of soil nitrogen cycling and new technological advances. A view held for many decades was that mineralization was the process that limited overall rates of nitrogen cycling in soils and that plants primarily assimilated NH_4^+ and NO_3^-, but it is now believed that overall soil nitrogen cycling is limited by the rate of depolymerization of amino acids from soil organic matter that occurs at localized microsites within the soil (Schimel & Bennett 2004). In addition to uptake of inorganic nitrogen, organic nitrogen uptake may occur either directly by plants or through mycorrhizae. For these reasons, and because soil nitrogen is dominated by a large, non-reactive recalcitrant pool, bulk soil values may provide little information on the $\delta^{15}N$ of nitrogen assimilated by organisms. Research in this area is being facilitated by advances in compound-specific isotope analysis that now allows measurement of the $\delta^{15}N$ of individual amino acids, because it is these individual amino acids in litter and soil that serve as the substrate for subsequent soil reactions. The $\delta^{15}N$ of individual compounds can vary greatly within plants; proteins are generally enriched in ^{15}N compared with secondary products such as chlorophyll, lipids, and amino sugars and differences among compounds can be as great as 20‰ (Werner & Schmidt 2002), and similar variation has been observed for amino acids in soils (Ostle et al. 1999; Bol et al. 2004).

Many early studies assumed that plant $\delta^{15}N$ could be used as an indicator of plant nitrogen source because it was presumed that the progressive products in the organic matter $\to NH_4^+ \to NO_3^-$ sequence would become increasingly depleted in ^{15}N. This assumption is often not correct because competing reactions may enrich the soil inorganic nitrogen pool (Figure 4.1). For example, Binkley et al. (1985) observed that soil NO_3^- may not differ or may even be enriched compared with NH_4^+. The $\delta^{15}N$ of soil inorganic nitrogen can also change rapidly over time because these pools are very labile (Figure 4.3); variation of 10–20‰ can occur over days or months (Feigin et al. 1974; Herman & Rundel 1989; Frank et al. 2004).

It is important to emphasize that isotopic measurements of soil inorganic nitrogen should be interpreted with caution and that values may be the product of artifacts during collection and purification rather than a soil biological or physical process. Collection of soil samples disturbs the system and this can alter $\delta^{15}N$ values (Högberg 1997), and many of the methods currently used to study soil inorganic nitrogen were developed for labeling studies and it is unclear whether they are appropriate at natural abundance levels. Robinson (2001) summarized the limitations of using methods that were developed for other applications. First, inorganic nitrogen is often isolated using diffusion methods developed for labeling studies. The pH of the solution containing NH_4^+ is raised causing NH_3 to volatilize where it is collected on an acidified disk. Robinson (2001) calculates that recovery must be >99 percent for maximum accuracy, and recovery of only 95 percent would result in an error of 3‰. Second, many of the isolation methods currently in use are not NH_4^+-specific and contamination by organic nitrogen is common. This could introduce substantial errors considering the large variation in soil organic compounds (Ostle et al. 1999). Robinson (2001) states that methods should

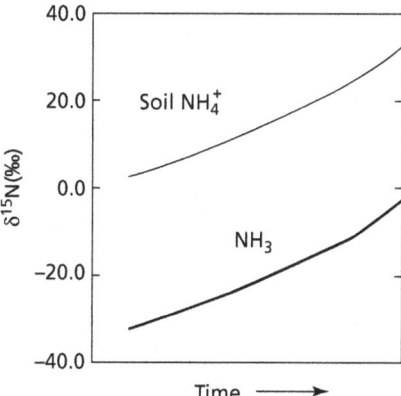

Figure 4.3 Predicted soil NH_4^+ and atmospheric NH_3 based on a Rayleigh distillation model using data from Frank et al. (2004). The model assumes a closed system over a 10-day period.

be used that have been specifically designed for natural abundance measurements to overcome the problems associated with diffusion and isolation. For NO_3^- these included NO_3^--specific dye-coupling (Johnston et al. 1999), ion exchange (Silva et al. 2000; Stickrod & Marshall 2000), and denitrifying bacteria (Sigman et al. 2001). Suitable methods are still being developed for NH_4^+.

Nitrogen loss

Volatilization

Relatively few studies have measured the $\delta^{15}N$ of inorganic reduced nitrogen compounds compared with oxidized forms. This is likely to change in the future because 70 percent of atmospheric NH_3 originates from volatilization of NH_4^+ deposited in the soil from domestic animals, fertilizers, and excrement (Schlesinger & Hartley 1992) leading to increased aerosol formation that can alter Earth's energy budget (Chapin et al. 2002). The observed discrimination with volatilization can be large because the formation of gaseous NH_3 from NH_4^+ involves several steps and each can discriminate against ^{15}N (Table 4.1). The first is the formation of NH_3 from NH_4^+. The isotope effect for this equilibrium reaction is ca. 20‰. NH_3 must then diffuse to the site volatilization, but the observed discrimination associated with this may be negligible (Shearer & Kohl 1986). Finally, there is volatilization of NH_3 into the atmosphere and the observed discrimination can be as great as 30‰ depending on the concentration gradient between the soil and atmosphere and the rate of removal as determined by atmospheric turbulence (Högberg 1997).

The $\delta^{15}N$ of NH_3 emitted from the soil can increase over time and often follows Rayleigh distillation kinetics (Figure 4.3). The increase is caused by the large observed discrimination with volatilization that enriches the remaining soil NH_4^+. The change over time can be significant over short periods of time; Frank et al. (2004) observed a 25‰ increase for NH_3 over a 10-day period following application of an artificial urine patch. The $\delta^{15}N$ of soil NH_4^+ was estimated over the same time period using the Rayleigh model and the predicted increase was from 0 on day 1 to almost 30‰ by day 10. The large observed discrimination with volatilization can have significant effects on total soil $\delta^{15}N$. Frank & Evans (1997) observed that grazed sites had soil $\delta^{15}N$ ca. 1‰ greater than sites where grazing had been excluded for 32 to 36 years, presumably due to microbial immobilization that retained NH_4^+ enriched by volatilization from urine patches in the ecosystem.

Nitrification and denitrification

Measuring the isotopic composition of oxidized nitrogen gases is becoming common in an attempt to constrain their global budgets because the N_2O from terrestrial ecosystems is depleted in ^{15}N and ^{18}O compared with N_2O with a marine origin (see Rahn & Wahlen 2000; Perez et al. 2001). Most of these studies have been associated with marine ecosystems even though soil emissions are estimated to be the largest global source to the atmosphere (Prather et al. 1995). It is difficult to estimate fractionation factors for the formation of N_2O and NO because both are the products of multiple soil transformations with different substrate $\delta^{15}N$ (Table 4.1). N_2O is a product of two reactions during nitrification (Wada & Ueda 1996; Stein & Yung 2003). The first is the oxidation of NH_2OH with NOH as a precursor; the second is the reduction of NO_2^- by nitrite reductase. N_2O is also an intermediate during the conversion of NO_3^- to N_2 during denitrification. Measuring the isotopic composition of NO is difficult because it is highly reactive and rapidly converted to organic and inorganic nitrates (Fehsenfeld et al. 1992).

The $\delta^{15}N$ of N_2O is variable over time because of the relative importance of nitrification versus denitrification and changes in the $\delta^{15}N$ of substrates (Perez et al. 2000, 2001; Tilsner et al. 2003; Toyoda et al. 2005). The fractionation factor for denitrification is less than that for nitrification (Barford et al. 1999; Yoshida 1988), so some investigators have attempted to separate the relative contribution of the two processes by assuming that N_2O with relatively greater $\delta^{15}N$ is the product of denitrification, while relatively low values are indicative of nitrification. Perez et al. (2000, 2001) observed that the $\delta^{15}N$ of N_2O produced in the first 5 days after fertilization was depleted 45–55‰ compared with the substrate presumably because nitrification was the dominant source. The depletion shifted to 10–30‰ following this as denitrification became more prevalent. A promising new approach to separate the processes responsible for N_2O production is measuring the $\delta^{15}N$ site preference of isotopomers. The site preference is defined as $\delta^{15}N^\alpha - \delta^{15}N^\beta$, where

N^{α} and N^{β} are the central and terminal nitrogen atoms in N_2O, respectively (Stein & Yung 2003). Schmidt et al. (2004) hypothesized that the site preferences should be determined by the population of microorganisms and the type of nitric oxide reductases and this is supported by recent experimental studies. Toyoda et al. (2005) compared two species of denitrifying bacteria and found that site preference was nearly constant and distinct over time, even though bulk $\delta^{15}N$ was highly variable. Site preference can also separate the relative importance of nitrification versus denitrification; Sutka et al. (2006) measured site preferences of 33‰ and 0‰ for the two respective processes even though the $\delta^{15}N$ of bulk N_2O changed over time.

Patterns of soil nitrogen isotope composition

Models

A consistent pattern across ecosystems is that soil $\delta^{15}N$ increases and nitrogen content decreases with depth, and the increase in $\delta^{15}N$ can be as great as 10‰ (Figure 4.4) (Shearer & Kohl 1986; Fry 1991; Högberg 1997). The mechanisms behind these corresponding changes were first addressed by Nadelhoffer & Fry (1988) and recent quantitative models by Brenner et al. (2001) and Baisden et al. (2002). Nadelhoffer & Fry (1988) hypothesized three possible mechanisms:

1 discrimination during decomposition;
2 differential preservation of components enriched in ^{15}N;
3 illuviation of ^{15}N enriched organic matter in deeper soil horizons.

Results from field and laboratory experiments by Nadelhoffer & Fry (1988) indicated that the most likely mechanisms for the increase with depth were inputs of litter at the soil surface that were isotopically lighter than organic matter and overall isotopic fractionation during microbial processing of organic matter during decomposition and nitrogen loss (Figure 4.4). This is supported by Kramer et al. (2003) who examined soil $\delta^{15}N$ in relation to the degree of humification by measuring the ratio of unsubstituted aliphatic carbon to oxygenated carbon and found that $\delta^{15}N$ increased with humification during microbial processing of organic matter. Unlike carbon, differential preservation was not an important factor because individual chemical components of litter had similar $\delta^{15}N$ values as whole leaf and root samples (Nadelhoffer & Fry 1988). The increase in $\delta^{15}N$ over time also occurred in long-term incubations in the absence of illuviation. Högberg et al. (1996, 1999) suggest a fourth mechanism; ectomycorrhizae become enriched during nitrogen uptake and this relatively recalcitrant material has a long turn-over time in the soil. Fractionation during nitrogen transfer from fungi to plants resulting in enrichment of the fungus has also been demonstrated by Hobbie et al. (1999, 2001) and Hobbie & Colpaert (2003).

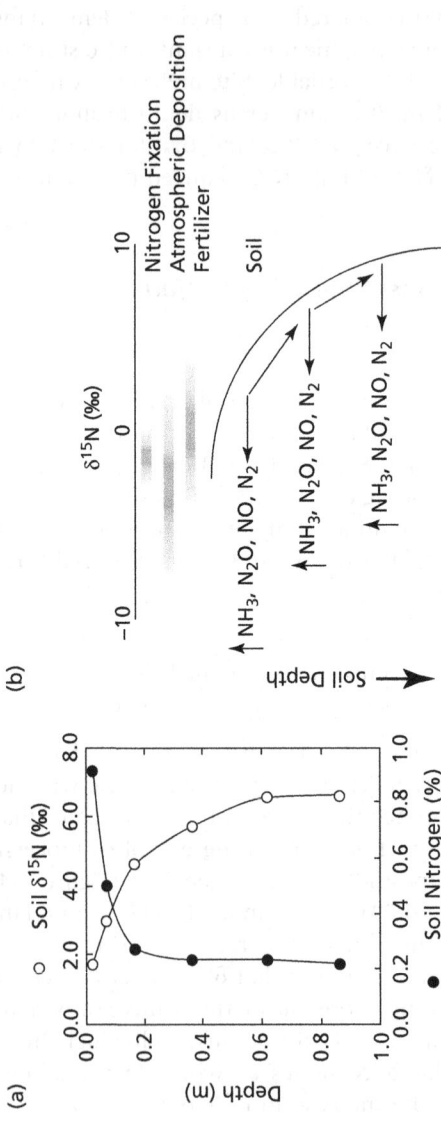

Figure 4.4 (Left panel) Soil $\delta^{15}N$ and nitrogen content of a typical soil profile (from Sperry et al. 2006). Closed symbols are soil nitrogen content, and open symbols are soil $\delta^{15}N$. (Right panel) Conceptual model based on Nadelhoffer & Fry (1988, 1994) proposing the mechanisms for the increase in soil $\delta^{15}N$ with depth. Horizontal arrows represent mineralization and nitrification; vertical arrows represent gaseous nitrogen loss. Fractionation during each transformation causes the remaining soil nitrogen to become progressively enriched in ^{15}N as nitrogen moves down the soil profile.

Quantitative models incorporating the basic principles of the conceptual model of Nadelhoffer & Fry (1988) have recently been developed using a chronosequence from 3 to 3,000 kyr in California grassland ecosystems. The first model is based on the $\delta^{15}N$ of nitrogen inputs into the soil and fractionation during nitrogen loss,

$$R_s = R_{total}/\alpha_{ex}$$

where R_s is the isotope ratio of soil organic matter, R_{total} is the weighted mean isotope ratio of nitrogen inputs, and α_{ex} is the apparent fractionation factor (Brenner et al. 2001). Soil $\delta^{15}N$ increased 2.7‰ along the chronosequence. In the absence of changes in R_{total}, Brenner et al. conclude that rates of soil nitrogen cycling increase with age, resulting in greater nitrogen loss, and that soil $\delta^{15}N$ therefore provides a quantitative measure of the openness of the nitrogen cycle in these ecosystems.

Ecosystem models of soil nitrogen and carbon dynamics recognize that there are several pools in the soil with different resident times (Parton et al. 1987). The most sophisticated isotope simulation effort to date is a multi-pool, multi-isotope model that considers input, loss, and transfer between soil pools that turn over at roughly annual (active), decadal (stabilized), and millennial (passive) rates at multiple depths (Baisden et al. 2002). The $\Delta^{14}C$ of samples collected from 1949, 1978, and 1998 were used to calculate transfer rates between the three pools and vertically within the profile. The model predicts that ^{15}N enrichment occurs primarily during the preservation of soil nitrogen in the passive pool, and to lesser extent in the stabilized pool.

Global patterns

Several recent studies have addressed broad patterns of soil $\delta^{15}N$ (Fry 1991; Austin & Vitousek 1998; Handley et al. 1999; Martinelli et al. 1999; Amundson et al. 2003; Aranibar et al. 2004). Fry (1991) measured soil $\delta^{15}N$ at 11 Long Term Ecological Research (LTER) sites in the USA and Puerto Rico that spanned a gradient from grasslands and deserts to forests and tundra. With the exception of desert sites, these relatively undisturbed locations generally had lower soil $\delta^{15}N$ than was previously observed in managed agricultural areas. Review of previous literature suggests that climate is the primary factor controlling soil $\delta^{15}N$ (Handley et al. 1999; Amundson et al. 2003). Handley et al. (1999) examined the correlation between soil $\delta^{15}N$ and mean annual temperature and precipitation, latitude, altitude, and soil pH and concluded that while soil $\delta^{15}N$ was correlated with rainfall and latitude, more data are required before we can describe general relationships. Amundson et al. (2003) reached similar conclusions; soil $\delta^{15}N$ decreases with decreasing mean annual temperature and increasing annual precipitation, but stated that understanding the mechanisms behind these patterns is still in its infancy. A general conclusion is that soils enriched in ^{15}N generally have more open nitrogen cycles with greater rates of nitrogen loss compared with inputs than more closed systems. The previously described model proposed by Brenner et al. (2001) illustrates this. Amundson et al. (2003) argued using this model that in general nitrogen inputs are restricted globally to values around 0‰, so the observed patterns are due solely to fractionation during nitrogen loss. However, direct measurements of soil $\delta^{15}N$ and gaseous nitrogen loss only partially support this hypothesis (Aranibar et al. 2004).

The observed correlations between temperature, precipitation, and soil $\delta^{15}N$ values are driven by greater soil $\delta^{15}N$ values in arid regions, but these regions also exhibit greater variability and can have values similar to more mesic and cooler ecosystems. An additional factor that may control soil $\delta^{15}N$ values in these regions is land-use change. A significant amount of the world's agriculture and grazing occur in arid areas and both factors can increase soil $\delta^{15}N$ (Frank & Evans 1997; Evans & Belnap 1999; Frank et al. 2004; Sperry et al. 2006). Comparisons of grazed and adjacent ungrazed sites demonstrate that grazing can increase soil $\delta^{15}N$ from 1 to 8‰ by removing sources of nitrogen input while nitrogen loss continues (Evans & Belnap 1999; Sperry et al. 2006), or by stimulating nitrogen cycling through the addition of more labile urine and feces causing greater volatilization (Frank & Evans 1997; Frank et al. 2004). Both mechanisms are consistent with models described previously, but it is clear that more mechanistic studies coupled with modeling efforts are needed to fully understand broad patterns of soil $\delta^{15}N$.

Conclusions

Interpretation of soil $\delta^{15}N$ has evolved over the past decades. Once thought of as a natural tracer it is now realized that $\delta^{15}N$ can serve as a valuable integrator of the complex transformations that occur in the soil. Modeling efforts such as those by Baisden et al. (2002) indicate that soil $\delta^{15}N$ provides valuable information on the balance between nitrogen inputs and losses in ecosystems. Recent contributions also demonstrate how $\delta^{15}N$ can be used to identify sources of important greenhouse gasses and their fate within the environment. Future research must address the role of organic nitrogen in the soil nitrogen cycle, and this will be facilitated by more sophisticated compound-specific isotope analyses as well as gas chromatography–mass spectrometry (GC–MS) systems. Significant advances in our knowledge of inorganic nitrogen compounds will be facilitated by dual-isotope analyses such as $\delta^{15}N$ and ^{18}O in NO_3^- and N_2O. A reliable method is also needed for NH_4^+, and microbial methods such as those developed for NO_3^- (Sigman et al. 2001) show great promise. It is now possible to make very rapid measurements of CO_2 using tunable diode lasers (Bowling et al. 2003), and future development of this technology should allow similar measurements of N_2O and NO. Further development of theoretical models coupled with empirical experiments using methods designed specifically for stable isotope analysis are required to further expand our knowledge of soil $\delta^{15}N$, but they also have the potential to expand our knowledge of the soil nitrogen cycle.

References

Amundson, R., Austin, A.T., Schuur, E.A.G., et al. (2003) Global patterns of the isotopic composition of soil and plant nitrogen. *Global Biogeochemical Cycles*, **17**, art. no.-1031.

Aranibar, J.N., Otter, L., Macko, S.A., et al. (2004) Nitrogen cycling in the soil–plant system along a precipitation gradient in the Kalahari sands. *Global Change Biology*, **10**, 359–373.

Austin, A.T. & Vitousek, P.M. (1998) Nutrient dynamics on a precipitation gradient in Hawai'i. *Oecologia*, **113**, 519–529.

Baisden, W.T., Amundson, R., Brenner, D.L., Cook, A.C., Kendall, C. & Harden, J.W. (2002) A multiisotope C and N modeling analysis of soil organic matter turnover and transport as a function of soil depth in a California annual grassland soil chronosequence. *Global Biogeochemical Cycles*, **16**, art. no.-1135.

Barford, C.C., Montoya, J.P., Altabet, M.A. & Mitchell, R. (1999) Steady-state nitrogen isotope effects of N_2 and N_2O production in *Paracoccus denitrificans*. *Applied and Environmental Microbiology*, **65**, 989–994.

Binkley, D., Sollins, P. & McGill, W.B. (1985) Natural abundance of ^{15}N as a tool for tracing Alder-fixed nitrogen. *Soil Science Society of America Journal*, **49**, 444–447.

Bol, R., Rockmann, T., Blackwell, M. & Yamulki, S. (2004) Influence of flooding on $\delta^{15}N$, $\delta^{18}O$, (1)$\delta^{15}N$ and (2)$\delta^{15}N$ signatures of N_2O released from estuarine soils – a laboratory experiment using tidal flooding chambers. *Rapid Communications in Mass Spectrometry*, **18**, 1561–1568.

Bowling, D.R., Sargent, S.D., Tanner, B.D. & Ehleringer, J.R. (2003) Tunable diode laser absorption spectroscopy for stable isotope studies of ecosystem–atmosphere CO_2 exchange. *Agricultural and Forest Meteorology*, **118**, 1–19.

Bragazza, L., Tahvanainen, T., Kutnar, L., et al. (2004) Nutritional constraints in ombrotrophic *Sphagnum* plants under increasing atmospheric nitrogen deposition in Europe. *New Phytologist*, **163**, 609–616.

Bragazza, L., Limpens, J., Gerdol, R., et al. (2005) Nitrogen concentration and $\delta^{15}N$ signature of ombrotrophic *Sphagnum* mosses at different N deposition levels in Europe. *Global Change Biology*, **11**, 106–114.

Brenner, D.L., Amundson, R., Baisden, W.T., Kendall, C. & Harden, J. (2001) Soil N and ^{15}N variation with time in a California annual grassland ecosystem. *Geochimica et Cosmochimica Acta*, **65**, 4171–4186.

Chapin III, F.S., Matson, P.A. & Mooney, H.A. (2002) *Principles of Ecosystem Ecology*. Springer-Verlag, New York.

Dawson, T.E. & Ehleringer, J.R. (1991) Streamside trees that do not use stream water. *Nature*, **350**, 335–337.

Ehleringer, J.R., Phillips, S.L., Schuster, W.S.F. & Sandquist, D.R. (1991) Differential utilization of summer rains by desert plants. *Oecologia*, **88**, 430–434.

Evans, R.D. & Belnap, J. (1999) Long-term consequences of disturbance on nitrogen dynamics in an arid ecosystem. *Ecology*, **80**, 150–160.

Evans, R.D. & Ehleringer, J.R. (1993) A break in the nitrogen cycle in arid lands? Evidence from $\delta^{15}N$ of soils. *Oecologia*, **94**, 314–317.

Evans, R.D. & Johansen, J.R. (1999) Microbiotic crusts and ecosystem processes, *Critical Reviews in Plant Sciences*, **18**, 183–225.

Farquhar, G.D., O'Leary, M.H. & Berry, J.A. (1982) On the relationship between carbon isotope discrimination and the intercellular carbon dioxide concentration in leaves. *Australian Journal of Plant Physiology*, **9**, 121–137.

Fehsenfeld, F.J., Calvert, J., Fall, R., et al. (1992) Emissions of volatile organic compounds from vegetation and the implications for atmospheric chemistry. *Global Biogeochemical Cycles*, **6**, 389–430.

Feigin, A., Kohl, D.H., Shearer, G. & Commoner, B. (1974) Variation in the natural nitroge^{15}N abundance in nitrate mineralized during incubation of several Illinois soils. *Soil Science Society of America Proceedings*, **38**, 90–96.

Flanagan, L.B., Comstock, J.P. & Ehleringer, J.R. (1991) Comparison of modelled and observed environmental influences on the stable oxygen and hydrogren isotope composition of leaf water in *Phaseolus vulgaris* L. *Plant Physiology*, **96**, 588–596.

Frank, D.A. & Evans, R.D. (1997) Effects of native grazers on grassland N cycling in Yellowstone National Park. *Ecology*, **78**, 2238–2248.

Frank, D.A., Evans, R.D. & Tracy, B.F. (2004) The role of ammonia volatilization in controlling the natural ^{15}N abundance of a grazed grassland. *Biogeochemistry*, **68**, 169–178.

Freyer, H.D. & Aly, A.I.M. (1974) ^{15}N variations in fertilizer nitrogen. *Journal of Environmental Quality*, **3**, 405–406.

Fry, B. (1991) Stable isotope diagrams of fresh-water food webs. *Ecology*, **72**, 2293–2297.

Galloway, J.N., Schlesinger, W.H., Levy, H., Michaels, A. & Schnoor, J.L. (1995) Nitrogen-fixation–anthropogenic enhancement–environmental response. *Global Biogeochemical Cycles*, **9**, 235–252.

Handley, L.L., Austin, A.T., Robinson, D., et al. (1999) The ^{15}N natural abundance ($\delta^{15}N$) of ecosystem samples reflects measures of water availability, *Australian Journal of Plant Physiology*, **26**, 185–199.

Heaton, T.H.E. (1986) Isotopic studies of nitrogen pollution in the hydrosphere and atmosphere: a review., *Chemical Geology*, **59**, 87–102.

Heaton, T.H.E. (1987) $^{15}N/^{14}N$ ratios of nitrate and ammonium in rain at Pretoria, South Africa., *Atmospheric Environment*, **21**, 843–852.

Herman, D.J. & Rundel, P.W. (1989) Nitrogen isotope fractionation in burned and unburned Chaparral soils. *Soil Science Society of America Journal*, **53**, 1229–1236.

Hobbie, E.A. & Colpaert, J.V. (2003) Nitrogen availability and colonization by mycorrhizal fungi correlate with nitrogen isotope patterns in plants, *New Phytologist*, **157**, 115–126.

Hobbie, E.A., Macko, S.A. & Shugart, H.H. (1999) Insights into nitrogen and carbon dynamics of ectomycorrhizal and saprotrophic fungi from isotopic evidence. *Oecologia*, **118**, 353–360.

Hobbie, E.A., Olszyk, D.M., Rygiewicz, P.T., Tingey, D.T. & Johnson, M.G. (2001) Foliar nitrogen concentrations and natural abundance of ^{15}N suggest nitrogen allocation patterns of Douglas-fir and mycorrhizal fungi during development in elevated carbon dioxide concentration and temperature. *Tree Physiology*, **21**, 1113–1122.

Högberg, P. (1997) Tansley Review No. 95: ^{15}N natural abundance in soil–plant systems. *New Phytologist*, **137**, 179–203.

Högberg, P., Högbom, L., Schinkel, H., Högberg, M., Johannisson, C. & Wallmark, H. (1996) ^{15}N abundance of surface soils, roots and mycorrhizas in profiles of European forest soils. *Oecologia*, **108**, 207–214.

Högberg, P., Högberg, M., Quist, M.E., Ekblad, A. & Näsholm, T. (1999) Nitrogen isotope fractionation during nitrogen uptake by ectomycorrhizal and non-mycorrhizal *Pinus sylvestris*. *New Phytologist*, **142**, 569–576.

Johnston, A.M., Scrimgeour, C.M., Henry, M.O. & Handley, L.L. (1999) Isolation of NO_3^--N as 1-phenylazo-2-naphthol (Sudan-1) for measurement of $\delta^{15}N$. *Rapid Communications in Mass Spectrometry*, **13**, 1531–1534.

Kramer, M.G., Sollins, P., Sletten, R.S. & Swart, P.K. (2003) N isotope fractionation and measures of organic matter alteration during decomposition. *Ecology*, **84**, 2021–2025.

Martinelli, L.A., Piccolo, M.C., Townsend, A.R., et al. (1999) Nitrogen stable isotopic composition of leaves and soil: Tropical versus temperate forests. *Biogeochemistry*, **46**, 45–65.

Nadelhoffer, K.J. & Fry, B. (1988) Controls on natural nitrogen-15 and carbon-13 abundances in forest soil organic matter. *Soil Science Society of America Journal*, **52**, 1633–1640.

Nadelhoffer, K.J. & Fry, B. (1994) Nitrogen isotope studies in forest ecosystems. In: *Stable Isotopes in Ecology and Enivronmental Studies* (Eds K. Lajtha & R.H. Michener), pp. 22–44. Blackwell Scientific Publications, Boston.

Ostle, N.J., Bol, R., Petzke, K.J. & Jarvis, S.C. (1999) Compound specific $\delta^{15}N$ values: amino acids in grassland and arable soils. *Soil Biology and Biochemistry*, **31**, 1751–1755.

Parton, W.J., Schimel, D.S., Cole, C.V. & Ojima, D.S. (1987) Analysis of factors controlling soil organic matter levels in Great Plains grasslands. *Soil Science Society of America Journal*, **51**, 1173–1179.

Perez, T., Trumbore, S.E., Tyler, S.C., Davidson, E.A., Keller, M. & de Camargo, P.B. (2000) Isotopic variability of N_2O emissions from tropical forest soils. *Global Biogeochemical Cycles*, **14**, 525–535.

Perez, T., Trumbore, S.E., Tyler, S.C., Matson, P.A., et al. (2001) Identifying the agricultural imprint on the global N_2O budget using stable isotopes. *Journal of Geophysical Research–Atmospheres*, **106**, 9869–9878.

Pichlmayer, F., Schoner, W., Seibert, P., Stichler, W. & Wagenbach, D. (1998) Stable isotope analysis for characterization of pollutants at high elevation alpine sites, *Atmospheric Environment*, **32**, 4075–4085.

Prather, M.J., Derwent, R., Ehhalt, D., Fraser, P., Sanhueza, E. & Zhou, X. (1995) Other trace gasses and atmospheric chemistry. In: *Climate Change 1994* (Eds J.T. Houghton, L.G. Filho, J. Bruce, et al.), pp. 73–126. Cambridge University Press, Cambridge.

Rahn, T. & Wahlen, M. (2000) A reassessment of the global isotopic budget of atmospheric nitrous oxide. *Global Biogeochemical Cycles*, **14**, 537–543.

Robinson, D. (2001) $\delta^{15}N$ as an integrator of the nitrogen cycle. *Trends in Ecology and Evolution*, **16**, 153–162.

Schimel, J.P. & Bennett, J. (2004) Nitrogen mineralization: Challenges of a changing paradigm. *Ecology*, **85**, 591–602.

Schlesinger, W.H. & Hartley, A.E. (1992) A global budget for atmospheric NH_3. *Biogeochemistry*, **15**, 191–211.

Schmidt, H.L., Werner, R.A., Yoshida, N. & Well, R. (2004) Is the isotopic composition of nitrous oxide an indicator for its origin from nitrification or denitrification? A theoretical approach from referred data and microbiological and enzyme kinetic aspects. *Rapid Communications in Mass Spectrometry*, **18**, 2036–2040.

Shearer, G. & Kohl, D.H. (1986) N_2-fixation in field settings: Estimations based on natural ^{15}N abundance. *Australian Journal of Plant Physiology*, **13**, 699–756.

Shearer, G. & Kohl, D.H. (1993) Natural abundance of ^{15}N: fractional contribution of two sources to a common sink and use of isotope discrimination. In: *Nitrogen Isotope Techniques* (Eds R. Knowles & T.H. Blackburn). pp. 89–125. Academic Press, New York.

Sigman, D.M., Casciotti, K.L., Andreani, M., Barford, C., Galanter, M. & Bohlke, J.K. (2001) A bacterial method for the nitrogen isotopic analysis of nitrate in seawater and freshwater. *Analytical Chemistry*, **73**, 4145–4153.

Silva, S.R., Kendall, C., Wilkison, D.H., Ziegler, A.C., Chang, C.C.Y. & Avanzino, R.J. (2000) A new method for collection of nitrate from fresh water and the analysis of nitrogen and oxygen isotope ratios. *Journal of Hydrology*, **228**, 22–36.

Sperry, L., Belnap, J. & Evans, R.D. (2006) *Bromus tectorum* invasion alters nitrogen dynamics in an undisturbed arid grassland ecosystem. *Ecology*, **87**, 603–615.

Steffen, W., Sanderson, A., Tyson, P.D., et al. (2004) *Global Change and the Earth System*. Springer-Verlag, New York.

Stein, L.Y. & Yung, Y.L. (2003) Production, isotopic composition, and atmospheric fate of biologically produced nitrous oxide. *Annual Review of Earth and Planetary Sciences*, **31**, 329–356.

Stickrod, R.D. & Marshall, J.D. (2000) On-line nitrate-δ^{15}N extracted from groundwater determined by continuous-flow elemental analyzer/isotope ratio mass spectrometry. *Rapid Communications in Mass Spectrometry*, **14**, 1266–1268.

Sutka, R.L., Ostrom, N.E., Ostrom, P.H., et al. (2006) Distinguishing nitrous oxide production from nitrification and denitrification on the basis of isotopomer abundances, *Applied and Environmental Microbiology*, **72**, 638–644.

Tilsner, J., Wrage, N., Lauf, J. & Gebauer, G. (2003) Emission of gaseous nitrogen oxides from an extensively managed grassland in NE Bavaria, Germany – II. Stable isotope natural abundance of N_2O. *Biogeochemistry*, **63**, 249–267.

Toyoda, S., Mutobe, H., Yamagishi, H., Yoshida, N. & Tanji, Y. (2005) Fractionation of N_2O isotopomers during production by denitrifier. *Soil Biology and Biochemistry*, **37**, 1535–1545.

Vitoria, L., Otero, N., Soler, A. & Canals, A. (2004) Fertilizer characterization: Isotopic data (N, S, O, C, and Sr). *Environmental Science and Technology*, **38**, 3254–3262.

Vitousek, P.M., Mooney, H.A., Lubchenco, J. & Melillo, J.M. (1997) Human domination of Earth's ecosystems. *Science*, **277**, 494–499.

Wada, E. & Ueda, S. (1996) Carbon, Nitrogen and Oxygen isotope ratios of CH_4 and N_2O in soil systems. In: *Mass Spectrometry of Soils* (Eds T. W. Boutton & S.-I. Yamasaki), pp. 177–204. Marcel Dekker: New York.

Werner, R.A. & Schmidt, H.L. (2002) The *in vivo* nitrogen isotope discrimination among organic plant compounds. *Phytochemistry*, **61**, 465–484.

Yoshida, N. (1988) ^{15}N-depleted N_2O as a product of nitrification. *Nature*, **335**, 528–529.

CHAPTER 5

Isotopic study of the biology of modern and fossil vertebrates

PAUL L. KOCH

Introduction

Naturally occurring variations in the stable isotope composition of fossil vertebrates have been studied since the late 1970s. Isotopic data from vertebrate fossils are sometimes used as proxies for environmental factors, such as temperature or precipitation, with biological processes viewed as annoying "vital" effects that must be filtered out to obtain a pure environmental signal. Yet knowledge of these biological factors, which include diet, digestive physiology, reproductive state, thermoregulatory or osmoregulatory physiology, habitat preference, and migration, deepen our understanding of the ecology and evolution of ancient vertebrates. Research on these biological issues was spearheaded by paleoanthropologists studying ancient humans and their kin (see reviews by van der Merwe 1982; DeNiro 1987; Schwarcz & Schoeninger 1991; Ambrose & Krigbaum 2003b), but has exploded in the last decade in animal ecology and physiology (see reviews by Gannes et al. 1998; Hobson 1999; Kelly 2000). The growing body of work on modern vertebrates offers a foundation for more nuanced paleobiological interpretations. In addition, technological advances have made isotopic analysis more routine and allowed study of new systems, new substrates, and extremely small samples.

Isotopic data are used in vertebrate paleobiology in two ways. Because of differences in their mass, isotopes of light elements (e.g., H, C, N, O, S, or Ca) are sorted (or fractionated) by chemical and physical processes (see Sulzman, this volume, pp. 1–21 for definitions and conventions for reporting isotopic data and fractionations). In some studies, the extent of isotopic fractionation is used to monitor the magnitude or rate of a process. For example, the process of evaporation is associated with preferential loss of water enriched in ^{16}O and ^{1}H. Experiments on rock doves showed that the δD value of body water is positively correlated with the fraction of water lost by evaporation (McKechnie et al. 2004). Alternately, isotopic differences among substances can serve as natural labels to trace the flow of these substances into vertebrates. This type of research can involve either light or more massive elements (e.g., Sr, Nd, Pb). For example, water and tissues in saguaros have unusual isotopic values relative to other resources available to desert animals. Wolf & Martínez del Rio (2003) exploited these differences to assess the impact of carbon and water from saguaros at the community level, showing that saguaros support a diverse guild of frugivorous, granivorous and insectivorous birds.

Here, I will explore recent advances in the study of isotopic variations in vertebrates, focusing on studies of fossils. Table 5.1 offers summary information about the isotopic systems to be discussed. After considering the vertebrate tissues that are found in the fossil record, I will introduce the main isotopic systems, focusing on physiological and environmental controls on isotopic values in vertebrate tissues. I will then briefly consider the reliability (or lack thereof) of vertebrate fossils as recorders of biogenic isotope compositions. Paleobiological applications are presented last, including examples of studies of diet, thermal physiology, reproductive biology, habitat preference, and migration. These examples are not meant to provide an exhaustive review; the literature on these subjects has become too vast to cover in a single paper.

Vertebrate tissues in the fossil record

Types of tissue

Vertebrate bodies are constructed from tissues with different macromolecular and elemental compositions, different styles of growth and turnover, and different potentials for post-mortem preservation (Table 5.2). Soft tissues such as skin, muscle, hair, and feathers contain protein and lipids, and well-preserved soft tissues persist for 10^3 to 10^4 years in unusual settings (e.g., mummification in dry environments, permafrost). Soft tissues are preserved in exceptional cases for up to 10^8 years, but often only as impressions or pseudomorphs composed of phosphate minerals or fossilized bacteria (Martill 1995; Briggs et al. 1997). A report of organic preservation in a Cretaceous dinosaur hints that the potential for excellent preservation in more typical depositional environments may have been underestimated (Schweitzer et al. 2005). Immunohistochemical and amino acid analyses suggest that traces of keratin (feather protein) survive in Mesozoic fossils (Schweitzer et al. 1999), but this material has not been isolated for isotopic analysis.

Mineralized tissues such as bone, tooth enamel and dentin, eggshell, and otoliths have much greater potential for preservation in deep time. Bone and tooth dentin and enamel are composites of mineral, protein, and lipid. The mineral is a highly substituted form of hydroxylapatite ($Ca_{10}[PO_4]_6[OH]_2$) I will call bioapatite. Bioapatite has a few weight percent carbonate substituting for hydroxyl and phosphate groups, and various cations (e.g, Sr, Pb) substituting for calcium (Simkiss & Wilbur 1989). Bioapatite readily adsorbs carbonate, rare earth elements, amino acids, and nucleic acids on crystal surfaces (Beshah et al. 1990; Hedges 2002; Trueman & Tuross 2002). Bone is com-

Table 5.1 Vertebrate isotope systems and their applications.

Element	Isotope	Fractional abundance	Standard for δ value calculation	Isotopic range in vertebrates*	Applications†
Hydrogen	^{1}H	0.999844	Standard Mean	−175 to +70‰	Migration, habitat use, diet, trophic
	^{2}H	0.000156	Ocean Water (SMOW)	Kelly et al. (2002) Wolf & del Rio (2000)	level, osmoregulatory physiology
Carbon	^{12}C	0.98889	Pee Dee	−60 to +5‰	Diet, digestive physiology, habitat
	^{13}C	0.01111	Belemnite Limestone (PDB)	Doucett et al. (2002) Kohn & Cerling (2002)	use, migration
Nitrogen	^{14}N	0.99634	Air	−30 to +30‰	Trophic level, diet, habitat use,
	^{15}N	0.00366		Hare et al. (1991)	migration, starvation, reproduction
Oxygen	^{16}O	0.99755	SMOW or PDB	0 to +35‰	Habitat use, migration, diet,
	^{17}O	0.00039		Kohn & Cerling	thermoregulation, osmoregulation
	^{18}O	0.00206		(2002)	
Sulfur	^{32}S	0.9493	Canyon Diablo	−15 to +20‰	Habitat use, migration, diet
	^{33}S	0.0076	Troilite	Lott et al. (2003)	
	^{34}S	0.0429			
	^{36}S	0.0002			
Calcium	^{40}Ca	0.96941	NIST 915a (CaF$_2$)	−3.1 to +1.8‰	Trophic level, habitat use, migration
	^{42}Ca	0.00647		Skulan & DePaolo	
	^{43}Ca	0.00135		(1999)	
	^{44}Ca	0.02086			
	^{46}Ca	0.00004			
	^{48}Ca	0.00187			
Strontium	^{84}Sr	0.0056	Bulk Earth (ε) or	0.7043–0.7583	Habitat use, migration, diet
	^{86}Sr	0.0986	Sea Water (δ)	Price et al. (2000)	
	^{87}Sr	0.0700		Nelson et al. (1986)	
	^{88}Sr	0.8258			

* The range of values is defined by data in cited papers.
† Study of migration is a potential application of all isotope systems.

Table 5.2 Summary information on materials used as substrate for isotopic analysis of vertebrates.

Tissue	Component	Signal window	Isotope systems	Preservation window (years)
Hair	Keratin	Accretion	H, C, N, O, S	10^4
Feather	Keratin	Accretion	H, C, N, O, S	10^4
Bone	Bioapatite	Years	CO_3–C, O	10^3 (10^6)
			PO_4–O	10^4 (10^8)
			Ca	10^7–10^8
			Sr, Nd, Pb	10^3
	Collagen	Years	H, C, N, O, S	10^5–10^6 (10^8)
	Lipid	Weeks–months	H, C	
Enamel	Bioapatite	Accretion	CO_3–C, O	10^8
			PO_4–O	10^8
			Ca	10^7–10^8
			Sr, Nd, Pb	10^7
Dentin	Bioapatite	Accretion	Same as bone	Same as bone
	Collagen	Accretion	H, C, N, O, S	Same as bone
Egg shell	Carbonate	Days–weeks	C, O	10^7–10^8
	Protein	Days–weeks	C, N	10^4
Otoliths	Carbonate	Accretion	C, O	10^7

Compound type	Signal window		Isotope systems	Preservation window
Amino acid	Depends on tissue type		H, C, N	10^5?
Cholesterol	Weeks–months		H, C	10^6?
Fatty acid	Weeks–months		H, C	10^4?

posed of tiny bioapatite crystals ($100 \times 20 \times 4$ nm) intergrown with an organic matrix (chiefly composed of the protein collagen) that comprises ca. 30 percent of its dry weight (Simkiss & Wilbur 1989). Enamel is much less porous than bone. It contains <5 weight percent organic matter (chiefly non-collagenous proteins) and has much larger crystals ($1000 \times 130 \times 30$ nm) with fewer substitutions (LeGeros 1991). The crystal size, organic content, and organic composition of tooth dentin resemble bone, whereas its porosity is intermediate between enamel and bone (Lowenstam & Weiner 1989).

Bird and crocodile eggshells are composed of tiny crystals secreted around a honeycomb of fibrous organic sheets. The crystalline portion of shells is almost entirely calcite. Mineral in bird eggshells occurs in three distinct layers that are covered externally by cuticle and anchored internally to a shell membrane (Simkiss & Wilbur 1989). The organic matrix comprises ca. 3 percent of the mass of bird eggshell and is largely protein (ca. 70%) with lesser amounts of carbohydrate and lipid (Burley & Vadehra 1989). A range of microstructures occurs in dinosaur eggshells, including some quite similar to those of bird eggshell (Mikhailov et al. 1996).

Otoliths are mineralized bodies in the vertebrate inner ear (Panella, 1980). Otolith mineralogy varies among vertebrates – bioapatite occurs in agnathans, aragonite occurs in jawed fish and amphibians, and calcite occurs in amniotes (Simkiss & Wilbur 1989). In teleosts, calcification occurs on a preformed organic matrix rich in non-collagenous proteins and mucopolysacchrides (Panella 1980).

Individual organic compounds

Most soft tissues decay rapidly, and even the organic matter in mineralized tissues degrades by various processes (Bada et al. 1999; Collins et al. 2002). Yet under favorable circumstances, individual organic molecules (or diagenetic products that can be directly related to biogenic molecules) can be isolated from fossil vertebrates by gas or liquid chromatography. Fatty acids occur in characteristic relative abundances in different organisms, as do amino acids in different types of protein (Tuross et al. 1988; Smith et al. 1997; Evershed et al. 1999). Abundance patterns can be used to trace sources and to ensure that the organic residues extracted from fossils are not exogenous (Tuross et al. 1988; Bada et al. 1999). Sometimes, particular molecules are highly specific to a class of organisms, making them especially reliable molecular substrates for isotopic analysis. For example, vertebrate collagen contains a relatively high abundance of the amino acid hydroxyproline, which is uncommon in other proteins from terrestrial organisms (Tuross et al. 1988). Similarly, the steroidal lipid cholesterol does not occur in plants, but does occur in relatively high concentrations in the bodies and bones of vertebrates (Stott et al. 1999).

If sufficient quantities of organic matter are available, the products of chromatographic separation can be collected and analyzed by standard, dual-inlet mass spectrometry, as was done in early studies of amino acids from bone collagen (Gaebler et al. 1966; Tuross et al. 1988; Hare et al. 1991) or with an elemental analyzer interfaced with a mass spectrometer. In many cases, however, only traces of organic compounds remain. By coupling a gas chromatograph (GC) to a combustion, pyrolysis, or reduction furnace, and then feeding the effluent to an isotope-ratio-monitoring mass spectrometer on a carrier gas stream, it is possible to separate minute quantities of individual organic molecules and measure their δD, $\delta^{13}C$, $\delta^{15}N$, or $\delta^{18}O$ values. Systems whereby organic molecules in the effluent of a liquid chromatograph are converted to CO_2 for carbon isotope analysis are available as well.

Three classes of compounds from vertebrates have been examined using these methods: fatty acids, sterols such as cholesterol, and amino acids. Lipids (e.g., fatty acids and sterols) are rich in carbon and hydrogen. Amino acids, in contrast, contain abundant carbon and nitrogen, and as well as hydrogen, oxygen, and sulfur. Early work on individual amino acids using dual-inlet mass spectrometry examined both $\delta^{13}C$ and $\delta^{15}N$ values (Tuross et al. 1988; Hare et al. 1991). Yet nearly all the papers on vertebrates using GC methods have examined only $\delta^{13}C$ values of individual compounds (e.g., Evershed et al. 1999; Fogel & Tuross 2003). There is a growing literature on $\delta^{15}N$ and δD values of individual molecules measured via GC methods from enriched-tracer studies in biomedical research (e.g., Scrimgeour et al. 1999; Metges et al. 2002), as well as work on the $\delta^{15}N$ values of individual amino acids in invertebrates (e.g., McClelland & Montoya 2002), but this approach has yet to impact ecological or paleontological studies of vertebrates.

Growth and turnover time

The time represented by an isotopic sample is dependent on the mode of growth and turnover time of each tissue, as well as the sampling strategy. Bone growth is complex, involving both ossification of cartilage and accretionary growth, which can be interrupted by lines of arrested growth (LAGs). Bone is also remodeled by dissolution and reprecipitation (Lowenstam & Weiner 1989). Remodeling is active in mammals, birds, and other rapidly growing vertebrates and in any bone under substantial load (Reid 1987; Chinsamy & Dodson 1995; Padian et al. 2004). A bulk sample of bone mineral or collagen thus contains material that may have formed over several years of growth, though it will be weighted towards periods of rapid growth. In portions of bone that show incremental features or LAGs, it is possible to obtain samples that represent a time series in the life of an animal, albeit one that is smoothed by bone turnover. In contrast, cholesterol in bone turns over more rapidly, and probably represents an average of at most a few months (Stott et al. 1999).

Dentin grows by accretion and experiences little post-depositional remodeling (Lowenstam & Weiner 1989). Dentin exhibits incremental laminations at a variety of temporal scales, from daily to annual (Carlson 1990), and sequential analysis of samples taken from these increments provides a time series of body chemistry (e.g., Koch et al. 1989). Enamel also bears incremental laminae indicating accretionary growth (striae of Retzius). Workers have tried to obtain sub-annual samples of body chemistry by collecting sequential samples from these increments (e.g., Bryant et al. 1996; Fricke & O'Neil 1996; Kohn et al. 1998; Sharp & Cerling 1998). Yet as noted by Fisher & Fox (1998), incremental laminae mark the front of organic matrix apposition, not the mineralization front. Complete enamel mineralization may lag organic matrix apposition by months, and the mineralization front need not parallel the organic apposition front (Balasse 2002; Hoppe et al. 2004; Zazzo et al. 2005). Thus even samples collected along incremental laminae will be time averaged. Passey & Cerling (2002) offered a mathematical model for dealing with this problem, and it may be minimized by examining enamel from ever-growing teeth (Stuart-Williams & Schwarcz 1997; Fox & Fisher 2004) or teeth that mature rapidly (Straight et al. 2004). In all mammals except those with ever-growing teeth, mineralization takes place early in the animal's life. In contrast, fish, amphibians, reptiles and some non-mammalian synapsids replace their teeth continuously throughout their lives.

Growth of feathers and hair may be continuous or episodic, but these tissues are typically replaced within a year or two. Eggshell crystallization occurs rapidly, lasting only 20 hours in chickens and perhaps 36 hours in ostriches (Burley & Vadehra 1989). Teleost otoliths exhibit incremental laminae formed at a variety of temporal scales (i.e. daily to lunar to annual) that are used to study age and growth rate (Panella 1980; Campana & Neilson 1985). Overall, we might expect that eggshell carbonate and proteins represent very short time intervals, that depending on sampling strategy hair and feather samples can represent weeks to months, and that otoliths offer a relatively complete record of body chemistry that can be microsampled for time series or bulk sampled for a life-time average.

Controls on the isotopic composition of vertebrate tissues

Controls on the isotopic composition of vertebrates have been reviewed repeatedly over the past two decades. Key reviews will be noted at the start of each section. The discussion in each section draws on these reviews, with citations only for more recent studies or topics not covered by the reviews.

Carbon isotopes

Key reviews of controls on carbon isotopes in vertebrates are van der Merwe (1982), Schoeninger (1985), DeNiro (1987), Schwarcz & Schoeninger (1991), Schoeninger & Moore (1992), Koch et al. (1994), Pate (1994), Koch (1998), Kelly (2000), Kohn & Cerling (2002), and Ambrose & Krigbaum (2003b).

Carbon in biominerals

Food is the source of carbon in the mineral and organic substrates of terrestrial vertebrate bones, teeth, and eggshells, yet each tissue differs in $\delta^{13}C$ value from diet by a characteristic amount. The fractionations among dissolved carbon dioxide (which is largely derived from oxidation of food), body fluid bicarbonate, and carbonate-bearing minerals at mammal and bird body temperatures are such that a ^{13}C-enrichment of 9–10‰ is expected between food and bioapatite or calcium carbonate. In terrestrial herbivores, the $\delta^{13}C$ value of bioapatite shows a strong 1:1 correlation with the $\delta^{13}C$ value of bulk diet, with a ^{13}C-enrichment of 9–11‰ for laboratory rodents and 12–14‰ for wild ungulates (Cerling & Harris 1999: Balasse 2002: Howland et al. 2003: Jim et al. 2004: Passey et al. 2005). Little is known about the diet-to-bioapatite fractionation in carnivores, but the value is thought to be ca. 9‰. The difference between bird eggshell carbonate and diet is quite large (14–16‰).

Differences in diet-to-mineral fractionation among terrestrial animals may have a number of sources, but modeling and laboratory experiments strongly suggest that greater degrees of enrichment occur in animals that obtain nutrients from microbial fermentation (Hedges 2003; Passey et al. 2005). Fermentation produces very ^{13}C-depleted CH_4, which escapes the body, and ^{13}C-enriched CO_2, which may diffuse from the gut to the blood stream, thereby labeling body fluid bicarbonate and mineral carbonate pools.

Diet-to-bioapatite fractionations in marine mammals are similar to those for terrestrial mammals (Clementz & Koch 2001). In aquatic ectotherms, carbon in bioapatite is derived from both respiration (i.e., food) and ambient water (Vennemann et al. 2001; Biasatti 2004). Carbon in otolith aragonite is also a mixture supplied by ambient water and respiration (Thorrold et al. 1997; Wurster & Patterson 2003). As a consequence, diet-to-mineral fractionations in these ectotherms are sensitive to any factor that alters the fluxes of metabolic versus ambient bicarbonate to body fluids (i.e., temperature, metabolic or growth rate, activity level, etc.).

Carbon in proteins and individual amino acids

Proteins are comprised of amino acids. Most amino acids have a central α-C atom to which is bonded (i) a hydrogen atom, (ii) a carboxyl group (—COOH), (iii) an amino group (—NH$_2$), and (iv) a distinct side chain, or R-group, which is often rich in carbon. The R-groups of essential amino acids (and non-essential amino acids whose sole precursors are essential) must originate from dietary protein. Non-essential amino acids may be ingested, or they may be assembled within the animal, and so may contain carbon from any dietary source (e.g., proteins, lipids, or carbohydrates) (Fogel et al. 1997). In collagen, ca. 20 percent of carbon atoms are essential and must be routed from dietary protein (Howland et al. 2003). Feeding experiments suggest that on relatively high protein diets, the δ^{13}C value of collagen is controlled by that of dietary protein, whereas on low protein diets, contributions from dietary carbohydrate and lipid are evident (Ambrose & Norr 1993; Tieszen & Fagre 1993; Howland et al. 2003; Jim et al. 2004). Associations between diet type and diet-to-tissue fractionation have also been detected in birds and fish (Bearhop et al. 2002; Pearson et al. 2003; Gaye-Siessegger et al. 2004).

Because different compound classes in diet can come from sources with different δ^{13}C values, the effects of carbon routing complicate dietary interpretations for omnivores if they are based on bulk collagen δ^{13}C values (Fogel & Tuross, 2003). For relatively committed herbivores or carnivores, in contrast, the picture is clearer. The bulk diet-to-collagen fractionation is 3–5‰

in mammals and birds. I am aware of no published controlled feeding experiments on fish or reptiles where collagen was analyzed, but data from ecological and paleoecological studies suggest a fractionation of similar magnitude (e.g., Sholto-Douglas et al. 1991; Ostrom et al. 1994; Dufour et al. 1999). In mammals, the diet-to-hair fractionation is 1–3‰ (Tieszen & Fagre 1993; Roth & Hobson 2000; O'Connell et al. 2001; Sponheimer et al. 2003a). In birds, the diet-to-feather fractionation is ca. 3–4‰ (Hobson & Clark 1992; Bearhop et al. 2002; Gaye-Siessegger et al. 2004), and the diet-to-eggshell matrix protein fractionation is ca. 2‰.

Studies of individual amino acids are shedding light on dietary routing and laying the groundwork for paleobiological analysis. Individual amino acids from collagen or muscle can differ greatly in $\delta^{13}C$ value, though consistent patterns of variation are emerging (Hare et al. 1991; Fogel et al. 1997). With respect to routing of dietary carbon, experiments on pigs with diets of variable protein content show that the $\delta^{13}C$ values of non-essential amino acids in collagen (e.g., glutamate and alanine) correlate well with the $\delta^{13}C$ value of bulk diet, whereas the $\delta^{13}C$ values of some important essential amino acids in collagen (e.g., leucine and phenylalanine) correlate with the same amino acid in diet without substantial isotopic fractionation (Howland et al. 2003). If these results hold for other taxa, analysis of glutamate or alanine could be used to assess the $\delta^{13}C$ value of bulk diet (i.e., calories), whereas analysis of leucine and phenylalanine could be used to assess the $\delta^{13}C$ value of dietary protein. The study by Fogel & Tuross (2003), which did not include feeding experiments, suggested substantial fractionations from plant diet-to-herbivore collagen for most essential amino acids, including phenylalanine. More experimental work is clearly needed, particularly for animals with different digestive physiologies.

Carbon in lipids, fatty acids, and cholesterol

The lipids within an animal's body are comprised of different fatty acids, some of which can be synthesized and some of which are essential and must be ingested. Lipids are ^{13}C-depleted relative to bulk diet and other body tissues (DeNiro & Epstein 1978). Few isotopic data have been reported on bulk lipids extracted from modern or fossil bones, perhaps because of concerns about differential degradation of different classes of lipids, differences between essential and non-essential lipids, and contamination (Koch et al. 2001; Collins et al. 2002). With the development of GC-C-IRMS, individual lipid molecules, especially cholesterol, have become targets for analysis.

Analyses of herbivorous mammals subjected to controlled feeding suggest that the $\delta^{13}C$ values of non-essential fatty acids and cholesterol are tightly correlated to the $\delta^{13}C$ value of bulk diet, and that the $\delta^{13}C$ values of essential fatty acids, such as linoleic acid, are directly related to the value of that fatty acid in diet without substantial isotopic fractionation (Stott et al. 1997, 1999;

Howland et al. 2003; Jim et al. 2004). Thus compound-specific isotope analysis of lipids may offer data on both bulk diet and dietary lipids.

Environmental controls on carbon isotopes

Carbon isotope differences among vertebrates largely reflect differences in the $\delta^{13}C$ values of primary producers at the base of the food web. In terrestrial ecosystems, the dominant control on the $\delta^{13}C$ value of plants is photosynthetic pathway. Basic physiological controls on the $\delta^{13}C$ values of plants using different pathways have been reviewed elsewhere (Ehleringer & Monson 1993) and will not be discussed here. The C3 pathway is the most common, occurring in all trees, most shrubs and herbs, and grasses in regions with a cool growing season. C3 plants have a mean $\delta^{13}C$ value of ca. −27‰ (range −22 to −35‰). C4 photosynthesis occurs in grasses from regions with a warm growing season, and in some sedges and dicots. C4 plants have higher $\delta^{13}C$ values (mean ca. −13‰, range −19 to −9‰). Crassulacean acid metabolism (CAM) is the least common pathway, occurring in succulent plants. CAM plants exhibit $\delta^{13}C$ values that range between the values for C3 and C4 plants.

There are strong abiotic influences on the distribution and isotopic composition of plants using different pathways. Among grasses, C4 species abundance and biomass increase with growing season temperature and wetness (Epstein et al. 1997). CAM plants and C4 dicots are most abundant in arid regions. Variations in light level, water and osmotic stress, nutrient levels, temperature, and P_{CO_2} produce predictable variations in the $\delta^{13}C$ values of C3 plants (reviewed by Tieszen 1991; Ehleringer & Monson 1993; Heaton 1999). C3 plants may also show large differences in $\delta^{13}C$ value related to plant functional type. In dense, closed-canopy forests, the $\delta^{13}C$ value of forest floor leaves may be ^{13}C depleted by up to 8‰ relative to leaves from the top of the canopy, owing to recycling of ^{13}C-depleted carbon dioxide and the effects of low irradiance. Because of their efficient method of carbon fixation, C4 plants show little environmental variability in $\delta^{13}C$ values (Ehleringer et al. 1997). CAM plants in arid regions have $\delta^{13}C$ values similar to those of C4 plants, whereas in wetter regions, CAM plants have values intermediate between those of C3 and C4 plants.

Marine primary producers exhibit strong spatial $\delta^{13}C$ gradients due to differences in (i) the rate of photosynthesis, (ii) the taxon and size of phytoplankton and bacteria fixing carbon, (iii) the $\delta^{13}C$ value of dissolved inorganic carbon, and (iv) water mass properties that influence factors i–iii (see discussion and references in Burton & Koch (1999) and Clementz & Koch (2001)). Primary producers are enriched in ^{12}C relative to starting substrates, and in areas of the ocean where primary and export production strip carbon from the ocean surface, primary producers have high $\delta^{13}C$ values. These conditions occur in highly productive regions (i.e., upwelling zones) and in oligotrophic waters, but not in regions where nutrient supply far exceeds photosynthetic demand, such as at high latitudes. Overall, primary producer $\delta^{13}C$ values increase from offshore to nearshore ecosystems, peaking in macrophytic ecosystems (i.e., kelp and seagrass beds). There are also strong gradients across different current systems (e.g., Rau et al. 1991). In estuarine and freshwater ecosystems, mean $\delta^{13}C$ values for primary producers are typically lower but much more variable than in marine systems.

Given these physiological and environmental controls, $\delta^{13}C$ values in vertebrates will vary with diet (photosynthetic pathway; marine vs. freshwater vs. terrestrial feeding), location, and ecosystem properties related to plant type and carbon cycling. They may shed light on digestive physiology and metabolic rate.

Nitrogen isotopes

Key reviews exploring nitrogen isotopes in vertebrates are Schoeninger (1985), DeNiro (1987), Ambrose (1991), Schwarcz & Schoeninger (1991), Schoeninger & Moore (1992), Koch et al. (1994), Pate (1994), Koch (1998), Kelly (2000), and Ambrose & Krigbaum (2003b).

Nitrogen in proteins and individual amino acids

Unlike carbon, which has multiple macromolecular dietary sources, nitrogen in animal protein is supplied almost entirely by dietary protein. A diet-to-tissue fractionation of ca. 3‰ has been observed or assumed in many studies. In general, most body and shell proteins have similar $\delta^{15}N$ values. This trophic level fractionation is thought to relate to excretion of urea and other nitrogenous wastes that are enriched in ^{14}N relative to body nitrogen pools (e.g., Parker et al. 2005), perhaps due to fractionation associated with deamination of glutamate (which yields one of the NH_2 groups in urea) or cleavage of arginine to yield urea (Fogel et al. 1997). Feeding experiments on mammals and birds have shown that the magnitude of fractionation increases with increasing protein content in the diet (Pearson et al. 2003; Sponheimer et al. 2003b, 2003c), and field observations suggest that fractionation is greater in mammals inhabiting arid habitats (Schwarcz et al. 1999). The causes of these variations in fractionation are debated, and variously attributed to changes in urea concentration, recycling of urea, or lack of nitrogen balance (Sponheimer et al. 2003c). Finally, for animals that are out of nitrogen balance, there is evidence that diet-to-tissue fractionation decreases for animals in anabolic states (growth) and increases for animals in catabolic states (fasting, starvation). Fuller et al. (2005) reviews these issues and offer compelling data on ^{15}N-enrichment in pregnant women with morning sickness.

To understand why trophic fractionation might vary with the protein content of diet, consider a simple open-system isotope mass balance model with the following properties:

1 a flux of dietary nitrogen into the body (F_d) with a value $\delta^{15}N_d$;
2 a single pool of body nitrogen with a value $\delta^{15}N_b$;
3 a flux of urea nitrogen out of the body (F_u) with a value $\delta^{15}N_u = \delta^{15}N_b + \varepsilon_{bu}$, where ε_{bu} is the net fractionation between body tissue and urea nitrogen associated with deamination and urea synthesis;
4 a flux of tissue nitrogen out of the body (chiefly fecal nitrogen, F_f) with a value $\delta^{15}N_f = \delta^{15}N_b$ (Figure 5.1).

At steady-state (i.e., with a body nitrogen pool of fixed size), input and output isotopic fluxes must be equal, so $F_d\delta^{15}N_d = F_u\delta^{15}N_u + F_f\delta^{15}N_f$. After conversion of actual fluxes to proportional fluxes (by dividing both sides of the equation by F_d), substitution of equations involving $\delta^{15}N_b$ for $\delta^{15}N_u$ and $\delta^{15}N_f$, and algebraic rearrangement, we obtain the following equation: $\delta^{15}N_b = \delta^{15}N_d - \varepsilon_{bu}X_u$, where X_u (= F_u/F_d) is the proportional flux of nitrogen lost as urea. Figure 5.1 plots $\delta^{15}N_b$ and $\delta^{15}N_u$ at different values of X_u with $\varepsilon_{bu} = -6‰$ and $\delta^{15}N_d = 0‰$.

For animals on diets rich in proteins, daily nitrogen intake far exceeds requirements for nitrogen balance. These animals catabolize the carbon skeletons of amino acids as fuel and shed the stripped amine groups as urea, leading to high proportional loss of body nitrogen as urea and high diet-to-tissue fractionations. Animals on low protein diets use most of their dietary nitrogen to build body protein, and consequently have a lower urea nitrogen flux, which reduces the diet-to-tissue fractionation. Models of greater complexity have been developed to investigate trophic fractionation of nitrogen isotopes (Schoeller 1999; Hedges & van Klinken 2000; Olive et al. 2003). The key points to note here are that fractionation of nitrogen isotopes between diet and tissue may vary among animals at steady-state, and that fractionation associated with urinary nitrogen loss may be a primary driver even when the $\delta^{15}N$ value of urea is not substantially lower than that of diet (*contra* Sponheimer et al. 2003c), as long as most nitrogen is lost via urine.

The controls on $\delta^{15}N$ values in individual amino acids are related to biosynthetic and catabolic pathways. The amine group is reversibly exchangeable by transamination for all amino acids except threonine and lysine, and so for all other amino acids, amine group nitrogen may be homogenized in the circulating amino acid pool. Several amino acids have nitrogen in their R-groups, and for essential amino acids, such as lysine and histidine, these must be supplied by diet. Most amino acids in collagen are enriched in ^{15}N relative to the same amino acid in diet, with the strongest enrichment for glutamate, which plays a central role in transamination reactions and the urea cycle (Hare et al. 1991; Fogel et al. 1997; see also McClelland & Montoya (2002) for a similar result involving invertebrates). Threonine in collagen shows very strong ^{15}N-depletion relative to threonine in diet, presumably because of an unusual fractionation associated with catabolism (Hare et al. 1991). Lysine and arginine also have low $\delta^{15}N$ values. For arginine, this may occur because it receives an NH_4^+ from deamination of glutamate. The fact

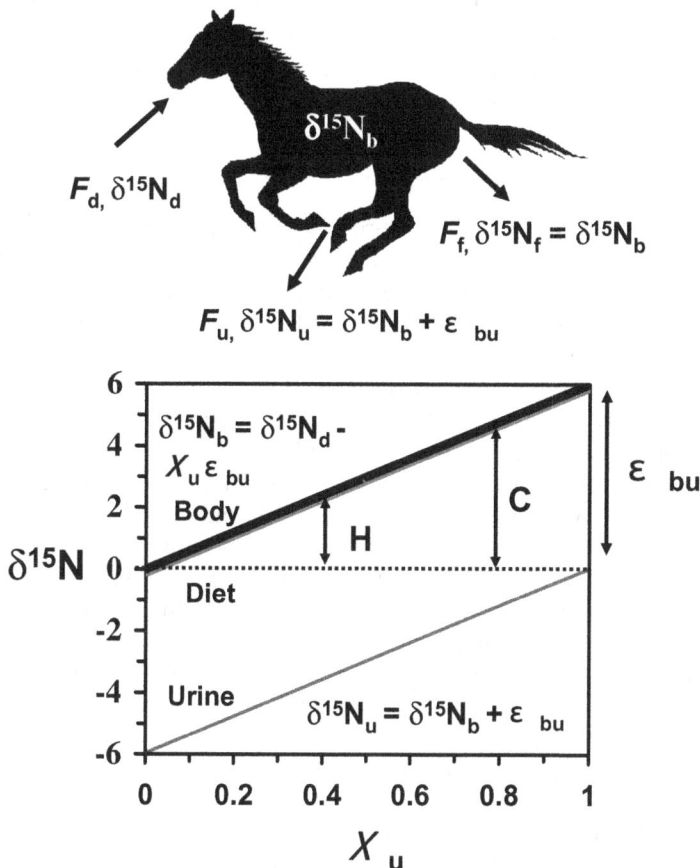

Figure 5.1 A steady-state nitrogen isotope mass balance model for mammals. The arrows into and out of the horse identify the major nitrogen fluxes and their isotopic compositions. The bivariate plot shows how the $\delta^{15}N$ values of body tissue (dark line) and urine (gray line) change as the proportion of body nitrogen lost as urine (X_u) changes, assuming that the main fractionation in this system occurs between body tissue and urine (ε_{bu}) and that the $\delta^{15}N$ value of diet is 0‰ (dotted line). Carnivores (C) consume protein for calories, and hence take in much more nitrogen than is needed to maintain mass balance. They shed excess nitrogen in ^{15}N-depleted urine and therefore show a relatively large ^{15}N-enrichment in body tissues relative to diet. In contrast, to maintain mass balance an herbivore (H) on a protein-poor diet must shed less nitrogen as waste/urine; it loses more nitrogen through unavoidable shedding of body protein as hair, milk, enzymes, and especially gut wall. As a consequence, herbivore body tissues exhibit a smaller ^{15}N-enrichment relative to diet. The abbreviations for all variables are defined in the text.

that nitrogen in lysine is entirely sourced from dietary lysine may buffer it from the ^{15}N-enrichment affecting most of the body amino acid pool (Hare et al. 1991; Fogel et al. 1997). As such, lysine is the best target for monitoring the δ^{15}N value of dietary protein.

Environmental controls on nitrogen isotopes

Controls on δ^{15}N values in primary producers at the base of food webs are complex. Nitrogen in most plants is taken up from soils, and soil and plant δ^{15}N values vary geographically depending on soil pH, moisture, and atmospheric nitrogen deposition (Nadelhoffer & Fry 1994; Högberg 1997). Most processes by which nitrogen is lost from soils (denitrification, ammonia volatilization, loss of dissolved species) lead to ^{15}N-enrichment of soil nitrogen, generating trends associated with soil age and maturity (Nadelhoffer & Fry 1994; Hobbie et al. 1998). The δ^{15}N value of foliar nitrogen is negatively correlated with rainfall abundance globally (Handley et al. 1999), though recent work in southern Africa suggests this relationship may be driven entirely by C3 plants (Swap et al. 2004). Plant δ^{15}N values are negatively correlated with soil moisture at a local scale as well (Evans & Ehleringer 1994). These correlations may reflect greater ^{15}N-enrichment in dry soils due to higher rates of nitrogen loss. The δ^{15}N value of plants that do not fix N_2 is higher in coastal regions, perhaps owing to deposition of marine nitrate (Heaton 1987). Finally, within sites, plants show consistent differences among growth forms related to differences in rooting depth, symbioses with nitrogen-fixing bacteria, and mycorrhizal associations. For example in boreal ecosystems, non-mycorrhizal plants (graminoids, clubmosses, forbs) have higher δ^{15}N values than mycorrhizal plants (chiefly trees and shrubs), mosses and lichens (Schulze et al. 1994; Nadelhoffer et al. 1996; Michelsen et al. 1998).

As with carbon in marine primary producers, there are strong δ^{15}N gradients in some regions due to differences in (i) the rate of nitrogen uptake, (ii) the extent of nitrate utilization or denitrification, (iii) the type of phytoplankton and bacteria fixing nitrogen, and (iv) water mass properties that influence factors i–iii (Michner & Schell 1994; Voss et al. 1996). In many regions of the ocean, primary producers remove all nitrogen from surface waters, so annually integrated production must have the same δ^{15}N value as the combined sources of nitrogen to the ocean surface. Spatial gradients may be produced by the differential mixing of these sources, particularly if deeper water that is ^{15}N-enriched due to denitrification is involved (e.g., along the eastern Pacific margin, Altabet et al. 1999). In regions where nitrogen utilization is not complete, the extent of ^{15}N-enrichment will correlate with degree of nitrogen utilization (e.g., from the Southern Ocean south to the sub-Antarctic, Altabet & François 2001). In any case, these strong spatial gradients in marine δ^{15}N values offer a natural tag of position for ecological and paleoecological studies (e.g., Schell et al. 1998, McClelland et al. 2003).

Given these physiological and environmental controls, it is clear that the $\delta^{15}N$ values of terrestrial vertebrates will vary with trophic level, marine vs. terrestrial feeding, factors affecting nitrogen balance (starvation, lactation, pregnancy, etc.), the type of plant food taken, rainfall abundance and location. In the ocean, all these factors except rainfall abundance affect vertebrate $\delta^{15}N$ values.

Sulfur isotopes

Key reviews exploring sulfur isotopes in vertebrates are Krouse (1989) and Richards et al. (2003).

Sulfur in proteins and biominerals

Sulfur in animals is primarily found in proteins (in the amino acids cysteine and methionine) and in bioapatite as sulfate (SO_4^{2-}) substituting for phosphate. Methionine is the most common sulfur-bearing amino acid in collagen (5 residues per 1000), whereas both cysteine (112 residues per 1000) and methionine (5 residues per 1000) are more abundant in hair. The proteinaceous organic matrix of otoliths contains 1–3 percent cysteine and methionine (Weber et al. 2002). Methionine is an essential amino acid and is the source of sulfur for cysteine biosynthesis.

In natural systems, the difference between potential foods and animal tissues is small, suggesting a small sulfur isotope fractionation. This inference has been verified by feeding experiments on bears and pigs, which found a small positive fractionation between diet and tissue (González-Martín et al. 2001; Felicetti et al. 2003). It is more difficult to assess the results of a feeding experiment on horses (Richards et al. 2003), because horses switched to low protein diets may not have fully equilibrated with the new diet.

Environmental controls on sulfur isotopes

Plants take up sulfur derived from (i) weathering of bedrock, which can vary widely in $\delta^{34}S$ value, (ii) wet atmospheric deposition (sea spray, acid rain), (iii) dry atmospheric deposition (SO_2 gas), and (iv) microbial processes in soils. As a consequence, the $\delta^{34}S$ value of terrestrial plants varies with location, with values ranging from −22 to +22‰ (Peterson & Fry 1987). In their study of grizzly bears, Felicetti et al. (2003) detected a large within-ecosystem difference in $\delta^{34}S$ value between pine nuts and all other plant and animal foods available to bears. They offered no explanation for the strong [34]S-enrichment in pine nuts, but it may relate to differences in rooting depth or soil properties near the edge of the tree line where whitebark pine occur. In rivers and lakes, differences in the extent of anaerobic sulfate reduction (which produces sulfate extremely depleted in [34]S) leads to a similarly wide range of $\delta^{34}S$ values (Peterson & Fry 1987). Sulfur in marine phytoplankton is relatively uniform, with a mean value of ca. 20‰.

Given the minor physiological impacts on $\delta^{34}S$ values, sulfur isotope data from vertebrates can be used to reconstruct marine vs. terrestrial vs. freshwater feeding (in the many settings where non-marine and marine values do not overlap), location, and type of plant consumed.

Oxygen isotopes

Key reviews of controls on oxygen isotopes in vertebrates are Schwarcz & Schoeninger (1991), Schoeninger & Moore (1992), Koch et al. (1994), Koch (1998), and Kohn & Cerling (2002).

Oxygen in biominerals

Oxygen in bioapatite phosphate and carbonate and in calcium carbonate has been used in isotopic studies of vertebrates. Some analytical methods extract all the oxygen in the bioapatite, including that in hydroxyl groups, into a combined pool (Kohn et al. 1996; Sharp & Cerling 1996), but to my knowledge, there are no studies that have isolated bioapatite hydroxyl oxygen for isotopic analysis. The $\delta^{18}O$ value of a biomineral depends on the temperature at which it forms and the $\delta^{18}O$ value of the body fluid from which it precipitates. For homeothermic mammals, there is a constant offset between the $\delta^{18}O$ value of body water and phosphate (ca. 18‰), and between the phosphate and carbonate components of bioapatite (ca. 8‰), close to values predicted for oxygen isotope equilibrium at body temperatures. Bird eggshell carbonate has values that range from near equilibrium with body water to values ^{18}O-enriched by 3‰ relative to equilibrium. For heterothermic animals (fish, reptiles, etc.), the oxygen isotope fractionation between body water and bone, tooth, or scale bioapatite phosphate and carbonate or otolith carbonate increases as temperature drops. Aquatic heterotherms form biominerals in isotopic equilibrium with ambient water at body temperature. For turtles, Barrick et al. (1999) noted that the fractionation between bone and environmental water (a chief source of body water oxygen) varies less than expected with environmental temperature, perhaps because turtles regulate their body temperatures within a narrow window behaviorally, or because bone growth occurs predominantly within a narrow thermal window.

Physiology affects the $\delta^{18}O$ value of body water by altering the magnitude of fluxes of oxygen into and out of the body, as well as fractionations associated with transport and/or transformation of oxygen-bearing compounds. Major oxygen fluxes into terrestrial mammals include drinking and diet water (>50%), which are not fractionated during uptake, and inhalation of atmospheric oxygen gas (ca. 25%) and water vapor (ca. 15%), which undergo isotopic fractionation during diffusion across the lung lining. Fluxes of oxygen out of the body include respired carbon dioxide (ca. 25%), water and organic matter in feces and urine (ca. 40%), and water lost during sweating, transcutaneous evaporation, and exhalation (ca. 35%). Oxygen in respired carbon dioxide and water lost by exhalation or transcutaneous evaporation are fractionated relative to body water. For aquatic mammals, ca. 98 percent of the oxygen flux into the animal may come from diffusion of water across the skin, and fluxes out of the body are fewer in number and less subject to strong fractionation because of high humidity. These physiological factors are relatively constant within species, leading to taxon-specific relationships between the $\delta^{18}O$ value of ingested water and body water that are some-what predictable from body size, taxonomy, or habitat, and that have been successfully approximated with mass balance models of varying degrees of complexity.

Oxygen in proteins

I am aware of only one paper and two abstracts examining oxygen isotope variations in vertebrate protein (deHart & Wooller 2004; Hobson et al. 2004; Wooller & O'Brien 2004). As such, basic physiological controls remain to be explored. One of the oxygen atoms in the carboxyl group of amino acids can be supplied by body water via hydrolysis of the peptide bonds that link amino acids in proteins. Oxygen also occurs in carboxyl, hydroxyl, and amide groups in the R-groups of some non-essential amino acids (e.g., serine, tyrosine, as-partate/asparagines, glutamate/glutamine, hydroxyproline) and one essential amino acid (threonine). Examination of the tricarboxylic acid cycle and amino acid biosynthetic pathways indicates that oxygen can enter these sites from food (largely carbohydrate), body water, phosphate, or oxygen gas (D.M. O'Brien, pers. comm.). Experiments on shrimp suggest that muscle protein is most strongly labeled by ambient water oxygen, not food oxygen (Epp et al. 2004). In general, animal proteins seem to be ^{18}O-enriched relative to drinking or ambient water.

Environmental controls on oxygen isotopes

Environmental factors that shift the $\delta^{18}O$ value of vertebrate body water will impact the $\delta^{18}O$ value of biominerals and protein. For terrestrial animals, in-gested drinking water is supplied by meteoric water. Meteoric water varies in $\delta^{18}O$ value geographically and temporally, with higher values in warm regions or seasons, and lower values in colder regions or seasons. Larger reservoirs of water will exhibit damped (or no) seasonal fluctuations. In ad-dition, evaporation will lead to progressive ^{18}O-enrichment of environmental water. Vertebrates also get substantial amounts of oxygen from water in food. Water in stems has a $\delta^{18}O$ value relatively close to that of meteoric water. In contrast, the water in leaves may be highly enriched in [18]O relative to meteoric water due to evapotranspiration, with increasing enrichment with decreasing relative humidity. Because of differences in water-use efficiency and anatomy, the $\delta^{18}O$ values of leaf water differ such that C3 dicots < C3 grass < C4 grass (Helliker & Ehleringer 2000). The $\delta^{18}O$ value in leaf water also varies with height in the canopy, with higher $\delta^{18}O$ values higher in the canopy. The flux of oxygen from food dry matter is relatively small, and the $\delta^{18}O$ value of atmospheric oxygen gas is globally homogeneous, so neither is a major contributor to environmental variability in vertebrate $\delta^{18}O$ values. Finally, as mentioned above, the oxygen flux in aquatic vertebrates is strongly dominated by environmental water. Consequently, aquatic vertebrate $\delta^{18}O$ values should vary spatially in freshwater systems, due to regional differences in meteoric water, but should be relatively homogeneous in marine vertebrates. Of course, biomineral from marine heterotherms will exhibit variance in $\delta^{18}O$ values related to differences in growth temperature (Thorrold et al. 1997; Vennemann et al. 2001).

In summary, vertebrate $\delta^{18}O$ values will vary with marine vs. freshwater vs. terrestrial habitat use, location, diet, and thermoregulatory and osmoregulatory physiology.

Hydrogen isotopes

Controls on hydrogen isotope variations in vertebrates are discussed briefly in Schwarcz & Schoeninger (1991).

Hydrogen isotopes in organic molecules

To my knowledge, hydrogen from the hydroxyl site in bioapatite has not been used for isotopic analysis. There is a small body of work on hydrogen isotope variations in bone collagen (Cormie et al. 1994a, 1994c; Birchall et al. 2005) and a growing literature on hydrogen isotopes in feathers, hair, and lipids (Estep & Dabrowski 1980; Chamberlain et al. 1997; Hobson & Wassenaar 1997; Hobson et al. 1999; Sharp et al. 2003; Cryan et al. 2004; deHart & Wooller 2004). Pioneering work by Estep & Dabrowski (1980) suggested that the δD value of body tissues was chiefly controlled by the δD value of food dry matter. Subsequent experiments revealed that organic molecules contain exchangeable hydrogen (bonded to oxygen or nitrogen), as well as non-exchangeable hydrogen bonded to carbon (Schimmelmann et al. 1993; Cormie et al. 1994b; Chamberlain et al. 1997; Hobson et al. 1999; Sharp et al. 2003). By equilibration with vapors of known composition, the δD value and amount of non-exchangeable hydrogen can be calculated. Lipids have nearly 100 percent non-exchangeable hydrogen, whereas the amount of non-exchangeable hydrogen in proteins ranges from 75 to 90 percent. Of the non-exchangeable hydrogen in feathers and hair protein, 18–32 percent comes from water, with the remainder coming from food

(Hobson et al. 1999; Sharp et al. 2003). In the case of collagen, ca. 25 percent of the non-exchangeable hydrogen is present in essential amino acids, and therefore must come from food (Birchall et al. 2005). There are two steps in the tricarboxylic acid cycle where hydrogen from water is added to carbon, and they occur immediately prior to the formation of two key intermediates in amino acid metabolism (α-ketogluterate and oxaloacetate), so labeling of non-exchangeable hydrogen in protein with hydrogen from water is not unexpected. Finally, Birchall et al. (2005) discovered a strong, trophic level hydrogen isotope fractionation in collagen from herbivore/omnivore to carnivore in both terrestrial and aquatic birds and mammals. The cause of this fractionation is unclear.

Environmental controls on hydrogen isotopes

Environmental controls on hydrogen isotopes are similar to those on oxygen isotopes, but generally produce signals of greater magnitude. δD values vary in meteoric water in space and time and evaporation leads to ^{2}H-enrichment of surface waters and leaf water, though of proportionally lower magnitude than for ^{18}O. Leaf water δD values differ depending on plant photosynthetic pathway in some settings, such that C3 < C4 < CAM, probably due to differences in evapotranspiration among the pathways (Ziegler 1989). Differences in leaf tissue δD values follow the same order, but are of much greater magnitude, suggesting differences in water-to-plant isotope fractionation among the pathways (Sternberg 1989; Ziegler 1989).

Vertebrate δD values are primarily used to study location and diet in modern animals, and it should be possible to assess marine vs. freshwater vs. terrestrial habitat use and trophic level.

Calcium isotopes

Measurement of $δ^{44}$Ca values has become more common with the development of the double-spike method by Skulan et al. (1997), but data are still sparse. Choice of a material for a calcium isotope reference standard is still in progress; a CaF_2 (NIST 915a) has been proposed (Hippler et al. 2003). Once a final decision is made, data from earlier papers that are standardized relative to different substances (e.g., carbonate rock, sea water) will need to be recalibrated. Calcium isotope systematics are reviewed by DePaolo (2004).

Calcium isotopes in biominerals

Calcium in terrestrial vertebrates is supplied chiefly by diet. In marine systems, ingestion of seawater might contribute to the body's calcium budget, though at least for carnivorous marine mammals, water is obtained chiefly from prey body fluids and metabolic water. Soft tissues have the same $δ^{44}$Ca value as diet or, in the case of marine invertebrates, as seawater. In contrast, biominerals are ^{44}Ca-depleted relative to diet by 1–1.5‰ (Skulan et al. 1997; Skulan & DePaolo 1999). Progressive ^{44}Ca-depletion with each trophic step has been demonstrated for both marine and terrestrial vertebrates, though this effect may depend on consumption of bone or shell (Clementz et al. 2003a). Inorganic synthesis of calcium carbonate and culturing experiments on foraminifera demonstrate that the fractionation is largely related to mineral precipitation, not an enzymatic or biological transport reaction, but there is disagreement about the physico-chemical mechanism producing fractionation (Gussone et al. 2003; Lemarchand et al. 2004). In addition, there is debate about the extent to which calcium isotope fractionation is sensitive to temperature, and the causes of this sensitivity (Marriott et al. 2004).

Environmental controls on calcium isotopes

Inorganic environmental variation in calcium isotope ratios is low. Most igneous rocks have a value ca. 0‰ (relative to the ultrapure calcite standard of Skulan et al. (1997)) and marine carbonate has a value of +1‰. As a consequence, δ^{44}Ca values in marine food webs are 1‰ higher than those in terrestrial food webs at a similar trophic level.

In summary, vertebrate δ^{44}Ca values can be used to reconstruct trophic level and marine vs. terrestrial resource use.

Strontium isotopes

Controls on strontium isotope variation in mammals were reviewed by Beard & Johnson (2000) and Kohn & Cerling (2002). Similar assumptions and processes apply for other high mass isotope systems in which a stable daughter isotope is produced by radioactive decay of a parent isotope (e.g., the U–Pb and Sm–Nd systems). The strontium isotope system has been applied the most widely to vertebrates, and so will be used as an example here. Finally, the δ notation is sometimes used to report strontium isotope data relative to modern seawater (^{87}Sr/^{86}Sr$_{\text{sea water}}$ = 0.7092) (Capo et al. 1998), or the ε notation is used, where relative to ε^{87}Sr = ([^{87}Sr/^{86}Sr$_{\text{sample}}$]/[^{87}Sr/^{86}Sr$_{\text{bulk earth}}$] − 1) × 10^4 and ^{87}Sr/^{86}Sr$_{\text{bulk earth}}$ = 0.7045. More often workers just report the ^{87}Sr/^{86}Sr ratio; we follow that convention here.

Strontium isotopes in biominerals

To date, there is no evidence for fractionation of strontium isotopes by biochemical processes, so the isotopic composition of a biomineral is assumed to be identical to that of the source of strontium and is passed up the food chain without modification.

Environmental controls on strontium isotopes

Continental rocks exhibit a large range in $^{87}Sr/^{86}Sr$ ratios that varies with rock type and age (average for rock type 0.702 to 0.716) (see Capo et al. (1998) for a review of controls on strontium isotopes in ecosystems). ^{87}Sr is produced by radioactive decay of ^{87}Rb. Older rocks with high initial Rb/Sr ratios (e.g., continental granites) have the highest $^{87}Sr/^{86}Sr$ ratios, whereas younger rocks, or rocks with low Rb concentrations (e.g., limestones, basalts) have lower ratios. Soil $^{87}Sr/^{86}Sr$ ratios are controlled by bedrock and by atmospheric deposition of strontium as dust and precipitation. Plants have $^{87}Sr/^{86}Sr$ ratios that match those of the soluble or available strontium in soils. Because of differences in rooting depth, and differences in the isotopic composition within soil weathering profiles, different plants within an ecosystem may have different $^{87}Sr/^{86}Sr$ ratios. The $^{87}Sr/^{86}Sr$ ratio of modern seawater (0.7092) is globally uniform because the residence time of marine strontium is much longer than the time required for oceanic mixing. Since the origin of vertebrates in the Cambrian, the seawater $^{87}Sr/^{86}Sr$ ratio has fluctuated between 0.7095 and 0.707 due to differences in continental weathering and hydrothermal alteration of oceanic basalt. Finally, $^{87}Sr/^{86}Sr$ ratios in estuaries are controlled by mixing. Strontium concentrations are much lower in freshwater than in seawater, so $^{87}Sr/^{86}Sr$ ratios of estuarine waters are quickly dominated by marine inputs (Bryant et al. 1995)

Strontium isotope data can be used to study location, marine vs. terrestrial vs. freshwater foraging, and perhaps diet if there are persistent strontium isotope differences among local plant types.

Preservation of biogenic isotope compositions by vertebrate fossils

Preservation of biogenic compositions is a prerequisite for most isotopic studies of fossil vertebrates. The subject has been contentious, with substantial disagreement among different research communities about the reliability of different tissues. In addition, preservation varies with mineralogy, tissue type, element, and depositional environment, so generalizations are hard to come by. The discussion below is based largely on my own research and reviews (Koch et al. 1994, 1997, 2001), as well as the recent review in Kohn & Cerling (2002).

Preservation of biogenic isotope compositions in biominerals

At least five post-mortem processes may lead to isotopic alteration of vertebrate biominerals.

1 Precipitation of secondary minerals in and around biogenic crystals. If diagenetic minerals have a different chemistry or crystallography than the biomineral, it may be possible to isolate unaltered mineral through pretreatment. In bones, infilling by secondary minerals may occur rapidly, even sub-aerially, and may be facilitated by the high porosity of bone and the rapid loss of the organic matrix (Nielsen-Marsh & Hedges 2000; Trueman et al. 2004). Furthermore, small poorly organized bone crystals undergo simultaneous dissolution and growth, such that large crystals grow at the expense of smaller ones. The new apatite may have isotope compositions unlike biogenic values due to incorporation of pore fluid ions at surface temperatures and it may be difficult or impossible to isolate bioapatite under such conditions.

2 Adsorption of ions at sites on the surface of crystals or in poorly organized hydration layers around crystals. This process affects even modern samples during preparation for isotopic analysis. Adsorbed ions may be released by leaching.

3 Solid-state exchange (i.e., diffusion of ions into the mineral lattice). For most crystallized materials, this process is too slow at Earth surface conditions to permit substantial alteration. The situation for bone is more complex, however.

4 Exchange of ions or atoms at lattice sites exposed on crystal surfaces. An exchange process of this sort may occur commonly in fossilized biominerals, and it may be facilitated by microbial activity (Zazzo et al. 2004a). For many minerals, surface-to-volume ratios are low, so this process cannot lead to wholesale resetting of biogenic isotope values. However, bone crystals are very small (only a few unit cells thick) with an extremely large surface area. Consequently surface exchange and even solid-state diffusion over very small distances could completely reset biogenic isotope values in bone mineral. It is not possible to strip away diagenetic mineral in this situation, so isotopic preservation is only possible if the ions in the fully exchanged bone came from the bone itself (not the sediments) and if there is minimal isotopic fractionation during exchange. This may explain why oxygen isotope analysis of fossil bone calcium and phosphate oxygen isotopes sometimes yields reliable biogenic isotope values.

5 Wholesale resetting by dissolution/reprecipitation or recrystallization. If the original biomineral is completely lost and replaced either by a similar mineral or an entirely different mineral, there is little hope that biogenic isotope values are preserved. Such wholesale resetting is typically obvious mineralogically (i.e., the transformation of aragonitic otoliths into calcite), crystallographically, or optically.

Kohn & Cerling (2002) discussed six types of tests used to assess isotopic preservation in bioapatite. They include: retention of expected levels of multi-sample isotopic variation within or among individuals; retention of expected isotopic differences among sympatric species; retention of expected isotopic differences among different tissues from the same specimen;

retention of expected isotopic differences among different ions from the same tissue; retention of original crystallinity; and retention of biogenic isotope values following deposition in a sedimentary environment with very different values.

Some generalizations have emerged about isotopic preservation in biominerals from these studies.

1 Enamel is much more likely to retain biogenic isotope values than bone or dentin. Bone $^{87}Sr/^{86}Sr$ ratios and $\delta^{18}O$ and $\delta^{13}C$ values from bone carbonate are often completely reset, even on Holocene–Pleistocene time scales (Hoppe et al. 2003). Enamel has been shown to carry biogenic values for these systems at least to the early Cenozoic, and perhaps to the Triassic (Botha et al. 2005), but even enamel is not completely closed to exchange and alteration. These conclusions from empirical studies of fossils are supported by experimental work showing extremely rapid exchange of bone carbonate and phosphate oxygen with pore solutions (Zazzo et al. 2004a). There may be settings where biogenic isotope values survive for these isotope systems if diagenetic processes impacting bone bioapatite lock in original values (Lee-Thorp & Sponheimer 2003). Likewise, bone phosphate sometimes retains biogenic $\delta^{18}O$ values. Still, for any isotopic study of bone bioapatite, preservation must be carefully demonstrated on a case-by-case basis; it cannot be assumed. In the very limited tests conducted to date, bone seems to carry biogenic $\delta^{44}Ca$ values in Cenozoic- and Cretaceous-aged specimens (Skulan et al. 1997; Clementz et al. 2003a).

2 With respect to oxygen in bioapatite, the long-standing assumption that phosphate is a more reliable substrate than carbonate may not be valid. This assumption is based on bond strength differences between the two ions. Experiments show that while phosphate is more resilient to inorganic isotope alteration than carbonate, when microbes are involved in the alteration process, phosphate oxygen alters much more rapidly than carbonate oxygen (Zazzo et al. 2004a). Zazzo et al. (2004b) offered a clever (albeit labor intensive) method for determining which type of alteration (if any) has occurred and for correcting back to biogenic oxygen isotope values in bioapatite.

3 At present, there are no crystallographic or chemical analyses that provide an unambiguous independent test of isotopic fidelity in bioapatite or eggshell calcite.

4 With respect to teleost otoliths, it has long been assumed that if original, aragonitic mineralogy is preserved, biogenic isotope values are as well. Reasonable isotopic data have been retained by aragonitic otoliths at least to the Jurassic (Patterson 1999).

5 With respect to eggshell carbonate, $\delta^{13}C$ values show mean values and trends congruent with those for enamel bioapatite and soil carbonate in specimens as old as the Miocene (Stern et al. 1994), and Cretaceous-aged shells yield plausible $\delta^{13}C$ values.

Preservation of biogenic isotope compositions in organic fossils

Post-mortem processes may alter isotopic values in proteins extracted from fossils. Since different amino acids in a protein have different isotopic compositions, if hydrolysis and amino acid loss are non-random, the isotopic composition of residual protein in bone may be shifted relative to that of unaltered protein. In addition, most tissues contain multiple proteins that may degrade at different rates, yet extraction protocols lump them as a single sample. For example, while collagen is the dominant protein in bone, non-collagenous proteins (albumin, osteocalcin) do occur. These non-collagenous proteins may have different isotopic compositions, thus isotopic trends may occur if diagenesis alters the proportions of proteins in bones. Bacterial or fungal proteins may be introduced during weathering and alter the isotopic composition of protein extracted from bones. Finally, amino acids and protein fragments may condense with exogenous organic matter (e.g., humic substances) during decomposition.

If chemical integrity can be demonstrated, then bulk protein from fossils is likely to yield biogenic isotope values. The simplest indicator is protein yield. Workers have noted anomalous isotope values when yield drops too low and developed rules-of-thumb based on these observations. The most common indicator is demonstration that the molar or atomic ratio of carbon to nitrogen (C/N) in the residue is biogenic or nearly so (Ambrose 1990). Quantitative amino acid analysis is the most robust indicator of chemical integrity (Tuross et al. 1988; Macko et al. 1999). If a bulk extract has biogenic amino acid abundances, then it is likely to have biogenic isotope values, as long as the materials subject to isotopic and amino acid analysis are identical. We note this last caveat because if humic substances are not removed from a bulk extract, even an extract with collagenous amino acid abundances, humics may contaminate the isotopic analysis but not the amino acid analysis. Finally, with respect to isotopic analysis of individual amino acids, the chief concern is contamination by sources from outside the fossil. Well-preserved collagen is common in many settings from Holocene-aged fossils, but becomes increasingly more rare with increasing age. Well-preserved collagen is present in Arctic fossils up to 100,000 years old, and is more sporadically present in older Pleistocene fossils. Claims of good collagen preservation in fossils much older than 100,000 years must be supported by stronger verification and testing.

Few studies have examined the state of preservation of lipids in fossil vertebrates. Most fatty acids are lost from bones quickly in sub-aerial and burial environments (Koch et al. 2001). In contrast, archaeological bone contains enough cholesterol for compound-specific isotope analysis (Stott et al. 1999), and cholesterol may be preserved in 100,000 year old Stellar sea cow bones (Clementz et al. 2003c). Cholesterol may degrade by reduction and oxidation, but if its presence can be demonstrated, then it was derived from a metazoan and will have biogenic isotope values. The chief concern is that bone cholesterol may be contaminated by cholesterol from saprotrophic organisms.

Paleobiological applications

Paleodietary reconstruction

Paleodietary reconstruction is the most common way that isotopes are used to study ancient vertebrates. The ability to discriminate between consumption of resources from C3 versus C4 food webs by carbon isotope analysis has allowed extensive study of the use of maize and other C4 plants by humans around the globe (van der Merwe 1982; Schwarcz & Schoeninger 1991). It has shown that sympatric camelids, equids, and proboscideans partitioned resources in Pleistocene North America (Connin et al. 1998; Koch et al. 1998; Feranec 2004a), as well as the surprising fact that many horses with high-crowned teeth were browsing, not grazing (MacFadden et al. 1999; Koch et al. 2004) (Table 5.3). Isotopic records from eggshells have revealed the impact of dietary preferences on differential survival of flightless birds in the late Pleistocene of Australia (Miller et al. 2005). Isotopic evidence has shown that australopithecines obtained up to 20 percent of their calories from C4 food webs. Either they ate C4 plant parts (perhaps underground storage organs) or they hunted or scavenged open country animals (Sponheimer & Lee-Thorp 2003). Finally, data from mammal teeth or bird eggshells have been used to examine climatically, atmospherically, or tectonically driven changes in the balance between C3 and C4 plants (MacFadden et al. 1994; Cerling et al. 1997; Koch et al. 2004; Miller et al. 2005). For example, eggshell isotope records have documented the rise to dominance of C4 plants in south Asia in the Miocene (Stern et al. 1994), and the lack of C4 dominance in the Miocene of Namibia (Segalen et al. 2002).

Isotopic studies of paleodiet in terrestrial C3-dominated ecosystems are more rare. In Eurasia, Beringia, and California, carbon and nitrogen isotope differences among Pleistocene herbivores seem to be related to differences in diet, though differences in digestive physiology and secular trends in environmental isotope values must be carefully evaluated as potential contributors to inter-specific differences (Bocherens et al. 1997; Iacumin et al. 1997; Drucker et al. 2003; Richards & Hedges 2003; Coltrain et al. 2004). Surprising results on C3-dominated Pleistocene Eurasian ecosystems come from the work of Bocherens, Drucker, and colleagues. They argued that extinct cave bears were highly herbivorous, whereas other co-occurring bears were more omnivorous (Bocherens et al. 1997). They also examined the diets of Neanderthals and early modern humans in Eurasia (Drucker & Bocherens 2004; Bocherens et al. 2005; Drucker & Henry-Gambier 2005). They argued that

Table 5.3 Carbon isotope values for tooth enamel bioapatite from Quaternary mammals. Data are for Rancholabrean- and Holocene-aged mammals, from Connin et al. (1998), Koch et al. (1998, 2004, unpublished data) and Feranec (2003), and are reported as the mean ± one standard deviation (in units of ‰ relative to VPDB), with the number of samples in parentheses.

		Texas	Florida	Missouri	$\delta^{13}C$†	%C4†
Cuvieronius	Gomphothere	−7.2 (1)	−6.1 (1)		2	97
Mammut	Mastodon	−10.5 ± 0.6 (25)	−11.0 ± 0.9 (41)	−11.4 ± 0.6 (37)	1	90
Mammuthus	Mammoth	−2.4 ± 1.4 (64)	−1.6 ± 1.8 (29)	−1.6 ± 0.4 (6)	0	83
Equus	Horse (inland)	−4.4 ± 1.8 (36)	−4.0 ± 3.2 (14)	−2.0 ± 0.8 (2)	−1	77
Equus	Horse (coastal)	−0.3 ± 0.4 (4)	−5.5 ± 2.7 (10)	n.a.	−2	70
Tapirus	Tapir	−11.2 ± 0.5 (7)	−12.9 ± 0.8 (9)	−12.8 ± 0.8 (2)	−3	63
Mylohyus	Pecarry	−9.9 ± 0.1 (2)	−10.3 ± 1.4 (5)		−4	57
Platygonus	Pecarry	−9.0 ± 0.2 (3)	−8.3 (1)		−5	50
Camelops	Camel	−4.8 ± 4.5 (17)			−6	43
Hemiauchenia	Llama		−8.5 ± 4.4 (25)		−8	30
Paleolama	Llama	−11.1 ± 0.8 (5)	−14.8 (1)		−9	23
Bison*	Bison	−1.0 ± 1.5 (22)	−0.7 ± 2.5 (8)	−0.7 ± 1.9 (2)	−10	17
Bootherium	Muskox			−11.3 ± 0.2 (5)	−11	10
Cervalces	Stag moose			−12.1 ± 0.3 (2)	−12	3
Odocoileus*	Deer	−12.2 ± 1.2 (17)	−13.6 ± 1.4 (15)	−14.4 ± 0.3 (2)	−13	−3
Casteroides	Giant beaver	−11.3 (1)				

* Genus survived in North America.

† Percent C4 estimates for given $\delta^{13}C$ values for mammalian bioapatite. Percent C4 was calculated from $\delta^{13}C$ values assuming a diet-to-apatite fractionation of 14‰ and a mixing model, assuming end-member $\delta^{13}C$ values of −26.5‰ for C3 plants and −11.5‰ for C4 plants in the late Pleistocene (Koch et al. 2004). Enamel $\delta^{13}C$ values below −12.5‰ yield negative percent C4 estimates, and potentially to feeding under a closed canopy forest.

Neanderthals were highly carnivorous, focusing on large open country herbivores, such as mammoths and woolly rhinoceros. Modern humans appeared in Europe around 45,000 years ago. Isotopic data indicate that they too were heavily reliant on large, open country herbivores, offering little support for the hypothesis that diet breadth increased in modern humans in Europe long before the Pleistocene–Holocene boundary (Drucker & Bocherens 2004; Drucker & Henry-Gambier 2005). Work on tropical Miocene and subtropical Eocene C3-dominated ecosystems has revealed resource partitioning among herbivores related to position in the canopy and plant functional group (Grimes et al. 2004; MacFadden & Higgins 2004), and comparisons of dinosaur eggshell isotope and trace element data with sedimentologic and paleosol carbonate data revealed the paleoenvironmental context of the late Cretaceous sites yielding eggshells (Cojan et al. 2003). Finally, Botha et al. (2005) have pushed back the temporal window on paleodietary research to the middle Triassic in their study of tooth enamel from non-mammalian cynodonts (*Cynognathus, Diademodon*). Carbon isotope data indicated that, as expected, both taxa fed in a C3-dominated ecosystem. Yet consistent differences between the taxa in both $\delta^{13}C$ and $\delta^{18}O$ values suggested resource or habitat partitioning.

There is a vast archaeological literature attempting to quantify the proportions of marine versus terrestrial foods in human diets using $\delta^{13}C$ and $\delta^{15}N$ values, and a growing body of work using $\delta^{34}S$ values and $^{87}Sr/^{86}Sr$ ratios (e.g., Sealy et al. 1991; Macko et al. 1999; Richards et al. 2003). In a non-archaeological application involving Pleistocene vertebrates, Chamberlain et al. (2005) used $\delta^{13}C$ and $\delta^{15}N$ data to show that a substantial fraction of the California condors from the La Brea tar pits consumed marine mammals (Figure 5.2). In deeper time, Clementz et al. (2003b) used $\delta^{13}C$ values to demonstrate that desmostylians, an extinct group related to elephants and sea cows, foraged on sea grass and kelp, whereas horses, proboscideans, and rhinos from the same marginal marine deposits foraged on C3 land plants.

There is a vast ecological literature examining diet and trophic relationships among modern marine vertebrates using $\delta^{13}C$ and $\delta^{15}N$ values, but work on the diets of ancient marine vertebrates is sparse. Clementz et al. (2003a) analyzed $\delta^{44}Ca$ values in Miocene marine toothed whales, seals, and desmostylians. As expected, high-trophic-level whales had the lowest values and fossil seals were reconstructed as mollusk-feeders (as suggested by morphological analysis), but calcium isotopes were unable to distinguish between herbivory and mollusk-consumption for desmostylians.

Despite the long-recognized promise of compound-specific isotope analysis, it is just now being used to answer paleodietary questions. Fogel & Tuross (2003) measured $\delta^{13}C$ values in essential versus non-essential amino acids from North American humans with and without corn in their diets. C4 carbon from corn labeled non-essential amino acids in humans, but essential amino acids reflected the C3 protein sources available to these people. Corr

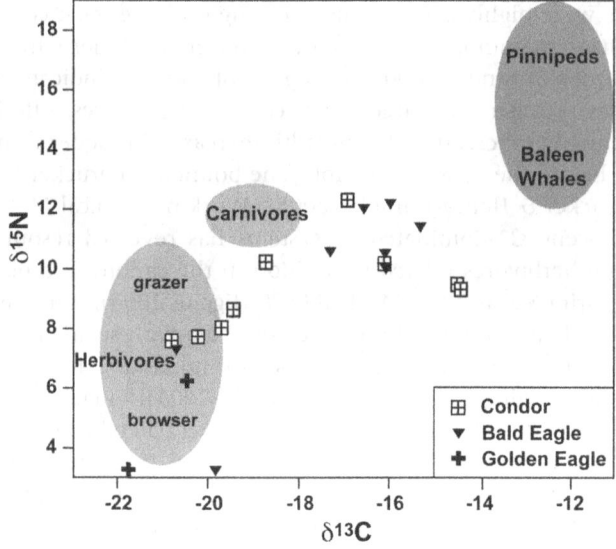

Figure 5.2 Isotopic reconstruction of the dietary preferences of Pleistocene birds from the La Brea tar pits. Estimates of the isotope composition of different dietary resources are provided by the labeled, gray oval fields. Isotope data from bone collagen for bald eagles, golden eagle, and California condors are plotted, after correcting these data to account for trophic level fractionation between diet and collagen. As expected golden eagle diets were strongly terrestrial, whereas bald eagle had a substantial amount of marine food in their diets. California condor diets were variable, but most are so enriched in both ^{13}C and ^{15}N that they must have contained a significant fraction of marine food. Even specialized predation on trapped carnivores (saber toothed cats and dire wolves), would not provide a dietary source with great enough heavy isotope enrichment to explain isotope values from condors.

et al. (2005) tackled the vexing problem of estimating the consumption of marine resources in arid coastal regions with C4 vegetation. In these settings, terrestrial foods have high δ^{15}N and δ^{13}C values that overlap values for marine protein. Corr et al. (2005) noted that the δ^{13}C value of the amino acid glycine is very ^{13}C-enriched in marine foodwebs. In contrast, the δ^{13}C value of phenylalanine (an essential amino acid) ultimately must track primary producers in marine or terrestrial ecosystems. They calculated a ratio of the δ^{13}C value of glycine to phenylalanine; this ratio was much higher in marine mammals than in terrestrial C3 or C4 feeders and was also high in Holocene humans from coastal South Africa thought to consume substantial amounts of marine protein. Finally, Clementz et al. (2003c) studied cholesterol in different fossil marine mammals; kelp consumption was apparent in δ^{13}C values from some fossil sea cows, but was absent from data for fossil whales.

Paleophysiology

The thermal physiology of dinosaurs has been debated for decades. Possible thermoregulatory strategies include:

1 high metabolic heat production (endothermy) with little variability in body temperature (homeothermy), as in extant birds and mammals;
2 low metabolic rates (ectothermy) with environmentally variable body temperatures (heterothermy), as in many living reptiles and amphibians;
3 mass homeothermy, where thermal inertia keeps body temperature higher and less variable than ambient temperature despite relatively low metabolic heat production;
4 behavioral homeothermy, where animals with low metabolic rates use behavioral traits (exercise, basking, shading, etc.) to maintain body temperatures within a narrower range than environmental temperatures.

Oxygen isotope data have been used to explore thermal physiology in two ways. Barrick & Showers (1994) noted that most endothermic homeotherms maintain temperatures within ±2°C across their entire bodies, whereas temperatures in the extremities of ectothermic heterotherms vary well beyond this range. Assuming that the $\delta^{18}O$ value of body water is constant across an individual, they argued that variations in bioapatite $\delta^{18}O$ values within an individual should reflect differences in body part temperatures. Their key working hypothesis was that an endotherm, with ≤4°C body temperature variability, should exhibit ≤1‰ variability in bone phosphate $\delta^{18}O$ values, whereas an ectothermic heterotherm should exhibit greater thermal and isotopic variability (Stoskopf et al. 2001). To date, they have examined large theropods (*Tyrannosaurus*, *Gigantosaurus*) and several ornithischians (small and large ceratopsians, a small hypsilophodont, and adult and juveniles of a large hadrosaur) (Barrick & Showers 1994, 1995, 1999; Barrick et al. 1996, 1998). Sample size limitations prevent rigorous statistical analysis, but most individuals exhibited within- and between-bone $\delta^{18}O$ variability near the limits of endothermic homeothermy. The preferred hypothesis of Barrick and colleagues is that many of the larger species of dinosaur were functionally homeothermic, but that they had lower metabolic rates than modern mammals or birds ("intermediate metabolism").

When dealing with ancient bone mineral, however, diagenetic homogenization is of great concern, as it would mimic the homeothermic pattern (Kolodny et al. 1996). Barrick & Showers (1994, 1995) measured an infrared index of crystallinity, which provided unambiguous evidence that fossil bones were recrystallized. Patterns of isotopic covariation among different phases (bioapatite phosphate, bioapatite carbonate, and diagenetic calcium carbonate) demonstrated that bioapatite carbonate $\delta^{13}C$ and $\delta^{18}O$ values were reset by recrystallization. Because $\delta^{18}O$ values of bioapatite phosphate and carbonate did not covary, Barrick & Showers (1994, 1995) argued that phosphate oxygen was not completely reset, though some degree of diagenetic homogenization cannot be precluded. The strongest evidence for isotopic preservation was their demonstration that a small ectothermic lizard had high $\delta^{18}O$ variability, while ornithischians from the same deposits did not (Barrick et al. 1996). If apparently primary isotopic differences are preserved across the skeleton of a small reptile, it is *ad hoc* to propose that fluid flow and exchange have erased differences across much larger dinosaur skeletons. It is more reasonable to conclude that the dinosaurs had little primary $\delta^{18}O$ (and presumably body temperature) variability. Perhaps $\delta^{18}O$ differences across these skeletons were preserved, despite recrystallization, because:

1 the bones were affected by inorganic, rather than bacterially mediated recrystallization;
2 surrounding sediments had little phosphate to exchange with bones;
3 the scale of phosphate diffusion in pore fluids was small, so that phosphate in recrystallized bone had an extremely local source.

If these conjectures are correct, retention of bone phosphate $\delta^{18}O$ values is likely to be site specific and should be carefully monitored on a site-by-site basis.

Fricke & Rogers (2000) assessed thermoregulatory physiology in a different fashion, through isotopic analysis of tooth enamel phosphate from sympatric taxa across a climatic gradient. For both endotherms and ectotherms, they predicted that body water (and hence bioapatite) $\delta^{18}O$ values would partially track meteoric water $\delta^{18}O$ values and drop with increasing latitude (i.e., with decreasing mean annual temperature). Yet oxygen isotope fractionation between body water and bioapatite increases as temperature drops. Therefore in tooth mineral from ectotherms, the drop in body water $\delta^{18}O$ values in cold regions should be offset by increased water-to-mineral fractionation. In endotherms, with a constant body temperature, this offsetting should not occur, and bioapatite $\delta^{18}O$ values should more closely mirror changes in meteoric water $\delta^{18}O$ values. They discovered that Cretaceous theropod dinosaurs showed a greater shift in $\delta^{18}O$ values with latitude than ectothermic crocodiles, implying a greater degree of homeothermy in theropods. In the absence of a comparison to an undisputed endotherm, it is impossible to assess whether theropods show intermediate metabolic rates, as suggested by Barrick and colleagues, or high, mammal/bird-like metabolic rates.

Little work has been done on isotopes as a monitor of osmoregulation and water-use efficiency. In modern mammals and birds, differences in water-use efficiency and evaporative water loss lead to substantial among-species, within-species, and within-individual $\delta^{18}O$ and δD variability (Bocherens et al. 1996; Kohn et al. 1996; McKechnie et al. 2004). Differences in mean and variability in $\delta^{18}O$ value have been used to assess the habitat preferences of fossil vertebrates and will be discussed below. I am not aware, however, of any studies that have used isotopic methods to study questions about the physiology of water use in ancient vertebrates. Similarly, study of modern

African herbivores has led to the conjecture that differences in urea concentrating mechanisms between obligate vs. non-obligate drinkers might contribute to differences in $\delta^{15}N$ values (Ambrose 1991). Yet this conjecture has never been verified experimentally, and alternate hypotheses exist to explain $\delta^{15}N$ differences among herbivores (Sealy et al. 1987). Again, I am aware of no attempts to use this approach to study the physiology of extinct vertebrates.

Despite evidence that nutritional stress, particularly protein deficient diets, might alter the $\delta^{15}N$ values of vertebrates, I am aware of no studies involving fossil vertebrates in which $\delta^{15}N$ values have been used to assess animal body condition, starvation, etc. Isotopic monitors of diet quality, especially the amount of protein in the diet, have been examined in concert with analyses of individual health and status (based on skeletal or dental metrics) and archaeological grave goods (e.g., Katzenberg et al. 1993; Ambrose et al. 2003). These studies shed light on the way that food availability and quality affect the health and demography of human populations, and how these impacts differ with gender, status and age.

Reproductive biology

Three aspects of reproductive biology have been studied using isotopic methods: nursing/weaning, lactation, and pregnancy. Only the first has been examined in fossil vertebrates. The diets of nursing mammals differ greatly from those of adults. As body water is ^{18}O-enriched relative to ingested water (Bryant & Froelich 1995; Kohn 1996), nursing young should have higher $\delta^{18}O$ values than adults. Very little information is available on calcium isotopes and nursing. A nursing porpoise and an adult porpoise had similar $\delta^{44}Ca$ values, suggesting that the maternal calcium pool used to synthesize milk is controlled by dietary calcium, not bone catabolism (Skulan et al. 1997). With respect to carbon and nitrogen, if lactating mothers catabolize their own tissues to produce milk, isotope values from nursing offspring should look like they are feeding at a trophic level higher than their mothers. For carbon isotopes, this prediction is complicated by the fact that milk is rich in lipids, and lipids are ^{13}C-depleted relative to proteins. Most nitrogen in milk occurs in protein, which is similar in $\delta^{15}N$ value to maternal body tissues, so nursing young should show a 3–5‰ trophic level enrichment relative to their mothers. This effect has been observed in a wide range of extant species (Fogel et al. 1989; Balasse et al. 2001; Jenkins et al. 2001; Polischuk et al. 2001). Thus, nursing offspring should have higher $\delta^{15}N$ values, but either lower or higher $\delta^{13}C$ values than their mothers, depending on the lipid content of the milk.

Nitrogen and carbon isotopes have been used to assess weaning age and the weaning process in prehistoric human populations (e.g., Fogel et al. 1989; Herring et al. 1998; Wright & Schwarcz 1999; Fuller et al. 2003). For example, isotopic and paleodemographic studies yielded the surprising result that weaning age and birth rates did not change in North America following the introduction of intensive agricultural production (Fogel et al. 1997; Schurr & Powell 2005). In a non-human example, Balasse & Tresset (2002) showed that Neolithic cattle in France were weaned at an earlier age than many modern cattle, perhaps because herders were reserving a greater fraction of the milk for human consumption. Likewise, Newsome et al. (in press) used $\delta^{15}N$ and $\delta^{13}C$ analysis to study the weaning age of the northern fur seal (*Callorhinus ursinus*), an eared seal that currently follows a strict weaning schedule (4 months) at its high-latitude rookery sites in the Bering Sea and elsewhere. In contrast, ancient northern fur seals from more temperate latitudes along the northeastern Pacific Rim weaned at >12 months, like nearly all other eared seals (Figure 5.3). In a deep-time study, Franz-Odendaal et al. (2003) used $\delta^{18}O$ values from tooth enamel to show that extinct

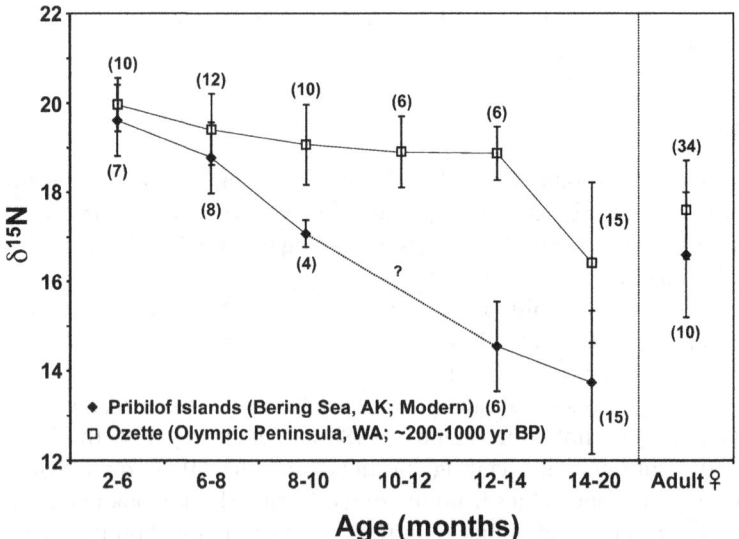

Figure 5.3 Nitrogen isotope evidence from bone collagen for a large change in weaning age for northern fur seals. $\delta^{15}N$ values (mean ± one standard deviation) are presented for animals in different age classes from a modern rookery in the Bering Sea (Pribilof Islands, filled symbol) and a late Holocene population from the Olympic Peninsula (Ozette, open symbol). Pribilof Island fur seals wean at 4 months; $\delta^{15}N$ values drop immediately (6–8 month age class) and are substantially lower for the 8–10 month old animals. In 12–14 month old animals, values are 5‰ lower than in unweaned pups, and are lower than values for adult females from the Pribilofs. Whereas $\delta^{15}N$ values are indistinguishable between Pribilof and Ozette populations for adult females and the youngest age class (2–6 month old animals), Ozette pups retain ^{15}N-enriched values for much longer than Pribilof pups. In Ozette seals, $\delta^{15}N$ values drop significantly only after 12–14 months, suggesting that this extinct population weaned at roughly this age. Numbers in parentheses indicate the number of specimens analyzed in each age class.

sivatheres (a fossil giraffid) from the Pliocene weaned at the same ontogenetic age as modern giraffes.

The isotopic consequences of lactation and pregnancy are rich areas for further study. While not designed to study patterns in mothers, the early work on human nursing did not uncover an isotopic effect in lactating women (Fogel et al. 1989, 1997). In contrast, a study of wild horses from Shackleford Island, NC, by Koch (1997) showed that lactating females had lower $\delta^{15}N$ values than other adults (males, non-lactating females) and used mass balance calculations to argue that ^{15}N-depletion is the expected result of the nitrogen balance perturbations associated with lactation in this herbivore. Fuller et al. (2004) reported $\delta^{15}N$ and $\delta^{13}C$ variations among pregnant human females. They found no significant effects of pregnancy on $\delta^{13}C$ values, but that $\delta^{15}N$ values dropped from conception to birth, and that the magnitude of the drop correlated to the birth weight of the baby as well as the amount of weight gained by the mother. The physiological mechanisms underlying these patterns are unknown, but likely relate to a proportionally reduced loss of ^{15}N-depleted urinary nitrogen as pregnant females achieve positive nitrogen balance. If these patterns associated with pregnancy and lactation are common among mammals, they offer the potential to study inter-birth interval, neonatal survival rate, and other critical aspects of reproductive biology, at least for Pleistocene and Holocene mammals with good organic preservation.

Habitat preference

For terrestrial animals, isotopic differences among taxa at a site or at different sites are sometimes interpreted as evidence for habitat partitioning or habitat preferences. The case is clearest where habitat is essentially congruent with diet. For example, many late Miocene to Recent fossil sites contain species with diets sourced from both C3 and C4 food webs. If we assume that these sites are not time-averaged, such localities must sample a habitat mosaic, with C3 feeders focusing on woodland/forest habitats, and C4 feeders focusing on grasslands. Cerling et al. (1999) discovered that from 5 to 1 Ma, most lineages of African and south Asian proboscideans (e.g., *Loxodonta, Elephas, Anancus, Stegotetrabelodon*, etc.) foraged in C4 grasslands; the only exceptions were C3-feeding deinotheres. Yet the two surviving modern genera (*Elephas* in Asia, *Loxodonta* in Africa), though opportunistic feeders, show a strong preference for C3 vegetation in forests and woodlands. The cause for this shift in diet and habitat is unclear, but it may relate to increased harassment by human hunters on grasslands. Among C3 feeders, extremely low $\delta^{13}C$ values have been viewed as evidence for foraging in dense forests below a closed canopy (Koch et al. 1998; Kohn et al. 2005; Palombo et al. 2005). Bocherens et al. (1996) showed that modern and fossil hippopotamus have lower $\delta^{18}O$ values than co-occurring terrestrial vertebrates, and speculated that this may reflect a reduced evaporative water flux due to daytime immersion or consumption of aquatic vegetation. MacFadden (1998) used this approach to test (and falsify) the hypothesis that the Miocene rhinoceros, *Teleoceras*, was aquatic.

Isotopic methods are excellent monitors of habitat preferences in aquatic vertebrates. For example, in fully marine settings, carbon isotope analysis revealed that earlier in the Holocene, northern fur seals foraged offshore, whereas harbor seals foraged close to shore, as these species do today (Burton et al. 2001). Similarity in tooth enamel $\delta^{18}O$ values in Jurassic pycnodont teleosts and sharks (*Asteracanthus*) from sites located at different paleodepths led Lécuyer et al. (2003) to conclude that these taxa lived in warm surface waters. Likewise, Billon-Bruyat et al. (2005) used bioapatite phosphate $\delta^{18}O$ values to estimate temperatures in order to illuminate the ecological and habitat preferences of fish, turtles, and crocodillians from Late Jurassic lithographic limestone deposits in Europe. They reconstructed plesiochelyid turtles as inhabitants of marine environments, making these the first known marine turtles, pre-dating chelonid sea turtles by ten million years.

For fossils from marginal marine deposits, it can be difficult to determine if co-occurring taxa are autochthonous marine species or if they represent a mixed assemblage of terrestrial, freshwater, estuarine, and marine species. Clementz & Koch (2001) analyzed tooth enamel from modern mammals across a gradient from terrestrial to open marine ecosystems. They found that $\delta^{13}C$ and $\delta^{18}O$ values differed among mammals from freshwater, estuarine, kelp, nearshore and offshore marine environments, but that they could not discriminate between terrestrial and some marine systems. The intra-population variance in $\delta^{18}O$ values was substantially lower in aquatic mammals than in terrestrial mammals, however, offering an independent means of discriminating among animals from all these habitats. Clementz et al. (2003b) used these isotopic proxies (as well as $^{87}Sr/^{86}Sr$ ratios) to demonstrate that Miocene desmostylians on the eastern Pacific margin were fully aquatic mammals foraging in estuaries and open marine systems. Tooth enamel $\delta^{13}C$ and $\delta^{18}O$ values suggested that the first sirenians (manatees and dugongs) consumed seagrass in shallow marine settings and estuaries (MacFadden et al. 2004a; Clementz et al. 2006). Later lineages diversified to consume macroalgae in open marine water, and then freshwater and terrestrial plants. The earliest known archaeocetes (primitive toothed whales) were also fully aquatic, but in contrast to sirenians, they fed in freshwater ecosystems, and only later invaded marine systems (Roe et al. 1998; Clementz et al. 2006) (Figure 5.4). Finally, Patterson (1999) measured $\delta^{13}C$ and $\delta^{18}O$ values from Jurassic teleost otoliths in Europe and was able to distinguish marine from estuarine taxa.

Migration

Any isotope system that varies spatially has the potential to provide information on animal movement or the movement of animals by other processes

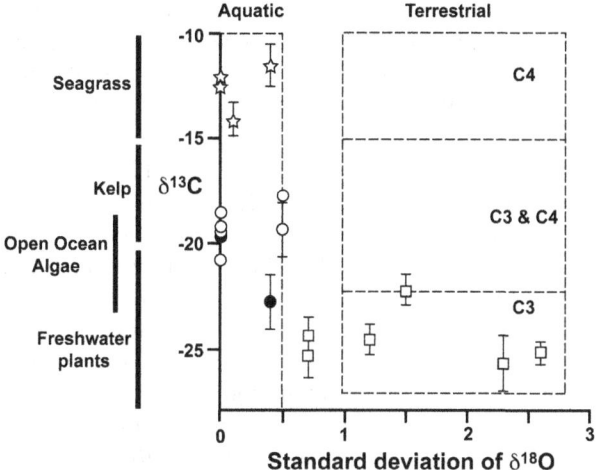

Figure 5.4 Isotopic evidence from tooth enamel carbonate for the paleoecology of Eocene sirenians (sea cows, stars), achaeocetes (archaic toothed whales, circles), and land mammals (squares) from Pakistan, northern Africa, and France. Each symbol represents data from a different species. Among archaeocetes the least derived group, the pakicetids, are marked with filled symbols. Mean $\delta^{13}C$ values (±1 standard deviation) have been converted to dietary isotope values by applying appropriate diet-to-apatite fractionations for carnivores and herbivores. The x axis reports the standard deviation for $\delta^{18}O$ values for fossil populations. For symbols plotting along the y axis, only single specimens were available, hence population $\delta^{18}O$ standard deviations could not be calculated. Finally, the fields for aquatic versus terrestrial adaptations (on the x axis), and the different types of terrestrial and aquatic ecosystems (along the y axis) are indicated by boxes and bars. As expected, land animals have highly variable $\delta^{18}O$ values, but $\delta^{13}C$ values indicating consumption of C3 land plants. The low $\delta^{18}O$ standard deviation of Eocene sea cows indicates they were fully aquatic, but very high $\delta^{13}C$ values indicate consumption of sea grass. Pakicetids have invariant $\delta^{18}O$ values, indicating fully aquatic lifestyes, as well as low mean $\delta^{18}O$ values (not plotted) and low $\delta^{13}C$ values, consistent with freshwater habitats. Finally, more derived cetaceans have mean $\delta^{13}C$ values and both mean (not plotted) and variance values for $\delta^{18}O$ that are consistent with life in marine habitats.

(natural and human predators, fluvial transport). There is a vast and growing literature in wildlife biology using isotope variations to study animal migration (Hobson 1999). The key to a successful study is to develop an isotopic map of the region over which animals or animal products might move. Because isotope values are so variable in soils, hydrologic systems, and plants, isotopic maps are often constructed using animals that have a small geographic range (e.g., rodents, rabbits, domestic pigs, etc.) (Sillen et al. 1998; Hoppe et al. 1999; Price et al. 2002; Budd et al. 2004; Hodell et al. 2004). A second key to success is the use of multiple isotopic or elemental tracers. Here, I will focus on studies of migration that use bulk tissue analysis, rather than studies based on isotopic time series from accreted tissues.

Isotopic research on human movement patterns, or the identification of the proportion of individuals in a skeletal population who are not local, is increasingly common (e.g., Price et al. 1994, 2000; Teschler-Nicola et al. 1999; Ezzo & Price 2002; Budd et al. 2004; White et al. 2004; Knudson et al. 2005; Wright, 2005). Interesting results include the recognition that an immigrant to an oasis along the Nile ca. 1750 yr BP had leprosy, perhaps indicating his exile (Dupras & Schwarcz 2001), that Anglo-Saxon (Scandinavian) immigrants to England from 1600 to 1400 yr BP included both sexes and all ages classes, not just a male military elite (Montgomery et al. 2005), and that first-generation slaves from tropical Africa were present in a late 18th/early 19th century burial ground in Cape Town, South Africa (Cox et al. 2001). Isotopic and other data have revealed the place of origin and lifetime movements of the Alpine iceman (Müller et al. 2003).

Studies of migration in other ancient terrestrial and marine vertebrates are much more rare. Hoppe et al. (1999) and Hoppe & Koch (2006) compared $^{87}Sr/^{86}Sr$ ratios, and $\delta^{13}C$ and $\delta^{18}O$ values in mastodons, mammoths, and co-occurring fauna from Pleistocene sites in Florida. They discovered that mastodons in northern Florida were making considerable migrations to the north to feed in forests on sediments sourced from the Appalachians, whereas mammoths across Florida were grazers on sediments derived from platform carbonates in peninsular Florida (Figure 5.5). The scale of movement was hundreds of miles, but not thousands. Hoppe (2004) used the same approach to study mammoth herd structure and movement, as well as the hypothesis that Paleoindians hunted mammoth family groups in the late Pleistocene. The lack of substantial isotopic differences in a suite of dinosaur eggshells led Cojan et al. (2003) to the conclusion that the dinosaurs that laid these eggs did not migrate over substantial distances prior to nesting. Finally, using pinnipeds that do not undertake large-scale migration, Burton & Koch (1999) showed that $\delta^{13}C$ and $\delta^{15}N$ differences at the base of northeast Pacific food webs cascade up to label top marine carnivores. Burton et al. (2001) and Newsome et al. (in press) verified that this same map applied in the Holocene, then used it to document that northern fur seals from along the California coast were not seasonal migrants from the Bering Sea (site of the current dominant rookery), but instead were sourced by mid-latitude rookeries.

Isotopic time series from accreted tissues

Samples collected in sequence from accreted tissues preserve a time series that can be used to explore aspects of animal biology that vary through ontogeny or with the seasons. These samples are typically collected by micromilling or laser ablation. A first step is to understand the pace and phasing of tissue accretion. Assuming that annual cycles in the $\delta^{18}O$ value of meteoric water are transmitted to body tissues via ingested water, annual cycles have been identified in fossil tusks and ever-growing teeth (Koch et al. 1989;

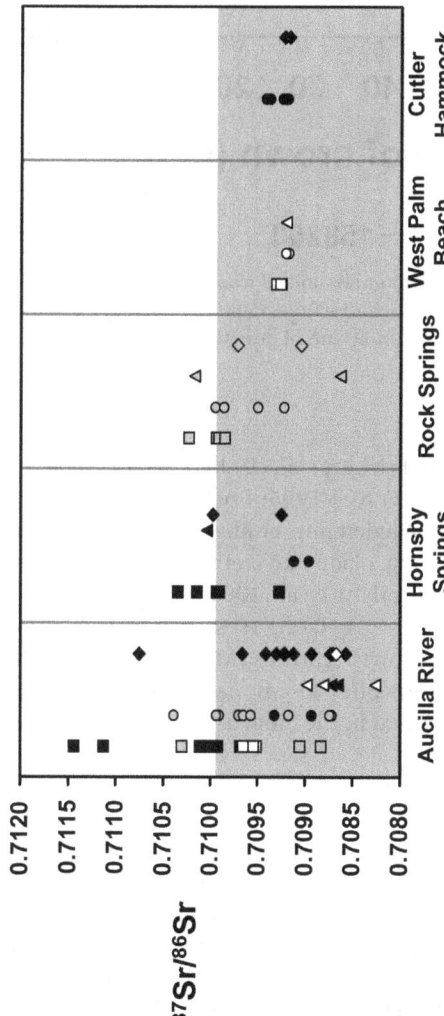

Figure 5.5 Strontium isotope values for bulk tooth enamel for Pleistocene mammals from Florida. Data are presented for mastodons (*Mammut,* squares), mammoths (*Mammuthus,* circles), tapirs (*Tapirus,* triangles), and deer (*Odocoileus,* diamonds). Solid symbols are post-glacial individuals (15,000 to 10,000 yr BP), open symbols are from the Last Glacial Maximum or earlier (15,000 to 70,000 yr BP), and gray-filled symbols are of indeterminate age (Rancholabrean). Aucilla River sites are on the Florida panhandle, Rock Springs and Hornsby Springs are in north-central Florida, and West Palm Beach and Cutler Hammock are in southern Florida. The gray shaded region indicates the range of values seen in different types of environmental samples from Florida, with values on the Plio-Pleistocene platform carbonates in southern and eastern coastal Florida similar to the modern ocean value of 0.7092. Higher isotope values in animals require inputs of strontium from sediments sourced by the Appalachians, to the north in Georgia. All taxa in southern Florida foraged on recent marine geological substrates. At northern sites (especially those along the Aucilla River), some individuals (especially mastodons) have ^{87}Sr/^{86}Sr ratios indicating that they foraged part of the year in the Appalachians or their foothills, several hundred miles from the sites where the animals died.

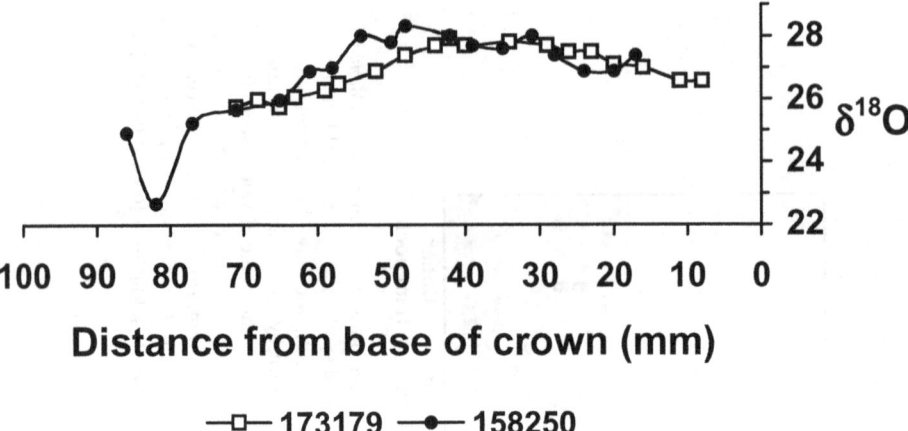

Figure 5.6 Oxygen isotope values for microsamples milled from the canines of two saber-toothed cats (*Smilodon fatalis*) from Rancho La Brea, California (Feranec 2004b). Approximately 1 year of growth is recorded in 80 mm of growth, indicating that the entire tooth crown formed in 18 months.

Stuart-Williams & Schwarcz 1997; Fricke et al. 1998b), enamel on tooth crowns (Bryant et al. 1996; Fricke & O'Neil 1996; Fricke et al. 1998a; Kohn et al. 1998; Sharp & Cerling 1998; Wiedemann et al. 1999; Franz-Odendaal et al. 2003; Botha et al. 2005), the long, blade like teeth of carnivorous mammals and reptiles (Feranec 2004b; Straight et al. 2004) (Figure 5.6), dental batteries in animals with continuously replaced teeth (Thomas & Carlson 2004), and bones with incremental growth features (MacFadden et al. 2004b; Tütken et al. 2004). These oscillations offer a seasonal chronometer against which to correlate other events recorded in the biogeochemistry and mineralogy of teeth and bones.

This approach has shown that incremental features in proboscidean tusks, dinosaur long bones, and shark vertebral centra are indeed annual, allowing estimates of growth rate and, in the case of proboscideans, season-of-death (Koch et al. 1989; MacFadden et al. 2004b; Tütken et al. 2004). Analyses of sheep tooth crowns revealed that there were two lambing seasons on the southwestern coast of South Africa ca. 2000 to 1000 yr BP (Balasse et al. 2003). Isotopic oscillations in time series from animals will be damped relative to environmental cycles due to reservoir effects in the body, bone turnover, and prolonged maturation of tooth enamel. This damping impacts paleoclimatic research (e.g., Dettman et al. 2001; Higgins & MacFadden 2004), but is of less concern if oscillations are used as a chronometer.

Oscillations in otolith $\delta^{18}O$ values, driven by changes in growth temperature and/or the $\delta^{18}O$ value of meteoric water, demarcate annual growth increments in both marine and continental settings (e.g., Patterson 1998; Weidman & Millner 2000). Most studies of isotopic time series from fossil otoliths have explored paleoclimatic, rather than paleobiologic, questions (e.g., Patterson 1998; Ivany et al. 2000; Wurster & Patterson 2001). An exception is the study by Wurster & Patterson (2003), which examined shifts in metabolic rate implied by changes in $\delta^{13}C$ values from core to rim in Holocene otoliths.

Migration is particularly amenable to analysis via isotopic time series. Koch et al. (1992) examined $^{87}Sr/^{86}Sr$ variations in the vertebral centra of modern and Neogene salmonids to assess their potential as a monitor of freshwater to marine migration. Bone turnover reduced the signal of migration in modern salmonids, and diagenesis completely erased it in fossil bone. Still, the approach can be successful if applied to unaltered fossil otoliths and is currently being developed to study the natal rivers of modern salmonids (Ingram & Weber 1999; Kennedy et al. 2000). An example is the study of $^{87}Sr/^{86}Sr$ ratios and $\delta^{13}C$ and $\delta^{18}O$ values from aragonitic otoliths of the teleost *Vorhisia vulpes* from late Cretaceous estuarine deposits in South Dakota, USA (Carpenter et al. 2003), which showed that fish spawned in brackish water, then migrated in their first year to open marine waters in the Western Interior Seaway. They remained at sea for 3 years before returning to estuaries to spawn and die. An example involving terrestrial vertebrates again comes from the work of Balasse et al. (2002), who measured the $^{87}Sr/^{86}Sr$ ratios and $\delta^{13}C$ and $\delta^{18}O$ values of the teeth of domestic sheep and cows on the southwestern coast of South Africa. They found regular oscillations in $\delta^{13}C$ values, indicating seasonal dietary change, but $^{87}Sr/^{86}Sr$ ratios offered no support for the hypothesis that these herders moved their animals seasonally between the coastal zone and inland regions. Schweissing & Grupe (2003) offered a different approach to time series construction. They examined multiple teeth from the same individual and constructed a time series of $^{87}Sr/^{86}Sr$ ratios by bulk sampling entire tooth crowns, and then estimating and modeling the time of crown formation. They identified at least three different sources for individuals in a late Roman site (ca. 1650–1550 yr BP) in Germany, confirming a hypothesis of population admixture as a result of Roman population policy.

Seasonal or ontogenetic dietary shifts are also amenable to study with isotopic time series, though few studies have been conducted to date. Koch et al. (1995) demonstrated the feasibility of sampling time series from molar roots and uncovered large changes in the diets of some modern elephants through $\delta^{13}C$ analysis. As discussed previously, Balasse et al. (2002) observed subtle $\delta^{13}C$ shifts in Holocene domestic sheep from the southwestern coast of South Africa. Sharp & Cerling (1998) collected $\delta^{13}C$ profiles from tooth enamel from several Pleistocene horses, and detected seasonal and unidirectional shifts in diet. In contrast, in the most comprehensive study of dietary time series, Fox & Fisher (2004) detected no seasonal shifts in the $\delta^{13}C$ values of tusk enamel from Miocene gomphotheres. Similarly, $\delta^{13}C$ data from individual therapsids showed small regular variations, but not in phase with variations in $\delta^{18}O$ values (Botha et al. 2005). Finally, Thomas & Carlson (2004) analyzed an ontogenetic series of teeth from a Cretaceous hadrosaur. They found moderately large (2–4‰) variations in $\delta^{13}C$ values, roughly in phase with $\delta^{18}O$ variations. However, the very high absolute $\delta^{13}C$ values, which were well outside the range for C3 plants, as well as clear evidence that diagenetically-altered apatite was ^{13}C-enriched, led them to be cautious in interpreting their data as solely reflecting diet.

Conclusions

The explosion of work on modern vertebrates has made life much easier for paleobiologists. New experimental data and extensive field observations increasingly offer a solid foundation for interpreting data from fossil vertebrates. Technological advances have made it possible to analyze large numbers of samples for paleobiological studies, they have opened up new isotope systems and/or new substrates (e.g., calcium in biominerals, oxygen in protein, hydrogen in individual molecules, laser sampling of biomineral oxygen). Many of these new developments have yet to be deployed to answer critical questions in palaebiology. The new methods that rely on organic tissues have not yet been applied in any systematic fashion to the study of Quaternary vertebrates and there is a potential to apply them to deep-time questions in fossils showing exceptional preservation. Studies using biominerals have largely focused on Cenozoic vertebrates, but there are enough successful studies of Mesozoic vertebrates to warrant further exploration. Finally, while paleodiet has received the lion's share of past research, future studies will hopefully fill out the paleobiology of vertebrates and explore physiology, reproduction, migration and habitat use.

A post-script on workshops and literature resources

Progress on the isotope paleobiology of vertebrates has been promoted by workshops that produced books and special issues of journals. The Advanced Seminars on Paleodietary Research provide a venue for exploration of major theoretical and analytical questions concerning chemical and isotopic approaches to hominid paleobiology. Each seminar has had a slightly different focus, emphasizing the most pressing questions facing workers. Experts from outside archaeology and paleoanthropology, including botanists, geochemists, soil scientists, paleontologists, physiologists, and calcified tissue biochemists, have contributed vital, multidisciplinary context. There have been six seminars and associated publications: 1986, Santa Fe, USA (Price 1989); 1989, Cape Town, South Africa (Sillen & Armelagos 1991); 1991, Bad Homburg, Germany (Lambert & Grupe 1993); 1993, Banff, Canada (Ambrose & Katzenberg 2000); 1997, Valbonne, France (Bocherens & Van Klinken 1999); and 2001, Santa Cruz, USA (Koch & Burton 2003). Taphonomy and diagenesis are also pressing concerns for biogeochemical study of vertebrates. The International Workshops on Bone Diagenesis have been the forum for exploring these issues. There have been four workshops and associated publications: 1988, Oxford, UK (Schwarcz et al. 1989); 1993, Oxford, UK (Hedges & Van Klinken 1995); 1996, Paris, France (Bocherens & Denys 1997–1998), and 2000, Albarracín, Spain (Fernández-Jalvo et al. 2002). Important volumes were produced following the Hal Krueger Memorial Symposium (Ambrose & Krigbaum 2003a) and the symposium on Incremental Growth in Vertebrate Skeletal Tissues (MacFadden 2004).

References

Altabet, M.A. & François, R. (2001) Nitrogen isotope biogeochemistry of the Antarctic Polar Frontal Zone at 170°W. *Deep-Sea Research Part II–Topical Studies in Oceanography*, **48**, 4247–4273.

Altabet, M.A., Pilskaln, C., Thunell, R., et al. (1999) The nitrogen isotope biogeochemistry of sinking particles from the margin of the Eastern North Pacific. *Deep-Sea Research Part I–Oceanographic Research Papers*, **46**, 655–679.

Ambrose, S.H. (1990) Preparation and characterization of bone and tooth collagen from isotopic analysis. *Journal of Archaeological Science*, **17**, 431–451.

Ambrose, S.H. (1991) Effects of diet, climate and physiology on nitrogen isotope abundances in terrestrial foodwebs. *Journal of Archaeological Science*, **18**, 293–317.

Ambrose, S.H. & Katzenberg, M.A. (eds.) (2000) *Biogeochemical Approaches to Paleodietary Research*. Kluwer Academic/Plenum Publishers, New York.

Ambrose, S.H. & Krigbaum, J. (2003a) Bone chemistry and bioarchaeology. *Journal of Anthropological Archaeology*, **22**, 191–192.

Ambrose, S.H. & Krigbaum, J. (2003b) Bone chemistry and bioarchaeology. *Journal of Anthropological Archaeology*, **22**, 193–199.

Ambrose, S.H. & Norr, L. (1993) Experimental evidence for the relationship of the carbon isotope ratios of whole diet and dietary protein to those of bone collagen and carbonate. In: *Prehistoric Human Bone: Archaeology at the Molecular Level*. (Eds J.B. Lambert & G. Grupe), pp. 1–37. Springer-Verlag, New York.

Ambrose, S.H., Buikstra, J. & Krueger, H.W. (2003) Status and gender differences in diet at Mound 72, Cahokia, revealed by isotopic analysis of bone. *Journal of Anthropological Archaeology*, **22**, 217–226.

Bada, J.L., Wang, X.Y.S. & Hamilton, H. (1999) Preservation of key biomolecules in the fossil record: current knowledge and future challenges. *Philosophical Transactions of the Royal Society of London Series B–Biological Sciences*, **354**, 77–86.

Balasse, M. (2002) Reconstructing dietary and environmental history from enamel isotopic analysis: Time resolution of intra-tooth sequential sampling. *International Journal of Osteoarchaeology*, **12**, 155–165.

Balasse, M. & Tresset, A. (2002) Early weaning of Neolithic domestic cattle (Bercy, France) revealed by intra-tooth variation in nitrogen isotope ratios. *Journal of Archaeological Science*, **29**, 853–859.

Balasse, M., Bocherens, H., Mariotti, A. & Ambrose, S.H. (2001) Detection of dietary changes by intra-tooth carbon and nitrogen isotopic analysis: An experimental study of dentine collagen of cattle (*Bos taurus*). *Journal of Archaeological Science*, **28**, 235–245.

Balasse, M., Ambrose, S.H., Smith, A.B. & Price, T.D. (2002) The seasonal mobility model for prehistoric herders in the South-western Cape of South Africa assessed by isotopic analysis of sheep tooth enamel. *Journal of Archaeological Science*, **29**, 917–932.

Balasse, M., Smith, A.B., Ambrose, S.H. & Leigh, S.R. (2003) Determining sheep birth seasonality by analysis of tooth enamel oxygen isotope ratios: The late stone age site of Kasteelberg (South Africa). *Journal of Archaeological Science*, **30**, 205–215.

Barrick, R.E. & Showers, W.J. (1994) Thermophysiology of *Tyrannosaurus rex* – evidence from oxygen isotopes. *Science*, **265**, 222–224.

Barrick, R.E. & Showers, W.J. (1995) Oxygen isotope variability in juvenile dinosaurs (*Hypacrosaurus*): Evidence for thermoregulation. *Paleobiology*, **21**, 552–560.

Barrick, R.E. & Showers, W.J. (1999) Thermophysiology and biology of *Giganotosaurus*; comparison with *Tyrannosaurus*. *Palaeontologia Electronica*, http://www.palaeo-electronica. org/1999_2/gigan/issue2_99.htm.

Barrick, R.E., Showers, W.J. & Fischer, A.G. (1996) Comparison of thermoregulation of four ornithischian dinosaurs and a varanid lizard from the Cretaceous Two Medicine formation: Evidence from oxygen isotopes. *Palaios*, **11**, 295–305.

Barrick, R.E., Stoskopf, M.K, Marcot, J.D, Russell, D.A. & Showers, W.J. (1998) The thermoregulatory functions of the *Triceratops* frill and horns: Heat flow measured with oxygen isotopes. *Journal of Vertebrate Paleontology*, **18**, 746–750.

Barrick, R.E., Fischer, A.G. & Showers, W.J. (1999) Oxygen isotopes from turtle bone: Applications for terrestrial paleoclimates? *Palaios*, **14**, 186–191.

Beard, B.L & Johnson, C.M. (2000) Strontium isotope composition of skeletal material can determine the birth place and geographic mobility of humans and animals. *Journal of Forensic Science*, **45**, 1049–1061.

Bearhop, S., Waldron, S., Votier, S.C. & Furness, R.W. (2002) Factors that influence assimilation rates and fractionation of nitrogen and carbon stable isotopes in avian blood and feathers. *Physiological and Biochemical Zoology*, **75**, 451–458.

Beshah, K., Rey, C., Glimcher, M., Schimizu, M. & Griffin, R.G. (1990) Solid state carbon-13 and proton NMR studies of carbonate-containing calcium phosphates and enamel. *Journal of Solid State Chemistry*, **84**, 71–81.

Biasatti, D.M. (2004) Stable carbon isotopic profiles of sea turtle humeri: implications for ecology and physiology. *Palaeogeography, Palaeoclimatology, Palaeoecology*, **206**, 203–216.

Billon-Bruyat, J.P., Lécuyer, C., Martineau, F. & Mazin, J.M. (2005) Oxygen isotope compositions of Late Jurassic vertebrate remains from lithographic limestones of western Europe: implications for the ecology of fish, turtles, and crocodillans. *Palaeogeography Palaeoclimatology Palaeoecology*, **216**, 359–375.

Birchall, J., O'Connell, T.C., Heaton, T.H.E. & Hedges, R.E.M. (2005) Hydrogen isotope ratios in animal body protein reflect trophic level. *Journal of Animal Ecology*, **74**(5), 877–881.

Bocherens, H. & Denys, C. (eds.) (1997–1998) *Bulletin de al Société géologique de France*, **168**, 485–490, 535–564, 751–781 and **169**, 101–114, 425–451.

Bocherens, H. & Van Klinken, G.J. (eds.) (1999) *Journal of Archaeological Science*, **26**, 593–728.

Bocherens, H., Koch, P.L., Mariotti, A., Geraads, D. & Jaeger, J.J. (1996) Isotopic biogeochemistry (^{13}C, ^{18}O) of mammalian enamel from African Pleistocene hominid sites. *Palaios*, **11**, 306–318.

Bocherens, H., Billiou, D., Patou-Mathis, M., Bonjean, D., Otte, M. & Mariotti, A. (1997) Paleobiological implications of the isotopic signatures (^{13}C, ^{15}N) of fossil mammal collagen in Scladina cave (Sclayn, Belgium). *Quaternary Research*, **48**, 370–380.

Bocherens, H., Drucker, D.G., Billiou, D., Patou-Mathis, M. & Vandermeersch, B. (2005) Isotopic evidence for diet and subsistence pattern of the Saint-Cesaire I Neanderthal: review and use of a multi-source mixing model. *Journal of Human Evolution*, **49**, 71–87.

Botha, J., Lee-Thorp, J. & Chinsamy, A. (2005) The palaeoecology of the non-mammalian cynodonts *Diademodon* and *Cynognathus* from the Karoo Basin of South Africa, using stable light isotope analysis. *Palaeogeography Palaeoclimatology Palaeoecology*, **223**, 303–316.

Briggs, D.E.G., Wilby, P.R, Perez Moreno, B.P., Sanz, J.L. & Fregenal Martinez, M. (1997) The mineralization of dinosaur soft tissue in the lower Cretaceous of Las Hoyas, Spain. *Journal of the Geological Society*, **154**, 587–588.

Bryant, J.D. & Froelich, P.N. (1995) A model of oxygen isotope fractionation in body water of large mammals. *Geochimica et Cosmochimica Acta*, **59**, 4523–4537.

Bryant, J.D., Froelich, P.N., Showers, W.J. & Genna, B.J. (1996) Biologic and climatic signals in the oxygen isotopic composition of Eocene–Oligocene equid enamel phosphate. *Palaeogeography Palaeoclimatology Palaeoecology*, **126**, 75–89.

Budd, P., Millard, A., Chenery, C., Lucy, S. & Roberts, C. (2004) Investigating population movement by stable isotope analysis: a report from Britain. *Antiquity*, **78**, 127–141.

Burley, R.W. & Vadehra, D.V. (1989) *The Avian Egg: Chemistry and Biology*. John Wiley & Sons, New York.

Burton, R.K. & Koch, P.L. (1999) Isotopic tracking of foraging and long-distance migration in northeastern Pacific pinnipeds. *Oecologia*, **119**, 578–585.

Burton, R.K., Snodgrass, J.J., Gifford-Gonzalez, D., Guilderson, T., Brown, T. & Koch, P.L. (2001) Holocene changes in the ecology of northern fur seals: Insights from stable isotopes and archaeofauna. *Oecologia*, **128**, 107–115.

Campana, S.E. & Neilson, J.D. (1985) Microstructure of fish otoliths. *Canadian Journal of Fisheries and Aquatic Science*, **42**, 1014–1032.

Capo, R.C., Stewart, B.W. & Chadwick, O.A. (1998) Strontium isotopes as tracers of ecosystem processes: theory and methods. *Geoderma*, **82**, 197–225.

Carlson, S.J. (1990) Vertebrate dental structures. In: *Skeletal Biomineralization: Patterns, Processes and Evolutionary Trends*, Vol. 1 (Ed. J.G. Carter), pp. 531–556. Van Nostrand Reinhold, New York.

Carpenter, S.J., Erickson, J.M. & Holland, F.D. (2003) Migration of a Late Cretaceous fish. *Nature*, **423**, 70–74.

Cerling, T.E. & Harris, J.M. (1999) Carbon isotope fractionation between diet and bioapatite in ungulate mammals and implications for ecological and paleoecological studies. *Oecologia*, **120**, 347–363.

Cerling, T.E., Harris, J.M., MacFadden, B.J., et al. (1997) Global vegetation change through the Miocene/Pliocene boundary. *Nature*, **389**, 153–158.

Cerling, T.E., Harris, J.M. & Leakey, M.G. (1999) Browsing and grazing in elephants: the isotope record of modern and fossil proboscideans. *Oecologia*, **120**, 364–374.

Chamberlain, C.P., Blum, J.D, Holmes, R.T., Feng, X.H., Sherry, T.W. & Graves, G.R. (1997) The use of isotope tracers for identifying populations of migratory birds. *Oecologia*, **109**, 132–141.

Chamberlain, C.P., Waldbauer, J.R., Fox-Dobbs, K., et al. (2005) Pleistocene to Recent dietary shifts in California condors: Implications for conservation strategies. *Proceedings of the National Academy of Sciences of the United States of America*, **102**, 16707–16711.

Chinsamy, A. & Dodson, P. (1995) Inside a dinosaur bone. *American Scientist*, **83**, 174–180.

Clementz, M.T. & Koch, P.L. (2001) Differentiating aquatic mammal habitat and foraging ecology with stable isotopes in tooth enamel. *Oecologia*, **129**, 461–472.

Clementz, M.T., Holden, P. & Koch, P.L. (2003a) Are calcium isotopes a reliable monitor of trophic level in marine settings? *International Journal of Osteoarchaeology*, **13**, 29–36.

Clementz, M.T., Hoppe, K.A. & Koch, P.L. (2003b) A paleoecological paradox: the habitat and dietary preferences of the extinct tethythere *Desmostylus*, inferred from stable isotope analysis. *Paleobiology*, **29**, 506–519.

Clementz, M.T., Jim, S., Koch, P.L. & Evershed, R.P. (2003c) Old lipids and the sea: Using cholesterol as a paleodietary proxy for extinct marine mammals. *Abstracts of Papers of the American Chemical Society*, **225**, U937.

Clementz, M.T., Goswami, A., Gingerich, P.D. & Koch, P.L. (2006) Isotopic records from early whales and sea cows: contrasting patterns of ecological transition. *Journal of Vertebrate Paleontology*, **26**(2), 355–370.

Cojan, I., Renard, M. & Emmanuel, L. (2003) Palaeoenvironmental reconstruction of dinosaur nesting sites based on a geochemical approach to eggshells and associated palaeosols (Maastrichtian, Provence Basin, France). *Palaeogeography Palaeoclimatology Palaeoecology*, **191**, 111–138.

Collins, M.J., Nielsen-Marsh, C.M., Hiller, J., et al. (2002) The survival of organic matter in bone: A review. *Archaeometry*, **44**, 383–394.

Coltrain, J.B., Harris, J.M., Cerling, T.E., et al. (2004) Rancho La Brea stable isotope biogeochemistry and its implications for the palaeoecology of late Pleistocene, coastal southern California. *Palaeogeography, Palaeoclimatology, Palaeoecology*, **205**, 199–219.

Connin, S.L., Betancourt, J. & Quade, J. (1998) Late Pleistocene C4 plant dominance and summer rainfall in the southwestern United States from isotopic study of herbivore teeth. *Quaternary Research*, **50**, 179–193.

Cormie, A.B., Luz, B. & Schwarcz, H.P. (1994a) Relationship between the hydrogen and oxygen isotopes of deer bone and their use in the estimation of relative humidity. *Geochimica et Cosmochimica Acta*, **58**, 3439–3449.

Cormie, A.B., Schwarcz, H.P. & Gray, J. (1994b) Determination of the hydrogen isotopic composition of bone collagen and correction for hydrogen exchange. *Geochimica et Cosmochimica Acta*, **58**, 365–375.

Cormie, A.B., Schwarcz, H.P. & Gray, J. (1994c) Relation between hydrogen isotopic ratios of bone collagen and rain. *Geochimica et Cosmochimica Acta*, **58**, 377–391.

Corr, L.T., Sealy, J.C., Horton, M.C. & Evershed, R.P. (2005) A novel marine dietary indicator utilising compound-specific bone collagen amino acid $\delta^{13}C$ values of ancient humans. *Journal of Archaeological Science*, **32**, 321–330.

Cox, G., Sealy, J., Schrire, C. & Morris, A. (2001) Stable carbon and nitrogen isotopic analyses of the underclass at the colonial Cape of Good Hope in the eighteenth and nineteenth centuries. *World Archaeology*, **33**, 73–97.

Cryan, P.M., Bogan, M.A., Rye, R.O., Landis, G.P. & Kester, C.L. (2004) Stable hydrogen isotope analysis of bat hair as evidence for seasonal molt and long-distance migration. *Journal of Mammalogy*, **85**, 995–1001.

DeHart, P.A.P. & Wooller, M.J. (2004) A multi-organismal isotopic study of north Pacific and Bering Sea marine mammals: responses to a changing environment. *Fourth International Conference on Applications of Stable Isotope Techniques to Ecologial Studies*, Wellington, New Zealand, 89 pp.

DeNiro, M.J. (1987) Stable isotopes and archaeology. *American Scientist*, **75**, 182–191.

DeNiro, M.J. & Epstein, S. (1978) Influence of diet on the distribution of carbon isotopes in animals. *Geochimica et Cosmochimica Acta*, **42**, 495–506.

DePaolo, D.J. (2004) Calcium isotopic variations produced by biological, kinetic, radiogenic and nucleosynthetic processes. *Geochemistry of Non-Traditional Stable Isotopes, Reviews in Mineralogy & Geochemistry*, **55**, 255–288.

Dettman, D.L., Kohn, M.J., Quade, J., Ryerson, F.J., Ojha, T.P. & Hamidullah, S. (2001) Seasonal stable isotope evidence for a strong Asian monsoon throughout the past 10.7 m.y. *Geology*, **29**, 31–34.

Doucett, R.R., Broders, H.G., Quinn, G. & Forbes, G.J. (2002) Those vain aerial ways: determining the relative importance of aquatic and terrestrial insects in the diets of bats. *Third International Conference on Applications of Stable Isotope Techniques to Ecologial Studies*, Flagstaff, Arizona, 55 pp.

Drucker, D. & Bocherens, H. (2004) Carbon and nitrogen stable isotopes as tracers of change in diet breadth during Middle and Upper Palaeolithic in Europe. *International Journal of Osteoarchaeology*, **14**, 162–177.

Drucker, D.G. & Henry-Gambier, D. (2005) Determination of the dietary habits of a Magdalenian woman from Saint-Germain-la-Riviere in southwestern France using stable isotopes. *Journal of Human Evolution*, **49**, 19–35.

Drucker, D.G., Bocherens, H. & Billiou, D. (2003) Evidence for shifting environmental conditions in Southwestern France from 33 000 to 15 000 years ago derived from ^{13}C and ^{15}N natural abundances in collagen of large herbivores. *Earth and Planetary Science Letters*, **216**, 163–173.

Dufour, E., Bocherens, H. & Mariotti, A. (1999) Palaeodietary implications of isotopic variability in Eurasian lacustrine fish. *Journal of Archaeological Science*, **26**, 617–627.

Dupras, T.L. & Schwarcz, H.P. (2001) Strangers in a strange land: Stale isotope evidence for human migration in the Dakhleh Oasis, Egypt. *Journal of Archaeological Science*, **28**, 1199–1208.

Ehleringer, J.R. & Monson, R.K. (1993) Evolutionary and ecological aspects of photosynthetic pathway variation. *Annual Review of Ecology and Systematics*, **24**, 411–439.

Ehleringer, J.R., Cerling, T.E. & Helliker, B.R. (1997) C4 photosynthesis, atmospheric CO_2, and climate. *Oecologia*, **112**, 285–299.

Epp, M., Howe, T., Conquest, L. & Wooller, M.J. (2004) Examining the relationships between δ^{18}O and δD of invertebrates, diet and water in an aquatic medium. *Fourth International Conference on Applications of Stable Isotope Techniques to Ecologial Studies*, Wellington, New Zealand, pp. 156.

Epstein, H.E., Lauenroth, W.K., Burke, I.C. & Coffin, D.P. (1997) Productivity patterns of C3 and C4 functional types in the U.S. Great Plains. *Ecology*, **78**, 722–731.

Estep, M.F. & Dabrowski, H. (1980) Tracing food webs with stable hydrogen isotopes. *Science*, **209**, 1537–1538.

Evans, R.D. & Ehleringer, J.R. (1994) Water and nitrogen dynamics in an arid woodland. *Oecologia*, **99**, 233–242.

Evershed, R.P., Dudd, S.N. & Charters, S. (1999) Lipids as carriers of anthropogenic signals from prehistory. *Philosophical Transactions of the Royal Society of London Series B–Biological Sciences*, **354**, 19–31.

Ezzo, J.A. & Price, T.D. (2002) Migration, regional reorganization, and spatial group composition at Grasshopper Pueblo, Arizona. *Journal of Archaeological Science*, **29**, 499–520.

Felicetti, L.A., Schwartz, C.C., Rye, R.O., et al. (2003) Use of sulfur and nitrogen stable isotopes to determine the importance of whitebark pine nuts to Yellowstone grizzly bears. *Canadian Journal of Zoology*, **81**, 763–770.

Feranec, R.S. (2003) Stable isotopes, hypsodonty, and the paleodiet of *Hemiauchenia* (Mammalia: Camelidae): a morphological specialization creating ecological generalization. *Paleobiology*, **29**(2), 230–242.

Feranec, R.S. (2004a) Geographic variation in the diet of hypsodont herbivores from the Rancholabrean of Florida. *Palaeogeography, Palaeoclimatology, Palaeoecology*, **207**, 359–369.

Feranec, R.S. (2004b) Isotopic evidence of saber-tooth development, growth rate, and diet from the adult canine of *Smilodon fatalis* from Rancho La Brea. *Palaeogeography, Palaeoclimatology, Palaeoecology*, **206**, 303–310.

Fernández-Jalvo, Y., Sánchez-Chillón, B. & Alcalá, L. (2002) The Fourth International Meeting on Bone Diagenesis – Summary and introduction to papers. *Archaeometry*, **44**, 315–318.

Fisher, D.C. & Fox, D.L. (1998) Oxygen isotopes in mammoth teeth: Sample design, mineralization patterns, and enamel–dentin comparisons. *Journal of Vertebrate Paleontology*, **18 (Suppl)**, 41–42A.

Fogel, M.L. & Tuross, N. (2003) Extending the limits of paleodietary studies of humans with compound specific carbon isotope analysis of amino acids. *Journal of Archaeological Science*, **30**, 535–545.

Fogel, M.L., Tuross, N. & Owsley, D.W. (1989) Nitrogen isotope tracers of human lactation in modern and archaeological populations. *Annual Report of the Director, Geophysical Laboratory, Carnegie Institute of Washington*, **1989**, 111–117.

Fogel, M.L., Tuross, N., Johnson, B.J. & Miller, G.H. (1997) Biogeochemical record of ancient humans. *Organic Geochemistry*, **27**, 275–287.

Fox, D.L. & Fisher, D.C. (2004) Dietary reconstruction of Miocene *Gomphotherium* (Mammalia, Proboscidea) from the Great Plains region, USA, based on the carbon isotope composition of tusk and molar enamel. *Palaeogeography, Palaeoclimatology, Palaeoecology*, **206**, 311–335.

Franz-Odendaal, T.A., Lee-Thorp, J.A. & Chinsamy, A. (2003) Insights from stable light isotopes on enamel defects and weaning in Pliocene herbivores. *Journal of Biosciences*, **28**, 765–773.

Fricke, H.C. & O'Neil, J.R. (1996) Inter- and intra-tooth variation in the oxygen isotope composition of mammalian tooth enamel phosphate: Implications for palaeoclimatological and palaeobiological research. *Palaeogeography, Palaeoclimatology, Palaeoecology*, **126**, 91–99.

Fricke, H.C. & Rogers, R.R. (2000) Multiple taxon-multiple locality approach to providing oxygen isotope evidence for warm-blooded theropod dinosaurs. *Geology*, **28**, 799–802.

Fricke, H.C., Clyde, W.C. & O'Neil, J.R. (1998a) Intra-tooth variations in $\delta^{18}O$ (PO_4) of mammalian tooth enamel as a record of seasonal variations in continental climate variables. *Geochimica et Cosmochimica Acta*, **62**, 1839–1850.

Fricke, H.C., Clyde, W.C., O'Neil, J.R. & Gingerich, P.D. (1998b) Evidence for rapid climate change in North America during the latest Paleocene thermal maximum: oxygen isotope compositions of biogenic phosphate from the Bighorn Basin (Wyoming). *Earth and Planetary Science Letters*, **160**, 193–208.

Fuller, B.T., Richards, M.P. & Mays, S.A. (2003) Stable carbon and nitrogen isotope variations in tooth dentine serial sections from Wharram Percy. *Journal of Archaeological Science*, **30**, 1673–1684.

Fuller, B.T., Fuller, J.L., Sage, N.E., Harris, D.A., O'Connell, T.C. & Hedges, R.E.M. (2004) Nitrogen balance and $\delta^{15}N$: why you're not what you eat during pregnancy. *Rapid Communications in Mass Spectrometry*, **18**, 2889–2896.

Fuller, B.T., Fuller, J.L., Sage, N.E., Harris, D.A., O'Connell, T.C. & Hedges, R.E.M. (2005) Nitrogen balance and $\delta^{15}N$: why you're not what you eat during nutritional stress. *Rapid Communications in Mass Spectrometry*, **19**, 2497–2506.

Gaebler, O.H., Trieste, G.V. & Vukmirovich, R. (1966) Isotope effects in metabolism of ^{14}N and ^{15}N from unlabeled dietary proteins. *Canadian Journal of Biochemistry*, **44**, 1249–1257.

Gannes, L.Z., Martínez del Rio, C. & Koch, P. (1998) Natural abundance variations in stable isotopes and their potential uses in animal physiological ecology. *Comparative Biochemistry and Physiology A–Molecular & Integrative Physiology*, **119**, 725–737.

Gaye-Siessegger, J., Focken, U., Abel, H. & Becker, K. (2004) Individual protein balance strongly influences $\delta^{15}N$ and $\delta^{13}C$ values in Nile tilapia, *Oreochromis niloticus*. *Naturwissenschaften*, **91**, 90–93.

González-Martín, I., González Peréz, C., Hernández Méndez, J. & Sánchez González, C. (2001) Differentiation of dietary regimene of Iberian swine by means of isotopic analysis of carbon and sulphur in hepatic tissue. *Meat Science*, **58**, 25–30.

Grimes, S.T., Collinson, M.E., Hooker, J.J., Mattey, D.P., Grassineau, N.V. & Lowry, D. (2004) Distinguishing the diets of coexisting fossil theridomyid and glirid rodents using carbon isotopes. *Palaeogeography, Palaeoclimatology, Palaeoecology*, **208**, 103–119.

Gussone, N., Eisenhauer, A., Heuser, A., et al. (2003) Model for kinetic effects on calcium isotope fractionation ($\delta^{44}Ca$) in inorganic aragonite and cultured planktonic foraminifera. *Geochimica et Cosmochimica Acta*, **67**, 1375–1382.

Handley, L.L., Austin, A.T., Robinson, D., et al. (1999) The ^{15}N natural abundance ($\delta^{15}N$) of ecosystem samples reflects measures of water availability. *Australian Journal of Plant Physiology*, **26**, 185–199.

Hare, P.E., Fogel, M.L., Stafford, T.W., Mitchell, A.D. & Hoering, T.C. (1991) The isotopic composition of carbon and nitrogen in individual amino acids isolated from modern and fossil proteins. *Journal of Archaeological Science*, **18**, 277–292.

Heaton, T.H.E. (1987) The $^{15}N/^{14}N$ ratios of plants in South Africa and Namibia: relationship to climate and coastal/saline environments. *Oecologia*, **74**, 236–246.

Heaton, T.H.E. (1999) Spatial, species, and temporal variations in the $^{13}C/^{12}C$ ratios of C3 plants: Implications for palaeodiet studies. *Journal of Archaeological Science*, **26**, 637–649.

Hedges, R.E.M. (2002) Bone diagenesis: An overview of processes. *Archaeometry*, **44**, 319–328.

Hedges, R.E.M. (2003) On bone collagen – Apatite–carbonate isotopic relationships. *International Journal of Osteoarchaeology*, **13**, 66–79.

Hedges, R.E.M. & Van Klinken, G.J. (eds) (1995) *Journal of Archaeological Science*, **22**, 147–340.

Hedges, R.E.M. & van Klinken, G.J. (2000) "Consider a spherical cow." – on modeling and diet. In: *Biogeochemical Approaches to Paleodietary Analysis* (Eds S.H. Ambrose & M.A. Katzenberg), pp. 211–241. Kluwer Academic/Plenum Publishers, New York.

Helliker, B.R. & Ehleringer, J.R. (2000) Establishing a grassland signature in veins: ^{18}O in the leaf water of C3 and C4 grasses. *Proceedings of the National Academy of Sciences of the United States of America*, **97**, 7894–7898.

Herring, D.A., Saunders, S.R. & Katzenberg, M.A. (1998) Investigating the weaning process in past populations. *American Journal of Physical Anthropology*, **105**, 425–439.

Higgins, P. & MacFadden, B.J. (2004) "Amount Effect" recorded in oxygen isotopes of Late Glacial horse (*Equus*) and bison (*Bison*) teeth from the Sonoran and Chihuahuan deserts, southwestern United States. *Palaeogeography, Palaeoclimatology, Palaeoecology*, **206**, 337–353.

Hippler, D., Schmitt, A.D., Gussone, N., et al. (2003) Calcium isotopic composition of various reference materials and seawater. *Geostandards Newsletter–the Journal of Geostandards and Geoanalysis*, **27**, 13–19.

Hobbie, E.A., Macko, S.A. & Shugart, H.H. (1998) Patterns in N dynamics and N isotopes during primary succession in Glacier Bay, Alaska. *Chemical Geology*, **152**, 3–11.

Hobson, K.A. (1999) Tracing origins and migration of wildlife using stable isotopes: A review. *Oecologia*, **120**, 314–326.

Hobson, K.A. & Clark, R.G. (1992) Assessing avian diets using stable isotopes II: factors influencing diet–tissue fractionation. *The Condor*, **94**, 189–197.

Hobson, K.A. & Wassenaar, L.I. (1997) Linking brooding and wintering grounds of neotropical migrant songbirds using stable hydrogen isotopic analysis of feathers. *Oecologia*, **109**, 142–148.

Hobson, K.A., Atwell, L. & Wassenaar, L.I. (1999) Influence of drinking water and diet on the stable hydrogen isotope ratios of animal tissues. *Proceedings of the National Academy of Sciences of the United States of America*, **96**, 8003–8006.

Hobson, K.A., Bowen, G.J., Wassenaar, L.I., Ferrand, Y. & Lormee, H. (2004) Using stable hydrogen and oxygen isotope measurements of feathers to infer geographical origins of migrating European birds. *Oecologia*, **141**, 477–488.

Hodell, D.A., Quinn, R.L., Brenner, M. & Kamenov, G. (2004) Spatial variation of strontium isotopes ($^{87}Sr/^{86}Sr$) in the Maya region: a tool for tracking ancient human migration. *Journal of Archaeological Science*, **31**, 585–601.

Högberg, P. (1997) ^{15}N natural abundance in soil–plant systems. *New Phytologist*, **137**, 179–203.

Hoppe, K.A. (2004) Late Pleistocene mammoth herd structure, migration patterns, and Clovis hunting strategies inferred from isotopic analyses of multiple death assemblages. *Paleobiology*, **30**, 129–145.

Hoppe, K.A. & Koch, P.L. (2006) The biogeochemistry of the Aucilla River Fauna. In: *First Floridians and Last Mastodons: The Page-Ladson Site in the Aucilla River* (Ed. S.D. Webb), pp. 379–401. Springer-Verlag, Dordrecht.

Hoppe, K.A., Koch, P.L., Carlson, R.W. & Webb, S.D. (1999) Tracking mammoths and mastodons: Reconstruction of migratory behavior using strontium isotope ratios. *Geology*, **27**, 439–442.

Hoppe, K.A., Koch, P.L. & Furutani, T.T. (2003) Assessing the preservation of biogenic strontium in fossil bones and tooth enamel. *International Journal of Osteoarchaeology*, **13**, 20–28.

Hoppe, K.A., Stover, S.M., Pascoe, J.R. & Amundson, R. (2004) Tooth enamel biomineralization in extant horses: implications for isotopic microsampling. *Palaeogeography Palaeoclimatology Palaeoecology*, **206**, 355–365.

Howland, M.R., Corr, L.T., Young, S.M.M., et al. (2003) Expression of the dietary isotope signal in the compound-specific $\delta^{13}C$ values of pig bone lipids and amino acids. *International Journal of Osteoarchaeology*, **13**, 54–65.

Iacumin, P., Bocherens, H. & Huertas, A.D. (1997) A stable isotope study of fossil mammal remains from the Paglicci cave, Southern Italy. N and C as palaeoenvironmental indicators. *Earth and Planetary Science Letters*, **148**, 349–357.

Ingram, B.L & Weber, P.K. (1999) Salmon origin in California's Sacramento–San Joaquin river system as determined by otolith strontium isotopic composition. *Geology*, **27**, 851–854.

Ivany, L.C., Patterson, W.P. & Lohmann, K.C. (2000) Cooler winters as a possible cause of mass extinction at the Eocene/Oligocene boundary. *Nature*, **407**, 887–890.

Jenkins, S.G., Partridge, S.T., Stephenson, T.R., Farley, S.D. & Robbins, C.T. (2001) Nitrogen and carbon isotope fractionation between mothers, neonates, and nursing offspring. *Oecologia*, **129**, 336–341.

Jim, S., Ambrose, S.H. & Evershed, R.P. (2004) Stable carbon isotopic evidence for differences in the dietary origin of bone cholesterol, collagen and apatite: Implications for their use in palaeodietary reconstruction. *Geochimica et Cosmochimica Acta*, **68**, 61–72.

Katzenberg, M.A., Saunders, S.R. & Fitzgerald, W.R. (1993) Age-differences in stable carbon and nitrogen isotope ratios in a population of prehistoric maize horticulturists. *American Journal of Physical Anthropology*, **90**, 267–281.

Kelly, J.F. (2000) Stable isotopes of carbon and nitrogen in the study of avian and mammalian trophic ecology. *Canadian Journal of Zoology*, **78**, 1–27.

Kelly, J.F., Atudorei, V., Sharp, Z.D. & Finch, D.M. (2002) Insights into Wilson's Warbler migration from analyses of hydrogen stable-isotope ratios. *Oecologia*, **130**, 216–221.

Kennedy, B.P., Blum, J.D, Folt, C.L. & Nislow, K.H. (2000) Using natural strontium isotopic signatures as fish markers: Methodology and application. *Canadian Journal of Fisheries and Aquatic Sciences*, **57**, 2280–2292.

Knudson, K.J., Tung, T.A., Nystrom, K.C., Price, T.D. & Fullagar, P.D. (2005) The origin of the Juch'uypampa Cave mummies: strontium isotope analysis of archaeological human remains from Bolivia. *Journal of Archaeological Science*, **32**, 903–913.

Koch, P.L. (1997) Nitrogen isotope ecology of carnivores and herbivores. *Journal of Vertebrate Paleontology*, **17 (Suppl.)**, 57A.

Koch, P.L. (1998) Isotopic reconstruction of past continental environments. *Annual Review of Earth and Planetary Sciences*, **26**, 573–613.

Koch, P.L. & Burton, J. (2003) Advanced Seminar on Paleodiet, University of California, Santa Cruz, September 2001 – Preface. *International Journal of Osteoarchaeology*, **13**, 1–2.

Koch, P.L., Fisher, D.C. & Dettman, D. (1989) Oxygen isotope variation in the tusks of extinct proboscideans – a measure of season of death and seasonality. *Geology*, **17**, 515–519.

Koch, P.L., Halliday, A.N., Walter, L.M., Stearley, R.F., Huston, T.J. & Smith, G.R. (1992) Sr isotopic composition of hydroxyapatite from recent and fossil salmon – the record of lifetime migration and diagenesis. *Earth and Planetary Science Letters*, **108**, 277–287.

Koch, P.L., Fogel, M.L. & Tuross, N. (1994) Tracing the diets of fossil animals using stable isotopes. In: Lajtha, K. & Michener, R.H. (eds.) *Stable Isotopes in Ecology and Environmental Science*. Blackwell Scientific Publications, Boston, pp. 63–92.

Koch, P.L., Heisinger, J., Moss, C., Carlson, R.W., Fogel, M.L. & Behrensmeyer, A.K. (1995) Isotopic tracking of change in diet and habitat use in African elephants. *Science*, **267**, 1340–1343.

Koch, P.L., Tuross, N. & Fogel, M.L. (1997) The effects of sample treatment and diagenesis on the isotopic integrity of carbonate in biogenic hydroxylapatite. *Journal of Archaeological Science*, **24**, 417–429.

Koch, P.L., Hoppe, K.A. & Webb, S.D. (1998) The isotopic ecology of late Pleistocene mammals in North America – Part 1. Florida. *Chemical Geology*, **152**, 119–138.

Koch, P.L., Behrensmeyer, A.K., Stott, A.W., Tuross, N., Evershed, R.P. & Fogel, M.L. (2001) The effects of weathering on the stable isotope composition of bones. *Ancient Biomolecules*, **3**, 117–134.

Koch, P.L., Diffenbaugh, N.S. & Hoppe, K.A. (2004) The effects of late Quaternary climate and pCO_2 change on C4 plant abundance in the south–central United States. *Palaeogeography Palaeoclimatology Palaeoecology*, **207**, 331–357.

Kohn, M.J. & Cerling, T.E. (2002) Stable isotope compositions of biological apatite. *Phosphates: Geochemical, Geobiological, and Materials Importance. Reviews in Mineralogy & Geochemistry*, **48**, 455–488.

Kohn, M.J., Schoeninger, M.J. & Valley, J.W. (1996) Herbivore tooth oxygen isotope compositions: effects of diet and physiology. *Geochimica et Cosmochimica Acta*, **60**, 3889–3896.

Kohn, M.J., Schoeninger, M.J. & Valley, J.W. (1998) Variability in oxygen isotope compositions of herbivore teeth: reflections of seasonality or developmental physiology? *Chemical Geology*, **152**, 97–112.

Kohn, M.J., McKay, M.P. & Knight, J.L. (2005) Dining in the Pleistocene – Who's on the menu? *Geology*, **33**, 649–652.

Kolodny, Y., Luz, B., Sander, M. & Clemens, W.A. (1996) Dinosaur bones: Fossils or pseudomorphs? The pitfalls of physiology reconstruction from apatitic fossils. *Palaeogeography, Palaeoclimatology, Palaeoecology*, **126**, 161–171.

Krouse, H.R. (1989) Sulfur isotope studies of the pedosphere and biosphere. In: *Stable Isotopes in Ecological Research* (Eds P.W. Rundel, J.R. Ehleringer & K.A. Nagy), pp. 424–444. Ecological Studies, Vol. 68, Springer-Verlag, New York.

Lambert, J.B. & Grupe, G. (1993) *Prehistoric Human Bone: Archaeology at the Molecular Level*. Springer-Verlag, Berlin.

Lécuyer, C., Picard, S., Garcia, J.P., Sheppard, S.M.F., Grandjean, P. & Dromart, G. (2003) Thermal evolution of Tethyan surface waters during the Middle–Late Jurassic: Evidence from $\delta^{18}O$ values of marine fish teeth. *Paleoceanography*, **18**, 1076, doi:10.1029/2002PA000863.

Lee-Thorp, J. & Sponheimer, M. (2003) Three case studies used to reassess the reliability of fossil bone and enamel isotope signals for paleodietary studies. *Journal of Anthropological Archaeology*, **22**, 208–216.

LeGeros, R.Z. (1991) *Calcium Phosphates in Oral Biology and Medicine*. Karger, Paris.

Lemarchand, D., Wasserburg, G.T. & Papanastassiou, D.A. (2004) Rate-controlled calcium isotope fractionation in synthetic calcite. *Geochimica et Cosmochimica Acta*, **68**, 4665–4678.

Lott, C.A., Meehan, T.D. & Heath, J.A. (2003) Estimating the latitudinal origins of migratory birds using hydrogen and sulfur stable isotopes in feathers: influence of marine prey base. *Oecologia*, **134**, 505–510.

Lowenstam, H.A. & Weiner, S. (1989) *On biomineralization*, New York, Oxford University Press.

MacFadden, B.J. (1998) Tale of two rhinos: Isotopic ecology, paleodiet, and niche differentiation of *Aphelops* and *Teleoceras* from the Florida Neogene. *Paleobiology*, **24**, 274–286.

MacFadden, B.J. (2004) Incremental growth in vertebrate skeletal tissues: paleobiological and paleoenviromental implications. *Palaeogeography, Palaeoclimatology, Palaeoecology*, **206**, 177–177.

MacFadden, B.J. & Higgins, P. (2004) Ancient ecology of 15-million-year-old browsing mammals within C3 plant communities from Panama. *Oecologia*, **140**, 169–182.

MacFadden, B.J., Wang, Y., Cerling, T.E. & Anaya, F. (1994) South American fossil mammals and carbon isotopes – A 25-million-year sequence from the Bolivian Andes. *Palaeogeography Palaeoclimatology Palaeoecology*, **107**, 257–268.

MacFadden, B.J., Solounias, N. & Cerling, T.E. (1999) Ancient diets, ecology, and extinction of 5-million-year-old horses from Florida. *Science*, **283**, 824–827.

MacFadden, B.J., Higgins, P., Clementz, M.T. & Jones, D.S. (2004a) Diets, habitat preferences, and niche differentiation of Cenozoic sirenians from Florida: evidence from stable isotopes. *Paleobiology*, **30**, 297–324.

MacFadden, B.J., Labs-Hochstein, J., Quitmyer, I. & Jones, D.S. (2004b) Incremental growth and diagenesis of skeletal parts of the lamnoid shark *Otodus obliquus* from the early Eocene (Ypresian) of Morocco. *Palaeogeography, Palaeoclimatology, Palaeoecology*, **206**, 179–192.

Macko, S.A., Engel, M.H., Andrusevich, V., Lubec, G., O'Connell, T.C. & Hedges, R.E.M. (1999) Documenting the diet in ancient human populations through stable isotope analysis of hair. *Philosophical Transactions of the Royal Society of London Series B–Biological Sciences*, **354**, 65–75.

Marriott, C.S., Henderson, G.M., Belshaw, N.S. & Tudhope, A.W. (2004) Temperature dependence of δ^7Li, $\delta^{44}Ca$ and Li/Ca during growth of calcium carbonate. *Earth and Planetary Science Letters*, **222**, 615–624.

Martill, D.M. (1995) An ichthyosaur with preserved soft tissue from the Sinemurian of southern England. *Palaeontology*, **38**, 897–903.

McClelland, J.W. & Montoya, J.P. (2002) Trophic relationships and the nitrogen isotopic composition of amino acids in plankton. *Ecology*, **83**, 2173–2180.

McClelland, J.W., Holl, C.M. & Montoya, J.P. (2003). Relating low d[15]N values of zooplankton to N2-fixation in the tropical North Atlantic: insights provided by stable isotope ratios of amino acids. *Deep-Sea Research I*, **50**, 849–861.

McKechnie, A.E., Wolf, B.O. & Martínez del Rio, C.M. (2004) Deuterium stable isotope ratios as tracers of water resource use: an experimental test with rock doves. *Oecologia*, **140**, 191–200.

Metges, C.C., Daenzer, M., Petzke, K.J. & Elsner, A. (2002) Low-abundance plasma and urinary ^{15}N urea enrichments analyzed by gas chromatography/combustion/isotope ratio mass spectrometry. *Journal of Mass Spectrometry*, **37**, 489–494.

Michelsen, A., Quarmby, C., Sleep, D. & Jonasson, S. (1998) Vascular plant ^{15}N natural abundance in heath and forest tundra ecosystems is closely correlated with presence and type of mycorrhizal fungi in roots. *Oecologia*, **115**, 406–418.

Michner, R.H. & Schell, D.M. (1994) Stable isotope ratios as tracers in marine aquatic food webs. In: *Stable Isotopes in Ecology and Environmental Science* (Eds K. Lajtha & R.H. Michener), pp. 138–157. Blackwell Scientific Publications, Boston.

Mikhailov, K.E., Bray, E.S. & Hirsch, K.E. (1996) Parataxonomy of fossil egg remains (veterovata): Principles and applications. *Journal of Vertebrate Paleontology*, **16**, 763–769.

Miller, G.H., Fogel, M.L., Magee, J.W., Gagan, M.K, Clarke, S.J. & Johnson, B.J. (2005) Ecosystem collapse in Pleistocene Australia and a human role in megafaunal extinction. *Science*, **309**, 287–290.

Montgomery, J., Evans, J.A., Powlesland, D. & Roberts, C.A. (2005) Continuity or colonization in Anglo-Saxon England? Isotope evidence for mobility, subsistence practice, and status at West Heslerton. *American Journal of Physical Anthropology*, **126**, 123–138.

Müller, W. Fricke, H., Halliday, A.N., McCulloch, M.T. & Wartho, J.A. (2003) Origin and migration of the Alpine Iceman. *Science*, **302**, 862–866.

Nadelhoffer, K.J. & Fry, B. (1994) Nitrogen isotopic studies in forest ecosystems. In: *Stable Isotopes in Ecology and Environmental Science* (Eds K. Lajtha & R. Michener), pp. 22–44. Blackwell Scientific Publications, Boston.

Nadelhoffer, K., Shaver, G., Fry, B., Giblin, A., Johnson, L. & McKane, R. (1996) ^{15}N natural abundances and N use by tundra plants. *Oecologia*, **107**, 386–394.

Nelson, B.K., DeNiro, M.J., Schoeninger, M.J., DePaolo, D.J. & Hare, P.E. (1986) Effects of diagenesis on strontium, carbon, nitrogen and oxygen concentration and isotopic composition of bone. *Geochimica et Cosmochimica Acta*, **50**, 1941–1949.

Newsome, S.D., Etnier, M.A., Gifford-Gonzalez, D., et al. (in press) The shifting baseline of northern fur seal (*Callorhinus ursinus*) ecology in the northeastern Pacific Ocean. *Proceedings of The National Academy of Sciences of the United States of America*, **104**.

Nielsen-Marsh, C.M. & Hedges, R.E.M. (2000) Patterns of diagenesis in bone I: The effects of site environments. *Journal of Archaeological Science*, **27**, 1139–1150.

O'Connell, T.C., Hedges, R.E.M., Healey, M.A. & Simpson, A.H.R. (2001) Isotopic comparison of hair, nail and bone: Modern analyses. *Journal of Archaeological Science*, **28**, 1247–1255.

Olive, P.J.W., Pinnegar, J.K., Polunin, N.V.C., Richards, G. & Welch, R. (2003) Isotope trophic-step fractionation: a dynamic equilibrium model. *Journal of Animal Ecology*, **72**, 608–617.

Ostrom, P.H., Zonneveld, J.P. & Robbins, L.L. (1994) Organic geochemistry of hard parts – Assessment of isotopic variability and indigeneity. *Palaeogeography, Palaeoclimatology, Palaeoecology*, **107**, 201–212.

Padian, K., Horner, J.R. & De Ricqles, A. (2004) Growth in small dinosaurs and pterosaurs: The evolution of archosaurian growth strategies. *Journal of Vertebrate Paleontology*, **24**, 555–571.

Palombo, M.R., Filippi, M.L., Iacumin, P., Longinelli, A., Barbieri, M. & Maras, A. (2005) Coupling tooth microwear and stable isotope analyses for palaeodiet reconstruction: the case study of late Middle Pleistocene *Elephas* (*Palaeoloxodon*) *antiquus* teeth from Central Italy (Rome area). *Quaternary International*, **126–28**, 153–170.

Panella, G. (1980) Growth patterns in fish sagittae. In: *Skeletal Growth of Aquatic Organisms* (Eds D.C. Rhoads & R.A. Lutz), pp. 519–560. Plenum, New York.

Parker, K.L., Barboza, P.S. & Stephenson, T.R. (2005) Protein conservation in female caribou (*Rangifer tarandus*): Effects of decreasing diet quality during winter. *Journal of Mammalogy*, **86**, 610–622.

Passey, B.H. & Cerling, T.E. (2002) Tooth enamel mineralization in ungulates: Implications for recovering a primary isotopic time-series. *Geochimica et Cosmochimica Acta*, **66**, 3225–3234.

Passey, B.H., Robinson, T.F., Ayliffe, L.K., et al. (2005) Carbon isotopic fractionation between diet, breath, and bioapatite in different mammals. *Journal of Archaeological Science*, **32**, 1459–1470.

Pate, F.D. (1994) Bone chemistry and paleodiet. *Journal of Archaeological Method and Theory*, **1**, 161–209.

Patterson, W.P. (1998) North American continental seasonality during the last millennium: high-resolution analysis of sagittal otoliths. *Palaeogeography, Palaeoclimatology, Palaeoecology*, **138**, 271–303.

Patterson, W.P. (1999) Oldest isotopically characterized fish otoliths provide insight to Jurassic continental climate of Europe. *Geology*, **27**, 199–202.

Pearson, S.F., Levey, D.J., Greenberg, C.H. & Martínez del Rio, C.M. (2003) Effects of elemental composition on the incorporation of dietary nitrogen and carbon isotopic signatures in an omnivorous songbird. *Oecologia*, **135**, 516–523.

Peterson, B.J. & Fry, B. (1987) Stable isotopes in ecosystem studies. *Annual Review of Ecology and Systematics*, **18**, 293–320.

Polischuk, S.C., Hobson, K.A. & Ramsay, M.A. (2001) Use of stable carbon and nitrogen isotopes to assess weaning and fasting in female polar bears and their cubs. *Canadian Journal of Zoology*, **79**, 499–511.

Price, T.D. (ed.) (1989) *The Chemistry of Prehistoric Human Bone*, Cambridge, Cambridge University Press.

Price, T.D., Grupe, G. & Schrotter, P. (1994) Reconstruction of migration patterns in the Bell Beaker Period by stable strontium isotope analysis. *Applied Geochemistry*, **9**, 413–417.

Price, T.D., Manzanilla, L. & Middleton, W.D. (2000) Immigration and the ancient city of Teothihuacan in Mexico: a study using strontium isotope ratios in human bone and teeth. *Journal of Archaeological Science*, **27**, 903–913.

Price, T.D., Burton, J.H. & Bentley, R.A. (2002) The characterization of biologically available strontium isotope ratios for the study of prehistoric migration. *Archaeometry*, **44**, 117–135.

Rau, G.H., Takahashi, T., DesMarais, D.J. & Sullivan, C.W. (1991) Particulate organic-matter $\delta^{13}C$ variations across the Drake Passage. *Journal of Geophysical Research-Oceans*, **96**, 15131–15135.

Reid, R.E.H. (1987) Bone and dinosaurian "endothermy". *Modern Geology*, **11**, 133–154.

Richards, M.P. & Hedges, R.E.M. (2003) Variations in bone collagen $\delta^{13}C$ and $\delta^{15}N$ values of fauna from Northwest Europe over the last 40 000 years. *Palaeogeography Palaeoclimatology Palaeoecology*, **193**, 261–267.

Richards, M.P., Fuller, B.T., Sponheimer, M., Robinson, T. & Ayliffe, L. (2003) Sulphur isotopes in palaeodietary studies: A review and results from a controlled feeding experiment. *International Journal of Osteoarchaeology*, **13**, 37–45.

Roe, L.J., Thewissen, J.G.M., Quade, J., et al. (1998) Isotopic approaches to understanding the terrestrial-to-marine transition of the earliest cetaceans. In: *The Emergence of Whales* (Ed. J.G.M. Thewissen), pp. 399–422. Plenum Press, New York.

Roth, J.D. & Hobson, K.A. (2000) Stable carbon and nitrogen isotopic fractionation between diet and tissue of captive red fox: Implications for dietary reconstruction. *Canadian Journal of Zoology*, **78**, 848–852.

Schell, D.M., Barnett, B.A. & Vinette, K.A. (1998) Carbon and nitrogen isotope ratios in zooplankton of the Bering, Chukchi and Beaufort seas. *Marine Ecology Progress Series*, **162**, 11–23.

Schimmelmann, A., Miller, R.F., Leavitt, S.W. (1993) Hydrogen isotopic exchange and stable isotope ratios in cellulose, wood, chitin, and amino compounds. In: *Climate Change in Continental Isotope Records* (eds. P.K. Swart, K.C. Lohmann, J. McKenzie, & S. Savin), pp. 367–374. Geophysical Monograph 68, American Geophysical Union: Washington, DC.

Schoeller, D.A. (1999) Isotope fractionation: Why aren't we what we eat? *Journal of Archaeological Science*, **26**, 667–673.

Schoeninger, M.J. (1985) Trophic level effects on $^{15}N/^{14}N$ and $^{13}C/^{12}C$ ratios in bone collagen and strontium levels in bone mineral. *Journal of Human Evolution*, **14**, 515–525.

Schoeninger, M.J. & Moore, K. (1992) Bone stable isotope studies in archaeology. *Journal of World Prehistory*, **6**, 247–296.

Schulze, E.D., Chapin, F.S. & Gebauer, G. (1994) Nitrogen nutrition and isotope differences among life forms at the northern treeline of Alaska. *Oecologia*, **100**, 406–412.

Schurr, M.R. & Powell, M.L. (2005) The role of changing childhood diets in the prehistoric evolution of food production: An isotopic assessment. *American Journal of Physical Anthropology*, **126**, 278–294.

Schwarcz, H.P. & Schoeninger, M.J. (1991) Stable isotope analyses in human nutritional ecology. *Yearbook of Physical Anthropology*, **34**, 283–321.

Schwarcz, H.P., Hedges, R.E.M. & Ivanovich, M. (1989) Editorial comments on the First International Workshop on Fossil Bone. *Applied Geochemistry*, **4**, 211–213.

Schwarcz, H.P., Dupras, T.L. & Fairgrieve, S.I. (1999) ^{15}N-enrichment in the Sahara: In search of a global relationship. *Journal of Archaeological Science*, **26**, 629–636.

Schweissing, M.M. & Grupe, G. (2003) Stable strontium isotopes in human teeth and bone: a key to migration events of the late Roman period in Bavaria. *Journal of Archaeological Science*, **30**, 1373–1383.

Schweitzer, M.H., Watt, J.A., Avci, R., et al. (1999) Keratin immunoreactivity in the Late Cretaceous bird *Rahonavis ostromi*. *Journal of Vertebrate Paleontology*, **19**, 712–722.

Schweitzer, M.H., Wittmeyer, J.L., Horner, J.R. & Toporski, J.K. (2005) Soft-tissue vessels and cellular preservation in *Tyrannosaurus rex*. *Science*, **307**, 1952–1955.

Scrimgeour, C.M., Begley, I.S. & Thomason, M.L. (1999) Measurement of deuterium incorporation into fatty acids by gas chromatography isotope ratio mass spectrometry. *Rapid Communications in Mass Spectrometry*, **13**, 271–274.

Sealy, J.C., van der Merwe, N.J., Lee-Thorp, J.A. & Lanham, J.L. (1987) Nitrogen isotopic ecology in southern Africa: implications for environmental and dietary tracing. *Geochimica et Cosmochimica Acta*, **51**, 2707–2717.

Sealy, J.C., van der Merwe, N.J., Sillen, A., Kruger, F.J. & Krueger, H.W. (1991) $^{87}Sr/^{86}Sr$ as a dietary indicator in modern and archaeological Bone. *Journal of Archaeological Science*, **18**, 399–416.

Segalen, L., Renard, M., Pickford, M., et al. (2002) Environmental and climatic evolution of the Namib Desert since the Middle Miocene: the contribution of carbon isotope ratios in ratite eggshells. *Comptes Rendus Geoscience*, **334**, 917–924.

Sharp, Z.D. & Cerling, T.E. (1996) A laser GC-IRMS technique for in situ stable isotope analyses of carbonates and phosphates. *Geochimica et Cosmochimica Acta*, **60**, 2909–2916.

Sharp, Z.D. & Cerling, T.E. (1998) Fossil isotope records of seasonal climate and ecology: Straight from the horse's mouth. *Geology*, **26**, 219–222.

Sharp, Z.D., Atudorei, V., Panarello, H.O., Fernandez, J. & Douthitt, C. (2003) Hydrogen isotope systematics of hair: archeological and forensic applications. *Journal of Archaeological Science*, **30**, 1709–1716.

Sholto-Douglas, A.D., Field, J.G., James, A.G. & van der Merwe, N.J. (1991) $^{13}C/^{12}C$ and $^{15}N/^{14}N$ isotope ratios in the southern Benguela Ecosystem – indicators of food web relationships among different size classes of plankton and pelagic fish – differences between fish muscle and bone collagen tissues. *Marine Ecology–Progress Series*, **78**, 23–31.

Sillen, A. & Armelagos, G. (1991) 2nd Advanced Seminar in Paleodietary Research – Introduction. *Journal of Archaeological Science*, **18**, 225–226.

Sillen, A., Hall, G., Richardson, S. & Armstrong, R. (1998) $^{87}Sr/^{86}Sr$ ratios in modern and fossil food-webs of the Sterkfontein Valley: Implications for early hominid habitat preference. *Geochimica et Cosmochimica Acta*, **62**, 2463–2473.

Simkiss, K. & Wilbur, K.M. (1989) *Biomineralization: Cell Biology and Mineral Deposition*. Academic Press, San Diego, CA.

Skulan, J. & DePaolo, D.J. (1999) Calcium isotope fractionation between soft and mineralized tissues as a monitor of calcium use in vertebrates. *Proceedings of the National Academy of Sciences of the United States of America*, **96**, 13709–13713.

Skulan, J., DePaolo, D.J. & Owens, T.L. (1997) Biological control of calcium isotopic abundances in the global calcium cycle. *Geochimica et Cosmochimica Acta*, **61**, 2505–2510.

Smith, S.J., Iverson, S.J. & Bowen, W.D. (1997) Fatty acid signatures and classification trees: new tools for investigating the foraging ecology of seals. *Canadian Journal of Fisheries and Aquatic Sciences*, **54**, 1377–1386.

Sponheimer, M. & Lee-Thorp, J.A. (2003) Differential resource utilization by extant great apes and australopithecines: towards solving the C4 conundrum. *Comparative Biochemistry and Physiology A–Molecular & Integrative Physiology*, **136**, 27–34.

Sponheimer, M., Robinson, T., Ayliffe, L., et al. (2003a) An experimental study of carbon-isotope fractionation between diet, hair, and feces of mammalian herbivores. *Canadian Journal of Zoology*, **81**, 871–876.

Sponheimer, M., Robinson, T., Ayliffe, L., et al. (2003b) Nitrogen isotopes in mammalian herbivores: Hair $\delta^{15}N$ values from a controlled feeding study. *International Journal of Osteoarchaeology*, **13**, 80–87.

Sponheimer, M., Robinson, T., Roeder, B.L et al. (2003c) An experimental study of nitrogen flux in llamas: is ^{14}N preferentially excreted? *Journal of Archaeological Science*, **30**, 1649–1655.

Stern, L.A., Johnson, G.D. & Chamberlain, C.P. (1994) Carbon isotope signature of environmental change found in fossil ratite eggshells from a south Asian Neogene sequence. *Geology*, **22**, 419–422.

Sternberg, L.D.L. (1989) Oxygen and hydrogen isotope ratios in plant cellulose: mechanisms and applications. In: *Stable Isotopes in Ecological Research* (Eds P.W. Rundel, J.R. Ehleringer & K.A. Nagy), pp. 124–141. Ecological Studies, Vol. 68, Springer-Verlag, New York.

Stoskopf, M.K, Barrick, R.E. & Showers, W.J. (2001) Oxygen isotope variability in bones of wild caught and constant temperature reared sub-adult American alligators. *Journal of Thermal Biology*, **26**, 183–191.

Stott, A.W., Davies, E., Evershed, R.P. & Tuross, N. (1997) Monitoring the routing of dietary and biosynthesised lipids through compound-specific stable isotope ($\delta^{13}C$) measurements at natural abundance. *Naturwissenschaften*, **84**, 82–86.

Stott, A.W., Evershed, R.P., Jim, S., et al. (1999) Cholesterol as a new source of palaeodietary information: Experimental approaches and archaeological applications. *Journal of Archaeological Science*, **26**, 705–716.

Straight, W.H., Barrick, R.E. & Eberth, D.A. (2004) Reflections of surface water, seasonality and climate in stable oxygen isotopes from tyrannosaurid tooth enamel. *Palaeogeography Palaeoclimatology Palaeoecology*, **206**, 239–256.

Stuart-Williams, H.L. & Schwarcz, H.P. (1997) Oxygen isotopic determination of climatic variation using phosphate from beaver bone, tooth enamel, and dentine. *Geochimica et Cosmochimica Acta*, **61**, 2539–2550.

Swap, R.J., Aranibar, J.N., Dowty, P.R, Gilhooly, W.P. & Macko, S.A. (2004) Natural abundance of ^{13}C and ^{15}N in C3 and C4 vegetation of southern Africa: patterns and implications. *Global Change Biology*, **10**, 350–358.

Teschler-Nicola, M., Gerold, F., Bujatti-Narbeshuber, M., et al. (1999) Evidence of genocide 7000 BP – Neolithic paradigm and geo-climatic reality. *Collegium Antropologicum*, **23**, 437–450.

Thomas, K.J.S. & Carlson, S.J. (2004) $\delta^{18}O$ and $\delta^{13}C$ isotopic analysis of an ontogenetic series of the hadrosaurid dinosaur *Edmontosaurus*: implications for physiology and ecology. *Palaeogeography Palaeoclimatology Palaeoecology*, **206**, 257–287.

Thorrold, S.R., Campana, S.E., Jones, C.M. & Swart, P.K. (1997) Factors determining $\delta^{13}C$ and $\delta^{18}O$ fractionation in aragonitic otoliths of marine fish. *Geochimica et Cosmochimica Acta*, **61**, 2909–2919.

Tieszen, L.L. (1991) Natural variations in the carbon isotope values of plants: implications for archaeology, ecology, and paleoecology. *Journal of Archaeological Science*, **18**, 227–248.

Tieszen, L.L. & Fagre, T. (1993) Effect of diet quality and composition on the isotopic composition of respiratory CO_2, bone collagen, bioapatite, and soft tissues. In: Lambert, J.B. & Grupe, G. (eds.) *Prehistoric Human Bone: Archaeology at the Molecular Level*. Springer-Verlag, New York, pp. 121–155.

Trueman, C.N. & Tuross, N. (2002) Trace elements in recent and fossil bone apatite. *Phosphates: Geochemical, Geobiological, and Materials Importance. Reviews in Mineralogy & Geochemistry*, **48**, 489–521.

Trueman, C.N.G., Behrensmeyer, A.K., Tuross, N. & Weiner, S. (2004) Mineralogical and compositional changes in bones exposed on soil surfaces in Amboseli National Park, Kenya: diagenetic mechanisms and the role of sediment pore fluids. *Journal of Archaeological Science*, **31**, 721–739.

Tuross, N., Fogel, M.L. & Hare, P.E. (1988) Variability in the preservation of the isotopic composition of collagen from fossil bone. *Geochimica et Cosmochimica Acta*, **52**, 929–935.

Tütken, T., Pfretzschner, H.U., Vennemann, T.W., Sun, G. & Wang, Y.D. (2004) Paleobiology and skeletochronology of Jurassic dinosaurs: implications from the histology and oxygen isotope compositions of bones. *Palaeogeography, Palaeoclimatology, Palaeoecology*, **206**, 217–238.

Van der Merwe, N.J. (1982) Carbon isotopes, photosynthesis, and archaeology. *American Scientist*, **70**, 596–606.

Vennemann, T.W., Hegner, E., Cliff, G. & Benz, G.W. (2001) Isotopic composition of recent shark teeth as a proxy for environmental conditions. *Geochimica et Cosmochimica Acta*, **65**, 1583–1599.

Voss, M., Altabet, M.A. & Bodungen, B.V. (1996) $\delta^{15}N$ in sedimenting particles as indicator of euphotic-zone processes. *Deep-Sea Research Part I–Oceanographic Research Papers*, **43**, 33–47.

Weber, P.K., Hutcheon, I.D., McKeegan, K.D. & Ingram, B.L. (2002) Otolith sulfur isotope method to reconstruct salmon (*Oncorhynchus tshawytscha*) life history. *Canadian Journal of Fisheries and Aquatic Sciences*, **59**, 587–591.

Weidman, C.R. & Millner, R. (2000) High-resolution stable isotope records from North Atlantic cod. *Fisheries Research*, **46**, 327–342.

White, C.D., Storey, R., Longstaffe, F.J. & Spence, M.W. (2004) Immigration, assimilation, and status in the ancient city of Teotihuacan: Stable isotopic evidence from Tlajinga 33. *Latin American Antiquity*, **15**, 176–198.

Wiedemann, F.B., Bocherens, H., Mariotti, A., Von Den Driesch, A. & Grupe, G. (1999) Methodological and archaeological implications of intra-tooth isotopic variations ($\delta^{13}C$, $\delta^{18}O$) in herbivores from Ain Ghazal (Jordan, Neolithic). *Journal of Archaeological Science*, **26**, 697–704.

Wolf, B.O. & Martínez del Rio, C. (2000) Use of saguaro fruit by white-winged doves: isotopic evidence of a tight ecological association. *Oecologia*, **124**, 536–543.

Wolf, B.O. & Martínez del Rio, C. (2003) How important are columnar cacti as sources of water and nutrients for desert consumers? A review. *Isotopes in Environmental and Health Studies*, **39**, 53–67.

Wooller, M.J. & O'Brien, D.M. (2004) Studying the Artic 'migration' of two university professors using stable oxygen and hydrogen isotope analyses. *Fourth International Conference on Applications of Stable Isotope Techniques to Ecologial Studies*, Wellington, New Zealand, p. 155.

Wright, L.E. (2005) Identifying immigrants to Tikal, Guatemala: Defining local variability in strontium isotope ratios of human tooth enamel. *Journal of Archaeological Science*, **32**, 555–566.

Wright, L.E. & Schwarcz, H.P. (1999) Correspondence between stable carbon, oxygen and nitrogen isotopes in human tooth enamel and dentine: Infant diets at Kaminaljuyu. *Journal of Archaeological Science*, **26**, 1159–1170.

Wurster, C.M. & Patterson, W.P. (2001) Late Holocene climate change for the eastern interior United States: evidence from high-resolution $\delta^{18}O$ values of sagittal otoliths. *Palaeogeography, Palaeoclimatology, Palaeoecology*, **170**, 81–100.

Wurster, C.M. & Patterson, W.P. (2003) Metabolic rate of late Holocene freshwater fish: evidence from $\delta^{13}C$ values of otoliths. *Paleobiology*, **29**, 492–505.

Zazzo, A., Lécuyer, C. & Mariotti, A. (2004a) Experimentally-controlled carbon and oxygen isotope exchange between bioapatites and water under inorganic and microbially-mediated conditions. *Geochimica et Cosmochimica Acta*, **68**, 1–12.

Zazzo, A., Lécuyer, C., Sheppard, S.M.F., Grandjean, P. & Mariotti, A. (2004b) Diagenesis and the reconstruction of paleoenvironments: A method to restore original $\delta^{18}O$ values of carbonate and phosphate from fossil tooth enamel. *Geochimica et Cosmochimica Acta*, **68**, 2245–2258.

Zazzo, A., Balasse, M., Patterson, W.P. & Patterson, P. (2005) High-resolution $\delta^{13}C$ intra-tooth profiles in bovine enamel: Implications for mineralization pattern and isotopic attenuation. *Geochimica et Cosmochimica Acta*, **69**, 3631–3642.

Ziegler, H. (1989) Hydrogen isotope fractionation in plant tissues. In: *Stable Isotopes in Ecological Research* (eds P.W. Rundel, J.R. Ehleringer & K.A. Nagy), pp. 105–123. Ecological Studies, Vol. 68, Springer-Verlag, New York.

CHAPTER 6

Isotopic tracking of migrant wildlife

KEITH A. HOBSON

Introduction

Animal movement, including dispersal and migration, is a phenomenon that captures the imagination of scientist and layperson alike. In the Northern Hemisphere, we are perhaps most familiar with the annual movements of birds and much research has been devoted to understanding proximate and ultimate mechanisms that trigger these impressive patterns in nature (Berthold et al. 2003). However, tracking animal movements, including those of insects, fish, and mammals, is also an endeavor important to both theoretical and conservation disciplines. Unfortunately, we have been severely restricted in our ability to follow vagile organisms using conventional techniques that rely on extrinsic markers. In addition, many organisms of interest are simply too small to hold satellite or radio transmitters for direct monitoring (reviewed by Hobson 2003). All studies that rely on the initial capture and ultimate recapture of an individual also are typically subject to sampling bias, especially for those species with widespread distributions, the so-called "needle in a haystack" problem. In contrast, animal tracking methods that rely on intrinsic markers are dependent only on the "recovery" sampling sites. Hobson (2005a) referred to this advantage as "every capture is a recapture". The measurement of naturally occurring stable isotopes in animal tissues represents one means of assaying intrinsic markers in animals in order to infer information on their origins. Few areas involving the application of stable isotope methods to ecological studies have seen such an explosive growth as the field of animal tracking, a consequence of the fundamental limitations of conventional methods involving extrinsic markers (Hobson 2005b).

Rubenstein & Hobson (2004) provided a review of the relative advantages of using intrinsic vs. extrinsic markers to track animal movements. Intrinsic biogeochemical markers include trace elements (Szép et al. 2003) as well as stable isotopes and each have advantages and disadvantages. As with trace elements, the use of stable isotopes is based on the fact that concentrations in animal tissues reflect those in their foodwebs and that spatial patterns, gradients, or changes in such markers exist in nature. Thus by making isotopic measurements in animal tissues and knowing how stable isotopic patterns in foodwebs change spatially, it is often possible to infer their origins. By combining our knowledge of the behavior of stable isotopes in animal tissues, the ecology and distribution of the species of interest, and the temporal and spatial patterns of isotopic signatures across foodwebs, it is at least theoretically possible to infer animal movements. In relatively simple systems, such as an organism moving into an isotopically distinct area from a few possible other isotopically distinct locations, it should be relatively straightforward to assign origin. In more complex systems or in situations where there is ambiguity in source isotopic profiles (i.e. more than one geographically distinct area with the same isotopic profile) then our success in using isotopic tracking will necessarily be compromised. The challenge, then, in the successful application of isotopic tracking of individual organisms is to combine several lines of evidence from the ecological to the biogeochemical in order to produce the best forensic tools possible.

Basic principles

There are three basic principles to using stable isotopes to infer geographical or biome origins of migratory animals. First, it is necessary to choose an appropriate tissue for analysis. Second, one must identify the appropriate isotopic landscapes or gradients appropriate to previous movements of the organism of interest. Finally, one needs to integrate all ecological information on the species (e.g. diet, distribution, physiology) relevant to narrowing down the range of possible isotopic information to allow inference of previous movements.

Choice of an appropriate animal tissue for analysis is an extremely important aspect of any application of stable isotopes in ecology. For animal tracking using metabolically active tissues, we are particularly concerned with the need to know the temporal window of integration of the tissue used. Similarly, for metabolically inert tissues such as hair, nails, and feathers, it is necessary to know when these tissues were formed and their rate of growth; information that is often elusive for many species and taxa. Luckily, elemental turnover rates in metabolically active tissues have been derived experimentally or can be deduced from allometric relationships and we currently have a reasonable idea of temporal periods of integration for useful tissues such as whole blood, blood cells or plasma (Hobson & Bairlein 2003; Evans-Ogden et al. 2004). Currently, we are less sure about the influence of metabolically demanding activities such as flight or extended work on elemental turnover rates during migration (Hobson 2005a). In addition, there have been some surprises. For example, Voigt et al. (2003) found exceptionally low carbon turnover rates in blood of nectarivorous bats despite their high metabolic rates. This may have been due to the differential way in which carbon is utilized for protein synthesis compared with nitrogen (Martinez del Rio & Wolf 2005) or that the experimental bats were not fed an appropriate diet (Mirón et al. 2006) Nonetheless, despite areas of needed research, we can currently make reasonable estimates for windows of integration for tissues of most warm-blooded animals at least to determine if the metabolically active tissue of interest will provide information of origin of the animal over the past days to months. With the exception of birds whose feather molt is often well known, less is typically known of windows of integration of metabolically inert tissues such as hair, baleen, and claw growth, but even these can be narrowed to reasonable periods of interest (Schell et al. 1998; Bearhop et al. 2003; Urton & Hobson 2005).

For migratory birds of the Northern Hemisphere that grow their feathers prior to southward migration and in areas close to their breeding grounds, it is possible to infer information about these molt origins if those feathers are later collected from individuals on or *en route* to their wintering grounds. Several European and some North American passerines also molt feathers on their wintering grounds providing the opportunity to infer winter origins in Africa or elsewhere (Møller & Hobson 2004; Yohannes et al. 2005). Some species have a partial prealternate moult that permits isotopic sampling representing both breeding and wintering grounds from the same individual (Mazerolle et al. 2005a). In cases of avian tracking using feathers, our isotopic information will be as useful as our confidence in molt chronology dictates and there is now considerable interest in using isotopic techniques to work out molt patterns in both well-known and more enigmatic species (Hobson et al. 2000a, Norris et al. 2004). Periods of growth of mammal hair is less well known and within populations can range over several months. Few researchers have measured growth rates and periods of replacement of claws and vibrissae in wild animals and these will require careful study in the future (e.g. Hirons et al. 2001; Bearhop et al. 2003). However, Bearhop et al. (2004, 2005) have provided some good examples of applications of the isotopic analyses of avian claws to answer questions related to general population delineation. Several marine mammals grow sequential teeth annuli and it is possible to examine the organic and inorganic components of these growth rings to reconstruct dietary and potentially provenance information throughout the lifetime of the animal (Hobson & Sease 1998). Similarly, otoliths and scales can be used to infer past diets and environments experienced by fish (McCarthy & Waldron 2000; Weber et al. 2002).

Most researchers have previously relied on the isotopic analysis of single tissues in order to reconstruct animal movements. However, it is often possible to infer powerful information from the analysis of two or more tissue types per individual. By comparing isotopic information in fast vs. slow elemental turnover tissues, it is possible to ascertain whether the organism has been in a local isotopic environment for a short or long period. High correlation between tissue types indicates long-term residency whereas a poor correlation can indicate only short-term residency. So, by contrasting blood plasma isotope values with the cellular portion, researchers can readily glean some information on residency time. In the case of poor knowledge of feather molt schedules in birds, it would be prudent to contrast isotopic values from different feather tracts known to be molted at different times. High correlation among feathers would indicate that birds did not move during molt, whereas poor correlation would indicate movement and/or dietary change during the moult period. Mazerolle & Hobson (2005) contrasted several tissue types (feathers, claws, blood) in migrating white-throated sparrows (*Zonotrichia albicollis*) in North America in order to decipher migratory connectivity between breeding and wintering grounds.

As in any scientific research involving animals, there can be no substitute for extensive knowledge of individual species ecology when considering the best forensic approach to establish animal movements. The most obvious consideration is the geographical range of the species so that isotopic alternates can be narrowed down only to those relevant. As we will see below, this approach is often made easier through the use of Geographic Information Systems (GIS) which allow the overlay of digital distribution information over isotopic landscape information. Since tissue isotopic values are ultimately determined by animal diet, knowledge of the nutritional ecology of species can be extremely important. For example, should diet differ isotopically among age and sex categories, this could influence the validity of inferences made about animal movements unless such information is considered explicitly (Rubenstein & Hobson 2004).

Without doubt, the most challenging aspect of tracking animal movements with stable isotopes is the establishment of clear and unambiguous isotopic landscapes through which an organism travels and derives nutrition. Such information will vary according to the scale of animal movement of interest, the choice of isotopes, and our ability to predict the stability of these patterns through time and space. As such, we are necessarily often restricted to fairly simple isotopic contrasts such as those provided by movements between terrestrial/freshwater and marine or estuarine habitats or between C3- and C4- or CAM-dominated plant communities. The recent application of continent-wide patterns of deuterium in precipitation is a recent exception in that it can potentially provide relatively fine-scale information for organisms moving across strong gradients in δD or which may move across elevations. The following are some examples of the use of various isotopic contrasts in nature to track animal movements. Rather than attempt an exhaustive list, only a few key examples will be considered and emphasis will be placed more on exploring the limitations and potential of each approach. The interested reader is urged to consult some of the previous good reviews on this and related topics (Hobson 1999a, 2005a; Kelly 2000; Rubenstein & Hobson 2004).

Marine systems

Arguably the most isotopically studied regions on Earth are the oceans. There has been extensive interest in using isotopic techniques to decipher aspects of marine foodwebs and sources of nutrients (Michener & Schell 1994). Nonetheless, the immensity of the world's oceans and the complexity of physical and chemical processes involved in isotopically labeling foodwebs and the animals that use them, results in a relatively poor understanding of the structure of useable marine isotopic patterns to track animal movements. Nonetheless, there are some patterns related to mean ocean temperature and salinity, patterns of upwelling and water flux, phytoplankton growth rates, and influence of terrestrial runoff due to major rivers that can be useful.

One of the earliest applications of stable isotope assays to infer animal movement was the measurement of $\delta^{18}O$ and $\delta^{13}C$ values in carbonates of barnacles retrieved from sea turtles and whales (Killingley 1980, Killingley & Lutcavage 1983). Since isotope abundance of both of these elements in carbonates are sensitive to water temperature during formation, sequential isotopic measurements across barnacles allowed a retrospective analysis of the water masses experienced by these marine mammals as they moved through them. These innovative studies suggested that instead of isotopically being what you eat, you can isotopically be what you move through.

Latitudinal gradients in zooplankton $\delta^{13}C$ values occur in both hemispheres (Rau et al. 1982) but such gradients have not yet been used to decipher animal movements directly (but see Burton & Koch 1999). Rather, known regional patterns in isotopic profiles of marine foodwebs seem to promise the best avenue of research. The most famous example of this is the isotopic measurement of the baleen plates of bowhead whales (*Balaena mysticetus*) that move between isotopically depleted regions of the Beaufort Sea to the more enriched foodwebs of the Bering and Chukchi seas (Schell et al. 1988). A similar investigation of baleen in southern right whales (*Eubalaena australis*) indicated their movement between two ocean regions of very different isotopic composition known as the Mid-Atlantic Convergence Zone just north of 40°S latitude (Best & Schell 1996, Figure 6.1). Since then, seabird researchers have similarly made use of this isotopic gradient to infer movements of far-ranging seabirds (Cherel et al. 2000). In addition, Antarctic foodwebs tend to be more depleted in ^{13}C and ^{15}N compared with more temperate regions of the southern oceans due to an absence of the essential nutrient iron, which restricts phytoplankton growth (Michener & Schell 1994).

Several species of diving duck in North America winter on both the north Pacific and Atlantic coasts and some breeding colonies consist of individuals adopting different migratory options. In an attempt to find a tool to differentiate between these migratory types (i.e. those wintering on east or west coasts) Mehl et al. (2004) found that for king eiders (*Somateria spectabilis*)

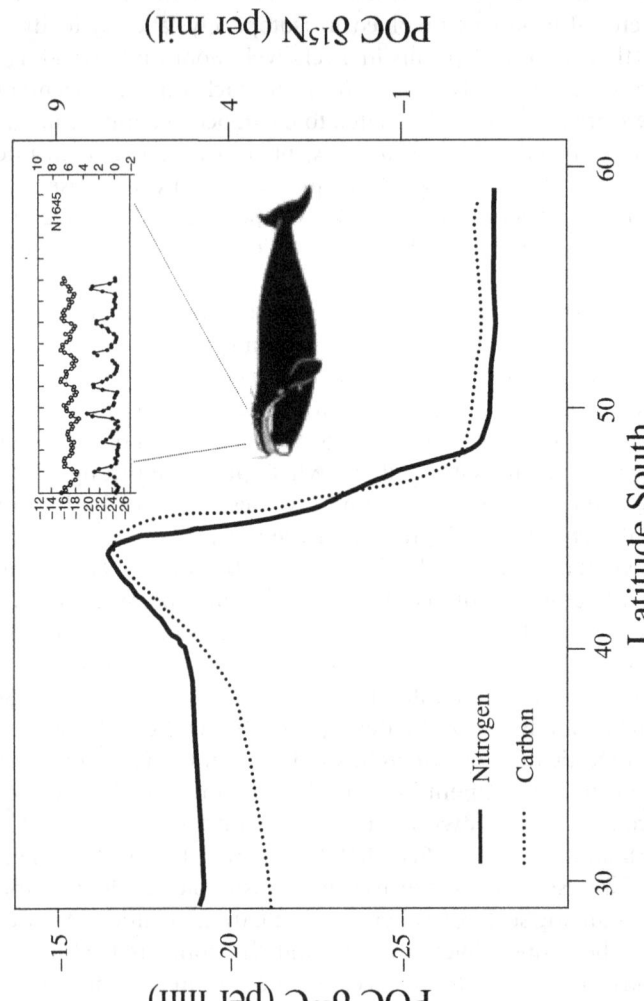

Figure 6.1 Isotopic pattern in marine particulate organic matter (POM) across the Subtropical Convergence in the southern Indian Ocean. This pattern has been used to decipher migration in the southern right whale (*Eubalaena australis*) using stable-carbon and nitrogen isotope measurements in baleen plates. (Figure is based on those presented in Best & Schell 1996.)

birds growing feathers on the western (Pacific) wintering grounds were typically enriched in ^{13}C and ^{15}N compared with those feathers grown off Greenland and assumed that this largely derived from the same principles driving the more regional phenomenon already discovered for the western Arctic population of the bowhead whale (Schell et al. 1998). It remains to be seen if such differences between populations are robust and reflect underlying isotopic patterns (see France et al. 1998) or are linked to more ephemeral patterns such as those expected from dietary differences between populations.

Population connectivity and the delineation of migratory routes are also possible in the case of marine species that might include populations that spend part of their annual cycle in terrestrial or freshwater biomes. This is the case with the long-tailed duck (*Clangula hyemalis*) that breeds across the high Arctic and, while most of the population winters in marine waters, some winter on the freshwater Great Lakes. Braune et al. (2005) used $\delta^{13}C$, $\delta^{15}N$, and $\delta^{34}S$ measurements of bone collagen from breeding populations across the species range to differentiate between western populations most likely to use the Great Lakes via an inland migratory route and eastern populations that moved more directly to marine coastal regions. Since several migratory seaducks use the Great Lakes of North America for staging and overwintering, such a freshwater isotopic signal is expected to be useful in delineating population structure and making migratory connections among sites for these species.

Like other biomes, marine systems are highly dynamic and it will be a challenge to identify isotopic patterns that remain consistently useful through time. For example, changes in oceanic circulation patterns such as the El Niño–Southern Oscillation (ENSO) and the North Atlantic Oscillation can potentially influence foodweb isotopic signatures due to changes in biotic and abiotic factors known to influence stable isotope abundance in primary productivity (Michener & Schell 1994). Ecologists interested in tracking marine organisms using stable isotopes will do well to examine available isotopic data and patterns available for particulate organic matter (POM) and zooplankton (Rau et al. 1982; Goericke & Fry 1994; Schell et al. 1998) and to promote the continued collection of such baseline oceanic data in areas of interest.

Terrestrial systems (excluding deuterium)

Some of the earliest research using stable isotopes elucidated photosynthetic pathways in plants and it is well established that $\delta^{13}C$ values in plant tissues can typically distinguish between C3 and C4 or between C3 and CAM strategies. Deuterium measurements are also useful for distinguishing between CAM and C4-based foodwebs. There is potential, then, to trace organisms moving between foodwebs dominated by different plant types. For example,

Alisauskas et al. (1998) readily identified Snow Geese (*Chen caerulescens*) from estuarine and C3 habitats that had recently moved into a spring staging area dominated by local birds using C4 corn forage. Caccamise et al. (2000) expanded the application of this approach to origins of gamebirds by using $\delta^{13}C$, $\delta^{15}N$, and $\delta^{34}S$ measurements of Canada geese (*Branta canadensis*) on their wintering grounds to distinguish between local and migratory individuals. In general, we might expect greater resolution using additional elements for isotopic measurements and this study is one of the few that have used three isotopes (but see Chamberlain et al. 1997; Yohannes et al. 2005).

In a classic study on the dependence of two species of doves on saguaro cactus (*Carnegiea gigantea*) in Arizona, Wolf & Martinez del Rio (2000) were able to track the use of cactus fruit pulp using $\delta^{13}C$ measurements of bird blood. As CAM plants, saguaro cactus represented a ^{13}C-enriched nutrient pulse against an otherwise C3-dominated foodweb. Interestingly, using δD measurements, these researchers were also able to trace the relative dependence of these birds on cactus as a source of water since cactus water was enriched in deuterium relative to local reservoir water.

In addition to photosynthetic pathway, other mechanisms influence plant isotopic composition. C3 plants can become enriched in ^{13}C as a result of water-use efficiency mechanisms. Marra et al. (1998) made use of this phenomenon to make connections between wintering habitat use by American redstarts (*Setophaga ruticilla*) in Jamaica and their arrival phenology on the breeding grounds in New Hampshire. Birds that occupied poorer quality scrub habitat were enriched in ^{13}C compared with those that occupied cooler and moister forest habitat and tended to arrive on the breeding grounds later. That study was the first to infer consequences of habitat use at one point in the annual cycle with subsequent events at another, an accomplishment virtually impossible without the use of intrinsic tracers such as stable isotopes. Chamberlain et al. (2000) similarly relied on $\delta^{13}C$ and $\delta^{15}N$ measurements of Scandinavian willow warblers (*Phylloscopus trochilis*) grown on African winter quarters to infer climate and subsequent geographic location of African origin. However, such measurements that infer xeric vs. mesic conditions are generally likely only to provide relatively crude insights to species' origins.

Factors other than climate *per se* can influence terrestrial plant isotopic composition. In the case of ^{13}C, agricultural land-use practices can alter the exposure of plants to sunlight by removing canopy forest. This involves potential enrichment due to water use efficiency mechanisms but also, in the case of previously dense forest, the removal of the canopy effect that tends to lower plant $\delta^{13}C$ values (Schlesser & Jayasekera 1985; Brooks et al. 1997). All things being equal, we might thus expect animal foodwebs to be more enriched in ^{13}C in agricultural areas vs. forested areas. Land-use practices and anthropogenic factors also influence plant $\delta^{15}N$ values. The use of animal-based and chemical fertilizers can cause enrichment in ^{15}N in foodwebs either directly or through ammonification, which favours the evaporative loss of isotopically lighter nitrogenous compounds (Nadelhoffer & Fry 1994). Tilling soils also exposes them to the atmosphere, leading to enrichment through the same process. As a result, agricultural landscapes tend to be more enriched in ^{15}N and this can provide a useful indicator of previous provenance for some wildlife (Hobson 1999b; Hebert & Wassenaar 2001).

There are a number of areas where the application of stable isotope methods to track movements of animals in terrestrial systems can be developed further. Recently, Still et al. (2003) reported on approximate global distributions of C3 and C4 plants and suggested that C4 plants predominantly occur in a band at around 10°N. However, such coarse-scale depictions of plant distributions will likely be of limited use in animal tracking studies. Instead, more continental-scale depictions of C4 distributions would be useful and could be augmented by GIS layers corresponding to distributions of C4 agricultural crops such as corn for those cases where species are known to depend on them (e.g. Wassenaar & Hobson 2000a). The measurement of isotopes of other elements such as Sr holds considerable promise and recently, Beard & Johnson (2000) provided a depiction of the expected distribution of δ^{87}Sr values in geological substrate across the USA. Similar depictions are required for other parts of North America and other continents. As well, a useful first step in evaluating the promise of this isotope in animal movement studies will be to measure δ^{87}Sr signatures in animal tissues grown across a known isotopic gradient.

Using deuterium patterns in precipitation

The relative abundance of deuterium (δD) in environmental waters is strongly influenced by kinetic processes associated with temperature, evaporation, and condensation (Craig 1961). Condensate is relatively enriched in deuterium compared with the vapour stage and the absolute water content of the atmosphere is dependent on temperature. So, with decreasing ambient temperature we see a general decrease in D content of vapor since the heavier water molecules are the first to condense. Schiegl (1970) summarized several effects that influence the deuterium content of precipitation. The **latitudinal effect** refers to the general decrease in precipitation D with increasing distance from the Equator due to the associated temperature gradient. The **altitude effect** similarly refers to a depletion of D in precipitation with altitude. The altitude effect is of the order of 1.6‰ 100 m^{-1} at low latitudes and 4.8‰ 100 m^{1} at high latitudes (Dansgaard 1964; Ziegler 1988). Deuterium content in precipitation also undergoes a **seasonal effect** at higher latitudes whereby summer precipitation tends to be more enriched in D than in winter. The **continental effect** refers to the phenomenon of air masses becoming more depleted in D as they move from the coast to interior locations. Finally, the **total precipitation** effect refers to the general depletion in precipitation D with increasing amount of rain fallout. The combination of these factors results in relatively robust patterns of deuterium in precipitation across continents. The primary source of data for such patterns is the global network of isotopes in precipitation (GNIP), administered by the International Atomic Energy Association and World Meteorological Organization (IAEA/WMO 2001).

For foodwebs involving migratory organisms, all hydrogen originates from plant water, in turn derived ultimately from precipitation or groundwater and from drinking water. As plants fix hydrogen in foodwebs, deuterium content of plant organic matter becomes depleted relative to precipitation. Although continental patterns of monthly or annual δD in precipitation have been known for decades (Sheppard et al. 1969; Taylor 1974), it was the investigations of Miller et al. (1988) on beetle exoskeletons and Cormie et al. (1994) on deer bone collagen who first demonstrated how growing-season average precipitation values could be transferred to higher-order herbivores, in turn allowing past climatic data to be derived from the isotopic analysis chitin and bones. Hobson & Wassenaar (1997) and Chamberlain et al. (1997) demonstrated that continental patterns in precipitation δD were transferred through foodwebs to feathers grown at known locations in North America. Importantly, a strong relationship was found between feather δD (δD_f) and the mean growing-season δD precipitation value (δD_p). The intercept of these relationships further provided an estimate of the discrimination factor between precipitation and feather formation – a value typically of the order of −25 to −30‰ for insectivorous passerines largely reflecting initial precipitation to plant isotopic discrimination (Wassenaar & Hobson 2000a). Further studies have largely confirmed that long-term average continental patterns of δD_p can be used to link tissue δD values of consumers with approximate latitude in North America (Meehan et al. 2001; Kelly et al. 2002; Figure 6.2).

Following on from a number of successful applications in North America, Bowen et al. (2005) examined δD_p patterns in other continents. That study revealed several other systems where the use of tissue δD values of migratory organisms could be used to infer origins. In particular, the European continent shows a general depletion in δD_p with latitude along a southwest to northeast gradient (Hobson et al. 2004). However, unlike North America, Hobson et al. (2004) found that there was equally good agreement between δD_f and both annual mean δD and mean growing season δD. South America shows vast regions of the Amazon basin with relatively constant δD_p values but the more temperate southern regions of Argentina and Chile show a useful stepwise depletion in δD_p with latitude (Bowen et al. 2005). That situation may well provide a means of tracking Austral migration (Jahn et al. 2004). The situation in Africa appears to be less encouraging because patterns of δD_p are highly dynamic from month to month. However, some attempts have been made to use measurements of δD_f to infer origins of sub-Saharan

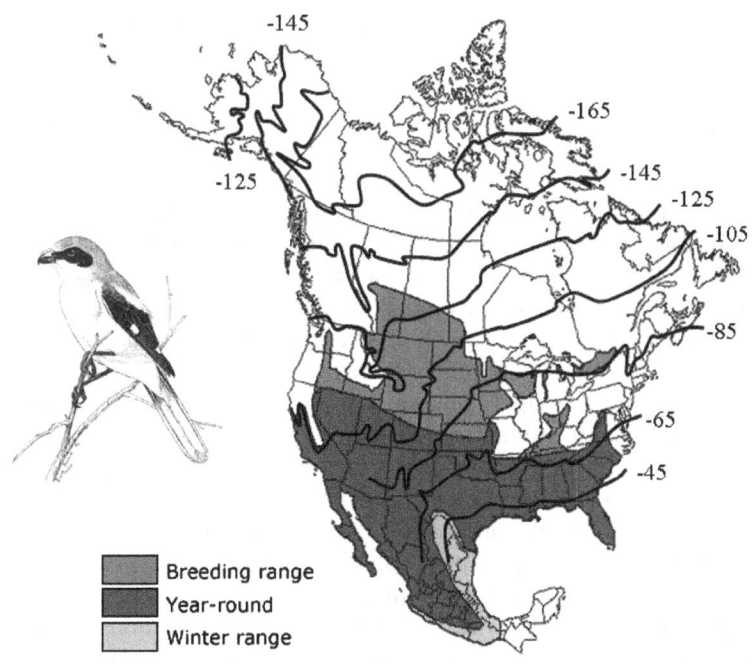

Figure 6.2 The range of the loggerhead shrike (*Lanius ludovicianus*) in North America in relation to the mean growing season average precipitation δD values. (Isotope basemap based on the altitude-corrected kriged dataset presented by Bowen et al. 2005.)

migrants from Europe (Møller & Hobson 2004; Pain et al. 2004; Yohannes et al. 2005) and of medium distance migrants within Europe (Bearhop et al. 2005).

An interesting application of the measurement of δD values in animal and plant tissues is the detection of altitudinal movements. That $δD_p$ values decrease with altitude is a well-known global phenomenon. Similarly, plants can also become enriched in ^{13}C with altitude, a result of physiological responses to changes in growing conditions (Körner et al. 1988, 1991; Graves et al. 2002). Hobson et al. (2003) demonstrated that δD and $δ^{13}C$ values of hummingbird feathers reflected altitude along a 4000 m gradient in the Ecuadorean Andes, clearly demonstrating the utility of the isotope approach to investigating altitudinal migration or movements in birds. To date, this aspect of isotopic research has been much underutilized. There are several species of tropical birds that are known to move altitudinally during their annual cycles. On the Hawaiian islands, viable populations of native avifauna are restricted to elevations above which they are not susceptible to avian malaria and there is great interest in tracking altitudinal movements of these birds to monitor their exposure (van Riper et al. 1986). In more temperate areas, the relative use of alpine vs. valley bottoms by bears and ungulates is of considerable conservation interest (Hobson et al. 2000b).

Some applications

One of the most successful applications of deuterium measurements to track animal migration was that of Wassenaar & Hobson (1998) who examined the North American origins of monarch butterflies (*Danauas plexippus*) at their winter roost sites in central Mexico. That study showed that each of the 13 discrete winter roost sites were panmictic, being composed of butterflies from throughout their breeding range. Moreover, the most important region of butterfly production was determined to be the corn belt of the northeastern USA (Hobson et al. 1999b). That study was important because it established a butterfly δD and $\delta^{13}C$ isotopic basemap for North American origins of butterflies corresponding to the same production year of the wintering butterflies later sampled in Mexico. Elementary students from throughout the breeding range of the eastern population of monarchs grew monarchs on milkweed plants exposed to natural precipitation. The resulting basemap showed the expected depletion in deuterium with latitude but also showed an unexpected pattern in butterfly wing $\delta^{13}C$ values. The combination of the two isotopes provided higher resolution than was possible using a single isotope. Most recently, this basemap was used to show that the increase in monarchs occurring in Cuba during the autumn was due to dispersing individuals originating in the northeast region of the USA (Dockx et al. 2004). Although several studies have successfully used the long-term deuterium basemap constructed from the GNIP dataset, the creation of organism-specific maps such as this enhances the level of information that can be gleaned. For example, Lott & Smith (2006) established an isotopic basemap for raptorial birds in North America based on museum and field collections. Interestingly, their analysis confirms that raptors show much more positive discrimination factors between δD_p and δD_f than have been recorded for insectivorous passerines.

Based on the analysis of museum specimens, there is good evidence that the pattern of δD_p for North America has remained reasonably robust over the past 100 years. However, current and pending climate change may alter the nature of these patterns toward new equilibria. Although the generation of new deuterium precipitation basemaps to track climate change will not be impossible, it does represent a huge undertaking that will take several years to achieve reasonable statistical confidence. In the meantime, individual research projects that are able to ground-truth animal tissue δD values for known locations may be able to calibrate models based on the long-term IAEA average.

The monarch butterfly study revealed just how powerful the stable isotope approach can be in solving questions that are simply intractable using conventional mark–recapture approaches. Similarly, good success in North American avian studies has been achieved by linking breeding and wintering populations. For example, Kelly et al. (2002) clearly showed that western North American populations of Wilson's warblers (*Wilsonia pusilla*.) perform a leapfrog migration to their wintering sites in Central America whereby more northern breeding populations winter further south than more southern breeders. Rubenstein et al. (2002) similarly showed that northern populations of black-throated blue warblers (*Dendroica caerulescens*) wintered in Cuba and the western Antilles whereas southern breeding populations wintered in the eastern Antilles. That study was significant since only the southern breeding populations appear to be declining and so this suggests that wintering ground factors may be important in these declines. Mazerolle & Hobson (2005) have used a number of tissues from the same bird to investigate aspects of migratory connectivity in white-throated sparrows. That species grows flight feathers on the breeding ground and head feathers on the wintering ground. By analyzing these feather types in addition to claws and blood in migrating birds, these authors delineated both breeding catchment areas in the Canadian boreal and wintering areas in the southeast USA. Moreover, it was demonstrated that male sparrows winter farther north than females, suggesting that stable isotope techniques will be valuable in investigating the phenomenon of differential migration in birds and other animals.

The ability to infer origin of migratory birds on their wintering grounds opens up the possibility of investigating how climate can effect bird populations. Dugger et al. (2004) combined long-term population data of neotropical migrant birds from North America that wintered on Puerto Rico. By narrowing down the origins of birds using δD measurements of feathers, they were able to investigate the role of rainfall on the breeding and wintering grounds as potential regulators of population. Similarly, Mazerolle et al. (2005a) investigated the effects of ENSO events on survival and production of young yellow warblers (*Dendroica petechia*) breeding in central Manitoba, Canada, and used the results of previous δD and molecular genetics measurements of this species on the wintering grounds to more precisely link winter weather events and population parameters. Other intriguing applications of tracking origins of migratory species using δD or other isotopic measurements is the geographical tracing of host parasites or disease organisms. For example, Smith et al. (2003) gained insight into the biogeography of avian blood hematozoa in sharp-shinned hawks using simultaneous blood parasite assays and feather isotope measurements of individuals captured during migration.

Analytical considerations

Unlike several other elements common to ecological isotopic investigations, hydrogen bound to oxygen and nitrogen forms weak bonds and is known to exchange with ambient water vapor in organic materials. Because ambient water vapor can change isotopic composition seasonally, it is important to take this exchangeability into account. Wassenaar & Hobson (2000b) described a method of steam equilibration for organic materials that provides a means of measuring δD of the non-exchangeable portion of animal materials. The advent of continuous flow isotope ratio mass spectrometry (CF-IRMS) for δD measurements using pyrolysis techniques, allows the calibration of unknown samples with standards of known non-exchangeable δD values (Wassenaar & Hobson 2003) but several laboratories continue to report whole δD values. Of course, for studies that measure all materials at one time, isotopic datasets are expected to be internally consistent. However, as the number of δD measurements of animal tissues grows, it would be extremely useful if various laboratories would conform to reporting only the non-exchangeable δD values so that large and consistent isotopic datasets can be constructed.

The exchangeability of hydrogen within organic matter also occurs during physiological processes within animals. Hobson et al. (1999a) showed experimentally that hydrogen in drinking water becomes incorporated into several avian tissues. If wild organisms of interest derive their food and water from the same source, then this is likely not to influence applications to tracking movement or migration. However, if drinking water is derived from hydrogen pools that differ substantially from dietary pools, then it will be difficult to associate tissue δD values with geographical origins at continental scales. For example, birds that feed in estuarine areas but travel to freshwater wetlands to drink experience very different sources of drinking water and diet. To some degree, datasets of primarily terrestrial migratory birds can be screened for those individuals with marine influences in their diets using also measurements of $\delta^{34}S$ (Lott et al. 2003).

Since the early revelation of good correlations between animal tissue δD values and δD_p, researchers have refined the presentation of the GNIP basemap for δD_p for North America to include elevation-corrected versions (Meehan et al. 2004; Bowen et al. 2005). These versions are now available in formats compatible with GIS software so that mean growing season δD_p values can be interpolated from the long-term kriged dataset. Bowen et al. (2005) also provide this information for Europe, Africa and South America. These digital datasets are extremely useful because they allow the depiction of the continental range corresponding to the isotope data of tissues. Mazerolle et al. (2005b) depicted their sparrow δD_f values as 95% confidence intervals associated with corresponding spatial coordinates in North America and so created linkages between possible breeding and wintering grounds.

The GIS approach to delineating spatially the distribution of tissue isotope values will go a long way to improving the kinds of information that can be gleaned from animal δD values. However, previous statistical inferences for determining animal origins have relied largely on the correlational relationship between tissue δD and δD_p and so have used continuous-response models for inferring geographic origin. More recently, there has been interest in using Bayesian or other discrete-response assignment tests that have the advantage of better dealing with uncertainty and bias in isotopic datasets without propagating error. This approach has largely been borrowed from the field of molecular genetics in order to establish a probability of association of an individual to population or region (Webster et al. 2002). The Bayesian approach is based on the use of conditional probabilities associated with the prior knowledge of stable isotope values associated with a given region. Royle & Rubenstein (2005) recently applied this approach to their black-throated blue warbler isotopic dataset and showed how this could work for three broad areas of the known breeding range. Wunder et al. (2005) showed that for cases where δD_f values could be associated with known locations of mountain plovers (*Charadrius montanus*) in the Rocky Mountains, the Bayesian approach provided much more useful information on assignment and more developments in this aspect of using stable isotopes to infer origins of wildlife are expected.

Conclusions

Ecologists interested in the development of isotopic techniques to track migratory organisms are to a large degree at the mercy of technological innovation. This is a rapidly developing field, and there are several innovative developments that have already moved the field forward. When Wassenaar & Hobson (1998) made their δD measurements of monarch butterflies, they used the zinc reduction approach involving offline preparation of each sample. Today, the automated measurement of δD in organic material using CF-IRMS has made such laborious and expensive analyses obsolete. Similar technological innovation in the realm of measuring stable isotope abundance for the heavier elements in organic material is now possible using inductively coupled plasma mass spectrometric (IC-PMS) techniques. For example, the routine measurement of strontium isotope ratios in feathers is now possible along with isotopes of numerous other elements. Once regional patterns in isotopes of such heavier elements are known, greater resolution in determining animal origins will undoubtedly follow. In addition to more elaborate isotopic assays of wild organisms, the combined use of isotopic and molecular techniques as well as trace element assays can provide further dimensions in multivariate space to be exploited (Clegg et al. 2003; Szép et al. 2003).

In addition to the search for more and more environmental markers that can be used in forensic applications to determine animal origins, there are still a number of areas requiring careful experimental study. For example, it is currently unknown which factors may be responsible for variation in isotopic discrimination between diet and metabolically active and inactive tissues of consumers (Pearson et al. 2003; Cherel et al. 2004). It would be particularly valuable to know why raptors tend to show much more enriched δD_f values compared with insectivorous passerines. Is this due to a trophic enrichment effect for deuterium in foodwebs or is it due to greater evapo-transpiration occurring in raptorial birds as a result of higher workloads for provisioning parents or longer nestling periods for neonates (McKechnie et al. 2004; Smith & Dufty 2005)? Isotopic turnover rates in tissues of actively migrating organisms have yet to be established and so we generally have a poor idea of the period of integration of isotopic signals during migration. Here, the use of stable isotope measurements of known individuals fitted with satellite transmitters or other markers and known to move between isotopically different landscapes could provide the best estimates. Ultimately, considerable progress will be made if studies on the physiology of migration can be integrated with isotopic measurements in captive and wild organisms.

References

Alisauskas, R.T., Klaas, E.E., Hobson, K.A. & Ankney, C.D. (1998) Stable-carbon isotopes support use of adventitious color to discern winter origins of lesser snow geese. *Journal of Field Ornithology*, **69**, 262–268.

Beard, B.L. & Johnson, C.M. (2000) Strontium isotope composition of skeletal material can determine the birth place and geographic mobility of humans and animals. *Journal of Forensic Science*, **45**, 1049–1061.

Bearhop, S., Furness, R.W., Hilton, G.M., Votier, S.C. & Waldron, S. (2003) A forensic approach to understanding diet and habitat use from stable isotope analysis of (avian) claw material. *Functional Ecology*, **17**, 270–275.

Bearhop, S., Hilton, G.M., Votier, S.C. & Waldron, S. (2004). Stable isotope ratios indicate that body condition in migrating passerines is influenced by winter habitat. *Proceedings of the Royal Society of London (Series B)*, **27**, 215–218.

Bearhop, S. Fiedler, W., Furness, R.W., et al. (2005). Assortative mating as a mechanism for rapid evolution of a migratory divide. *Science*, **310**, 502–504.

Berthold, P., Gwinner, E. & Sonnenschein, E. (eds) (2003) *Avian Migration*. Springer-Verlag, Berlin.

Best, P.B. & Schel, D.M. (1996) Stable isotopes in southern right whale (*Eubalaena australis*) baleen as indicators of seasonal movements, feeding and growth. *Marine Biology*, **124**, 483–494.

Bowen, G.J., Wassenaar, L.I. & Hobson, K.A. (2005) Application of stable hydrogen and oxygen isotopes to wildlife forensic investigations at global scales. *Oecologia*, **143**, 337–348.

Braune, B.M., Hobson, K.A. & Malone, B.J. (2005) Using stable isotope and contaminant profiles to discriminate populations of long-tailed duck breeding in the Canadian Arctic. *Science of the Total Environment*, **346**,156–168.

Brooks, J.R., Flanagan, L.B., Buchmann, N. & Ehleringer, J.R. (1997) Carbon isotope composition of boreal plants: Functional grouping of life forms. *Oecologia*, **110**, 301–311.

Burton, R.K. & Koch, P.L. (1999) Isotopic tracking of foraging and long distance migration in northeastern Pacific pinnipeds. *Oecologia*, **119**, 578–585.

Caccamise, D.F., Reed, L.M., Castelli, P.M., Wainright, S. & Nichols, T.C. (2000) Distinguishing migrant and resident Canada geese using stable isotope analysis. *Journal of Wildlife Management*, **64**, 1084–1091.

Chamberlain, C.P., Blum, J.D., Holmes, R.T., Feng, X., Sherry, T.W. & Graves, G.R. (1997) The use of isotope tracers for identifying populations of migratory birds *Oecologia*, **109**, 132–141.

Chamberlain, C.P., Bensch, S., Feng, X., Akesson, S. & Andersson, T. (2000) Stable isotopes examined across a migratory divide in Scandinavian willow warblers (*Phylloscopus trochilus trochilus* and *Phlloscopus trochilus acredula*) reflect African winter quarters. *Proceedings of the Royal Society of London (Series B)*, **267**, 43–48.

Cherel, Y., Hobson, K.A. & Weimerskirch, H. (2000) Using stable-isotope analysis of feathers to distinguish moulting and breeding origins of seabirds. *Oecologia*, **122**, 155–162.

Cherel, Y., Hobson, K.A. & Hassani, S. (2004) Isotopic discrimination between diet and blood and feathers of captive penguins: implications for dietary studies in the wild. *Physiological and Biochemical Zoology*, **78**, 106–115.

Clegg, S.M., Kelly, J.F., Kimura, M. & Smith, T.B. (2003) Combining genetic markers and stable isotopes to reveal population connectivity and migration patterns in a Neotropical migrant, Wilson's Warbler (*Wilsonia pusilla*). *Molecular Ecology*, **12**, 819–830.

Cormie, A.B., Schwarcz, H.P. & Gray, J. (1994) Relation between hydrogen isotopic ratios of bone collagen and rain. *Geochimica et Cosmochimica Acta*, **58**, 377–391.

Craig, H. (1961) Isotopic variations in meteoric water. *Science*, **133**, 1702–1703.

Dansgaard, W. (1964) Stable isotopes in precipitation. *Tellus*, **16**, 436–468.

Dockx, C., Brower, L.P., Wassenaar, L.I. & Hobson, K.A. (2004) Do North American Monarch Butterflies migrate to Cuba? Insights from combined isotope and chemical tracer techniques. *Ecological Applications*, **14**, 1106–1114.

Dugger, K.M., Faaborg, J., Arendt, W.J. & Hobson, K.A. (2004) Understanding survival and abundance of overwintering warblers: does rainfall matter? *Condor*, **106**, 744–760.

Evans-Ogden, L.J., Hobson, K.A. & Lank, D.B. (2004) Blood isotopic ($\delta^{13}C$ and $\delta^{15}N$) turnover and diet-tissue fractionation factors in captive Dunlin. *Auk*, **121**, 170–177.

France, R., Loret, J., Mathews R. & Springer, J. (1998) Longitudinal variation in zooplankton $\delta^{13}C$ through the Northwest Passage: inference for incorporation of sea-ice POM into pelagic foodwebs. *Polar Biology*, **20**, 335–341.

Goericke, R. & Fry, B. (1994) Variations in marine $\delta^{13}C$ with latitude, temperature, and dissolved CO_2 in the world ocean. *Global Biogeochemical Cycles*, **8**, 85–90.

Graves, G.R., Romanek, C.S. & Navarro, A.R. (2002) Stable isotope signature of philopatry and dispersal in a migratory songbird. *Proceedings of the National Academy of Science*, **99**, 8096–8100.

Hebert, C. & Wassenaar, L.I. (2001). Stable nitrogen isotopes in waterfowl feathers reflect agricultural land use in western Canada. *Environmental Science and Technology*, **35**, 3482–3487.

Hirons, A.C., Schell, D.M. & St. Aubin, D.J. (2001) Growth rate of vibrissae of harbour seals (*Phoca vitulina*) and Steller sea lions (*Eumatopias jubatus*). *Canadian. Journal of Zoology*, **79**, 1053–1061.

Hobson, K.A. (1999a) Tracing origins and migration of wildlife using stable isotopes: a review. *Oecologia*, **120**, 314–326.

Hobson, K.A. (1999b) Stable-carbon and nitrogen isotope ratios of songbird feathers grown in two terrestrial biomes: implications for evaluating trophic relationships and breeding origins. *Condor*, **101**, 799–805.

Hobson, K.A. (2003). Making Migratory Connections with Stable Isotopes. In: *Avian Migration* (Eds P. Berthold, E. Gwinner & E. Sonnenschein), pp. 379–392. Springer-Verlag, Berlin.

Hobson, K.A. (2005a) Flying fingerprints: Making connections with stable isotopes and trace elements. In: *Birds of Two Worlds: The Ecology and Evolution of Migratory Birds* (Eds R. Greenberg & P.P. Marra), pp. 235–246. Johns Hopkins University Press, Washington, DC.

Hobson, K.A. (2005b) Stable isotopes and the determination of avian migratory connectivity and seasonal interactions. *Auk*, **122**, 1037–1048.

Hobson, K.A. & Bairlein, F. (2003) Isotopic discrimination and turnover in captive Garden Warblers (*Sylvia borin*): implications for delineating dietary and migratory associations in wild passerines. *Canadian Journal of Zoology*, **81**, 1630–1635.

Hobson, K.A. & Sease, J. (1998) Stable isotope analysis of tooth annuli reveal temporal dietary records: an example using Steller sea lions. *Marine Mammal Science*, **14**, 116–129.

Hobson, K.A. & Wassenaar, L.I. (1997) Linking breeding and wintering grounds of neotropical migrant songbirds using stable hydrogen isotopic analysis of feathers. *Oecologia*, **109**, 142–148.

Hobson, K.A., Atwell, L. & Wassenaar, L.I. (1999a) Influence of drinking water and diet on the stable-hydrogen isotope ratios of animal tissues. *Proceedings of the National Academy of Science*, **96**, 8003–8006.

Hobson, K.A., Wassenaar, L.I. & Taylor, O.R. (1999b) Stable isotopes (δD and δ^{13}C) are geographic indicators of natal origins of monarch butterflies in eastern North America. *Oecologia*, **120**, 397–404.

Hobson, K.A. Brua, R.B., Hohman, W.L. & Wassenaar, L.I. (2000a) Low frequency of "double molt" of remiges in ruddy ducks revealed by stable isotopes: implications for tracking migratory waterfowl. *Auk*, **117**, 129–135.

Hobson, K.A., McLellan, B.N. & Woods, J. (2000b) Using stable-carbon (δ^{13}C) and nitrogen (δ^{15}N) isotopes to infer trophic relationships among black and grizzly bears in Upper Columbia River Basin, British Columbia. *Canadian Journal of Zoology*, **78**, 1332–1339.

Hobson, K.A., Wassenaar, L.I., Milá, B., Lovette, I., Dingle, C. & Smith, T.B. (2003) Stable isotopes as indicators of altitudinal distributions and movements in an Ecuadorean hummingbird community. *Oecologia*, **136**, 302–308.

Hobson, K.A., Bowen, G., Wassenaar, L.I., Ferrand, Y. & Lormee, H. (2004) Using stable hydrogen isotope measurements of feathers to infer geographical origins of migrating European birds. *Oecologia*, **141**, 477–488.

IAEA/WMO (2001) *Global Network for Isotopes in Precipitation, the GNIP Database.* International Atomic Energy Agency/World Meteorogical Organization, Geneva.

Jahn, A.E., Levey, D.J. & Smith, K.G. (2004) Reflections across hemispheres: A system-wide approach to new world bird migration. Auk, **121**, 1005–1013.

Kelly, J.F. (2000) Stable isotopes of carbon and nitrogen in the study of avian and mammalian trophic ecology. *Canadian Journal of Zoology*, **78**, 1–27.

Kelly, J.F., Atudorei, V. Sharp, Z.D. & Finch, D.M. (2002) Insights into Wilson's Warbler migration from analyses of hydrogen stable-isotope ratios. *Oecologia*, **130**, 216–221.

Killingley, J.S. (1980) Migrations of California gray whales tracked by oxygen-18 variations in their epizoic barnacles. *Science*, **207**, 759–760.

Killingley, J.S. & Lutcavage, M. (1983) Loggerhead turtle movements reconstructed from ^{18}O and ^{13}C profiles from commensal barnacle shells. *Estuarine Coastal Shelf Science*, **16**, 345–349.

Körner, C., Farquhar, G.D. & Roksandic, Z. (1988) A global survey of carbon isotope discrimination plants from high altitude. *Oecologia*, **74**, 623–632.

Körner, C., Farquhar, G.D. & Wong, S.C. (1991) Carbon isotope discrimination by plants follows latitudinal and altitudinal trends. *Oecologia*, **74**, 623–632.

Lott, C.A. & Smith, J.P. (2006) A GIS approach to estimating the origins of migratory raptors in North America using hydrogen stable isotope ratios in feathers. *Auk*, **123**, 822–835.

Lott, C.A., Meehan, T.D. & Heath , J.A. (2003) Estimating the latitudinal origins of migratory birds using stable hydrogen and sulfur isotopes in feathers: influences of marine prey base. *Oecologia*, **134**, 505–510.

Marra, P.P., Hobson, K.A. & Holmes, R.T. (1998). Linking winter and summer events in a migratory bird using stable carbon isotopes. *Science*, **282**, 1884–1886.

Martinez del Rio, C. & Wolf, B.O. (2005) Mass balance models for animal isotope ecology. In: *Physiological and Ecological Adaptations to Feeding in Vertebrates* (Eds J.M. Starck & T. Wang,), pp 141–174. Science Publishers, Enfield, NH.

Mazerolle, D. & Hobson, K.A. (2005) Estimating origins of short-distance migrant songbirds in North America: Contrasting inferences from hydrogen isotope measurements of feathers, claws, and blood. *Condor*, **107**, 280–288.

Mazerolle, D., Hobson, K.A. & Wassenaar, L.I. (2005a) Combining stable isotope and band-encounter analyses to delineate migratory patterns and catchment areas of white-throated sparrows at a migration monitoring station. *Oecologia*, **144**, 541–549.

Mazerolle, D.F., Dufour, K., Hobson, K.A. & den Haan, H. (2005b) Effects of large-scale climatic fluctuations on survival and productivity of a Neotropical migrant songbird. *Journal of Avian Biology*, **36**, 155–163.

McCarthy, I.D. & Waldron, S. (2000) Identifying migratory *Salmo trutta* using carbon and nitrogen stable isotope ratios. *Rapid Communications in Mass Spectrometry*, **14**, 1325–1331.

McKechnie. A.E., Wolf, B.O. & Martinez del Rio, C. (2004) Deuterium, stable isotope ratios as tracers of water resources use: an experimental test with rock doves. *Oecologia*, **140**, 191–200.

Meehan, T.D., Lott, C.A., Sharp, Z.D., et al. (2001) Using hydrogen isotope geochemistry to estimate the natal latitudes of immature Cooper's hawks migrating through the Florida Keys. *Condor*, **103**, 11–20.

Meehan, T.D., Giermakowski, J.T. & Cryan, P.M. (2004) A GIS-based model of stable hydrogen isotope ratios in North American growing season precipitation for use in animal movement studies. *Isotopes in Environment and Health Studies*, **40**, 291–300.

Mehl, K.R., Alisauskas, R.T., Hobson, K. A. & Kellett, D. K. (2004) To winter east or west? Heterogeneity in winter site philopatry in a central Arctic population of King Eiders. *Condor*, **106**, 241–251.

Michener, R.H. & Schell, D.M. (1994) Stable isotope ratios as tracers in marine aquatic food webs. In: *Stable Isotopes in Ecology and Environmental Science* (Eds K. Lajtha & R.H. Michener), pp. 138–157. Blackwell Scientific, Oxford.

Miller, R.F., Fritz, P. & Morgan, A.V. (1988) Climatic implications of D/H ratios in beetle chitin. *Palaeogeography, Palaeoclimatology, Plaeoecology*, **66**, 277–288.

Mirón, M.L., Gerardo Herrera, M., Nicte Ramírez, P. & Hobson, K.A. (2006) Carbon and nitrogen turnover rates and trophic fractionation in whole blood in a new world nectarivorous bat. *Journal of Experimental Biology*, **209**, 541–548.

Møller, A.P. & Hobson, K.A. (2004) Heterogeneity in stable isotope profiles predicts coexistence of two populations of barn swallows *Hirundo rustica* differing in morphology and reproductive performance. *Proceedings of the Royal Society of London*, **271**, 1355–1362.

Nadelhoffer, K.J. & Fry, B. (1994) Nitrogen isotopes studies in forest ecosystems. In: *Stable Isotopes in Ecology and Environmental Science* (Eds K. Lajtha & R.H. Michener), pp. 22–44. Blackwell Scientific, Oxford.

Norris, D.R., Marra, P.P., Montgomerie, R.M., Kyser, T.K. & Ratcliffe, L. (2004) Reproductive effort, molting latitude, and feather color in a migratory songbird. *Science*, **306**, 2249–2250.

Pain, D.J., Green, R.E., Giebing, B., et al. (2004). Using stable isotopes to investigate migratory connectivity of the globally threatened aquatic warbler *Acrocephalus paludicola*. *Oecologia*, **138**, 168–174.

Pearson, D.F., Levey, D.J., Greenberg, C.H. & Martinez del Rio, C. (2003) Effects of elemental composition on the incorporation of dietary nitrogen and carbon isotopic signatures in an omnivorous songbird. *Oecologia*, **135**, 516–523.

Rau, G.H., Sweeney, R.E. & Kaplan, I.R. (1982) Plankton ^{13}C:^{12}C ratio changes with latitude: differences between northern and southern oceans. *Deep-Sea Research*, **29**, 1035–1039.

Royle, J.A. & Rubenstein, D.R. (2005) The role of species abundance in determining the breeding origins of migratory birds using stable isotopes. *Ecological Applications*, **14**, 1780–1788.

Rubenstein, D.R. & Hobson, K.A. (2004) From birds to butterflies: animal movement patterns and stable isotopes. *Trends in Ecology and Evolution*, **19**, 256–263.

Rubenstein, D.R., Chamberlain, C.P., Holmes, R.T., et al. (2002). Linking breeding and wintering ranges of a migratory songbird using stable isotopes. *Science*, **295**, 1062–1065.

Schell, D.M., Saupe, S.M. & Haubenstock, N. (1988) Natural isotope abundances in bowhead whale (*Balaena mysticetus*) baleen: markers of aging and habitat usage. In: *Stable Isotopes in Ecological Research* (eds P.W. Rundel, J.R Ehleringer & K.A. Nagy), pp. 261–269. Springer Verlag, New York.

Schell, D. M., Barnett, B.A. & Vinette, K. (1998) Carbon and nitrogen isotope ratios in zooplankton of the Bering, Chukchi and Beaufort Seas. *Marine Ecology Progress Series*, **162**, 11–23.

Schiegl, W.G. 1970. *Natural deuterium in biogenic materials. Influence of environment and geophysical applications*. PhD thesis, University of South Africa, Pretoria.

Schleser, G.H. & R. Jayasekera (1985) δ^{13}C-variations of leaves in forests as an indication of reassimilated CO_2 from the soil. *Oecologia*, **65**, 536–542.

Sheppard, S.M.F., Nielsen, R.L. & Taylor, H.P. (1969) O and H isotope ratios of clay minerals from porphyry copper deposits. *Economic Geology*, **64**, 755–777.

Smith, A.D. & Dufty, A.M. Jr. (2005). Variation in the stable-hydrogen isotope composition of Northern Goshawk feathers: relevance to the study of migratory origins. *Condor*, **107**, 547–558.

Smith, R.B., Greiner, E.C. & Wolf., B.O. (2003) Migratory movements of sharp-shinned hawks *Accipiter striatus* captured in New Mexico in relation to prevalence, intensity, and biogeography of avian hematozoa. *Auk*, **121**, 837–846.

Still, C.J., Berry, J.A., Collatz, G.J. & DeFries, R.S. (2003) Global distribution of C3 and C4 vegetation: carbon cycle implications. *Global Biogeochemical Cycles*, **17**, 1006–1029.

Szép, T., Møller, A.P., Vallner, J., Kovacs, B. & Norman, D. (2003) Use of trace elements in feathers of sand martin *Riparia riparia* for identifying moulting areas. *Journal of Avian Biology*, **34**, 307–320.

Taylor, H.P. Jr. (1974) An application of oxygen and hydrogen isotope studies to problems of hydrothermal alteration and ore deposition. *Economic Geology*, **69**, 843–883.

Urton, E.J.M. & Hobson, K.A. (2005) Intrapopulation variation in gray wolf isotope (δ^{15}N and δ^{13}C) profiles: implications for the ecology of individuals. *Oecologia*, **14**, 317–326.

Van Riper, C., van Riper, S.G., Goff, M.L. & Laird, M. (1986) The epizootiology and ecological significance of malaria in Hawaiian landbirds. *Ecological Monographs*, **56**, 327–344.

Voigt, C.C., Matt, F., Michener, R. & Kunz, T.H.. (2003) Low turnover of carbon isotopes in tissues of two nectar-feeding bat species. *Journal of Experimental Biology*, **206**, 1419–1427.

Wassenaar, L.I. & Hobson, K.A. (1998) Natal origins of migratory Monarch Butterflies at wintering colonies in Mexico: New isotopic evidence. *Proceedings of the National Academy of Sciences*, **95**, 15436–15439.

Wassenaar, L.I. & Hobson, K.A. (2000a) Stable-carbon and hydrogen isotope ratios reveal breeding origins of red-winged blackbirds. *Ecological Applications*, **10**, 911–916.

Wassenaar, L.I. & Hobson, K.A. (2000b) Improved method for determining the stable-hydrogen isotopic composition (δD) of complex organic materials of environmental interest. *Environmental Science and Technology*, **34**, 2354–2360.

Wassenaar, L.I. & Hobson, K.A. (2003) Comparative equilibration and online technique for determination of non-exchangeable hydrogen of keratins for use in animal migration studies. *Isotopes in Environmental and Health Studies*, **39**, 1–7.

Weber, P.K., Hutcheon, I.D., Mckeegan, K.D. & Ingram, B.L. (2002) Otolith sulfur isotope method to reconstruct salmon (*Oncorhynchos tschawytscha*) life history. *Canadian Journal of Fisheries and Aquatic Sciences*, **59**, 587–591.

Webster, M.S., Marra, P.P., Haig, S.M., Bensch, S. & Holmes, R.T. (2002) Links between worlds: unraveling migratory connectivity. *Trends in Ecology and Evolution*, **17**, 76–83.

Wolf, B. & Martinez del Rio, C. (2000) Use of saguaro fruit by white-winged doves: isotopic evidence of a tight ecological association. *Oecologia*, **124**, 536–543.

Wunder, M.B., Kester, C.L., Knopf, F.L. & Rye, R.O. (2005) A test of geographic assignment using isotope tracers in feathers of known origin. *Oecologia*, **144**, 607–617.

Yohannes, E., Hobson, K.A., Pearson, D., Wassenaar, L.I. & Biebach, H. (2005) Stable isotope analyses of feathers help identify autumn stopover sites of three long-distance migrants in northeastern Africa. *Journal of Avian Biology*, **36**, 235–241.

Ziegler, H. (1988). Hydrogen isotope fractionation in plant tissues. In: *Stable Isotopes in Ecological Research* (Eds P.W. Rundel, J.R. Ehleringer & K.A. Nagy), pp. 105–123. Springer-Verlag, New York.

CHAPTER 7

Natural abundance of ^{15}N in marine planktonic ecosystems

JOSEPH P. MONTOYA

Introduction

The availability of nitrogen plays a central role in regulating biological productivity in many marine environments. The distribution of the stable isotopes of nitrogen within marine ecosystems can provide critical insights into the sources of N supporting production and the pathways and mechanisms of movement of nitrogen through those ecosystems. This biogeochemical approach generally requires little manipulation of the system being sampled and integrates over space and time in a manner that nicely complements experimental studies using ^{15}N-labeled substrates, which typically involve short-term measurements in small, isolated volumes. Despite the potential for using ^{15}N natural abundance data to study important nitrogen cycle processes, the complexity of the marine nitrogen cycle and the potential influence of multiple processes on the isotopic composition of most individual pools of nitrogen require a judicious choice of research questions if stable isotope studies are to provide unambiguous insights into the behavior of marine systems.

Significant variations in the natural abundance of ^{15}N in marine organisms were first documented in the 1950s (Hoering 1955), but the earliest focused studies of nitrogen isotopic abundance in marine systems were carried out beginning in the mid-1960s (Miyake & Wada 1967; Wada et al. 1975: Wada & Hattori 1976; Minagawa & Wada 1984). These and subsequent studies of nitrogen isotope systematics suggested that variations in the natural abundance of ^{15}N in marine systems could be used as an indicator of the sources of nitrogen supporting the growth of marine organisms and as an index to the trophic position of marine animals. These two general themes continue to motivate the bulk of the ongoing work on the stable isotopes of nitrogen in marine systems. In this paper, I provide a brief review of selected applications of nitrogen isotope measurements to oceanographic studies and a prospectus of some outstanding research questions that remain to be addressed.

Background

Isotopic fractionation

In general, the biological processing of nitrogen in the ocean is accompanied by significant kinetic isotopic fractionation, which in turn generates predictable patterns of distribution of ^{15}N among biologically active pools of nitrogen. Kinetic fractionation fundamentally reflects the different reaction rates for molecules containing the two isotopes of nitrogen and is typically expressed as a fractionation factor, α:

$$\alpha = {}^{14}k/{}^{15}k \tag{7.1}$$

where ^{14}k and ^{15}k are the rate constants for reaction of molecules containing the light and heavy isotopes, respectively. With this convention for expressing isotopic fractionation, "normal" kinetic fractionation, which discriminates against the heavy isotope, has a fractionation factor greater than unity. Note that the inverse convention is also used by some authors (e.g. Mariotti et al. 1981), requiring careful attention when comparing fractionation factors reported by different authors.

The relative difference in reaction rates between isotopic species is generally on the order of parts per thousand (‰) or parts per hundred (%), so the isotopic discrimination factor, ε, provides a more convenient way to express the magnitude of fractionation:

$$\varepsilon = (\alpha - 1) \times 1000 \tag{7.2}$$

To a very good approximation for samples not artificially enriched in ^{15}N, the discrimination factor is equal to the instantaneous difference in δ^{15}N between the substrate and product of a reaction:

$$\varepsilon \approx \delta^{15}N_{substrate} - \delta^{15}N_{product} \tag{7.3}$$

as long as residual substrate remains and is undergoing reaction. As a reaction progresses in a closed system, both the residual substrate and the product formed will become progressively enriched in ^{15}N, following a typical Rayleigh fractionation trajectory (Figure 7.1). Conservation of mass and isotopes requires that the δ^{15}N of the combined substrate and product pool remain constant throughout the course of reaction. For a closed system, this also implies that complete conversion of substrate to product will leave no measurable isotopic imprint in the system since the accumulated product will have exactly the same δ^{15}N as the initial pool of substrate even if the reaction itself discriminates strongly between isotopes (Figure 7.1). The effect of isotopic fractionation is thus observable only under conditions of partial consumption of the available substrate pool by the fractionating reaction.

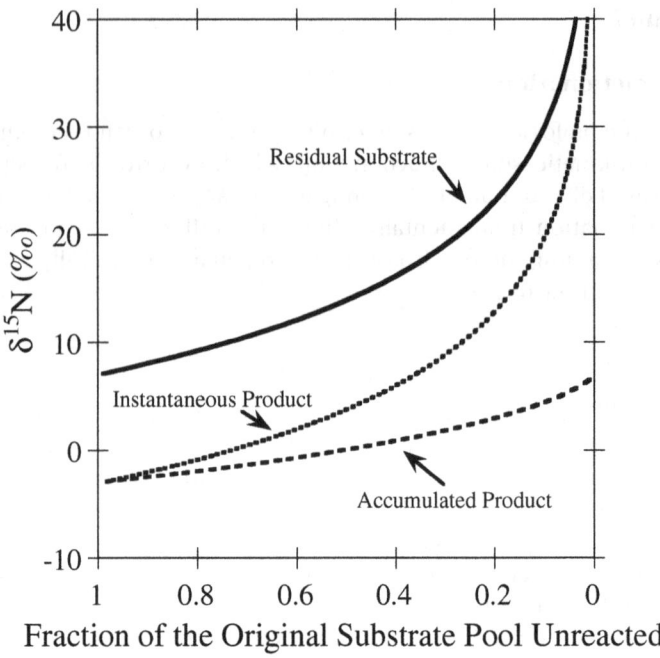

Figure 7.1 Trajectory of $\delta^{15}N$ of the substrate and product of a reaction with an isotope discrimination facter (ε) of 10‰. As reaction proceeds from left to right, the $\delta^{15}N$ of the residual substrate and accumulated produce pool both increase, as does the $\delta^{15}N$ of the product formed at any instant.

Isotopic variation in marine nitrogen

Biological processes that move nitrogen in and out of the ocean play a key role in determining the natural abundance of ^{15}N in marine systems (Table 7.1). In ncarshorc and contincntal shclf waters, terrigenous runoff may be a significant source of nutrients, but the impact of runoff decreases dramatically away from the shoreline (Galloway et al. 1996; Michaels et al. 1996; Nixon et al. 1996). Terrestrial organic matter is often depleted in ^{15}N ($\delta^{15}N$ = ca. −1 to 3‰) relative to marine nitrogen, although dissolved inorganic nitrogen derived from sewage and farm wastes may be significantly enriched in ^{15}N as a result of volatilization and microbial processing of the nitrogen in solution (Sweeney et al. 1978, 1980; Rau et al. 1981; Heaton 1986; Van Dover et al. 1992). Atmospheric deposition may make a significant contribution to the nitrogen budget of surface waters, but this effect is also largest in areas relatively near continental land masses with significant populations and industrial activity or regions downwind of significant continental dust sources (Prospero et al. 1996). Atmospheric sources of N derived from

Table 7.1 Summary of recent published estimates of the oceanic nitrogen budget.

Oceanic nitrogen budget term	Oceanic nitrogen budget ($TgNyr^{-1}$)	
	Codispoti et al. (2001)	Gruber (2004)
Inputs:		
atmospheric deposition	86	50
river input (total)	76	80
benthic N_2-fixation	15	15
water column N_2-fixation	110	120
Total Inputs	287	265
Outputs:		
sedimentary denitrification	300	180
water column denitrification	150	50
sedimentation	25	25
N_2O emissions	6	4
organic export	1	1
Total Outputs	482	260

anthropogenic sources generally have a low δ^{15}N (−5 to 2‰; Paerl & Fogel 1994; Russell et al. 1998), again providing an isotopic contrast to average marine nitrogen.

The dominant source of nitrogen to support primary production in the open ocean is generally assumed to be NO_3^-, which is present at relatively high concentrations (tens of μmol kg^{-1}) below the surface mixed layer. Most efforts at ocean-scale mass and isotope budgets (e.g., Brandes & Devol 1997, 2002) therefore focus on oceanic NO_3^- as an integrator of fluxes in and out of the ocean. Despite the importance of NO_3^- in supporting marine primary production, a growing body of biological and geochemical evidence points to N_2 fixation as a dominant local source of combined nitrogen supporting production in open ocean ecosystems (Gruber 2004; Montoya et al. 2004; Capone et al. 2005).

From an isotopic standpoint, all of the different sources of N (runoff, atmospheric deposition, subsurface NO_3^-, and N_2 fixation) may alter the baseline δ^{15}N of the marine biota, though their impact clearly varies both laterally and vertically through the water column. Beyond the variations imposed by the isotopic composition of different sources of N, biological processes dominate the movement of nitrogen through marine ecosystems and can lead to marked variations in the δ^{15}N of different biologically active pools of N. For example, the isotopic fractionations associated with N_2-fixation, denitrification and, to some extent, nitrification (Figure 7.2) can all affect the overall isotopic composition of marine nitrogen. In addition to creating spatial heterogeneity in the δ^{15}N of marine systems, these processes,

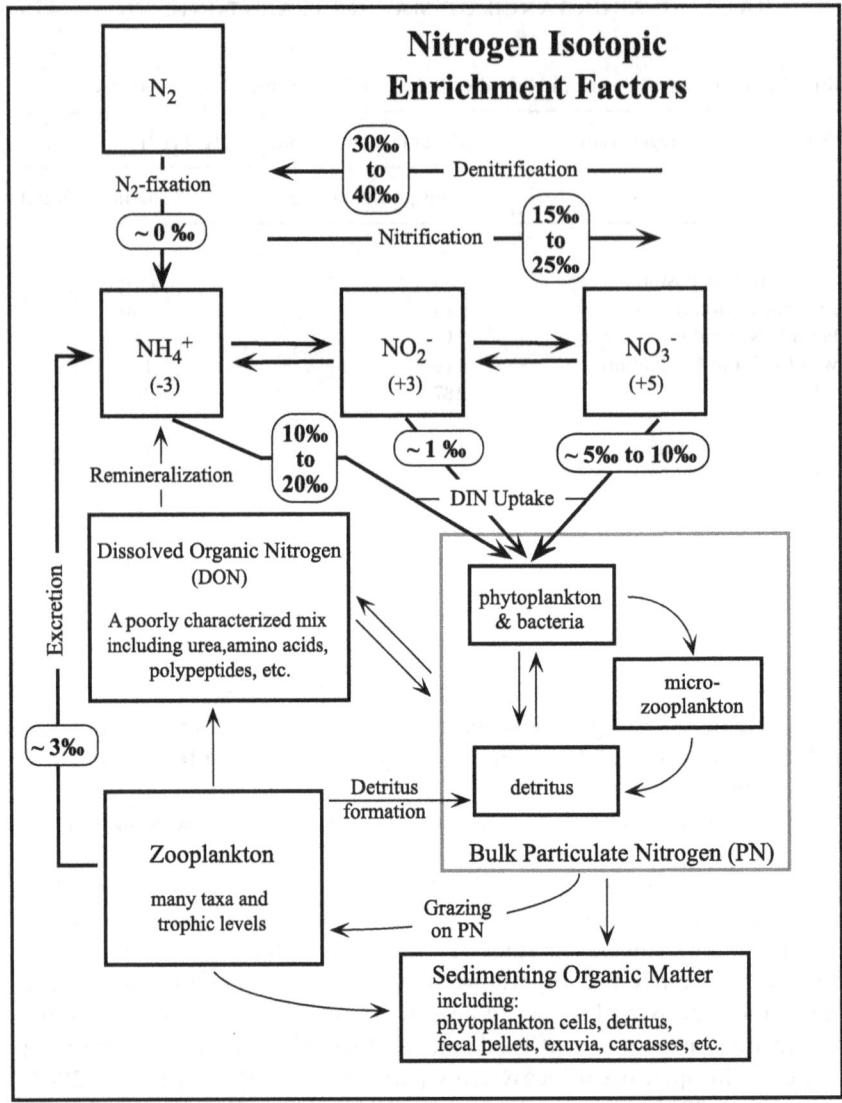

Figure 7.2 Schematic overview of the marine nitrogen cycle and the magnitude of isotopic fractionation associated with various reactions. Each box represents one major functional pool of nitrogen in the ocean. Some of the important components of heterogeneous pools are listed beneath the name of the pool, as is the oxidation state of the dissolved inorganic forms. Arrows depict the major biological transformations of nitrogen. Typical values of the isotopic enrichment factor (ε) are given for reactions that have been characterized isotopically. Estimates of ε were drawn from the available literature on N_2-fixation and the $\delta^{15}N$ of diazotrophs (Hoering & Ford 1960; Delwiche & Steyn 1970; Macko et al. 1987; Carpenter et al. 1997; Montoya et al. 2002), denitrification (Delwiche & Steyn 1970; Miyake & Wada 1971; Cline & Kaplan 1975; Wada 1980; Mariotti et al. 1981, 1982; McCready et al. 1983; Barford et al. 1999; Voss et al. 2001), nitrification (Delwiche & Steyn 1970; Miyake & Wada 1971; Mariotti et al. 1981; Yoshida 1988), NO_3^- uptake (Wada & Hattori 1978; Pennock et al. 1988; Montoya & McCarthy 1995; Pennock et al. 1996; Waser et al. 1998a, 1998b; Needoba et al. 2003; Needoba & Harrison 2004), NO_2^- uptake (Wada & Hattori 1978; Wada 1980), NH_4^+ uptake (Wada & Hattori 1978; Wada 1980; Pennock et al. 1988; Cifuentes et al. 1989; Montoya et al. 1991), and zooplankton excretion (Checkley & Miller 1989).

over time, act to set the average isotopic composition of the global pool of marine nitrogen. Recent attempts to compile an overall nitrogen budget for the world ocean have produced quite divergent results (Table 7.1), and the overall flux of nitrogen in and out of the ocean, as well as the degree to which the oceanic nitrogen budget approaches steady state, are all areas of continuing research and debate (Codispoti et al. 2001; Gruber 2004). In any case, all current evidence indicates that the marine nitrogen cycle is quite dynamic and the overall magnitude of the fluxes of nitrogen into and out of the ocean are large enough to turn over the oceanic pool of NO_3^- on a time scale of about 1.5 kyr (Codispoti et al. 2001), which is similar to the time scale of the deep thermohaline circulation.

δ^{15}N of marine dissolved inorganic nitrogen

With the exception of N_2-fixing prokaryotes, all marine autotrophs require combined nitrogen (i.e., NO_3^-, NO_2^-, or NH_4^+) and are unable to use molecular N_2 as a substrate for growth. As a result, the isotopic composition of dissolved inorganic nitrogen (DIN) acts as a master variable controlling the isotopic composition of marine plankton (Figure 7.3). Deepwater NO_3^- is the largest pool of combined nitrogen in the ocean, and its isotopic composition is affected by a variety of processes and inputs. On a global scale, the δ^{15}N of NO_3^- is sensitive to the relative importance of N_2-fixation and denitrification (Brandes & Devol 2002), which respectively add and remove N from this pool. Measurements of the δ^{15}N of NO_3^- from oxic deep waters typically range between 3‰ and 6‰, with a global average of about 4.8‰ (Liu & Kaplan 1989; Sigman et al. 1997, 2000).

The δ^{15}N of NO_3^- is significantly higher in and around suboxic water masses such as the Arabian Sea and the Eastern Tropical Pacific (Cline & Kaplan 1975; Liu & Kaplan 1989; Brandes et al. 1998; Voss et al. 2001; Montoya & Voss, 2006). In these midwater oxygen-minimum zones, denitrifying bacteria use NO_3^- as an electron acceptor to support heterotrophic growth, reducing it largely to N_2. Both field and laboratory experiments have shown that denitrification discriminates strongly (20–30‰) against ^{15}N (Cline & Kaplan 1975; Brandes et al. 1998; Barford et al. 1999; Voss et al. 2001). In the major pelagic oxygen minimum zones, denitrification consumes only part of the available pool of NO_3^-, resulting in significant enrichment of ^{15}N in the residual NO_3^- (δ^{15}N = 15–18‰). Note that the isotopic impact of pelagic denitrification differs fundamentally from that of sedimentary denitrification, which typically goes to completion within the sediment and therefore results in no expression of the isotope effect associated with the denitrification process itself (Brandes & Devol 1997, 2002). In other words, sedimentary denitrification is effectively invisible from an isotopic standpoint because it removes combined nitrogen from the ocean without altering the isotopic composition of the dissolved inorganic nitrogen left behind.

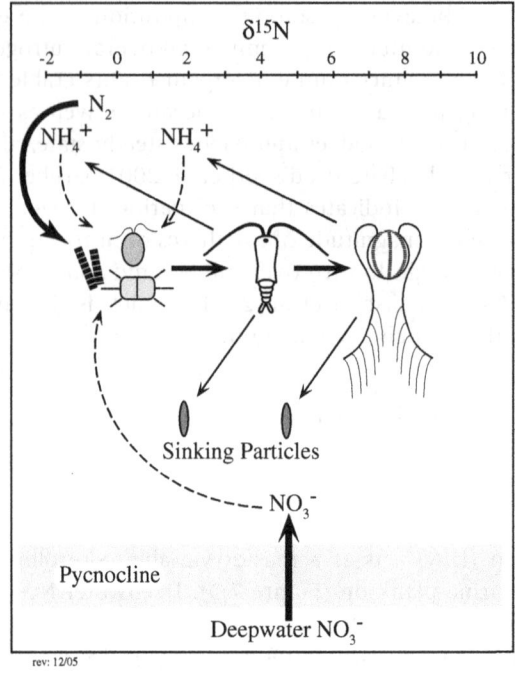

Figure 7.3 Schematic depicting the interactions among uptake of inorganic nitrogen (NO_3^-, NH_4^+, and N_2) by phytoplankton, trophic transfers, recycling, and export processes in controlling the $\delta^{15}N$ of plankton. Broken arrows represent phytoplankton uptake of upwelled NO_3^- ($\delta^{15}N = 4.8$‰) and remineralized NH_4^+ ($\delta^{15}N = -1$ to 3‰). N_2 fixation, which generates organic matter with a low $\delta^{15}N$, and upwelling of NO_3^- are shown with solid heavy arrows. Within the mixed layer, medium solid arrows represent trophic processes, and light solid arrows represent losses of ingested nitrogen to dissolved and solid excreta. For each of these processes, the transfer of nitrogen shown by an arrow will tend to bring the $\delta^{15}N$ of the target pool nearer that of the source pool of nitrogen. Note that the loss of sinking particles in combination with the uptake of NH_4^+ within the mixed layer will tend to remove ^{15}N preferentially from the mixed layer. The mean $\delta^{15}N$ of the organic nitrogen exported from the upper water column will equal the weighted average of the inputs; in this case, the $\delta^{15}N$ of the sinking particles is intermediate between the $\delta^{15}N$ expected in a system dominated by diazotrophs and one supported exclusively by inputs of NO_3^- across the pycnocline.

In the euphotic zone, phytoplankton assimilation of NO_3^- can have a measurable impact on the $\delta^{15}N$ of the residual NO_3^-. Phytoplankton express significant discrimination against $^{15}NO_3^-$ during uptake (Montoya & McCarthy 1995; Pennock et al. 1996; Waser et al. 1998a, 1998b; Needoba et al. 2003; Needoba & Harrison 2004), which can lead to significant alteration of the $\delta^{15}N$ of the residual NO_3^-. This is particularly important at the base of the mixed layer of waters where the nitracline shows significant overlap with the lower portion of the euphotic zone (Sigman et al. 1997, 1999). In culture studies, the isotopic discrimination factor for NO_3^- uptake varies between taxa and with growth conditions, but a typical value is about 5‰ (Montoya & McCarthy 1995; Waser et al. 1998a, 1998b; Needoba et al. 2003).

Although most marine autotrophs require combined nitrogen (e.g., NO_3^-, NH_4^+) as a substrate for growth, a wide variety of prokaryotes are able to reduce and assimilate N_2 into biomass, a process known as N_2 fixation, or diazotrophy. N_2 fixation produces organic matter that is depleted in ^{15}N relative to deepwater NO_3^-. Thus, in contrast to denitrification, N_2 fixation lowers the δ^{15}N of NO_3^- while adding to the oceanic pool of combined nitrogen. For example, the δ^{15}N of NO_3^- in the upper thermocline at the Bermuda Atlantic Time Series Station averages 2.8‰ (Knap et al. 2005), or roughly 2‰ lower than the oceanic average. Note that the process of N_2 fixation itself does not directly alter the oceanic NO_3^- pool; instead, the organic matter formed by N_2-fixing organisms is ultimately remineralized and contributes to the subsurface pool of NO_3^- through nitrification of the NH_4^+ produced during remineralization. Nitrification is thus a critical step in the conversion of organic matter to NO_3^-, and a potential source of ^{15}N-depleted NO_3^- in the water column (Ostrom et al. 1997; Sutka et al. 2004) since the oxidation of NH_4^+ is accompanied by significant isotopic fractionation (Mariotti et al. 1981). In the upper and middle water column of oligotrophic waters, however, the nitrification process itself is unlikely to have an impact on the δ^{15}N of NO_3^- since little or no residual NH_4^+ accumulates in the water column. Unless an appreciable fraction of the NH_4^+ produced by remineralization remains in solution unconsumed by nitrifiers (i.e., the nitrification reaction fails to go to completion), nitrification simply transfers the isotopic signature of remineralized NH_4^+ into the NO_3^- pool.

Given the size of the oceanic pool of NO_3^- and the central role of NO_3^- as a nutrient, electron acceptor, and metabolic end product, the isotopic composition of NO_3^- naturally provides an integrative measure of the biological sinks, sources, and transformations of nitrogen within a water column. In regions sampled to date, the δ^{15}N of subsurface NO_3^- typically falls in a range of 4–5‰ (Sigman et al. 1996, 1999; Knap et al. 2005), with significantly higher values (15–20‰) in and around pelagic oxygen minimun zones (Cline & Kaplan 1975; Brandes et al. 1998; Barford et al. 1999; Voss et al. 2001) and lower values (2.8‰) in the upper thermocline of oligotrophic waters such as the Sargasso Sea (Knap et al. 2005). The impact of regional processes such as denitrification in an oxygen minimum zone may also propagate laterally for substantial distances through advection (Liu & Kaplan 1989; Montoya & Voss 2006). As discussed below, both denitrification and N_2 fixation generate isotopic signatures that can propagate into other components of the ecosystem, including organisms in the upper ocean (Voss et al. 2001; Montoya et al. 2002; McClelland et al. 2003; Montoya & Voss, 2006) and benthic sediments (Altabet et al. 1995; Ganeshram et al. 1995). In effect, nitrogen isotopic discrimination creates isotopic variations and perturbations that can potentially be used as natural, *in situ* tracer experiments to study nitrogen cycle processes as well as physical transport in oceanic systems.

$\delta^{15}N$ of marine organic nitrogen

The $\delta^{15}N$ of marine organic matter shows a number of broad spatial patterns. In oligotrophic waters such as the Sargasso Sea or the North Pacific Subtropical Gyre, organic matter in the upper ocean is generally depleted in ^{15}N, with particle $\delta^{15}N$ values ranging as low as −2‰ and zooplankton $\delta^{15}N$ ranging down to near 0‰ (Wada & Hattori 1976; Saino & Hattori 1980; Montoya et al. 1992, 2002; Dore et al. 2002). This is one of the most robust patterns of nitrogen isotopic variation in marine systems and is now generally attributed to the influence of N_2-fixing organisms (diazotrophs). N_2 fixation is accompanied by only weak discrimination against ^{15}N, producing organic matter that is slightly depleted in ^{15}N relative to tropospheric N_2.

The best known oceanic diazotroph, the colonial nonheterocystous cyanobacterium *Trichodesmium*, typically has a $\delta^{15}N$ of about −1 to −2‰ (Carpenter et al. 1997; Montoya et al. 2002), while the heterocystous *Richelia* that lives endosymbiotically within diatoms has a $\delta^{15}N$ of about −1‰ (Montoya et al. 2002). Although *Trichodesmium* makes a substantial contribution to the nitrogen budget of both the North Atlantic (Capone et al. 2005) and the North Pacific (Karl et al. 1996, 1997; Dore et al. 2002), diazotrophy is broadly distributed among the prokaryotes and a diverse array of free-living and endosymbiotic cyanobacteria, archaea, and bacteria are capable of utilizing N_2 as a substrate for growth (Zehr et al. 1998). Diazotrophs symbiotic in diatoms in particular can occur in dense blooms and are at times a very important component of the phytoplankton community. Such associations can make a very large contribution to the local nitrogen budget (Carpenter et al. 1999), but their overall contribution to global oceanic N_2 fixation remains poorly constrained.

Recent field experiments and surveys in the Pacific have demonstrated that small, presumably unicellular, diazotrophic cyanobacteria occur throughout the mixed layer in subtropical waters across the basin (Zehr et al. 2001; Montoya et al. 2004; Church et al. 2005). Furthermore, experimental rate measurements show that these small diazotrophs make a large contribution to the pelagic nitrogen budget (Montoya et al. 2004) in the Pacific and perhaps in the North Atlantic (Falcon et al. 2004; Mills et al. 2004; Voss et al. 2004). Although the diversity and broad distribution of unicellular diazotrophs have been demonstrated using molecular techniques (Zehr et al. 2001; Church et al. 2005), these organisms have not yet been cultured. As a result, their $\delta^{15}N$, the factors that control their N_2 fixation activity, and their overall impact on the marine nitrogen budget are all active areas of investigation.

Natural abundance and experimental rate measurements both provide strong evidence for substantial N_2 fixation in the open ocean, but low $\delta^{15}N$

values in marine organic matter may also arise through other mechanisms. Significant isotopic fractionation occurs during the movement of nitrogen through marine food webs, and field and experimental studies consistently show a "trophic" enrichment in animal tissues, with a typical δ^{15}N difference of 2–4‰ between an animal's tissues and its food (DeNiro & Epstein 1981; Minagawa & Wada 1984; Mullin et al. 1984; Wada et al. 1987; Fry 1988; Montoya et al. 1990, 1992, 2002). The biochemical and physiological mechanisms leading to this consistent isotopic enrichment in heterotrophs are not fully understood (see below), but it is clear that catabolic and excretory processes play a central role in enriching animals in ^{15}N relative to their food (Bada et al. 1989; Gannes et al. 1997).

For an animal at steady state, the mean δ^{15}N of the nitrogen entering as food and leaving as dissolved and solid excreta must be equal. Relatively few measurements of these isotopic fluxes are available for marine animals. Fecal material typically has a δ^{15}N similar to or slightly higher than that of the animal's food (Checkley & Entzeroth 1985; Altabet & Small 1990; Montoya et al. 1992), an observation that is consistent with the limited opportunities for isotopic fractionation during bulk digestion and nutrient absorption by animals. Zooplankton and fish are primarily ammonotelic (i.e., excrete NH_4^+ as their major nitrogenous waste product), so deamination reactions play a critical role in their whole-body nitrogen isotope budget. Isotopic discrimination in connection with deamination will produce NH_4^+ that is depleted in ^{15}N relative to the substrate being catabolized (Macko et al. 1986; Bada et al. 1989). For ammonotelic animals, this will in turn lead to a preferential loss of $^{14}NH_4^+$ and a progressive enrichment of the residual pool of amino nitrogen within an animal's body (Montoya et al. 2002).

At the ecosystem level, the net effect of these trophic processes is to retain ^{14}N preferentially within the upper water column through tight recycling of ^{15}N-depleted NH_4^+ excreted *in situ* by zooplankton (Mullin et al. 1984; Checkley & Miller 1989). At the same time, rapidly sinking fecal pellets and detritus act to preferentially transport ^{15}N to the deep ocean (Checkley & Entzeroth 1985; Altabet & Small 1990; Montoya et al. 1992). As a result, animal feeding and excretory processes will tend to lower the average δ^{15}N of organic matter in the upper water column, particularly in oligotrophic waters where remineralized N supports a large fraction of total primary production (Checkley & Miller 1989). Although this trophic mechanism may contribute to the low δ^{15}N typical of the oligotrophic open ocean, a number of lines of evidence clearly point to N_2 fixation as the primary source of the low δ^{15}N of these waters.

1 In general, the δ^{15}N of mixed-layer plankton is inversely correlated with the biomass of diazotrophs in the water column (Wada & Hattori 1976, 1991; Macko et al. 1984; Minagawa & Wada 1984; Saino & Hattori 1987; Capone et al. 1998, 2005; Carpenter et al. 1999; Dore et al. 2002; Montoya et al. 2002).

2 Zooplankton biomass in the tropical and subtropical Atlantic has too low a $\delta^{15}N$ to produce an export flux that can account for the low $\delta^{15}N$ of mixed-layer particles (Montoya et al. 2002).

3 The isotopic composition of individual amino acids isolated from zooplankton collected in the tropical and subtropical Atlantic (McClelland et al. 2003) show no change in overall trophic structure across the basin and a clear change in the $\delta^{15}N$ of primary producers that correlates well with the abundance of *Trichodesmium*.

4 The $\delta^{15}N$ of NO_3^- in the upper thermocline of the Sargasso Sea is 2‰ lower than the global average (Knap et al. 2005), reflecting the input of diazotrophic nitrogen through remineralization and nitrification.

5 Independently of the isotopic evidence, basin- and global-scale studies of nutrient stoichiometry (i.e., N*) and new production both require a major input of new N to oligotrophic waters (Gruber & Sarmiento 1997; Deutsch et al. 2001; Lee et al. 2002). N_2 fixation appears to be the only process that can add nitrogen to the upper water column in quantities sufficient to match the geochemical budgets.

Source delineation and isotope budgets

An obvious application of ^{15}N natural abundance measurements is in resolving the contribution of isotopically distinct nitrogen sources to the elemental budget of an ecosystem. For example, a number of studies have exploited differences in $\delta^{15}N$ to trace the movement of sewage-derived N into estuarine and nearshore benthic communities. In general, terrestrial organic matter and particulate nitrogen in sewage are both depleted in ^{15}N relative to marine nitrogen (Sweeney et al. 1980; Rau et al. 1981; Heaton 1986; Van Dover et al. 1992; Tucker et al. 1999). Studies of several nearshore ecosystems have demonstrated that this distinctive isotopic signature propagates into the food web and can be used to trace the movement of the solid component of sewage nitrogen into marine organisms (Rau et al. 1981; Spies et al. 1989; Van Dover et al. 1992; McClelland & Valiela 1998a, 1998b; Tucker et al. 1999). At a qualitative level, this isotopic tracing simply involves comparison of the $\delta^{15}N$ of representative organisms with the $\delta^{15}N$ of the potential sources of nitrogen to support growth.

In contrast to particulate solids, the dissolved pools (primarily NH_4^+ and NO_3^-) in sewage often have a high $\delta^{15}N$ as a result of microbial activity (nitrification and denitrification) and volatilization of gaseous NH_3 in sewage treatment plants (Heaton 1986; Schlacher et al. 2005). The degree of ^{15}N enrichment in these dissolved pools depends strongly on the details of sewage plant construction and operation (Sheats 2000), but the net result is that the dominant component of the total nitrogen pool in sewage generally has a $\delta^{15}N$ significantly higher than the $\delta^{15}N$ of marine pools of inorganic nitrogen. Here again, this isotopic contrast can be exploited to follow the movement of dissolved inorganic nitrogen from sewage into various pools including sediments, primary producers, plankton, and fish (Voss & Struck 1997; McClelland & Valiela 1998a; Sheats 2000; Schlacher et al. 2005).

On a much broader spatial scale, the low δ^{15}N of oligotrophic oceanic waters such as the subtropical Pacific and the Sargasso Sea is attributable to the input of new nitrogen to these ecosystems through N_2 fixation (see above). In these systems, ^{15}N measurements provide an *in situ* measure of the contribution of N_2 fixation to the water column nitrogen budget. Marine diazotrophs produce organic matter with a δ^{15}N of about −1 to −2‰ (Carpenter et al. 1997; Montoya et al. 2002), values that contrast strongly with the δ^{15}N of deepwater NO_3^-, which averages about 4.8‰ globally (Liu & Kaplan 1989; Sigman et al. 1997, 2000). This isotopic contrast allows the use of simple mixing models to quantify the relative importance of upwelled NO_3^- and N_2 fixation in supporting biological production. This approach integrates over time scales defined by the turnover of organic matter in the mixed layer and NO_3^- in the upper thermocline. On these time scales, the contribution of N_2 fixation to the particle pool, can be calculated with a simple mixing model (Montoya et al. 2002):

$$\% \text{ Diazotroph N} = 100 \times \left(\frac{\delta^{15}N_{\text{particles}} - \delta^{15}NO_3^-}{\delta^{15}N_{\text{diazotroph}} - \delta^{15}NO_3^-} \right) \tag{7.4}$$

where $\delta^{15}N_{\text{particles}}$ represents the δ^{15}N of bulk suspended particles, $\delta^{15}N_{\text{diazotroph}}$ = −2‰, and $\delta^{15}NO_3^-$ = 4.8‰. This calculation is sensitive to the choice of end-members, but the values used here are conservative in terms of the calculation of the diazotroph contribution to the nitrogen budget (Montoya et al. 2002). This approach has recently been used over a broad span of the North Atlantic to calculate that an average of 36% of the upper water column nitrogen budget can be accounted for through N_2 fixation (Capone et al. 2005).

Animal fractionation and food web processes

The δ^{15}N of animal tissues is frequently used as an indicator of trophic position in marine food webs. Although the scaling of δ^{15}N with trophic position is not fully understood at a biochemical level, the excretory loss of NH_4^+ is accompanied by significant isotopic fractionation and appears to be the primary factor contributing to the isotopic enrichment of an animal's tissues relative to its food (Bada et al. 1989; Gannes et al. 1997). From this perspective, the δ^{15}N of marine zooplankton might be better viewed as an index to the efficiency of nitrogen transfer through a food web rather than as a marker of trophic level itself. A simple calculation of the effects of isotopic discrimination on the $\delta^{15}N$ of animals in a food chain can be used to illustrate this point.

A complete isotope budget is not available for any marine animal, but the available data suggest that the fractionation associated with the assimilation of food nitrogen is small, while significant fractionation accompanies the excretory loss of nitrogenous wastes (see above). Marine zooplankton are primarily ammonotelic (i.e., their primary nitrogenous waste product is NH_4^+) and most rely upon diffusion of NH_4^+ to remove wastes from the body. This similarity of excretory mechanism across taxa may contribute to the relatively simple pattern of variation in animal $\delta^{15}N$ in marine ecosystems. To my knowledge, the only direct measurements of the $\delta^{15}N$ of the NH_4^+ excreted by marine zooplankton are those reported by Checkley & Miller (1989) for a variety of marine zooplankton collected in the Pacific. In this study, the $\delta^{15}N$ of zooplankton and the $\delta^{15}N$ of their excreta were strongly correlated, and $\delta^{15}NH_4^+$ was lower than animal $\delta^{15}N$ by an average of 2.7‰ (Checkley & Miller 1989). One experimental detail is important in interpreting this result: Checkley & Miller (1989) incubated their zooplankton in filtered seawater to remove the confounding effects of phytoplankton uptake of the excreted NH_4^+. Under these conditions, the pool of amino acids catabolized by the zooplankton would be drawn largely from the animal's tissues, minimizing the influence of any food consumed before the start of the experiment on the isotopic composition of the excreted NH_4^+. In other words, the NH_4^+ excreted by marine zooplankton was 2.7‰ depleted in ^{15}N relative to the animal's tissues, which was the primary substrate for deamination under the experimental conditions. Note that the substrate pool for deamination would be weighted toward the food consumed by an actively feeding animal. Under these conditions, an animal will excrete $NH4^+$ with a $\delta^{15}N$ lower than that of the food consumed and substantially lower than the $\delta^{15}N$ of the animal's tissues (Frazer et al. 1997).

These experimental results now allow us to estimate the excretory losses required to account for the typical trophic enrichment seen in marine communities. If excretory losses are treated as a first-order, irreversible process, then a simple Rayleigh fractionation model can be applied to calculate the extent of reaction necessary to account for the isotopic difference between an animal's food and its tissues:

$$\delta^{15}N_{animal} = \delta^{15}N_{food} - \varepsilon \ln f \tag{7.5}$$

which can be rearranged to give an expression for f:

$$f = \exp[(\delta^{15}N_{food} - \delta^{15}N_{animal})/\varepsilon] \tag{7.6}$$

where ε is the isotopic enrichment factor for the excretion process and f is the fraction of nitrogen assimilated that remains in the animal's body, that is the net growth efficiency of the animal. Using a value of $\varepsilon = 2.7‰$ (Checkley & Miller 1989) and a trophic enrichment ($\delta^{15}N_{animal} - \delta^{15}N_{food}$) of 3.5‰, we can use equation 7.6 to estimate a net growth efficiency of 27% for a typical marine zooplankter. Although Checkley & Miller found that ε was remarkably similar for a variety of zooplankton, it is important to note that this estimate of growth efficiency is sensitive to the value of ε used. For a constant trophic enrichment of 3.5‰, a 0.5‰ increase in ε increases the estimated growth efficiency to 33% while a 0.5‰ decrease in ε reduces the calculated efficiency to 20%. Finally, it is worth noting that a typical "trophic" enrichment of roughly 3.5‰ suggests that many marine zooplankton grow with about the same overall efficiency with respect to nitrogen. In contrast, microzooplankton may exhibit a much smaller "trophic" effect, implying that these animals have a much higher net growth efficiency than macrozooplankton. For example, rotifers grown in large volume culture show a trophic shift in $\delta^{15}N$ of 1.7‰ (McClelland & Montoya 2002), which implies a growth efficiency of 53%.

These simple calculations suggest that the $\delta^{15}N$ of marine plankton may provide a simple index to the efficiency of nitrogen transfer within a marine food web. In combination with ancillary measurements of assimilation efficiency, the isotopically based estimates of net growth efficiency can be used to produce estimates of gross-growth efficiency from animal samples collected at sea with minimal handling and experimental intervention. At the very least, the difference between the $\delta^{15}N$ of an animal and the $\delta^{15}N$ of the primary producers can be used to provide an upper limit for the overall efficiency of transfer of nitrogen through the food web to that animal.

Isotopic transients in marine systems

Although nitrogen isotopes provide an integrative measure of the sources of nitrogen supporting production in a community, a number of relatively short-lived events can produce significant changes in the distribution of ^{15}N within an ecosystem. The best single example is the isotopic transient that accompanies a phytoplankton bloom supported by NO_3^- (Wu et al. 1999; Holmes et al. 2002) At the start of a bloom, isotopic fractionation results in the production of organic matter depleted in ^{15}N relative to the available NO_3^- (Figures 7.1 & 7.4). As the bloom progresses, the preferential removal of $^{14}NO_3^-$ causes a steady increase in the $\delta^{15}N$ of the residual NO_3^- pool, which results in turn in a steady increase in the $\delta^{15}N$ of the suspended particles (PN) formed (Figures 7.1 & 7.4). In effect, this process creates an isotopic transient that can be exploited as a natural, *in situ* tracer of the biomass formed during the bloom (Figure 7.4). Note that this temporal change is exactly analogous to the spatial trends in $\delta^{15}N$ in regions where advection results in consumption of NO_3^- and an increase in $\delta^{15}NO_3^-$ with greater distance from an upwelling

(a)

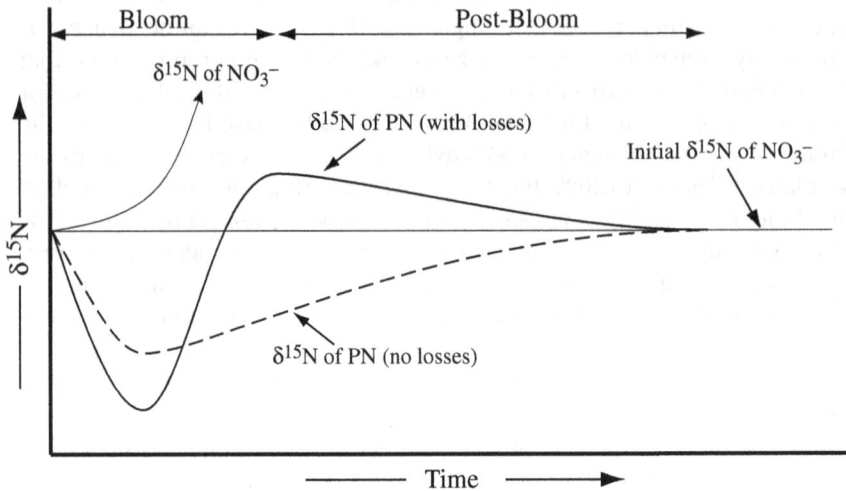

(b)

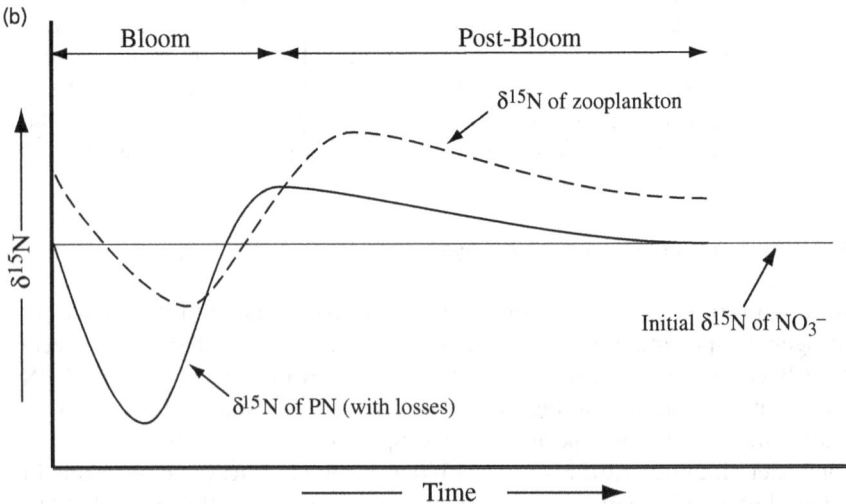

Figure 7.4 Schematic showing the impact of a bloom on the $\delta^{15}N$ of suspended particles (PN) and zooplankton. (a) Biological processes can create a distinct pattern of $\delta^{15}N$ variation with time as the initial pool of substrate (here NO_3^-) is consumed. Early in the bloom, isotopic fractionation during NO_3^- uptake by phytoplankton produces PN with a low $\delta^{15}N$. As the bloom progresses, the $\delta^{15}N$ of the residual NO_3^- increases, leading in turn to an increase in the $\delta^{15}N$ of PN formed. If the bloom is rapid with little material lost through sedimentation or grazing, the $\delta^{15}N$ of PN will converge on the $\delta^{15}N$ of the initial pool of NO_3^- available to support growth (dashed line). If significant losses occur through grazing or sedimentation, the $\delta^{15}N$ of PN may overshoot and exceed the initial $\delta^{15}N$ of NO_3^- (solid line). (b) The isotopic perturbation associated with a bloom can propagate into the food web. Zooplankton nitrogen turns over more slowly than PN, so the magnitude of the isotopic perturbation will decrease with trophic position, and the temporal pattern of variation in $\delta^{15}N$ may be offset from that of the suspended particles.

center or oceanographic front (Altabet & François 1994; Holmes et al. 1996a, 1996b).

Intensive studies in estuarine and nearshore systems (Cifuentes et al. 1988, 1989; Goering et al. 1990; Montoya et al. 1991) as well as the open ocean (Altabet et al. 1991; Nakatsuka et al. 1992; Wu et al. 1999; Dore et al. 2002; Holmes et al. 2002) have documented significant changes in the δ^{15}N of phytoplankton and sedimenting organic matter in the course of a bloom on time scales ranging from days to months. In each of these cases, the temporal changes in the δ^{15}N of mixed-layer PN reflected the time course of change in δ^{15}N of inorganic nitrogen as a result of isotopic fractionation during uptake by primary producers.

Temporal shifts in the isotopic composition at the base of the food web will propagate to the zooplankton and higher trophic levels of the ecosystem (Figure 7.4). Studies in Auke Bay (Alaska) and Chesapeake Bay have documented significant changes in the δ^{15}N of zooplankton on a time scale of days to weeks in the course of a phytoplankton bloom (Goering et al. 1990; Montoya et al. 1991). Although such rapid isotopic changes in animal communities are not well documented for oceanic communities, Montoya et al. (1992) provide circumstantial evidence that the δ^{15}N of a wide variety of zooplankton changes significantly during and after the spring bloom in the Gulf Stream Region.

The movement of an isotopic signal through a food web provides a natural tracer for quantifying the dynamics of nitrogen flow through an ecosystem, though it is important to note that the nitrogen present in an animal's body at any instant represents an integration of the animal's feeding and excretory processes over some finite time interval. This has the effect of buffering the animal's isotopic composition against rapid changes, ensuring that its δ^{15}N will always change more slowly than that of its food (Figure 7.4). In effect, animals act as low-pass filters with respect to isotopic changes in their food, and the greatest isotopic responses will appear among small, short-lived herbivores. Animals feeding at a higher trophic position as well as larger animals with a relatively long turnover time for body nitrogen should show little response to temporal changes in the δ^{15}N of phytoplankton. Thus, different components of the zooplankton community are sensitive to isotopic variation on different time scales and may provide an index to patterns of nutrient cycling over periods ranging from the hours or days to the months preceding the time of sampling.

Compound-specific nitrogen isotope analyses

The isotopic composition of bulk organic matter (e.g., suspended particles, zooplankton, or fish) has been used in diverse ecosystems as an index to the sources and processes affecting the biologically active pools of nitrogen. Food web and source studies, however, generally require the measurement of multiple pools to quantify relationships between nitrogen sources and animals. A great deal of additional information on the biological pathways and processing of nitrogen can be obtained from the isotopic composition of specific biomolecules isolated from individual organisms or detrital samples. This sort of compound-specific isotope ratio analysis can provide very specific information about the diet and physiology of organisms (Hare et al. 1991; Uhle et al. 1997; Fantle et al. 1999; McClelland & Montoya 2002), but until fairly recently, the large sample size needed for biochemical separations and isotopic analysis placed a major constraint on the range of compounds that could be characterized isotopically. This is a particularly challenging problem for nitrogen isotope studies since nitrogen is much less abundant than carbon in organic matter. With the development of continuous-flow gas chromatography/combustion/isotope ratio mass spectrometry (CF-GC/C/IRMS) techniques, however, compound-specific $\delta^{15}N$ measurements are now feasible for a wide range of marine samples (Macko & Uhle 1997; McClelland & Montoya 2002).

To date, amino acids have been the primary focus of work on compound-specific $\delta^{15}N$ analysis (Metges et al. 1996; Macko & Uhle 1997; Metges & Petzke 1997; McClelland & Montoya 2002; McClelland et al. 2003; Pakhomov et al. 2004; Schmidt et al. 2004). The analytical protocols, though labor intensive and time consuming, are now reasonably well established. In the first analytical step, amino acids have to be extracted from a biological sample, typically by hydrolysis at low pH. After purification of the extract, the amino acids must be derivatized to allow separation by gas chromatography. In one approach used in a number of recent studies, amino acids are derivatized to produce N-pivaloyl-i-propyl-amino acid esters (Metges et al. 1996; Metges & Petzke 1997; McClelland & Montoya 2002; McClelland et al. 2003; Pakhomov et al. 2004; Schmidt et al. 2004). This derivatization scheme allows analysis of 18 common amino acids as well as an internal standard (α-aminoadipic acid). Both glutamine and asparagine are deaminated and converted to glutamic acid and aspartic acid, respectively. Tryptophan, cystine/cysteine, and arginine are not compatible with this method.

A number of intriguing patterns have already emerged from studies of the $\delta^{15}N$ of individual amino acids in plankton. First, the $\delta^{15}N$ of individual amino acids can differ significantly from the $\delta^{15}N$ of the bulk organic matter where they were obtained. Second, laboratory feeding experiments and comparisons among different size fractions of zooplankton collected in the North Atlantic have revealed that the increase in bulk organic matter $\delta^{15}N$ with trophic position is associated with a wide range of shifts in $\delta^{15}N$ of amino acids: some amino acids have the same $\delta^{15}N$ in food and consumer tissues while other amino acids show a very large trophic shift (McClelland & Montoya 2002). The trophic shifts of individual amino acids are generally consistent for different experiments and sample sets, and two amino acids show particularly robust behavior: the δ^{15}N value of phenylalanine is unchanged between food source and consumer tissues, while the δ^{15}N of glutamic acid (+glutamine) consistently increases by about 7‰ between food and consumer (McClelland & Montoya 2002). The differential response of these amino acids to trophic processing provides a record of both the nitrogen sources supporting primary production at the base of the food web and the trophic position of an individual consumer. These two pieces of information are recorded by the δ^{15}N of phenylalanine and the difference in δ^{15}N of glutamic acid/glutamine and phenylalanine, or $\Delta\delta^{15}$N(glu-phe), respectively (McClelland & Montoya 2002).

Compound-specific measurement of δ^{15}N is a comparatively new area of research and relatively few investigators have applied this technique to marine systems. The work to date provides a good foundation for interpreting compound-specific δ^{15}N data and recent studies have used such measurements to explore spatial and temporal variations in the nitrogen-supporting primary production in the oligotrophic North Atlantic (McClelland et al. 2003) as well as the Southern Ocean (Pakhomov et al. 2004). Note that this approach allows us to sample the isotopic composition of autotrophs indirectly by analyzing the isotopic composition of heterotroph tissues, making compound-specific measurements particularly valuable in settings where the primary producers are difficult or impossible to obtain in sufficient quantity for isotopic analysis. In the same way, differences in the δ^{15}N of amino acids within an animal, in particular the parameter $\Delta\delta^{15}$N(glu-phe) (McClelland & Montoya 2002), can provide novel insight into the trophic structure or, as argued above, the efficiency of nitrogen transfer through an ecosystem, based on analyses of animal samples alone (McClelland et al. 2003).

Conclusions

Our current knowledge of the isotopic systematics of the marine nitrogen cycle gives us a useful natural tracer for studying the movement of nitrogen into and within oceanic ecosystems. The isotopic composition of a biologically active pool of nitrogen is an integrated measure of the processes affecting that pool over periods comparable to its turnover time, and we now have a general understanding of the nature and typical patterns of interaction of such processes. Nonetheless, it is worth noting that a number of fundamental aspects of nitrogen isotope biogeochemistry remain poorly understood and invite additional investigation. For example, the patterns and mechanisms of isotopic fractionation by animals are not yet well understood. A complete isotope budget has not been measured for any animal and would represent an important contribution to the quantitative use of δ^{15}N measurements in studies of plankton trophic biology and physiology. Similarly, compound-specific isotopic analysis is an emerging technology and a better understanding of the physiological and biochemical processes that generate differences in $\delta^{15}N$ among amino acids is clearly needed.

Samples for ^{15}N analysis typically can be collected with minimal handling at sea and often require little preparation before isotope ratio analysis with an automated mass spectrometer, potentially making stable isotope measurements accessible to a broad community of investigators. Although ^{15}N natural abundance measurements are not yet a routine part of large oceanographic field programs, they provide a powerful complement to traditional experimental studies of the marine nitrogen cycle using ^{15}N-labeled substrates and are especially useful in studying nitrogen-cycle processes acting on larger spatial and longer temporal scales than can be studied readily with standard shipboard experiments. The current challenge is to continue to develop and refine the analytical tools for carrying out stable isotope abundance measurements on small organic samples as well as specific biomolecules. Taken together, these stable isotope approaches may ultimately provide the best single tool for studying the dynamics of the nitrogen cycle on both short and long time scales.

Acknowledgment

This work was supported in part by NSF grants OCE-0425583 and OCE-0425610.

References

Altabet, M.A. & François, R. (1994) Sedimentary nitrogen isotopic ratio records surface ocean nitrate utilization. *Global Biogeochemical Cycles*, **8**, 103–116.

Altabet, M.A. & Small, L.F. (1990) Nitrogen isotopic ratios in fecal pellets produced by marine zooplankton. *Geochimica et Cosmochimica Acta*, **54**, 155–163.

Altabet, M.A., Deuser, W.G., Honjo, S. & Stienen, C. (1991) Seasonal and depth-related changes in the source of sinking particles in the North Atlantic. *Nature*, **354**, 136–139.

Altabet, M.A., François, R., Murray, D.W. & Preil, W.L. (1995) Climate-related variations in denitrification in the Arabian Sea from sediment $^{15}N/^{14}N$ ratios. *Nature*, **373**, 506–509.

Bada, J.L., Schoeninger, M.J. & Schimmelmann, A. (1989) Isotopic fractionation during peptide bond hydrolysis. *Geochimica et Cosmochimica Acta*, **53**, 3337–3341.

Barford, C.C., Montoya, J.P., Altabet, M.A. & Mitchell, R. (1999) Steady state nitrogen isotope effects of N_2 and N_2O production in *Paracoccus denitrificans*. *Applied and Environmental Microbiology*, **65**, 989–994.

Brandes, J.A. & Devol, A.H. (1997) Isotopic fractionation of oxygen and nitrogen in coastal marine sediments. *Geochimica et Cosmochimica Acta*, **61**, 1793–1801.

Brandes, J.A. & Devol, A.H. (2002) A global marine fixed nitrogen isotopic budget: Implications for Holocene nitrogen cycling. *Global Biogeochemical Cycles*, **16**, 1120, doi:10.1029/2001GB001856.

Brandes, J.A., Devol, A.H., Yoshinari, T., Jayakumar, D.A. & Naqvi, S.W.A. (1998) Isotopic composition of nitrate in the central Arabian Sea and eastern tropical North Pacific: A tracer for mixing and nitrogen cycles. *Limnology and Oceanography*, **43**, 1680–1689.

Capone, D.G., Subramanian, A., Montoya, J.P., et al. (1998) An extensive bloom of the N_2-fixing cyanobacterium, *Trichodesmium erythraeum*, in the central Arabian Sea during the spring intermonsoon. *Marine Ecology Progress Series*, **172**, 281–292.

Capone, D.G., Burns, J.A., Montoya, J.P., et al. (2005) Nitrogen fixation by *Trichodesmium* spp.: an important source of new nitrogen to the tropical and subtropical North Atlantic Ocean. *Global Biogeochemical Cycles*, **19**, GB2024, doi:10.1029/2004GB002331.

Carpenter, E.J., Harvey, H.R., Fry, B. & Capone, D.G. (1997) Biogeochemical tracers of the marine cyanobacterium *Trichodesmium*. *Deep-Sea Research*, **44**, 27–38.

Carpenter, E.J., Montoya, J.P., Burns, J., Mulholland, M., Subramanian, A. & Capone, D.G. (1999) Extensive bloom of a N_2-fixing symbiotic association (*Hemiaulis hauckii* and *Richelia intracellularis*) in the tropical Atlantic Ocean. *Marine Ecology Progress Series*, **188**, 273–283.

Checkley, D.M. & Entzeroth, L.C. (1985) Elemental and isotopic fractionation of carbon and nitrogen by marine, planktonic copepods and implications to the marine nitrogen cycle. *Journal of Plankton Research*, **7**, 553–568.

Checkley, D.M., Jr. & Miller, C.A. (1989) Nitrogen isotope fractionation by oceanic zooplankton. *Deep-Sea Research*, **36**, 1449–1456.

Church, M.J., Jenkins, B.D., Karl, D.M. & Zehr, J.P. (2005) Vertical distributions of nitrogen-fixing phylotypes at Station ALOHA in the oligotrophic North Pacific Ocean. *Aquatic Microbial Ecology*, **38**, 3–14.

Cifuentes, L.A., Sharp, J.H. & Fogel, M.L. (1988) Stable carbon and nitrogen isotope biogeochemistry in the Delaware estuary. *Limnology and Oceanography*, **33**, 1102–1115.

Cifuentes, L.A., Fogel, M.L., Pennock, J.R. & Sharp, J.H. (1989) Biogeochemical factors that influence the stable nitrogen isotope ratio of dissolved ammonium in the Delaware Estuary. *Geochimica et Cosmochimica Acta*, **53**, 2713–2721.

Cline, J.D. & Kaplan, I.R. (1975) Isotopic fractionation of dissolved nitrate during denitrification in the eastern tropical North Pacific Ocean. *Marine Chemistry*, **3**, 271–299.

Codispoti, L.A., Brandes, J.A., Christensen, J.P., et al. (2001) The oceanic fixed nitrogen and nitrous budgets: moving targets as we enter the Anthropocene? *Scientia Marina*, **65**, 85–105.

Delwiche, C.C. & Steyn, P.L. (1970) Nitrogen isotope fractionation in soils and microbial reactions. *Environmental Science and Technology*, **4**, 929–935.

DeNiro, M.J. & Epstein, S. (1981) Influence of diet on the distribution of nitrogen isotopes in animals. *Geochimica et Cosmochimica Acta*, **45**, 341–351.

Deutsch, C., Gruber, N., Key, R.M. & Sarmiento, J.L. (2001) Denitrification and N_2 fixation in the Pacific Ocean. *Global Biogeochemical Cycles*, **15**, 483–506.

Dore, J.E., Brum, J.R., Tupas, L. & Karl, D.M. (2002) Seasonal and interannual variability in sources of nitrogen supporting export in the oligotrophic subtropical North Pacific Ocean. *Limnology and Oceanography*, **47**, 1595–1607.

Falcon, L.I., Carpenter, E.J., Cipriano, F., Bergman, B. & Capone, D.G. (2004) N_2 fixation by unicellular bacterioplankton from the Atlantic and Pacific Oceans: Phylogeny and in situ rates. *Applied and Environmental Microbiology*, **70**, 765–770.

Fantle, M.S., Dittel, A.I., Schwalm, S.M., Epifanio, C.E. & Fogel, M.L. (1999) A food web analysis of the juvenile blue crab, *Callinectes sapidus*, using stable isotopes in whole animals and individual amino acids. *Oecologia*, **120**, 416–426.

Frazer, T.K., Ross, R.M., Quetin, L.B. & Montoya, J.P. (1997) Turnover of carbon and nitrogen during growth of larval krill, *Euphausia superba*: a stable isotope approach. *Journal of Experimental Marine Biology and Ecology*, **212**, 259–275.

Fry, B. (1988) Food web structure on Georges Bank from stable C, N, and S isotopic compositions. *Limnology and Oceanography*, **33**, 1182–1190.

Galloway, J.N., Howarth, R.W., Michaels, A.F., Nixon, S.W., Prospero, J.M. & Dentener, F.J. (1996) Nitrogen and phosphorus budgets of the North Atlantic Ocean and its watershed. *Biogeochemistry*, **35**, 3–25.

Ganeshram, R.S., Pedersen, T.F., Calvert, S.E. & Murray, J.W. (1995) Large changes in oceanic nutrient inventories from glacial to interglacial periods. *Nature*, **376**, 755–758.

Gannes, L.Z., O'brien, D.M. & Martinez Del Rio, C. (1997) Stable isotopes in animal ecology: Assumptions, caveats, and a call for more laboratory experiments. *Ecology*, **78**, 1271–1276.

Goering, J., Alexander, V. & Haubenstock, N. (1990) Seasonal variability of stable carbon and nitrogen isotopic ratios of organisms in a North Pacific bay. *Estuarine, Coastal and Shelf Science*, **30**, 239–260.

Gruber, N. (2004) The dynamics of the marine nitrogen cycle and its influence on atmospheric CO_2. In: *Carbon–Climate Interactions* (Eds M. Follows & T. Oguz), pp. 97–148. Kluwer Academic, Dordrecht.

Gruber, N. & Sarmiento, J.L. (1997) Global patterns of marine nitrogen fixation and denitrification. *Global Biogeochemical Cycles*, **11**, 235–266.

Hare, P.E., Fogel, M.L., Stafford, T.W., Jr., Mitchell, A.D. & Hoering, T.C. (1991) The isotopic composition of carbon and nitrogen in individual amino acids isolated from modern and fossil proteins. *Journal of Archaeological Science*, **18**, 277–292.

Heaton, T.H.E. (1986) Isotopic studies of nitrogen pollution in the hydrosphere and atmosphere: a review. *Chemical Geology (Isotope Geoscience Section)*, **59**, 87–102.

Hoering, T. (1955) Variations of nitrogen-15 abundance in naturally occurring substances. *Science*, **122**, 1233–1234.

Hoering, T.C. & Ford, H.T. (1960) The isotope effect in the fixation of nitrogen by *Azotobacter*. *Journal of the American Chemical Society*, **82**, 376–378.

Holmes, M.E., Muller, P.J., Schneider, R.R., Segl, M., Patzold, J. & Wefer, G. (1996a) Stable nitrogen isotopes in Angola Basin surface sediments. *Marine Geology*, **134**, 1–12.

Holmes, M.E., Müller, P.J., Schneider, R.R., Segl, M., Pötzold, J. & Wefer, G. (1996b) Stable nitrogen isotopes in Angola Basin surface sediments. *Marine Geology*, **134**, 1–12.

Holmes, M.E., Lauvik, G., Fischer, G., Segl, M., Ruhland, G. & Wefer, G. (2002) Seasonal variability of $\delta^{15}N$ in sinking particles in the Benguela Upwelling Region. *Deep-Sea Research Part I*, **49**, 377–394.

Karl, D.M., Christian, J.R., Dore, J.E., et al. (1996) Seasonal and interannual variability in primary production and particle flux at Station ALOHA. *Deep-Sea Research Part II*, **43**, 539–568.

Karl, D., Letelier, R., Tupas, L., Dore, J., Christian, J. & Hebel, D. (1997) The role of nitrogen fixation in biogeochemical cycling in the subtropical North Pacific Ocean. *Nature*, **388**, 533–538.

Knap, A.N., Sigman, D.M. & Lipschultz, F. (2005) N isotopic composition of dissolved organic nitrogen and nitrate at the Bermuda Atlantic Time-series Study site. *Global Biogeochemical Cycles*, **19**, doi:10.1029/2004GB002320,doi:10.1029/2004GB002320.

Lee, K., Karl, D.M., Wanninkhof, R.H. & Zhang, J.Z. (2002) Global estimates of net carbon production in the nitrate-depleted tropical and subtropical oceans. *Geophysical Research Letters*, **29**, doi:10.1029/2001GLO14198.

Liu, K.-K. & Kaplan, I.R. (1989) The eastern tropical Pacific as a source of ^{15}N-enriched nitrate in seawater off southern California. *Limnology and Oceanography*, **34**, 820–830.

Macko, S.A. & Uhle, M.E. (1997) Stable nitrogen isotope analysis of amino acid enantiomers by gas chromatography/combustion/isotope ratio mass spectrometry. *Analytical Chemistry*, **69**, 926–929.

Macko, S.A., Entzeroth, L. & Parker, P.L. (1984) Regional differences in nitrogen and carbon isotopes on the continental shelf of the Gulf of Mexico. *Naturwissenschaften*, **71**, 374–375.

Macko, S.A., Fogel Estep, M.L., Engel, M.H. & Hare, P.E. (1986) Kinetic fractionation of stable nitrogen isotopes during amino acid transamination. *Geochimica et Cosmochimica Acta*, **50**, 2143–2146.

Macko, S.A., Fogel, M.L., Hare, P.E. & Hoering, T.C. (1987) Isotopic fractionation of nitrogen and carbon in the synthesis of amino acids by microorganisms. *Chemical Geology (Isotope Geoscience Section)*, **65**, 79–92.

Mariotti, A., Germon, J.C., Hubert, P., et al. (1981) Experimental determination of nitrogen kinetic isotope fractionation: some principles; illustration for the denitrification and nitrification processes. *Plant and Soil*, **62**, 413–430.

Mariotti, A., Germon, J.C., Leclerc, A., Catroux, G. & Letolle, R. (1982) Experimental determination of kinetic isotopic fractionation of nitrogen isotopes during denitrification. In: *Stable Isotopes* (Eds H.-L. Schmidt, H. Forstel & K. Heinzinger), pp. 459–464. Elsevier, Amsterdam.

McClelland, J.W. & Montoya, J.P. (2002) Trophic relationships and the nitrogen isotopic composition of amino acids in plankton. *Ecology*, **83**, 2173–2180.

McClelland, J.W. & Valiela, I. (1998a) Changes in food web structure under the influence of increased anthropogenic nitrogen inputs to estuaries. *Marine Ecology Progress Series*, **168**, 259–271.

McClelland, J.W. & Valiela, I. (1998b) Linking nitrogen in estuarine producers to land-derived sources. *Limnology and Oceanography*, **43**, 577–585.

McClelland, J.W., Holl, C.M. & Montoya, J.P. (2003) Nitrogen sources to zooplankton in the Tropical North Atlantic: Stable isotope ratios of amino acids identify strong coupling to N_2-fixation. *Deep-Sea Research Part I*, **50**, 849–861.

McCready, R.G.L., Gould, W.D. & Barendregt, R.W. (1983) Nitrogen isotope fractionation during the reduction of NO_3 to NH_4 by *Desulfovibrio* sp. Canadian *Journal of Microbiology*, **29**, 231–234.

Metges, C.C. & Petzke, K.J. (1997) Measurement of ^{15}N/^{14}N isotopic composition in individual plasma free amino acids of human adults at natural abundance by gas chromatography-combustion isotope ratio mass spectrometry. *Analytical Biochemistry*, **247**, 158–164.

Metges, C.C., Petzke, K.-J. & Hennig, U. (1996) Gas Chromatography/combustion/isotope ratio mass spectrometric comparison of N-acetyl- and N-pivaloyl amino acid esters to measure ^{15}N isotopic abundances in physiological samples: a pilot study on amino acid synthesis in the upper gastro-intestinal tract of minipigs. *Journal of Mass Spectrometry*, **31**, 367–376.

Michaels, A.F., Olson, D., Sarmiento, J., et al. (1996) Inputs, losses and transformation of nitrogen and phosphorus in the deep North Atlantic Ocean. *Biogeochemistry*, **35**, 181–226.

Mills, M.M., Ridame, C., Davey, M., La Roche, J. & Geider, R.J. (2004) Iron and phosphorus co–limit nitrogen fixation in the eastern tropical North Atlantic. *Nature*, **429**, 292–294.

Minagawa, M. & Wada, E. (1984) Stepwise enrichment of ^{15}N along food chains: further evidence and the relation between $\delta^{15}N$ and animal age. *Geochimica et Cosmochimica Acta*, **48**, 1135–1140.

Miyake, Y. & Wada, E. (1967) The abundance ratio of $^{15}N/^{14}N$ in marine environments. *Records of Oceanographic Works in Japan*, **9**, 37–53.

Miyake, Y. & Wada, E. (1971) The isotope effect on the nitrogen in biochemical oxidation–reduction reactions. *Records of Oceanographic Works in Japan*, **11**, 1–6.

Montoya, J.P. & McCarthy, J.J. (1995) Nitrogen isotope fractionation during nitrate uptake by marine phytoplankton in continuous culture. *Journal of Plankton Research*, **17**, 439–464.

Montoya, J.P. & Voss, M. (2006) Nitrogen cycling in suboxic waters: Isotopic signatures of nitrogen transformations in the Arabian Sea Oxygen Minimum Zone. In: *Past and Present Water Column Anoxia* (Ed. L. Neretin), pp. 259–281. Springer-Verlag, Dordrecht.

Montoya, J.P., Horrigan, S.G. & McCarthy, J.J. (1990) Natural abundance of ^{15}N in particulate nitrogen and zooplankton in the Chesapeake Bay. *Marine Ecology Progress Series*, **65**, 35–61.

Montoya, J.P., Horrigan, S.G. & McCarthy, J.J. (1991) Rapid, storm-induced changes in the natural abundance of ^{15}N in a planktonic ecosystem. *Geochimica et Cosmochimica Acta*, **55**, 3627–3638.

Montoya, J.P., Wiebe, P.H. & McCarthy, J.J. (1992) Natural abundance of ^{15}N in particulate nitrogen and zooplankton in the Gulf Stream region and Warm-Core Ring 86A. *Deep-Sea Research*, **39**(Supplement 1), S363–S392.

Montoya, J.P., Carpenter, E.J. & Capone, D.G. (2002) Nitrogen-fixation and nitrogen isotope abundances in zooplankton of the oligotrophic North Atlantic. *Limnology and Oceanography*, **47**, 1617–1628.

Montoya, J.P., Holl, C.M., Zehr, J.P., Hansen, A., Villareal, T.A. & Capone, D.G. (2004) High rates of N_2-fixation by unicellular diazotrophs in the oligotrophic Pacific. *Nature*, **430**, 1027–1031.

Mullin, M.M., Rau, G.H. & Eppley, R.W. (1984) Stable nitrogen isotopes in zooplankton: some geographic and temporal variations in the North Pacific. *Limnology and Oceanography*, **29**, 1267–1273.

Nakatsuka, T., Handa, N., Wada, E. & Wong, C.S. (1992) The dynamic changes of stable isotope ratios of carbon and nitrogen in suspended and sedimented particulate organic matter during a phytoplankton bloom. *Journal of Marine Research*, **50**, 267–296.

Needoba, J.A. & Harrison, P.J. (2004) Influence of low light and a light:dark cycle on NO_3^- uptake, intracellular NO_3^-, and nitrogen isotope fractionation by marine phytoplankton. *Journal of Phycology*, **40**, 505–516.

Needoba, J.A., Waser, N.A.D., Harrison, P.J. & Calvert, S.E. (2003) Nitrogen isotope fractionation by 12 species of marine phytoplankton during growth on nitrate. *Marine Ecology Progress Series*, **255**, 81–91.

Nixon, S.W., Ammerman, J.W., Atkinson, L.P., et al. (1996) The fate of nitrogen and phosphorus at the land–sea margin of the North Atlantic Ocean. *Biogeochemistry*, **35**, 141–180.

Ostrom, N.E., Macko, S.A., Deibel, D. & Thompson, R.J. (1997) Seasonal variation in the stable carbon and nitrogen isotope biogeochemistry of a coastal cold environment. *Geochimica et Cosmochimica Acta*, **61**, 2929–2942.

Paerl, H.W. & Fogel, M.L. (1994) Isotopic characterization of atmospheric nitrogen inputs as sources of enhanced primary production in coastal Atlantic Ocean waters. *Marine Biology*, **119**, 635–645.

Pakhomov, E.A., Mcclelland, J., Bernard, K., Kaehler, S. & Montoya, J.P. (2004) Spatial and temporal shifts in stable isotope values of the bottom-dwelling shrimp *Nauticaris marionis* at the sub-Antarctic archipelago. *Marine Biology*, **144**, 317–325.

Pennock, J.R., Sharp, J.H., Ludlam, J., Velinsky, D.J. & Fogel, M.L. (1988) Isotopic fractionation of nitrogen during the uptake of NH_4^+ and NO_3^- by *Skeletonema costatum*. *Eos*, **69**, 1098.

Pennock, J.R., Velinsky, D.J., Ludlam, J.M., Sharp, J.H. & Fogel, M.L. (1996) Isotopic fractionation of ammonium and nitrate during uptake by *Skeletonema costatum*: Implications for delta N-15 dynamics under bloom conditions. *Limnology and Oceanography*, **41**, 451–459.

Prospero, J.M., Barrett, K., Church, T., et al. (1996) Atmospheric deposition of nutrients to the North Atlantic Basin. *Biogeochemistry*, **35**, 27–73.

Rau, G.H., Sweeney, R.E., Kaplan, I.R., Mearns, A.J. & Young, D.R. (1981) Differences in animal ¹³C, ¹⁵N, and D abundance between a polluted and an unpolluted coastal site: likely indicators of sewage uptake by a marine food web. *Estuarine, Coastal and Shelf Science*, **13**, 701–707.

Russell, K.M., Galloway, J.N., Macko, S.A., Moody, J.L. & Scudlark, J.R. (1998) Sources of nitrogen in wet deposition ot the Chesapeake Bay region. *Atmospheric Envionment*, **32**, 2453–2465.

Saino, T. & Hattori, A. (1980) ¹⁵N natural abundance in oceanic suspended particulate matter. *Nature*, **283**, 752–754.

Saino, T. & Hattori, A. (1987) Geographical variation of the water column distribution of suspended particulate organic nitrogen and its ¹⁵N natural abundance in the Pacific and its marginal seas. *Deep-Sea Research*, **34**, 807–827.

Schlacher, T.A., Liddell, B., Gaston, T.F. & Schlacher-Hoenlinger, M. (2005) Fish track wastewater pollution to estuaries. *Oecologia*, **144**, 570–584.

Schmidt, K., McClelland, J., Mente, E., Montoya, J.P., Atkinson, A. & Voss, M. (2004) Trophic-level interpretation based on $\delta^{15}N$ values: Implications of tissue-specific fractionation and amino acid composition. *Marine Ecology Progress Series*, **266**, 43–58.

Sheats, N. (2000) *The use of stable isotopes to define the extent of incorporation of sewage nitrogen into aquatic foodwebs and to discern differences in habitat suitability within a single estuary*. Dissertation, Department of Earth and Planetary Sciences, Harvard University. Cambridge, MA, 278 pp.

Sigman, D.M., Altabet, M.A., Michener, R., McCorkle, D.C. & François, R. (1996) A new method for the nitrogen isotopic analysis of oceanic nitrate and first results from the Southern Ocean. *Eos*, **76**, 143.

Sigman, D.M., Altabet, M.A., Michener, R., McCorkle, D.C., Fry, B. & Holmes, R.M. (1997) Natural abundance-level measurment of the nitrogen isotopic composition of oceanic nitrate: an adaptation of the ammonia diffusion method. *Marine Chemistry*, **57**, 227–242.

Sigman, D.M., Altabet, M.A., McCorkle, D.C., François, R. & Fisher, G. (1999) The $\delta^{15}N$ of nitrate in the Southern Ocean: Comsumption of nitrate in surface waters. *Global Biogeochemical Cycles*, **13**, 1149–1166.

Sigman, D.M., Altabet, M.A., Mccorkle, D.C., François, R. & Fischer, G. (2000) The $\delta^{15}N$ of nitrate in the Southern Ocean: Nitrogen cycling and circulation in the ocean interior. *Journal of Geophysical Research*, **105**, 19599–19614.

Spies, R.B., Kruger, H., Ireland, R. & Rice, D.W., Jr. (1989) Stable isotope ratios and contaminant concentrations in a sewage-distorted food web. *Marine Ecology Progress Series*, **54**, 157–170.

Sutka, R.L., Ostrom, N.E., Ostrom, P.H. & Phanikumar, M.S. (2004) Stable nitrogen isotope dynamics of dissolved nitrate in a transect from the North Pacific Subtropical Gyre to the Eastern Tropical North Pacific. *Geochimica et Cosmochimica Acta*, **68**, 517–527.

Sweeney, R.E., Liu, K.K. & Kaplan, I.R. (1978) Oceanic nitrogen isotopes and their uses in determing the sources of sedimentary nitrogen. In: *Stable Isotopes in the Earth Sciences* (Ed. B. W. Robinson), pp 9–26. Bulletin 220, Department of Scientific and Industrial Research, Wellington.

Sweeney, R.E., Kalil, E.K. & Kaplan, I.R. (1980) Characterisation of domestic and industrial sewage in southern California coastal sediments using nitrogen, carbon, sulphur, and uranium tracers. *Marine Environmental Research*, **3**, 225–243.

Tucker, J.N., Sheats, N., Giblin, A.E., Hopkinson, C.S. & Montoya, J.P. (1999) Using stable isotopes to trace sewage derived material through Boston Harbor and Massachusetts Bay. *Marine and Environmental Research*, **48**, 353–375.

Uhle, M.E., Macko, S.A., Spero, H.J., Engel, M.H. & Lea, D.W. (1997) Sources of carbon and nitrogen in modern planktonic foraminifera: the role of algal symbionts as determined by bulk compound specific stable isotopic analysis. *Organic Geochemistry*, **27**, 103–113.

Van Dover, C.L., Grassle, J.F., Fry, B., Garritt, R.H. & Starczak, V.R. (1992) Stable isotope evidence for entry of sewage-derived organic material into a deep-sea food web. *Nature*, **360**, 153–155.

Voss, M. & Struck, U. (1997) Stable nitrogen and carbon isotopes as indicator of eutrophication of the Oder river (Baltic sea). *Marine Chemistry*, **59**, 35–49.

Voss, M., Dippner, J. & Montoya, J.P. (2001) Nitrogen isotope patterns in the oxygen deficient waters of the Eastern Tropical North Pacific (ETNP). *Deep-Sea Research*, **48**, 1905–1921.

Voss, M., Croot, P., Lochte, K., Mills, M. & Peeken, I. (2004) Patterns of nitrogen fixation along 10°N in the tropical Atlantic. *Geophysical Research Letters*, **31**, doi:10.1029/2004GL020127.

Wada, E. (1980) Nitrogen isotope fractionation and its significance in biogeochemical processes occurring in marine environments. In: *Isotope Marine Chemistry* (Eds E. Goldberg, Y. Horibe & K. Saruhashi), pp. 375–398. Uchida Rokakuho, Tokyo.

Wada, E. & Hattori, A. (1976) Natural abundance of ^{15}N in particulate organic matter in the North Pacific Ocean. *Geochimica et Cosmochimica Acta*, **40**, 249–251.

Wada, E. & Hattori, A. (1978) Nitrogen isotope effects in the assimilation of inorganic nitrogenous compounds. *Geomicrobiology Journal*, **1**, 85–101.

Wada, E. & Hattori, A. (1991) *Nitrogen in the Sea: Forms, Abundances, and Rate Processes*. CRC Press, Boca Raton, FL.

Wada, E., Kadonaga, T. & Matsuo, S. (1975) ^{15}N abundance in nitrogen of naturally occurring substances and global assessment of denitrification from isotopic viewpoint. *Geochemical Journal*, **9**, 139–148.

Wada, E., Terazaki, M., Kobaya, Y. & Nemoto, T. (1987) ^{15}N and ^{13}C abundances in the Antarctic Ocean with emphasis on biogeochemical structure of the food web. *Deep-Sea Research*, **34**, 829–841.

Waser, N.A., Yin, K., Yu, Z., et al. (1998a) Nitrogen isotope fractionation during nitrate, ammonium and urea uptake by marine diatoms and coccolithophores under various conditions of N availability. *Marine Ecology Progress Series*, **169**, 29–41.

Waser, N.A.D., Harrison, P.J., Nielsen, B., Calvert, S.E. & Turpin, D.H. (1998b) Nitrogen isotope fractionation during the uptake and assimilation of nitrate, nitrite, ammonium, and urea by a marine diatom. *Limnology and Oceanography*, **43**, 215–224.

Wu, J.P., Calvert, S.E. & Wong, C.S. (1999) Carbon and nitrogen isotope ratios in sediment-ing particulate organic matter at an upwelling site off Vancouver Island. *Estuarine Coastal and Shelf Science*, **48**, 193–203.

Yoshida, N. (1988) ^{15}N-depleted N_2O as a product of nitrification. *Nature*, **335**, 528–529.

Zehr, J.P., Mellon, M.T. & Zani, S. (1998) New nitrogen-fixing microorganisms detected in oligotrophic oceans by amplification of nitrogenase (nifH) genes. *Applied and Environmental Microbiology*, **64**, 3444–3450.

Zehr, J.P., Waterbury, J.B., Turner, P.J., et al. (2001) New nitrogen-fixing unicellular cyanobacteria discovered in the North Pacific Central Gyre. *Nature*, **412**, 635–638.

CHAPTER 8

Stable isotope studies in marine chemoautotrophically based ecosystems: An update

CINDY LEE VAN DOVER

Introduction

When deep-sea hydrothermal vents were discovered in the late 1970s, stable isotope compositions of vent invertebrates outside the expected range for organisms whose nutrition is based on photosynthetically derived organic carbon suggested that primary production at vents was autochthonous and chemoautotrophic (Rau & Hedges 1979; Rau 1981a, 1981b; Fry et al. 1983). During the decades that followed, stable isotope techniques have been used extensively as tools for understanding trophic relationships in chemoautotrophic ecosystems (reviewed in Conway et al. 1994; Van Dover 2000). This update highlights several themes that prevail in recent studies that use stable isotope techniques as a means of understanding energy metabolism and transfer in chemoautotrophic ecosystems (sulfide and methane-based systems), including documentation of archael–bacterial syntrophy in methane-rich sediments, studies of ontogenetic and spatial variations in stable isotope compositions within species to identify important autecological variables, and continuing surveys of patterns of bulk stable isotope compositions in vent and seep communities to generate preliminary assessments of carbon sources and trophic interactions. Carbon and nitrogen isotope analyses and interpretations prevail in most of this work. Sulfur isotopes have the potential to be very useful wherever sulfide oxidation or sulfate reduction are important microbial processes, but their use has continued to be limited due to the larger sample sizes and more complicated analytical procedures required.

Definition of chemoautotrophic ecosystems

Chemoautotrophically based marine ecosystems (herein referred to as chemoautotrophic ecosystems) are now known to occur in diverse settings extending through all depths of the ocean, from coastal waters into the deepest regions of subduction zones. In aerobic microbial chemotrophic production, reduced compounds [typically hydrogen sulfide (used by thiotrophs) and/or methane (used by methanotrophs)] serve as electron donors, and oxygen serves as the electron acceptor. In anaerobic microbial chemoautotrophic production, H_2 is typically the electron donor, with CO_2 (consumed by methanogens), sulfate or elemental sulfur (consumed by sulfur/sulfate reducers), or nitrate (consumed by denitrifiers) serving as electron acceptors. Discussion in this review is for the most part restricted to those chemoautotrophic ecosystems in the deep sea that support a large biomass of invertebrates, i.e., hydrothermal vents and cold seeps.

Hydrothermal vents at mid-ocean ridge volcanic settings (reviewed in Van Dover 2000) are chemoautotrophic ecosystems where sources of reduced compounds (especially hydrogen sulfide) result from degassing of magma chambers and from water-rock (basalt) interactions at elevated temperatures (1200°C). Vents are also common on back-arc spreading centers, where water may react with oceanic and continental crust as well as with pelagic sediments. Less well known are hydrothermal vents associated with volcanic seamounts (e.g., Loihi Seamount; Hilton et al. 1998). Warm- or hot-water vents are also known in coastal settings, where the rock component of the reaction may be continental rather than oceanic, where reaction temperatures may be well below 1200°C, and where interaction with sediments is an important factor. The most recently discovered form of hydrothermal venting in the deep sea may be promoted by tectonic rather than volcanic activity, and has been attributed to exothermic reactions of cold mantle rock exposed to seawater (serpentinization), resulting in alkaline, methane- and sulfide-enriched fluids (Kelley et al. 2001). Given that source inorganic carbon, nitrogen, and sulfur isotopic compositions vary among each of these settings, there is potential for a range of isotopic compositions in organisms from these different types of hydrothermal vents, even in organisms relying on the same microbial metabolism.

In contrast to vents, cold seeps (reviewed in Sibuet & Olu 1998) lack any significant thermal anomaly. Like hydrothermal vents, seeps occur in diverse settings. Methane hydrates of tectonically passive continental margins, where methane accumulated from anaerobic microbial degradation of sedimentary organic material (methanogenesis) may be delivered through conduits in the overlying sediments to the seafloor, are among the most ubiquitous of seep environments (Kvenvolden & Lorenson 2001). Methane-hydrate seep communities are often associated with relatively buoyant salt diapers that deform the overlying sediment and result in destabilization of the subsurface gas hydrates, as in the case of the Blake Ridge seep site (Taylor et al. 2000). Seeps also commonly occur in tectonically active continental margins, where sediments are squeezed, resulting in migration of methane-rich pore waters to the seafloor (e.g., Lewis & Marshall 1996; Olu et al. 1996). Where methane-rich pore-water fluids in anaerobic sediments are exposed to seawater sulfate (i.e., the upper 10 cm of the seabed), methane oxidation is typically coupled to sulfate reduction to yield CO_2 and H_2S, the inorganic compounds that are prerequisites for aerobic chemoautotrophy.

Chemoautotrophic ecosystems also occur in association with bulk organic enrichments on the seafloor. Lipid-rich whale skeletons are one example of such an environment; wood falls and bulk deposition of macroalgae (e.g., kelp) are others. As at seeps, reduced sulfur compounds used in chemoautotrophic primary production are generated at whale falls by microbial sulfate reduction (reviewed by Smith & Baco 2003).

Isotopic compositions of inorganic sources

Carbon

Carbon dioxide and methane are the inorganic and C1 (single-carbon organic) carbon sources of interest in chemoauototrophic ecosystems. $\delta^{13}C$ values [vs. PDB (Pee Dee Belemnite)] of carbon dioxide in seawater varies globally by less than 3‰, centered around 0.5‰ (Lynch-Stieglitz et al. 1995). CO_2 degassing from submarine mid-ocean ridges derives from carbon present in mid-ocean ridge basalt, which has a $\delta^{13}C$ value of −5.2 ± 0.7‰ (Marty & Zimmerman 1999). A comprehensive review of carbon-isotope systematics for methane may be found in Whiticar (1999), who used C and H isotopes to define compositional fields of methane for different methanogenic pathways (Figure 8.1). While H isotope systematics are useful in resolving methane sources, they have so far not been exploited in ecological studies of marine chemoautotrophic ecosystems.

Bacterially derived methane (often called "biogenic" methane) has carbon-isotope compositions that vary from −110‰ to −50‰ (Whiticar 1999). This range is subdivided into methane formed in freshwater (−70‰ to −50‰; bacterial, methyl-type fermentation) and methane formed in marine sedimentary systems (<−100‰ to ca. −50‰; bacterial carbonate reduction) (Whiticar 1999). Thermogenic methane, i.e., methane generated from organic matter or kerogen that has undergone heating, has a non-bacterial origin but is ultimately derived from biological processes. Thermogenic methane is typically enriched in ^{13}C relative to bacterial CH_4, and ranges from −50‰ to −20‰ (Whiticar 1999).

Methane can also be of mantle origin in volcanic systems, with carbon isotopic compositions ranging from −12‰ to 0‰ (Whiticar 1999). Recent experimental studies of the carbon isotopic composition of methane formed from dissolved bicarbonate under hydrothermal conditions suggest the potential for kinetic isotopic fractionation yielding carbon isotope values of −53.6‰ to −19.1‰, i.e., overlapping with values for microbially derived CH_4 (Horita & Berndt 1999). Methane collected from hydrothermal vents on the Mid-Atlantic Ridge, East Pacific Rise, and southern Juan de Fuca Ridge has carbon isotopic compositions ranging from −24.1‰ to −8‰ (Table 8.1),

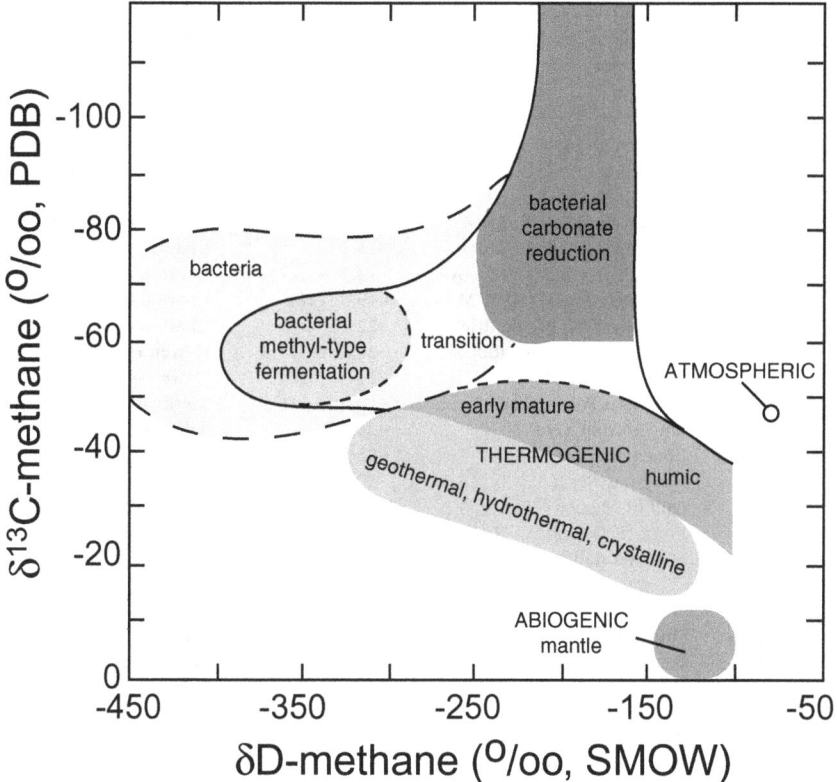

Figure 8.1 Carbon and hydrogen isotopic signatures of methane sources. Thermogenic methane includes methane thermogenically released from keogens (early-mature to humic) and other sources (geothermal, hydrothermal, crystalline). Bacterial methane includes methane formed in freshwater (methyl-type fermentation) and in marine environments (carbonate reduction). (Redrawn from Whiticar 1999.)

consistent with kinetic isotopic fractionation during abiotic processes, although possible contributions from a deep biosphere cannot be excluded (Charlou et al. 2002). Contributions of shallow or deep subsurface organic material to methane production in hydrothermal systems has been inferred from ^{12}C-enriched methane carbon-isotope values found in vent fluids of Endeavour and Guaymas Basin vents (−55‰ to −43‰, Table 8.1; Lilley et al. 1983; Welhan & Lupton 1987). CO_2 in predominantly thermogenic methane settings (e.g., Guaymas Basin, Gulf of Mexico) has a $\delta^{13}C$ composition closer to that of seawater than CO_2 in predominantly biogenic methane settings, where $\delta^{13}C$ values are more depleted in ^{13}C due to the production of

Table 8.1 Partial list of carbon isotopic compositions of inorganic carbon sources (methane and carbon dioxide). MAR = Mid-Atlantic Ridge.

Source	Location	$\delta^{13}C$ (‰)	Reference
Vent methane	Logatchev (14°45′N, MAR)	−14.3	Lein & Sagalevich 2000
	Logatchev (14°45′Ń, MAR)	−13.6	Charlou et al. 2002
	TAG (26°N, MAR)	−9.5, −8.0	Charlou et al. 2002
	Broken Spur (29°N, MAR)	−19, −18	Charlou et al. 2002
	Rainbow (36°14′N, MAR)	−15.8	Charlou et al. 2002
	Lucky Strike (37°17′N, MAR)	−13.7, −12.7	Charlou et al. 2002
	Menez Gwen (37°50′N, MAR)	−19.6, −18.8	Charlou et al. 2002
	Southern East Pacific Rise	−22	Charlou et al. 1996
	South Juan de Fuca Ridge	−20.8 to −17.8	Cowen et al. 2002
	21N, East Pacific Rise	−16	Cowen et al. 2002
	13N, East Pacific Rise	−19.5 to −16	Merlivat et al. 1987
	Lihir Harbor vent	−24.1	Schmidt et al. 2002
	Endeavour Ridge	−55 to −48.4	Cowen et al. 2002
Thermogenic methane	Guaymas Basin	−50.8 to −43.2	Cowen et al. 2002
	Gulf of Mexico	−46.3	Sassen et al. 1999
Laboratory abiogenic methane	From HCO_3 and H_2; Ni–Fe alloy catalyst	−53.6 to −46.0 (200°C)	Horita & Berndt 1999 Horita & Berndt 1999
		−33.6 to −19 (300°C)	
Seep biogenic methane	Unaltered biogenic methane	−70	Whalen 1993
	Eel River Basin	−65 to −35; average −50.4 ($n = 28$)	Paull, in Orphan et al. 2001
	Oregon Margin	−65	Suess & Whiticar 1989
	Eel River (CA) microbial mats	−60	Richnow & Ziebis, in Levin & Michener 2002
	Eel River (CA) *C. Pacifica* beds	−47	Richnow & Ziebis, in Levin & Michener 2002
	Eel River (CA) *C. pacifica* beds	−49.7	Orphan et al. 2002
	New Ireland Fore arc basin (mix of biogenic and thermogenic)	−51 to −58	Schmidt et al. 2002
	Pockmark, Norwegian coastline (xx m)	−29.2 to −38.3	Hovland & Risk 2003
	Gulf of Mexico	−48.7	Fisher et al. 2000
	Black Sea	−68.3 to −62.4	Michaelis et al. 2002
	Håkon Mosby Mud Volcano	−60.6 to −59.2	Damm & Budéus 2003
	Nankai Trough	−90	Tsunogai et al. 2002
	Makran accretionary prism (Arabian Sea)	−77.8	Von Rad et al. 1996
CO_2	Seawater	~0	
	Hydrothermal vent fluid	−7	Craig et al. 1980
	Guaymas Basin porewater CO_2 where thermogenic processes dominate	−13.3 to −8.5	Teske et al. 2002
	Gulf of Mexico CO_2 where thermogenic processes dominate	−2.5	Sassen et al. 1999
	Gulf of Mexico CO_2 where biogenic processes dominate	−29.8 to −21	Sassen et al. 1999
	Nankai Trough where biogenic processes dominate	<−40	Tsunogai et al. 2002

Table 8.2 Isotopic discriminations.

	^{12}C enrichment (‰)	Reference
CO_2 to CH_4 (marine methanogenesis)*	49–95	Whiticar 1999
CH_4 to organic $C^†$ (marine methanotrophy)	5–30	Whiticar 1999
CH_4 to anaerobic lipid biomarkers	30–50	Tables 8.1 & 8.4
CO_2 to organic C (marine chemoautotrophy)		
RubisCO form I	22–30	Guy et al. 1993
RubisCO form II	18–23	Roeske & O'Leary 1985, Guy et al. 1993, Robinson et al. 2003

* Fractionation factor decreases with increasing temperature (Whiticar 1999) and rate of methanogenesis (Zyakun 1992).
† Plus up to 10‰ discrimination against ^{13}C during lipid synthesis (Summons et al. 1994).

^{13}C-depleted CO_2 during microbial oxidation of methane (Sassen et al. 1999; Teske et al. 2002; Table 8.1).

Fractionation of carbon isotopes during methanotrophy and chemoautotrophy

In addition to documentation of the isotopic composition of plausible inorganic or C1 carbon substrates, interpretation of $\delta^{13}C$ values of field collected primary producers in chemosynthetic ecosystems requires an understanding of fractionation effects introduced during assimilation of the source inorganic carbon (Table 8.2). From *in vitro* experiments, such as those briefly mentioned below for selected methanotrophic and chemoautotrophic enzyme systems, theoretical $\delta^{13}C$ values can be calculated by applying measured fractionation effects to the $\delta^{13}C$ value of a source CO_2. $\delta^{13}C$ values obtained from field samples are typically less fractionated than predicted by enzymatic fractionation factors alone, due to other processes (including diffusion and carbon limitation) that reduce the degree of isotopic discrimination (Robinson et al. 2003).

Growth of aerobic methanotrophs in open culture systems with a constant supply of methane results in bulk ^{13}C-fractionation depletion factors as high as 31‰, with a further 10‰ depletion taking place during lipid synthesis (Summons et al. 1994). Maximum isotopic fractionation is associated with type I and type X methanotrophs, which use the ribulose monophosphate pathway for carbon assimilation, and with the p-methane monooxygenase enzyme system (Jahnke et al. 1999). In type II methanotrophs, which use a serine pathway, assimilated carbon derives from a mix of methane and carbon dioxide, and the degree of fractionation is dependent on the relative availability of CO_2 and CH_4, as well as other factors (especially the type of methane monooxygenase; Jahnke et al. 1999).

CO_2-fixation during chemoautotrophy by microbial endosymbionts typically involves the Calvin cycle and the CO_2-fixing enzyme ribulose 1,5-bisphosphate carboxylase/oxygenase (RubisCO; Cavanaugh 1994). Form I and form II RubisCOs catalyze the same reactions, but form I RubisCOs generally have higher affinities and specificity factors and function more efficiently under low CO_2 and high O_2 conditions (Horken & Tabita 1999). Form I RubisCO discriminates more effectively against [13]C (fractionation factor = 22–30‰; Guy et al. 1993) than form II RubisCO (fractionation factor = 18–23‰; Roeske & O'Leary 1984; Guy et al. 1993).

Nitrogen

Atmospheric and dissolved seawater nitrogen (N_2) have $\delta^{15}N$ (vs. air) values of 0‰, whereas oceanic nitrate has an average $\delta^{15}N$ value between +4‰ and +5‰ (Sigman et al. 1997). Under certain conditions at hydrothermal vents and seeps, ammonium may be present (Von Damm et al. 1985, 1998; Lilley et al. 1993) or inferred (e.g., MacAvoy et al. 2002). Volatile nitrogen (N_2) derived from mid-ocean ridge basalts presumably has a $\delta^{15}N$ signature of $-3.3 \pm 1.0‰$ (Marty & Zimmerman 1999). Direct measures of source inorganic nitrogen isotopic ratios are scarce or non-existent in studies of chemoautotrophic ecosystems, although bulk isotopic compositions of bacterial and invertebrate biomass repeatedly document a ^{14}N-enriched food-web, presumably dependent on a local source of nitrogen (rather than oceanic nitrate).

Sulfur

Seawater sulfate has a $\delta^{34}S$ [vs. CDT (Canyon Diablo Troilite)] value of +20‰ (Bottrell & Raiswell 2000). Microbial sulfate reduction in sediments favors the lighter, [32]S isotope, yielding [32]S-enriched sulfide and [32]S-depleted residual sulfate, but the fractionation factor depends on the degree to which the system is open or closed and on the rate of reduction (Habicht & Canfield 1997); the resulting sulfide may have $\delta^{34}S$ values from $-50‰$ to +50‰ (Bottrell & Raiswell 2000). Dissolved H_2S in end-member hydrothermal-vent fluids typically has a $\delta^{34}S$ value of +1‰ to +8‰, i.e., enriched in [32]S relative to seawater sulfate (Shanks et al. 1995, and references cited therein).

Isotopic tracing of carbon at methane seeps

Biogenic methane (and other low-molecular-weight gases) derived from microbial degradation of organic material via methanogenesis can be trapped in a lattice of hydrogen-bonded water molecules to form methane (gas) hydrates. The biogenic origin of the hydrated methane is inferred from the carbon and hydrogen isotopic systematics of the methane, as in Figure 8.1. Methane hydrates are stable at low temperatures (<10°C) and high pressures (>3 MPa), which means that they occur naturally in methane-rich seafloor sediments. Hydrate deposits can be detected using remote sensing (i.e., seismic profiling of sediments to reveal the Bottom Simulating Reflector (BSR), which is indicative of subsurface hydrates) and are a potential exploitable energy resource, containing two to five times the amount of methane in conventional gas sources (Kvenvolden 1999). Methane is often readily available at the sediment–water interface in hydrate settings, with delivery resulting from various processes, including steady-state deposition/dissolution of methane hydrates, deformation and elevation of hydrate-bearing sediments by diapiric processes, and by tectonic processes (faulting, fissuring) that create conduits for migration of methane from free-gas zones beneath the hydrate cap. Methane hydrates in deep water may support localized chemoautotrophic communities; examples include Hydrate Ridge on the Cascadia convergent margin (northeast Pacific), dominated by vesicomyid clams and mats of filamentous bacteria (Sahling et al. 2002), and Blake Ridge (northwest Atlantic, off the Carolina coast), dominated by bathymodiolid mussels and vesicomyid clams (Van Dover et al. 2003). Invertebrates making up the bulk of the biomass at these seeps depend on primary production by sulfide-oxidizing microbial endosymbionts (vesicomyid clams and certain bathymodiolin mussel species) or on a combination of sulfide- and methane-oxidizing endosymbionts (other bathymodiolin mussel species).

Evidence for archaeal–bacterial syntrophy in methane-rich sediments

Recent studies using lipid biomarkers, molecular probes, and isotope techniques have led to an appreciation of how the inverse relationship between sulfate reduction and methane oxidation in the upper layers of marine sediments, where sulfate becomes depleted downward as methane becomes depleted upward (Figure 8.2a; Borowski et al. 1999), can be microbially mediated (DeLong 2000). Reverse methanogenesis by archaeal methanogens (Hoehler et al. 1994) consumes methane under anaerobic conditions according to the reaction:

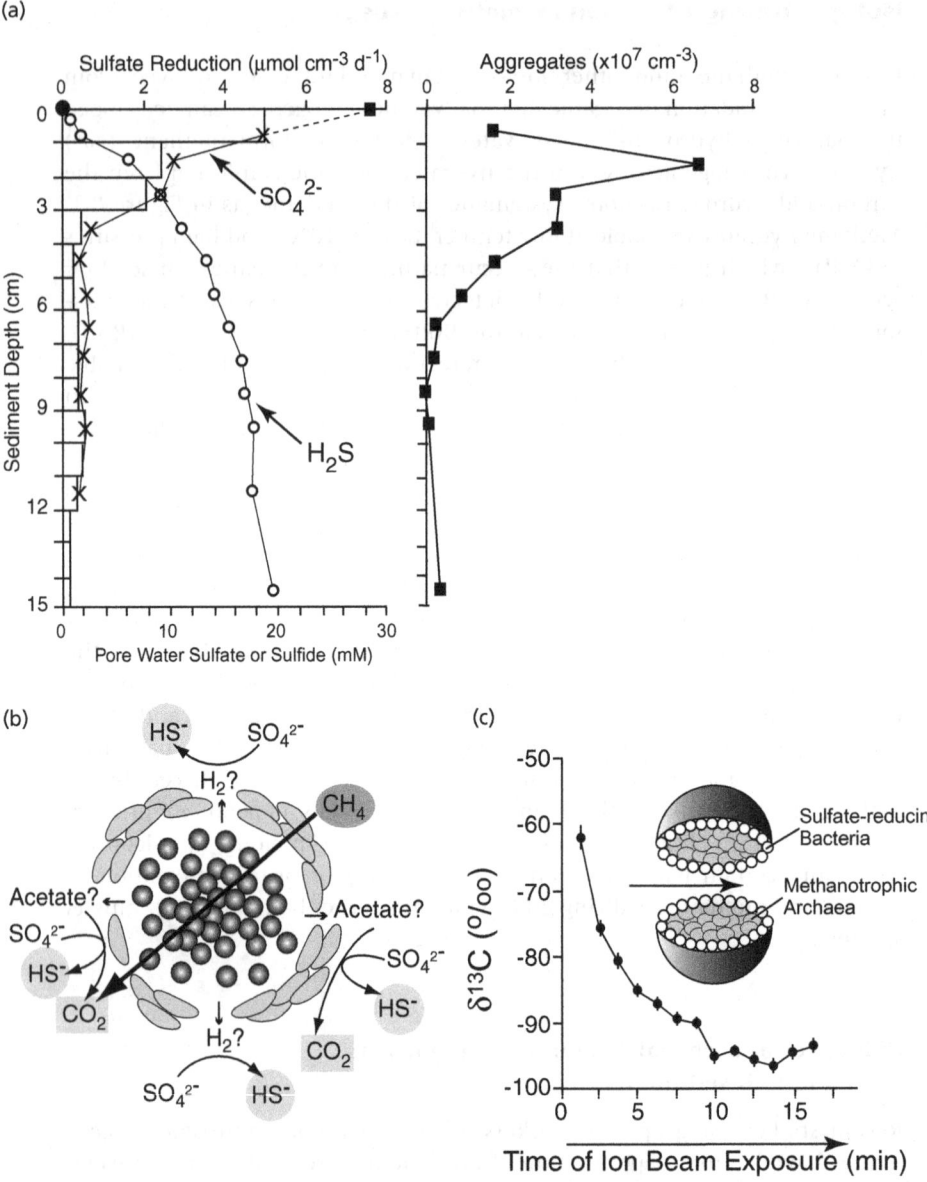

Figure 8.2 (a) Depth profiles of sulfate concentration, sulfide concentration, sulfate reduction rates, and abundance of methanotrophic archaea and sulfate-reducing bacteria aggregates (consortia) in methane-rich sediments. (Redrawn from Boetius et al. 2000.) (b) Spatial arrangement and potential metabolic couplings of the microbial consortium reported by Boetius et al. (2000). Cells in the center of the consortia are methane-consuming archaea; cells on the periphery are sulfate-reducing bacteria. (Redrawn from DeLong 2000.) (c) $\delta^{13}C$ (‰) profile (ion microprobe technique) vs. time through a 10-μm aggregate. (Redrawn from Orphan et al. 2001.)

$$CH_4 + 2H_2O \rightarrow CO_2 + 4H_2$$

This reaction is favored when H_2 is removed by hydrogen-oxidizing sulfate-reducers,

$$H^+ + 4H_2 + SO_4^{2-} \rightarrow HS^- + 4H_2O$$

and by CO_2 removal through carbonate deposition (Zehnder & Brock 1979). The overall reaction of such a microbial consortium is

$$CH_4 + SO_4^{2-} \rightarrow HCO^{3-} + HS^- + H_2O$$

Microorganisms responsible for these reactions have not been isolated and cultured, but their existence has been unquestionably demonstrated using multiple diagnostic techniques. Evidence for anaerobic methanotrophic archaea in marine sediments with high concentrations of biogenic methane in the pore waters include the presence of archaeal ether lipid biomarkers (e.g., archaeol, crocetane, and hydroxyarchaeols) depleted in ^{13}C ($\delta^{13}C$ as low as -130‰; e.g., Elvert et al. 2000; Hinrichs et al. 1999; Thiel et al. 1999; Pancost et al. 2000; Table 8.3). ^{13}C depletion of the archaeal biomarkers implicate methane as an energy resource rather than as a metabolic product in these organisms (Elvert et al. 2000; Hinrichs et al. 1999). Bacterially derived, ^{13}C-depleted ($\delta^{13}C = -111\text{‰}$ to -88‰) glycerol ethers and fatty acids extracted together with ^{13}C-depleted archaeal ether lipids ($\delta^{13}C$ as low as -128‰) from methane seeps in the Eel River and Santa Barbara Basins, implicate a syntrophic relationship (Hinrichs et al. 2000, Pancost et al. 2000). Boetius et al. (2000) used fluorescence $in\ situ$ hybridization to distinguish archaea and bacteria in naturally occurring aggregations in the upper 3–5 cm of mud cores from the Eel River Basin, revealing a central core of archaea surrounded by an outer layer of sulfate-reducing bacteria (Figure 8.2b). $In\ situ$ fluorescence hybridization coupled with secondary ion mass spectrometry was used to determine the carbon isotope composition of different cell types. As the ion beam progressed from the outer layer of sulfate reducers to the inner core, $\delta^{13}C$ values became increasingly depleted in ^{13}C (from $<-100\text{‰}$ to ca. -60‰; Figure 8.2c), indicating assimilation of isotopically light methane into archaeal cells (Orphan et al. 2001). Biomarker studies demonstrate that $CH_4/H_2/SO_4^{2-}$-based archaeal/bacterial syntrophic metabolism is not exclusive to methane-hydrate deposits; these consortia occur in a variety of settings, including gas seeps in the Black Sea (Thiel et al. 2001; Michaelis et al. 2002), mud volcanoes in the Mediterranean Sea (Werne et al. 2002), ancient sediments (Thiel et al. 1999), and anoxic waters (Schouten et al. 2001). ^{13}C-depleted archaeal biomarkers are also reported from hydrothermally active sediments in Guaymas Basin (Teske et al. 2002), where the methane source is thermogenic ($\delta^{13}C = -51\text{‰}$ to -43‰) and the corresponding archaeal lipids

Table 8.3 Carbon isotopic composition of archaea and archaeal and bacterial biomarkers in sediments. ANME-1,2 = anaerobic methane oxidizing phyologenetic groups based on 16S rRNA genes; PMI = 2,6,10,15,19-pentamethylicosane.

	Location	Biomarker	$\delta^{13}C$ (‰)	Reference
Archaeal	Aleutian trench	Crocetane	−130.3 to −124.6	Elvert et al. 2000
		PMI	−107 to −71	Elvert et al. 2000
	Kazan Mud Volcano	sn-3-hydroxy archaeol	−107 ± 4.5 sd	Werne et al. 2002
	(Mediterranean Sea)	Archaeol	−95 ± 8.8 sd	Werne et al. 2002
	Napoli, Amsterdam Mud Volcanoes (Mediterranean Sea)	Archaeol, sn-3-hydroxyarchaeol, crocetane	−66.1 to −99.0	Aloisi et al. 2002
	Hydrate Ridge	Crocetane	−117.9 to −107.6	Elvert et al. 1999
		Crocetane	−124	Boetius et al. 2000
		PMI	−123.8 to −101.7	Elvert et al. 1999
		sn-3-hydroxyarchaeol	−133	Boetius et al. 2000
		Archaeol	−114	Boetius et al. 2000
	Black Sea microbial mat	PMI	−75	Thiel et al. 2001
	Black Sea	Crocetane	−94.7 ± 0.7	Michaelis et al. 2002
		Archaeol	−87.9 ± 0.6	Michaelis et al. 2002
		sn-2-hydroxyarchaeol	−90.0 ± 1.0	Michaelis et al. 2002
		PMI	−95.6 ± 0.1	Michaelis et al. 2002
	Mediterranean Ridge	Archaeol	−85	Pancost et al. 2000
		PMI	−77	Pancost et al. 2000
	Santa Barbara Basin	Archaeol	−119 to −100	Hinrichs et al. 2000
		Crocetane	−119	Hinrichs et al. 2000
		sn-2-hydroxyarchaeol	−128 to −101	Hinrichs et al. 2000
		PMI	−129 to −76	Hinrichs et al. 2000
	Eel River Basin	ANME-1	to −83; average −75	Orphan et al. 2002
		Archaeol	−100	Hinrichs et al. 1999
		Archaeol	−104.1	Orphan et al. 2001
		ANME-2	to −92	Orphan et al. 2002
		sn-2-hydroxyarchaeol	−110, −105	Hinrichs et al. 1999
		sn-2-hydroxyarchaeol	−107.6	Orphan et al. 2001
	Guaymas Basin*	Archaeol	−81 to −58	Teske et al. 2002
		sn-2-hydroxyarchaeol	−89 to −70	Teske et al. 2002
Bacterial	Kazan Mud Volcano (Mediterranean Sea)	Diether 1†	−83 ± 11.4 sd	Werne et al. 2002
		Diether 2†	−78 ± 13.1 sd	Werne et al. 2002
		Diploptene‡	−57	Werne et al. 2002
		Diplopterol‡	−59	Werne et al. 2002
	Hydrate Ridge	Iso-C_{15} fatty acids†	−63	Boetius et al. 2000
		Anteiso-C_{15} fatty acids†	−75	Boetius et al. 2000
	Aleutian Trench	Diploptene‡	−74.4 to −72.7	Elvert et al. 2000
	Mediterranean Ridge	Diplopterol‡	−54	Pancost et al. 2000
	Napoli, Amsterdam Mud Volcanoes (Mediterranean Sea)	Alkyl diethers†	−83.6 to −51.9	Aloisi et al. 2002
	Eel River	n-$C_{14:0}$†	−69.1	Orphan et al. 2001
		n-$C_{16:1\ (\omega7)}$†	−62.8	Orphan et al. 2001
		n-$C_{16:1\ (\omega5)}$†	−76.1	Orphan et al. 2001
		10-Me-$C_{16:0}$†	−69.4	Orphan et al. 2001

* Thermogenic methane source (Teske et al. 2002).
† Biomarkers for sulfate reduction (Werne et al. 2002).
‡ Biomarkers for aerobic methane oxidizers (Werne et al. 2002).

($\delta^{13}C = -89‰$ to $-58‰$) and bacterial biomarkers are not as ^{13}C-depleted as those where biogenic methane is the source (Table 8.3). In addition to consortia of anaerobic archaeal methanotrophs and sulfate reducing bacteria, other microbial groups, including single cells, monospecific aggregates, or multispecies consortia are capable of anaerobic methane consumption (Orphan et al. 2002).

Microbially mediated carbonate deposition in methane-rich systems

Authigenic (formed in place) carbonate deposits are ubiquitous in submarine, methane-rich environments, including methane-hydrate seeps, mud volcanoes, and other settings along passive and active continental margins (e.g., Kulm & Suess 1990; Sakai et al. 1992; Stakes et al. 1999; Aloisi et al. 2000, 2002; Greinert et al. 2002; Van Dover et al. 2003; Carson et al. 2003) and in shallow and deep hydrothermal settings (Schmidt et al. 2002; Canet et al. 2003). Carbonate precipitation results from super-saturation of pore fluids with bicarbonate ions due to anaerobic oxidation of methane; the high energy involved in inorganic oxidation of methane implicated microbially mediated oxidation in carbonate formation (Carson et al. 2003), and carbon isotope compositions of the carbonates and of organic extracts from the carbonates confirm this view. Carbonate precipitation occurs under anaerobic conditions that favor the production of HCO_3^- and carbonate precipitation during methane oxidation, rather than aerobic conditions where methane oxidation produces CO_2 and lowers the pH, favoring carbonate dissolution (Aloisi et al. 2002). Carbonate deposits display various morphologies (slabs, chimneys, mounds) and mineralogies (calcite, aragonite, dolomite), and are regarded as a fossil record of bacterial activity (e.g., Cavagna et al. 1999; Peckmann et al. 2002). Sources of carbon in pore fluids (and thus available for carbonate precipitation) include biogenic methane ($\delta^{13}C < -65‰$), thermogenic methane ($\delta^{13}C = -50$ to $-30‰$), sedimentary organic carbon ($\delta^{13}C = -20‰$), and marine carbonate ($\delta^{13}C = 0‰$). Diagnostic markers of chemoautotrophic settings include ^{13}C-depleted $\delta^{13}C_{inorganic}$ values in carbonates reflecting a methane source of carbon (Table 8.4), and organic extracts from carbonates with extremely ^{13}C-depleted $\delta^{13}C$ values for biomarkers of methanotrophic bacteria [crocetane, archaeol, pentamethylicosanes (PMIs); $\delta^{13}C$ to $-96‰$] that match those described above from methane-rich sediments (Peckmann et al. 1999; Thiel et al. 2001; Pancost et al. 2001; Aloisi et al. 2002). Carbonate $\delta^{13}C$ values more positive than $-25‰$ probably represent a mixture of methane-derived CO_2, CO_2 generated from heterotrophic degradation of organic matter, and bottom-water dissolved CO_2. Relative contributions of CO_2-carbon from different CO_2 sources can vary within carbonates from a geographically restricted area (Sassen et al. 1993; Stakes et al. 1999, and references therein).

Table 8.4 Carbonate-carbon isotope compositions.

	Location	$\delta^{13}C$ (‰)	Reference
Seeps	Mediterranean mud volcanoes	−28.9 to −24.8	Aloisi et al. 2002
	Aleutian Trench	−45.4 to −48.7	Elvert et al. 2000
	Derugin Basin, Sea of Okhotsk	−40 to −43.5	Greinert et al. 2002
	Monterey Bay, California (west of San Gregorio Fault Zone)	−35 to −56	Stakes et al. 1999
	Monterey Bay, California (San Gregorio Fault Zone)	−7 to −26	Stakes et al. 1999
	Makran accretinary prism (Arabian Sea)	<−40	Von Rad et al. 1996
Hydrothermal systems	Nankai Trough	−30 to −50	Sakai et al. 1992
	Shallow-water (10 m) hot spring (85°C)	To −39.2	Canet et al. 2003
Fossil carbonates	Lincoln Creek Formation (WA):		
	clear and yellow aragonite	−51.3 to −26.4	Peckmann et al. 2002
	brownish calcite	5.2 to 6.6	Peckmann et al. 2002
	microspar	−4.2 to −1.6	Peckmann et al. 2002
	yellow calcite spar	0.1 to 6.7	Peckmann et al. 2002
	Monferrato, Italy (Tertiary)	−20 to −40	Cavagna et al. 1999

Characterization of trophic relationships in methane-rich environments

Sources of carbon in food webs of shallow-water methane seeps

There is developing evidence that microbial primary production (non-symbiotic) at shallow-water methane seeps (generally <500 m) need not have a strong influence on the associated invertebrate community composition or nutrition (Levin et al. 2000). Photosynthetically derived organic material may be more readily available to heterotrophic macrofaunal and megafaunal invertebrates than chemoautotrophically derived organic carbon in shallow-water environments, although trophic specialists on bacterial autotrophic carbon and opportunistic predators of chemoautotrophic symbioses are known. Methane pockmarks in the North Sea at 150 m, for example, support thyasirid clams that host chemoautotrophic endosymbionts and have $\delta^{13}C$ values of −35‰ to −31‰, consistent with dependence on sulfide oxidation, but other macrofaunal species analyzed from the site (cnidarians, nemerteans, polychaetes, gastropods, other bivalve mollusks, enteropneusts, echinoids, and fish) have $\delta^{13}C$ values −23‰ to −16‰ (Dando et al. 1991), in the range expected for benthos feeding heterotrophically on plankton-derived organic carbon. A pogonophoran tubeworm, *Siboglinum poseidoni*, from Skagerrak methane seeps (280–340 m) is so far the only known siboglinid polychaete to host methanotrophic endosymbionts (Schmaljohann & Flügel 1987), with a correspondingly ^{13}C-depleted δ^{13}C value of –78‰ (Schmaljohann et al. 1990). A sigalionid polychaete (*Leanira* sp.) from this same Skaggerak seep had a δ^{13}C value of –20.0‰, consistent with nutrition based on sedimentary organic matter at the site rather than chemoautotrophic production (Schmaljohann et al. 1990). In the Eel River Basin on the northern California slope (ca. 500 m), methane seeps support clams (*Calyptogena pacifica*) that host endosymbiotic sulfide-oxidizing bacteria (Kennicutt et al. 1989). Of 28 infaunal species examined from slope (ca. 500 m) and shelf (18–30 m) methane seeps off the northern California slope, only one macrofaunal species had a δ^{13}C value far outside the range expected for heterotrophs feeding on plankton-derived organic carbon (Levin et al. 2000). This one species, a dorvilleid polychaete (δ^{13}C = –33.5‰), is presumed to rely on a chemoautotrophic food source, derived either from symbiotic or free-living bacteria (Levin et al. 2000). An opportunistic predatory gastropod had δ^{13}C values (to –27.5‰) that reflected a partial diet of *C. pacifica* (δ^{13}C = –36‰; Kennicutt et al. 1989). Additional reports of non-symbiotic components of shallow seep food webs deriving most of their nutrition from photosynthetically derived carbon may be cited (e.g., methane seeps off Oregon; Juhl & Taghon, 1993). Other seep sites nearly as shallow as the Eel River slope seeps of Levin et al. (2000) support heterotrophic species with ^{13}C-depleted δ^{13}C values indicative of reliance on chemoautotrophic production. Methane seeps between 500 and 600 m in the Gulf of Mexico, for example, support a variety of macrofaunal invertebrates with δ^{13}C compositions <–50‰; there is even evidence for export of some of this chemoautotrophically derived organic material into the surrounding deep-sea environment via large, mobile benthic predators (e.g., fish, sea stars, predatory gastropods; MacAvoy et al. 2002).

Partitioning of organic carbon in deep-sea methane seeps: Are seep food webs dominated by microbial methanotrophy or sulfide oxidation?

Sulfide and methane are delivered to the sediment–seawater interface in seep environments, where they are available as resources for aerobic microbial oxidation. Levin & Michener (2002) used a two-component, linear mixing model (Fry & Sherr 1984) to estimate the relative contribution of methane-derived organic carbon in the nutrition of infaunal macro-invertebrates. Because there were three possible carbon sources (from methane oxidation, from sulfide oxidation, and from phytoplankton-derived carbon), Levin & Michener (2002) calculated maximal [$F_m = (\delta_i - \delta_{POC})/(\delta_m - \delta_{POC})$] and minimal [$F_m = (\delta_i - \delta_{SOB})/(\delta_m - \delta_{SOB})$] possible contributions of methane-derived carbon (F_m), where δ_i, δ_m, δ_{POC}, and δ_{SOB} are the $\delta^{13}C$ values of infauna, methane, particulate organic carbon, and sulfide-oxidizing bacteria, respectively. For seeps in the abyssal Gulf of Alaska (4400 m), where the flux of photosynthetically derived organic carbon is lower than on the upper slope, the contribution of methane-derived carbon to macrofaunal invertebrate organic carbon was estimated to be 30% to 50% on average. Some of these Gulf of Alaska seep species (dorvilleid and nereid polychaetes, gammarid amphipods, anemones; $\delta^{13}C = -90‰$ to $-63‰$) may be microvores specializing on ^{13}C-depleted organic carbon produced by anaerobic archaea or aerobic bacteria that oxidize methane, while other seep macrofaunal species (syllid and trichobranchid polychaetes, sipunculids) had $\delta^{13}C$ values suggesting reliance on photosynthetically derived organic material (Levin & Michener 2002). Methane-derived carbon was less important for macrofauna at other Pacific seeps (Table 8.5), with lowest estimated contributions of 0–5% in microbial-mat-dominated seeps of the Eel River Basin (500 m). Levin & Michener (2002) propose that macrofaunal assemblages associated with seeps lie on a continuum from reliance mainly on chemoautotrophic production to reliance mainly on photosynthetically derived carbon, and that differences in seepage and fluid flow regimes among habitat types within a seep site (e.g., bacterial mat, clam bed, tubeworms) may be key determinants of the position of a macrofaunal assemblage along this continuum.

At the Blake Ridge methane-hydrate site, invertebrate biomass was dominated by bathymodiolin mussels, with both methane-oxidizing and sulfide-oxidizing bacteria in their gills and, to a lesser degree, by vesicomyid clams shown to host sulfide-oxidizing bacteria in their gills (Van Dover et al. 2003). Assuming that photosynthetically derived organic carbon is unimportant to the bulk nutrition of the mussels (Fisher 1990), the importance of methane-derived carbon to the mussel nutrition (F_m) can be determined using a two-

Table 8.5 Contributions of methane-derived organic material to Pacific seep macrofaunal invertebrates (Levin & Michener 2002).

Seep site and microhabitat	Estimated average % contribution
Gulf of Alaska pogonophoran fields	32–51
Gulf of Alaska clam (*Calyptogena phaseoliformis*) beds	12–40
Gulf of Alaska microbial mats	20–44
Hydrate Ridge (Oregon margin) microbial mats	20–44
Hydrate Ridge (Oregon margin) clam (*Calyptogena pacifica*) beds	0–27
Eel River (California slope) clam (*Calyptogena pacifica*) beds	0–27
Eel River (California slope) microbial mats	0–5

source mixing model $F_m = (\delta_{mussel} - \delta_{SOB})/(\delta_{MOB} - \delta_{SOB})$, where δ_{mussel}, δ_{SOB}, and δ_{MOB} are the $\delta^{13}C$ values of mussel tissue, sulfide-oxidizing bacteria, and methane-oxidizing bacteria, respectively. Assuming also that the $\delta^{13}C$ value of methanotrophically generated biomass matches that of the source methane (–67.8‰; Paull et al. 1995) and the $\delta^{13}C$ value of thiotrophically generated biomass is that of the vesiocomyid clams (–36‰), Blake Ridge mussels ($\delta^{13}C$ = –56‰) derive approximately 60% of their organic carbon from their methanotrophic endosymbionts and approximately 40% of their carbon from their sulfide-oxidizing symbionts (Van Dover et al. 2003). A substantial amount of the biomass at the Blake Ridge site (40% of the mussel organic carbon plus 100% of the clam organic carbon) is thus dependent on endosymbiont oxidation of sulfide generated by microbial sulfate reduction. Measures of $\delta^{34}S$ in mussel tissues ($\delta^{34}S_{gills}$ = +13‰; $\delta^{34}S_{mantle}$ = +8‰) compared with clam tissues ($\delta^{34}S_{gills}$ = –16‰; $\delta^{34}S_{mantle}$ = –12‰) are consistent with the view of differing importance of sulfide in mussels vs. clams (Van Dover et al. 2003). Mussel sulfur isotope composition suggests a greater dependence on seawater sulfate ($\delta^{34}S$ = +20‰) than on sulfide; sulfide generated by microbial sulfate reduction (where $\delta^{34}S$ can be as low as –50‰) is more important as a sulfur source in clams. Sulfate reduction is evidently less important in biomass production at mussel-dominated Gulf of Mexico seeps, where only methanotrophs are found in the gills and the carbon isotopic composition of the tissues matches that of the source methane ($\delta^{13}C_{mussel}$ = –40.6‰; $\delta^{13}C_{methane}$ = –41.2‰; Brooks et al. 1987).

A food-web model for the Blake Ridge methane-hydrate seep

As in other ecosystems, carbon, nitrogen, and sulfur stable isotope data (Figure 8.3a & b) may be used in conjunction with other information to constrain trophic relationships among invertebrates (Figure 8.3c) at methane seep sites. The basic rules apply regarding the relative fidelity of $\delta^{13}C$ (and $\delta^{34}S$) values and a ca. 3–4‰ enrichment in $\delta^{15}N$ between consumers and the organisms they consume (Conway et al. 1994). The food-web model developed for the Blake Ridge chemotrophic ecosystem includes assignments of species to trophic levels as well as predictions (indicated by italics and dashed box outlines) of trophic levels (i.e., of $\delta^{15}N$), since several important species were not analyzed. The food-web model for Blake Ridge is fairly typical of food webs in general within chemoautotrophic systems, including the characteristically ^{15}N-depleted primary production, but there were two surprises. The first surprise was that nematodes, which were abundant in sediments associated with mussel beds at Blake Ridge, proved to have the greatest ^{15}N-enrichment, as if they were top consumers in the system. Given no evidence for a predatory mode of feeding, the nematodes were inferred to be decomposers within

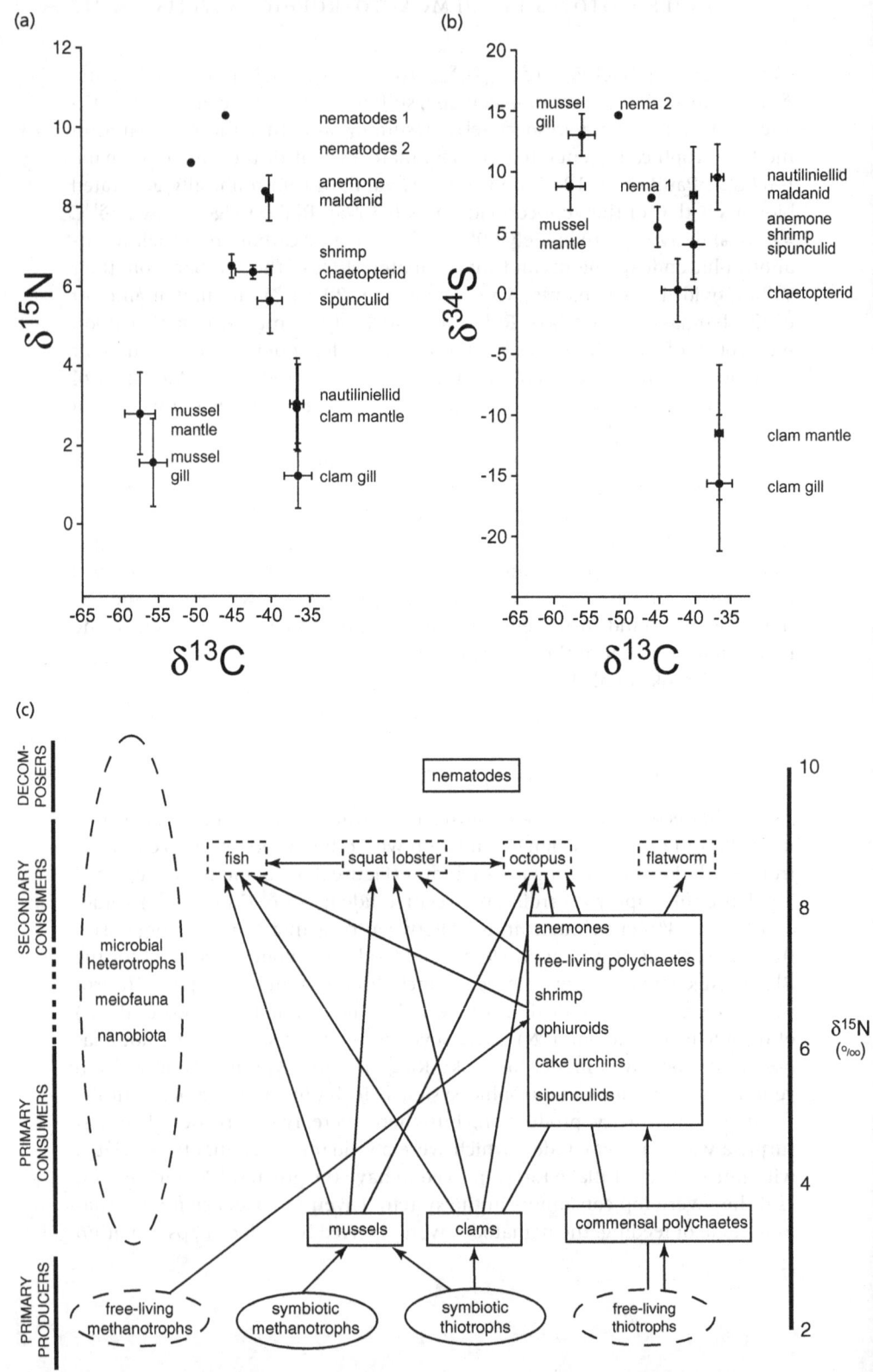

the system (Van Dover et al. 2003). The second surprise derives from the use of sulfur isotopes to infer the diet of the nautiliniellid polychaete, which lives in the mantle cavity of the clams. Carbon and nitrogen isotope data alone do not discriminate between two possible sources of food, namely clam tissues and some other organic source (e.g., free-living sulfide-oxidizing bacteria; Figure 8.3a). But sulfur isotopes of the polychaete are so much more enriched in ^{34}S than the clam tissues ($\delta^{34}S_{polychaete} = +9‰$; $\delta^{34}S_{clam\ gill} = -15‰$; Figure 8.3b) that an alternate source of nutrition must be invoked (Van Dover et al. 2003).

Whale falls

Dead whales sinking to the seafloor deliver massive pulses of labile organic matter. After the flesh of the whale is consumed by mobile scavengers (fish, amphipods, isopods, brittle stars, etc.), the lipid-rich skeleton supports microbial sulfate reduction, which delivers sufficient sulfide to the local environment to sustain sulfide-oxiding bacteria at the base of a chemotrophic food web (Figure 8.4; Deming et al. 1997). At adult whale skeletons in deep water (ca. 1000m or deeper), after the flesh is consumed, sources of organic material available to consumers include lipids within bones, free-living heterotrophic and sulfide-oxidizing bacteria, endosymbiotic sulfide-oxidizing bacteria, and photosynthetically derived organic material. Carbon isotope data from diverse microbial samples and invertebrate tissues (Figure 8.5) provide no evidence for generation of significant amounts of bacterial methane (Smith & Baco 2003), despite the organic enrichment that whale falls represent. Several invertebrate species colonizing whale falls (including vesicomyid clams) had $\delta^{13}C$ values in the range of $-36.5‰$ to $-29.6‰$, with the most ^{15}N-depleted $\delta^{15}N$ values (-0.9–$4.0‰$), consistent with isotopic compositions of invertebrates hosting sulfide-oxidizing bacterial endosymbionts (Conway et al. 1994). Nitrogen isotope values for dominant

---

Figure 8.3 Stable isotope composition of selected Blake Ridge invertebrates. (a) $\delta^{15}N$ vs. $\delta^{13}C$ (‰). (b) $\delta^{34}S$ vs. $\delta^{13}C$ (‰). (c) Food-web model for the Blake Ridge Diapir mussel and clam beds. Italicized entries and dashed envelopes indicate taxa for which isotopic data are not available. Positions in the food web for these taxa are inferred from behavior, distribution, and feeding characteristics of the taxa in other habitats. All other taxa are positioned vertically according to their nitrogen isotopic ratios. Dashed lines beneath primary consumer and secondary consumer legends represent a mixotrophic region. Note that nitrogen values for the symbiotic and free-living autotrophs are approximated by analyses of bivalve gills that include a mix of symbiont and host organic material. (From Van Dover et al. 2003.)

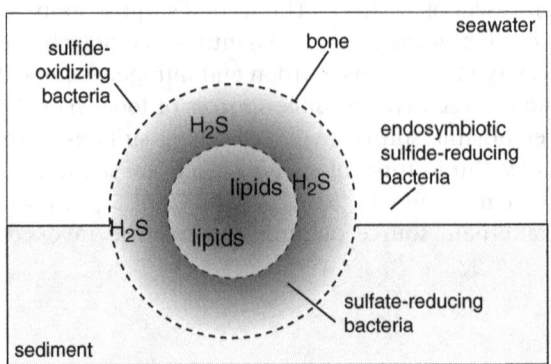

Figure 8.4 Schematic cross-section of a whale vertebra on the seafloor. Decompositional processes include: (i) diffusion of seawater sulfate into the bone; (ii) sulfate-reduction and oxidation of lipids from the core of the bone by anaerobic bacteria; (iii) diffusion of microbially generated sulfide outward; (iv) microbial sulfide oxidation and autotrophic production on surfaces (bacterial mats) or in symbiotic associations (e.g., with clams). (From Smith & Baco 2003.)

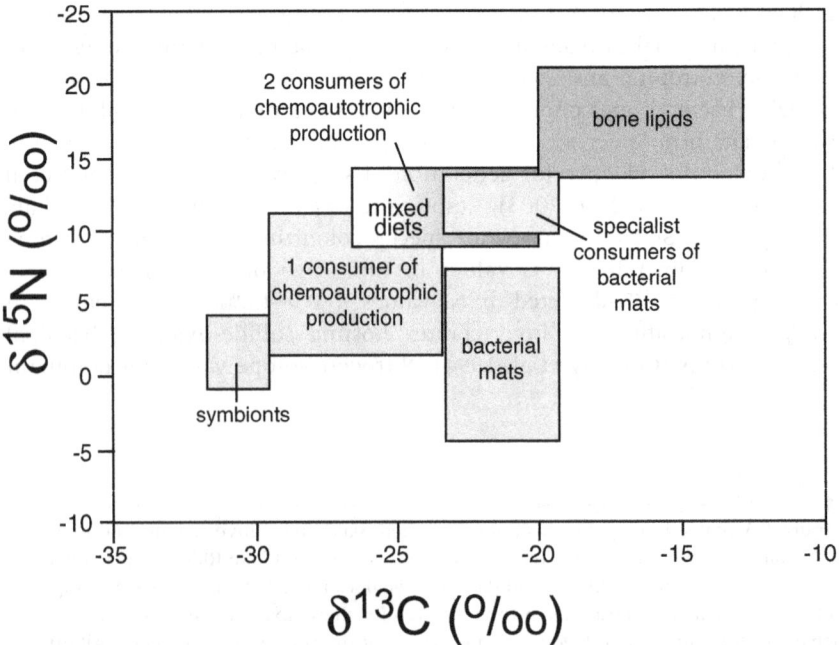

Figure 8.5 Carbon and nitrogen isotope compositions of food-web components at whale skeletons. Diets of invertebrates selectively deriving their nutrition from endosymbiotic sulfide-oxidizing bacteria (primary consumers, labeled as "1") have the most negative carbon isotopic compositions. Secondary consumers (or grazers on free-living microorganisms) are also relatively depleted in ^{13}C (labeled as "2"). Bacterial mats have $\delta^{15}N$ values suggesting that they are primary producers. Specialist consumers of bacterial mats have $\delta^{15}N$ values enriched in ^{15}N relative to the mats. (Data from Smith & Baco 2003.)

invertebrate primary consumer species (e.g., small gastropods and poly-chaetes) are relatively depleted in ^{15}N (as they are at seeps and vents), suggesting a minor role for photosynthetically derived organic material in whale-fall food webs (Figure 8.5). Secondary consumers or scavengers (in-cluding amphipods and squat lobsters) have carbon and nitrogen isotopic compositions that reflect a mixed diet that includes photo- and chemoau-totrophically derived organic material. Resident invertebrates on smaller, juvenile whale skeletons were not dependent on a sulfide-based food web, but instead had carbon and nitrogen isotopic compositions consistent with nutrition based on the whale lipids (Smith & Baco 2003). Juvenile whale skeletons are more poorly calcified than those of the adults; the minimal degree of calcification and small size are inferred to preclude sustained sulfide production of sufficient duration to support a chemoautotrophic food web (Smith & Baco 2003).

Hydrothermal vents

Isotopic tools continue to be important in ecological studies at vents. Recent applications include studies investigating dietary shifts during the life history of invertebrate species or in response to competition with congeneric species, and studies of site-specific variations correlated with biological and environ-mental variables. Stable isotope data continue to provide a first-order synop-sis of trophic interactions within newly discovered or poorly studied vent ecosystems and have proved useful in identifying invertebrate species likely to host endosymbiotic, chemoautotrophic microorganisms. Before exploring some of these ecological applications of isotope techniques, a brief account of an important advance in our understanding of factors underlying isotopic compositions of vent invertebrates is presented.

Form I and form II RubisCOs and the dichotomous $\delta^{13}C$ values of vent invertebrates hosting symbiotic chemoautotrophs

Early surveys of stable isotopic compositions of vent invertebrates hosting symbiotic, chemoautotrophic bacteria identified two fundamental groups (Conway et al. 1994): those invertebrates with $\delta^{13}C$ values of ca. $-34‰$ (bivalve mussels and clams), and those with $\delta^{13}C$ values centered at ca. $-13‰$ (tubeworms, swarming shrimp). Several explanations were put for-ward to account for a $20‰$ dichotomy between these two groups of organ-isms. Since the microbial symbionts of the two groups of organisms presumably use the same substrate CO_2, source variations in carbon isotopic composition have been dismissed as a likely factor (Robinson & Cavanaugh 1995). One explanation invoked a physiological CO_2-transport limitation in tubeworms and consequent limitation of expression of any enzymatic fractionation effect (Rau 1981a; Fisher et al. 1990); another suggested that a C4 pathway of carbon fixation, where there is no isotopic discrimination during initial CO_2 fixation by phosphenol pyruvate carboxylase, was operative in tubeworms (Felbeck 1985). Robinson & Cavanaugh (1995) suggested yet another possibility, namely that the symbiotic microorganisms in the two groups use different forms of RubisCO that have different kinetic fractionation effects. They showed that tubeworms (*Riftia pachyptila* and *Ridgeia piscesae*) express only form II RubisCO, while bivalves from vents (*Bathymodiolus thermophilus*) and from a shallow-water sulfidic environment (*Solemya velum*) express only form I RubisCO (Robinson & Cavanaugh 1995). Kinetic isotope effects of form II RubisCO from symbionts of *Riftia pachyptila* have subsequently been determined to be smaller (maximum discrimination against $^{13}C = 22‰$) than kinetic isotope effects of form II RubisCO (maximum discrimination against $^{13}C = 30‰$; Robinson et al. 2003). Even accounting for this kinetic isotope effect during symbiont carbon fixation and the $\delta^{13}C$ composition of the source CO_2 (ca. −7‰; Childress et al. 1993), tubeworm tissues are still 10–17‰ more enriched in ^{13}C than expected. Thus, while kinetic isotope effects associated with form I and form II RubisCOs contribute to some of the dichotomy of carbon isotope values observed, other factors are involved. CO_2-limitation, generated by diffusion and transport barriers, remains a plausible process contributing to the ^{13}C-enriched $\delta^{13}C$ values of tubeworms and other vent invertebrate species (Robinson et al. 2003).

Ontogenetic variations in diet at hydrothermal vents

One of the best-documented examples of a dietary shift during development in a vent invertebrate is that observed between juveniles and adults of the shrimp, *Rimicaris exoculata*, which lives in dense populations on the sides of black smoker chimneys of Mid-Atlantic Ridge and Indian Ocean vents. Juvenile *R. exoculata* live adjacent to, but segregated from, the adult populations and are readily differentiated from the adults by their smaller size and orange-red coloration. These characters, together with other anatomical differences, resulted in placement of the juveniles in a different species (Martin et al. 1997), and even a different genus (Vereschaka 1996), than the adults. Molecular studies, however, convincingly demonstrated that the "new species" was in fact the juvenile stage of *R. exoculata* (Creasey et al. 1996), Shank et al. (1998).

A large (6–7‰) and systematic shift to more positive $\delta^{13}C$ values with increasing body size has been documented for *R. exoculata* populations in the Atlantic (Figure 8.6; Polz et al. 1998, Colaçao et al. 2002a) and in the Indian Ocean (Van Dover 2002). A corresponding decrease in $\delta^{15}N$ values (by 3–4‰) with increasing body size has also been documented for *R.*

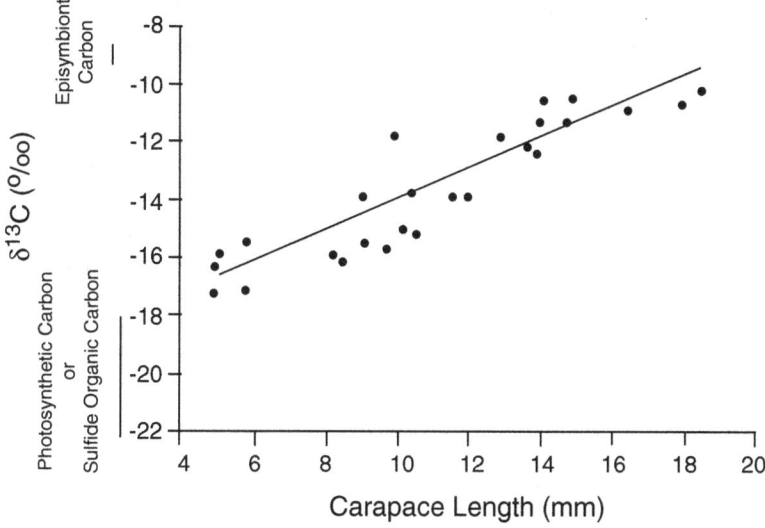

Figure 8.6 Ontogenetic shift in diet: Relationship between carapace length of shrimp (*Rimicaris exoculata*) and $\delta^{13}C$. Ranges of carbon isotope values for organic carbon in chimney sulfides, organic carbon from surface-derived photosynthetic production, and shrimp episymbiotic bacteria are shown on the vertical axis. (Redrawn from Polz et al. 1998.)

exoculata (Van Dover 2002). An abundance of fatty acids originating in the euphotic zone [20:5(n-3) and 22:6(n-3); Pond et al. 1997] was observed in larval and post-larval *R. exoculata*, and the fatty acid profiles of juvenile *R. exoculata* suggest that they rely on photosynthetically derived organic material (Allen-Copley et al. 1998). Fatty-acid profiles and compound-specific $\delta^{13}C$ compositions indicate that the adult shrimp likely derive the bulk of their nutrition from the episymbiotic bacteria that cover the surface of the carapace of the branchial chamber, although carbon isotopic data suggest that some essential compounds not produced by bacteria, specifically shrimp cholesterol (which cannot be created *de novo* in the shrimp), are likely derived from the oceanic photic zone (Rieley et al. 1999). ^{15}N-depleted bulk nitrogen isotopic compositions of adult shrimp tissues relative to juvenile tissues (Van Dover 2002) are consistent with a trophic-level shift between juvenile and adult stages. The resulting view is that larval and juvenile *Rimicaris exoculata* spend all (larvae) or a good portion (juvenile) of their time in the water column away from hydrothermal vents, where they feed on photosynthetically derived carbon (Allen Copley et al. 1998). When the shrimp take up life at a black smoker, they switch to a diet based on che-moautotrophic production by episymbiotic bacteria, but with some essential nutrients traceable to photosynthetic production (Rieley et al. 1999). A similar reliance of juveniles on photosynthetically derived organic material and a shift to chemoautotrophically produced organic material has also been reported for the shrimp, *Mirocaris fortunata*, from other Mid-Atlantic Ridge hydrothermal vents (Pond et al. 1997).

An ontogenetic shift in diet has also been reported for a bythograeid crab (*Austinograea rodriguezensis*) from hydrothermal vents in the Indian Ocean (Van Dover 2002). $\delta^{15}N$ values are enriched in ^{15}N in larger crabs (from 6‰ in the smallest crab analyzed to 12‰ in adult crabs), suggesting a step increase in trophic level as the crabs become able to procure large prey items. Together with a corresponding shift toward ^{13}C-enriched $\delta^{13}C$ values (from −16‰ in juveniles to −13‰ in adults), the bulk isotope data are consistent with an adult diet of shrimp (*R. exoculata*), upon which crabs have been observed to prey (Van Dover 2002).

Interspecific competition and dietary shifts

Sympatric coexistence of congeneric species is often assumed to imply some degree of dietary niche partitioning that may be relaxed when species live allopatrically (e.g., Schoener 1974). Differences in stable isotopic composi-tions can be useful in identifying potential niche partitioning vs. competitive interactions among congeners, a strategy that Levesque et al. (2003) used to investigate resource partitioning among three species of alvinellid polychaetes from hydrothermal vents on the Juan de Fuca Ridge. All three alvinellid species (*Paralvinella palmiformis*, *P. sulfincola*, and *P. pandorae*) have similar feeding appendages that permit them to engage in deposit- and/or suspen-sion-feeding. Where *P. palmiformis* and *P. sulfincola* or *P. sulfincola* and *P. pan-dorae* co-occurred, their pair-wise carbon and nitrogen isotopic compositions were significantly different, indicating reliance on different food resources. In contrast, isotopic compositions of *P. palmiformis* and *P. pandorae* could not be distinguished, suggesting the potential for interspecific competition between these two species (Levesque et al. 2003).

Site-specific variations in isotopic compositions within vent mussels hosting dual symbionts

An inferred advantage of dual symbioses (i.e., the simultaneous occurrence of aerobic methanotrophic and sulfide-oxidizing bacteria) observed in gill bacteriocytes of some species of mussels from chemotrophic systems is that it confers to the host a greater environmental tolerance and degree of nutri-tional flexibility (Distel et al. 1995, Fiala-Médioni et al. 2002). Where sulfide-

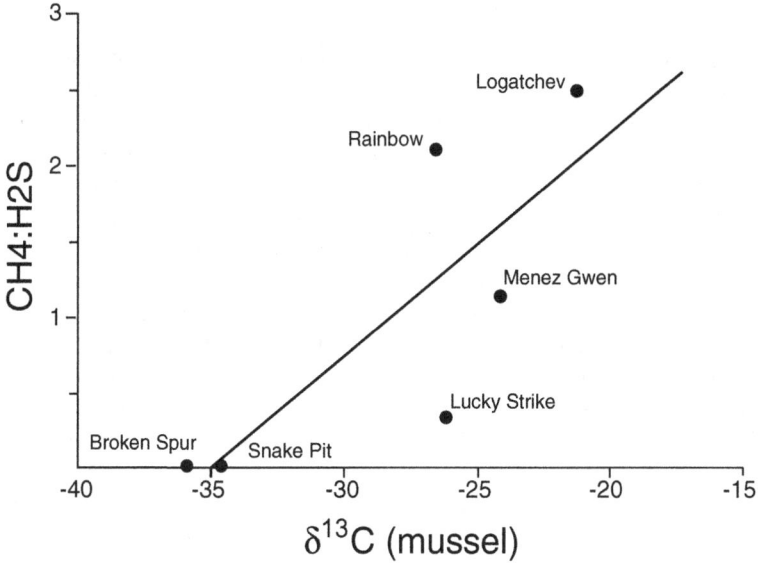

Figure 8.7 Relationship between methane:sulfide in end-member (black-smoker) hydrothermal fluids and $\delta^{13}C$ values for mussels from Mid-Atlantic Ridge hydrothermal vents. (From Colaço et al. 2002b.)

oxidizers and methanotrophs produce isotopically distinct organic carbon, the bulk carbon isotopic composition of the host should reflect the relative importance of each autotrophic source. The first report of site-specific variations in carbon and nitrogen isotopic compositions of mussels with dual symbionts was for *Bathymodiolus azoricus* from the Lucky Strike vent field (Trask & Van Dover 1999). The mussels, collected from two chemically distinct sites (Von Damm et al. 1998), segregated into two isotopic groups: those inferred to rely primarily on microbial sulfide oxidation ($\delta^{13}C = -30.7‰$, $\delta^{15}N = -10.5‰$) and those inferred to rely primarily on microbial methanotrophy ($\delta^{13}C = -21.3‰$, $\delta^{15}N = -4.5‰$; Trask & Van Dover 1999). These site-specific differences were observed independently by Colaço et al. (2002a), and a positive relationship between the ratio of methane to hydrogen sulfide in end-member (black smoker) fluid composition and the $\delta^{13}C$ composition of the mussel tissue has been documented (Figure 8.7; Colaço et al. 2002b) Mussels relying on microbial methanotrophy had twice as many methanotrophic bacteria for a given area of gill bacteriocyte compared with mussels relying on microbial sulfide oxidation (Trask & Van Dover 1999). Additional observations of site-specific variations in isotopic compositions and symbiont populations are reported by Fiala-Médioni et al. (2002) for *B. azoricus*.

Vent food webs: trends and anomalies

Stable isotope methods continue to be valuable in preliminary assessments of trophic interactions within vent food webs, including those at Marianas (western Pacific) and 21°N (East Pacific Rise) vents (Van Dover & Fry 1989), Juan de Fuca vents (Southward et al. 1994; Van Dover & Fry 1994), Galapagos (Galapagos Spreading Center) vents (Fisher et al. 1994), Indian Ocean vents (Van Dover 2002), and Mid-Atlantic Ridge vents (Vereshchaka et al. 2000; Colaço et al. 2002a). Cross-plots of carbon and nitrogen compositions of vent species within vent fields show a consistent pattern of increasing $\delta^{15}N$ with increasing $\delta^{13}C$, regardless of the geographic setting (Figure 8.8; Van Dover 2002). A satisfactory explanation for this pattern is lacking, although Van Dover (2002) suggests that there may be an as yet unidentified primary producer within these vent fields that has $\delta^{13}C$ values of ca. −10‰ to −15‰ and $\delta^{15}N$ values of ca. −1‰ and 5‰, or that there may be much more than a 1‰ shift in $\delta^{13}C$ in favor of ^{13}C with trophic level in vent food webs.

Carbon and nitrogen compositions of vent organisms are sufficiently predictable that stable isotope surveys can be used as a rapid means to identify organisms that fall outside the expected range of values and that thus warrant further study if an understanding of their nutrition is to be gained. An example is an undescribed species of scaly-footed gastropod from the Kairei vent field (Central Indian Ridge; Warén et al. 2003). Gills of this gastropod had a ^{15}N-depleted $\delta^{15}N$ composition (3.1‰) relative to other members of the vent community (Van Dover et al. 2001; Van Dover 2002), which was consistent with a reliance on endosymbiotic bacteria (Conway et al. 1994). Despite isotopic support for endosymbionts, tests for enzymes diagnostic of autotrophic carbon fixation in the gills were negative and examination of gill tissues using electron microscopy failed to detect bacteria (Van Dover et al. 2001). The gastropods were subsequently shown to have an unusually enlarged esophageal gland that housed bacteriocytes and sulfide-oxidizing bacteria (Goffredi et al. 2004).

Conclusions

Stable isotope analyses in studies of chemosynthetic ecosystems continue to be extremely important in guiding our understanding of microbial-resource interactions and all levels of trophic processes, from investigations of the autecology of a species, to between-species interactions, to synoptic views of carbon flow within entire communities. Results from carbon isotope studies of archaeal-bacterial consortia and biomarkers reveal the potential for significant insight from isotopic data at a microscopic scale to global phenomena associated with chemoautotrophic ecosystems. There remain surprising gaps in our understanding of isotope systematics at vents and seeps. For example, the nitrogen cycle continues to remain a mystery, with little authoritative

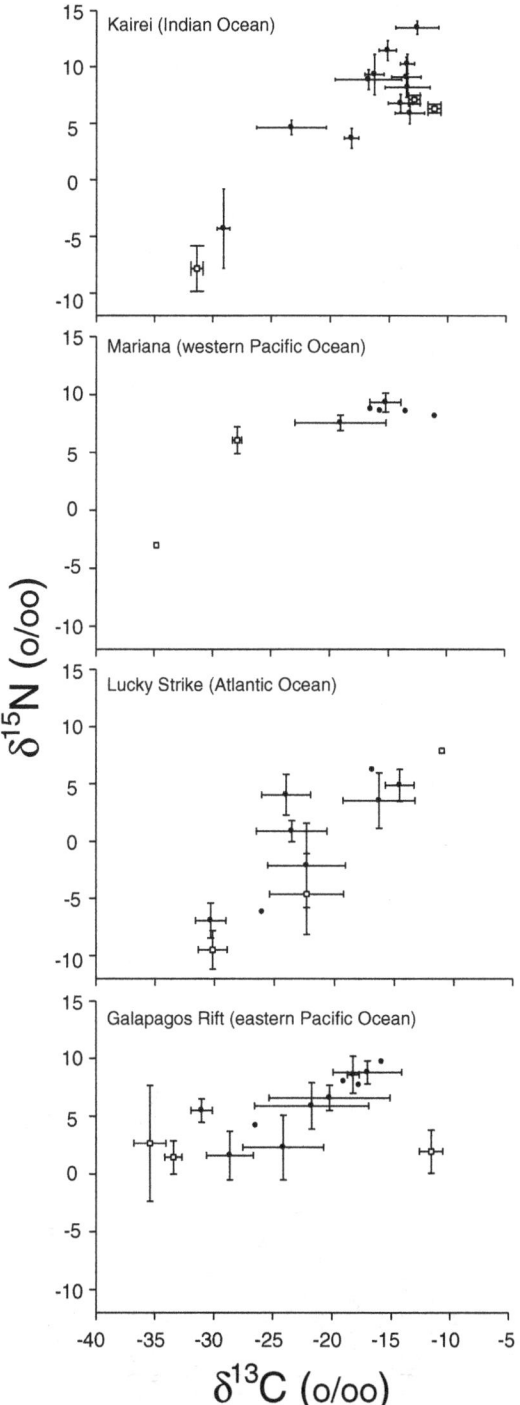

Figure 8.8 Comparison of $\delta^{13}C$ and $\delta^{15}N$ values (‰; mean ± SD for >2 individuals) for biogeographically disjunct hydrothermal vent communities. Data from Table 8.6. Open squares = species with endo- or episymbionts.

Table 8.6 Comparison of stable-isotope compositions of invertebrates from eastern Pacific (Galapagos Rift), western Pacific (Mariana Back-Arc), Indian (Kairei), and Atlantic (Lucky Strike) Ocean hydrothermal vents. Data are ordered by decreasing. ^{13}C values to highlight similarities and differences in carbon resources among sites and species. S = symbiont-bearing (epi- or endosymbiont); P = primary consumer (suspension feeder or grazer); O = omnivore or secondary consumer. Isotopic data are from Fisher et al. (1994), Van Dover & Fry (1989), Colaço et al. (2002a, 2002b), and Van Dover (2002). For bivalve species with gill symbionts, gill isotopic compositions are reported, except for *Bathymodiolus azoricus* from Lucky Strike where gill and muscle tissue were not significantly different and were reported as pooled data by Colaço et al. (2002b).

^{13}C	Indian Ocean	^{15}N		Marianas	^{15}N		Galapagos	^{15}N		Lucky Strike	^{15}N	
−11	*Alviniconcha* n. sp.	6.3	S	*Paralvinella* sp	8.2	P	*Riftia pachyptila*	2.0	S	*Rimicaris exoculata*	7.9	S
−12	*Branchinotogluma* sp.	13.5	O									
	Rimicaris aff. *exoculata*	7.1	S									
−13	*Phymorhynchus* sp.	5.9	O	Limpet	8.6	P						
	Archinome cf. *rosacea*	8.2	O									
	Marianactis cf. *bythios*	10.3	O									
	Austinograea n. sp.	9.1	O									
−14	*Lepetodrilus* sp.	6.8	P							*Chorocaris chacei*	4.9	O
−15	Nemertean	11.4	O	*Austinograea williamsi*	9.3	O	Polynoid	9.7	O			
				Marianactis bythios	8.6	O						
−16	*Neolepas* n. sp.	9.3	P	*Chorocaris vandoverae*	8.8	O				*Mirocaris fortunata*	3.5	P
	Amphisamytha n. sp.	8.9	O							Anemone	6.2	O
−17							*Bythograea thermydron*	8.8	O			
							Munidopsis subsquamosa	7.7	O			
−18	Scaly-footed vetigastropod	3.7	P				*Neomphalus fretterae*	8.6	P			

−19				*Neolepas* sp.	7.6	P	*Bythites hollisi*	8.0	O	
−20							*Lepetodrilus elevatus*	6.6	P	
−21							*Alvinocaris lusca*	5.9	P	
−22										*Bathymodiolus azoricus*	−4.6	S
										Branchipolynoe seepensis	−2.1	P
−23	*Desbruyeresia* sp.	4.6	S or P							*Amathys lutzi*	0.9	P
−24							*Ventiella sulfuris*	2.3	P	*Sericosura* sp.	4.1	P
−25										
−26							*Eulepetopsis vitrea*	4.2	P	*Phymorhynchus* sp.	−6.2	O
−27										
−28				*Alviniconcha hessleri*	6.1	S	*Archinome rosacea*	1.6	O	
−29	*Branchipolynoe* n. sp.	−4.3	P							
−30										*Bathymodiolus azoricus*	−9.5	S
−31	*Bathymodiolus* aff. *brevior*	−7.8	S				*Phymorhynchus* sp.	5.5	O	*Branchipolynoe seepensis*	6.9	P
−32										
−33							*Calyptogena magnifica*	1.4	S	
−34				*Bathymodiolus brevior*	−3.0	S				
−35							*Bathymodiolus thermophilus*	2.7	S	

evidence to explain the consistent ^{15}N depletion observed in vent and seep food webs. Isotopic analysis of sulfur cycles is also full of potential for tremendous insight, especially given systematic variations in source inorganic sulfide that could be used to advantage in parsing the sulfur systems at vents and seeps. Given the relative inaccessibility of many marine chemoautotrophic ecosystems, isotopic methods will continue to be among the most useful of tools for tracing the flow of nutrients and energy.

References

Allen-Copley, C.E., Tyler, P.A. & Varney, M.S. (1998) Lipid profiles of hydrothermal vent shrimps. *Cahiers de Biologie Marine*, **39**, 229–231.

Aloisi, G., Pierre, C., Rouchy, J.-M., Foucher, J.-P., Woodside, J. & MEDINAUT Scientific Party. (2000) Methane-related authigenic carbonates of eastern Mediterranean Sea mud volcanoes and their possible relation to gas hydrate destabilization. *Earth and Planetary Science Letters*, **184**, 321–338.

Aloisi, G., Bouloubassi, I., Heijs, S.K., et al. (2002) CH_4-consuming microorganisms and the formation of carbonate crusts at cold seeps. *Earth and Planetary Science Letters*, **203**, 195–203.

Boetius, A., Ravenschlag, K., Schubert, C.J., et al. (2000) A marine microbial consortium apparently mediating anaerobic oxidation of methane. *Nature*, **407**, 623–626.

Borowski, W.S., Paull, C.K. & Ussler III, W. (1999) Global and local variations of interstitial sulfate gradients in deep-water, continental margin sediments: Sensitivity to underlying methane and gas hydrates. *Marine Geology*, **159**, 131–154.

Bottrell, S.H. & Raiswell, R. (2000) Sulphur isotopes and microbial sulphur cycling in sediments. In: *Microbial Sediments* (eds R.E. Riding & S.M. Awramik), pp. 96–104. Springer-Verlag, Berlin.

Brooks, J.M., Kennicutt II, M.C., Fisher, C.R., et al. (1987) Deep-sea hydrocarbon seep communities: evidence for energy and nutritional carbon sources. *Science*, **238**, 1138–1142.

Canet, C., Prol-Ledesma, R.M., Melgarejo, J.-C. & Reyes, A. (2003) Methane-related carbonates formed at submarine hydrothermal springs: a new setting for microbially derived carbonates? *Marine Geology*, **199**, 245–261.

Carson, B., Kastner, M., Bartlett, D., Jaeger, J., Jannasch, H. & Weinstein, Y. (2003) Implications of carbnon flux from the Cascadia accretionary prism: results from long-term, in situ measurements at ODP Site 892B. *Marine Geology*, **198**, 159–180.

Cavagna, S., Clari, P. & Martire, L. (1999) The role of bacteria in the formation of cold-seep carbonates: geological evidence from Monferrato (Tertiary, NW Italy). *Sedimentary Geology*, **126**, 253–270.

Cavanaugh, C.M. (1994) Microbial symbiosis: Patterns of diversity in the marine environment. *American Zoologist*, **34**, 79–89.

Charlou, J.-L., Fouquet, Y., Donval, J.P., et al. (1996) Mineral and gas chemistry of hydrothermal fluids on an unltrafast sprading ridge: East Pacific Rise, 17° to 19°S (*Naudur* cruise, 1993) phase separation processes controlled by volcanic and tectonic activity. *Journal of Geophysical Research*, **101**, 15899–15919.

Charlou, J.L., Donval, J.P., Fouquet, Y., Jean-Baptiste, P. & Holm, N. (2002) Geochemistry of high H_2 and CH_4 vent fluids issuing from ultramafic rocks at the Rainbow hydrothermal field (36°14′N, MAR). *Chemical Geology*, **191**, 345–359.

Childress, J.J., Lee, R.W., Sanders, N.K., et al. (1993) Inorganic carbon uptake in hydrothermal vent tubeworms facilitated by high environmental pCO_2. *Nature*, **362**, 147–149.

Colaço, A., Dehairs, F. & Desbruyères, D. (2002a) Nutritional relations of deep-sea hydrothermal fields at the Mid-Atlantic Ridge: A stable isotope approach. *Deep-Sea Research, Part I*, **49**, 395–412.

Colaço, A, Dehairs, F.& Desbruyeres, D. (2002b) $\delta^{13}C$ signature of hydrothermal mussels is related with the end-member fluid concentrations of H_2S and CH_4 at the Mid-Atlantic Ridge hydrothermal vent fields. *Cahiers de Biologie Marine*, **43**, 259–262.

Conway, N., Kennicutt, M.C. & Van Dover, C.L. (1994) Stable isotopes, microbial symbioses and microbial physiology. In: *Stable Isotopes in Ecology* (Eds K. Lajtha & R. Michener), pp. 158–188. Blackwell Scientific Publications, Oxford.

Cowen, J.P., Wen, X. & Popp, B.N. (2002) Methane in aging hydrothermal plumes. *Geochimica et Cosmochimica Acta*, **66**, 3563–3571.

Craig, H., Welhan, J.A., Kim, K., Poreda, R. & Lupton, J.E. (1980) Geochemical studies of the 21 degrees N EPR hydrothermal fluids. *Eos (Transactions of the American Geophysical Union)*, **61**, 992.

Creasey, S., Rogers, A.D. & Tyler, P.A. (1996) Genetic comparison of two populations of the deep-sea vent shrimp *Rimicaris exoculata* (Decapoda: Bresiliidae) from the Mid-Atlantic Ridge. *Marine Biology*, **125**, 473–482.

Damm, E. & Budéus, G. (2003) Fate of vent-derived methane in seawater above the Håkon Mosby mud volcano (Norwegian Sea). *Marine Chemistry*, **82**, 1–11.

Dando, P.R., Austen, M.C., Burke, R.A., Jr., et al. (1991) Ecology of a North Sea pockmark with an active methane seep. *Marine Ecology Progress Series*, **70**, 49–63.

DeLong, E.F. (2000) Resolving a methane mystery. *Nature*, **407**, 577–578.

Deming, J., Reysenbach, A.-L., Macko, S.A. & Smith, C.R. (1997) The microbial diversity at a whale fall on the seafloor: bone-colonizing mats and animal-associated symbionts. *Microscopy Research and Technique*, **37**, 162–170.

Distel, D.L., Lee, H.K.-W. & Cavanaugh, C.M. (1995) Intracellular coexistence of methano- and thioautotrophic bacteria in a hydrothermal vent mussel. *Proceedings of the National Academy of Sciences*, **92**, 9598–9602.

Elvert, M., Suess, E. & Whiticar, M.J. (1999) Anaerobic methane oxidation associated with marine gas hydrates: superlight C-isotopes from saturated and unsaturated C_{20} and C_{25} irregular isoprenoids. *Naturwissenschaften*, **86**, 295–300.

Elvert, M., Suess, E., Greinert, J. & Whiticar, M.J. (2000) Archaea mediating anaerobic methane oxidation in deep-sea sediments at cold seeps of the eastern Aleutian subduction zone. *Organic Geochemistry*, **31**, 1175–1187.

Felbeck, H. (1985) Carbon dioxide fixation in the hydrothermal vent tubeworm *Riftia pachyptila* (Jones). *Physiological Zoology*, **58**, 272–281.

Fiala-Médioni, A., McKiness, Z.P., Dando, P., et al. (2002) Ultrastructural, biochemical, and immunological characterization of two populations of the mytilids mussel *Bathymodiolus azoricus* from the Mid-Atlantic Ridge: evidence for a dual symbiosis. *Environmental Microbiology*, **8**(11), 1902–1912.

Fisher, C.R. (1990) Chemoautotrophic and methanotrophic symbioses in marine invertebrates. *Reviews in Aquatic Science*, **2**, 399–436.

Fisher, C.R., Kennicutt, M.C. & Brooks, J.M. (1990) Stable carbon isotopic evidence for carbon limitation in hydrothermal vent vestimentiferans. *Science*, **247**, 1094–1096.

Fisher, C.R., Childress, J.J., Macko, S.A. & Brooks, J.M. (1994) Nutritional interactions in Galapagos Rift hydrothermal vent communities: Inferences from stable carbon and nitrogen isotope analyses. *Marine Ecology Progress Series*, **103**, 45–55.

Fisher, C.R., MacDonald, I.R., Sassen, R., et al. (2000) Methane ice worms: *Hesiocaeca methanicola* colonizing fossil fuel reserves. *Naturwissenschaften*, **87**, 184–187.

Fry, B. & Sherr, E.B. (1984) $\delta^{13}C$ measurements as indicators of carbon flow in marine and freshwater ecosystems. *Contributions to Marine Science*, **27**, 13–47.

Fry, B., Gest, H. & Hayes, J.M. (1983) Sulphur isotopic composition of deep-sea hydrothermal vent animals. *Nature*, **306**, 51–52.

Goffredi, S.K., Warén, A., Orphan V.J., Van Dover, C.L. & Vrijenhoek, R.C. (2004) Novel forms of structural integration between microbes and a hydrothermal vent gastropod from the Indian Ocean. *Applied and Environmental Microbiology*, **70**(5), 3082–3090.

Greinert, J., Bollwerk, S.M., Derkachev, A., Bohrmann, G. & Suess, E. (2002) Massive barite deposits and carbonate mineralization in the Derugin Basin, Sea of Okhotsk: precipitation processes at cold seep sites. *Earth and Planetary Science Letters*, **203**, 165–180.

Guy, R.D., Fogel, M.L. & Berry, J.A. (1993) Photosynthetic fractionation of the stable isotopes of oxygen and carbon. *Plant Physiology*, **101**, 37–47.

Habicht, K.S. & Canfield, D.E. (1997) Sulfur isotope fractionation during bacterial sulfate, reduction in organic-rich sediments. *Geochimica et Cosmochimica Acta*, **61**, 5351–5361.

Hilton, D.R., McMurtry, G.M. & Goff, F. (1998) Large variations in vent fluid $CO_2/{}^3He$ ratios signal rapid changes in magma chemistry at Loihi Seamount, Hawaii. *Nature*, **396**, 359–362.

Hinrichs, K.-U., Hayes, J.M., Sylva, S.P., Brewer, P.G. & DeLong, E. (1999) Methane-consuming archaebacteria in marine sediments. *Nature*, **398**, 802–805.

Hinrichs, K-U., Summons, R.E., Orphan, V., Sylva, S.P. & Hayes, J.M. (2000) Molecular and isotopic analysis of anaerobic methane-oxidizing communities in marine sediments. *Organic Geochemistry*, **31**, 1685–1701.

Hoehler, T.M., Alperin, M.J., Albert, D.B. & Martens, C.S. (1994) Field and laboratory studies of methane oxidation in an anoxic marine sediment; evidence for a methanogen-sulfate reducer consortium. *Global Biogeochemical Cycles*, **8**, 451–463.

Horita, J. & Berndt, M. (1999) Abiogenic methane formation and isotopic fractionation under hydrothermal conditions. *Science*, **285**, 1055–1057.

Horken, K.M. & Tabita, F.R. (1999) Closely related form I ribulose bisphosphate/oxygenase molecules that possess different CO_2/O_2 substrate specificities. *Archives of Biochemistry and Biophysics*, **361**, 183–194.

Hovland, M. & Risk, M. (2003) Do Norwegian deep-water coral reefs rely on seeping fluids? *Marine Geology*, **198**, 83–96.

Jahnke, L.L., Summons, R.E., Hope, J.M. & DesMarais, D.J. (1999) Carbon isotopic fractionation in lipids from methanotrophic bacteria II: The effects of physiology and environmental parameters on the biosynthesis and isotopic signatures of biomarkers. *Geochimica et Cosmochimica Acta*, **63**, 79–93.

Juhl, A. & Taghon, G. (1993) Biology of an active methane seep on the Oregon continental shelf. *Marine Ecology Progress Series*, **102**, 287–294.

Kelley, D.S., Karson, J.A., Blackman, D.K., et al. (2001) An off-axis hydrothermal vent field near the Mid-Atlantic Ridge at 30 degree N. *Nature*, **412**, 145–149.

Kennicutt, M.C. II, Brooks, J.M., Bidigare, R.R., McDonald, S.J., Adkison, D.L. & Macko, S.A. (1989) An upper slope "cold" seep community: northern California. *Limnology and Oceanography*, **34**, 635–640.

Kulm, L.D. & Suess, E. (1990) Relationship between carbonate deposits and fluid venting; Oregon accretionary prism. *Journal of Geophysical Research*, **B95**, 8899–8915.

Kvenvolden, K.A. (1999) Potential effects of gas hydrate on human welfare. Proc. Nat. Acad. Sci. **96**, 3420–3426.

Kvenvolden, K.A. & Lorenson, T.D. (2001) Global occurrences of gas hydrate. In: *The Proceedings of the Eleventh International Offshore and Polar Engineering Conference* (Eds J.S. Chung, M. Sayed, H. Saeki, & T. Setoguchi), pp. 462–467.

Lein, A.Y. & Sagalevich, A.M. (2000) Smokers of the Rainbow field – an area of a large abiogenic methane synthesis. *Priroda*, **8**, 44–53. [In Russian.]

Levesque, C., Juniper, S.K. & Marcus, J. (2003) Food resource partitioning and competition among alvinellid polychaetes of Juan de Fuca Ridge hydrothermal vents. *Marine Ecology Progress Series*, **246**, 173–182.

Levin, L.A. & Michener, R.H. (2002) Isotopic evidence for chemosynthesis-based nutrition of macrobenthos: The lightness of being at Pacific methane seeps. *Limnology and Oceanography*, **47**, 1336–1345.

Levin, L.A., James, D.W., Martin, C.M., Rathburn, A.E., Harris, L.H. & Michener, R.H. (2000) Do methane seeps support distinct macrofaunal assemblages? Observations on community structure and nutrition from the northern California slope and shelf. *Marine Ecology Progress Series*, **208**, 21–39.

Lewis, K.B. & Marshall, B.A. (1996) Seep faunas and other indicators of methane-rich dewatering on New Zealand convergent margins. *New Zealand Journal of Geology and Geophysics*, **39**, 181–200.

Lilley, M.D., Butterfield, D.A., Olson, E.J., Lupton, J.E., Mackos, S.A. & McDuff, R.E. (1993) Anomalous CH_4 and NH_4^+ concentrations at an unsedimented mid-ocean ridge hydrothermal system. *Nature*, **364**, 45–47.

Lynch-Stieglitz, J., Stocker, T.F., Broecker, W.S. & Fairbanks, R.G. (1995) The influence of air–sea exchange on the isotopic composition of oceanic carbon: observations and modeling. *Global Biogeochemical Cycles*, **9**, 653–665.

MacAvoy, S.E., Carney, R.S., Fisher, C.R. & Macko, S.A. (2002) Use of chemosynthetic biomass by large, mobile, benthic predators in the Gulf of Mexico. *Marine Ecology Progress Series*, **225**, 65–78.

Martin, J., Signorovitch, J. & Patel, H. (1997) A new species of *Rimicaris* (Crustacea: Decapoda: Bresiliidae) from the Snake Pit hydrothermal vent field on the Mid-Atlantic Ridge. *Proceedings of the Biological Society of Washington*, **110**, 399–411.

Marty, B. & Zimmerman, L. (1999) Volatiles (He, C, N, Ar) in mid-ocean ridge basalts: Assessment of shallow-level fractionation and characterization of source composition. *Geochimica et Cosmochimica Acta*, **63**, 3619–3633.

Merlivat, L., Pineau, F. & Javoy, M. (1987) Hydrothermal vent waters at 13°N on the East Pacific Rise: Isotopic composition and gas concentration. *Earth and Planetary Science Letters*, **84**, 100–108.

Michaelis, W., Seifert, R., Nauhaus, K., et al. (2002) Microbial reefs in the Black Sea fueled by anaerobic oxidation of methane. *Science*, **297**, 1013–1015.

Olu, K., Duperret, A., Sibuet, M., Foucher, J.-P. & Fiala-Medioni, A. (1996) Structure and distribution of cold seep communities along the Peruvian active margin: Relationship to geological and fluid patterns. *Marine Ecology Progress Series*, **132**, 109–125.

Orphan, V.J., House, C.H., Hinrichs, K.-U., McKeegan, K.D. & DeLong, E.F. (2001) Methane-consuming Archaea revealed by directly coupled isotopic and phylogenetic analysis. *Science*, **293**, 484–487.

Orphan, V.J., House, C.H., Hinrichs, K.-U., McKeegan, K.D. & DeLong, E. (2002) Multiple archaeal groups mediate methane oxidation in anoxic cold seep sediments. *Proceedings of the National Academy of Sciences*, **99**, 7663–7668.

Pancost, R.D., Sinninghe Damsté, de Lint, S., van der Maarel, M.J.E.C., Gottschal, J.C. & MEDINAUT Shipboard Scientific Party (2000) Biomarker evidence for widespread anaerobic methane oxidation in Mediterranean sediments by a consortium of methanogenic archaea and bacteria. *Applied and Environmental Microbiology*, **66**, 1126–1132.

Pancost, R.D., Bouloubassi, I., Aloisi, G., Sinninghe Damsté, J.S. & MEDINAUT Shipboard Scientific Party. (2001) Three series of non-isoprenoidal dialkyl glycerol diethers in cold-seep carbonate crusts. *Organic Geochemistry*, **32**, 695–707.

Paull, C.K., Ussler III, W., Borowski, W. & Spiess, F. (1995) Methane-rich plumes on the Carolina continental rise: associations with gas hydrates. *Geology*, **23**, 89–92.

Peckmann, J., Thiel, V., Michaelis, W., Clari, P., Gaillard, C., Martire, L. & Reitner, R. (1999) Cold seep deposits of Beauvoisin (Oxfordian; southeastern France) and Marmorito (Miocene; northern Italy): microbially induced authigenic carbonates. *International Journal of Earth Sciences*, **88**, 60–75.

Peckmann, J., Goedert, J.L., Thiel, V., Michaelis, W. & Reitner, J. (2002) A comprehensive approach to the study of methane-seep deposits from the Lincoln Creek Formation, western Washington State, USA. *Sedimentology*, **49**, 855–873.

Polz, M.F., Robinson, J.J., Cavanaugh, C.M. & Van Dover, C.L. (1998) Trophic ecology of massive shrimp aggregations at a Mid-Atlantic Ridge hydrothermal site. *Limnology and Oceanography*, **43**, 1631–1638.

Pond, D.W., Segonzac, M., Bell, M.V, Dixon, D.R., Fallick, A.E. & Sargent, J.R. (1997) Lipid and lipid carbon stable isotope composition of the hydrothermal vent shrimp *Mirocaris fortunata*: Evidence for nutritional dependence on photosynthetically fixed carbon. *Marine Ecology Progress Series*, **157**, 221–231.

Rau, G.H. (1981a) Hydrothermal vent clam and vent tubeworm $^{13}C/^{12}C$: Further evidence of a non-photosynthetic food source. *Science*, **213**, 338–339.

Rau, G.H. (1981b) Low $^{15}N/^{14}N$ in hydrothermal vent animals: ecological implications. *Nature*, **289**, 484–485.

Rau, G.H. & Hedges, J.I. (1979) Carbon-13 depletion in a hydrothermal vent mussel: suggestion of a chemosynthetic food source. *Science*, **203**, 648–649.

Rieley, G., Van Dover, C.L., Hedrick, D.B. & Eglinton, G. (1999) Trophic ecology of *Rimicaris exoculata*: a combined lipid abundance/stable isotope approach. *Marine Biology*, **133**, 495–499.

Robinson, J.J. & Cavanaugh, C.M. (1995) Expression of form I and form II ribulose-1,5-bisphosphate carboxylase/oxygenase (Rubsico) in chemoautotrophic symbioses: Implications for the interpretation of stable carbon isotope ratios. *Limnology and Oceanography*, **40**, 1496–1502.

Robinson, J.J., Scott, K.M., Swanson, S.T., et al. (2003) Kinetic isotope effect and characterization of form II RubisCO from the chemoautotrophic endosymbionts of the hydrothermal vent tubeworm *Riftia pachyptila*. *Limnology and Oceanography*, **48**, 48–54.

Roeske, C.A. & O'Leary, M.H. (1984) Carbon isotope effects on the enzyme-catalyzed carboxylation of ribulose bisphosphate. *Biochemistry*, **23**, 6275–6284.

Sahling, H., Rickert, D., Lee, R.W., Linke, P. & Suess, E. (2002) Macrofaunal community structure and sulfide flux at gas hydrate deposits from the Cascadia convergent margin, NE Pacific. *Marine Ecology Progress Series*, **231**, 121–138.

Sakai, H., Gamo, T., Ogawa, Y. & Boulegue, J. (1992) Stable isotopic ratios and origins of carbonates associated with cold seepage at the eastern Nankai Trough. *Earth and Planetary Science Letters*, **109**, 391–404.

Sassen, R., Roberts, H., Aharon, P., Larkin, J., Chinn, E.W. & Carney, R. (1993) Chemosynthetic bacterial mats at cold hydrocarbon seeps, Gulf of Mexico continental slope. *Organic Geochemistry*, **20**, 77–89.

Sassen, R., Joye, S., Sweet, S.T., DeFreitas, D.A., Milkov, A.V. & MacDonald, I.R. (1999) Thermogenic gas hydrates and hydrocarbon gases in complex chemosynthetic communities, Gulf of Mexico continental slope. *Organic Geochemistry*, **30**(7), 485–497.

Schmaljohann, R. & Flügel, H.J. (1987) Methane-oxidizing bacteria in Pogonophora. *Sarsia*, **72**, 91–98.

Schmaljohann, R., Faber, E., Whiticar, M.J. & Dando, P.R. (1990) Co-existence of methane- and sulfur-based endosymbioses between bacteria and invertebrates at a site in the Skagerrak. *Marine Ecology Progress Series*, **61**, 119–124.

Schmidt, M., Botz, R., Winn, K., Stoffers, P., Thiessen, O. & Herzig, P. (2002) Seeping hydrocarbons and related carbonate mineralisations in sediments south of Lihir Island (New Ireland fore arc basin, Papua New Guinea). *Chemical Geology*, **186**, 249–264.

Schoener, T.W. (1974) Resource partitioning in ecological communities. *Science*, **185**, 27–39.

Schouten, S., Wakeham, S.G. & Sinninghe Damsté, J.S. (2001) Anaerobic methane oxidation by archaea in the euxinic water column of the Black Sea. *Organic Geochemistry*, **32**, 1277–1281.

Shank, T.M., Lutz, R.A. & Vrijenhoek, R.C. (1998) Molecular systematics of shrimp (Decapoda: Bresiliidae) from deep-sea hydrothermal vents: Enigmatic "small orange" shrimp from the Mid-Atlantic Ridge are juvenile *Rimicaris exoculata*. *Molecular Marine Biology and Biotechnology*, **7**, 88–96.

Shanks, W.C. III, Böhlke, J.K. & Seal, R.R. II. (1995) Stable isotopes in mid-ocean ridge hydrothermal systems: Interactions between fluids, minerals, and organisms. In: *Seafloor Hydrothermal Processes: Physical, Chemical, Biological, and Geological Interactions* (Eds S.E. Humphris, R.A. Zierenberg, L.S. Mullineaux & R.E. Thompson), pp. 194–221. Geophysical Monograph 91, American Geophysical Union, Washington, DC.

Sibuet, M. & Olu, K. (1998) Biogeography, biodiversity and fluid dependence of deep-sea cold-seep communities at active and passive margins. *Deep-Sea Research, Part II*, **45**, 517–567.

Sigman, D.M., Altabet, M.A., Michener, R., McCorkle, D.C., Fry, B. & Holmes, R.M. (1997) Natural abundance-level measurement of the nitrogen isotopic composition of oceanic nitrate; an adaptation of the ammonia diffusion method. *Marine Chemistry*, **57**, 227–242.

Smith, C.R. & Baco, A.R. (2003) Ecology of whale falls at the deep-sea floor. *Oceanography and Marine Biology: an Annual Review*, **41**, 311–354.

Southward, A.J., Southward, E.C., Spiro, B., Rau, G.H. & Tunnicliffe, V. (1994) $^{13}C/^{12}C$ organisms from Juan de Fuca Ridge hydrothermal vents: A guide to carbon and food sources. *Journal of the Marine Biological Association of the United Kingdom*, **74**, 265–278.

Stakes, D.S., Orange, D., Paduan, J.B., Salamy, K.A. & Maher, N. (1999) Cold-seeps and authigenic carbonate formation in Monterey Bay, California. *Marine Geology*, **159**, 93–109.

Suess, E. & Whiticar, M.J. (1989) Methane-derived CO_2 in pore waters fluids expelled from the Oregon subduction zone. *Palaeogeography, Palaeoclimatology, Palaeoecology*, **71**, 119–136.

Summons, R.E., Jahnke, L.L. & Roksandic, Z. (1994) Carbon isotopic fractionation in lipids from methanotrophic bacteria: relevance for interpretation of the geochemical record of biomarkers. *Geochimica et Cosmochimica Acta*, **58**, 2853–2863.

Taylor, M.H., Dillon, W.P. & Pecher, I.A. (2000) Trapping and migration of methane associated with the gas hydrate stability zone at the Blake Ridge diapir; new insights from seismic data. *Marine Geology*, **164**, 79–89.

Teske, A., Hinrichs, K.-U., Edgcomb, V., et al. (2002) Microbial diversity of hydrothermal sediments in the Guaymas Basin: Evidence for anaerobic methanotrophic communities. *Applied and Environmental Microbiology*, **68**, 1994–2007.

Thiel, V., Peckmann, J., Seifert, R., Wehrung, P., Reitner, J. & Michaelis, W. (1999) Highly isotopically depleted isoprenoids: Molecular markers for ancient methane venting. *Geochimica et Cosmochimica Acta*, **63**, 3959–3966.

Thiel, V., Peckmann, J., Richnow, H., Luth, U., Reitner, J. & Michaelis, W. (2001) Molecular signals for anaerobic methane oxidation in Black Sea seep carbonates and microbial mats. *Marine Chemistry*, **73**, 97–112.

Trask, J.L. & Van Dover, C.L. (1999) Site-specific and ontogenetic variations in nutrition of mussels (*Bathymodiolus* sp.) from the Lucky Strike hydrothermal vent field, Mid-Atlantic Ridge. *Limnology and Oceanography*, **44**, 334–343.

Tsunogai, U., Yoshida, N. & Gamo, T. (2002) Carbon isotopic evidence of methane oxidation through sulfate reduction in sediment beneath cold seep vents on the seafloor at Nankai Trough. *Marine Geology*, **187**, 145–160.

Van Dover, C.L. (2000) *The Ecology of Hydrothermal Vents*. Princeton University Press, Princeton, NJ.

Van Dover, C.L. (2002) Trophic relationships among invertebrates at the Kairei hydrothermal vent field (Central Indian Ridge). *Marine Biology*, **141**, 761–772 + electronic supplement.

Van Dover, C.L. & Fry, B. (1989) Stable isotopic compositions of hydrothermal vent organisms. *Marine Biology*, **98**, 209–216.

Van Dover, C.L. & Fry, B. (1994) Microorganisms as food resources at deep-sea hydrothermal vents. *Limnology and Oceanography*, **39**, 51–57.

Van Dover, C.L., Humphris, S.E., Fornari, D., et al. (2001) Biogeography and ecological setting of Indian Ocean hydrothermal vents. *Science*, **294**, 818–823.

Van Dover, C.L., Aharon, P., Bernhard, J.M., et al. (2003) Blake Ridge methane seeps: characterization of a soft-sediment, chemosynthetically based ecosystem. *Deep-Sea Research, Part I*, **50**, 281–300.

Vereshchaka, A.L. (1996) A new genus and species of caridean shrimp (Crustacea: Decapoda: Alvinocarididae) from north Atlantic hydrothermal vents. *Journal of the Marine Biology Association, UK*, **76**, 951–961.

Vereschaka, A.L., Vinogradov, G.M., Lein, A. Yu., Dalton, S. & Dehairs, F. (2000) Carbon and nitrogen isotopic composition of the fauna from the Broken Spur hydrothermal vent field. *Marine Biology*, **136**, 11–17.

Von Damm, K.L., Edmond, J., Measures, C.I. & Grant, B. (1985) Chemistry of submarine hydrothermal solutions at Guaymas Basin, Gulf of California. *Geochimica et Cosmochimica Acta*, **49**, 2221–2237.

Von Damm, K.L., Bray, A.M., Buttermore, L.G. & Oosting, S.E. (1998) The geochemical controls on vent fluids from the Lucky Strike vent field, Mid-Atlantic Ridge. *Earth and Planetary Science Letters*, **160**, 521–536.

Von Rad, U., Rosch, H., Berner, U., Geyh, M., Marchig, V. & Schulz, H. (1996) Authigenic carbonates derived from oxidized methane vented from the Makran accretionary prism off Pakistan. *Marine Geology*, **136**, 55–77.

Warén, A., Bengtson, S., Goffredi, S.K. & Van Dover, C.L. (2003) A hot-vent gastropod with iron-sulphide dermal sclerites. *Science*, **302**, 1007.

Welhan, J.A. & Lupton, J.E. (1987) Light hydrocarbon gases in Guaymas Basin hydrothermal fluids: thermogenic verss abiogenic origin. *American Association of Petroleum Geologists Bulletin*, **71**, 215–223.

Werne, J.P., Baas, M. & Sinninghe Damsté, J.S. (2002) Molecular isotopic tracing of carbon flow and trophic relationships in a methane-supported benthic microbial community. *Limnology and Oceanography*, **47**, 1694–1701.

Whalen, M. (1993) The global methane cycle. *Annual Reviews in Earth and Planetary Science*, **21**, 407–426.

Whiticar, M.J. (1999) Carbon and hydrogen isotope systematics of bacterial formation and oxidation of methane. *Chemical Geology*, **161**, 291–314.

Zehnder, A.J.B. & Brock, T.D. (1979) Methane formation and methane oxidation by methanogenic bacteria. *Journal of Bacteriology*, **137**, 420–432.

Zyakun, A.M. (1992) Isotopes and their possible use as biomarkers of microbial products. In: *Bacterial Gas* (Ed. R. Vially), pp. 27–46. Editions Technip, Paris.

CHAPTER 9

Stable isotope ratios as tracers in marine food webs: An update

ROBERT H. MICHENER AND LES KAUFMAN

Introduction

Scientists concerned with organic matter flow and food web structures in aquatic systems have increasingly realized the potential for using stable isotopes as both natural tracers and as a means of characterizing trophic structure. Since the first edition of this book, there has been a marked expansion of published studies using stable isotopes to better understand the dynamics of marine ecosystems. Here we discuss how natural abundance stable isotopes vary within various marine ecosystems and how they have been used in feeding experiments both in the laboratory and the field. Stable isotopes also have been used to predict and track the flows and concentration of anthropogenic contaminants and thus reveal the efficacy of management regimes erected to mitigate these impacts. We conclude by summarizing studies of food webs by habitat, including marine, estuarine/nearshore systems, and salt-marsh systems. We also look at how stable isotopes have been applied to marine conservation and management.

Methods of assessing food webs

Traditional approaches

Non-stable isotope approaches to food web analysis include gut contents analysis, fecal analysis, direct observation both in the field and laboratory, radiotracer, immunological, and fatty acid applications (Smith et al. 1979; Beviss-Challinor & Field 1982; Rounick & Winterbourn 1986; Hopkins 1987; Sondergaard et al. 1988; Warren 1989; Kioboe et al. 1990; Båmstedt et al. 2000; Trites 2001). These methods have helped resolve food web structure, yet each has its drawbacks. Analysis of gut contents involves collecting and some time spent on dissecting a broad range of individuals to determine what each member of the food web has recently eaten. Employing few tools or equipment, a food web structure can be constructed. However, some organisms digest their prey rapidly, making identification difficult, whereas others do not, confounding between-species comparisons (Feller et al. 1979).

However long digestion may take, gut contents are at best a snapshot of what an individual ate the previous day, and is not always a good reflection of what it might have been eating on a regular basis. There is also the potential for overestimation of prey items with hard body parts, due to differential digestion of different types of prey within a species. There is also a need for large sample sizes required to overcome variability in gut fullness and feeding plasticity. Moreover, gut morphology varies considerably among species, challenging what one means by "gut" contents, given that it often includes material that is never assimilated (e.g., bones, scales, teeth). In addition, gut content analysis requires the researcher to have a good taxonomic knowledge of nearly all the macroorganisms present. In aquatic ecosystems, predators at all levels of the food web can be remarkably undiscriminating in what they eat.

Predator–prey relationships have also been studied through behavioral work in the laboratory, such as tests of feeding preference, but obtaining an adequate sample of prey items may be difficult and laboratory studies produce at best a highly imperfect prediction of species relationships in the field (Ockelmann & Vahl 1970; Feller et al. 1979; Boyd et al. 1984). Other traditional methods used both in the field and the laboratory include food or prey exclusion or inclusion, monitoring predator/prey abundances, and the posting of pieces of prey items that can be recovered to signify a predation event (Gerlach et al. 1976; Arntz 1977).

The use of radio-labeled prey items in food web analysis involves adding a radiotracer (e.g., ^{14}C or ^{3}H) to a food source or prey species and following the label through the food chain (Marples 1966; Smith et al. 1979; Beviss-Challinor & Field 1982; Pearcy & Stuiver 1983; Hessen et al. 1990). Smith et al. (1979) used this technique to partially characterize a coral reef and subtropical estuary involving demersal zooplankton. As with other methods, radio-labeling has its disadvantages: recovery of a statistically significant number of labeled species can be difficult, the need to secure permits to use radioactive isotopes, and the need to use high dosages of isotope to overcome dilution, especially in the case of tritium. Sophisticated models are also often needed to extract feeding relations from time courses of label appearance and loss.

Immunological methods have also been used to characterize aquatic food webs (Boreham & Ohiagu 1978; Feller et al. 1979, 1985). These methods involve developing antisera from whole organism extracts, followed by double immunodiffusion precipitin tests of antiserum specificity. Antisera are usually taxon-specific and thus can be used to trace trophic relationships. This method is appealing for investigating organisms whose gut contents cannot be identified visually. Feller et al. (1985) used this technique to investigate deep sea food web structure, where changes in water pressure frequently deform organisms and make gut contents difficult to identify. However, this method is limited to the specificities and number of antisera developed and is strictly qualitative. For systems with a large number of species, it would be prohibitively expensive and time-consuming to explore all possible antisera. For further discussion and review, see Boreham & Ohiagu (1978). Another molecular approach involves the use of new DNA identification techniques. These are revolutionizing gut contents studies. A big advantage is that species relations are identified and preserved; isotopes track element flows much less directly than species interactions. Along similar lines, investigators have enjoyed some success using ratios of fatty acids in prey and tissues (Meier-Augenstein 2002; Kainz et al. 2004; Evershed et al., this volume, pp. 480–540). Fatty acid analysis has the potential to be a major competitor to isotopic analysis, and probably will be coupled with isotopic analysis in many cases.

Stable isotope analysis

Stable isotope analysis has proven its value as an alternate, complementary, and in some cases, better tool for food web analysis in marine systems. The collection technique is simple and straightforward, and with the development of continuous flow systems, a broad survey of organisms in a food web system can be performed within reasonable limits on time and funds. Stable isotopes record both source and trophic level information, with sulfur and carbon isotopes strongest for source information and nitrogen isotopes recording trophic information. Oxygen is also of possible interest in future food studies, to the extent that oxygen isotopes record dietary rather than source water information. Further details on theory, methods, and collection strategies can be found in the Introduction to this volume and in Sulzman, this volume, pp. 1–21. To interpret the data, we need to next look at how carbon, nitrogen, hydrogen, and sulfur stable isotopes circulate within biological systems.

Carbon

Carbon isotopic compositions of animals reflect their diet within about 1‰ (Haines 1976a; DeNiro & Epstein 1978; Fry et al. 1978a; Haines & Montague 1979; Teeri & Schoeller 1979; Rau 1980; Rau & Anderson 1981; Fry & Arnold 1982; Tieszen et al. 1983; Checkley & Entzeroth 1985; Peterson & Fry 1987; Figure 9.1). Overall, there is a small (0.5–1‰) enrichment (i.e. enriched in the rare heavier isotope of ^{13}C) in the animal relative to its diet. This can also be referred to as the "trophic enrichment factor" (TEF): $\delta_{animal} - \delta_{food} = \Delta^{13}C_{TEF}$, $\Delta^{15}N_{TEF}$, and so on, depending on the isotope in question. There are several possible biological processes that could contribute to this enrichment: (i) preferential loss of ^{12}C during respiration, (ii) preferential uptake of ^{13}C enriched compounds during digestion and/or assimilation, or (iii) metabolic

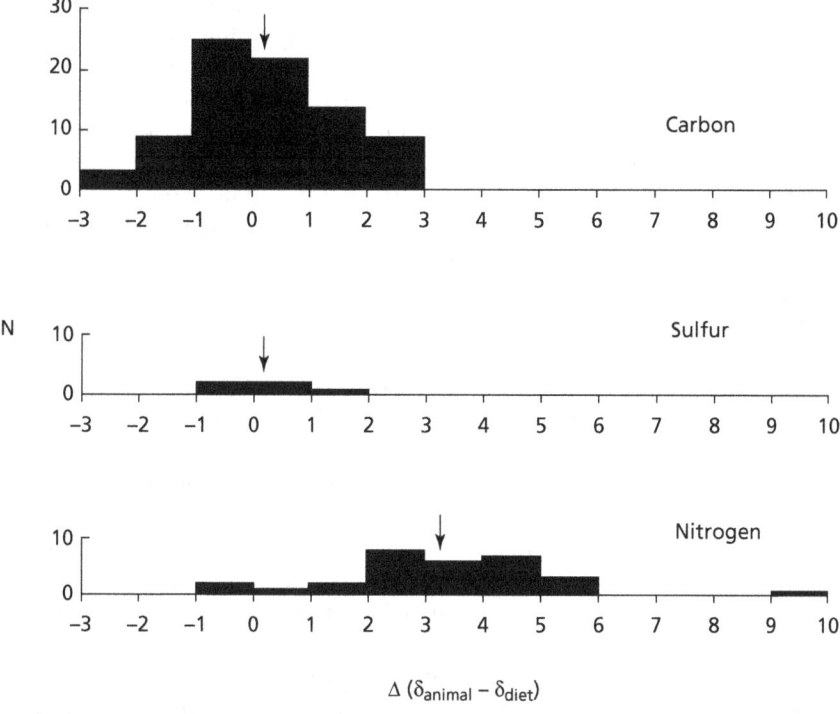

Figure 9.1 Trophic enrichment factors (TEF's) between organisms and diet for carbon, sulfur, and nitrogen isotopes. (From Peterson & Fry 1987.)

fractionation during synthesis of different tissue types (DeNiro & Epstein 1978; Rau et al. 1983; Tieszen et al. 1983; Fry et al. 1984). More recently, research on isotopic routing has been suggested, where an animal's tissue reflects the biochemical component used to make it (Focken & Becker 1998; Vander Zanden & Rasmussen 2001). Ambrose & Norr (1993) found that consumers ingesting a high protein vs. a low protein diet can have a different trophic level fractionation. Organisms on a high protein diet show carbon isotopic values reflecting the protein component, whereas organisms on a low protein diet tended to reflect the bulk isotopic composition of the food source. Thus, knowledge of food quality is needed to fully understand the importance of isotopic routing.

Currently, there is no general synthetic model to precisely predict enrichment levels, although recent analyses by Fry (2006) indicate that trophic enrichments should be linearly related to net growth efficiency; at 100% efficiency enrichment vs. diet is small and largest at 0% efficiency. These model results derive from simple mass balance considerations, some-times highlighted in earlier studies. For example, DeNiro & Epstein (1978) investigated the grasshopper *Melanoplus* on a wheat diet and found that respired CO_2 was depleted and whole body composition and feces were enriched, relative to the diet. Mass balance showed an approximate 1‰ enrichment. Stephenson et al. (1986) examined lobsters and oysters in the laboratory and found a significant enrichment of the animals vs. their diet and suggested a selective assimilation of compounds from the diet. They also found an inverse relationship between the calorific value of the diet vs. the $(\delta^{13}C_{animal} - \delta^{13}C_{diet})$ of the lobster tissue. Lobsters fed a high calorie, higher fat diet will produce tissues with a higher lipid content and lower $\delta^{13}C$ values. This agrees with studies showing that lipids are isotopically lighter than other biochemical fractions (DeNiro & Epstein 1977; Parker 1964; Pinnegar & Polunin 1999).

This conservative transfer of carbon isotopic compositions (<1‰) to the animal from the diet can be useful in tracing food webs in systems where food sources show large differences in $\delta^{13}C$ values, such as C3 vs. C4 plants, marine vs. terrestrial systems, or nearshore vs. offshore systems (Haines 1976b; Fry et al. 1977, 1978a; DeNiro & Epstein 1978; Rau 1981a; Schoeninger & DeNiro 1984; Koch, this volume, pp. 99–154). However, investigators must also be aware of isotopic variations among different tissues within an organism. This can reflect differences in lipid or amino acid composition and concentration, variation in turnover time among different tissues, and different rates of tissue turnover when an organism selectively feeds. For example, Pinnegar & Polunin (1999) found differences in stable isotopic values among white muscle, red muscle, liver and heart tissue of juvenile rainbow trout. White muscle was found to be less variable than the other trout tissues. One must also consider time averaging, where food sources change temporally. This can have consequences for which tissue type is used (O'Reilly et al. 2002), and may add a confounding factor to seasonal studies. These issues are especially important when working in environments with multiple, disparate sources of carbon (e.g. estuaries, and benthic assemblages of mixed deposit and suspension feeders.)

The isotopic composition of tissues can change over time, reflecting changes in diet, with the more metabolically active tissues turning over more quickly. For example, in a dietary study of gerbils, Tieszen et al. (1983) switched the diet from a C4 corn to a C3 wheat and analyzed the major tissues for ^{13}C. They found that the ^{13}C enrichment for the individual tissues decreased from hair > brain > muscle > liver > fat. Other laboratory studies (Teeri & Schoeller 1979; Rau & Anderson 1981; Fry & Arnold 1982; Macko et al. 1982; Voigt et al. 2003) have shown that organisms fed an isotopically distinct diet will approach the dietary value as the organism grows and tissue turns over.

Ontogenetic changes, especially in estuarine fish species, can have a significant affect on the interpretation of the food web. The question of changes in isotopic ratios based on either biomass gain or metabolic tissue turnover rates is an area that has mostly focused on larvae or juvenile organisms (Bosley et al. 2002; Herzka 2005). Several studies have found that isotopic changes primarily reflect increases in body mass (Hesslein et al. 1993; Jennings et al. 2001b; Gaye-Siessegger et al. 2004; Herzka 2005), with an equation giving turnover in terms of both time and growth developed by Hesslein et al. (1993).

Juvenile animals that migrate offshore from estuaries will tend to change their isotopic composition as they incorporate a new diet. In a field study, Fry (1983) measured the isotopic compositions of juvenile shrimp feeding in south Texas grass flats. Juveniles collected at the beginning of the study had isotopic values of $\delta^{13}C = -11$ to $-14‰$, and 6–8‰ for nitrogen and sulfur. As the shrimp grew and migrated offshore, the isotopic values converged toward offshore values of ca. $-16‰$, $+11.5‰$ and $+16‰$ for C, N, and S, respectively. A study that investigated benthic flatfish found that as the fish grew, their diet switched from riverine particulate organic matter (POM) to offshore marine POM (Darnaude 2005).

Depending on tissue turnover rates, $\delta^{13}C$ values will be biased towards the most recent feeding patterns. In ecosystem studies, unless the sampling protocol includes excellent replication both within and between study sites, it is difficult to determine if the isotopic compositions of mobile animals reflect local feeding or food from other sources with different isotopic compositions (Fry & Arnold 1982; Wainright & Fry 1994; Wainright et al. 1996; Sullivan et al. 2006). Given these problems and different isotopic compositions of tissues, it would seem prudent to either sample the entire organism to get an integrated isotopic value, to sample several tissue types covering a range of tissue turnover times, or consistently sample a particular tissue – usually tissue from the same muscles in comparable creatures – as this is likely to show the least bias due to biochemical heterogeneity (e.g., lipid composition). Sampling several individuals of a smaller species would also help eliminate variations within individuals, since animals of the same species fed on the same diet can vary up to 2‰ (DeNiro & Epstein 1978; Voigt et al. 2003). The problem with pooling, however, is the information that is potentially lost; for example, individual variation due to feeding plasticity, which represents one of the most interesting patterns that could be revealed. This variability is also useful for measuring niche parameters (Bearhop et al. 2004).

Despite the seeming difficulties, there are several excellent studies of food web systems using $\delta^{13}C$ measurements (Black & Bender 1976; Haines 1976a; Fry et al. 1978a; Fry & Parker 1979; Haines & Montague 1979; Rau et al. 1983; Peterson et al. 1985; Gu et al. 2001). Each study was characterized by primary food sources that were isotopically distinct, and again, direct relationships were revealed between the organisms and their diets. Rau et al. (1983) found roughly a 1.1‰ ^{13}C enrichment per trophic level, and the species were selected by their clearly identifiable trophic status and where only one tissue – muscle – was analyzed.

Nitrogen

When looking at the $\delta^{15}N$ of organisms and their diet, important dietary relationships can be found (Gaebler et al. 1966; Steele & Daniel 1978; DeNiro & Epstein 1981a; Rau 1981b; Macko et al. 1982; Checkley & Entzeroth 1985; Peterson & Fry 1987; France 1995; Adams & Sterner 2000; Vander Zanden & Rasmussen 2001; Vanderklift & Ponsard 2003). This is not surprising because stable nitrogen isotopes, while influenced by source, can offer an especially strong signal for the mean effective trophic level of individuals for the period of time equal to the turnover time for nitrogen atoms in the sampled tissue. Thus, $\delta^{15}N$ can record both source and TEF. Because there are no "empty stomachs" in this approach, it offers a consistent assessment of at least some aspects of recent feeding behavior and how it varies among individuals within and between populations. As with carbon, $\delta^{15}N$ in the organism reflects the $\delta^{15}N$ of the diet, but in most cases the whole animal is enriched in $\delta^{15}N$ relative to the diet (DeNiro & Epstein 1981a). In laboratory and field studies of two marine amphipods, Macko et al. (1982) showed −0.3‰ and +2.3‰ TEFs, regardless of the food source. DeNiro & Epstein (1981b) saw similar fractionations for two insect species.

When enrichment occurs, a preferential excretion of ^{15}N-depleted nitrogen occurs, usually in the form of urea and ammonia (Minigawa & Wada 1984). ^{15}N retention varies according to species, diet, and nutritional stress (Hobson 1991; Koch, this volume, pp. 99–154 and Hobson, this volume, pp. 155–175). Isotopic analysis of cattle and their diets found that urine was depleted in ^{15}N relative to diet as well as to blood, feces, and milk (Steele & Daniel 1978).

When the laboratory mouse *Mus musculus* was fed isotopically distinct diets, various tissues measured were enriched relative to the diet, with $\delta^{15}N$ increasing from kidney to hair to liver to brain (DeNiro & Epstein 1981a). Part of these differences in tissue $\delta^{15}N$ may reflect isotope fractionation during amino acid transamination and deamination (Gaebler et al. 1966; Macko et al. 1986; Vander Zanden & Rasmussen 2001). Heavier amine groups appear to be preferentially metabolized and excreted, leaving isotopically light metabolites. Thus, during assimilation there is isotopic discrimination, resulting in excreted nitrogen that is isotopically light (Gannes et al. 1998; Pinnegar & Polunin 1999; Adams & Sterner 2000; Schmidt et al. 2003). As with carbon, analysis of several tissue types, whole organism, or within-tissue comparisons of $\delta^{15}N$ should be performed when comparing animal and diet $\delta^{15}N$.

The quality of food and the method of excretion by the consumer can have profound affects on the degree of enrichment, which can be a confounding factor in determining food web relationships. For example, the higher $C:N$ ratios in the diet can increase the ^{15}N enrichments in the consumer, similar to that seen in starving organisms (Adams & Sterner 2000; Vanderklift & Ponsard 2003; Koch, this volume, pp. 99–154). Organisms excreting primarily ammonium had lower levels of enrichment than organisms excreting urea. That being said, Pearson et al. (2003) presented evidence that in birds, the reverse is true: ^{15}N enrichment increases as the concentration of N in the diet rises. On a high N diet, animals lose more N as waste (e.g. urea and ammonia), whereas on a low N diet they lose more N as body tissue. It is possible that the quality rather than the quantity of protein is the deciding factor. These examples illustrate the complexity of diet–consumer relationships, pointing out the need for more controlled feeding studies (Gannes et al. 1998; Robbins et al. 2005).

Although some studies have shown ^{15}N variability in trophic enrichment (Ruess et al. 2004; MacNeil et al. 2005), the original study by DeNiro & Epstein (1981a) has proven to be remarkably robust (Post 2002). Field studies have shown an average of 3.2‰ enrichment in animal $\delta^{15}N$ vs. diet for a wide range of species (Figure 9.1), which is reflected as a trophic level effect in food web studies. Minigawa & Wada (1984) found a ^{15}N enrichment of $+3.4 \pm 1.1$‰ per trophic level, independent of habitat. A survey of bone collagen by Schoeninger & DeNiro (1984) for 66 species of vertebrates resulted in an average 3‰ enrichment per trophic level. Hobson (1991) showed that dietary information can be acquired on both short- and long-term intervals in birds through the use of different tissue types, and suggested that nutritional stress was important and must be considered along with diet. In another study of Arctic marine food webs, Hobson & Welch (1992) reported a trophic enrichment in $\delta^{15}N$. Rau (1981b), in a study of hydrothermal vent animals, also observed an increase in $\delta^{15}N$ as a function of the presumed trophic level.

Nitrogen isotopes can also help determine spatial variation when two distinct nutrient sources are available. Gaston & Suthers (2004), in a 2-year study of planktivorous fish on a rocky reef found that the influence of sewage effluent could be seen, with liver and muscle recording short-term and long-term integration of light POM, respectively. Moving offshore, two studies looking at various species of sharks in pelagic systems found that stable isotopes were excellent predictors of trophic position (Estrada et al. 2003; Domi et al. 2005). Still, given the variability observed in some studies, there is a need for controlled laboratory studies to investigate the relationship between consumers and diet (Gorokhova & Hansson 1999; Adams & Sterner 2000; Gaye-Siessegger et al. 2004). Laboratory experiments investigating TEFs should probably use a known and isotopically uniform diet that is fairly natural and supports good growth. Additionally, animals should be grown for extended periods (preferably for more than one generation), e.g. Tieszen et al. (1983), with mass balance summaries of all inputs and outputs to really understand TEF mechanisms and magnitudes.

Hydrogen and sulfur

There appears to be little or no enrichment in [34]S at different trophic levels or in animal diets (Figure 9.1; Mekhtiyeva et al. 1976; Peterson & Howarth 1983). However, the isotopic difference between seawater sulfate and sulfides (~21‰ vs. ca. −10‰) makes sulfur useful in distinguishing benthic vs. pelagic producers and marsh plants vs. phytoplankton in estuarine studies (Fry et al. 1982; Peterson & Howarth 1987). In combination with carbon isotopes, sulfur can be used to distinguish among primary producers in marine food web systems (Connolly et al. 2004; Oakes & Connolly 2004). Benthic systems and marsh plants tend to be richer than pelagic or more aerobic sources in sulfur derived from sulfides, and thus reflect a lighter $\delta^{34}S$ signal.

Hydrogen also appears to show no enrichment of whole animal δD vs. diet (Estep & Dabrowski 1980; Macko et al. 1983). Although exchangeable hydrogen dilutes the signal between organic tissue and water (DeNiro & Epstein 1981b; Hobson, this volume, pp. 155–175), it appears that the δD of diet is reflected in the animal. This may make hydrogen useful in tracing organisms that feed in freshwater vs. marine systems and tracing terrestrial vs. marine food webs (Fry & Sherr 1984). Hydrogen also may be useful as a tracer for anadromous fishes or migratory marine organisms (see also Hobson, this volume, pp. 155–175).

Phytoplankton and particulate organic carbon

This section covers carbon isotopes in phytoplankton and particulate organic carbon (POC), and the biological and physical factors that cause fractionation. More extensive reviews can be found in Fry & Sherr (1984) and Johnston & Kennedy (1998). When investigating food webs, it is important to first look at the carbon isotopic composition in the primary producers, since the isotopic label derived from these organisms will be transferred to successive trophic levels. Few measurements of phytoplankton have been published; typically researchers measure POC, assuming that the bulk of POC is derived from phytoplankton.

In natural oceanic populations of phytoplankton, there is a large range of isotopic values within and between geographic regions (Degens et al. 1968; Deuser et al. 1968; Rau et al. 1983, 1990; Gearing et al. 1984). Several factors may contribute to this variability, discussed in further detail below:

1 the isotopic composition of the dissolved inorganic carbon pool (DIC) may vary with temperature and location;
2 isotopic discrimination may be related to the morphology of the particular species of phytoplankton;

3 there may be isotopic discrimination by the carboxylating enzyme involved in CO_2 fixation;
4 growth rates of phytoplankton can affect their carbon isotopic composition.

Early ideas concerning the effects of temperature on isotopic composition of phytoplankton (Sackett et al. 1965; Eadie & Jeffrey 1973) have been refined as the role of carbon dioxide pools in marine water became more evident. Although the DIC pool in the open ocean is fairly uniform at 0‰ (Kroopnick 1985; Sherr 1982; Sackett & Moore 1966), estuarine environments with freshwater input will vary, since DIC values in freshwater can range from −5 to −10‰. As most phytoplankton obtain carbon from the free carbon dioxide pool, and this pool size increases with decreasing temperature, temperature both directly and indirectly contributes to observed variation in $\delta^{13}C$. Calculated $\delta^{13}C$ values of aqueous CO_2 vary from −6‰ at 25°C to −12‰ at 0°C (Mook et al. 1974) when the $\delta^{13}C$ of DIC is 0‰.

Gearing et al. (1984) sampled different plankton sizes and found that $\delta^{13}C$ varied with species and size, ranging from −20.3‰ for diatoms to −22.2‰ for nanoplankton. Rau et al. (1990) also noted isotopic differences in both $\delta^{13}C$ and $\delta^{15}N$ with size in suspended POM. These differences were thought to reflect microbial breakdown and fractionation of the POM, and a trophic-level effect in the smallest size fraction. The smallest fractions were thought to consist of nano- and picoplankton, which may be a significant low trophic level component (Rau et al. 1990). By contrast, a study of size fractions of POC in Martha's Vineyard Sound (Woods Hole, MA) found little or no correlation of $\delta^{13}C$ vs. size (Wainwright 1990). One possible explanation was a problem in obtaining pure samples of diatoms, since laboratory cultures have shown size-related differences. Thus, sampling protocols should be chosen carefully and one must be cognizant that phytoplankton species size and composition may affect $\delta^{13}C$.

Another variable affecting $\delta^{13}C$ in phytoplankton is the metabolic pathway of photosynthesis. As discussed by Marshall et al. (this volume, pp. 22–60), the two primary pathways of photosynthesis utilize either the RuBP carboxylase enzyme (isotopic discrimination −23 to −41‰, C3 pathway) or the PEP carboxylase enzyme (isotopic discrimination −0.5 to −3.6‰, C4 pathway). There is little evidence that any phytoplankton utilize C4 pathways; however, they do concentrate HCO_3^- (Tortell et al. 1997), which may eventually result in C4-like $\delta^{13}C$ values. Wong & Sackett (1978) found that marine phytoplankton species differ in their metabolic pathways, leading to a range of $\Delta(\delta^{13}C$ algae vs. $HCO_3^-)$ of −22.1 to −35.5‰ for 17 species in laboratory cultures. As they cautioned, this was a controlled situation and optimal growth conditions were not achieved for all species. Within cells, different macromolecular fractions have been shown to have different $\delta^{13}C$ values. Cellulose fractions are more depleted in $\delta^{13}C$ than other carbohydrates (Degens et al, 1968), with lipids being the most ^{13}C-depleted (Parker 1964).

Thus, variations in lipid content may account for some of the observed isotopic differences.

Rau et al. (1982) synthesized the then-available information on latitudinal gradients from north to south in $\delta^{13}C$ of plankton and found a much steeper isotopic gradient in the Southern Hemisphere. Rau et al. (1990) later ascribed the poleward depletion in ^{13}C to the increasing size of the free carbon dioxide pool in seawater. This could not explain, however, the large differences between Arctic and Antarctic carbon isotopic values. Antarctic phytoplankton as a whole also has very low $\delta^{15}N$ relative to Arctic phytoplankton. This probably reflects slow growth rates despite the higher concentrations of nutrients and carbon in the Antarctic. Ultimate control of $\delta^{13}C$ and $\delta^{15}N$ in the southernmost marine waters is most likely due to some other environmental factor such as light or a trace element (such as iron) limitation (Ostrom et al. 1997).

Understanding variations in $\delta^{13}C$ of POC are important when defining a pelagic food web. For example, natural gradients among POC and phytoplankton species $\delta^{13}C$ provide a powerful means of defining habitat usage and acquiring insight into the natural history of migratory fauna. Saupe et al. (1989) described the gradient in $\delta^{13}C$ of zooplankton across the Alaskan Beaufort, Chukchi and Bering seas, and Schell et al. (1989a, 1989b) used this gradient to identify critical feeding habitats and seasonal feeding cycles in bowhead whales (*Balaena mysticetus*). Oscillations in $\delta^{13}C$ derived from feeding in differing summer and winter habitats were recorded in the baleen plates of the whales from both the Pacific and Atlantic Oceans. These oscillations provided a means of age determination in the whales and defined regions essential for feeding (Schell 1987).

With the development of fine-scale drilling and milling devices, as well as with a decrease in sample size requirements, it is now possible to use fish otoliths as tracers of fish migration and geographic origin. Lenanton et al. (2003) used $\delta^{13}C$ and $\delta^{18}O$ of whitefish otoliths to determine their geographical source in Western Australia, as whitefish are an important food source for little penguins (*Eudyptes minor*). Analysis of otolith bands in Atlantic cod made it possible for researchers to record the movements of these fish as juveniles and to record their movement as adults to deeper waters on the Scotian Shelf, North Atlantic (Jamieson et al. 2004).

Phytoplankton and particulate organic nitrogen

As with carbon, nitrogen isotope ratios in phytoplankton are affected by the abundance and various forms of inorganic nitrogenous nutrients. Particulate organic nitrogen (PON) plays an important role in the vertical transport of material out of the euphotic zone. In terms of POM, there are two types of organic matter: rapidly sinking particles (usually made up of fecal pellets and marine "snow"; i.e., detrital precipitation from surface to deeper waters) and slowly sinking particles, which are mostly decomposed and remineralized within the euphotic zone (Saino & Hattori 1987). C/N ratios of POM increase with depth, implying that N is more rapidly lost than C during degradation (Gordon 1977; Tanoue & Handa 1979; Saino & Hattori 1980). It is thus important to determine how this PON is cycled, since ultimately the $\delta^{15}N$ of this nitrogen will determine the $\delta^{15}N$ of the phytoplankton (Owens 1987; Wada & Hattori 1991; see also Montoya, this volume, pp. 176–201).

In the euphotic zone, the forms of inorganic nitrogen important to phytoplankton include N_2 gas, ammonia and nitrate (Wada et al. 1975). Regenerated nitrogen is recycled nitrogen within the euphotic zone and typically refers to ammonia uptake and release and to a lesser extent low molecular weight organic nitrogen (e.g. amino acids and urea) (Dugdale & Goering 1967). "New" nitrogen refers to the influx of nitrate from deeper waters, nitrogen fixation, and atmospheric washout of fixed dinitrogen (Saino & Hattori 1987). In order to maintain a nitrogen balance, the loss of nitrogen from the downward flux of particles out of the euphotic zone must be balanced by an input of this new nitrogen to the system (Eppley & Peterson 1979).

Nitrogen fixation and atmospheric washout of fixed nitrogen appear to introduce a small fraction of new nitrogen to pelagic systems (Eppley et al. 1973; Wada et al. 1975; Mullin et al. 1984; Altabet 1989; Montoya, this volume, pp. 176–201), although for certain species this input of additional N may be very helpful (Martinez et al. 1983). This input will introduce nitrogen with a $\delta^{15}N$ value close to 0‰, since nitrogen fixation has a small fractionation factor (e.g., laboratory cultures of *Azotobacter* and *Anabaena cylindrica*, $\alpha = 0.996–1.009$; Wada et al. 1975). *Trichodesmium*, a nitrogen fixer, has a lower $\delta^{15}N$ value than non-N_2-fixing phytoplankton (Owens 1987). Oxidized nitrogen and ammonia in rainfall have also been shown to have low $\delta^{15}N$ (Hoering 1957; Wada et al. 1975). In estuarine systems, inputs of nitrogen with low $\delta^{15}N$ material via river water become increasingly important to the overall nitrogen budget (Wada et al. 1975; Mariotti et al. 1984; Owens 1985; Fry 2002).

The most important source of nitrogen to the oceanic euphotic zone is in the form of upwelled nitrate (Altabet & McCarthy 1985; Altabet 1989; Liu & Kaplan 1989). Nitrate with high $\delta^{15}N$ is left during denitrification, which has a large fractionation factor (α up to 1.04). Nitrate with high $\delta^{15}N$ often occurs in oxygen-depleted water (Cline & Kaplan 1975; Liu & Kaplan 1989). Subsequent vertical transport will introduce this enriched nitrate into the water column to be taken up by phytoplankton and in turn give a higher $\delta^{15}N$ value for marine organisms (Wada et al. 1975; Table 9.1 and see Figure 9.3). This leads to the possibility of using nitrogen isotope ratios to follow upwelling. Values for $\delta^{15}N$ nitrate in different geographic regions can be found in Table 9.1 and Figure 9.2.

Table 9.1 ^{15}N values for nitrate and particulate organic nitrogen (PON).

Location*	δ^{15}N nitrate (‰)	δ^{15}N PON (‰)	Reference
Northeastern North Pacific	5.1–7.0		Miyake & Wada 1967
Seawater	5.8 ± 1.6		Wada et al. 1975
Denitrifying zones, ETNP	Up to 19		Cline & Kaplan 1975
Denitrifying zones, ETSP	Up to 13		Liu et al. 1987
Northern Atlantic >200 m	4.8–6.8		Liu & Kaplan 1989
Denitrifying, ETNP 200–500 m	12–18		
ETNP >500 m	ca. 6.5		
Central North Pacific	5–6		
North Pacific		Mean −4.9	Wada et al. 1975
Southern California Bight		6.5–12.1	Sweeney & Kaplan 1980a, 1980b
North Pacific		−1.7 to +9.7	Wada & Hattori 1978
North Pacific		3.4–7.2	Miyake & Wada 1967
Northeast Indian Ocean, surface		1.4 ± 0.8	Saino & Hattori 1980
Northeast Indian Ocean to 500 m		2.9–13.0	
Northeast Indian Ocean >500 m		ca. 13	

* ETNP, Eastern Tropical North Pacific; ETSP, Eastern Tropical South Pacific.

As with POC, it is difficult to obtain clean samples of phytoplankton-PON for analysis. Future work may isolate chlorophyll for δ^{15}N specific to phytoplankton (Sachs et al. 1999). But at present, PON samples usually include a mix of phytoplankton, detritus, microzooplankton, and bacteria (Altabet & McCarthy 1985). As noted above, nitrogen available in the euphotic zone is recycled, which may result in isotopic fractionation. The dissolved recycled nitrogen primarily is in the form of ammonia and urea, present in small quantities relative to the total PON. This pool is totally recycled and does not change the δ^{15}N of the bulk PON even if the different components differ in δ^{15}N (Altabet & McCarthy 1985). Variations in δ^{15}N of PON within the euphotic zone appear to reflect both the influx of nitrate and loss through sinking particles.

Particulate organic nitrogen below the euphotic zone increases in δ^{15}N (Saino & Hattori 1980, 1987; Altabet & McCarthy 1986; Altabet, 1989). Saino & Hattori (1987) found approximately a 6‰ enrichment in PON with depth at each station they sampled, which they ascribed to biological degradation of the sinking PON. Altabet & McCarthy (1985) hypothesized that the change with depth is due to (i) selective degradation of different chemical fractions, and (ii) the sinking rates of the particles will be different, therefore more

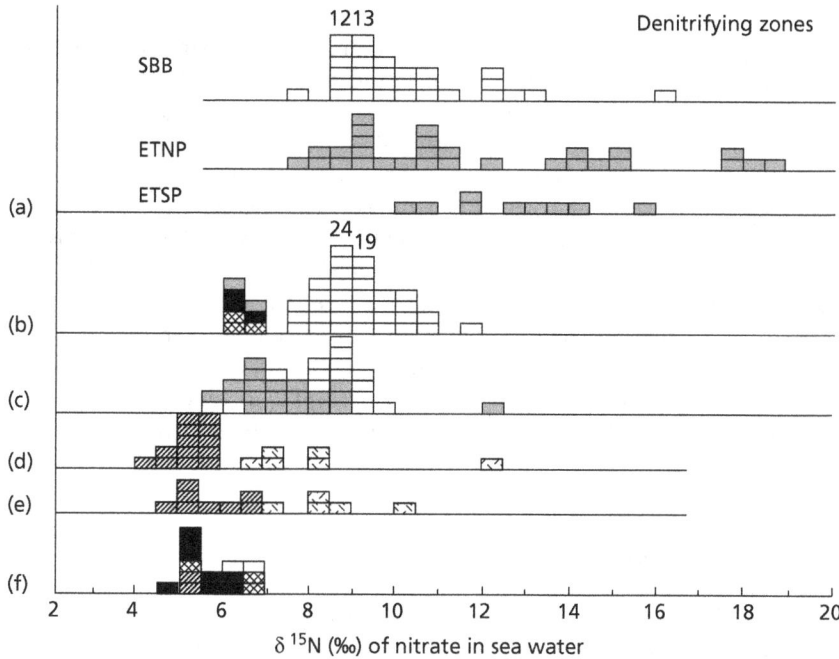

Figure 9.2 Nitrate ^{15}N values measured in seawater below 200 m. (a) SBB, Santa Barbara Basin, California; ETNP, Eastern Tropical North Pacific; ETSP, Eastern Tropical South Pacific. (b) Pacific subsurface water (200–500 m); (c) Pacific intermediate water (500–1500 m); (d) Atlantic subsurface water (200–500 m); (e) Atlantic intermediate water (500–1500 m); (f) deep waters (>1500 m). ⊠ = Northwestern North Pacific; ■ = Central North Pacific; □ = California; ⊞ = Northern North Atlantic; ▨ = Eastern Pacific; ⊠ = Tropical North Atlantic. (Reproduced from figure 3 in Liu & Kaplan 1989, Copyright (2007) by the American Society of Limnology and Oceanography, Inc.)

reworking will occur in slower-settling particles and the refractory components will increase in δ^{15}N. Altabet (1989) proposed that the differences in δ^{15}N between the PON within and below the euphotic zone indicated the average number of trophic steps between phytoplankton and sinking particulate organic matter, and could be used as indicators of vertical change.

Numerous profiles of PON-δ^{15}N show a sharp decrease, and then increase with depth in the euphotic zone (e.g. Saino & Hattori 1980). The δ^{15}N minimum is usually accompanied by a decrease in nitrate concentrations. Phytoplankton has been shown to fractionate ^{15}N during assimilation of nitrate (Wada & Hattori 1978), and this profile in δ^{15}N may result from preferential uptake of $^{14}NO_3$ by phytoplankton (Altabet et al. 1986; Liu & Kaplan 1989). In the euphotic zone of the Sargasso Sea under stratified conditions, Altabet (1989) found a δ^{15}N minimum, PON maxima, and the top of the nitracline all occurring at approximately the same depth. Nitrate reduction and uptake was thought to outweigh the sinking of PON enriched in ^{15}N to produce the δ^{15}N minima. It has been noted that suspended particulate matter in the euphotic zone has usually been found to be more ^{15}N -enriched in oligotrophic than eutrophic seas (Saino & Hattori 1980; Minigawa & Wada 1984; Checkley & Entzeroth 1985). Eutrophic systems are often characterized by nitrate uptake, whereas oligotrophic systems depend on recycled nitrogen uptake, especially ammonia which is generally depleted in ^{15}N (e.g. δ^{15}N of $NH_4 = -3.5‰$; Miyaki & Wada 1967). Checkley & Entzeroth (1985) proposed that a major source of this remineralized nitrogen comes from excretion by pelagic heterotrophs (especially copepods). Since excreta was shown to be depleted in ^{15}N, sinking feces would be enriched, leading to an overall ^{15}N-depletion in the remaining PON within the euphotic zone of oligotrophic systems. Any comprehensive survey should attempt to include δ^{15}N measurements of nitrate and PON to develop a complete picture of the food web under study.

Marine food webs

A number of studies have used stable isotopes to determine marine and estuarine food web structure (Table 9.2). In the past, studies used carbon isotopes, but investigators have increasingly used isotopes of C, N, S, and in some cases O. Some investigators (e.g. Mariotti et al. 1984) found that use of a single isotope tracer can produce ambiguous results and be of little help in defining complex estuarine systems. The following studies are roughly divided into Antarctic/Arctic pelagic systems, offshore systems, nearshore/estuarine systems and salt marsh systems. Finally, we examine how stable isotopes have been used in marine systems, from a marine conservation perspective.

Antarctic/Arctic systems

Studies using stable isotopes in open water Arctic/sub-Arctic and Antarctic systems usually show clear trophic structures (McConnaughey & McRoy 1979a, 1979b; Minagawa & Wada 1984; Wada 1987; Wada et al. 1987, 1991). In most cases there is one primary food source (phytoplankton) or at most two (phytoplankton and macrophytes) with distinct isotopic values that can be easily traced in the food web. McConnaughey & McRoy (1979a) found a significant enrichment in δ^{13}C with increasing trophic level in the Bering Sea, and estimated 1.5‰ enrichment per trophic level. All animals were enriched relative to the phytoplankton (δ^{13}C ca. $-20‰$). They also noted that benthic organisms were typically more enriched than pelagic animals. This was attributed to possible bacterial and meiofaunal reworking of the food source, or to shorter food chains with greater fractionation. The enrichment was

Table 9.2 Marine and estuarine food web studies utilizing stable isotopes. References are listed in approximate order of discussion in the text.

System	Isotopes used	Reference
Bering Sea	Carbon	McConnaughey & McRoy 1979a
Beaufort Sea	Carbon	Schell 1987
Beaufort, Chuckchi, Bering Seas	Carbon	Saupe et al. 1989
Bering Sea	Nitrogen	Minagawa & Wada 1984
Antarctic Ocean	Carbon, nitrogen	Wada 1987
Antarctic Ocean	Carbon, nitrogen	Wada et al. 1987
Antarctic Ocean	Carbon, nitrogen	Schmidt et al. 2003
Barrow Strait–Lancaster Sound	Carbon, nitrogen	Hobson & Welch 1992
Beaufort–Chukchi Seas	Carbon, nitrogen	Hoekstra et al. 2003
Puget Sound, Washington	Carbon	Simenstad & Wissmar 1985
Nova Scotia, Canada	Carbon	Stephenson et al. 1986
East China Sea	Nitrogen	Minagawa & Wada 1984
Baltic Sea	Nitrogen	Hansson et al. 1997
George's Bank	Carbon, nitrogen, sulfur	Fry 1988
South California Bight	Carbon	Rau et al. 1983
East Tropical Pacific	Carbon	Rau et al. 1983
Monaco	Carbon, nitrogen	Rau et al. 1990
Gulf of Mexico	Carbon	Fry et al. 1984
Gulf of Mexico	Carbon, nitrogen, sulfur	Fry 1983
Gulf of Mexico	Carbon	Thayer et al. 1983
Seto Inland Sea, Japan	Carbon, nitrogen	Takai et al. 2002
Galicia, northwest Spain	Carbon, nitrogen	Bode et al. 2003
Ria Formosa, Portugal	Carbon, nitrogen, sulfur	Machás & Santos 1999
Kariega Estuary, South Africa	Carbon	Paterson & Whitfield 1997
St Lawrence River, Canada	Carbon, nitrogen	Martineau & Vincent 2004
Nearshore Gulf of Mexico	Carbon	Fry & Parker 1979
Kitakami River Estuary, Honshu Island, Japan	Carbon, nitrogen	Doi et al. 2005
Mont Saint Michel, Northwest France	Carbon, nitrogen	Créach et al. 1997
South African estuaries	Carbon, nitrogen	Perissinotto et al. 2003
Stefansson Sound, Alaska	Carbon	Dunton & Schell 1987
Scheldt Estuary, North Sea	Carbon, nitrogen	Guelinckx et al. 2006
Tampa Bay, Florida	Carbon	Conkright & Sackett 1986
Narragansett Bay, Rhode Island	Carbon	Gearing et al. 1984
Stagnone di Marsala, western Sicily	Carbon, nitrogen	Vizzini et al. 2002
Western Australia	Carbon, nitrogen	Smit et al. 2006
Gulf St Vincent, South Australia	Carbon	Connolly et al. 2005
St Louis Bay, Mississippi	Carbon	Hackney & Haines 1980
Redfish Bay, Texas	Carbon	Parker 1964
Torres Strait, Australia	Carbon	Fry et al. 1983
Scheldt Estuary, North Sea	Carbon	Mariotti et al. 1984
Freshwater, Estuarine Systems	Nitrogen	Minagawa & Wada 1984
Otcuchi River Estuary	Carbon, nitrogen	Wada 1987
Camargue Biosphere Reserve	Carbon, nitrogen	Persic et al. 2004
Tijuana Estuary, San Dieguito Lagoon	Carbon, nitrogen	Kwak & Zedler 1997
Seagrass Meadow, Texas	Carbon	Fry et al. 1977
Seagrass Meadow, Texas	Carbon	Fry & Parker 1979
Narragansett Bay, Rhode Island	Carbon, nitrogen	Pruell et al. 2005
Sapelo Island, Georgia	Carbon	Haines 1976a
Sapelo Island, Georgia	Carbon	Haines & Montague 1979
Sapelo Island, Georgia	Carbon, nitrogen, sulfur	Peterson & Howarth 1987
Sipewissett Marsh, Massachusetts	Carbon, sulfur	Peterson et al. 1985

still evident when the isotope data were corrected for lipid content. The enrichment was speculatively thought to be due to respiration of light $^{12}CO_2$ (McConnaughey & McRoy 1979a).

The Antarctic Ocean is characterized by phytoplankton and consumers largely depleted in ^{13}C and ^{15}N (Wada 1987; Wada et al. 1987). High nitrate concentrations, high pCO_2 concentrations combined with low light intensities, and perhaps limitations of trace elements may contribute to the slow growth rates. No clear pattern of trophic enrichment in $\delta^{13}C$ was noted. Low $\delta^{13}C$ and $\delta^{15}N$ values in the animals (in some cases lower than in phytoplankton) may also reflect seasonal variations in growth conditions of the phytoplankton. Again, an average ^{15}N enrichment of 3.3‰ per trophic level was observed. A controlled feeding study of marine copepods found that fecal pellets were depleted in ^{13}C by 6–11‰ and in ^{15}N by 0.7–9‰, and the authors speculated that this may contribute to the depleted values seen in POM in high-latitude marine systems (Tamelander et al. 2006). Schmidt et al. (2003) investigated the relationship of krill and copepods in the oceans around Antarctica. They noted that the feeding behavior of the krill focused more in pelagic areas, rather than feeding within the ice. They also suggested the krill did not rapidly equilibrate with their diet, which could present a confounding factor when looking at trophic relationships within the Southern Ocean.

An analysis of the High Arctic food web of the Barrow Strait and Lancaster Sound region by Hobson & Welch (1992) spanned five trophic levels. Higher trophic levels were characterized by little change in $\delta^{13}C$ but $\delta^{15}N$ increased an average of 3.8‰ per trophic level. This study confirmed the observed importance of Arctic cod *Boreogadus saida* in the transfer of energy from lower trophic levels to top consumers. Trophic relationships are also important when investigating organochlorine contaminants (Hoekstra et al. 2003). A comprehensive study of the nearshore marine food web of the southern Beaufort-Chukchi Seas, using both ^{13}C and ^{15}N, found that the contaminants were biomagnified with increasing trophic level.

Offshore temperate systems

Offshore ecosystems in temperate oceans also show heavy isotope enrichment in food webs (Rau et al. 1983, 1990; Fry et al. 1984; Minigawa & Wada 1984; Fry, 1988). In open water systems where the primary food source is limited to phytoplankton, a relatively clear food web can be traced using carbon and nitrogen isotopic values. For example, both the Gulf of Mexico and George's Bank had similar $\delta^{13}C$ values (−21.7 and −21.3‰, respectively) for POC, and organisms at higher trophic levels were enriched relative to the phytoplankton (Fry et al. 1984; Fry 1988). In the Gulf of Mexico there was an increase in $\delta^{13}C$ from POC to zooplankton to benthic crustaceans.

Shipboard experiments confirmed the 0.5 to 1‰ enrichment between successive trophic levels noted by DeNiro & Epstein (1978).

Carbon isotope ratios do not always show a clear picture of food web structure even with a single primary food source. Rau et al. (1983) analyzed whole muscle tissue from each organism in a pelagic food web. All fish isotopic compositions were enriched relative to the POC (Williams & Gordon 1970; Rau et al. 1982), although ^{13}C enrichments did not always correspond to trophic level increases. For example, skipjack and yellowfin tuna had the same isotopic values but are known to feed at different trophic levels (Fry & Sherr 1984).

Nitrogen isotopic compositions in offshore systems also showed a ^{15}N increase per trophic level, and reflected the source phytoplankton. In the East China Sea, a system dominated by nitrogen-fixing blue-green algae, δ^{15}N values were lower than other systems, reflecting the initial phytoplankton depletion in ^{15}N (−0.55‰; Minagawa & Wada 1984). Still, the average enrichment per trophic level was 3.4 ± 1.1‰. Nitrogen isotopes are also quite useful when assessing fish migrations where there are food sources such as anthropogenic inputs from sewage that have distinctive isotope values. Hansson et al. (1997) looked at three distinct Baltic Sea areas to determine their pelagic food web structure, and found that the average trophic level fractionation was 2.4‰.

Given the difficulties of using a single isotopic tracer, Fry (1988) illustrated the benefit of multiple isotopes in his study of George's Bank. Sulfur isotopes showed little change among all organisms (δ^{34}S = +15.6 to +17.7‰) and were similar to seawater sulfate values (+20‰). Carbon and nitrogen isotopes were heavier with increasing trophic level, although δ^{13}C increases were not as consistent as δ^{15}N increases. This was thought to be due to variations in phytoplankton isotopic values; scallop analyses showed a gradient in increasing δ^{13}C from deep to shallow water, indicating a possible gradient in phytoplankton ^{13}C, which was confirmed in later studies (Fry & Wainright 1991). Nitrogen isotopes showed a consistent enrichment of 3–4‰ per trophic level and overall trophic structure estimated from δ^{15}N data agreed well with fisheries production models (Fry 1988).

The dual isotope (carbon and nitrogen) approach also has proven to be useful in pelagic studies. In the western Seto Inland Sea, Japan, Takai et al. (2002) looked at the coupling between POM and demersal fish. They determined that pelagic primary production was not as important as expected, and that benthic primary production is the main driver in this system. In contrast, nutrients in an upwelling ecosystem of Galacia (northwest Spain) are the primary driver of primary production (Bode et al. 2003). Particulate organic matter was enriched in both ^{13}C and ^{15}N, which was transferred up through several trophic levels. However, the results also illustrated how omnivory can be a confounding factor when trying to determine food web relationships.

Nearshore/estuarine systems

In studying estuarine systems, it is important to focus on variables that could affect the food source at the base of the food web. These include terrestrial vs. marine inputs to the system, seasonality, the importance of macrophytes, and taxonomic changes in phytoplankton populations. Owing to the similarity of nitrogen isotopic compositions of food sources in nearshore systems, researchers in general have concentrated on carbon isotopes.

Organic inputs to estuarine systems (Fry & Sherr 1984) can include C3 terrestrial plant material ($\delta^{13}C$ = −23 to −30‰), seagrasses (−3 to −15‰; −26‰ in some species growing in low-salinity reaches), macroalgae (−8 to −27‰), C3 marsh plants (−23 to −26‰), C4 marsh plants (−12 to −14‰), benthic algae (−10 to −20‰), and marine phytoplankton (−18 to −24‰). Most systems generally have multiple inputs, making data interpretation difficult. One way to resolve this is to compare estuarine organisms with offshore organisms (Fry & Parker 1979; Fry 1983; Fry & Sherr 1984; Clementz & Koch 2001). Benthic macrophytes and seagrasses are usually enriched relative to phytoplankton and epiphytes, so that as the carbon from these isotopically heavier sources is incorporated into the food chain, differences can be seen. In south Texas, Fry & Parker (1979) found that organisms collected in seagrass flats were consistently heavier than samples collected in the Gulf of Mexico (−8.3 to −14.5‰ vs. −15.0 to −19.0‰), and concluded that benthic plants were the source of the heavier $\delta^{13}C$ values.

Other estuarine systems are less clear and can have several primary producers supplying carbon to higher trophic levels. For example, the Torres Strait (Australia) exemplifies this complexity (Fry et al. 1983). Offshore benthic systems were influenced by phytoplankton, whereas inshore systems reflected ^{13}C-enriched benthic algae and seagrasses. Motile organisms did not have isotopic values corresponding to one primary producer and reflected access to multiple food sources. In the above example, a multiple isotope study may have provided better insight into this food web. For example, in the case of a tidal lagoon, carbon, nitrogen, and sulfur isotopes were used to determine the sources of organic matter to consumers (Machás & Santos 1999). Although POM variation was low, by using all three isotopes the investigators found that upland plants, benthic plants, and phytoplankton were important food sources along a gradient from near the entrance to the lagoon to an inner station. Upland plants were more important in the inner lagoon, whereas phytoplankton dominated as a food source near the entrance to open water. In a South Africa estuary, the question of the importance of macrophytes as a food source for consumers was studied using carbon isotopes (Paterson & Whitfield 1997); it was found that enriched *Spartina maritima* and *Zostera capensis* were dominant food sources for the littoral community. As the sampling transect moved into the channel, those organisms consumed more depleted phytoplankton, terrestrial detritus and C4 macrophytes, than microalgae and other C3 macrophytes that were found to be less important to primary consumers.

In systems limited to two food sources, the use of stable isotopes can provide more insight into defining the structure of food webs. In attempting to resolve the influence of terrestrial carbon to the Gulf of Mexico, Thayer et al. (1983) found that although dissolved organic carbon (DOC) and particulates $< 0.45\,\mu m$ approached terrestrial $\delta^{13}C$ values (-24.0 and $-24.6\permil$) and phytoplankton averaged $-22.7\permil$, zooplankton and larval fish isotopic compositions indicated that their carbon was derived from marine phytoplankton. In a similar study, Martineau & Vincent (2004) found that terrestrial POM was readily available but was not used at the freshwater transition zone in an upper estuary. Carbon isotopic values indicated zooplankton utilized phytoplankton. Macrophytes also may provide an important source of carbon to coastal ecosystems. Stephenson et al. (1986) studied a system in Nova Scotia dominated by two species of macrophytes. Their group determined that marine phytoplankton was lighter in ^{13}C than the macrophytes, and thus was the dominant source of carbon to this food web system.

Benthic diatoms have increasingly been found to be an important source of food to estuarine consumers. Doi et al. (2005) analyzed ocypodid (ghost) crabs in a spatial gradient from upstream to downstream in the Kitakami River estuary. Upstream, benthic diatoms (ca. $-19\permil$) were the dominant food source, and further downstream sediment organic matter dominated (ca. $-24\permil$). In another study, the gut contents and stable isotopes were analyzed (Créach et al. 1997) for a variety of consumers. In some cases, benthic diatoms dominated as a food source, and in others plant detritus was important. Finally, many of South Africa's estuaries are closed to the open ocean during the dry season. This can eliminate phytoplankton as a source of organic matter to zooplankton, causing them to switch to alternate food sources. Perissinotto et al. (2003) looked at three different estuaries. By using $\delta^{13}C$ and $\delta^{15}N$ values, they were able to define three separate food sources: POM, detritus, and benthic microalgae. For further studies, see the review by Peterson (1999).

Arctic systems are typified by more limited carbon sources and lend themselves to clearer definition with stable isotope ratios. Dunton & Schell (1987) compared seasonal feeding in sessile and motile invertebrates in Sefansson Sound near Prudhoe Bay, Alaska. Here the two sources of energy were phytoplankton and the kelp *Laminaria solidungula*. During the darkness of winter, only macrophyte carbon was available and the shift to macrophyte-based diets by consumers was readily evident. In the North Sea Schelde estuary, researchers were able to use carbon isotopes in conjunction with stomach contents to trace the migration of herring and sprat (Guelinckx et al. 2006). The food source varied both spatially and temporally, which was reflected in the fish muscle and gut contents.

In some systems seasonal effects on carbon isotope ratios have been reported. Since estuarine systems usually experience an input of freshwater, seasonal changes in the $\delta^{13}C$ of DIC may be incorporated by the phytoplankton. Conkright & Sackett (1986) found that coastal marine phytoplankton, POC, and bivalves differed in their $\delta^{13}C$ values between the dry and rainy seasons, with all samples lighter during the wet season. This was thought to be due to either terrestrial organics entering the food web or lighter DIC from freshwater input being incorporated by the phytoplankton. A similar situation was observed in Puget Sound, Washington (Simenstad & Wissmar 1985). Depletions of up to 8‰ were noted in autotrophs, DOC and some herbivores in estuarine and nearshore habitats during the winter. Nearshore waters were also depleted in the winter, and it was thought that freshwater DIC influenced the isotopic values of the autotrophs. Longer-lived and secondary consumers did not show a seasonal trend, due to long-term integration of isotopic values.

Other seasonal effects may be due to changes in phytoplankton populations. Gearing et al. (1984) noted that Narragansett Bay, Rhode Island has distinct phytoplankton populations of differing isotopic compositions. Nanoplankton were dominant in the summer and ^{13}C-enriched diatom blooms occurred in winter–spring. Zooplankton also showed a seasonal trend in $\delta^{13}C$ values and were 0.5–0.6‰ enriched relative to the phytoplankton. Larval fish did not show a seasonal trend; it was postulated that they selectively used the diatoms or there was a large difference in fractionation between fish and phytoplankton. All consumers and predators in the water column and benthos were enriched relative to the phytoplankton, indicating no terrestrial influence. Increasing $\delta^{13}C$ followed a trend based on the organism's presumed trophic position in the food web.

Needless to say, the complexity of estuarine systems can be challenging. A case in point is the San Francisco Bay system (Cloern et al. 2002), where the primary producers (from phytoplankton to upland plants) spanned 10 groups. Variation in ^{13}C and ^{15}N was high within each group (5–10‰), making it difficult to develop a comprehensive picture of the estuary using only stable isotopes. In this case, molecular biomarkers were used in addition to stable isotopes. Sampling at higher trophic levels may be necessary to see the emergent isotope averaging in this sort of system, and sometimes a multiple isotope approach does work. Kwak & Zedler (1997) used carbon, sulfur, and nitrogen isotopes to study the Tijuana Estuary and San Dieguito Lagoon. They were clearly able to distinguish primary producers by the carbon and sulfur isotopes; nitrogen isotopes were used to distinguish three to four trophic levels. When all three isotopes were used in conjunction with two- and three-source mixing models, Kwak & Zedler (1997) found that the Tijuana Estuary had a more complex food web than San Dieguito. Each system is different, and sometimes it is best to take a few pilot samples to determine the best approach. One also needs to look at both temporal and spatial variability when studying estuaries.

Salt marsh systems

Although much of the early work on salt marshes was done in the USA, more work is being done in Mediterranean and tropical seagrass systems (Vizzini et al. 2002; Lepoint et al. 2004; Persic et al. 2004; Smit et al. 2006). Seagrass systems can be valuable indicators of anthropogenic inputs and nitrogen isotopes have shown their utility, as they reflect nitrogen loads to both salt marsh systems and estuaries (McClelland et al. 1997; Valiela et al. 2000). In a series of salt marshes along a nutrient gradient in Narragansett Bay, Rhode Island, Pruell et al. (2005) found that nitrogen isotopes increased in primary producers as the levels of nutrients from primary consumers increased.

Most salt marshes of the USA are fringed by stands of *Spartina alterniflora*, which may add a significant amount of carbon to the consumers of the ecosystem (Fry & Sherr 1984). *Spartina* $\delta^{13}C$ values generally range from -12 to $-14‰$, i.e. ^{13}C-enriched relative to most terrestrial C3 plants and phytoplankton. When stands of marsh plants are largely monospecific, it appears that the associated invertebrates reflect that carbon isotopic composition (Haines 1976a, 1967b). However, physical processes within the system as well as biological processes can blur the distinctions between producer and consumer. At Sapelo Island, Georgia, the isotopic values of the invertebrates that were sampled reflected those of the foods they ingested (Haines & Montague 1979). Again, where distinct stands were evident and had differing isotopic compositions, organisms such as mud snails that stayed within a given area reflected the source carbon. More mobile species, such as foraging crabs, tended to deviate from the marsh plant $\delta^{13}C$ values and were thought to feed on a mixture of benthic diatoms and *Spartina*.

Connolly et al. (2005) found that seagrasses in Australia dominated the organic matter in transects taken from seagrass meadows into the open estuary, and that the animals sampled had isotopic compositions similar to the seagrasses, regardless of where they were found. In another study in Australia, the seagrass *Posidonia sinuosa* is the dominant seagrass. Smit et al. (2006) examined its role as a food source for food webs within the marshes. Although *P. sinuosa* appeared to be a major carbon source, it apparently is not assimilated into the food web. Nitrogen isotopic values of the various food sources, including *P. sinuosa* and macroalgae, did not vary significantly, making their use limited, although they did track trophic enrichment quite well. Carbon values did differ, with the seagrass values of $-11‰$ and macroalgae ranging from -16.6 to $-31.7‰$. The consumers were found to be similar to macroalgae. From this study one can see that if $\delta^{15}N$ is constant across sources, then it is of no use as a source indicator. However, in this case the use of $\delta^{15}N$ as a trophic level indicator is maximal. The ideal case is to use $\delta^{13}C$ or $\delta^{34}S$ as a source indicator and $\delta^{15}N$ as a trophic level indicator. A similar situation was observed in the Mediterranean (Vizzini et al. 2002), where stands of *Posidonia oceanica* dominate the salt marshes. Here, epiphytic algae found on the seagrass leaves were an important food source for benthic consumers as the leaves broke down into detritus. Pelagic consumers utilized phytoplankton as their food source, which was then transferred into the pelagic food web.

Hackney & Haines (1980) studied two adjacent marshes in St Louis Bay, Mississippi which had different nearly monospecific stands of marsh grass. One marsh was a C4-dominated system with *Spartina alterniflora* ($\delta^{13}C$ = −12.4‰), the other was a C3-dominated system of *Juncus roemerianus* ($\delta^{13}C$ = −26.2‰). A range of organisms sampled from both systems had similar carbon isotopic compositions of −20 to −26‰, indicating a possible mixing of material from both marshes or mixing with unsampled carbon sources. This ecosystem also had very negative values for the filter-feeding bivalves, suggesting that either a significant quantity of terrestrial organic matter or that very negative estuarine algae was entering the system. Hackney & Haines (1980) concluded that in contrast to Georgia marshes, terrestrial C3 plant material played a major role in this Mississippi marsh. This study indicated a minor role of C4 *Spartina* in the marsh food web.

Peterson & Howarth (1987) completed an extensive survey of organic matter flow in the salt marsh and estuarine waters of Sapelo Island, Georgia. In this study, a combination of carbon, nitrogen, and sulfur isotopes were used. Sulfur isotopes were especially important, since *Spartina* stands utilize S-depleted sulfides which can be as much as 30–40‰ lower than $\delta^{34}S$ values for sulfate. Sulfur input to terrestrial systems comes primarily from precipitation ($\delta^{34}S$ = 2–8‰) whereas marine systems utilize seawater sulfate ($\delta^{34}S$ = 21‰). Analysis of a broad range of organisms showed that the terrestrial organic matter input was not important to this system. Here, the combination of carbon and sulfur isotopes (Figure 9.3) illustrates this pattern. The dominant source of organic matter to the systems was a mixture of algae and *Spartina*. Nitrogen isotopes tended to show trophic level enrichment and reflected the mixed diet (Figure 9.4). A study based on a transect of ribbed mussels in Sippewissett Marsh, Massachusetts came to a similar conclusion on the importance of *Spartina* in marsh food webs (Peterson et al. 1985). These two studies further illustrate the efficacy of employing a multiple isotope approach to resolve food web systems.

An important factor that should be considered in using isotopes is the signal-to-noise ratio of the sources, defined as the difference in the mean isotope values of the food sources versus the sum of the standard deviations of the mean values of the food sources. When ecologists use isotopes in mixing models, resolution will be limited by the separation between end-member sources. This can vary with the isotope measured. For example, in the Sapelo Island system, upland C3 plants and *Spartina* had a large ^{13}C

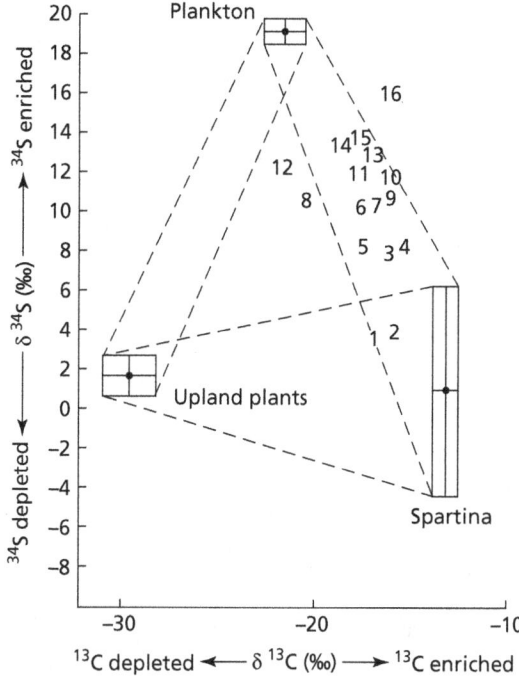

Figure 9.3 Illustration of the utility of using multiple isotope studies to resolve food web questions. Plot of $\delta^{34}S$ vs. $\delta^{13}C$ in relation to the mean of potential food sources and marsh consumers. Numbers represent consumers. (Reproduced from figure 8 in Peterson & Howarth 1987, Copyright (2007) by the American Society of Limnology and Oceanography, Inc.)

isotopic difference (16‰) and low standard deviations, leading to a signal:noise ratio of 8.6 (Peterson & Howarth 1987). However, ^{34}S isotope values were very similar and had a signal:noise ratio of 0.1. Resolution will also be limited by the sampling density; in a system with a 1‰ separation between sources, 50 samples may be necessary, whereas a system with a 10‰ separation may only need three to five samples. Working in systems with high signal: noise ratios is optimal when applying the multiple isotope approach and will help reduce the number of samples required for successful answers.

Stable isotopes in marine conservation biology

Most of the applications cited above have relevance for both basic research and management in the conservation of marine habitats, species, and commercial resources. The use of stable isotopes is now a routine procedure for

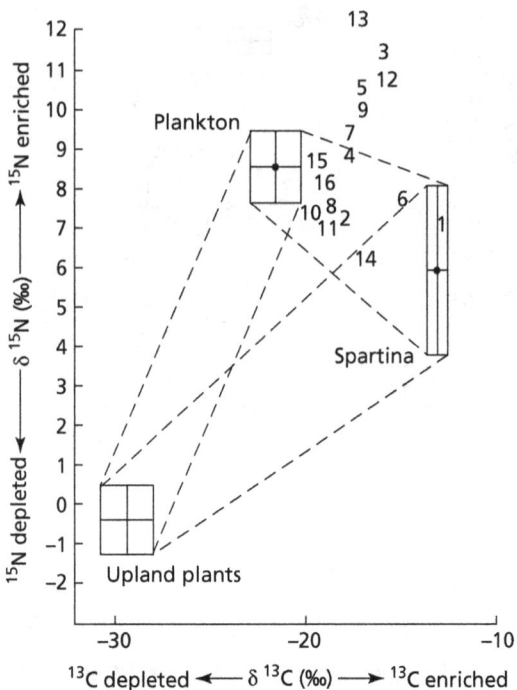

Figure 9.4 Carbon and nitrogen isotope ratios of food sources vs. consumers for a salt marsh. Numbers represent consumers. Note the enrichment in ^{15}N for consumers vs. food sources. (Reproduced from figure 10 in Peterson & Howarth 1987, Copyright (2007) by the American Society of Limnology and Oceanography, Inc.)

tracking contaminants, characterizing food webs and describing how they may be altered by resource extraction, measuring habitat connectivity, and deciphering ambient temperatures at which organisms lived during the courses of their lifetimes, and even far back into the past (e.g., Thorrold et al. 1997; Patterson 1998; Wilkinson & Ivany 2002; Surge & Walker 2005; Wurster et al. 2005). Stable isotope data can be rich in spatial and temporal information of the sort required by managers. However, practitioners must also be wary of the contribution to isotopic values by variables other than those of primary interest in any given situation (e.g., MacNeil et al. (2006) and Suzuki et al. (2005) for tissue variation in fish, Hall-Aspland et al. (2005) on seasonal variation in leopard seals). However, with attention to the appropriate caveats, stable isotope data can play a major role in conservation decision making and resource management.

One of the most important problems in fisheries is matching the spatial scale of management to that of the biology of exploited species, a process still in its infancy. To this end, stable isotopes can provide critical information on movements and stock structure (e.g., Gao & Beamish 1999; Cunjak et al. 2005) as well as about the prey base for species of commercial or conservation interest (Lenanton et al. 2003). Another key concern is the sourcing of carbon and nitrogen to higher tiers of the food web, and in particular, the role of marine macrophytes. Marine macrophytes are obviously important in providing habitat for a host of structurally dependent organisms, but their role in the food web is more complicated. Marine macrophytes are dominant features of many littoral communities, but they tend to be heavily armed against herbivory via various mechanical and chemical defense mechanisms, so the assumption that they are nutritionally important is open to question. Indeed, some marine habitats, particularly at high latitude, are characterized by luxuriant accumulations of fleshy macroalgae and seagrasses showing little evidence of consumption by large organisms such as fishes, as compared with, say, plants and insects on land. This does not prevent detritus from such plants providing an important source of carbon or nitrogen via the detritivore food chain. A few examples of isotopic study of carbon and nitrogen sourcing from macrophytes have already been mentioned. Investigators examining the contribution of seagrass to consumers have arrived at a range of different answers, from seagrass is a major contributor (Smit et al. 2006), to the conclusion that seagrass, despite its biomass and abundance, makes a negligible contribution as compared with benthic algae (Smit et al. 2006) or other sources. In a South African estuary, Paterson & Whitfield (1997) found littoral organisms to show ^{13}C-enrichment from marsh grass and marine benthic algae, while tidal channel animals had a low $\delta^{13}C$, indicating dependence upon pelagic primary producers for their carbon. Kwak & Zedler (1997) found a strong signal from both *Spartina* and benthic algae as well as a tight linkage between the marsh and the tidal channels. They did something else interesting. They profiled $\delta^{13}C$, $\delta^{15}N$, and $\delta^{34}S$ and considered variability in this profile as a measure of trophic complexity.

Similarly, in assessing the functional importance of mangrove forests – one of several justifications for their preservation – it is important to know if the leaves and detritus from mangrove forests do or do not nourish the marine community bustling about their trunks and prop roots (Haines & Montague 1979; Rodelli et al. 1984; Stoner & Zimmerman 1988; Harrigan et al. 1989; Twilley 1995; Primavera 1996). Also, do mangrove forests nourish adjacent seagrass beds and coral reefs? This is an area of active research with varied conclusions (e.g. Mancera & Twilley 2003; Fry & Ewel 2003).

It is also useful to know when and how, during the life of an animal, different dietary constituents become important. Despite a general awareness of the variation in diet exhibited by marine organisms, particularly fishes, there is a tendency among ecologists to regard herbivores as particularly narrowly specialized, perhaps due to the extreme morphological and physiological modifications necessary to deal effectively with marine plants. Indeed, many fishes begin life as carnivores, and only later switch to a plant-based diet once the necessary morphology has developed. Frequently overlooked is the fact that an ability to eat plants does not preclude the continued consumption of animals. Halfbeaks (relatives of needlefishes and flying fishes) are somewhat unusual for being pelagic herbivores. Carseldine & Tibbetts (2005) demonstrated that adult halfbeaks fully equipped with herbivorous capabilities, rely lifelong on animal matter for their nitrogen supply, though shifting with adulthood to floating seagrass as a primary source of carbon.

Widespread coral reef degradation has drawn attention to factors responsible for assuring coral reef health. In particular, coral reef ecologists must understand the relationship between reef-building corals and their symbiotic dinoflagellates, since elevated sea surface temperatures and emergent pathogens are now often triggering the expulsion of these algae from their host corals, in a process known as bleaching. A key question in this respect is whether hermatypic (reef building) corals derive their nutrition primarily from their symbiotic dinoflagellates (zooxanthellae), by preying on zooplankton, by filtering small particles from the water column, or by absorbing DOM (e.g. Ward-Paige et al. 2005). The answer would appear to be "all of the above" but varying in accordance with local conditions, species, and other factors that are not well understood. Filling in this picture through clever use of stable isotopes could help in predicting and profiling the species-specific impacts of coral bleaching as well as help to explain why some species are easily killed by bleaching, while others are not.

No-take marine protected areas have been proposed as a major instrument for marine conservation, but their utility depends upon MPAs (Marine Protected Areas) being properly designed to include all essential components of a marine ecosystem. Often this could mean a need to protect adjacent or functionally connected habitats. Temperate zone nearshore marine environments may be subsidized by terrestrial plants, marine algae, or marine angiosperms. Fortunately these taxonomic and functional groups of primary producers fractionate stable isotopes differently, and thus offer traceable signals in adjacent food webs. It is generally assumed that coastal marine communities are subsidized by terrestrial sources of carbon and nitrogen. Testing this hypothesis was one of the earliest uses of stable isotopes in marine ecology (e.g. Haines & Montague 1979). It has since been backed up by several studies (Paterson & Whitfield 1997), and in particular those that rely on isotopically characteristic sewage or other forms of anthropogenically enriched terrestrial isotopic signals as tracers (Spies et al. 1989; Van Dover et al. 1992; Hansson et al. 1997; Tewfik 2005; Martinetto et al. 2006). Sewage is commonly assumed to be $\delta^{13}C$ enriched due to the prevalence of C4 plants on land, and $\delta^{15}N$ enriched as well. In sewage, the ^{15}N of PON is low, but dissolved inorganic nitrogen (DIN) is high (after ammonia is converted to nitrate and the nitrate is denitrified). Figure 9.5 shows the results of a survey of fish collected from three Australian estuaries, with increasing contributions

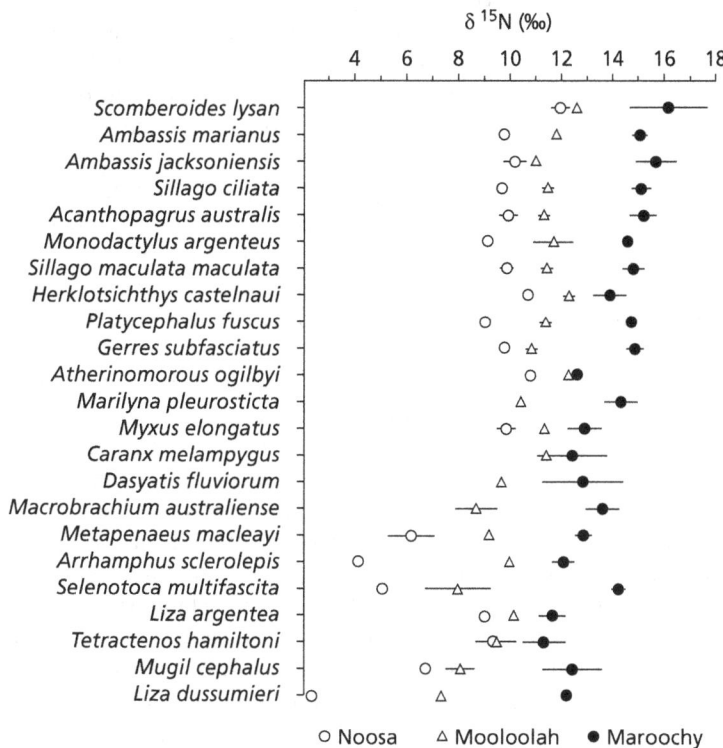

Figure 9.5 Inter-estuarine comparison of fishes and prawn collected from three estuaries for increasing anthropogenic inputs of sewage. (Reproduced with permission from Schlacher et al. 2005.)

of sewage. In this case, there is a clear enrichment in ^{15}N across all species (Schlacher et al. 2005). However, this should not be taken for granted. Gaston & Suthers (2004) studied a common zooplanktivore known to marine aquarists as the "stripey" at sewage-impacted and control sites. In their case the sewage was ^{15}N depleted, not enriched, possibly due to the degree of effluent treatment at the treatment plant. They also found that the signal of sewage entering the food web was strong enough to be visible even with sporadic upwelling events, and that these environmental signals were most evident in tissues such as the liver, with its fast turnover time. Although there is not room here for lengthy discussion, there is quite a bit of literature about salmon-subsidies to terrestrial systems, and marine-derived nutrients (Bilby et al. 1996; Jacoby et al. 1999; Ben-David et al. 2004).

Another common assumption about habitat connectivity is that nutrients flow from the watershed into estuaries and from there out and down the shelf (e.g. Wissel & Fry 2005). Odum (1968) was perhaps the first to propose that organic matter from coastal wetlands could be a major source of nourishment for nearshore marine communities, and as already discussed, there is ample evidence to this effect from stable isotope studies. The strength of these conclusions is often limited by signal:noise ratios and overlapping source isotope values. The careful reader should evaluate these studies in light of recently published error analyses (Phillips & Greg 2003). The story, however, is not always that simple. Connolly et al. (2005) used stable isotopes to trace carbon from subtidal seagrass to shallower mudflats up to hundreds of meters away and up-slope; perhaps a reminder that both tides and animals move in both directions. The geography of marine parks and other forms of area management must obviously take this into account, but often fail to.

As for any trophic study, stable isotope profiles and variability may be as or more informative than the individual raw signals (Persic et al. 2004). We already mentioned the work of Kwak & Zedler (1997). Schwartz et al. (2006) used within-lake $\delta^{15}N$ variance as a measure of niche breadth in haplochromine cichlid species, and found the introduced Nile perch to have simplified trophic structure in those lakes where it occurred. What is needed is a better understanding of both the causes and functional consequences of trophic variation. Along these lines, we must better understand the nature and meaning of the individual variation in behavior that contributes to the apparent noise in trophic structure (Bolnick et al. 2003). Also, when disparate locations are compared, it is crucial to correct for differences in source signals. Cabana & Rasmussen (1996) found high variation in $\delta^{15}N$ among 40 Canadian lakes, but most of this vanished once contamination from human sewage had been corrected for.

One of the more spectacular applications of stable isotopes to marine conservation is to use them to peer into the lives of mysterious, rarely seen and little-understood creatures so that their conservation needs and priorities can be mapped out. Estrada et al. (2003) used $\delta^{15}N$ to measure and compare the trophic positions of sharks, as well as to reconstruct their movements across distance and habitats. In similar work, Domi et al. (2005) demonstrated the advantages of using multiple methods; in their case, combining data on stable isotopes and heavy metal contaminants. MacNeil et al. (2005) used these methods to investigate diet-switching in big sharks. Ruiz-Cooley et al. (2004) studied the relationship between sperm whales and jumbo squid. Cherel & Hobson (2005) reconstructed the trophic relationships of an entire community of pelagic cephalopods, including the famed giant and colossal squids (the world's two most massive invertebrates), by subjecting squid beaks in predators' stomachs to stable isotopic scrutiny. Large, charismatic free-ranging marine animals are very difficult to study and killing them to learn about them is often out of the question. Securing tissue samples, however, may not be nearly so difficult. For this reason stable isotope analysis, particularly when coupled to other methods, can be among the most powerful tools available in the study of seabird and marine mammal biology (e.g., Cherel et al. 2005; Herman et al. 2005). Ramsay & Hobson (1991) used $\delta^{15}N$ values to show that polar bears derive nearly all of their nutrition from marine sources despite spending about one-third of their lives on land. This is an important conservation finding, given the predicted shrinking of the Arctic ice cap that currently supports most marine feeding by polar bears. Stable isotopes have also opened the door on our understanding of unusual or spatially isolated habitats, such as seamounts (Duineveld et al. 2004), pack-ice (Lovvorn et al. 2005), and deep-sea vents (MacAvoy et al. 2005; Levin et al. 2000; Levin & Michener 2002; Van Dover, this volume, pp. 202–237). Without this kind of insight, conservation planning must be based on guesswork, a situation in which conservation will always lose to short-term economic interests.

Perhaps the greatest potential for new uses of stable isotopes in conservation lies in the development of new diagnostics for the managing of fishery impacts. The persistence of ecosystems and their constituent species requires and reflects that predator–prey relationships remain within certain boundary conditions. Deviations beyond this envelope of conditions should theoretically be detectable through shifts in stable isotope values and profiles, making this an easier and cheaper ecosystem indicator than sorting through thousands of fish gut contents under a microscope. Changes in food web structure and shifts in the position of any species or life stage in the food web is one of the key indicators that fishery scientists are looking toward in the development of a new science of ecosystem-based management (EBM). Ecosystem-based management is in the process of being invented, but it is already clear that it will require a multitude of new biological reference points, including some that allow us to measure and diagnose the state of marine food webs. The size frequency spectrum of marine organisms and stomach contents analysis are two primitive but effective ways to assess anthropogenic alterations to marine food webs. If large-scale fishery-independent sampling is being conducted anyway, these are the traditional and direct ways to measure food web deformation. But they do have limitations. Large-scale fishery-independent sampling is rather a luxury of rich developed nations. Size structure data are extremely useful (Jennings et al. 2001a; Bozec et al. 2005; Graham et al. 2005; Pauly & Watson 2005) but obtaining them requires a well-established capability for fisheries-independent stock assessment. Stomach contents analysis only reveals what some animals have been eating in the previous few hours, plus it is consumptive of time and space, requires noxious fixatives and extraordinary expertise, and is expensive when conducted on a large scale. The traditional approach also precludes longitudinal and nondestructive sampling. It is simply inappropriate if the goal is a better understanding of relatively small, fragile coastal marine habitats, or the currently few and mostly minute protected areas that we are so interested in as conservation and reference sites.

But consider stable isotope analysis. It requires only that a tiny bit of sample tissue be collected, usually less than 10 mg. The isotope results are efficient, offer a signal integrated over the tissue turnover time (weeks for muscle tissue), does not require a full stomach and so can be obtained from every individual sampled, tissue can often be obtained without killing larger organisms, and affords cross-comparability independent of the idiosyncratic natural histories of marine organisms. The question is whether it tells us what we actually need to know.

The use of stable isotope values as biological indicators in EBM is in its infancy, and neither reference points nor theory have yet been developed that would allow stable isotopes to be used routinely as management tools. However, this is clearly a direction worth pursuing, as a few have been. Jennings et al. (2001b) examined the relationship between body size and $\delta^{15}N$ values across species in a European food fish community (Figure 9.6). They found that while the maximum size of species did not correlate well to mean $\delta^{15}N$ values, the size spectrum of individuals, independent of species, did correlate very well to their isotopic profiles. In other words, even among large species one can find such functional modes as detritivores and plankti-vores, but in general big mouths eat big things, at least occasionally. Based on this and other work (e.g. Jennings & Blanchard 2004; Bozec et al. 2005), the fish community size spectrum plus isotope profiles hold considerable promise as an EBM indicator and may be particularly sensitive to the effects of fishing. The trophic structure of infaunal marine benthic communities, however, may prove surprisingly resilient to fishing effects, with the array of

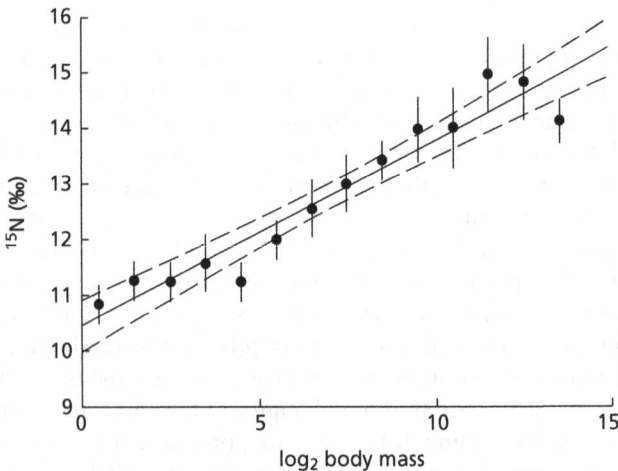

Figure 9.6 Relationship between $\delta^{15}N$ of white muscle tissue and size class for a North Sea fish community. (Reproduced with permission from Jennings et al. 2001b.)

species most vulnerable to fishing giving way to resistant species that fulfill the same functional roles (Jennings et al. 2001b).

Stable isotope signals offer a wealth of information about exploited and endangered marine species that may be unavailable by any other means. This may be especially critical in aquatic environments, where the ecological phase shifts of greatest concern can occur with stunning abruptness, and have sobering impacts on the economy of all dependent species, and most especially humans. We currently live in an era of strong overfishing, and analyses of archived specimens may be necessary and helpful for reconstructing and conserving marine fisheries.

Conclusions

Stable isotope analysis (SIA) offers an effective natural tracer approach for following energy and nutrient flows in ecosystems. Although many studies now illustrate the fundamentals of SIA, limitations and caveats are often not fully appreciated. These are becoming more evident. Through a synthesis of past work several generalizations can be made.

1 Carbon isotope ratios reflect the primary production important to the ecosystem energy flow. The transfer of carbon isotope ratios is essentially conservative between trophic levels although minor ^{13}C-enrichment ($<1‰$) may occur. This allows allocation of energy sources in the cases where major inputs are limited to two and the sources have distinctive isotope values.

2 An advantage of the isotope tracer approach is that organisms from all trophic levels can be analyzed with relative ease, leading to quick estimates of how organic matter flows throughout an entire food web. This SIA monitoring can measure food web shifts in time and space and so help evaluate management practices.

3 Within estuarine and coastal ecosystems carbon isotope values of primary producers are altered by seasonal and environmental changes. Depletion in the ^{13}C content of DIC through increased respiratory inputs or reductions in primary producer growth rates by decreasing light, nutrients, or trace element concentrations will decrease the $\delta^{13}C$ of fixed carbon. Large cell sizes and fast growth rates produce ^{13}C-enriched fixed carbon.

4 Carbon isotope tracing of food webs in pelagic systems must take into account seasonal, geographic, and multi-year changes in the $\delta^{13}C$ of primary producers. The $\delta^{13}C$ values of migratory fauna can reflect feeding in distant areas where isotope ratios are different from the environment in which they were sampled. Seasonal changes in lipid content will alter bulk $\delta^{13}C$ but such changes are usually small and easily corrected (Sweeting et al. 2006).

5 Nitrogen isotope ratios reflect trophic status in most ecosystem food webs and to a lesser extent, nutritional status. Food deprivation leads to enrichment of ^{15}N in body tissues. Overall $\delta^{15}N$ in fauna usually reflects the nitrogen isotope ratios in the primary producers with a gain of ca. 3–4‰ per trophic level.

6 When possible, laboratory dietary studies are encouraged to confirm producer–consumer linkages and to determine if "you are what you eat" truly holds. Laboratory experiments should carefully budget inputs and outputs using natural diets and stocking densities to arrive at realistic estimates of field trophic enrichment factors.

7 Multiple element isotope ratio studies often can provide much better insight into ecosystem processes than single element studies. Multiple source inputs can sometimes be separated, with a combined tracer approach often much better than using just one isotope tracer.

8 Stable isotopes offer a wealth of information about exploited and endangered marine species, and properly applied, hold the potential to revolutionize both our knowledge of life in the sea, and our ability to protect it from our tendency to unwittingly destroy it.

Acknowledgments

This work was partially supported by the Gordon and Betty Moore Foundation and their support to the Marine Management Area Science Program of Conservation International. This work is also supported in part by grants from the National Marine Fisheries Service for work on the food web effects of the western Gulf of Maine Closed Area.

References

Adams, T.S. & Sterner, R.W. (2000) The effect of dietary nitrogen content on trophic level ^{15}N enrichment. *Limnology and Oceanography*, **45**, 601–607.

Altabet, M.A. (1989) A time-series study of the vertical structure of nitrogen and particle dynamics in the Sargasso Sea. *Limnology and Oceanography*, **34**, 1185–1201.

Altabet, M.A. & McCarthy, J.J. (1985) Temporal and spatial variations in the natural abundance of ^{15}N on PON from a warm-core ring. *Deep-Sea Research*, **32**, 755–772.

Altabet, M.A. & McCarthy, J.J. (1986) Vertical patterns in ^{15}N natural abundance in PON from the surface waters of several warm core rings and the Sargasso Sea. *Journal of Marine Research*, **44**, 185–201.

Altabet, M.A., Robinson, A.R. & Walstad, L.J. (1986) A model for the vertical flux of nitrogen in the upper ocean: simulating the alteration of isotopic ratios. *Journal of Marine Research*, **44**, 203–225.

Ambrose, S.H. & Norr, L. (1993) Experimental evidence for the relationship of the carbon isotope ratios of whole diet and dietary protein to those of bone collagen and carbonate. In: *Prehistoric Human Bone: Archaeology at the Molecular Level* (Eds J.B. Lambert & G. Grupe), pp. 1–37. Springer-Verlag, New York.

Arntz, W.E. (1977) Results and problems of an "unsuccessful" benthos cage predation experiment (western Baltic). In: *Biology of Benthic Organisms* (Eds B.F. Keegan, P.O. Ceidigh & P.J.S. Boaden), pp. 31–44. Pergammon Press, Oxford.

Båmstedt, U., Gifford, D.J., Irigoien, X., Atkinson, A. & Roman, M. (2000) Feeding. In: *ICES zooplankton Methodological Manual* (Eds R. Harris, P. Weibe, J. Lenz, H.R. Skjødal & H. Huntley), pp. 297–399. Academic Press, New York.

Bearhop, S., Adams, C.E., Waldron, S., Fuller, R.A. & Macleod, H. (2004) Determining trophic niche width: a novel approach using stable isotope analysis. *Ecology*, **73**, 1007–1012.

Ben-David, M., Titus, K. & Beier, L.R. (2004) Consumption of salmon by Alaskan brown bears: a trade-off between nutritional requirements and the risk of infanticide? *Oecologia*, **138**, 465–474.

Beviss-Challinor, M.H. & Field, J.G. (1982) Analysis of a benthic community food web using isotopically labeled potential food. *Marine Ecology Progress Series*, **9**, 223–230.

Bilby, R.E., Fransen, B.R. & Bisson, P.A. (1996) Incorporation of nitrogen and carbon from spawning salmon into the trophic system of small streams: evidence from stable isotopes. *Canadian Journal of Fisheries and Aquatic Sciences*, **53**, 164–173.

Black, C.C. & Bender, M.M. (1976) ^{13}C values in marine organisms from the Great Barrier Reef. *Australian Journal of Plant Physiology*, **3**, 25–32.

Bode, A., Carrera, P. & Lens, S. (2003) The pelagic foodweb in the upwelling ecosystem of Galicia during spring: natural abundance of stable carbon and nitrogen isotopes. *ICES Journal for Marine Science*, **60**, 11–12.

Bolnick, D.I., Svanback, R., Fordyce, J.A., et al. (2003) The ecology of individuals: incidence and implications of individual specialization. *The American Naturalist*, **161**, 1–28.

Boreham, P.F.L. & Ohiagu, C.E. (1978) The use of serology in evaluating invertebrate predator-prey relationships: a review. *Bulletin of Entomological Research*, **68**, 171–194.

Bosley, K.L., Chambers, R.C. & Wainright, S.C. (2002) Estimating turnover rates of carbon and nitrogen in recently metamorphosed winter flounder *Pseudopleuronectes americanus* with stable isotopes. *Marine Ecology Progress Series*, **236**, 233–240.

Boyd, C.M., Heyraud, M. & Boyd, C.N. (1984) Feeding of the Antarctic krill *Euphausia superba*. *Journal of Crustacean Biology*, **4**, 123–141.

Bozec, Y-M., Kulbicki, M., Chassot, E. & Gascuel, D. (2005) Trophic signature of coral reef fish assemblages: Towards a potential indicator of ecosystem disturbance. *Aquatic Living Resources*, **18**, 103–109.

Cabana, G. & Rasmussen, J.B. (1996) Comparison of aquatic food chains using nitrogen isotopes. *Proceedings of the National Academy of Sciences*, **93**, 10844–10847.

Carseldine, L. & Tibbetts, I.R. (2005) Dietary analysis of the herbivorous hemiramphid *Hyporhamphus regularis ardelio*: an isotopic approach. *Journal of Fish Biology*, **66**, 1589–1600.

Checkley, D.M. & Entzeroth, L.C. (1985) Elemental and isotopic fractionation of carbon and nitrogen by marine, planktonic copepods and implications to the marine nitrogen cycle. *Journal of Planktonic Research*, **7**, 553–568.

Cherel, Y. & Hobson, K.A. (2005) Stable isotopes, beaks and predators: a new tool to study the trophic ecology of cephalopods, including giant and colossal squids. *Proceedings of Biological Science*, **272**, 1601–1607.

Cherel, Y., Hobson, K.A. & Weimerskirch, H. (2005) Using stable isotopes to study resource acquisition and allocation in procellariiform seabirds. *Oecologia*, **145**, 533–540.

Clementz, M.T. & Koch, P.L. (2001) Differentiating aquatic mammal habitat and foraging ecology with stable isotopes in tooth enamel. *Oecologia*, **129**, 461–472.

Cline, J.D. & Kaplan, I.R. (1975) Isotopic fractionation of dissolved nitrate during denitrification in the eastern tropical North Pacific Ocean. *Marine Chemistry*, **3**, 271–299.

Cloern, J.E., Canuel. E.A. & Harris, D. (2002) Stable carbon and nitrogen isotope composition of aquatic and terrestrial plants of the San Francisco Bay estuarine system. *Limnology and Oceanography*, **47**, 713–729.

Conkright, M.E. & Sackett, W.M. (1986) A stable carbon isotope evaluation of the contribution of terriginous carbon to the marine food web in Bayboro Harbor, Tampa Bay, Florida. *Contributions to Marine Science*, **29**, 131–139.

Connolly, R.M., Gorman, D. & Guest, M.A. (2005) Movement of carbon among estuarine habitats and its assimilation by invertebrates. *Oecologia*, **144**, 684–691.

Créach, V., Schricke, M.T., Bertru, G. & Mariotti, A. (1997) Stable isotope and gut analysis to determine feeding relationships in saltmarsh macroconsumers. *Estuarine Coastal and Shelf Science*, **44**, 599–611.

Cunjak, R.A., Roussel, J.-M., Gray, M.A., et al. (2005) Using stable isotope analysis with telemetry or mark-recapture data to identify fish movement and foraging. *Oecologia*, **144**, 636–646.

Darnaude, A.M. (2005) Fish ecology and terrestrial carbon use in coastal areas: implications for marine fish production. *Journal of Animal Ecology*, **74**, 864–876.

Degens, E.T., Behrendt, M., Gotthardt, B. & Reppmann, E. (1968) Metabolic fractionation of carbon isotopes in marine plankton-II. Data on samples collected off the coasts of Peru and Ecuador. *Deep-Sea Research*, **15**, 11–20.

DeNiro, M.J. & Epstein, S. (1977) Mechanism of carbon isotope fractionation associated with lipid synthesis. *Science*, **197**, 261–263.

DeNiro, M.J. & Epstein, S. (1978) Influence of diet on the distribution of carbon isotopes in animals. *Geochimica et Cosmochimica Acta*, **42**, 495–506.

DeNiro, M.J. & Epstein, S. (1981a) Influence of diet on the distribution of nitrogen isotopes in animals. *Geochimica et Cosmochimica Acta*, **45**, 1885–1894.

DeNiro, M.J. & Epstein, S. (1981b) Hydrogen isotope ratios of mouse tissues are influenced by a variety of factors other than diet. *Science*, **214**, 1374–1376.

Deuser, W.G., Degens, E.T. & Guillard, R.R.L. (1968) Carbon isotope relationships between plankton and sea water. *Geochimica et Cosmochimica Acta*, **32**, 657–660.

Doi, H., Matsumasa, M., Toya, T., et al. (2005) Spatial shifts in food sources for macrozoobenthos in an estuarine ecosystem: carbon and nitrogen stable isotope analysis. *Estuarine Coastal and Shelf Science*, **64**, 316–322.

Domi, N., Bouquegneau, J.M. & Das, K. (2005) Feeding ecology of five commercial shark species of the Celtic Sea through stable isotope and trace metal analysis. *Marine Environmental Research*, **60**, 551–569.

Dugdale, V.A. & Goering, J.J. (1967) Uptake of new and regenerated forms of nitrogen in primary productivity. *Limnology and Oceanography*, **12**, 196–206.

Duineveld, G.C.A., Lavaleye, M.S.S. & Berghuis, E.M. (2004) Particle flux and food supply to a seamount cold-water coral community (Galicia Bank, NW Spain). *Marine Ecology Progress Series*, **277**, 13–23.

Dunton, K.H. & Schell, D.M. (1987) Dependence of consumers on macroalgal (*Laminaria solidungula*) carbon in an arctic kelp community: [13]C evidence. *Marine Biology*, **93**, 615–625.

Eadie, B.J. & Jeffrey, L.M. (1973) δ^{13}C analyses of oceanic particulate organic matter. *Marine Chemistry*, **1**, 199–209.

Eppley, R.W. & Peterson, B.J. (1979) Particulate organic matter flux and planktonic new production in the deep ocean. *Nature (London)*, **282**, 671–680.

Eppley, R.W., Renger, E.H., Venrick, E.L. & Mullin, M.M. (1973) A study of plankton dynamics and nutrient cycling in the central gyre of the North Pacific Ocean. *Limnology and Oceanography*, **18**, 534–551.

Estep, M.F. & Dabrowski, H. (1980) Tracing food webs with stable hydrogen isotopes. *Science*, **209**, 1537–1538.

Estrada, J.A., Rice, A.N., Lutcavage, M.E. & Skomal, G.B. (2003) Predicting trophic position in sharks of the northwest Atlantic Ocean using stable isotope analysis. *Journal of Marine Biological Assessment, UK*, **83**, 1347–1350.

Feller, R.J., Taghon, G.L., Gallagher, E.D., Kenny, G.E. & Jumars, P.A. (1979) Immunological methods for food web analysis in a soft-bottom benthic community. *Marine Biology*, **54**, 61–74.

Feller, R.J., Zargursky, G. & Day, E.A. (1985) Deep-sea food web analysis using cross-reacting antisera. *Deep-Sea Research*, **4**, 485–497.

Focken, U. & Becker, K. (1998) Metabolic fractionation of stable carbon isotopes: implications of different proximate compositions for studies of the aquatic food webs using $\delta^{13}C$ data. *Oecologia*, **115**, 337–343.

France, R.L. (1995) Stable nitrogen isotopes in fish: literature synthesis on the influence of Ecotonal coupling. *Estuarine Coastal and Shelf Science*, **41**, 737–742.

Fry, B. (1983) Fish and shrimp migrations in the northern Gulf of Mexico analyzed using stable C, N, and S isotope ratios. *Fisheries Bulletin*, **81**, 789–801.

Fry, B. (1988) Food web structure on Georges Bank from stable C, N, and S isotopic compositions. *Limnology and Oceanography*, **33**, 1182–1190.

Fry, B. (2002) Conservative mixing of stable isotopes across estuarine salinity gradients: A conceptual framework for monitoring watershed influences on downstream fisheries production. *Estuaries*, **25**, 264–271.

Fry, B. (2006) *Stable Isotope Ecology*. Springer-Verlag, New York, 308 pp.

Fry, B. & Arnold, C. (1982) Rapid $^{13}C/^{12}C$ turnover during growth of brown shrimp (*Penaeus aztecus*). *Ocecologia, Berlin*, **54**, 200–204.

Fry, B. & Ewel, K.C. (2003) Using stable isotopes in mangrove fisheries research – review and outlook. *Isotopes in Environmental and Health Studies*, **39**, 191–196.

Fry, B. & Parker, P.L. (1979) Animal diet in Texas seagrass meadows: $\delta^{13}C$ evidence for the importance of benthic plants. *Estuarine and Coastal Marine Science*, **8**, 499–509.

Fry, B. & Sherr, E.B. (1984) ^{13}C measurements as indicators of carbon flow in marine and freshwater ecosystems. *Contributions to Marine Science*, **27**, 13–47.

Fry, B. & Wainright, S.C. (1991) Diatom sources of ^{13}C-rich carbon in marine food webs. *Marine Ecology Progress Series*, **76**, 149–157.

Fry, B., Scalan, R.S. & Parker, P.L. (1977) Stable carbon isotope evidence for two sources of organic matter in coastal sediments: seagrasses and plankton. *Geochimica et Cosmochimica Acta*, **41**, 1875–1877.

Fry, B., Jeng, W., Scanlan, R.S., Parker, P.L. & Baccus, J. (1978a) $\delta^{13}C$ food web analysis of a Texas sand dune community. *Geochimica et Cosmochimica Acta*, **42**, 1299–1302.

Fry, B., Joern, A. & Parker, P.L. (1978b) Grasshopper food web analysis: use of carbon isotope ratios to examine feeding relationships among terrestrial herbivores. *Ecology*, **59**, 498–506.

Fry, B., Scalan, R.S., Winters, K. & Parker, P.L. (1982) Sulphur uptake by salt grasses, mangroves and seagrasses in anaerobic sediments. *Geochimica et Cosmochimica Acta*, **46**, 1121–1124.

Fry, B., Scanlan, R.S. & Parker, P.L. (1983) $^{13}C/^{12}C$ ratios in marine food webs of the Torres Strait, Queensland. *Australian Journal of Marine Freshwater Reseach*, **34**, 707–716.

Fry, B., Anderson, R.K., Entzeroth, L., Byrd, J.L. & Parker, P.L. (1984) ^{13}C enrichment and oceanic food web structure in the northwestern Gulf of Mexico. *Contributions Marine Science*, **27**, 49–63.

Gaebler, O.H., Vitti, T.G. & Vukmirovich, R. (1966) Isotope effects in metabolism of ^{15}N and ^{14}N from unlabeled dietary proteins. *Cananadian Journal of Biochemistry*, **44**, 1149–1257.

Gannes, L.Z., Martínez del Rio, C. & Koch, P. (1998) Natural abundance variations in stable isotopes and their potential uses in animal physiological ecology. *Comparative Biochemistry and Physiology A-Molecular & Integrative Physiology*, **119**, 725–737.

Gao, Y.W. & Beamish, R.J. (1999) Isotopic composition of otoliths as a chemical tracer in population identification of sockeye salmon (*Oncorhynchus nerka*). *Canadian Journal of Fisheries and Aquatic Sciences*, **56**, 2062–2068.

Gaston, T.F. & Suthers, I.M. (2004) Spatial variation in $\delta^{13}C$ and $\delta^{15}N$ of liver, muscle and bone in a rocky reef planktivorous fish: the relative contribution of sewage. *Journal of Experimental Marine Biology and Ecology*, **304**, 17–33.

Gaye-Siessegger, J., Focken, U., Muetzel, S., Abel, H. & Becher, K. (2004) Feeding level and individual metabolic rate affect $\delta^{13}C$ and $\delta^{15}N$ values in carp: implications for food web studies. *Oecologia*, **138**, 175–183.

Gearing, J.N., Gearing, P.J., Rudnick, D.T., Requejo, A.G. & Hutchins, M.J. (1984) Isotopic variability of organic carbon in a phytoplankton-based temperate estuary. *Geochimica et Cosmochimica Acta*, **48**, 1089–1098.

Gerlach, S.A., Ekstrom, D.K. & Eckhardt, P.H. (1976) Filter feeding in the hermit crab *Pagurus bernhardus*. *Oecologia (Berlin)*, **24**, 257–265.

Gordon, D.C., Jr. (1977) Variability of particulate organic carbon and nitrogen along the Halifax-Bermuda section. *Deep-Sea Research*, **24**, 257–270.

Gorokhova, E. & Hansson, S. (1999) An experimental study on variation in stable carbon and nitrogen isotope fractionation during growth of *Mysisi mixta* and *Neomysis integer*. *Canadian Journal of Fisheries and Aquatic Science*, **56**, 2203–2210.

Graham, N.A.J., Dulvy, N.K., Jennings, S. & Polunin, N.V.C. (2005) Size spectra as indicators of fishing effects on coral reef fish assemblages. *Coral Reefs*, **24**, 118–124.

Gu, B., Schell, D.M., Frazer, T., Hoyer, M. & Chapman, F.A. (2001) Stable carbon isotope evidence for reduced feeding of Gulf of Mexico sturgeon during their prolonged river residence period. *Estuarine Coastal and Shelf Science*, **53**, 275–280.

Guelinckx, J., Maes, J., De Brabandere, L., Dehairs, F. & Ollevier, F. (2006) Migration dynamics of clupeoids in the Schelde estuary: a stable isotope approach. *Estuarine Coastal and Shelf Science*, **66**, 612–623.

Hackney, C.T. & Haines, E.B. (1980) Stable carbon isotope composition of Fauna and organic matter collected in a Mississippi estuary. *Estuarine and Coastal Marine Science*, **10**, 703–708.

Haines, E.B. (1976a) Stable carbon isotope ratios in the biota, soils and tidal water of a Georgia salt marsh. *Estuarine and Coastal Shelf Science*, **4**, 609–616.

Haines, E.B. (1976b) Relation between the stable carbon isotope composition of fiddler crabs, plants & soils in a salt marsh. *Limnology and Oceanography*, **21**, 880–883.

Haines, E.B. & Montague, C.L. (1979) Food sources of estuarine invertebrates analyzed using $^{13}C/^{12}C$ ratios. *Ecology*, **60**, 48–56.

Hall-Aspland, S.A., Rogers, T.L. & Canfield, R.B. (2005) Stable carbon and nitrogen isotope analysis reveals seasonal variation in the diet of leopard seals. *Marine Ecology Progress Series*, **305**, 249–259.

Hansson, S., Hobbie, J.E., Elmgren, R., Larsson, U., Fry, B. & Johansson, S. (1997) The stable nitrogen isotope ratio as a marker of food-web interactions and fish migration. *Ecology*, **78**, 2249–2257.

Harrigan, P., Zieman, J.C. & Macko, S.A. (1989) The base of nutritional support for the gray snapper (*Lutjanus griseus*): An evaluation based on a combined stomach content and stable isotope analysis. *Bulletin of Marine Science*, **44**, 65–77.

Herman, D.P., Burrows, D.G., Wade, P.R., et al. (2005) Feeding ecology of eastern North Pacific killer whales *Orcinus orca* from fatty acid, stable isotope, and organochlorine analyses of blubber biopsies. *Marine Ecology Progress Series*, **302**, 275–291.

Herzka, S.Z. (2005) Assessing conectivity of estuarine fishes based on stable isotope ratio analysis. *Estuarine Coastal and Shelf Science*, **64**, 58–69.

Hessen, D.O., Andersen, T. & Lyche, A. (1990) Carbon metabolism in a humic lake: pool sizes and cycling through zooplankton. *Limnology and Oceanography*, **35**, 84–99.

Hesslein, R.H., Hallard, K.A. & Ramlal, P. (1993) Replacement of sulfur, carbon and nitrogen in tissue of growing Broad Whitefish (*Coregonus nasus*) in response to a change in diet traced by $\delta^{34}S$, $\delta^{13}C$ and $\delta^{15}N$. *Canadian Journal of Fisheries and Aquatic Sciences*, **50**, 2071–2076.

Hobson, K.A. (1991) Use of stable carbon and nitrogen isotope analysis in seabird dietary studies. Ph. D. Dissertation, Dept. of Biology, University of Saskatchewan, Saskatoon.

Hobson, K.A. & Welch, H.E. (1992) Determination of trophic relationships within a high arctic marine food web using $\delta^{13}C$ and $\delta^{15}N$ analysis. *Marine Ecology Progress Series*, **84**, 9–18.

Hoering, T. (1957) The isotopic composition of the ammonia and nitrate ion in rain. *Geochimica et Cosmochimica Acta*, **12**, 97–102.

Hoekstra, P.F., O'Hara, T.M., Fisk, A.T., Borga, K., Solomon, K.R. & Muir, D.C.G. (2003) Trophic transfer of persistent organochlorine contaminants (OCs) within and Arctic marine food web from the southern Beaufort–Chukchi Seas. *Environmental Pollution*, **124**, 509–522.

Hopkins, T.L. (1987) Midwater food web in McMurdo Sound, Ross Sea, Antarctica. *Marine Biology*, **96**, 93–106.

Jacoby, M.E., Hilderbrand, G.V., Servheen, C., et al. (1999) Trophic relations of brown and black bears in several western north american ecosystems. *Journal of Wildlife Management*, **63**, 921–929.

Jamieson, R.E., Schwarcz, H. & Brattey, J. (2004) Carbon isotopic records from the otoliths of Atlantic cod (*Gadus morhua*) from eastern Newfoundland, Canada. *Fisheries Research*, **68**, 83–97.

Jennings, S. & Blanchard, J.L. (2004) Fish abundance with no fishing: predictions based on macroecological theory. *Journal of Animal Ecology*, **73**, 632–642.

Jennings, S., Dinmore, T.A., Duplisea, D.E., Warr, K.J. & Lancaster, J.E. (2001a) Trawling disturbance can modify benthic production processes. *Journal of Animal Ecology*, **70**, 459–475.

Jennings, S., Pinnegar, J.K., Polunin, N.V.C. & Boon, T. (2001b) Weak cross-species relationships between body size and trophic level belie powerful size-based trophic structuring in fish communities. *Journal of Animal Ecology*, **70**, 934–944.

Johnston, A.M. & Kennedy, H. (1998) Carbon stable isotope fractionation in marine systems. In: *Stable Isotopes: Integration of Biological, Ecological and Geochemical Processes* (Ed. H. Griffith), pp. 239–256. Bios Scientific Publishers, Oxford.

Kainz, M., Arts, M.T. & Mazumder, A. (2004) Essential fatty acids in the planktonic food web and their ecological role for higher trophic levels. *Limnology and Oceanography*, **49**, 1784–1793.

Kioboe, T., Kaas, H., Kruse, B., Mohlenberg, F., Tiselius, P. & Aertebjerg, G. (1990) The structure of the pelagic food web in relation to water column structure in the Skagerrak. *Marine Ecology Progress Series*, **59**, 19–32.

Kroopnick, P.M. (1985) The distribution of ^{13}C of TCO_2 in the world oceans. *Deep-Sea Research*, **32**, 57–84.

Kwak, T.J. & Zedler, J.B. (1997) Food web analysis of southern California coastal wetland using multiple stable isotopes. *Oecologia*, **110**, 262–277.

Lenanton, R.C.J., Valesini, F., Bastow, T.P., Nowara, G.B., Edmonds, J.S. & Connard, M.N. (2003) The use of stable isotope ratios in whitebait otolish carbonate to identify the source of prey for Western Australian penguins. *Journal of Experimental Marine Biology and Ecology*, **291**, 17–27.

Lepoint, G., Dauby, P. & Gobert, S. (2004) Applications of C and N stable isotopes to ecological and environmental studies in sea grass ecosystems. *Marine Pollution Bulletin*, **49**, 887–891.

Levin, L.A. & Michener, R. (2002) Isotopic evidence for chemosynthesis-based nutrition on macrobenthos: The lightness of being at Pacific methane seeps. *Limnology and Oceanography*, **47**, 1336–1345.

Levin, L.A., James, D.W., Martin, C.M., Rathburn, A.E., Harris, L.H. & Michener, R.H. (2000) Do methane seeps support distinct macrofaunal assemblages? Observations on community structure and nutrition from the northern California slope and shelf. *Marine Ecology Progress Series*, **208**, 21–39.

Liu, K.K. & Kaplan, I.R. (1989) The eastern tropical Pacific as a source of ^{15}N-enriched nitrate in seawater off southern California. *Limnology and Oceanography*, **34**, 820–830.

Liu, K.K., Shaw, P.T. & Kaplan, I.R. (1987) Modeling of nitrogen isotopic variation of nitrate within the denitrifying zone in the eastern tropical South Pacific. *Eos*, **68**, 1714.

Lovvorn, J.R., Cooper, L.W., Brooks, M.L., De Ruyck, C.C., Bump, J.K. & Grebmeier, J.M. (2005) Organic matter pathways to zooplankton and benthos under pack ice in late winter and open water in late summer in the north-central Bering Sea. *Marine Ecology Progress Series*, **291**, 135–150.

MacAvoy, S.E., Fisher, C.R., Carney, R.S. & Macko, S.A. (2005) Nutritional associations among fauna at hydrocarbon seep communities in the Gulf of Mexico. *Marine Ecology Progress Series*, **292**, 51–60.

Machás, R. & Santos, R. (1999) Sources of organic matter in Ria Formosa revealed by stable isotope analysis. *Acta Oceologia*, **20**, 463–469.

Macko, S.A., Lee, W.Y. & Parker, P.L. (1982) Nitrogen and carbon isotope fractionation by two species of marine amphipods: laboratory and field studies. *Journal of Experimental Marine Biology, Ecology*, **63**, 145–149.

Macko, S.A., Estep, M.L.F. & Lee, W.Y. (1983) Stable hydrogen isotope analysis of food webs on laboratory and field populations of marine amphipods. *Journal of Experimental Marine Biology and Ecology*, **72**, 243–249.

Macko, S.A., Fogel Estep, M.L., Engel, M.H. & Hare, P.E. (1986) Kinetic fractionation of stable nitrogen isotopes during amino acid transamination. *Geochimica et Cosmochimica Acta*, **50**, 2143–2146.

MacNeil, M.A., Skomal, G.B. & Fisk, A.T. (2005) Stable isotopes from multiple tissues reveal diet switching in sharks. *Marine Ecology Progress Series*, **302**, 199–206.

MacNeil, M.A., Drouillard, K.G. & Fisk, A.T. (2006) Variable uptake and elimination of stable nitrogen isotopes between tissues in fish. *Canadian Journal of Fisheries and Aquatic Sciences*, **63**, 345–353.

Mancera, J.E. & Twilley, R.R. (2003) Testing the energy signature hypothesis of mangroves: using carbon isotope ratio analysis (poster). Presented at *Florida Coastal Everglades All Scientists Meeting*, 5 January, Miami, FL.

Mariotti, A., Lancelot, C. & Billen, G. (1984) Natural isotopic composition of nitrogen as a tracer of origin for suspended organic matter in the Scheldt estuary. *Geochimica et Cosmochimica Acta*, **48**, 549–555.

Marples, T.G. (1966) A radionuclide tracer study of arthropod food chains in a *Spartina* salt march ecosystem. *Ecology*, **47**, 270–277.

Martineau, C. & Vincent, W.F. (2004) Primary consumers and particulate organic matter: Isotopic evidence of strong selectivity in the estuarine transition zone. *Limnology and Oceanography*, **49**, 1679–1686.

Martinetto, P., Teichberg, M. & Valiela, I. (2006) Coupling of estuarine benthic and pelagic food webs to land-derived nitrogen sources in Waquoit Bay, Massachusetts, USA. *Marine Ecology Progress Series*, **307**, 37–48.

Martinez, L., Silver, M.W., King, J.M. & Alldredge, A.L. (1983) Nitrogen fixation by floating diatom mats: a source of new nitrogen to oligotrophic ocean waters. *Science*, **221**, 152–154.

McClelland, J.W., Valiela, I. & Michener, R.H. (1997) Nitrogen-stable isotope signatures in estuarine food webs: a record of increasing urbanization in coastal watersheds. *Limnology and Oceanography*, **42**, 930–937.

McConnaughey, T. & McRoy, C.P. (1979a) Food web structure and the fractionation of carbon isotopes in the Bering Sea. *Marine Biology*, **53**, 257–262.

McConnaughey, T. & McRoy, C.P. (1979b) ^{13}C label identifies eelgrass (*Zostera marina*) carbon in an Alaskan estuarine food web. *Marine Biology*, **53**, 263–269.

Meier-Augenstein, W. (2002) Stable isotope analysis of fatty acids by gas chromatography-isotope ratio mass spectrometry. *Analytica Chimica Acta*, **465**, 63–79.

Mekhtiyeva, V.L., Pankina, R.G. & Gavrilov, Y.Y. (1976) Distribution and isotopic compositions of forms of sulfur in water, animals and plants. *Geokhimiya*, **9**, 1419–1426.

Minagawa, M. & Wada, E. (1984) Stepwise enrichment of δ^{15}N along food chains: further evidence and the relation between δ^{15}N and animal age. *Geochimica et Cosmochimica Acta*, **48**, 1135–1140.

Minagawa, M., Winter, D.A. & Kaplan, I.R. (1984) Comparison of Kjeldahl and combustion methods for measurement of nitrogen isotope ratios in organic matter. *Analytical Chemistry*, **56**, 1859–1861.

Miyake, Y. & Wada, E. (1967) The abundance ratio of ^{15}N/^{14}N in marine environments. *Records of Oceanographic Works, Japan*, **9**, 37–53.

Mook, W.G., Bommerson, J.C. & Staverman, W.H. (1974) Carbon isotope fractionation between dissolved bicarbonate and gaseous carbon dioxide. *Earth and Planetary Science Letters*, **22**, 139–149.

Mullin, M.M., Rau, G.H. & Eppley, R.W. (1984) Stable nitrogen isotopes in zooplankton: some geographic and temporal variations in the North Pacific. *Limnology and Oceanography*, **29**, 1267–1273.

Oakes, J.M. & Connolly, R.M. (2004) Causes of sulfur isotope variability in the seagrass, *Zostera capricorni. Journal of Experimental Marine Biology and Ecology*, **302**, 153–164.

Ockelmann, K.W. & Vahl, O. (1970) On the biology of the polychaete *Glycera alba*, especially its burrowing and feeding. *Ophelia*, **8**, 275–294.

Odum, E.P. (1968) A research challenge: evaluating the productivity of coastal and estuarine water. In: *Proceedings of the Second Sea Grant Conference*, October, University of Rhode Island, pp. 63–64.

O'Reilly, C.M., Hecky, R.E., Cohen, A.S. & Plisnier, P.-D. (2002) Interpreting stable isotopes in food webs: Recognizing the role of time averaging at different trophic levels. *Limnology and Oceanography*, **47**, 306–309.

Ostrom, N.E., Stephen, M., Deibel, A.D. & Thompson, R.J. (1997) Seasonal variation in the stable carbon and nitrogen isotope boigeochemistry of a coastal cold ocean environment. *Geochimica et Cosmochimica Atca*, **61**, 2929–2942.

Owens, N.J.P. (1985) Variations in the natural abundance of ^{15}N in estuarine suspended particulate matter: a specific indicator of biological processing. *Estuarine and Coastal Marine Science*, **20**, 505–510.

Owens, N.J.P. (1987) Natural variations in ^{15}N in the marine environment. *Advances in Marine Biology*, **24**, 389–451.

Parker, P.L. (1964) The biogeochemistry of the stable isotopes of carbon in a marine bay. *Geochimica et Cosmochimica Acta*, **28**, 1155–1164.

Paterson, A.W. & Whitfield, A.K. (1997) A stable carbon isotope study of the food-web in a freshwater-deprived South African estuary, with particular emphasis on the ichthyofauna. *Estuarine Coastal and Shelf Science*, **45**, 705–715.

Patterson, W.P. (1998) North American continental seasonality during the last millennium: high-resolution analysis of sagittal otoliths. *Palaeogeography, Palaeoclimatology, Palaeoecology*, **138**, 271–303.

Pauly, D. & Watson, R. (2005) Background and interpretation of the "Marine Trophic Index" as a measure of biodiversity. *Philosophical Transactions of The Royal Society: Biological Sciences*, **360**, 415–423.

Pearcy, W.G. & Stuiver, M. (1983) Vertical trasport of carbon-14 into deep-sea food webs. *Deep-Sea Research*, **30**, 427–440.

Pearson, S.F., Levey, D.J., Greenberg, C.H. & del Rio, C.M. (2003) Effects of elemental composition on the incorporation of dietary nitrogen and carbon isotopic signatures in an omnivorous songbird. *Oecologia*, **135**, 516–523.

Persic, A., Roche, H. & Ramade, F. (2004) Stable carbon and nitrogen isotopes quantitative structural assessment of dominant species from the Vaccares Lagoon trophic web (Camargue Bioshpere Reserve, France). *Estuarine Coastal and Shelf Science*, **60**, 261–272.

Perissinotto, R., Nozais, C., Kibirige, I. & Andandraj, A. (2003) Planktonic food webs and bethic-pelagic coupling in three South African temporarily-open estuaries. *Acta Oceologia*, **24**, s307–s316.

Peterson, B.J. (1999) Stable isotopes as tracers of organic matter input and transfer in benthic food webs: a review. *Acta Oceologia*, **20**, 479–487.

Peterson, B.J. & Fry, B. (1987) Stable Isotopes in Ecosystem Studies. *Annual Review of Ecological Systems*, **18**, 293–320.

Peterson, B.J. & Howarth, R.W. (1983) Sulfur and carbon isotopes as tracers of organic matter flow in salt marshes. *Estuaries*, **6**, 305.

Peterson, B.J. & Howarth, R.W. (1987) Sulfur, carbon and nitrogen isotopes used to trace organic matter flow in the salt-marsh estuaries of Sapelo Island, Georgia. *Limnology and Oceanography*, **32**, 1195–1213.

Peterson, B.J., Howarth, R.W. & Garitt, R.H. (1985) Multiple stable isotopes used to trace the flow of organic matter in estuarine food webs. *Science*, **227**, 1361–1363.

Phillips, D.L. & Gregg, J.W. (2003) Source partitioning using stable isotopes: coping with too many sources. *Oecologia*, **136**, 261–269.

Pinnegar, J.K. & Polunin, N.V.C. (1999) Differential fractionation of $\delta^{13}C$ and $\delta^{15}N$ among fish tissues: implications for the study of trophic interactions. *Functional Ecology*, **13**, 225–231.

Post, D.M. (2002) Using stable isotopes to estimate trophic position: models, methods and assumptions. *Ecology*, **83**, 703–718.

Primavera, J.H. (1996) Stable carbon and nitrogen isotope ratios of Penaeid juveniles and primary producers in a riverine mangrove in Guimaras, Philippines. *Bulletin of Marine Science*, **58**, 675–683.

Pruell, R.J., Taplin, B.K., Lake, J.L. & Jayaraman, S. (2005) Nitrogen isotope ratios in estuarine biota collected along a nutrient gradient in Narragansett Bay, Rhode Island, USA. *Marine Pollution Bulletin*, **52**, 612–620.

Ramsay, M.A. & Hobson, K.A. (1991) Polar bears make little use of terrestrial food webs: evidence from stable-carbon isotope analysis. *Oecologia*, **86**, 598–600.

Rau, G.H. (1980) $^{13}C/^{12}C$ variation in subalpine lake aquatic insects: food source implications. *Canadian Journal of Fisheries and Aquatic Sciences*, **37**, 742–745.

Rau, G.H. (1981a) Hydrothermal vent clam and tube worm $^{13}C/^{12}C$: further evidence of nonphotosynthetic food sources. *Science*, **213**, 338–340.

Rau, G.H. (1981b) Low $^{15}N/^{14}N$ in hydrothermal vent animals: ecological implications. *Science*, **289**, 284–285.

Rau, G.H. & Anderson, N.H. (1981) Use of $^{13}C/^{12}C$ to trace dissolved and particulate organic matter utilization by populations of an aquatic invertebrate. *Oecologia*, **48**, 19–21.

Rau, G.H., Sweeney, R.E. & Kaplan, I.R. (1982) Plankton $^{13}C:^{12}C$ ratio changes with latitude: differences between northern and southern oceans. *Deep-Sea Research*, **29**, 1035–1039.

Rau, G.H., Mearns, A.J., Young, D.R., Olson, R.J., Schafer, H.A. & Kaplan, I.R. (1983) Animal $^{13}C/^{12}C$ correlates with trophic level in pelagic food webs. *Ecology*, **64**, 1314–1318.

Rau, G.H., Teyssie, J.L., Rassoulzadegan, F. & Fowler, S.W. (1990) $^{13}C/^{12}C$ and $^{15}N/^{14}N$ variations among size-fractionated marine particles: implications for their origin and trophic relationships. *Marine Ecology Progress Series*, **59**, 33–38.

Robbins, C.T., Felicetti, L.A. & Sponheimer, M. (2005) The effect of dietary protein on nitrogen isotope discrimination in mammals and birds. *Oecologia*, **144**, 534–540.

Rodelli, M.R., Gearing, J.N., Gearing, P.J., Marshall, N. & Sasekumar, A. (1984) Stable isotope ratio as a tracer of mangrove carbon in Malaysian ecosystems. *Oecologia*, **61**, 326–333.

Rounick, J.S. & Winterbourn, M.J. (1986) Stable carbon isotopes and carbon flow in ecosystems. *Bioscience*, **36**, 171–177.

Ruess, L., Haggblom, M., Langel, R. & Scheu, S. (2004) Nitrogen isotope ratios and fatty acid composition as indicators of animal diets in belowground systems. *Oecologia*, **139**, 336–346.

Ruiz-Cooley, R.I., Gendron, D., Aguiniga, S., Mesnick, S. & Carriquiry, J.D. (2004) Trophic relationships between sperm whales and jumbo squid using stable isotopes of C and N. *Marine Ecology Progress Series*, **277**, 275–283.

Sachs, J.P., Repeta, D.J. & Goericke, R. (1999) Nitrogen and carbon isotopic ratios of chlorophyll from marine phytoplankton. *Geochimica et Cosmochimica Acta*, **63**, 1431–1441.

Sackett, W.M. & Moore, W.S. (1966) Isotopic variations of dissolved inorganic carbon. *Chemical Geology*, **1**, 323–328.

Sackett, W.M., Eckelmann, W.R. & Bender, M.L. (1965) Temperature dependence of carbon isotope composition in marine plankton and sediments. *Science*, **148**, 235–237.

Saino, T. & Hattori, A. (1980) ^{15}N natural abundance in oceanic suspended particulate matter. *Nature*, **283**, 752–754.

Saino, T. & Hattori, A. (1987) Geographical variation of the water column distribution of suspended particulate opganic nitrogen and its ^{15}N natural abundance in the Pacific and its marginal seas. *Deep-Sea Research*, **34**, 807–827.

Saupe, S.M., Schell, D.M. & Griffiths, W. (1989) Carbon isotope ratio gradients in western arctic zooplankton. *Marine Biology*, **103**, 427–433.

Schell, D.M. (1987) Bowhead whale feeding: allocation of regional habitat importance based on stable isotope abundances. In: *Importance of the Eastern Alaskan Beaufort Sea to*

Feeding Bowhead Whales, 1985–1986 (Ed. W.J. Richardson), pp. 369–415. Report 87-0037, U.S. Minerals Management Service. NTIS PB88-150271/AF.

Schell, D.M., Saupe, S.M. & Haubenstodck, N. (1989a) Bowhead whale (*Balaena mysticetus*) growth and feeding as estimated by ^{13}C techniques. *Marine Biology*, **103**, 433–443.

Schell, D.M., Saupe, S.M. & Haubenstock, N. (1989b) Natural isotope abundances in bowhead whale (*Balaena mysticetus*) baleen: markers of aging and habitat usage. *Ecological Studies*, **68**, 260–269.

Schlacher, T.A., Liddell, B., Gaston, T.F. & Schlacher-Hoenlinger, M. (2005) Fish track wastewater pollution to estuaries. *Oecologia*, **144**, 570–584.

Schmidt, K., Atkinson, A., Stubing, D., McClelland, J.W., Montoya, J.P. & Voss, M. (2003) Trophic relationships among Southern Ocean copepods and krill: Some uses and limitations of a stable isotope approach. *Limnology and Oceanography*, **48**, 277–289.

Schoeninger, M.J. & DeNiro, M.J. (1984) Nitrogen and carbon isotopic composition of bone collogen from marine and terrestrial animals. *Geochimica et Cosmochimica Acta*, **48**, 625–639.

Schwartz, J.D.M., Pallin, M.J., Michener, R.H., Mbabazi, D. & Kaufman, L.S. (2006) Effects of Nile perch, *Lates niloticus*, on functional and specific fish diversity in Uganda's Lake Kyoga system. *African Journal of Ecology*, **44**, 145–156.

Sherr, E.B. (1982) Carbon isotope composition of organic seston and sediments in a Georgia salt march estuary. *Geochimica et Cosmochimica Acta*, **46**, 1227–1232.

Simenstad, C.A. & Wissmar, R.C. (1985) ^{13}C evidence of the origins and fates of organic carbon in estuarine and nearshore food webs. *Marine Ecology Progress Series*, **22**, 141–152.

Smit, A.J., Brearley, A., Hyndes, G.A., Lavery, P.S. & Walker, D.I. (2006) δ^{15}N and δ^{13}C analysis of a *Ponsidonia sinuosa* sea grass bed. *Aquatic Botany*, **84**, 277–282.

Smith, D.F., Bulleid, N.C., Campbell, R., et al. (1979) Marine food-web analysis: an experimental study of demersal zooplankton using isotopically labeled prey species. *Marine Biology*, **54**, 49–59.

Sondergaard, M., Riemann, B., Jensen, L.M., et al. (1988) Pelagic food web processes in an oligotrophic lake. *Hydrobiologia*, **164**, 271–286.

Spies, R.B., Kruger, H., Ireland, R. & Rice, D.W. (1989) Stable isotope ratios and contaminant concentrations in a sewage distorted food web. *Marine Ecology Progress Series*, **54**, 157–170.

Steele, K.W. & Daniel, R.M. (1978) Fractionation of nitrogen isotopes by animals: a further complication to the use of variations in the natural abundance of ^{15}N for tracer studies. *Journal of Agricultural Science*, **90**, 7–9.

Stephenson, R.L., Tann, F.C. & Mann, K.H. (1986) Use of stable carbon isotope ratios to compare plant material and potential consumers in a seagrass bed and a kelp bed in Nova Scotia, Canada. *Marine Ecology Progress Series*, **30**, 1–7.

Stoner, A.W. & Zimmerman, R.J. (1988) Food pathways associated with penaeid shrimps in a mangrove-fringed estuary. *Fisheries Bulletin*, **86**, 543–551.

Sullivan, J.C., Buscetta, K.J., Michener, R.H., Whitaker, J.J.O., Finnerty, J.R. & Kunz, T.H. (2006) Models developed from δ^{13}C and δ^{15}N of skin tissue indicate non-specific hapitat use by the big brown bat (*Eptisicus fuscus*). *Ecoscience*, **13**, 11–22.

Surge, D. & Walker, K.J. (2005) Oxygen isotope composition of modern and archaeological otoliths from the estuarine hardhead catfish (*Ariopsis felis*) and their potential to record low-latitude climate change. *Palaeogeography, Palaeoclimatology, Palaeoecology*, **228**, 179–191.

Suzuki, K.W., Kasai, A., Nakayama, K. & Tanaka, M. (2005) Differential isotopic enrichment and half-life among tissues in Japanese temperate bass (*Lateolabrax japonicus*) juveniles:

implications for analyzing migration. *Canadian Journal of Fisheries and Aquatic Sciences*, **62**, 671–678.

Sweeney, R.E. & Kaplan, I.R. (1980a) Tracing flocculent industrial and domestic sewage transport on San Pedro shelf, southern California, by nitrogen and sulfur isotope ratios. *Marine Environmental Research*, **3**, 214–224.

Sweeney, R.E. & Kaplan, I.R. (1980b) Natural abundances of ^{15}N as a source indicator for near-shore marine sedimentary and dissolved nitrogen. *Marine Chemistry*, **9**, 81–94.

Sweeting, C.J., Polunin, N.V.C. & Jennings, S. (2006) Effects of chemical lipid extraction and arithmetic lipid correction on stable isotope ratios of fish tissues. *Rapid Communications in Mass Spectrometry*, **20**, 595–601.

Takai, N., Mishima, Y., Yoroza, A. & Hoshika, A. (2002) Carbon sources for demersal fish in the western Seto Inland Sea, Japan, examined by δ^{13}C and δ^{15}N analyses. *Limnology and Oceanography*, **47**, 730–741.

Tamelander, T., Soreide, J.E., Hop, H. & Carroll, M.L. (2006) Fractionation of stable isotopes in the arctic marine copepod Calanus glacialis: effects on the isotopic composition of marine particulate organic matter. *Journal of Experimental Marine Biology and Ecology*, **333**, 231–240.

Tanoue, E. & Handa, N. (1979) Distribution of particulate organic carbon and nitrogen in the Bering Sea and the northern North Pacific Ocean. *Journal of the Oceanographic Society of Japan*, **35**, 47–62.

Teeri, J.A. & Schoeller, D.A. (1979) δ^{13}C values of an herbivore and the ratio of C_3 to C_4 plant carbon in its diet. *Oecologia*, **39**, 197–200.

Tewfik, A., Rasmussen, J.B. & McCann, K.S. (2005) Anthropogenic enrichment alters a marine benthic food web. *Ecology*, **86**, 2726–2736.

Thayer, G.W., Govoni, J.J. & Connally, D.W. (1983) Stable carbon isotope ratios of the planktonic food web in the northern Gulf of Mexico. *Bulletin of Marine Science*, **33**, 247–256.

Thorrold, S.R., Campana, S.E., Jones, C.M. & Swart, P.K. (1997) Factors determining δ^{13}C and δ^{18}O fractionation in aragonitic otoliths of marine fish. *Geochimica et Cosmochimica Acta*, **61**, 2909–2919.

Tieszen, L.L., Boutton, T.W., Tesdahl, K.G. & Slade, N.A. (1983) Fractionation and turnover of stable carbon isotopes in animal tissues: implications for ^{13}C analysis of diet. *Oecologia*, **57**, 32–37.

Tortell, P.D., Reinfelder, J.R. & Morel, F.M.M. (1997) Active uptake of bicarbonate by diatoms. *Nature*, **390**, 243–244.

Trites, A.W. (2001) Food webs in the ocean: who eats whom and how much? *Reykjavik Conference on Responsible Fisheries in the Marine Ecosystem*, 1–4 October.

Twilley, R.R. (1995) Properties of mangrove ecosystems related to the energy signature of coastal environments. In: *Maximum Power: the Ideas and Applications of H.T. Odum* (Ed. C. Hall), pp. 43–62. The University Press of Colorado, Boulder, CO.

Valiela, I., Geist, M., McClelland, J. & Tomasky, G. (2000) Nitrogen loading from watersheds to estuaries: Verification of the Waquoit Bay Nitrogen Loading Model. *Biogeochemistry*, **49**, 277–293.

Van Dover, C.L., Grassle, J.F., Fry, B., Garritt, R.H. & Starczak, V.R. (1992) Stable isotope evidence for entry of sewage-derived organic material into a deep-sea food web. *Nature* **360**, 153–155.

Vanderklift, M.A. & Ponsard, S. (2003) Source of variation in consumer-diet δ^{15}N enrichment: a meta-analysis. *Oecologia*, **136**, 169–182.

Vander Zanden, M.J. & Rasmussen, J.B. (2001) Variation in ^{15}N and ^{13}C trophic fractionation; implications for aquatic food web studies. *Limnology and Oceanography*, **46**, 1–27.

Vizzini, S., Sara, G., Michener, R. & Mazzola, A. (2002) The role and contribution of the seagrass *Posidonia oceanica* (L.) delile organic matter for sencondary consumers as revealed by carbon and nitrogen stable isotope analysis. *Acta Oceologia*, **23**, 277–285.

Voigt, C.C., Matt, F., Michener, R. & Kunz, T.H. (2003) Low turnover rates of carbon isotopes in tissues of two nectar feeding bats. *The Journal of Experimental Biology*, **206**, 1419–1427.

Wada, E. (1987) ^{15}N and ^{13}C abundances in marine environments with emphasis on biogeochemical structure of food web. *Isotopenpraxis*, **23**, 639–646.

Wada, E. & Hattori, A. (1978) Nitrogen isotope effects in the assimilation of inorganic nitrogenous compounds by marine diatoms. *Geomicrobiology Journal*, **1**, 85–101.

Wada, E. & Hattori, A. (1991) *Nitrogen in the Sea: Forms, Abundance, and Rate Processes*. CRC Press, Boca Raton, FL.

Wada, E., Kadonaga, T. & Matsuo, S. (1975) ^{15}N abundance in nitrogen of naturally occurring substances and global assessment of denitrification from isotopic viewpoint. *Geochemical Society of Japan*, **9**, 139–148.

Wada, E., Terazaki, M., Kabaya, Y. & Nemoto, T. (1987) ^{15}N and ^{13}C abundances in the Antarctic Ocean with emphasis on the biogeochemical structure of the food web. *Deep-Sea Research*, **34**, 829–841.

Wada, E., Mizutani, H. & Minagawa, M. (1991) The use of stable isotopes for food web analysis. *Critical Reviews in Food Science and Nutrition*, **30**, 361–371.

Wainwright, S.C. (1990) Sediment-to-water fluxes of particulate material and microbes by resuspension and their contribution to the planktonic food web. *Marine Ecology Progress Series*, **62**, 271–281.

Wainright, S.C. & Fry, B. (1994) Seasonal variation of the stable isotopic compositions of coastal marine phytoplankton from Woods Hole, Massachusetts and Georges Bank. *Estuaries*, **17**, 552–560.

Wainright, S.C., Fuller, C.M., Michener, R.H. & Richards, R.A. (1996) Spatial variation of growth rate and trophic position of juvenile striped bass (*Morone saxitalis*) in the Delaware River. *Canadian Journal of Fisheries and Aquatic Sciences*, **53**, 685–692.

Ward-Paige, C.A., Risk, M.J. & Sherwood, O.A. (2005) Reconstruction of nitrogen sources on coral reefs: δ^{15}N and δ^{13}C in gorgonians from Florida Reef Tract. *Marine Ecology Progress Series*, **296**, 155–163.

Warren, P.H. (1989) Spatial and temporal variation in the structure of a freshwater food web. *Oikos*, **55**, 299–311.

Wilkinson, B.H. & Ivany, L.C. (2002) Paleoclimatic inference from stable isotope profiles of accretionary biogenic hardparts – a quantitative approach to the evaluation of incomplete data. *Palaeogeography, Palaeoclimatology, Palaeoecology*, **185**, 95–114.

Williams, P.M. & Gordon, L.I. (1970) [13]C:[12]C ratios in dissolved and particulate organic matter in the sea. *Deep-Sea Research*, **17**, 19–27.

Wissel, B. & Fry, B. (2005) Tracing Mississippi River influences in estuarine food webs of coastal Louisiana. *Oecologia*, **144**, 659–672.

Wong, W.W. & Sackett, W. (1978) Fractionation of stable carbon isotopes by marine phytoplankton. *Geochimica et Cosmochimica Acta*, **42**, 1809–1815.

Wurster, C.M., Patterson, W.P., Stewart, D.J., Bowlby, J.N. & Stewart, T.J. (2005) Thermal histories, stress, and metabolic rates of chinook salmon (*Oncorhynchus tshawytscha*) in Lake Ontario: evidence from intra-otolith stable isotope analyses. *Canadian Journal of Fisheries and Aquatic Sciences*, **62**, 700–713.

CHAPTER 10

Stable isotope tracing of temporal and spatial variability in organic matter sources to freshwater ecosystems

JACQUES C. FINLAY AND CAROL KENDALL

Introduction

The recent expansion of natural abundance stable isotope methods has rapidly increased our understanding of freshwater food web ecology. Investigators now routinely use stable isotopes to answer questions related to plant and animal ecophysiology, trophic structure, and energy pathways within freshwater ecosystems and at their interfaces with marine and terrestrial ecosystems. Stable isotope studies are particularly useful in aquatic settings because of limited opportunities for direct observations, high degree of spatial complexity and diverse potential sources of nutrients, organic matter, and prey.

All applications of natural abundance methods to trace nutrient and energy sources and trophic interactions depend on variation in isotope ratios in organisms and their environment. While natural abundance stable isotope techniques have been applied to food web research for over 25 years (i.e., Fry et al. 1978; Rounick et al. 1982), the increased availability of automated preparation systems for analyses of carbon (C), nitrogen (N), and sulfur (S) stable isotopes has led to a recent, rapid increase in data. These data have allowed greater exploration of isotopic variation in the environment, leading to many new applications in ecological studies. Moreover, increased knowledge of isotopic variability at the base of food webs, especially in aquatic ecosystems, provides an opportunity to re-evaluate poorly tested assumptions and improve older, established methods.

Isotopic variation of basal resources (detritus and primary producers at the base of food webs) and prey determine if, how, and when stable isotope techniques may be applied. Overlapping or variable isotopic compositions may make natural abundance methods ineffective (e.g., Phillips & Gregg 2003). However, if recognized, such isotopic variability may sometimes prove to be a useful natural signal that may actually enhance the power of isotope methods. Because isotope methods are still rapidly evolving, however, there is currently limited ability to predict when and where natural abundance tracers will be effective.

Excellent introductions to stable isotope approaches in freshwaters are available elsewhere (see Peterson & Fry 1987; Hershey & Peterson 1996). In this chapter, we explore use of natural abundance stable isotopes (primarily $\delta^{13}C$ and $\delta^{15}N$, and to a lesser extent, $\delta^{34}S$) in ecosystems, with an emphasis on nutrient and energy flow at the base of aquatic food webs. Our focus is on freshwaters, especially rivers, and their interface with surrounding watersheds, although the underlying principles apply to many other ecosystems. Below, we review the environmental basis for variation in $\delta^{13}C$, $\delta^{15}N$, and $\delta^{34}S$ of organic matter at the base of river food webs, and explore broad patterns, biogeochemical predictors, and applications of natural abundance measurements in river environments.

Overview of river food webs and stable isotope approaches

Food webs in rivers have high spatial and temporal complexity compared with most other ecosystems. Fluvial ecosystems are characterized by strong longitudinal gradients in environmental conditions from headwaters to mouth, heterogeneity at all spatial scales, and considerable temporal variation.

Two features of rivers are particularly important to the study of food webs. First, drainage networks contain both allochthonous organic matter, derived primarily from terrestrial sources, and autochthonous organic matter derived from production within rivers. Five major types of living and dead organic matter (OM) are available to stream consumers (Figure 10.1), including two allochthonous sources (terrestrial plant detritus and soils), two autochthonous sources (aquatic macrophytes and algae), and aquatic heterotrophic bacteria and fungi, that may rely on a mix of the other sources.

The River Continuum Concept (RCC; Figure 10.2) predicts that the relative contribution of these sources to total organic carbon varies with stream size (Vannote et al. 1980). In this model, allochthonous terrestrial carbon is expected to be important for food webs in small forest streams because overhanging trees block sunlight, greatly reducing algal productivity, and adding large amounts of detritus to streams. With increasing stream size, algal productivity (i.e. periphyton in small streams, transitioning to planktonic forms in large rivers) becomes a more important source of carbon as canopy shading decreases. As turbidity increases in large or disturbed rivers, light limits autotrophic production again; hence, allochthonous forms of carbon are expected to dominate energy flow in food webs in large rivers as well as in headwater streams. However, several recent studies have shown that some large rivers (e.g., the Mississippi River) have appreciable amounts of algal productivity (Kendall et al., 2001; Wissel & Fry, 2005; Delong & Thorp 2006). While the RCC describes broad patterns in the relative abundance of

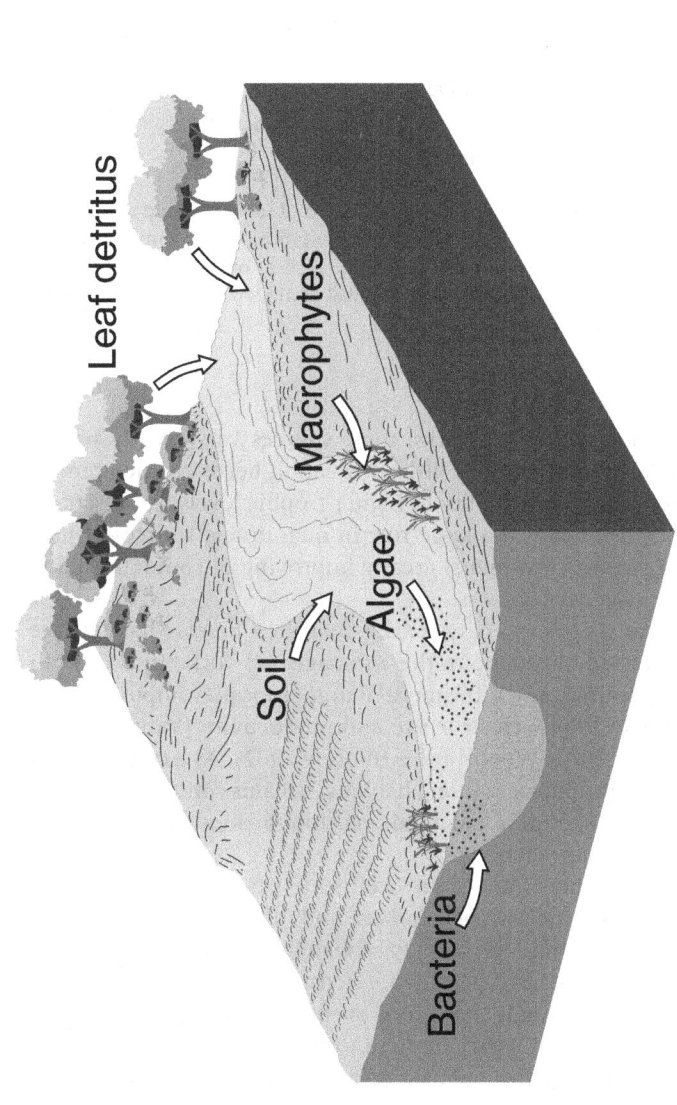

Figure 10.1 The five main sources of organic matter to stream ecosystems. "Bacteria" = both benthic and planktonic heterotrophs, and "algae" = benthic and planktonic algae and cyanobacteria.

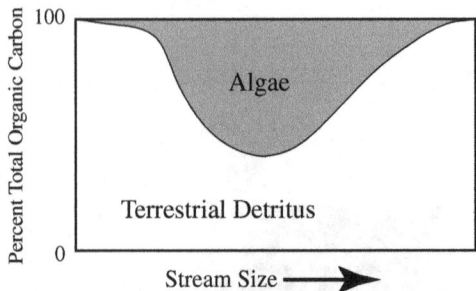

Figure 10.2 General predictions of the relative contribution of terrestrial and autotrophic organic matter to rivers, as inferred from the River Continuum Model of Vannote et al. (1980). Some recent publications have found that a large percent of the seston in large rivers in temperate climates is algal derived. (e.g., Kendall et al. 2001; Wissel & Fry 2005; Delong & Thorp 2006.)

organic matter sources in river networks, these sources vary greatly widely in nutritional quality; hence, there is little correlation between the amount of organic resources and its availability to higher trophic levels. As described later, stable isotopes have played a large role in quantifying sources of production supporting stream consumers, greatly improving our understanding of energy sources to river consumers.

A second key feature of river ecosystems is that their diverse habitats are hydrologically connected throughout the drainage network. Hence, both fluvial transport of resources and movement of organisms play important and often definitive roles in food web structure and productivity (Vannote et al. 1980; Power & Dietrich 2002; Woodward & Hildrew 2002). As a consequence of the variable hydrologic linkages among habitats, streams contain a complex mixture of organic matter types, and food web interactions are difficult to constrain by traditional approaches.

Natural abundance stable isotopes are well suited to studying processes and interactions in river food webs. Carbon and nitrogen stable isotope ratios of organic matter sources in rivers vary widely, providing the basis of the stable isotope tools that have provided extensive insight about transfer of energy and nutrients through these food webs. When key assumptions are satisfied, $\delta^{13}C$, $\delta^{15}N$, and $\delta^{34}S$ may be used to trace the source of productivity fueling food webs (i.e. the habitat, resource type, or in some cases, the specific taxa), the origin of organisms, and the trophic position of consumers. Such information is essential for understanding the dynamics of food webs as well as for detecting responses to environmental and human-driven change.

As discussed below in detail, $\delta^{13}C$ values are useful because of the wide range of $\delta^{13}C$ of algae at the base of food webs. The other major source

Table 10.1 Potential locations and mechanisms for stable isotope variation in organic matter sources in rivers and riparian zones.

Habitats or sources	Stable isotope	Primary mechanism	Conditions or locations observed	Examples
Longitudinal				
Pool–riffle	$\delta^{13}C$, $\delta^{15}N$	Fractionation	High algal growth, low CO_2	Finlay et al. (1999); Trudeau & Rasmussen (2003)
River confluence	$\delta^{13}C$, $\delta^{15}N$, $\delta^{34}S$	Source	Mixing of chemically distinct waters	Figure 10.8
Upstream–downstream	$\delta^{13}C$, $\delta^{15}N$	Source and fractionation	Small or spring fed streams, point source nutrient inputs	Kennedy et al. (2005)
Within site Terrestrial–aquatic	$\delta^{13}C$, $\delta^{15}N$, $\delta^{34}S$	Source and fractionation	Widespread	Finlay (2001)
Benthic–planktonic	$\delta^{13}C$	Fractionation	Rivers with attached and planktonic algae	Debruyn & Rasmussen (2002); Delong & Thorp (2006)
River–riparian interface	$\delta^{13}C$, $\delta^{15}N$, $\delta^{34}S$	Source and fractionation	Large streams, rivers	Bastow et al. (2002); Kato et al. (2004)
Marine–river	$\delta^{13}C$, $\delta^{15}N$, $\delta^{34}S$	Source	Coastal streams an rivers	Chaloner et al. (2002); MacAvoy et al. (2000)

of energy in river food webs, terrestrial detritus, has a much better constrained $\delta^{13}C$, so that these two carbon sources are often isotopically distinct (Finlay 2001). Recent literature reviews confirm that there are very minor changes in $\delta^{13}C$ during trophic transfer of organic carbon following fixation by plants (Vander Zanden & Rasmussen 2001; McCutchan et al. 2003) and minimal changes during decomposition. These features often make $\delta^{13}C$ the most effective tracer of organic C sources and energy flow in aquatic ecosystems.

In rivers, investigators have identified a growing number of ways that variation in $\delta^{13}C$ may be used to distinguish carbon sources in food webs (Table 10.1). Variation in $\delta^{13}C$ of sources has been applied toward basic research to study energy sources for riparian and aquatic consumers (e.g. Finlay 2001; Bastow et al. 2002; Finlay et al. 2002; Huryn et al. 2002; McCutchan & Lewis 2002), parasitism (Doucett et al. 1999a), and origins of juvenile progeny (e.g. Doucett et al. 1999b; McCarthy & Waldron 2000). $\delta^{13}C$ have also been used to examine applied issues in river food webs, including land-use change (e.g. Hicks 1997; England & Rosemond 2004), invasive species (e.g. Kennedy et al. 2005), contaminant transfer (e.g. Berglund et al. 2005), and eutrophication (e.g. Debruyn & Rasmussen 2002).

Nitrogen stable isotopes are a powerful tracer of the nitrogen cycle and food web interactions in freshwater ecosystems. $\delta^{15}N$ natural abundance techniques are primarily used to study nitrogen sources and cycling (see Kendall et al., this volume, pp. 375–449), as an integrative measure of food chain length (Finlay et al. 2002; Jepsen & Winemiller 2002), and to quantify energy or nutrient sources in food webs when differences in $\delta^{15}N$ among organic matter or prey sources are very large (Table 10.1). $\delta^{15}N$ may be used to estimate trophic position because there is a consistent increase in $\delta^{15}N$ of consumers that is related to trophic position in food webs. Reviews of controlled feeding studies and comparisons with gut content analyses confirm that $\delta^{15}N$ yield quantitative information regarding trophic position of consumers (Vander Zanden et al. 1997), although fractionation is variable and poorly understood for basal consumers that have N-poor diets (see McCutchan et al. 2003; Vanderklift & Ponsard 2003). However, $\delta^{15}N$ offers many advantages over alternate techniques to measure trophic structure in ecosystems. In particular, isotope-based estimates of trophic position are spatially and temporally integrated, in contrast to observational approaches. However, when prey $\delta^{15}N$ vary among habitats or energy sources, measurements of $\delta^{13}C$ or other tracers are necessary to accurately estimate trophic position from $\delta^{15}N$ (see Post 2002).

As with $\delta^{13}C$, $\delta^{34}S$ values are useful for determining food sources because of the wide range in the $\delta^{34}S$ of organic matter at the base of food webs, especially those of coastal environments or ecosystems with strong gradients in redox conditions, and because of the apparent minimal fractionation during trophic transfer of S. Relatively little is known about trophic fractionation, however. A recent review of controlled-diet studies found that organisms fed high-protein diets had higher $\delta^{34}S$ values than when fed low-protein diets, with fractionation ranging from −0.5 to +2‰ (McCutchan et al. 2003).

As briefly summarized here, natural abundance stable isotope studies are increasingly integrated into food web studies in rivers and elsewhere. However, use of these techniques depends on key issues such as distinct end-members, and limited or well characterized temporal and spatial variation of potential organic matter or prey sources. Thus, the extent of isotopic variation at the base of the food web determines the usefulness of these tools for studies of both energy and material flow, and trophic structure. While predictive understanding of such variation is limited in rivers and other aquatic ecosystems, this situation is changing. Below, we describe spatial and temporal variations in the isotopic compositions of organic matter at the base of food webs, synthesize recent progress in understanding mechanisms driving these patterns, and provide recent examples of new or improved applications of stable isotope methods to basic and applied research in rivers.

Table 10.2 Typical compositional values of major organic matter sources; the ranges of observed values are in parentheses. The data are gleaned from the references cited in text. For a much more detailed list of C:N values of potential contributors to stream seston, see Rostad et al. (1997) and Sterner & Elser (2002).

Organic matter source	$\delta^{13}C$ (‰)	$\delta^{15}N$ (‰)	$\delta^{34}S$ (‰)	C:N (at.)
Heterotrophic bacteria	Similar to substrate	−15 to +20	−15 to +20	4 to 8
Freshwater autotrophs		−15 to +20	−10 to +33	5 to 12
Periphyton	−35 to −18 (−47 to −8)			5 to 8
Phytoplankton	−32 to −23 (−42 to −19)			
Macrophytes*	−27 to −20	−15 to +20	−10 to +33	10 to >50
Soil organic matter		0 to +5	0 to +5	8 to >25
C3	−27 (−32 to −22)		(−30 to +35)	
C4	−13 (−16 to −9)			
Terrestrial plants		−3 to +7	0 to +5	15 to >50
C3	−27 (−32 to −22)	(−10 to +10)	(−10 to +20)	
C4	−13 (−16 to −9)			

* Excluding bryophytes.

Stable isotope ratios of organic matter sources in stream ecosystems

The main types of organic matter in rivers shown in Figure 10.1 have widely variable $\delta^{13}C$, $\delta^{15}N$, and $\delta^{34}S$ values (Table 10.2). Examination of the ranges of $\delta^{13}C$ and $\delta^{15}N$ in rivers (Figure 10.3) shows substantial overlap among the various sources. However, isotopic ranges for organic matter sources are much smaller in a specific river than found in a global literature survey, and sources often have distinct compositions. For example, the typical range of $\delta^{13}C$ and $\delta^{15}N$ values of organic matter sources in the San Joaquin River, a major agricultural river in California (USA), shows that algae (phytoplankton) often have distinct isotope ratios compared with other sources (Figure 10.4). Moreover, other tracers, in particular elemental ratios, also vary among sources (Figures 10.3 & 10.4) and may be used to complement isotope approaches for determining the sources of organic matter. For example, the C:N of seston (suspended particulate organic matter – POM) is usually more useful than $\delta^{13}C$ and/or $\delta^{15}N$ for determining the dominant source of organic matter to major rivers (Kendall et al. 2001), and lower C:N values generally correlate with higher quality of organic matter for consumers.

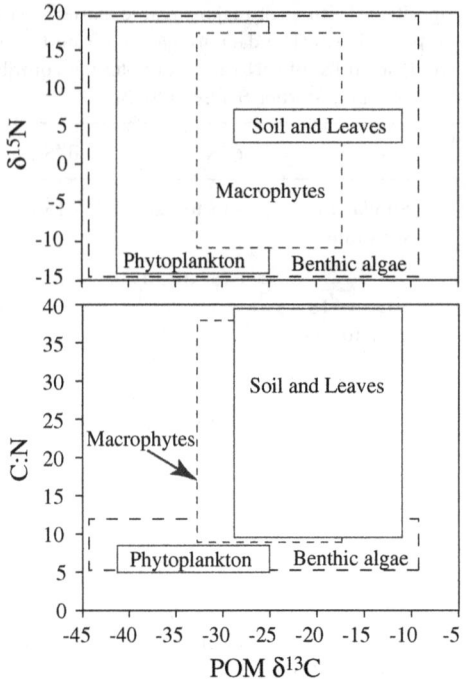

Figure 10.3 Typical ranges in $\delta^{15}N$, $\delta^{13}C$, and C:N (atomic) values of different particulate organic matter (POM) sources to rivers, based on a literature survey.

Understanding the environmental and physiological controls of stable isotope ratios in food webs is a first step for effective use of these tracers in rivers. Natural isotopic variation in the plant sources of organic materials is determined by two general factors:

1 the isotope ratio of inorganic **source** elements (i.e., C, N, S), either of dissolved species (such as NO_3^-, HCO_3^-, or SO_4^{-2}) or gaseous compounds (such as CO_2, N_2, or H_2S);
2 their subsequent **fractionation** during assimilation (uptake) by terrestrial or aquatic plants.

A well-known model of carbon isotope variation in terrestrial plants (Farquhar et al. 1982) provides a useful framework for discussing controls of isotopic variation. The model states that $\delta^{13}C$ of plants is described by the following equation:

$$\delta^{13}C_{CO2} - a - (b - a)c_i/c_e$$

where a is the discrimination due to slower diffusion of $^{13}CO_2$, b is the discrimination against $^{13}CO_2$ by Rubisco (ca. 27‰), c_i is the intercellular CO_2

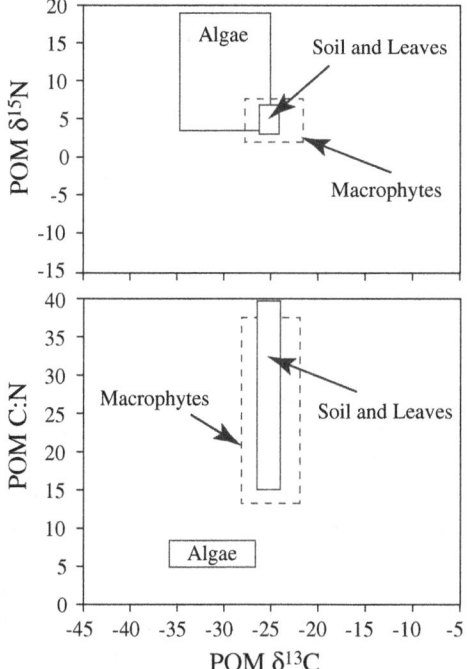

Figure 10.4 Typical compositions of different particulate organic matter (POM; seston) sources to the San Joaquin River, California, USA. Note that the ranges for different organic matter sources are usually much less than shown in Figure 10.3.

concentration ($\approx$ growth rate), and c_e is the external CO_2 concentration. The discrimination factor b is equivalent to the more commonly used term ε, or fractionation factor, and is almost always >0. Similar principles apply to understanding variation in plant $\delta^{15}N$ and $\delta^{34}S$ because they also are influenced by both the isotope ratio of the inorganic source and fractionation during uptake, but far less is known about these isotopes in freshwater ecosystems.

The influence of source and fractionation effects varies widely within and between terrestrial and aquatic ecosystems. Because both environments are significant sources of organic matter to river and riparian food webs (Figure 10.1), we must consider the controls on the isotopic compositions of food sources in both terrestrial and freshwater ecosystems. Below we briefly review isotopic variation within the five major sources of organic matter in rivers, with greater emphasis on aquatic producers because of larger, but less well known, isotopic variation in freshwaters.

We also consider briefly C:N ratios because of their utility in describing the source and quality of organic matter in ecosytems. C:N ratios can be reported as either atomic (at.) or mass (wt.) ratios. Mass ratios (the normal output of mass spectrometer analysis of %C and %N in OM) can be converted to atomic ratios by multiplying by 14/12. We use atomic ratios in this chapter.

Terrestrial plants

Detailed reviews of controls of terrestrial plant stable isotope ratios may be found elsewhere (Dawson et al. 2002 and references therein) and are only briefly described here. A more thorough discussion of this topic is found in Garten et al., this volume, pp. 61–82 and Evans, this volume, pp. 83–98.

Controls on $\delta^{13}C$ of plants

For $\delta^{13}C$, terrestrial plants fall into two main categories, based on different photosynthetic pathways for uptake of carbon: C4 plants, of which corn and some grasses are the most important members, and C3 plants, which include deciduous and coniferous trees. The average $\delta^{13}C$ value of C3 plants is around −27‰ (Bender 1968; Smith & Epstein 1971; Troughton 1972), with a total range of about 15‰. In semi-arid regions, water limitation promotes use of the C4 photosynthetic pathway, which enhances water-use efficiency and lowers isotopic fractionation (as discussed more thoroughly in Marshall et al., this volume, pp. 22–60 and Evans, this volume, pp. 83–98). As a consequence, the average $\delta^{13}C$ value of C4 plants is about −13‰, with a smaller range of values (Figure 10.3). The $\delta^{13}C$ values of C4 plants are higher than most other organic sources to rivers, except for attached algae and macrophytes under some conditions (Hillaire-Marcel 1986).

Terrestrial plants use a single form of inorganic carbon (CO_2) derived from a large, well-mixed atmospheric reservoir ($\delta^{13}C$ of −8‰); thus, source effects on $\delta^{13}C$ are relatively minor. Isotopic variation for terrestrial plants is thus related to fractionation during CO_2 assimilation, and fractionation is primarily related to plant growth rates and water use. The variation that arises for C3 and C4 plants is generally small compared with freshwater autotrophs (aquatic photosynthetic plants), although the presence of both C3 and C4 plants in a watershed expands the potential $\delta^{13}C$ range of the terrestrial end-member substantially (Figure 10.3).

Controls on $\delta^{15}N$ of plants

Nitrogen metabolism also distinguishes plants into two general categories: N-fixing plants (e.g., legumes and certain grasses, notably alfalfa) that assimilate atmospheric N_2, and non-N-fixing plants that use only other forms of plant-available N. These forms include inorganic N (NH_4^+, NO_3^-) as well as organic N (e.g. amino acids), that may have a wide range in $\delta^{15}N$ values depending on environmental conditions (Figure 10.3). However, most terrestrial plants have $\delta^{15}N$ values in the range of −6 to +5‰ (Fry 1991). Plants fixing atmospheric N_2 have a more restricted $\delta^{15}N$ range of about -3 to $+1‰$ (Fogel & Cifuentes, 1993), close to the $\delta^{15}N$ value of atmospheric N.

Most investigations have concluded that there is negligible fractionation during terrestrial plant uptake (assimilation) in most natural N-limited systems (Nadelhoffer & Fry 1994; Högberg 1997); nevertheless, tree tissues and litter typically have slightly lower $\delta^{15}N$ values than the source of nitrogen. Under higher nutrient conditions, preferential uptake of ^{14}N by plants results in a few per mil fractionation between plants and dissolved inorganic N (DIN). Many processes (e.g., volatilization, nitrification, and denitrification) can alter the $\delta^{15}N$ values of the ammonium or nitrate that is ultimately utilized by the plant (Handley et al. 1999), as will be discussed in more detail below. As a result of these complications, plants with $\delta^{15}N$ values $< -10‰$ or $> +10‰$ are not unusual, especially when they are growing near streams or in wetlands where $\delta^{15}N_{DIN}$ is high because of denitrification or eutrophic conditions.

Controls on $\delta^{34}S$ of plants

Sulfur isotopes ($\delta^{34}S$) are less commonly used than those of either C or N for determining sources of organic matter, primarily because the analytics are more complicated but also because much less is understood about sources of variability in ecosystems. Nevertheless, they have been used when a dual (C, N) isotope approach has proved unsuccessful (Peterson et al. 1985; Loneragan et al. 1997; Connolly et al. 2003), especially in marine or coastal environments where plants are likely to be isotopically labeled with the high $\delta^{34}S$ of marine sulfate (about $+20‰$).

Terrestrial plants and soils have $\delta^{34}S$ values that average $+2‰$ (Table 10.2), but soils have a much larger range (Krouse et al. 1991). The wide range of $\delta^{34}S$ values for soils and terrestrial plants has been attributed to atmospheric inputs, local mineral sources and reduction processes in anaerobic environments (Krouse et al. 1991). In areas with low atmospheric inputs of sulfur, the $\delta^{34}S$ values of vegetation generally reflect those of the soil sulfate. However, in areas with high atmospheric S inputs, the $\delta^{34}S$ value of vegetation may also reflect direct incorporation of S as sulfur dioxide through the stomata as well as possible enrichment in ^{34}S due to biogenic emission of reduced sulfur gases depleted in ^{34}S (Krouse et al. 1991; Wadleigh & Blake 1999). An intriguing study of the $\delta^{34}S$ of lichens across the island of Newfoundland (Canada), which showed an $11‰$ gradient in $\delta^{34}S$ related to distance from the coast and proximity to local S emission sources, beautifully illustrates how the spatial distribution of atmosphere $\delta^{34}S$ sources is reflected in the $\delta^{34}S$ of plants (Wadleigh & Blake 1999).

C:N values

Terrestrial plant species vary widely in their nutritional quality for growth of stream consumers. C:N provides a general indication of their quality for heterotrophs. The $C:N$ of fresh and senesced leaves vary widely, with a mean value around 20 (Sterner & Elser 2002). Litter with low $C:N$ can be a significant food source to local stream consumers. However, OM with high $C:N$ (e.g., woody debris, soil organic matter, and litter of nutrient-limited plants) is unlikely to be very nutritive to consumers without substantial microbial processing. In general, the $C:N$ of terrestrial litter exceed those of aquatic production, and there is insufficient N to meet growth demands of many stream heterotrophs. However, decomposers colonize leaves and immobilize N, thus decreasing bulk $C:N$ and improving its nutritional quality for invertebrate consumers (see 'Microbial heterotrophs', below).

Soil organic matter

The isotopic composition of soil organic matter largely reflects the isotopic compositions of the plants growing on them. Soil organic matter $\delta^{13}C$ values of about $-27‰$ and $-13‰$ are expected in areas dominated by C3 and C4 plants, respectively (Boutton 1996). Most soils have $\delta^{15}N$ values of +2 to +5‰ (Broadbent et al. 1980). On a regional and global scale, soil and plant $\delta^{15}N$ values systematically decrease with increasing mean annual precipitation and decreasing mean annual temperature (Amundson et al. 2003). Most soils have organic matter $\delta^{34}S$ values in the range of 0 to +5‰. The $\delta^{34}S$ of the organic matter is strongly affected by the $\delta^{34}S$ of atmospheric inputs. For example, organic matter in soils in New Zealand has $\delta^{34}S$ values approaching that of seawater (ca. +20‰), suggesting that the oceanic spray is the primary sulfur source (Kusakabe et al. 1976). Some soils in Alberta (Canada) and California (USA) with low $\delta^{34}S$ values (ca. $-30‰$) have been encountered and it has been suggested that the sulfur was derived from the weathering of sulfide minerals and/or organic sulfur in shales (Krouse et al. 1991).

Median $C:N$ ratios of organic matter in the top 15 cm of arable soils range from 10 to 12, with most ratios in the range of 8 to 25 (Brady 1990; Aitkenhead & McDowell 2000). The ratios are higher in humid than arid areas, and higher in colder than warmer areas (Brady 1990). Most soil microorganisms have ratios between 4 and 9 (Brady 1990; Rostad et al. 1997). The much lower $C:N$ values of soils compared with terrestrial plants reflect the cycling of plant material during decomposition (Brady 1990).

Aquatic plants

Macrophytes and algae have a wide range of $\delta^{13}C$ and $\delta^{15}N$ values (Figure 10.3); the same is true of $\delta^{34}S$ but much less data are available. These ranges are so large and have such extensive overlap with terrestrial plants and soils that one might wonder how contributions from different sources of organic matter can be distinguished. The ranges of autotroph $\delta^{13}C$ and $\delta^{15}N$ values from any specific river, however, are substantially smaller than the total literature range (Figure 10.4), often making quantitative separation with two or more tracers feasible. In addition, macrophytes (and bryophytes) have restricted distributions in river networks, and are usually quantitatively unimportant in fueling food webs because they are inedible for most herbivores.

The isotopic compositions of aquatic plants, and the organic matter derived from them, are more variable and less predictable than terrestrial plants and detritus (Figure 10.3), mainly because of (i) the large variation in the concentrations and stable isotope ratios of dissolved inorganic carbon (DIC), nitrogen (DIN), and sulfur (DIS) in freshwater systems, and (ii) the physiological diversity of aquatic autotrophs. The effects of some of the physical and biogeochemical processes that cause the large variations in the isotopic compositions of the dissolved inorganic species, and ultimately lead to variability in the isotopic compositions of aquatic plants, are illustrated in schematic form in Figure 10.5.

Because of their complexity, the *controls* on the carbon, nitrogen, and sulfur isotope biogeochemistry of aquatic plants are described separately below, in greater detail than for the other sources considered in this chapter.

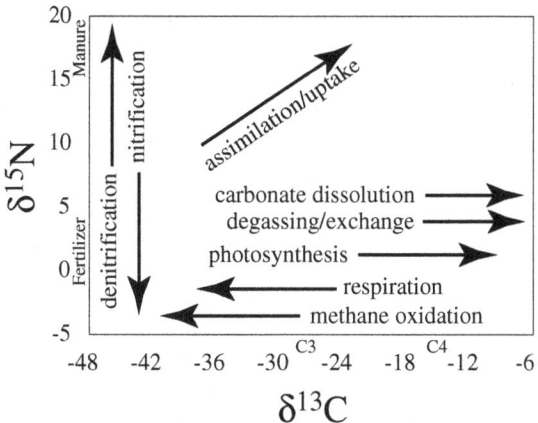

Figure 10.5 Conceptual model showing the main biogeochemical processes that control the $\delta^{13}C_{DIC}$ and the $\delta^{15}N_{NO_3}$, and consequently the $\delta^{13}C$ and $\delta^{15}N$ of aquatic plants and particulate organic matter (POM). The arrows indicate the usual effect of an increased amount of the specified process on the $\delta^{13}C_{DIC}$ and/or $\delta^{15}N_{NO_3}$, the $\delta^{13}C$ and/or $\delta^{15}N$ of the aquatic plants growing in the ecosystem, and ultimately the food webs based on these plants. For example, increased amounts of NO_3^- formed by nitrification of NH_4^+ probably causes decreases in $\delta^{15}N_{NO_3}$ (but usually minimal affect on $\delta^{13}C_{DIC}$), and assimilation causes significant increases in both $\delta^{13}C_{DIC}$ and $\delta^{15}N_{NO_3}$. The approximate $\delta^{13}C$ and $\delta^{15}N$ values of important C and N sources are also shown. (e.g., C3 plants and nitrate from manure, respectively.)

Furthermore, each section is subdivided into separate discussions of "source effects" and "fractionation effects." The section on source effects is intended to provide an overview of:

1 the biogeochemistry and isotopic compositions of the major dissolved inorganic species (e.g., HCO_3^-, NO_3^-, SO_4^{2+}) that affect the isotopic compositions of aquatic plants;
2 how different watershed and in-stream processes affect the isotopic compositions of the dissolved species;
3 the typical ranges of fractionation factors for dissolved species caused by these processes;
4 how these processes vary seasonally.

The section on fractionation effects focuses only on fractionations during assimilation (uptake), not on fractionations that only control the isotopic compositions of the dissolved species.

Controls on $\delta^{13}C$ of aquatic plants and DIC

Reported values for $\delta^{13}C$ of freshwater aquatic plants and algae range from about −47 to −8‰, with values typically falling in the range of −30 to −20‰ (LaZerte & Szalados 1982; Hamilton & Lewis 1992; Angradi 1993, 1994; Schlacher & Wooldridge 1996; Thorp et al. 1998; Cloern et al. 2002; Finlay 2004). Macrophytes have a more restricted $\delta^{13}C$ range, perhaps mostly because of their limited geographic distribution within drainage networks, i.e., lentic (still-water) habitats of larger rivers. Planktonic (free living, suspended) and periphytic (attached) algae are widely distributed in drainage networks, with periphyton dominating small streams and rivers, and planktonic forms more important in larger rivers. Where these forms co-occur, attached forms should have higher $\delta^{13}C$ than planktonic forms as discussed in detail below.

Source effects

In contrast to terrestrial plants, aquatic autotrophs derive inorganic carbon from DIC. Depending on the pH of the stream, the DIC may be comprised mostly of aqueous CO_2 (at pH < 6.4), HCO_3^- (pH 6.4–10.3), or CO_3^{-2} (pH > 10.3). The $\delta^{13}C$ of DIC has an observed range of ca. 30‰ due to variation in sources and subsequent physical (e.g. mixing) and biological (e.g. respiration, photosynthesis) processes, as discussed below. However, $\delta^{13}C_{DIC}$ values of −12 to −8‰ are most commonly observed in temperate rivers (Mook & Tan 1991; Kendall 1993; Bullen & Kendall 1998; Finlay 2003).

Variation in the sources and sinks for DIC control $\delta^{13}C_{DIC}$ in rivers. The main sources of DIC in fresh waters are atmospheric CO_2, carbonate rock dissolution, and respiration, and each has a distinct $\delta^{13}C$. The DIC produced by carbonic acid (i.e., H_2CO_3) dissolution of marine carbonates ($\delta^{13}C$ values = ca. 0‰) generally produces $\delta^{13}C$ values in the range of −15 to −5‰ depending mainly on whether the source of the C in the carbonic acid is C3 or C4 plants. Addition of respired CO_2, which has a $\delta^{13}C$ value similar to the organic carbon substrate (e.g., $-27‰$ for C3 plants) lowers $\delta^{13}C_{DIC}$; respiration has a large effect when DIC is low and less influence when DIC is high (Fry & Sherr, 1984). The main sinks affecting DIC in rivers are photosynthetic uptake by plants, degassing of CO_2 to the atmosphere, recharge to groundwater, and carbonate precipitation.

The $\delta^{13}C_{DIC}$ is affected by both in-stream and watershed-scale processes. The main watershed-scale processes that affect the $\delta^{13}C_{DIC}$ are:

1 dominance of C3 vs. C4 plants in the watershed;
2 presence of carbonate minerals in the bedrock and soil;
3 the relative proportions of groundwater vs. surface runoff contributing to streamflow.

As summarized in Figure 10.5, the main in-stream processes that affect $\delta^{13}C_{DIC}$ in rivers are:

1 degassing and CO_2 exchange with the atmosphere;
2 dissolution/precipitation of carbonate minerals in the stream;
3 discrimination during photosynthesis, which leaves the residual DIC pool enriched in ^{13}C;
4 respiration.

Other processes, including oxidation of methane produced in anoxic stream sediments, soils, or bedrock may also affect the $\delta^{13}C_{DIC}$. The relative strength of these processes affecting stream $\delta^{13}C_{DIC}$ varies spatially with stream size and productivity (Finlay 2003), and with geology and hydrology (Bullen & Kendall 1998). The acidity of the stream, which is a function of both in-stream and watershed-scale processes, also has a profound affect on $\delta^{13}C_{DIC}$ because this controls whether the dominant form of DIC is CO_2, HCO_3^-, or CO_3^{-2}.

Streams and rivers typically show large seasonal changes in $\delta^{13}C_{DIC}$, caused by both variation in contributions of groundwater vs. soil runoff to the river and changes in stream metabolism. For example, Figure 10.6 shows seasonal variation in discharge, alkalinity ($\approx$ DIC), and $\delta^{13}C_{DIC}$ in a small stream in Maryland, USA. Hydrology strongly affects the $\delta^{13}C_{DIC}$, with the oscillations in $\delta^{13}C$ during storm events reflecting shifts in the relative proportion of flow from soil water vs. groundwater flowpaths contributing to discharge. The broad seasonal changes in alkalinity (Figure 10.6c) reflect differences in the proportions of DIC derived from calcite and soil CO_2 (Figure 10.6d). Riverine CO_2 concentrations often decrease while the $\delta^{13}C_{DIC}$ increases during the summer because of photosynthesis, whereas CO_2 concentrations often increase and $\delta^{13}C_{DIC}$ decrease during the late fall as photosynthesis declines and in-stream decay and respiration increases (Kendall 1993; Atekwana & Krishnamurthy 1998; Finlay 2003). Episodic algal blooms in rivers can cause the $\delta^{13}C$ of the seston to oscillate as the DIC pool is drawn down by photosynthesis and then replenished by respiration and transport of DIC from upstream (Figure 10.7).

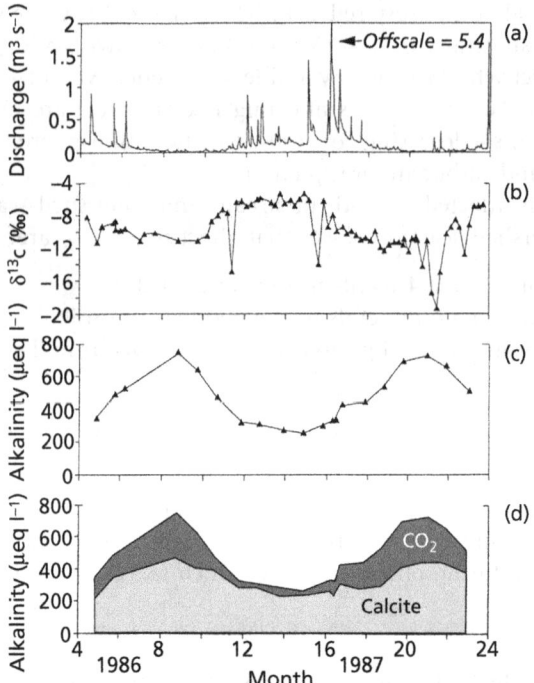

Figure 10.6 Seasonal changes at Hunting Creek in the Catoctin Mountains, MD, 1986–87. (a) Discharge. (b) $\delta^{13}C$ of stream dissolved inorganic carbon (DIC) collected weekly. (c) Alkalinity. (d) Estimation of the relative contributions of carbon from CO_2 and calcite to stream alkalinity using $\delta^{13}C$ of calcite = −5‰ and $\delta^{13}C$ of CO_2 = −21‰; shaded areas show the relative proportions of carbon sources. (Modified from Kendall 1993.)

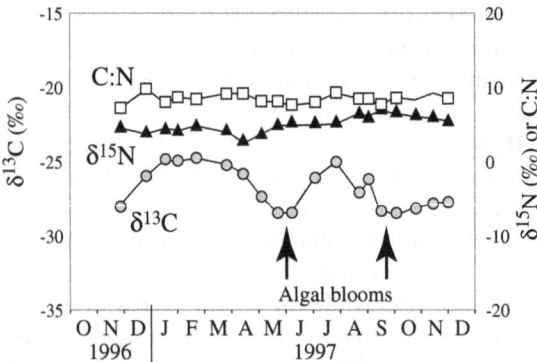

Figure 10.7 Temporal changes in the $\delta^{13}C$, $\delta^{15}N$, and C:N (at.) of particulate organic matter (POM) in the Yazoo River (USA) due to successive algal blooms. The average C:N is 8.5, indicating that the POM is largely algal; the almost constant C:N shows that the proportion of terrestrial vs. algal POM shows little seasonal variability. The oscillations in $\delta^{13}C$ while the $\delta^{15}N$ remains almost constant reflect the relative effects of the small dissolved inorganic carbon (DIC) and large NO_3^- pools on the fractionations caused by assimilation. (Modified from Kendall et al. 2001.)

An additional important source of variation in algal $\delta^{13}C$ is that aquatic plants commonly use two species of DIC (i.e. dissolved CO_2 and HCO_3^-). The $\delta^{13}C$ of these species are controlled by temperature-dependent equilibrium reactions, with the $\delta^{13}C$ of aqueous CO_2 being 6–11‰ lower than the $\delta^{13}C$ of HCO_3^- for temperatures of 35–0°C, respectively (Mook et al. 1974). Although rigorously studied for only a few taxa, many lotic (flowing water) autotrophs, including diatoms and green algae, apparently switch from CO_2 to HCO_3^- when CO_2 supply is insufficient (Raven & Beardall 1981). Some exceptions include cyanobacteria that have specialized inorganic carbon pumps to concentrate DIC, resulting in very little fractionation (Goericke et al. 1994), and red algae and bryophytes that are obligate CO_2 users (Raven & Beardall 1981; Glime & Vitt 1984). Thus, $\delta^{13}C_{DIC}$ and differences in the form of DIC used by autotrophs can contribute considerably to variation in the $\delta^{13}C$ of autotrophs in freshwaters.

Fractionation effects in aquatic plants
Fractionation during carbon uptake and assimilation also varies more widely in aquatic compared with terrestrial plants (Keeley & Sandquist 1992), contributing to the wide range of observed algal $\delta^{13}C$ in streams and lakes. The isotopic fractionations during photosynthesis are dependent on several factors, including aqueous CO_2 concentration and diffusion rate, and plant growth rate (Fry & Sherr 1984; Fogel & Cifuentes 1993; Hecky & Hesslein 1995; Laws et al. 1995). In general, when the availablility of dissolved CO_2 is high and/or when the growth rate is low (i.e., little CO_2 is required), the fractionation between the $\delta^{13}C_{DIC}$ and plants is greatest. However, when there is a scarcity of CO_2 and/or when growth rates are high, plants discriminate less against $^{13}CO_2$ and, as described above, may shift to use of HCO_3^- (Fry & Sherr 1984; Mariotti et al. 1984; Fogel & Cifuentes 1993; Laws et al. 1995; Barth et al. 1998). This mechanism may explain the typical increase in algal $\delta^{13}C$ during spring and summer when primary production is highest and the concentration of CO_2 is lowest because of uptake. An additional mechanism affecting oscillations in $\delta^{13}C$ of algae during successive algal blooms is that as photosynthesis consumes more and more of the DIC pool, the residual DIC becomes progressively ^{13}C-enriched due to fractionation, resulting in a corresponding increase in the $\delta^{13}C$ of algae (Figure 10.7).

Although there is a large range of growth rates for algae in rivers due to variable light and nutrient limitation, much more variation in ε appears to be related to effects of CO_2 supply (Finlay 2004). CO_2 availability is strongly influenced by the physical constraints of gas diffusion in aquatic ecosystems. CO_2 diffuses four times more slowly through water than air, and this has two important consequences for DIC use and isotope fractionation by freshwater autotrophs. First, the thickness of the diffusive (boundary) layer around aquatic plants varies according to water velocity or turbulence. Thick boundary layers lead to a reduced "apparent fractionation" because $^{13}CO_2$ that is discriminated against is effectively trapped within the boundary layer and partially assimilated before it can diffuse away, resulting in algae with higher $\delta^{13}C$. While the actual kinetic isotope fractionation factor for assimilation may be unaffected by diffusion, comparison of the $\delta^{13}C$ of the algae and $\delta^{13}C_{DIC}$ of water (beyond the boundary layer) shows a smaller difference in their $\delta^{13}C$ values than when the boundary layer is thinner. This effect has been consistently demonstrated in lakes and open canopied streams with low CO_2 concentrations (e.g., Hecky & Hesslein 1995; Finlay et al. 1999).

A second consequence of slow CO_2 diffusion is that the dissolved CO_2 pool (and consequently the entire DIC pool) equilibrates very slowly with the atmosphere, resulting in large deviations from atmospheric concentrations when rates of photosynthesis and respiration are not balanced, or when groundwater inputs of DIC are high. CO_2 concentrations in rivers thus range from levels well below atmospheric saturation to as much as 20 fold higher or more (Duarte & Agusti 1998; Finlay 2003). Therefore, in contrast to terrestrial plants, isotopic fractionation of DIC by algae often is influenced by multiple factors in aquatic environments.

Controls on $\delta^{15}N$ of aquatic plants and DIN

Values for $\delta^{15}N$ of freshwater aquatic plants range from about -15 to $+20‰$ (Hamilton & Lewis, 1992; Angradi 1993, 1994; Thorp et al. 1998; Cloern et al. 2002; and unpublished U.S. Geological Survey data from the Everglades, Florida). The extreme $\delta^{15}N$ values are often associated with human disturbances of various kinds, with $\delta^{15}N$ values in the range of -1 to $+7‰$ perhaps typical of undisturbed riverine and marsh ecosystems. In a study of >1000 aquatic plants in the Everglades, algae (periphyton and epiphyton) had the same average $\delta^{15}N$ as macrophytes ($+2‰$), and both had $\delta^{15}N$ values as high as $+15‰$; however, macrophytes (e.g. lilypads) had values as low as $-13‰$ whereas the lowest algae $\delta^{15}N$ values were $-7‰$ (unpublished U.S. Geological Survey data). The same principles of source and fractionation effects described above for $\delta^{13}C$ also apply to $\delta^{15}N$. The $\delta^{15}N$ of aquatic plants is controlled by the type of DIN utilized, its $\delta^{15}N$ value, and the fractionations associated with discrimination against ^{15}N during uptake of N that may vary by plant species and environmental conditions. As for $\delta^{13}C$, fractionations between the $\delta^{15}N$ of DIN and the plant are greatest when the pool sizes are large and/or growth rates low, and smallest when DIN is scarce and/or growth rates are high.

Controls of $\delta^{15}N$ of basal resources in food webs have been less well studied than for $\delta^{13}C$ in aquatic ecosystems, in part because of the complexity of the nitrogen cycle, and in part because of the analytical challenges in measuring the $\delta^{15}N_{DIN}$ and of organic matter compounds with low %N. Recent improvements in natural abundance methods for analyzing NH_4^+ and NO_3^- are rapidly increasing the available data (Sigman et al. 1997, 2001; Holmes et al. 1998; Sebilo et al. 2004), but there are still few studies of the effect of different DIN species on the $\delta^{15}N$ of plants and detritus in rivers.

Source effects

The main N sources for non-fixing aquatic plants are NH_4^+ and NO_3^- (i.e. DIN). The $\delta^{15}N_{DIN}$ range widely in the environment (see Kendall et al., this volume, pp. 375–449), from typical low values (−10 to +5‰) for N derived from N fixation or fossil fuel combustion to over +30‰ for N derived from animal waste or that has undergone intensive denitrification (Heaton 1986). Variation in the sources and sinks for DIN control $\delta^{15}N_{DIN}$ in rivers. The main sources of DIN in fresh waters are atmospheric deposition, mineralization of organic matter (from soil, terrestrial plant detritus, sediments, and N-fixing aquatic plants), and fertilizer and animal waste. The main sinks are assimilation, degassing of N_2 and other gases to the atmosphere via denitrification and nitrification pathways, and recharge to groundwater.

Like $\delta^{13}C_{DIC}$, $\delta^{15}N_{DIN}$ are affected by both in-stream and watershed-scale processes. In addition, the $\delta^{15}N_{DIN}$ and the $\delta^{15}N$ of plants are both strongly affected by the form of the DIN (i.e. NH_4^+ and NO_3^-). The first control is briefly described here, and explored much more thoroughly in Kendall et al., this volume, pp. 375–449. The main watershed-scale influences on $\delta^{15}N_{DIN}$ are: (i) land use and extent of human disturbance in the watershed and airshed; (ii) the relative proportions of groundwater vs. surface runoff contributing to streamflow; (iii) the prevalence of N-fixing plants (such as alders and legumes) in the watershed; (iv) the presence of reducing conditions in groundwater, the riparian and hyporheic zones, and sediments; and (v) climate (e.g., temperature, rain amount).

The main processes that affect $\delta^{15}N_{DIN}$ in rivers are (i) assimilation, (ii) nitrification, (iii) denitrification, and (iv) mineralization. Redox conditions in the stream, a function of both in-stream and watershed-scale processes, also have a profound affect on $\delta^{15}N_{DIN}$ because it controls whether the dominant form of DIN is NO_3^-, NO_2^-, or NH_4^+. All of these processes influence $\delta^{15}N_{DIN}$ because the residual reactants in the water column, sediments, or soils become enriched in ^{15}N (Figure 10.5). The fractionations associated with microbial N transformation that control, in part, $\delta^{15}N_{DIN}$ are discussed here, while plant fractionation of $\delta^{15}N_{DIN}$ during N uptake is discussed in the following section.

The $\delta^{15}N_{DIN}$ is strongly affected by watershed-scale processes and cycling of N in streams. Mineralization usually causes only a small fractionation (ca. 1‰) between organic matter and ammonium. The extent of fractionation during nitrification is dependent on the size of the substrate pool (reservoir). In N-limited systems, the fractionation associated with nitrification is usually minimal and depends on which step is rate determining. Because the oxidation of nitrite to nitrate is generally rapid in natural systems, this is often not the rate determining step, and most of the N fractionation is probably caused by the slow oxidation of ammonium. Uptake of NO_3^- by algae may also be a major control on in-stream variation in $\delta^{15}N_{NO_3}$, particularly in eutrophic, nitrate-rich rivers.

In pristine basins, DIN from watersheds is derived mainly through mineralization of organic matter and subsequent nitrification in soils and groundwater; nitrate is the main form of DIN transported to streams. In such streams, transport of dissolved organic N (DON) may equal or exceed DIN inputs to streams; much less is known about $\delta^{15}N_{DON}$. The $\delta^{15}N$ of NO_3^- leaving terrestrial ecosystems is often low in such settings (Kendall et al., this volume, pp. 375–449). However, given that it is estimated that half the nitrogen added to the biosphere currently is from anthropogenic sources (Vitousek et al. 1997), it is likely that much of the N supplied by watersheds to river systems is from agricultural fertilizer applications, animal feed lots, atmospheric deposition, and sewage. In many situations, nitrate from fertilizer can be distinguished from nitrate from animal waste because most synthetic fertilizers are generated from air nitrogen and therefore possess $\delta^{15}N$ values within a few per mil of zero, whereas nitrate derived from waste usually has $\delta^{15}N$ values between +10‰ and +20‰ (Heaton 1986; Kendall 1998; Kendall et al., this volume, pp. 375–449).

Biologically mediated reduction of nitrate to N_2 and other gases (i.e. denitrification) in low-O_2 groundwater, soils, and sediments can have a large influence on $\delta^{15}N_{DIN}$ moving from land to water, described in greater detail in Kendall et al., this volume, pp. 375–449. Briefly, denitrification in groundwater and riparian zones causes the $\delta^{15}N$ of the residual nitrate to increase exponentially as nitrate concentrations decrease. For example, denitrification of fertilizer nitrate ($\delta^{15}N$ of +0‰) can yield residual nitrate with much higher $\delta^{15}N$ values (e.g., +15 to +30‰) that are within the range of compositions expected for nitrate from manure or septic-tank sources. Measured enrichment factors (apparent fractionations) associated with denitrification range from −40 to −5‰ (Kendall 1998).

Denitrification in the water column in rivers is rare but common in benthic sediments. Benthic denitrification has been shown to cause small fractionations (ranging from −1.5 to −3.6‰) in the $\delta^{15}N_{NO_3}$ in the overlying waters. These fractionations are much smaller than expected for groundwater denitrification because nitrate diffusion through the water–sediment interface, which causes minimal fractionation, is the rate determining step (Sebilo et al. 2003; Lehman et al. 2004). The isotopic fractionation caused by diffusive transport of DIN may influence algal $\delta^{15}N$, and small effects have been observed under experimental conditions (Macleod & Barton 1998; Trudeau & Rasmussen 2003).

The final source effects we consider are isotopic differences between the two main forms of DIN available to aquatic plants, NH_4^+ and NO_3^-. NH_4^+ is usually the more preferred form of DIN, so isotopic differences between these forms or their relative availability may have a large influence on plant $\delta^{15}N$. Unlike DIC, the speciation of NH_4^+ and NO_3^- is not controlled by equilibrium chemical reactions; there is no a priori reason for their $\delta^{15}N$ values to be similar or related. Instead, the relative proportion of these species is strongly affected by redox conditions, with NO_3^- the more abundant species in well-oxygenated waters and NH_4^+ and NO_2^- more abundant in low-O_2 waters. The concentration of NO_2^- can be more than 10 times that of NO_3^- in highly eutrophic waters (e.g., organic-rich sediments or animal waste lagoons). Many studies have observed large and variable differences in the $\delta^{15}N$ values of NH_4^+ and NO_3^-, particularly in disturbed or eutrophic ecosystems (Kendall 1998); rapid and complete cycling of N may result in relatively consistent steady-state $\delta^{15}N$ values among the various N pools.

Fractionation effects in aquatic plants

As for carbon, aquatic plant fractionation of $\delta^{15}N$ is also influenced by the supply relative to demand for DIN, but far less is known about the fractionations for freshwater algae than for terrestrial or marine plants. Under nitrogen-limited conditions, uptake of DIN by terrestrial and aquatic plants causes little or no fractionation of $\delta^{15}N_{DIN}$ but this situation changes rapidly in the presence of excess N (Fogel & Cifuentes 1993; Pennock et al 1996; Casciotti et al. 2002; Granger et al. 2004; Needoba et al. 2004). A conceptual model developed in the Great Lakes (USA) describes that the $\delta^{15}N$ of seston is controlled by the balance between NH_4^+ uptake and degradative processes that increase the $\delta^{15}N$, and NO_3^- uptake that decreases the $\delta^{15}N$ of seston (McCusker et al. 1999).

The $\delta^{15}N$ of algae in rivers in the USA is generally about 4–5‰ lower than the $\delta^{15}N$ of the associated NO_3^- (Battaglin et al. 2001a, 2001b; Kratzer et al. 2004). Data from a longitudinal sampling of the San Joaquin River (USA) illustrates this pattern (Figure 10.8). Seston samples collected in the riverine part of this transect were > 90% algae. In this section, NH_4^+ concentrations were too low to analyze for $\delta^{15}N$, and NO_3^- is believed to be the main source of N for algal growth. The downstream increase in NO_3^- concentration, $\delta^{15}N_{NO3}$, and $\delta^{15}N_{POM}$ in the riverine section are thought to reflect increases in groundwater (with higher $\delta^{15}N$ values suggestive of animal waste) downstream (Kratzer et al. 2004).

Controls on $\delta^{34}S$ of aquatic plants and DIS

As described for terrestrial plants, $\delta^{34}S_{DIS}$ may be influenced by atmospheric or local mineral sources, and the isotopic fractionations associated with S cycling that alter the $\delta^{34}S$ of sulfate or sulfide. The $\delta^{34}S$ of aquatic plants are controlled primarily by the $\delta^{34}S$ of DIS, and are little affected by plant fractionation. A survey of the $\delta^{34}S$ of aquatic plants from the Everglades

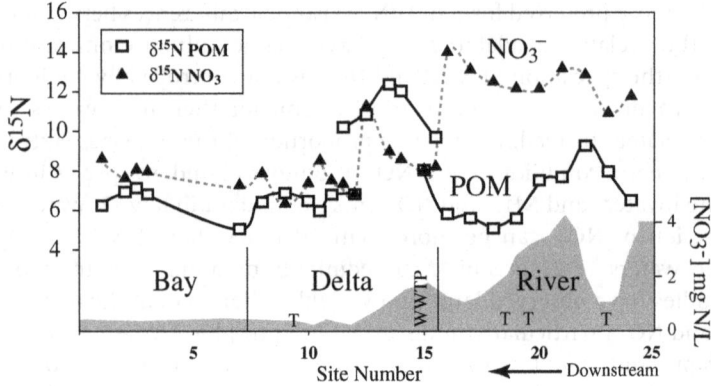

Figure 10.8 Spatial changes in nitrate concentrations, nitrate $\delta^{15}N$, and suspended particulate organic matter (POM; which is predominantly algal) $\delta^{15}N$ in the San Joaquin River, CA, due to downstream changes in inputs of NO_3^- from groundwater, waste-water treatment plants (WWTPs), and other inputs. The river, delta, and bay reaches, as well as the locations of major tributaries (T) and WWTPs are marked. The transect spans a distance of >100 km, from the confluence with Mud Slough (site 25) to the Golden Gate Bridge (site 1). The drop in NO_3^- at site 23 and the slightly downstream peaks in $\delta^{15}N$ are a result of a large algal bloom at the confluence with the Merced River; the $\delta^{13}C$ values of POM also oscillate near the confluence owing to algal blooms due to mixing of waters. (C. Kendall, unpublished U.S. Geological Survey data.)

found values of +5 to +33‰ (Kendall, unpublished U.S. Geological Survey data), similar to the range of $\delta^{34}S_{SO4}$ observed in another study in the Everglades (Bates et al. 2002). The highest plant $\delta^{34}S$ values were found in anoxic areas affected by sulfate reduction (Kendall et al. 2000). More commonly, anoxic areas in streams and bogs result in organic matter with lower than normal $\delta^{34}S$ (Trust & Fry 1992). For example, mangroves can have $\delta^{34}S$ values as low as −20‰ (Fry et al. 1982).

Aquatic plants are less commonly analyzed for $\delta^{34}S$ than for either $\delta^{13}C$ or $\delta^{15}N$ because the $\delta^{34}S$ analytics are more complicated, many aquatic plants and seston have low %S, and also much less is understood about sources of S variability in ecosystems. However, part of the explanation for the scarcity of $\delta^{34}S$ data for organics is the fear (mostly unfounded) that analysis will cause contamination of the mass spectrometer. Recent improvements in natural abundance methods for analyzing organic samples are rapidly increasing the available data (e.g., Fry et al. 2002; Yun et al. 2005), but there are still very few studies of the effect of different DIS species on the $\delta^{34}S$ of plants and detritus in rivers. Despite these problems, $\delta^{34}S$ of aquatic plants has been used when a dual (C, N) isotope approach has proved unsuccessful (Peterson et al. 1985; Loneragan et al. 1997; Connolly et al. 2003), especially in marine or coastal environments where plants are likely to be isotopically labeled with the high $\delta^{34}S$ of marine sulfate (about +20‰).

Source effects

A literature review of $\delta^{34}S$ values in aquatic ecosystems found that the $\delta^{34}S_{SO_4}$ in lakes ranged from +3.5 to +87‰, with the low values found in deep waters, and the $\delta^{34}S_{H_2S}$ ranged from −32 to −11‰ (Nriagu et al. 1991). The $\delta^{34}S_{SO_4}$ in rivers also show a wide range (−20 to >+30‰) with mean values for rivers in North America, Europe, and globally of +4, +6, and +7‰, respectively (Nriagu et al. 1991). Marine sulfate has a $\delta^{34}S$ of ca. +20‰, and atmospheric $\delta^{34}S_{SO_4}$ values are typically in the range of −5 to +25‰ (Wadleigh et al. 1996; Krouse & Mayer 2000).

Like $\delta^{13}C_{DIC}$ and $\delta^{15}N_{DIN,}$ the $\delta^{34}S_{DIS}$ is affected by both in-stream and watershed processes. In addition, the $\delta^{34}S_{DIS}$ and the $\delta^{34}S$ of plants are both strongly affected by the form of the DIS (i.e. SO_4^{2-} or H_2S). The main watershed characteristics that affect $\delta^{34}S_{DIS}$ are: (i) land use and extent of human disturbance in the watershed and airshed; (ii) the presence of sulfate on soil exchange sites or sulfide minerals in soils, sediments, or bedrock; (iii) the relative proportions of groundwater vs. surface runoff contributing to stream-flow; and (iv) the presence of reducing conditions in groundwater, the riparian and hyporheic zones, and sediments.

The main processes that affect $\delta^{34}S_{DIS}$ in rivers are (i) sulfate reduction and (ii) sulfide oxidation. Redox and pH conditions in the stream, which are a product of both in-stream and watershed-scale processes, also have a profound affect on $\delta^{34}S_{DIS}$ because this controls whether the dominant form of DIS is SO_4^{2-} or H_2S. For good discussions of watershed and biogeochemical controls on the $\delta^{34}S$ of sulfate and sulfides, see Mitchell et al. (1998) and Trust & Fry (1992), respectively.

As described for terrestrial plants, $\delta^{34}S_{DIS}$ may be influenced by the isotopic fractionations associated with S cycling that alter the $\delta^{34}S$ of sulfate or sulfide. Bacterial reduction of sulfate is the primary source of the variability of $\delta^{34}S$ observed in freshwater systems because sulfide oxidation involves minimal fractionation of S. During sulfate reduction, the bacteria produce H_2S gas that has a $\delta^{34}S$ value ca. 25‰ lower than the sulfate source (Clark & Fritz 1997). Consequently, the residual pool of sulfate becomes progressively enriched in ^{34}S. Thus, long-term anaerobic conditions ought to be reflected in the $\delta^{34}S$ values of sulfate, and consequently $\delta^{34}S$ of plants, that are significantly higher than that of the original sulfate source.

Sulfate-reducing bacteria use dissolved sulfate as an electron acceptor during the oxidation of organic matter. Since this process is accompanied by oxidation of organic matter, there is a corresponding shift in $\delta^{13}C_{DIC}$ toward that of the organic source due to the production of CO_2. Under conditions of low concentrations of reactive organic matter, re-oxidation of mineral sulfides can lead to constant recycling of the dissolved sulfur pool. In such cases, increases in $\delta^{34}S$ do not occur in conjunction with reaction progress since the sulfide concentration in solution is held approximately constant (Fry 1986; Spence et al. 2001).

Fractionation effects in aquatic plants
Sulfate is the dominant DIS species used by plants, especially in environments where sulfide concentrations are low. However, in environments where sulfide concentrations are significant, H_2S inhibits the uptake of sulfate (Brunold & Erismann 1975). In most cases, assimilatory sulfate reduction does not lead to significant fractionations; for aquatic plants, typical fractionations range from 0 to 3‰ (Nriagu et al. 1991). The unusually low $\delta^{34}S$ values of mangroves have been attributed to assimilation of H_2S by plant roots in the mud (Fry et al. 1982).

Microbial heterotrophs

The last organic matter "source" for rivers we consider, heterotrophic bacteria and fungi, are potential consumers of the four other types of organic matter that we have considered (Figure 10.1). However, because of their ubiquity and potential to modify isotope ratios of other sources, we consider them separately from other consumers in food webs.

The $\delta^{13}C$ of heterotrophic bacteria growing on organic carbon substrates are typically the same or slightly higher than the $\delta^{13}C$ of the organic matter used (Coffin et al. 1989). However, $\delta^{15}N$ relationships are considerably more complicated. Since N availability often limits terrestrial plant growth, terrestrial detritus inputs to freshwaters initially have high C:N ratios (Sterner & Elser 2002), with insufficient N to meet microbial growth demands. Subsequent microbial immobilization of DIN by bacteria can cause significant changes in the $\delta^{15}N$ of bacteria, as well any organic particles they are adhered to, if the $\delta^{15}N$ of DIN used by bacteria are different from the $\delta^{15}N$ of the original organic matter (Caraco et al. 1998). Such effects on $\delta^{15}N$ would be minimal where microbial incorporation of N does not occur (e.g. low C:N detritus). However, terrestrial plant litter has high C:N and frequently has lower $\delta^{15}N$ than aquatic DIN, leading to conditions that favor changes in bulk $\delta^{15}N$ during decomposition of terrestrial organic matter in aquatic ecosystems.

To illustrate this point, we compare $\delta^{15}N$ of oak leaves decomposing in a large eutrophic river (Hudson River, NY) and a small pristine stream (Fox Creek, CA; Figure 10.9). In Fox Creek, $\delta^{15}N_{DIN}$ is apparently similar to that of leaves, so that minor temporal changes are observed during decomposition. In contrast, $\delta^{15}N_{DIN}$ in the nitrate-rich Hudson is much higher than $\delta^{15}N$ of riparian plants, leading to rapid increases in the $\delta^{15}N$ of decomposing leaves.

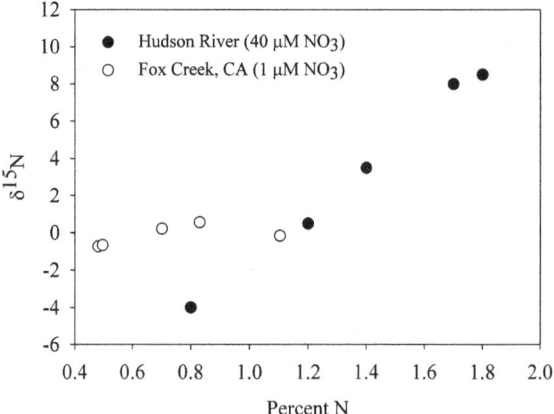

Figure 10.9 Change in δ¹⁵N in terrestrial leaf litter (*Quercus* spp. for both sites) in a pristine stream (Fox Creek, CA; Finlay, unpublished data) and eutrophic river (Hudson River, NY; Caraco et al. 1998) during *in situ* decomposition and microbial incorporation of external N from river water.

There is little known about the $\delta^{15}N$ and $\delta^{34}S$ of bacteria in natural settings, mainly because of the difficulty of isolating pure samples. A recent study of different fractions of organic matter in the Ohio River (USA) showed that the colloidal ($<0.2\,\mu m$) fraction, which contains a large proportion of heterotrophic bacteria, had $\delta^{13}C$ and $\delta^{15}N$ values of about $-30‰$ ($\pm2‰$) and ca. $0‰$ ($\pm2‰$), respectively. These values were ca. $2‰$ lower than the values for terrestrial OM and 7–$12‰$ lower than the $\delta^{15}N$ of both algal and detrital OM (Delong & Thorp 2006). Low $\delta^{15}N$ values have previously been reported for bacteria. For example, Wissel & Fry (2005) suggested that the low $\delta^{15}N$ of seston (to $-6‰$) collected in the winter from river sites near the mouth of the Mississippi River (USA) were due to large inputs of bacteria. This explanation was consistent with the low C:N (ca. 5.5) and the low chlorophyll levels. This might be the explanation for $\delta^{15}N$ values seen at other large river sites in the USA, where changes in seston $\delta^{15}N$ values as low as $-16‰$ have been observed (Kendall et al. 2001; unpublished U.S. Geological Survey data for these same sites). Bacteria growing on organic-rich bottom sediments may be affected by the low $\delta^{34}S$ of H_2S produced by sulfate reduction in such reducing environments.

C:N ratios

Data for C:N (at.) ratios of freshwater aquatic vegetation exclusive of phytoplankton are sparse. The reported range of C:N ratio for freshwater phytoplankton is about 5–8, averaging close to the Redfield ratio of 6.6 for marine phytoplankton (Redfield 1958; LaZerte 1983; Harris 1986). Thorp et al. (1998) reported C:N values ranging from about 8 to 10 for benthic algae and from about 11 to 12 for aquatic macrophytes in the Ohio River (USA). Hamilton & Lewis (1992) report C:N ratios of different size-fractions of algal-derived seston from the Orinoco River floodplain (Venezuela) between about 6 and 9. The higher and more variable C:N ratios of lacustrine plankton compared with marine plankton appear to be caused by variations in N and P nutrient limitations in the lakes (Hecky et al. 1993). Riverine microorganisms from the Mississippi River (USA) have C:N values in the range of 5–15 (Rostad et al. 1997). Seston and algae samples collected from some 70 small-river sites in the upper Mississippi River basin during a dry period in August 1998 (unpublished U.S. Geological Survey data, C. Kendall) showed the following average C:N values: seston was 9.4 (±1.8, $n = 68$), handpicked *Spirogyra* and *Cladophora* was 11.7 (±2.3, $n = 68$), and periphyton on woody snags was 14.6 (±3.5, $n = 42$). Various submerged macrophytes ($n = 490$) and periphyton ($n = 640$) from the Everglades (USA) have C:N ratios with median values of 22 and 16, respectively; C:N values for old, woody, and/or partially degraded Everglades aquatics can have C:N values > 80 (unpublished U.S. Geological Survey data, C. Kendall). A study of ca. 900 plants in the fresh, brackish, and salt water in the San Francisco Bay ecosystem showed C:N (wt.) ratios of 10 to >100 (Cloern et al. 2002). In summary, there can be a wide range in C:N ratio for various types of aquatic plants (Figure 10.3), but generally plankton has C:N values (5–8) that are lower than periphyton and macrophytes (10–30), and the C:N values of terrestrial and aquatic plants show considerable overlap.

C, N, and S isotopic variability and its applications in river ecology

Variation in stable isotope ratios among organic materials in rivers provides the basis of many important applications in rivers (Table 10.1). As we have seen, stable isotope ratios are highly varied in rivers and often influenced by multiple biogeochemical, physical, and physiological processes throughout watersheds. When well understood and predictable, this variation is useful; however, isotope tracers may be much less efficiently used if the signals are unexpectedly variable, or if source separation is ultimately not possible. Environmental conditions and resource type and abundance show substantial seasonal and spatial variability. These changes directly influence $\delta^{13}C$, $\delta^{15}N$, and $\delta^{34}S$ in resources and consumers, and thus determine when, where, and how stable isotopes may be applied in food web studies. Below we identify the major spatial and temporal patterns in stable isotope ratios in river food webs, identify predictive relationships between controlling variables, and show how and where this variation can serve as a useful tracer for the study of river ecology.

Spatial patterns and applications of stable isotope tracers

Recent studies in watersheds ranging from meters to basin scales have shown three types of strong spatial patterns in the isotopic compositions of organic matter sources that are helpful in the planning and design of stable isotope studies: longitudinal, water velocity gradients, and anthropogenic. These patterns reflect spatial variability in the $\delta^{13}C$, $\delta^{15}N$, and $\delta^{34}S$ of dissolved inorganic species, which is integrated into the isotopic compositions of organisms living in the river.

Longitudinal effects on $\delta^{13}C$

The percent of the organic C that is derived from allochthonous (terrestrial) sources vs. autotrophic (in stream) sources varies systematically from headwaters to mid-size streams to major rivers (Figure 10.2; Vannote et al. 1980). $\delta^{13}C$ can help determine the origin of organic matter in river networks and its contribution to production in food webs, but only when terrestrial detritus and autothonous OM have distinct $\delta^{13}C$ values. The terrestrial detritus that often dominates organic matter pools in rivers has a well-constrained mean $\delta^{13}C$ value of $-28.2 \pm 0.2‰$ ($\pm$ standard error; C3 plants only), with no consistent trend with stream size (Figure 10.10; Finlay 2001) because the aquatic detrital pool integrates terrestrial plant $\delta^{13}C$ through time and space. In contrast, many rivers show strong longitudinal (i.e., downstream) changes in the $\delta^{13}C$ of DIC, algae, seston, and consumers. For example, in small temperate watersheds, algal $\delta^{13}C$ are highly varied, from as low as $-47‰$ in headwater streams and springs to values as high as $-8‰$ in moderate-size rivers (Figure 10.10).

Longitudinal changes in algal $\delta^{13}C$ are due to combined effects of physical and biogeochemical processes in streams and their watersheds. Examples of physical processes that can cause longitudinal changes in $\delta^{13}C_{DIC}$ and consequently in $\delta^{13}C$ of algae include variable percentages of water in the stream derived from groundwater vs. soil water with watershed scale (Bullen & Kendall 1998), and degassing of CO_2 from springs and groundwater that cause a gradual increase in $\delta^{13}C_{DIC}$ (Kendall & Doctor 2004; Finlay 2004; Doctor et al. in press). The main biological processes causing longitudinal changes in $\delta^{13}C_{DIC}$ and algal $\delta^{13}C$ are the balance of total ecosystem respiration and photosynthesis (Figure 10.5). These physical and biological processes may be occurring simultaneously along the stream reach.

In temperate forested rivers, strong longitudinal patterns in algal $\delta^{13}C$ are caused by downstream changes in $\delta^{13}C_{DIC}$ and algal fractionation, which appear to be strongly linked to stream CO_2 (Finlay 2004). In a study within a

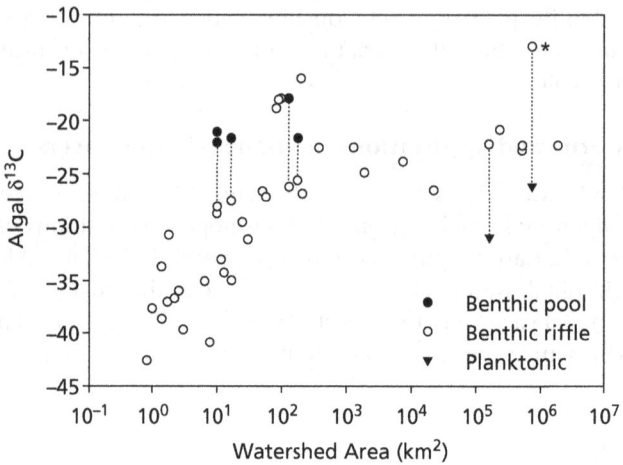

Figure 10.10 Spatial variation in $\delta^{13}C$ of autotrophic organic carbon sources to temperate river and riparian food webs during summer. $\delta^{13}C$ data representing the most edible algal form (usually epilithic diatoms or periphyton) within a site were used for the plot. Considerable variation exists between autotrophs within many of these sites, but most of these forms (bryophytes, rhodophyta, and filamentous chlorophytes) are inedible or edible only by specialist grazers. In some cases, data for herbivores with primarily algal diets were used instead of direct measurements. Data for other biomes (e.g. alpine, arctic, desert) were not included. Dotted lines indicate data for pool and riffle or benthic and planktonic algae pairs within a location. The * indicates data from the same river (St Lawrence, Canada) but different sampling sites and dates. (Data are updated and expanded from Finlay (2001) to include additional sources: Delong & Thorp (2006), Barnard et al. (2006), Finlay (2004), Huryn et al. (2002), deBruyn & Rasmussen (2002), McCutchan & Lewis (2002), Delong et al. (2001), and Thorp et al. (1998).)

single watershed, Finlay (2004) found that 90% of variation in benthic algal $\delta^{13}C$ in riffles was explained by CO_2 concentration since it directly affects fractionation. However, algal $\delta^{13}C$ also showed correlations with $\delta^{13}C_{DIC}$ and algal growth rates. Thus, measurements of CO_2 or its correlates (e.g. pH, O_2) may provide a predictive tool for identification of useful spatial variation in $\delta^{13}C$.

Longitudinal patterns in CO_2 concentrations in river networks appear to control $\delta^{13}C$ differences between algae and terrestrial OM. In the headwater streams, CO_2 is high due to inputs from soil and groundwater, and $\delta^{13}C_{DIC}$ and algal growth rates are low. These conditions result in algal $\delta^{13}C$ values that are lower than the $\delta^{13}C$ of terrestrial detritus. With increasing stream size, algal $\delta^{13}C$ increase because of the combined effects of several processes, including lower fractionation due to low CO_2 and high algal growth rates and biomass level, and higher $\delta^{13}C_{DIC}$ due to uptake of DIC and perhaps exchange with atmospheric CO_2 and degassing. In large rivers, algal $\delta^{13}C$ might then decrease if the rivers become strongly heterotrophic, as observed in the Amazon River (Araujo-Lima et al. 1986). However, some recent publications have found that a large percent of the seston in large rivers in the USA is algal derived (e.g., Kendall et al. 2001; Wissel & Fry 2005; Delong & Thorp 2006).

Algal $\delta^{13}C$ are thus most different from those of terrestrial detritus in headwater streams, lentic habitats of small rivers, and some large, heterotrophic rivers (Figure 10.10; Araujo-Lima et al. 1986; Finlay 2004). Overlap between algal and terrestrial $\delta^{13}C$ is most often observed in fast flowing, well-mixed riffles of small rivers (Figure 10.10). It is important to note, however, that substantial variation may exist for a given stream size. In addition, considerable $\delta^{13}C$ variation exists between autotroph taxa; if diverse autotroph growth forms are present, and if specialist herbivores that consume macroalgae, bryophytes, or macrophytes are important to the food web under consideration, these sources must be accounted for.

Longitudinal gradients in $\delta^{13}C_{DIC}$ and consequently algal $\delta^{13}C$ can also be caused by physical, biogeochemical, or anthropogenic factors that create variations in chemical conditions over short spatial scales. Some examples may include springs or groundwater upwelling zones (e.g., Rounick & James 1984; Hoffer-French & Herman 1989; Pentecost 1995; Finlay 2003, 2004; Kennedy et al. 2005); and junctions with other rivers, lakes, and impoundments that contribute nutrients and cause algal blooms. For example, Angradi (1993) interpreted variations in $\delta^{13}C$ and $\delta^{15}N$ of seston collected at intervals downstream from a dam on a Rocky Mountain (USA) river to reflect changes in the proportions of local source (e.g. riverine) materials that were rapidly overprinting the reservoir (e.g lacustrine) signature.

Water velocity gradient effects on $\delta^{13}C$

Downstream changes in stream channel morphology caused by the combined effects of differences in bedrock geology, basin slope, and sediment transport create alternating pools and riffles that have very different mean water velocities. These changes in stream gradient can have large differences in algal fractionation between habitats, as discussed below, and may also influence $\delta^{13}C_{DIC}$ by affecting groundwater–surface-water interactions. Newly introduced subsurface water usually has a lower $\delta^{13}C_{DIC}$ than the streamwater because of inputs of respiratory CO_2 (see Figure 10.6). However, the effect of the groundwater inputs may be masked by increases in $\delta^{13}C_{DIC}$ caused by rapid degassing of this CO_2-rich groundwater in the turbulence of the riffle (Doctor et al. in press).

Local water velocity conditions influence algal $\delta^{13}C$ through a negative relationship between flow rate and algal fractionation of $\delta^{13}C$. Slower waters

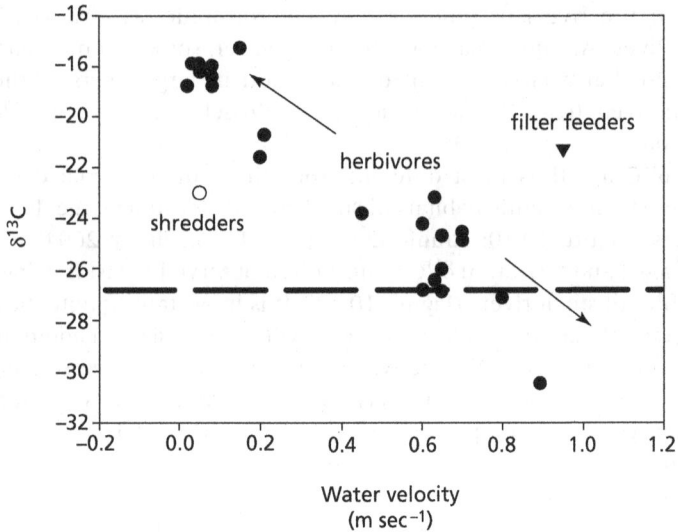

Figure 10.11 Relationship of water velocity with primary invertebrate functional feeding groups in the South Fork Eel River, CA. Herbivores included both collector-gathers and scraper taxa; herbivore $\delta^{13}C$ were strongly correlated with algal $\delta^{13}C$ at the site. The dashed line represents mean $\delta^{13}C$ for terrestrial organic carbon in the river. (Modified from Finlay et al. 2002.)

in pools have thicker benthic boundary layers, dominated by molecular diffusion, than the faster waters in riffles. The thicker diffusive layers inhibit diffusion of CO_2 and consequently cause more build-up of $^{13}CO_2$ near benthic algae in pools than found in riffles. The higher $\delta^{13}C_{DIC}$ in pools causes $\delta^{13}C$ of algae to be higher in pools than in riffles (Figure 10.11). This pattern is more likely observed in streams with pool–riffle geomorphology, where contrasts in water velocity are large, and where CO_2 supply is low and/or photosynthesis rates are high, than in deeper streams or ones with a more constant stream gradient. Thus water velocity effects are not obvious in small shaded streams where CO_2 concentrations are high and algal growth low (Macleod & Barton 1998; Finlay et al. 1999).

The higher $\delta^{13}C$ of algae in pools compared with riffles can impart a characteristic isotopic signature to a patch or habitat, thus providing a useful tool for food web analyses. Specifically, water velocity effects spatially label diet sources of consumers over a short distance along a stream. This label can then be traced through food webs, thus providing new information on the source and scale of resource use by herbivores (e.g., Finlay et al. 2002). For example, in the South Fork Eel River (USA), herbivorous invertebrates clearly rely on local sources of algal production, while presumed leaf-eaters (shredders) were found to consume an equal mix of leaves and algae within their pool habitats (Figure 10.11). Diet sources of filter-feeding taxa could not be completely resolved with $\delta^{13}C$ because there are three potential sources of suspended particles available to them: suspended terrestrial detritus, and algae transported from both pools and riffles.

The presence of multiple organic matter sources within a site with similar isotope ratios, or isotopic variation over small spatial scales, creates conditions where a single isotope diet tracer is not sufficient to address the full complexity of trophic interactions. Such conditions appear to be most common in large rivers (Delong & Thorp 2006), and potentially at junctions with other rivers, lakes and impoundments (Table 10.1). In some cases, careful separation and isotopic analysis of mixed organic materials can help determine the $\delta^{13}C$ and/or $\delta^{15}N$ of individual sources (Hamilton et al. 2005; Delong & Thorp 2006). Other measurements, such as gut content analyses or C:N ratios, may also help eliminate some potential sources for consumers. However, for situations where there are more than two OM sources present, the best solution to resolving all sources is to use a second tracer, most often $\delta^{15}N$, that shows good isotopic separation among the potential sources to rivers.

Anthropogenic and other spatial effects on $\delta^{15}N$ of algae and consumers

Natural variation and human alterations for the N cycle sometimes create large differences in $\delta^{15}N$ among sources. Of the potential causes of spatial variation in $\delta^{15}N$ in streams, human perturbations of the nitrogen cycle are probably the main cause of large-scale spatial changes in biota $\delta^{15}N$ in watersheds.

Longitudinal changes in nitrate concentrations and $\delta^{15}N_{NO_3}$ often result in longitudinal changes in the $\delta^{15}N$ of algae. There are many potential causes of longitudinal gradients in $\delta^{15}N$ in streams, including anthropogenic contaminants (e.g. fertilizer, animal waste, emissions from power plants), proximity to the ocean (e.g., marine-derived nutrients), redox chemistry (e.g. denitrification, ammonium vs. nitrate as the dominant DIN form), and site-specific natural labeling of DIN pools. In particular, the $\delta^{15}N$ of aquatic biota can be affected by inputs of water from waste water treatment plants (WWTPs) and combined sewage overflows (CSOs) in urban areas, or confined animal feeding operations (CAFOs) and agricultural runoff in farming areas.

Several of these causes of longitudinal gradients in $\delta^{15}N$ are observed in data from a transect along the San Joaquin River (USA). Figure 10.8 shows spatial variation in the $\delta^{15}N$ of nitrate and seston (largely algae) in the San Joaquin River, from the headwaters, through the deltaic wetlands, and to the mouth of the San Francisco Bay. In the riverine part where nitrate concentrations are high, probably because of inputs from agricultural activities, the $\delta^{15}N$ of seston is 4–5‰ lower than $\delta^{15}N_{NO_3}$. In contrast, in the bay part where nitrate concentrations are low and DIN may be limiting photosynthesis

(Wankel et al. 2006), there is little difference between the $\delta^{15}N$ of nitrate and seston. The relationship between nitrate and seston $\delta^{15}N$ is more complex in the delta area because of tidal mixing, multiple channels in the delta, and ammonium inputs to the river from insufficiently treated (oxidized) waste water. Nitrification of the ammonium causes decreases in nitrate $\delta^{15}N$; the subsequent uptake of the now ^{15}N-enriched residual ammonium by algae, causes increases in the $\delta^{15}N$ of the seston.

Sites with higher N concentrations or loads typically have much higher $\delta^{15}N$ compared with pristine watersheds (Mayer et al. 2002). Several watershed studies have shown that the high $\delta^{15}N$ value of DIN derived from sewage (+10‰ to +25‰; Heaton 1986) can be traced in aquatic food webs influenced by urban development (e.g., Cabana & Rasmussen 1996; McClelland & Valiela 1998; Lake et al. 2001). Other studies have shown that the low $\delta^{15}N$ of seston derived from domestic wastes can result in whole food webs having lower $\delta^{15}N$ in impacted sites than in non-impacted sites (e.g., Van Dover et al. 1992; Tucker et al. 1999; DeBruyn & Rasmussen 2002). In agricultural areas where animal waste is often used as a fertilizer, $\delta^{15}N$ in aquatic organisms have been shown to be positively correlated with the percent of agriculture in the watershed (Harrington et al. 1998; Hebert & Wassenaar 2001; Udy & Bunn 2001; Anderson & Cabana 2005). For example, a detailed analysis of longitudinal changes in the $\delta^{15}N$ values of primary consumers (and organisms of higher trophic levels) in nested watersheds in the St Lawrence River Basin (Canada) showed that the $\delta^{15}N$ closely tracked spatial differences in land use, and that the percent of the watershed devoted to agricultural explained 69% of the total variation in $\delta^{15}N$ (Anderson & Cabana, 2005). These increases in the $\delta^{15}N$ of organisms and DIN with increasing N loads, watershed size, and percent agricultural land use in the watershed may be caused by increased amounts of denitrification in the groundwater flowing into the stream. Kinetic fractionation during denitrification can cause substantial increases in the $\delta^{15}N$ of the residual nitrate in streams (see Kendall et al., this volume, pp. 375–449). As a consequence, autotroph $\delta^{15}N$ varies with factors such as land use or amount of sewage input, and not watershed area *per se*.

Other environmental effects can cause spatial variations in $\delta^{15}N$ in rivers. For example, marine organic matter and organisms often have higher $\delta^{15}N$ values than oligotrophic freshwater ecosystems (France 1995; Bilby et al. 1996), creating a useful isotope separation for detecting marine-derived nutrients or prey in freshwater food webs. In addition, terrestrial organic matter appears to be often ^{15}N-depleted relative to aquatic autotrophs (e.g. Harrington et al. 1998; Mulholland et al. 2000; Delong & Thorp 2006), perhaps due to intensive denitrification in riparian and aquatic sediments (Ostrom et al. 2002). At sites where there are large $\delta^{15}N$ differences between autotrophic and detrital food web components, $\delta^{15}N$ are proving useful for source separation, especially when used with advanced statistical techniques (e.g., Delong & Thorp 2006). Finally, local changes in redox chemistry due to processes occurring in dams or wetlands can also cause spatial changes in $\delta^{15}N$ of biota (Angradi 1993; Kendall et al. 2000, 2001). Much less is known about the behavior of $\delta^{15}N$ at the base of food webs, however, suggesting some caution in its application particularly under highly eutrophic conditions, as discussed below.

Spatial effects on $\delta^{34}S$ of biota

There is much less known about spatial distributions in the $\delta^{34}S$ of aquatic resources and organisms. We identify three potential causes of longitudinal gradients in $\delta^{34}S$ in streams: proximity to the ocean, bedrock geology, and redox chemistry. Sulfur isotopes are an effective tracer of organic matter at the land–ocean margin (Connolly et al. 2003) because ^{34}S-enriched sulfate can be transported up to hundreds of kilometers inland. For example, the $\delta^{34}S$ of marine sulfate and vegetation near the ocean are ca. +20‰ but decreases to +6‰ over ca. 100 km (Wadleigh et al. 1996; Wadleigh & Blake 1999). Geology may influence $\delta^{34}S$ where rivers cross geologic units with large concentrations of sulfide minerals. Oxidation of the sulfide minerals can cause decreases in the $\delta^{34}S$ of riverine sulfate, and presumably aquatic plants, as observed in the McKenzie River system (Hitchon & Krouse 1972). Last, reducing conditions typical of wetlands, bogs, and the deep waters of some lakes lead to low $\delta^{34}S$ values because sulfate reduction produces H_2S with low $\delta^{34}S$ values, which then may be assimilated into plants (Nriagu et al. 1991).

Along a river reach with variable contributions of organic matter derived from wetlands, the $\delta^{34}S$ of stream organic matter and consequently consumers would be expected to vary. For example, in the Everglades the $\delta^{34}S$ of sulfate in canals and marshes increases as agricultural sulfate from near Lake Okechobee is progressively reduced as water flows to the south (Bates et al. 2002). In contrast to the normal $\delta^{34}S$ pattern expected for redox gradients (e.g., lower $\delta^{34}S$ of biota with increasing amount of sulfate reduction to ^{34}S-depleted H_2S), the $\delta^{34}S$ of algae and fish increase with increasing extent of eutrophication and sulfate reduction in the Everglades (Kendall et al. 2000).

Probably the main use of $\delta^{34}S$ in food web studies is differentiating terrestrial versus marine sources of organic matter in near coastal environments, because of the large contrast in $\delta^{34}S$ values at the land–ocean margin (Connolly et al. 2003). There are two main types of studies: ones that use $\delta^{34}S$ to determine terrestrial plant contributions to coastal ecosystems (e.g., Peterson & Howarth 1987; Peterson & Fry 1987), and ones that use $\delta^{34}S$ to determine marine-derived nutrient contributions to near-coast watersheds (e.g., MacAvoy et al. 1998). Because of the extremely sharp gradient in $\delta^{34}S$ values between marine and terrestrial ecosystems, such studies have a high likelihood of producing quantitative results. When two isotope tracers are employed, the combination of $\delta^{34}S$ and $\delta^{13}C$ separates more producers than other isotope tracer combination (Connolly et al. 2003), despite the high within-producer variability of $\delta^{34}S$ in coastal areas.

Temporal patterns and applications of stable isotope tracers

Implicit in the preceding discussion of source identification is the assumption that consumers are in approximate "isotopic equilibrium" with their food sources. This assumption was not widely tested in many previous applications of stable isotope methods, but temporal variation is increasingly evident in time series of $\delta^{13}C$ and $\delta^{15}N$ in consumers and resources (Kendall et al. 2001; McCutchan & Lewis 2001). Temporal variation in the $\delta^{13}C$, $\delta^{15}N$, and $\delta^{34}S$ of dissolved species and consequently algae has the potential to greatly influence isotope-based inferences of food web relations (e.g. O'Reilly 2002), and should ideally be assessed prior to study design since it determines sampling frequency and the nature of other data necessary to interpret results. Thus, prediction of temporal variation and methods for dealing with it are important for the use of stable isotopes in food web studies.

Seasonal variation in $\delta^{13}C$ of algae

For reasons discussed previously, terrestrial plants show limited temporal variation in $\delta^{13}C$ relative to aquatic plants (e.g. Garten & Taylor 1992; Leffler & Evans 1999; McCutchan & Lewis 2002). While some minor variation exists among terrestrial detrital fractions during decomposition in streams (Hicks & Laboyrie 1999; Finlay 2001), allochthonous detritus has a comparatively constant $\delta^{13}C$ relative to autotrophic $\delta^{13}C$.

In contrast, autotrophs in streams and wetlands can show considerable seasonality in $\delta^{13}C$. In small unproductive streams, $\delta^{13}C_{DIC}$ values often change seasonally in response to watershed processes. In particular, seasonal variation in DIC sources to the stream (e.g., DIC derived from soil respiration vs. weathering reactions), causes higher $\delta^{13}C_{DIC}$ in the winter when DIC concentrations are low, and lower $\delta^{13}C_{DIC}$ in the summer when DIC concentrations are higher (Figure 10.6; Kendall 1993; Bullen & Kendall 1998). The algal $\delta^{13}C$ track the seasonality in $\delta^{13}C_{DIC}$ and CO_2 concentrations (Finlay 2004).

In larger streams and rivers, biological effects (e.g., the effects of algal productivity and/or respiration on CO_2 concentrations and $\delta^{13}C_{DIC}$), can overprint the effects of seasonality in watershed influences on $\delta^{13}C_{DIC}$. For example, when algal productivity is the main control on CO_2 or even DIC concentration, maximum algal $\delta^{13}C$ are typically observed during summer baseflow when productivity is greatest and CO_2 concentrations are lowest (McCutchan & Lewis 2001; Finlay 2004). Successive algal blooms can cause multiple maxima in the $\delta^{13}C$ of algae, as shown in Figure 10.7, where there is a 5‰ oscillation in the $\delta^{13}C$ of predominately algal POM in the Yazoo River (USA) due to algal blooms. Alternately, if respiration dominates ecosystem metabolism, such as in streams with high detrital inputs, algal $\delta^{13}C$ values tend to be lower in the summer when respiration rates are highest (Figure 10.6).

The effect of this variation on stable isotope studies depends on whether it occurs as gradual (i.e. seasonal) environmental changes or due to more stochastic processes such as floods. Seasonal changes in environmental conditions may alter factors influencing algal $\delta^{13}C$ (e.g. productivity, DIC availability, and $\delta^{13}C_{DIC}$) slowly, producing predicable patterns and long periods of stable signals under baseflow conditions (McCutchan & Lewis 2002; Finlay 2004). This type of situation is more amenable to straightforward use of stable isotope methods. If consumer growth and $\delta^{13}C$ are measured, temporal variation in resource $\delta^{13}C$ may be incorporated into biomass models (McCutchan & Lewis 2002). However, the fast turnover of autotroph biomass causes rapid isotopic responses to environmental changes that affect $\delta^{13}C$. Thus, under eutrophic conditions or when discharge is highly varied (e.g. Singer et al. 2005), frequent sampling for both $\delta^{13}C$ and growth of consumers is necessary to account for temporal changes in autotrophic $\delta^{13}C$.

Given the strong effect of discharge on parameters influencing $\delta^{13}C$ in streams (e.g. nutrients, water velocity, DIC), flow data should provide a useful indicator of temporal variation, and inform sampling decisions and overall study design. For example, Kendall (1993) found a strong inverse relation between discharge and DIC, and between $\delta^{13}C_{DIC}$ and discharge (Figure 10.6). Finlay (2004) also found strong inverse relationships between DIC and discharge, and discharge determined much of the temporal variation in algal $\delta^{13}C$ observed in streams and rivers of the watershed. In a eutrophic stream with thick periphyton, Singer et al. (2005) found that flow history was a better predictor than instantaneous water velocity in explaining spatial and temporal variation in algal $\delta^{13}C$. This finding makes sense considering that algal $\delta^{13}C$ integrate the effects of temporal and spatial variations in local conditions, whereas instantaneous velocity does not.

Seasonal variability in $\delta^{15}N$ of algae

Streams commonly show seasonal variability in $\delta^{15}N_{DIN}$ due to temporal changes in DIN sources and their $\delta^{15}N$ values, flow, and in-stream biogeochemical processes (see Kendall et al., this volume, pp. 375–449). Temporal variation of algae $\delta^{15}N$ is less well known. In their study of four watersheds of differing elevations and sizes on North St Vrain Creek in the Rocky Mountains (USA), McCutchan & Lewis (2002) found that only one site (the montane site) showed seasonality in the $\delta^{15}N$ of algae, with $\delta^{15}N$ values that were highest just before snowmelt and decreased over the summer. Several large rivers in the USA showed minor seasonal variations in the $\delta^{15}N$ of algae-dominated POM (Kendall et al. 2001). Subsequent multi-year data from these same 40 river sites shows that many rivers showed significant seasonal variability, with $\delta^{15}N$ oscillations frequently correlated with discharge changes (C. Kendall, unpublished U.S. Geological Survey data for these same rivers, 1998–2005). In some cases, DIN concentrations were so high, and DIN sources sufficiently constant, that even massive algal blooms did not cause enough drawdown of the DIN pool to cause changes in the algal $\delta^{15}N$ (Figure 10.7).

Investigations of temporal changes in the $\delta^{13}C$ and $\delta^{15}N$ of algae in marshes in the Everglades (USA) have shown that the $\delta^{13}C$ values show a strong inverse relation to $\delta^{15}N$ values; spatial data also show this same inverse relation. There are several coupled reactions that could cause this pervasive temporal and spatial pattern: respiration and denitrification, respiration and mineralization, and methane formation and then oxidation and denitrification. When water levels are low in the marshes, the $\delta^{13}C$ of algae are lower than when the water levels are higher, consistent with respiration effects exceeding atmospheric exchange effects in the dry season. The coincidence of low $\delta^{13}C$, high $\delta^{15}N$, and high $\delta^{34}S$ values with areas of increased eutrophication supports the theory that seasonal and spatial changes redox chemistry is a major control on the isotopic compositions of biota in this environment (Figure 10.12; Kendall et al. 2000).

Temporal changes in macrophytes related to growth and senescence cycles

Riparian plants can show significant seasonal variations in $\delta^{13}C$, $\delta^{15}N$, and C:N. In a study of wetland plants in the San Francisco Estuary (USA), the most common seasonal pattern was low C:N of foliage in spring when leaves/shoots first emerged, a gradual increase in C:N during the growth season, and a rapid increase in C:N in the fall when the new biomass died (Cloern

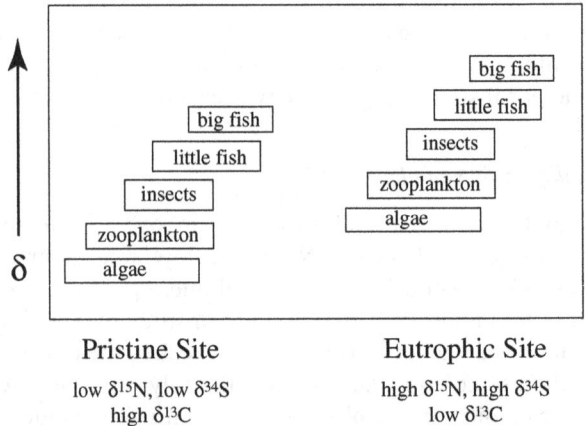

Figure 10.12 Cartoon showing how the $\delta^{13}C$, $\delta^{15}N$, and $\delta^{34}S$ of primary producers at oligotrophic (pristine) vs. eutrophic wetlands sites in the Everglades (USA) affect the isotopic compositions of the entire food web. These site-specific shifts in δ values are due to different suites of biogeochemical reactions that affect the δ values at the base of the food webs.

et al. 2002); however, while some species (e.g., *Salix, Typha*) showed strong seasonality in C:N, others (e.g., *Spartina*) did not. Different species also showed different $\delta^{13}C$ and $\delta^{15}N$ patterns, with the isotopic compositions of specific species of plants declining, increasing, or showing no consistent pattern during the growth season (Cloern et al. 2002). There was no consistent relation between $\delta^{15}N$ and C:N of plants during the growth–death cycle; high C:N dead biomass showed high $\delta^{15}N$ values for *Spartina* and low $\delta^{15}N$ for *Typha*. These species-specific patterns of seasonal variability in wetland plants are similar to those reported in a study of terrestrial vegetation by Handley & Scrimgeour (1997).

Temporal changes in isotopic composition caused by degradative processes

Once terrestrial detritus enters streams, its $\delta^{15}N$ may be changed substantially during decomposition as N is immobilized onto litter. The degree of modification during decomposition should depend on the amount of N immobilized and on the contrast between the $\delta^{15}N$ of DIN and the terrestrial plants. High $\delta^{15}N_{NO_3}$ is commonly observed when NO_3 concentrations are high; hence, the strongest effects should be expected in eutrophic streams. This is illustrated by a comparison of $\delta^{15}N$ of decomposing oak leaves in high and low NO_3^- ecosystems (the Hudson River, NY, and Fox Creek, CA, respectively). Work of Caraco et al. (1998) shows that changes in $\delta^{15}N$ during decomposition may be substantial; however, such effects are more subtle under low NO_3^- conditions, as observed for the oligotrophic stream (Figure 10.9). Overall, these observations suggest that use of $\delta^{15}N$ as a source tracer will be least complicated in undisturbed, low DIN streams and lakes. However, separation of aquatic vs. terrestrial sources is usually greater in human-impacted rivers.

Temporal changes in seston composition

Nutrient transport and food webs in larger rivers are often dominated by suspended organic matter so they are considered separately from the preceding sections which dealt primarily with benthic OM and organisms. Few studies in major rivers have had sufficient temporal or spatial coverage for adequate assessment of seasonal or hydrologic changes in seston composition. However, detailed studies in the Amazon River (Brazil) have found seston compositions to be nearly constant over substantial time periods, distances, hydrologic fluctuations, and size fractions (Hedges et al. 1986; Quay et al. 1992). In contrast, seston in the St Lawrence River (Canada) was dominated by *in situ* photosynthesis during warm seasons, and terrestrial detritus during colder periods and storm surges (Barth et al. 1998). Seston in the Sanaga River (Cameroon) and Congo River also varied with discharge, with high proportions of seston derived from C4 plants in savannas during high discharge periods, and high proportions of seston derived from C3 plants from the riverbanks during low discharge periods (Mariotti et al. 1991; Bird et al. 1998). A study of monthly seston samples from 40 sites on major rivers in the USA found that plankton is the dominant source of seston at many sites, with lower $\delta^{13}C$ and higher percentages of plankton during algal blooms in the spring and summer and downstream of reservoirs (Kendall et al. 2001). Particulate organic matter with unusually low C:N and low $\delta^{15}N$ values at a Mississippi River site was attributed to a bacterial bloom in the winter (Wissel & Fry 2005). These and other studies involving seston are just beginning to elucidate the origin, transport, and cycling of seston in major rivers, and the role it plays in riverine food webs.

Diel changes in algal $\delta^{13}C$ and $\delta^{15}N$

Daily cycles of temperature, redox potential, photosynthesis, and respiration cause cyclic changes in many important aquatic constituents including pH, DIC, conductivity, dissolved gases such as O_2 and CO_2, major elements, and trace metals. Very little is known about diel variations in the isotopic compositions of dissolved constituents in streams. However, recent studies show substantial diel changes in $\delta^{18}O_{O_2}$ and $\delta^{13}C_{DIC}$ that may eventually provide new insight into biogeochemistry of freshwaters. For example, in a recent study in a small stream in Montana (USA), there was a 14‰ change in $\delta^{18}O_{O_2}$ and 2‰ change in $\delta^{13}C_{DIC}$ (Parker et al. 2005), with minimum $\delta^{18}O$ values found at mid-day and minimum $\delta^{13}C$ values found in at dawn. A recent multi-isotope investigation of diel changes in the San Joaquin River (USA) showed that the $\delta^{13}C$ and $\delta^{15}N$ of algae, the $\delta^{15}N$ and $\delta^{18}O$ of nitrate, and $\delta^{13}C_{DIC}$ varied by about 2‰ while the $\delta^{18}O_{O_2}$ varied by >10‰. Recent work by Venkiteswaran et al. (in press) has shown that the temporal "shape" of the diel $\delta^{18}O$ curve can be explained by a small number of parameters, including photosynthesis to respiration ratio, uptake fractionation, temperature, and gas exchange rate. Hence, the natural diel change in the $\delta^{18}O$ of O_2 (and probably the isotopic compositions of the other constituents) provides a powerful biogeochemical "signal" in aquatic systems.

Isotopic indicators of ecosystem processes

In the preceding sections, we have mainly considered isotopes as tracers of organic matter, nutrients, or trophic interactions. Because stable isotopes of inorganic and organic materials are intricately linked to biogeochemical cycles, they are increasingly useful as indictors or predictors of various types of ecosystem processes. For example, algal $\delta^{13}C$ values appear to be strongly linked to the overall ecosystem metabolism of rivers or lakes (e.g. Finlay 2004). Metabolic processes such as photosynthesis and respiration strongly indicate or influence the parameters that control algal $\delta^{13}C$ such as algal growth rates, CO_2, and $\delta^{13}C_{DIC}$. Thus, algal $\delta^{13}C$ could potentially be used to provide an integrated measure of the carbon balance of aquatic ecosystems (Finlay 2004).

Nitrogen isotopes in food webs appear to provide robust, integrated measurements of nitrogen loading and cycling in aquatic ecosystems. For example, $\delta^{15}N$ of estuarine plants and river consumers are well-predicted from N loads and land use (Cole et al. 2004; Anderson & Cabana 2005). Anderson & Cabana (2005) conclude that in the absence of important point-source N inputs, $\delta^{15}N$ in aquatic organisms appear to increase in a predictable pattern, following the increasing amounts of agricultural N inputs. The cause of the relationship between $\delta^{15}N$ of biota and agricultural land use or N loads is a subject of some debate, since it could be caused by increased denitrification, N cycling, or use of manure and/or septic waste in larger watersheds.

Stable isotopes may be used to detect "hot spots" of intense biogeochemical activity (McClain et al. 2003) in aquatic ecosystems and at their interfaces with terrestrial environments. Eutrophic or hypoxic environments may cause large decreases in $\delta^{13}C_{DIC}$ because of methane formation and subsequent oxidation, increases in $\delta^{15}N_{NO_3}$ because of denitrification, and either decreases or increases in $\delta^{34}S_{SO_4}$ because of sulfate reduction – with consequent incorporation of these biogeochemical signatures of redox conditions into local biota (Figure 10.12).

Probably the best example of use of stable isotope indicators to detect environmental change is for detecting effects of sewage. Sewage inputs can cause local "hot spots" of labile, organic matter with distinct isotopic signatures compared with other sources. For example, a study of municipal sewage and pulp-mill inputs to rivers in Canada showed that the waste material contained materials with $\delta^{15}N$ and $\delta^{34}S$ values that were distinct from upstream background conditions in the receiving environment (Wayland & Hobson 2001). In another sewage waste study on tributaries of the St Lawrence River (Canada), point-source N inputs from human waste lowered the $\delta^{15}N$ of biota collected below the points of discharge at the scale of kilometers (Anderson & Cabana 2006). When domestic waste was diverted from the stream, the $\delta^{15}N$ of small consumers changed by >4‰ over the next year, showing that streams impacted by large point-source urban discharges can show rapid recovery when the discharge is entirely eliminated. The study suggested that biogeochemical indicators such as stable isotopes may reveal recovering changes in the ecosystem nutrient metabolism, even before a complete taxonomic recovery of the community is observed (Anderson & Cabana 2006).

Alternates for complex and variable situations: other approaches and isotopic tracers

Ongoing research on variability in $\delta^{13}C$, $\delta^{15}N$, and $\delta^{34}S$ of dissolved species, plants, and consumers will undoubtedly continue to refine their use. Given the widely varied environmental conditions present throughout drainage networks, it is clear that stable isotope values reported in one field study cannot always be applied to the next without thorough examination of the biogeochemical conditions. A careful attempt must be made to sample the spatial and temporal variability in the potential sources and in the organisms of interest, at scales relevant to the questions or organisms present.

Under many conditions, overlap of isotope signals or high temporal variation decrease the efficacy of isotopes as tracers of organic matter or trophic position. Improvements in statistical analyses of mixing models offer increasingly sophisticated means to separate sources in situations of variable or partially overlapping isotope ratios and to make use of multiple tracers (Phillips & Koch 2002; Phillips & Gregg 2003; Benstead et al. 2006).

High variability in the isotopic compositions of primary sources may sometimes be dealt with by using consumers to integrate small-scale spatial and temporal variation. For example, as we have seen, $\delta^{15}N$ of plants, microbes, and detritus are highly varied. To address this variability, investigators may use the $\delta^{15}N$ of a basal consumer instead of the primary resource to base (i.e., normalize) trophic position calculations (Cabana & Rasmussen 1996; Post 2002) or to assess the contribution of a specific habitat (i.e. not the specific resource type) to a predator's diet (e.g. Vander Zanden & Vadeboncoeur 2002). Since the diet of primary consumers integrates temporal and spatial variability in the isotopic compositions of resources, using primary consumers removes a considerable amount of environmental noise. Furthermore, as illustrated in Figure 10.12, use of primary consumers to normalize subsequent calculations of trophic position is advantageous since it allows comparisons of food web structure across ecosystems that may have very different $\delta^{13}C$, $\delta^{15}N$, and $\delta^{34}S$ at the base of the food web due to differences in environmental conditions and nutrient cycling (Cabana & Rasmussen 1996; Post 2002).

In many cases, however, additional tracers are necessary to fully resolve food webs isotopically. Fortunately, a number of alternate isotope tracers are available and are increasingly used to enhance natural abundance techniques. For example, purposeful enrichment of the aquatic environment with ^{15}N-DIN, ^{13}C-DIC, and ^{13}C-DOC is relatively inexpensive, and such isotopic labeling techniques are increasingly used to study N and C cycling and energy flow in food webs. Isotope enrichment is expensive compared with natural abundance approaches, and equilibration between food sources may not occur, requiring sophisticated modeling approaches to interpret food web results. However, these approaches may be effectively coupled with natural abundance measurements to gain new insight into stream food web ecology (e.g. Mulholland et al. 2000; Hamilton et al. 2004).

Other stable isotope tracers are less well developed but have the potential to resolve issues associated with limitations of $\delta^{13}C$ and $\delta^{15}N$. As noted, applications of $\delta^{34}S$ are rapidly increasing. Strontium isotopes ($^{87}Sr/^{86}Sr$) have been found to be extremely useful for studies of bird migration because the different Sr isotopic ratios of different geologic materials become incorporated in the plants growing on these materials (e.g. Chamberlain et al. 1997). The $^{87}Sr/^{86}Sr$ values are transferred up the food chain from the soil-exchange pool to leaves and then to small consumers without modification (Blum et al. 2000). The oxygen ($\delta^{18}O$) and hydrogen (δ^2H) isotopes of butterfly wings and bird feathers have been found extremely useful for tracking migration paths (Wassenaar & Hobson 1998; Hobson et al. 1999) because of the strong environmental gradients in the $\delta^{18}O$ and δ^2H of surface waters (Kendall & Coplen 2001) and precipitation (Dutton et al. 2005) in North America. Furthermore, δ^2H and radiocarbon (^{14}C) difference among diet sources for consumers observed decades ago (e.g., Estep & Dabrowski (1980) and Broecker & Walton (1959) respectively) are only now being more thoroughly explored for their potential as routine tracers. Finally, compound-specific isotope ratio mass spectrometry (Evershed et al., this volume, pp. 480–540) has the potential for much more accurate determinations of food web relations than possible with the more conventional analysis of bulk solid samples by tracing specific organic molecules from diet to consumer.

Stable isotope techniques are always best used in combination with other methods and complementary data. This is particularly true when it is not possible to unequivocally separate sources or estimate trophic position in food webs with stable isotopes alone. For example, gut content analyses may allow an investigator to eliminate potential sources to consumers, and allow greater use of isotope techniques. Cation ratios (Sr/Ca and Ba/Ca) were found to decrease at each successive trophic level, suggesting that these ratios can be used to identify the trophic level at which an organism is primarily feeding (Blum et al. 2000). The measurement of resource availability, in terms of edibility or abundance, may also help constrain the number and variability of end members that must be considered for use of stable isotope tracers.

Conclusions

Spatial and temporal isotopic variation among organic matter sources is pervasive and is much less well understood compared with other factors that affect use of natural abundance approaches, such as trophic fractionation. Such variation determines when, where, and how isotope techniques may be applied in food web studies. Thus, assessment of the inherent environmental variability in organic matter or prey sources should be the first step in any attempt to use stable isotope methods in food web studies. Accurate assessments of extent of source variation is especially critical for studies of energy flow in food webs, but also influences use of $\delta^{13}C$ and $\delta^{15}N$ to migration of animals as well as trophic position estimation, since $\delta^{15}N$ often vary by habitat or organic matter source.

Physical and chemical conditions of rivers are characterized by spatial gradients, patchiness at all scales (e.g. Vannote et al. 1980; Frissell et al. 1986), and temporal variability. Natural abundance stable isotope tools are increasingly used to study basic and applied aspects of river food webs. Isotope tools are particularly useful because the physical and biogeochemical diversity of rivers creates variation in isotope ratios that is useful to test hypotheses and examine global change effects. Isotopic variation often complicates the simplest applications of these methods (e.g. sampling on one habitat or date to characterize a river food web); however, this variation can be used as a natural "signal" for carefully designed studies. As spatial and temporal complexity of food webs increase, measurements of multiple source tracers, consumer growth rates, movement rates of predators and prey, and modeling approaches are increasingly necessary. Recognition of sources of variation affecting stable isotope ratios enhances the usefulness and efficiency of isotope techniques to understand freshwater food web ecology.

Acknowledgments

JCF was supported while preparing this manuscript by the National Science Foundation via support to the National Center for Earth-surface Dynamics (STC EAR 0120914) and via a Division of Environmental Biology grant (DEB 0315990). CK was supported by the National Research Program of the U.S. Geological Survey.

References

Aitkenhead, J.A. & McDowell, W.H. (2000) Soil C:N ratio as a predictor of annual riverine DOC flux at local and global scales. *Global Biogeochemical Cycles*, **14**, 128–138.

Amundson, A., Austin, A.T., Schuur, E.A.G., et al. (2003). Global patterns of the isotopic composition of soil and plant nitrogen. *Global Biogeochemical Cycles*, **17**, 31-1–31-10.

Anderson, C. & Cabana, G. (2005) Delta N-15 in riverine food webs: effects of N inputs from agricultural watersheds. *Canadian Journal of Fisheries and Aquatic Sciences*, **62**, 333–340.

Anderson, C. & Cabana, G. (2006) Does $\delta^{15}N$ in river food webs reflect the intensity and origin of N loads from the watershed? *Science of the Total Environment*, **367**, 968–978.

Angradi, T.R. (1993) Stable carbon and nitrogen isotope analysis of seston in a regulated rocky mountain river, USA. *Regulated Rivers: Research and Management*, **8**, 251–270.

Angradi, T.R. (1994) Trophic linkages in the lower Colorado River: multiple stable isotope evidence. *Journal of the North American Benthological Society*, **13**, 479–495.

Araujo-Lima, C.A.R.M., Forsberg, B.R., Victoria, R.L. & Martinelli, L. (1986) Energy sources for detritivorous fishes in the Amazon. *Science*, **234**, 1256–1259.

Atekwana, E.A. & Krishnamurthy, R.V. (1998) Seasonal variations of dissolved inorganic carbon and $\delta^{13}C$ of surface waters: application of a modified gas evolution technique. *Journal of Hydrology*, **205**, 265–278.

Barnard, C., Martineau, C., Frenette, J.J., Dodson, J.J. & Vincent, W.F. (2006) Trophic position of zebra mussel veligers and their use of dissolved organic carbon. *Limnology and Oceanography*, **51**, 1473–1484.

Barth, J.A.C., Veizer, J. & Mayer, B. (1998) Origin of particulate organic carbon in the upper St. Lawrence: isotopic constraints. *Earth and Planetary Science Letters*, **162**, 111–121.

Bastow, J.L., Sabo, J.L., Finlay, J.C. & Power, M.E. (2002) A basal aquatic-terrestrial trophic link in rivers: algal subsidies via shore-dwelling grasshoppers. *Oecologia*, **131**, 261–268.

Bates, A.L., Orem, W.H., Harvey, J.W. & Spiker, E.C. (2002) Tracing sources of sulfur in the Florida Everglades. *Journal of Environmental Quality*, **31**, 287–299.

Battaglin, W.A., Kendall, C., Chang, C.C.Y., et al. (2001a) Chemical and isotopic composition of organic and inorganic samples from the Mississippi River and its tributaries, 1997–98. *U.S.Geological Survey Water Resources Investigation Report*, **01-4095**, 57 pp.

Battaglin, W.A., Kendall, C., Chang, C.C.Y., et al. (2001b) Chemical and isotopic evidence of nitrogen transformation in the Mississippi River, 1997–98. *Hydrological Processes*, **15**, 1285–1300.

Bender, M.M. (1968) Mass spectrometric studies of carbon-13 variations in corn and other grasses. *Radiocarbon*, **10**, 468–472.

Benstead, J.P., March, J.G., Fry, B., Ewel, K.C. et al. (2006) Testing isosource: Stable isotope analysis of a tropical fishery with diverse organic matter sources. *Ecology*, **87**, 326–333.

Berglund, O., Nystrom, P. & Larsson, P. (2005) Persistent organic pollutants in river food webs: influence of trophic position and degree of heterotrophy. *Canadian Journal of Fisheries and Aquatic Sciences*, **62**, 2021–2032.

Bilby, R.E., Fransen, B.R. & Bisson, P.A. (1996) Incorporation of nitrogen and carbon from spawning coho salmon into the trophic system of small streams: evidence from stable isotopes. *Canadian Journal of Fisheries and Aquatic Sciences*, **53**, 164–173.

Bird, M.I., Giresse, P. & Ngos, S. (1998) A seasonal cycle in the carbon-isotopic composition of organic carbon in the Sanaga River, Cameroon. *Limnology and Oceanography*, **43**, 143–146.

Blum, J.D., Taliaferro, E.H., Weisse, M.T. & Holmes, R.T. (2000) Changes in Sr/Ca, Ba/Ca and $^{87}Sr/^{86}Sr$ ratios between trophic levels in two forest ecosystems in the northeastern U.S.A. *Biogeochemistry*, **49**, 87–101.

Boutton, R.W. (1996) Stable carbon isotope ratios of soil organic matter and their use as indicators of vegetation and climate change. In: *Mass Spectrometry of Soils* (Eds T.W. Boutton & S. Yamasaki), pp. 47–82. Marcel Dekker, New York.

Brady, N.C. (1990) *The Nature and Properties of Soils*. Macmillan, New York, 621 pp.

Broadbent, F.E., Rauschkolb, R.S., Lewis, K.A. & Chang, G.Y. (1980) Spatial variability in nitrogen-15 and total nitrogen in some virgin and cultivated soils. *Soil Science Society of America Journal*, **44**, 524–527.

Broecker, W.S. & Walton, A. (1959) The geochemistry of C14 in fresh-water systems. *Geochemica et Cosmochemica Acta*, **16**, 15–38.

Brunold, C. & Erismann, K.H. (1975) H_2S as sulfur sources in *Lemna minor* L.: direct incorporation in to cystein and inhibition of sulfate assimilation. *Experientia*, **31**, 508–510.

Bullen, T.D. & Kendall, C. (1998) Tracing of weathering reactions and water flowpaths: a multi-isotope approach. In: *Isotope Tracers in Catchment Hydrology* (Eds C. Kendall & J.J. McDonnell), pp. 611–646. Elsevier, Amsterdam.

Cabana, G. & Rasmussen, J.B. (1996) Comparison of aquatic food chains using nitrogen isotopes. *Proceedings of the National Academy of Sciences*, **93**, 10844–10847.

Caraco, N.F., Lampman, G., Cole, J.J., Limburg, K.E., Pace, M.L. & Fischer, D. (1998) Microbial assimilation of DIN in a nitrogen rich estuary: implications for food quality and isotope studies. *Marine Ecology – Progress Series*, **167**, 59–71.

Casciotti, K.L., Sigman, D.M., Hastings, M.G., et al. (2002) Measurement of the oxygen isotopic composition of nitrate in seawater and freshwater using the denitrifier method. *Analytical Chemistry*, **74**, 4905–4912.

Chaloner, D.T., Martin, K.M., Wipfli, M.S., et al. (2002) Marine carbon and nitrogen in southeastern Alaska stream food webs: evidence from artificial and natural streams. *Canadian Journal of Fisheries and Aquatic Science*, **59**, 1257–1265.

Chamberlain, C.P., Blum, J.D., Holmes, R.T., et al. (1997) The use of isotope tracers for identifying populations of migratory birds. *Oecologia*, **109**, 132–141.

Clark, I.D. & Fritz, P. (1997) *Environmental Isotopes in Hydrogeology*. CRC Press, Boca raton, FL.

Cloern, J.E., Canuel, E.A. & Harris, D. (2002) Stable carbon and nitrogen isotope composition of aquatic and terrestrial plants of the San Francisco Bay estuarine system. *Limnology and Oceanography*, **47**, 713–729.

Coffin, R.B., Fry, B., Peterson, B.J. & Wright, R.T. (1989) Carbon isotopic compsitions of estuarine bacteria. *Limnology and Oceanography*, **34**, 1305–1310.

Cole, M.L., Valiela, I., Kroeger, K.D., et al. (2004) Assessment of a delta N-15 isotopic method to indicate anthropogenic eutrophication in aquatic ecosystems. *Journal of Environmental Quality*, **33**, 124–132.

Connolly, R.M., Guest, M.A., Melville, A.J. & Oakes, J.M. (2003) Sulfur stable isotopes separate producers in marine food-web analysis. *Oecologia*, **138**, 161–167.

Dawson, T.E., Mambelli, S., Plamboeck, A.H., et al. (2002) Stable isotopes in plant ecology. *Annual Review of Ecology and Systematics*, **33**, 507–559.

DeBruyn, A.M H. & Rasmussen, J.B. (2002) Quantifying assimilation of sewage-derived organic matter by riverine benthos. *Ecological Applications*, **12**, 511–520.

Delong, M.D. & Thorp, J.H. (2006) Significance of instream autotrophs in trophic dynamics of the Upper Mississippi River. *Oecologia*, **147**, 76–85.

Delong, M.D., Thorp, J.H., Greenwood, K.S. & Miller, M.C. (2001) Responses of consumers and food resources to a high magnitude, unpredicted flood in the upper Mississippi River basin. *Regulated Rivers-Research and Management*, **17**, 217–234.

Doctor, D.H., Kendall, C., Sebestyen, S.D., et al. (in press) Carbon isotope fractionation of dissolved inorganic carbon (DIC) due to outgassing of CO2 from a headwater stream. *Hydrological Processes*.

Doucett, R.R., Giberson, D.J. & Power, G. (1999a) Parasitic association of *Nanocladius* (Diptera : Chironomidae) and *Pteronarcys biloba* (Plecoptera : Pteronarcyidae): insights from stable-isotope analysis. *Journal of the North American Benthological Society*, **18**, 514–523.

Doucett, R.R., Hooper, W. & Power, G. (1999b) Identification of anadromous and nonanadromous adult brook trout and their progeny in the Tabusintac River, New Brunswick, by means of multiple-stable-isotope analysis. *Transactions of the American Fisheries Society*, **128**, 278–288.

Duarte, C.M. & Agusti, S. (1998) The CO_2 balance of unproductive aquatic ecosystems. *Science*, **281**, 234–235.

Dutton, A., Wilkinson, B.H., Welker, J.M., et al. (2005) Spatial distribution and seasonal variation in $^{18}O/^{16}O$ of modern precipitation and river water across the conterminous USA. *Hydrological Processes*, **19**(20), 4121–4146.

England, L.E. & Rosemond, A.D. (2004) Small reductions in forest cover weaken terrestrial-aquatic linkages in headwater streams. *Freshwater Biology*, **49**, 721–734.

Estep, M.F. & Dabrowski, H. (1980) Tracing food webs with stable hydrogen isotopes. *Science*, **209**, 1537–1538.

Farquhar, G.D., O'Leary, M.H. & Berry, J.A. (1982) On the relationship between carbon isotope discrimination and the intercellular carbon dioxide concentration in leaves. *Australian Journal of Plant Physiology*, **9**, 121–137.

Finlay, J.C. (2001) Stable-carbon-isotope ratios of river biota: Implications for energy flow in lotic food webs. *Ecology*, **82**, 1052–1064.

Finlay, J.C. (2003) Controls of streamwater dissolved inorganic carbon dynamics in a forested watershed. *Biogeochemistry*, **62**, 231–252.

Finlay, J.C. (2004) Patterns and controls of lotic algal stable carbon isotope ratios. *Limnology and Oceanography*, **49**, 850–861.

Finlay, J.C., Power, M.E. & Cabana, G. (1999) Effects of water velocity on algal carbon isotope ratios: Implications for river food web studies. *Limnology and Oceanography*, **44**, 1198–1203.

Finlay, J.C., Khandwala, S. & Power, M.E. (2002) Spatial scales of carbon flow in a river food web. *Ecology*, **83**, 1845–1859.

Fogel, M.L. & Cifuentes, L.A. (1993) Isotope fractionation during primary production. In: *Organic Geochemistry* (Eds M.H. Engel & S.A. Macko), pp. 73–94. Plenum Press, New York.

France, R. (1995) Stable nitrogen isotopes in fish: Literature synthesis on the influence of ecotonal coupling. *Estuarine Coastal and Shelf Science*, **41**, 737–742.

Frissell, C.A., Liss, W.J., Warren, C.E. & Hurley, M.D. (1986) A hierarchical framework for stream habitat classification: viewing streams in a watershed context. *Environmenal Management*, **10**, 199–214.

Fry, B. (1986) Stable sulphur isotopic distributions and sulphate reduction in lake-sediments of the Adirondack Mountains, New York. *Biogeochemistry*, **2**(4), 329–343.

Fry, B. (1991) Stable isotope diagrams of freshwater foodwebs. *Ecology*, **72**, 2293–2297.

Fry, B. & Sherr, E.B. (1984) ^{13}C measurements as indicators of carbon flow in marine and freshwater ecosystems. *Contributions to Marine Science*, **27**, 13–47.

Fry, B., Joern, A. & Parker, P.L. (1978) Grasshopper food web analysis: use of carbon isotope ratios to examine feeding relationships among terrestrial herbivores. *Ecology*, **59**, 498–506.

Fry, B., Scalan, R.S., Winters, K. & Parker, P.L. (1982) Sulphur uptake by salt grasses, mangroves and seagrasses in anaerobic sediments. *Geochimica et Cosmochimica Acta*, **46**, 1121–1124.

Fry, B., Silva, S.R., Kendall, C. & Anderson, R.K. (2002) Oxygen isotope corrections for online δ^{34}S analysis. *Rapid Communications in Mass Spectrometry*, **16**, 854–858.

Garten, C.T., Jr. & Taylor, G.E., Jr. (1992) Foliar δ^{13}C within a temperate deciduous forest: spatial, temporal and species sources of variation. *Oecologia*, **90**, 1–7.

Glime, J.M. & Vitt, D.H. (1984) The physiological adaptations of aquatic Musci. *Lindbergia*, **10**, 41–52.

Goericke, R., Montoya, J.P. & Fry, B. (1994) Physiology of isotope fractionation in algae and cyanobacteria. In: *Stable Isotopes in Ecology and Environmental Science* (Eds K. Lajtha & B. Michener), pp. 199–233. Blackwell Scientific Publications, Oxford.

Granger, J., Sigman, D.M., Needoba, J.A. & Harrison, P.J. (2004) Coupled nitrogen and oxygen isotope fractionation of nitrate during assimilation by cultures of marine phytoplankton. *Limnology and Oceanography*, **49**(5), 1763–1773.

Hamilton, S.K. & Lewis, Jr., W.M. (1992) Stable carbon and nitrogen isotopes in algae and detritus from the Orinoco River floodplain, Venezuela. *Geochimica et Cosmochimica Acta*, **56**, 4237–4246.

Hamilton, S.K., Tank, J.L., Raikow, D.E., Siler, E.R., et al. (2004) The role of instream vs allochthonous N in stream food webs: modeling the results of an isotope addition experiment. *Journal of the North American Benthological Society*, 23, 429–448.

Hamilton, S.K., Sippel, S.J. & Bunn, S.E. (2005) Separation of algae from detritus for stable isotope or ecological stoichiometry studies using density fractionation in colloidal silica. *Limnology and Oceanography, Methods*, 3, 149–157.

Handley, L.L. & Scrimgeour, C.M. (1997) Terrestrial plant ecology and N-15 natural abundance: The present limits to interpretation for uncultivated systems with original data from a Scottish old field. *Advances in Ecological Research*, 27, 133–212.

Handley, L.L., Austin, A.T., Robinson, D., et al. (1999) The ^{15}N natural abundance (δ^{15}N) of ecosystem samples reflects measures of water availability. *Australian Journal of Plant Physiology*, 26, 185–199.

Harrington, R.R., Kennedy, B.P., Chamberlain, C.P., et al. (1998) ^{15}N enrichment in agricultural catchments: Field patterns and applications to tracking Atlantic salmon (Salmo. Salar). *Chemical Geology*, 147, 281–294.

Harris, G.P. (1986) *Phytoplankton Ecology*. Chapman and Hall, London.

Heaton, T.H.E. (1986) Isotopic studies of nitrogen pollution in the hydrosphere and atmosphere: a review. *Chemical Geology*, 56, 87–102.

Hebert, C.E. & Wassenaar, L.I. (2001) Stable nitrogen isotopes in waterfowl feathers reflect agricultural land use in western Canada. *Environmental Science and Technology*, 35, 3482–3487.

Hecky, R.E. & Hesslein, R.H. (1995) Contributions of benthic algae to lake food webs as revealed by stable isotope analysis. *Journal of the North American Benthological Society*, 14, 631–653.

Hecky, R.E., Campbell, P. & Hendzel, L.L. (1993) The stoichiometry of carbon, nitrogen, and phosphorus in particulate matter of lakes and oceans. *Limnology and Oceanography*, 38, 709–724.

Hedges, J.L., Clark, W.A., Quay, P.D., et al. (1986) Compositions and fluxes of particulate organic material in the Amazon River. *Limnology and Oceanography*, 31, 717–738.

Hershey, A.E. & Peterson, B.J. (1996) Stream food webs. In: *Methods in Stream Ecology* (Eds G.A. Lamberti & F.R. Hauer), pp. 511–530. Academic Press, New York.

Hicks, B.J. (1997) Food webs in forest and pasture streams in the Waikato region, New Zealand: A study based on analyses of stable isotopes of carbon and nitrogen, and fish gut contents. *New Zealand Journal of Marine and Freshwater Research*, 31, 651–664.

Hicks, B.J. & Laboyrie, J.L. (1999) Preliminary estimates of mass-loss rates, changes in stable isotope composition, and invertebrate colonisation of evergreen and deciduous leaves in a Waikato, New Zealand, stream. *New Zealand Journal of Marine and Freshwater Research*, 33, 221–232.

Hillaire-Marcel, G. (1986) Isotopes and food. In: *Handbook of Environmental Isotope Geochemistry*, Vol. 2 (Eds P. Fritz & J.Ch. Fontes), pp. 507–548. Elsevier Science, Amsterdam.

Hitchon, B. & Krouse, H.R. (1972) Hydrogeochemistry of the surface waters of the Mackenzie River drainage basin, Canada – III. Stable isotopes of oxygen, carbon and sulphur. *Geochimica et Cosmochimica Acta*, 36, 1337–1357.

Hobson, K.A., Wassenaar, L.I. & Taylor, O.R. (1999) Stable isotopes (δD and δ^{13}C) are geographic indicators of natal origins of monarch butterflies in eastern North America. *Oecologia*, 120, 397–404.

Hoffer-French, K.J. & Herman, J.S. (1989) Evaluation of hydrological and biological influences on CO_2 fluxes from a karst stream. *Journal of Hydrology*, 108, 189–212.

Högberg, P. (1997) ^{15}N natural abundance in soil-plant systems. *New Phytologist*, 137, 179–203.

Holmes, R., McCleland, M.J., Sigman, W., et al. (1998) Measuring ^{15}N–NH$_4^+$ in marine, estuarine and fresh waters: An adaptation of the ammonia diffusion method for samples with low ammonium concentrations. *Marine Chemistry*, **60**, 235–243.

Huryn, A.D., Riley, R.H., Young, R.G., et al. (2002) Natural-abundance stable C and N isotopes indicate weak upstream-downstream linkage of food webs in a grassland river. *Archiv für Hydrobiologie*, **153**, 177–196.

Jepsen, D.B. & Winemiller, K.O. (2002) Structure of tropical river food webs revealed by stable isotope ratios. *Oikos*, **96**, 46–55.

Kato, C., Iwata, T. & Wada, E. (2004) Prey use by web-building spiders: stable isotope analyses of trophic flow at a forest-stream ecotone. *Ecological Research*, **19**, 633–643.

Keeley, J.E. & Sandquist, D.R. (1992) Carbon: freshwater plants. *Plant, Cell and Environment*, **15**, 1021–1035.

Kendall, C. (1993) *Impact of isotopic heterogeneity in shallow systems on stormflow generation*. PhD dissertation, University of Maryland, College Park, 310 pp.

Kendall, C. (1998) Tracing nitrogen sources and cycling in catchments. In: *Isotope Tracers in Catchment Hydrology* (Eds C. Kendall & J.J. McDonnell), pp. 519–576. Elsevier, Amsterdam.

Kendall, C. & Coplen, T.B. (2001) Distribution of oxygen-18 and deuterium in river waters across the United States. *Hydrological Processes*, **15**, 1363–1393.

Kendall, C. & Doctor, D.H. (2004) Stable isotope applications in hydrologic studies. In: *Treatise on Geochemistry*, Vol. 5, *Surface and Ground Water, Weathering, and Soils* (Ed. J.I. Drever), pp. 319–364. Elsevier, Amsterdam.

Kendall, C., Silva, S.R., Chang, C.Y., et al. (2000) Spatial changes in redox conditions and foodweb relations at low and high nutrient sites in the Everglades. *U.S. Geological Survey Open File Report*, **00-449**, 61–63.

Kendall, C., Silva, S.R. & Kelly, V.J. (2001) Carbon and nitrogen isotopic compositions of particulate organic matter in four large river systems across the United States. *Hydrological Processes*, **15**, 1301–1346.

Kennedy, T.A., Finlay, J.C. & Hobbie, S.E. (2005) Eradication of invasive *Tamarix ramosissima* along a desert stream increases native fish density. *Ecological Applications*, **15**, 2072–2083.

Kratzer, C.R., Dileanis, P.D., Zamora, C., et al. (2004) *Sources and transport of nutrients, organic carbon, and chlorophyll-a in the San Joaquin River upstream of Vernalis, California, during summer and fall, 2000 and 2001*. WRI 03-4127, U.S. Geological Survey. Online: http://water.usgs.gov/pubs/wri/wri034127/.

Krouse, H.R. & Mayer, B. (2000) Sulphur and oxygen isotopes in sulphate. In: *Environmental Tracers in Subsurface Hydrology* (Eds P.G. Cook & A.L.Herczeg), pp. 195–231. Kluwer Academic Publishers, Boston.

Krouse, H.R., Legge, A. & Brown, H.M. (1984) Sulphur gas emissions in the boreal forest: the West Whitecourt case study V: stable sulfur isotopes. *Water, Air and Soil Pollution*, **22**, 321–347.

Krouse, H.R., Stewart, J.W.B. & Grinenko, V.A. (1991) Pedosphere and biosphere. In: *Stable Isotopes: Natural and Anthropogenic Sulphur in the Environment* (Eds. H.R. Krouse and V.A. Grinenko), pp. 267–306. SCOPE 43, John Wiley and Sons, Chichester.

Kusakbe, M., Rafer, J.D., Stout, J. & Colle, T.W. (1976) Isotopic ratios of sulphur extracted from some plants, soils and related materials. *New Zealand Journal of Science*, **19**, 433–440.

Lake, J.L., McKinney, R.A., Osterman, F.A., et al. (2001) Stable nitrogen isotopes as indicators of anthropogenic activities in small freshwater systems. *Canadian Journal of Fisheries and Aquatic Science*, **58**, 870–878.

Laws, E.A., Popp, B.N., Bidigare, R.R., et al. (1995) Dependence of phytoplankton isotopic composition on growth rate and $[CO_2]_{aq}$: theoretical considerations and experimental results. *Geochimica et Cosmochimica Acta*, **59**, 1131–1138.

LaZerte, B.D. (1983) Stable carbon isotope ratios: Implications for the source sediment carbon and for phytoplankton carbon assimilation in Lake Memphremagog, Quebec. *Canadian Journal of Fisheries and Aquatic Science*, **40**, 1658–1666.

LaZerte, B.D. & Szalados, J.E. (1982) Stable carbon isotope ratio of submerged freshwater macrophytes. *Limnology and Oceanography*, **27**(3), 413–418.

Leffler, A.J. & Evans, A.S. (1999) Variation in carbon isotope composition among years in the riparian tree *Populus fremontii*. *Oecologia*, **119**, 311–319.

Lehmann, M.F., Sigman, D.M. & Berelson, W.M. (2004) Coupling the $^{15}N/^{14}N$ and $^{18}O/^{16}O$ of nitrate as a constraint on benthic nitrogen cycling. *Marine Chemistry*, **88**, 1–20.

Loneragan, N.R., Bunn, S.E. & Kellaway, D.M. (1997) Are mangroves and seagrasses sources of organic carbon for penaeid prawns in a tropical Australian estuary? A multiple stable-isotope study. *Marine Biology*, **130**(2), 289–300.

MacAvoy, S.E., Macko, S.A. & Garman, G.C. (1998) Tracing marine biomass into tidal freshwater ecosystems using stable sulfur isotopes. *Naturwissenschaften*, **85**, 544–546.

MacAvoy S.E., Macko S.A., McIninch S.P. & Garman G.C. (2000) Marine nutrient contributions to freshwater apex predators. *Oecologia*, **122**(4), 568–573.

Macleod, N.A. & Barton, D.R. (1998) Effects of light intensity, water velocity, and species composition on carbon and nitrogen stable isotope ratios in periphyton. *Canadian Journal of Fisheries and Aquatic Sciences*, **55**, 1919–1925.

Mariotti, A., Lancelot, C. & Billen, G. (1984) Natural isotopic composition of nitrogen as a tracer of origin for suspended organic matter in the Scheldt estuary. *Geochimica et Cosmochimica Acta*, **48**, 549–555.

Mariotti, A., Gadel, F., Giresse, P. & Mouzeo, K. (1991) Carbon isotope composition and geochemistry of particulate organic matter in the Congo River (Central Africa): Application to the study of Quaternary sediments off the mouth of the river. *Chemical Geology*, **86**, 345–357.

Mayer, B., Boyer, E.W., Goodale, C., et al. (2002) Sources of nitrate in rivers draining sixteen watersheds in the northeastern US: Isotopic constraints, *Biogeochemistry*, **57**, 171–197.

McCarthy, I.D. & Waldron, S. (2000) Identifying migratory *Salmo trutta* using carbon and nitrogen stable isotope ratios. *Rapid Communications in Mass Spectrometry*, **14**, 1325–1331.

McClain, M.E., Boyer, E.W., Dent, C.L., et al. (2003) Biogeochemical hot spots and hot moments at the interface of terrestrial and aquatic ecosystems. *Ecosystems*, **6**, 301–312.

McClelland, J.W. & Valiela, I. (1998) Linking nitrogen in estuarine producers to land-derived sources. *Limnology and Oceanography*, **43**, 577–585.

McCusker, E.M., Ostrom, P.H., Ostrom N.E., et al. (1999) Seasonal variation in the biogeochemical cycling of seston in Grand Traverse Bay, Lake Michigan. *Organic Geochemistry*, **30**, 1543–1557.

McCutchan, J.H. & Lewis, W.M. (2001) Seasonal variation in stable isotope ratios of stream algae. *Verhandlungen Internationale Vereinigung für Theoretische und Angewandte Limnologie*, **27**, 3304–3307.

McCutchan, J.H. & Lewis, W.M. (2002) Relative importance of carbon sources for macroinvertebrates in a Rocky Mountain stream. *Limnology and Oceanography*, **47**, 742–752.

McCutchan, J.H., Lewis, W.M., Kendall, C. & McGrath, C.C. (2003) Variation in trophic shift for stable isotope ratios of carbon, nitrogen, and sulfur. *Oikos*, **102**, 378–390.

Mitchell, M.J, Krouse, H.R., Mayer, B., et al. (1998) Use of stable isotopes in evaluating sulfur biogeochemistry of forest ecosystems. In: *Isotope Tracers in Catchment Hydrology* (Eds C. Kendall & J.J. McDonnell), pp. 489–518. Elsevier, Amsterdam.

Mook, W.G. & Tan, F.C. (1991) Stable carbon isotopes in rivers and estuaries. In: *Biogeochemistry of Major World Rivers* (eds E. Degens, S. Kempe & J. Richey), pp. 245–264. J. Wiley & Sons, Chichester.

Mook, W.G., Bommerson, J.C. & Staverman, W.H. (1974) Carbon isotope fractionation between dissolved bicarbonate and gaseous carbon dioxide. *Earth and Planetary Science Letters*, **22**, 169–176.

Mulholland, P.J., Tank, J.L., Sanzone, D.M., et al. (2000) Food resources of stream macroinvertebrates determined by natural-abundance stable C and N isotopes and a N-15 tracer addition. *Journal of the North American Benthological Society*, **19**, 145–157.

Nadelhoffer, K.J. & Fry, B. (1994) Nitrogen isotope studies in forest ecosystems. In: *Stable Isotopes in Ecology and Environmental Science* (Eds K. Lajtha & R.M. Michener), pp. 22–44. Blackwell Scientific Publishers, Oxford.

Needoba, J.A., Sigman, D.M. & Harrison, P.J. (2004) The mechanism of isotope fractionation during algal nitrate assimilation as illuminated by the N-15/N-14 of intracellular nitrate. *Journal of Phycology*, **40**, 517–522.

Nriagu, J.O., Rees, C.E., Mekhtiyeva, V.L., et al. (1991) Hydrosphere. In: *Stable Isotopes: Natural and Anthropogenic Sulphur in the Environment*, SCOPE 43 (Scientific Committee on Problems of the Environment) (Eds H.R. Krouse & V. Grinenko), pp. 177–265. J. Wiley & Sons, Chichester.

O'Reilly, C.M., Hecky, R.E., Cohen, A.S. & Plisnier, P.D. (2002) Interpreting stable isotopes in food webs: Recognizing the role of time averaging at different trophic levels. *Limnology and Oceanography*, **47**, 306–309.

Ostrom, N.E., Hedin, L.O., Von Fischer, J.C. & Robertson, G.P. (2002) Nitrogen transformations and NO_3 removal at a soil–stream interface: A stable isotope approach. *Ecological Applications*, **12**, 1027–1043.

Parker, S.R., Poulson, S.R., Gammons, C.H. & DeGrandpre, M.D. (2005) Biogeochemical controls on diel cycling of stable isotopes of dissolved O2 and dissolved inorganic carbon in the Big Hole River, Montana. *Environmental Science and Technology*, **39**, 7134–7140.

Pennock, J., Velinsky, D., Ludlam, J., et al. (1996) Isotopic fractionation of ammonium and nitrate during uptake by Skeletonema costatum: Implications for $\delta^{15}N$ dynamics under bloom conditions. *Limnology and Oceanography*, **41**(3), 451–459.

Pentecost, A. (1995) Geochemistry of carbon dioxide in six travertine-depositing waters of Italy. *Journal of Hydrology*, **167**, 263–278.

Peterson, B.J. & Fry, B. (1987) Stable isotopes in ecosystem studies. *Annual Review of Ecology and Systematics*, **18**, 293–320.

Peterson, B.J. & Howarth, R.W. (1987) Sulfur, carbon, and nitrogen isotopes used to trace organic matter flow in the salt-marsh estuaries of Sapelo Island, Georgia. *Limnology and Oceanography*, **32**, 1195–1213.

Peterson, B.J., Howarth, R.W. & Garrett, R.H. (1985) Multiple stable isotopes used to trace the flow of organic matter in estuarine food webs. *Science*, **227**, 1361–1363.

Phillips, D.L. & Gregg, J.W. (2003) Source partitioning using stable isotopes: coping with too many sources. *Oecologia*, **136**, 261–269.

Phillips, D.L. & Koch, P.L. (2002) Incorporating concentration dependence in stable isotope mixing models. *Oecologia*, **130**, 114–125.

Post, D.M. (2002) Using stable isotopes to estimate trophic position: Models, methods, and assumptions. *Ecology*, **83**, 703–718.

Power, M.E. & Dietrich, W.E. (2002) Food webs in river networks. *Ecological Research*, **17**, 451–471.

Quay, P.D., Wilbur, D.O., Richey, J.E., et al. (1992) Carbon cycling in the Amazon River: Implications from the ^{13}C compositions of particles and solutes. *Limnology and Oceanography*, **37**, 857–871.

Raven, J.A. & Beardall, J. (1981) Carbon dioxide as the exogenous inorganic carbon source for Batrachospermum and Lemanea. *British Phycological Journal*, **16**, 165–175.

Redfield, A.C. (1958) The biological control of chemical factors in the environment. *American Scientist*, **46**, 205–221.

Rostad, C.E., Leenheer, J.A. & Daniel, S.R. (1997) Organic carbon and nitrogen content associated with colloids and suspended particules from the Mississippi River and some of its tributaries. *Environmental Science and Technology*, **31**, 3218–3225.

Rounick, J.S. & James, M.R. (1984) Geothermal and cold springs faunas: inorganic carbon sources affect isotopic values. *Limnology and Oceanography*, **29**, 386–389.

Rounick, J.S., Winterbourn, M.J. & Lyon, G.L. (1982) Differential utilization of allochthonous and autochthonous inputs by aquatic invertebrates in some New Zealand streams: a stable carbon isotope study. *Oikos*, **39**, 191–198.

Sebilo, M., Billen, G., Grably, M. & Mariotti, A. (2003) Isotopic composition of nitrate-nitrogen as a marker of riparian and benthic denitrification at the scale of the whole Seine River system. *Biogeochemistry*, **63**, 35–51.

Sebilo, M., Mayer, B., Grably, M., et al. (2004) The use of the "ammonium diffusion" method for $\delta^{15}N$–NH_4^+ and $\delta^{15}N$–NO_3^- measurements: comparison with other techniques. *Environmental Chemistry*, **1**, 99–103.

Sigman, D.M., Altabet, M.A., Micheneer, R., et al. (1997) Natural abundance-level measurements of the nitrogen isotopic composition of oceanic nitrate: an adaptation of the ammonia diffusion method. *Marine Chemistry*, **57**, 227–242.

Sigman, D.M., Casciotti, K.L., Andreani, M., et al. (2001) A bacterial method for the nitrogen isotopic analyses of nitrate in seawater and freshwater. *Analytical Chemistry*, **73**, 4145–4153.

Singer, G.A., Panzenbock, M., Weigelhofer, G., et al. (2005) Flow history explains temporal and spatial variation of carbon fractionation in stream periphyton. *Limnology and Oceanography*, **50**, 706–712.

Schlacher, T.A. & Wooldridge, T.H. (1996) Origin and trophic importance of detritus – evidence from stable isotopes in the benthos of a small, temperate estuary. *Oecologia*, **106**, 382–388.

Smith, B.N. & Epstein, S. (1971) Two categories of $^{13}C/^{12}C$ ratios for higher plants. *Plant Physiology*, **47**, 380–384.

Spence, M.J., Bottrell, S.H., Thornton, S.F. & Lerner, D.N. (2001) Isotopic modelling of the significance of bacterial sulphate reduction for phenol attenuation in a contaminated aquifer. *Journal of Contaminant Hydrology*, **53**, 285–304.

Sterner, R.W. & Elser, J.J. (2002) *Ecological Stoichiometry: The Biology of Elements from Molecules to the Biosphere*. Princeton University Press, Princeton, NJ.

Thorp, J.H., Delong, M.D., Greenwood, K.S. & Casper, A.F. (1998) Isotopic analysis of three food web theories in constricted and floodplain regions of a large river. *Oecologia*, **117**, 551–563.

Troughton, J.H. (1972) Carbon isotopic fractionation in plants. *Proceedings of the 8th Conference on Radiocarbon Dating*, Lower Hutt, 18–25 October, Vol. 2, 39–57.

Trudeau, W. & Rasmussen, J.B. (2003) The effect of water velocity on stable carbon and nitrogen isotope signatures of periphyton. *Limnology and Oceanography*, **48**, 2194–2199.

Trust, B.A. & Fry, B. (1992) Stable sulphur isotopes in plants: a review. *Plant, Cell, and Environment*, **15**, 1105–1110.

Tucker, J., Sheats, N., Giblin, A.E., et al. (1999) Using stable isotopes to trace sewage-derived material through Boston Harbor and Massachusetts Bay. *Marine Environmental Research*, **48**, 353–375.

Udy, J.W. & Bunn, S.E. (2001) Elevated $\delta^{15}N$ values in aquatic plants from cleared catchments: why? Marine and Freshwater Research, **52**, 347–351.

Vander Zanden, M.J. & Rasmussen, J.B. (1999) Primary consumer $\delta^{13}C$ and $\delta^{15}N$ and the trophic position of aquatic consumers. Ecology, 80, 1395–1404.

Vander Zanden, M.J. & Rasmussen, J.B. (2001) Variation in delta N-15 and delta C-13 trophic fractionation: Implications for aquatic food web studies. *Limnology and Oceanography*, **46**, 2061–2066.

Vander Zanden, M.J. & Vadeboncoeur, Y. (2002) Fishes as integrators of benthic and pelagic food webs in lakes. *Ecology*, **83**(8), 2152–2161.

Vander Zanden, M.J., Cabana, G. & Rasmussen, J.B. (1997) Comparing trophic position of freshwater fish calculated using stable nitrogen isotope ratios (delta N-15) and literature dietary data. *Canadian Journal of Fisheries and Aquatic Sciences*, **54**, 1142–1158.

Vanderklift, M.A. & Ponsard, S. (2003) Sources of variation in consumer-diet delta N-15 enrichment: a meta-analysis. *Oecologia*, **136**, 169–182.

Van Dover, C.L., Grassle, J.F., Fry, B., et al. (1992) Stable isotope evidence for entry of sewage-derived organic material into a deep-sea food web. *Nature*, **360**, 153–155.

Vannote, R.L., Minshall, G.W., Cummins, K.W., et al. (1980) The river continuum concept. *Canadian Journal of Fisheries and Aquatic Sciences*, **37**, 130–137.

Venkiteswaran, J.J., Wassenaar, L.I. & Schiff, S.L. (In press) Dynamics of dissolved O_2 isotopic ratios: a transient model to quantify primary production, community respiration, and air-water exchange in aquatic ecosystems. *Oecologia*.

Vitousek, P.M., Mooney, H.A., Lubchenco, J. & Melillo, J. (1997) Human domination of Earth's ecosystems. *Science*, **277**, 494–499.

Wadleigh, M.A. & Blake, D.M. (1999) Tracing sources of atmospheric sulphur using epiphytic lichens. *Environmental Pollution*, **106**, 265–271.

Wadleigh, M.A., Schwarcz, H.P. & Kramer, J.R. (1996) Isotopic evidence for the origin of sulphate in coastal rain. *Tellus*, **48B**, 44–59.

Wankel, S.D., Kendall, C., Francis, C.A. & Paytan, A. (2006) Nitrogen sources and cycling in the San Francisco Bay Estuary: A nitrate dual isotopic composition approach. *Limnology and Oceanography*, **51**(4), 1654–1664.

Wassenaar, L.I. & Hobson, K.A. (1998) Natal origins of migratory monarch butterflies at wintering colonies in Mexico: New isotopic evidence. *Proceedings of the National Academy of Sciences, USA*, **95**, 15436–15439.

Wayland, M. & Hobson, K.A. (2001) Stable carbon, nitrogen, and sulfur isotope ratios in riparian food webs on rivers receiving sewage and pulp-mill effluents. *Canadian Journal of Zoology*, **79**, 5–15.

Wissel, B. & Fry, B. (2005) Sources of particulate matter in the Mississippi River, USA. *Large Rivers. Archiv für Hydrobiologie, Supplementband*, **15**(1–4), 105–118.

Woodward, G. & Hildrew, A.G. (2002) Food web structure in riverine landscapes. *Freshwater Biology*, **47**, 777–798.

Yun, M., Mayer, B. & Taylor, S.W. (2005) $\delta^{34}S$ measurements on organic materials by continuous flow isotope ratio mass spectrometry. *Rapid Communications in Mass Spectrometry*, **19**, 1429–1436.

CHAPTER 11

Stable isotope tracers in watershed hydrology

KEVIN McGUIRE AND JEFF McDONNELL

Introduction

Watershed hydrology is a field of study that concerns itself with questions of where water goes when it rains, what flowpaths the water takes to the stream, and how long water resides in the watershed. Even though these questions seem basic and water-focused, they often form the underpinning for questions of water availability, biogeochemical cycling, microbial production, and other ecological processes that depend on the water cycle. While stable isotopes of water (e.g., $^1H_2^{18}O$ and $^1H^2H^{16}O$) have been used to study global-scale water cycling since the early 1950s (Epstein & Mayeda 1953; Craig 1961; Dansgaard 1964), they were not used for watershed-scale problems of water source, flowpath, and age until the 1970s (Dinçer et al. 1970). Sklash & Farvolden (1979) were among the first hydrologists to quantify the composition of stream water and its temporal and geographical sources using water isotopes in small watersheds. Since then, watershed-scale stable isotope hydrology has blossomed (Kendall & McDonnell 1998), and today, stable isotopes are a standard tool for helping hydrologists understand the basic functioning of watersheds. More importantly, stable isotope tracing and analysis forms an important link between hydrological and ecological processes at the watershed scale where knowledge of flow path, water source, and age inform many water-mediated ecological processes.

This chapter shows how an understanding of watershed hydrology is fundamental to watershed ecology. We also show that rather basic isotope techniques can help to better understand water quality, sustainability, land-use change effects, nutrient cycling, and general terrestrial and aquatic system interactions. We first review basic concepts in watershed and stable isotope hydrology, and then present some isotope-based approaches relevant to the hydrology–ecology interface.

Basic concepts in watershed hydrology

Our introduction to the basic concepts in watershed hydrology is to provide readers with a background for understanding hydrological systems so that cross-disciplinary linkages are realized. For more advanced material pertinent to isotope hydrology, the reader may wish to consult one of several good books and book chapters written on this topic by Gat & Gonfiantini (1981), Sklash (1990), Coplen (1993), Coplen et al. (2000), Clark & Fritz (1997), Kendall & McDonnell (1998), and Buttle & McDonnell (2004). Here, we restrict ourselves to an overview of stable isotope techniques in small watersheds, which we define as 10^{-2} to $10^{2}\,km^{2}$. Our overview of watershed hydrology is from a process-oriented perspective, i.e., focused on physical and functional relationships to the generation of streamflow (the drainage of water to streams). More detailed treatment of this topic can be found in Anderson & Burt (1990), Bonell (1998), Buttle (1998), Dunne & Leopold (1978), and Ward & Robinson (2000).

The water balance

Watersheds are hydrologic systems where inputs and outputs of water, sediment and nutrients are cycled within topographically restricted landscape units (Dunne & Leopold 1978). As such, the watershed serves as the control volume where mass is conserved according to the following water balance equation:

$$\frac{dS}{dt} = I - O = P - Q - ET \tag{11.1}$$

where dS/dt are changes in water storage within the watershed, I are watershed inputs, equivalent to P (precipitation), and O are the watershed outputs. Variables Q and ET are the streamflow discharge (runoff) and the evapotranspiration, respectively. This equation can be further simplified when looking at long-term averages, since changes in the volume of stored water (dS/dt) are typically small compared with the remaining terms; thus, dS/dt can be neglected. While equation 11.1 illustrates the most simple of conceptual hydrologic frameworks, the dynamic terms on the right-hand side of the equation can be difficult to quantify or understand in detail. This is especially the case for the transfer between terms (i.e., flow pathways), and is where isotope tracers have been most useful.

Streamflow generation processes

Water flow pathways control many ecological processes, biochemical transformations, exchange reactions, and mineral weathering rates. For example, stream nutrient dynamics are often very sensitive lateral flow paths through shallow organic mats or other zones where water may mobilize or flush labile constituents. Flow paths determine largely the geochemical evolution along the flow gradient and the contact time in the subsurface (or residence time) has much control on the translocation of weatherable products in the soil and bedrock.

Equation 11.1 shows that precipitation is balanced by the sum of stream discharge and evapotranspiration. Thus, the proportion of precipitation that contributes to streamflow is what remains after considering several losses, including the evaporation of intercepted precipitation by the vegetation canopy and ground cover (e.g., litter), evaporation from the soil, and transpiration. Transpiration (i.e., passive water loss through plant stomata driven by climatic forces) is generally assumed to be minimal during storm events, since vapor pressure deficits are low and leaf surfaces are wet (Penman 1963). During wet canopy conditions, transpiration reduction is partly compensated by the evaporation of intercepted precipitation (Stewart 1977; Klaassen 2001). However, transpiration exerts significant control on antecedent soil moisture conditions by plant extraction of water in the rooting zone as described in Marshall et al., this volume, pp. 22–60. The net precipitation remaining after these loss terms are removed may be delivered to the stream through a variety flow pathways as shown in Figure 11.1.

Channel precipitation

The most rapid precipitation contribution to streamflow is from precipitation that falls directly onto the channel or near-stream saturated areas, which can become incorporated directly and immediately into streamflow (channel precipitation) (Figure 11.1). Under most conditions, this term is generally small, since stream channels represent 1–2% of the total watershed area. However, as channels and saturated areas (where near-stream groundwater tables rise to and intersect the soil surface) expand during storms or seasonally, this contribution can increase and have major impacts on the chemical dilution of stream water. Channel precipitation can account for approximately 30% of stormflow in some watersheds and is typically highest (as a percent of total runoff) for low antecedent wetness conditions and low storm intensities (Crayosky et al. 1999) where runoff response ratios are low (i.e., where runoff divided by total storm precipitation is low).

Overland flow

Once the net precipitation reaches the soil surface it will move vertically into the soil at a rate less than the infiltration capacity and (under certain conditions) contribute to streamflow as a subsurface flow source (Figure 11.1). If the rainfall intensity exceeds the infiltration capacity of the soil, surface ponding will fill small depressions, which eventually connect to form rill-like sheets of overland flow (Smith & Goodrich 2005). Overland flow will continue and contribute as surface runoff as long as infiltration capacity is exceeded as the water moves over downslope soils; otherwise it infiltrates and becomes one of the subsurface flow paths shown on the right side of the Figure 11.1. This process was first described by Horton (1933) and is now termed Hortonian or infiltration-excess overland flow (although recent

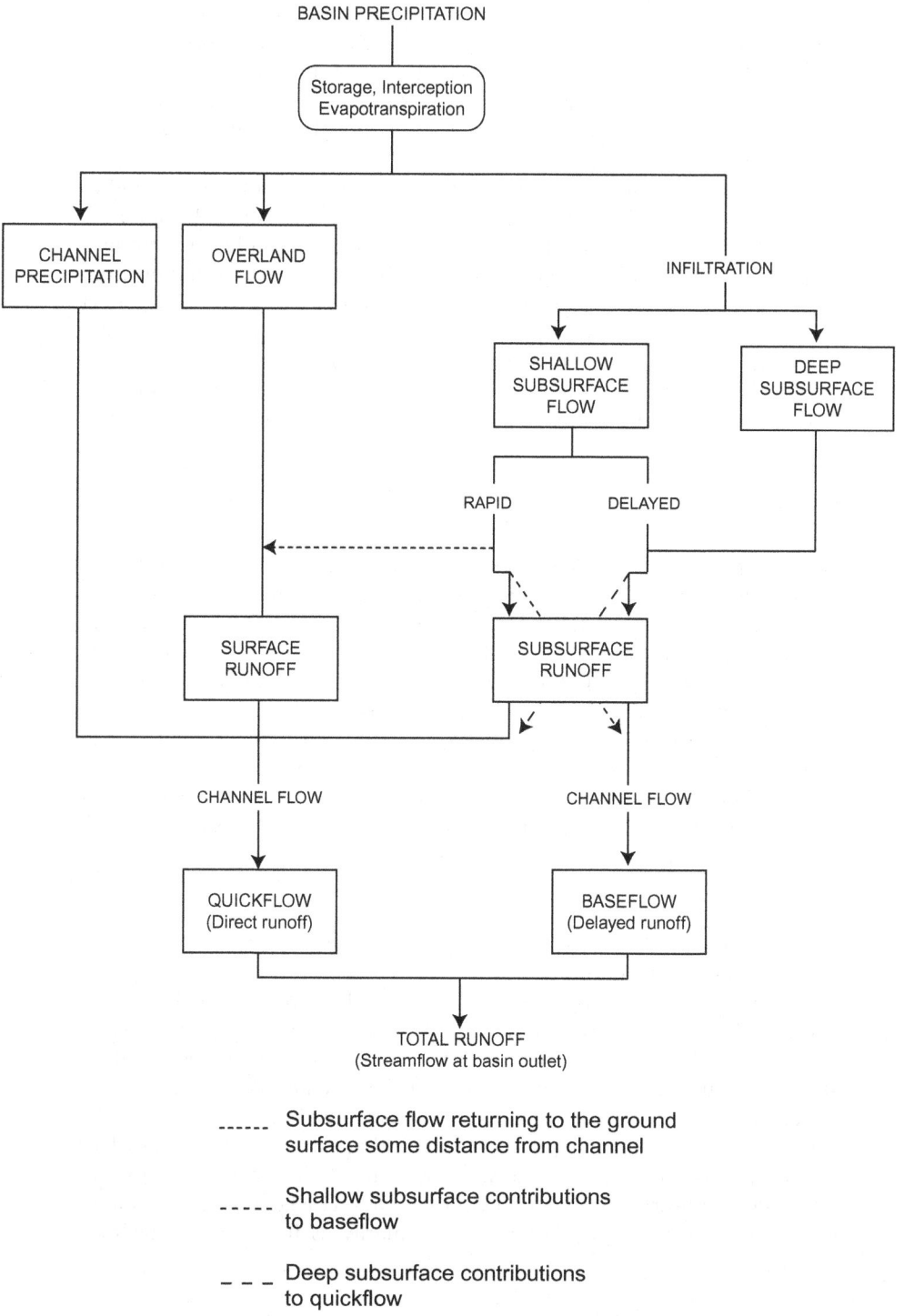

Figure 11.1 A representation of flow paths through a watershed from precipitation (rainfall and snowmelt) to streamflow at the basin outlet. (Modified after Buttle (1998) and Ward & Robinson (2000).)

papers have clarified Horton's perception of this and other runoff generating processes – see Beven (2004)). In most undisturbed forested ecosystems, the precipitation rate (e.g., a 25-yr return period storm for the southern USA is about $10\,cm\,h^{-1}$ for a 1-h duration) rarely exceeds the infiltration capacity of soil (e.g. $>20\,cm\,h^{-1}$) and therefore, the dominant flow paths are generally subsurface.

Subsurface flow may return to the surface and contribute to overland as groundwater exfiltration or seepage (Dunne & Black 1970; Eshleman et al. 1993). This is an overland-flow-producing effect but, unlike infiltration-excess overland flow (saturation from above), it is surface saturation from below. The area on the hillslope where this occurs will also receive direct precipitation onto pre-saturated areas developed from shallow water tables emerging at the soil surface, and together, return flow and direct precipitation onto saturated areas are termed saturation-excess overland flow (see (Anderson & Burt 1990) for detailed treatment and review of this process).

Subsurface flow

Subsurface flow processes are often considerably slower, more tortuous, and more difficult to discern than overland flow processes. First, we consider two major mechanisms that deliver subsurface water to streams: transport and displacement. Transport is defined as the movement of water according the pore water velocity field (Freeze & Cherry 1979). Therefore the physical processes of advection, diffusion, and dispersion affect water transport. Displacement, on the other hand, is much faster than actual flow velocities and can be characterized as the pressure propagation of precipitation rates through the saturated zone, which affects the discharge rate (Horton & Hawkins 1965; Beven 1989; Rasmussen et al. 2000). An example of the displacement or piston process is illustrated by considering water that enters a garden hose is not the same water that immediately exits the hose at the opposite end. Thus, in the watershed context, new rainfall may displace water to the stream, which previously had been stored in the soil mantle (Zimmermann et al. 1966).

We broadly separate subsurface flow into shallow and deep processes and consider the deep subsurface flow to be largely the groundwater flow component, i.e., saturated zone flow. However, saturated flow can also occur as shallow subsurface flow, sometimes called subsurface stormflow (e.g., perched water tables at the soil–bedrock interface or at some impeding horizon in the soil profile) (see the following section). We distinguish the shallow subsurface and deep subsurface flow processes based on depth and regional extent, where deep subsurface flow is thought to occur over a larger regional aquifer system. Groundwater flow is regarded primarily as flow through bedrock and/or confined to lowland areas (i.e., near stream) of a watershed such that it mimics the general topographic form of the drainage basin. Under driven conditions (i.e., during precipitation), groundwater may respond rapidly and contribute to streamflow via the piston-displacement mechanism, which is represented by the large dashed line within the subsurface runoff box in Figure 11.1. During non-driven periods, groundwater flow through bedrock, soils, and the near-stream zone sustains low flows through the dry season. Since water at depth, where soil permeability is often some orders of magnitude lower than the surface soil horizons, can only move slowly through connected pore space, outflow from groundwater may lag behind the precipitation episode by days, weeks, or even years.

In undisturbed forested watersheds in upland terrain, shallow subsurface flow often dominates the stream stormflow response. The specific processes that give rise to this component vary with climate, soils, and geology. At any particular location in the watershed, the initiation of shallow subsurface flow is highly dependent upon antecedent moisture conditions. Therefore, evapotranspiration (largely transpiration) exerts a major control on the generation of subsurface flow processes, mainly by establishing the initial moisture deficit necessary to overcome by removing water from the rooting zone. Shallow subsurface flow, which is often termed subsurface stormflow or throughflow, is very threshold dependent and describes the lateral movement (i.e., downslope) of water in the soil profile within the time frame of a storm hydrograph. There are numerous mechanisms ascribed to the formation of subsurface stormflow; however, in most cases, it represents a 'quickflow' pathway, meaning that it rapidly contributes to the formation of the hydrograph rise.

Shallow subsurface flow

Shallow subsurface flow processes have perplexed hydrologists since the early work of Hursh (1936, 1944) and to some extent are still ignored as a contribution to the storm event response. As mentioned above, overland flow is rarely observed in undisturbed upland watersheds; thus, hydrographs are largely composed of rapid subsurface flow and saturation-excess sources. The challenge has been in explaining how subsurface flow can so rapidly cause a streamflow response when measured soil matrix hydraulic conductivity data often contradict seemingly high soil water velocities. Observations have shown that two major processes give rise to rapid subsurface flow: (i) the rapid displacement of water stored in the watershed prior to the onset of precipitation, and (ii) preferential flow, mainly in the form of macropore flow. The displacement flow process, termed translatory flow by Hewlett & Hibbert (1967), suggests that streams can respond to rainfall inputs even though individual water molecules only travel centimeters or meters per day. This process is most effective when soils are at or near saturation and is assisted by the frequently observed decrease in saturated hydraulic conductivity with depth in soil profiles (Taha et al. 1997; Buttle 1998). However, recent studies have indicated that pressure propagation in unsaturated soils causes a similar response (Torres et al. 1998; Williams et al. 2002) by the thickening of water films around soil particles and resulting in a water flux pulse as saturated conditions are approached (Hewlett & Hibbert 1967).

Water percolating vertically through the soil may encounter permeability decreases with depth (generally, the hydraulic conductivity decreases exponentially) that can cause localized areas of transient saturation (or near-saturation). When this happens in steeply sloping terrain, the gravitational component of the soil water potential causes flow vectors to move in a lateral direction, which might only occur briefly during storm events (Weyman 1973; Harr 1977; Torres et al. 1998). Lateral flow will increase as the soil approaches saturation because the hydraulic conductivity increases nearly exponentially with degree of saturation. As the saturated layer (i.e., perched water table) develops and extends upward in the soil profile into more transmissive soils, an additional water flux increase is often observed called the transmissivity feedback (Kendall et al. 1999; Bishop et al. 2004; Laudon et al. 2004). The development of lateral flow and transient saturation also is assisted by flow convergence in topographic and bedrock hollows (Beven 1978; Tsuboyama et al. 2000), along bedrock surfaces (Freer et al. 2002), and adjacent to bedrock exfiltration zones (Anderson et al. 1997; Uchida et al. 2003).

A rapid conversion from near-saturation (e.g., capillary fringe) to saturation can also occur in the soil profile when large inputs from rainfall or snowmelt combine with low effective porosity soils, yielding a disproportionately large and rapid rise in the water table (Abdul & Gillham 1984; Gillham 1984; Ragan 1968; Sklash & Farvolden 1979). This response occurs typically at the toe of the hillslope or near-stream zone and resembles a groundwater ridge or mound. The groundwater ridge induces locally steepened hydraulic gradients, which enhances groundwater discharge to the stream, and some studies have shown that the gradient on the other side of the mound is reversed back toward the hillslope (Bates et al. 2000; Burt et al. 2002). However, the applicability of the groundwater ridging mechanism has been questioned for soils with little capillary fringe development (i.e., coarse textured soils) (McDonnell & Buttle 1998). Soils that do develop a significant capillary fringe tend to have low saturated hydraulic conductivity, which conflicts with the hypothesis that it rapidly contributes to stormflow generation (Cloke et al. 2006).

Rapid flow through non-capillary soil pores (i.e., macropores) caused by root channels, animal burrows, cracks/fissures, or simply coarse textured or aggregated soils, is also frequently evoked as a major subsurface stormflow mechanism, especially in forested watersheds (Mosley 1979; Beven & Germann

1982; McDonnell 1990). Flow through macropores is conditional on saturation of the surrounding soil matrix or flow through the macropores exceeding the rate of loss to the surrounding matrix. Macropore flow and other preferential flow processes produced by wetting front instability (i.e., fingering) in unsaturated soils (Hill & Parlange 1972; Hillel 1998), cause accelerated movement of water to depth often bypassing portions of the soil matrix that can ultimately trigger a rapid conversion to saturated conditions at depth in the soil profile, where effective porosity is low compared with shallow soils (McDonnell 1990; Buttle & Turcotte 1999). Subsequently, the location of macropores and soil pipes (Jones 1971) that occur near the bedrock interface can enhance lateral drainage from hillslopes (Uchida et al. 2001).

Contributing source areas

The temporal and spatial nature of the aforementioned streamflow generation processes changes in response to antecedent moisture, precipitation intensity, and season, which are reflected by the varying extent of surface saturated areas produced in the watershed. This concept, which was introduced in the USA by Hewlett (1961) and simultaneously by Cappus (1960) in France and Tsukamoto (1961) in Japan, remains the major theoretical paradigm of streamflow generation. Saturated areas present an opportunity for the rapid conversion of rainfall to streamflow and thus are considered the primary contributing source area in a watershed. However, it is important to note that even though saturated areas expand and contract reflecting the storm response, those areas are not necessarily the only sources that actively contribute to stormflow (Ambroise 2004). Disjunct areas of the watershed must be hydrologically connected to organized drainage for some period of time to be considered a contributing source area. Connectivity may occur via surface saturated area development (Burt & Butcher 1985; Grayson et al. 1997), water table development (Stieglitz et al. 2003; Tromp-van Meerveld & McDonnell 2006), or by the generation of subsurface flow networks (Sidle et al. 2001). Often hydrologic connectivity is threshold driven such that a specific soil moisture state is needed prior to activating runoff from an area within the watershed (Bazemore et al. 1994; Grayson et al. 1997; Sidle et al. 2000; McGlynn & McDonnell 2003). Many observations have indicated that hillslope connections to near-stream zones also operate as thresholds requiring specific antecedent conditions prior to activation (McDonnell et al. 1998; Freer et al. 2002; McGlynn & McDonnell 2003). Recent work indicates that the threshold is not necessarily controlled by moisture status alone, but the depth to bedrock depressions, which fill to form transient saturated zones that connect and flow downslope depending upon event size and bedrock topography (Buttle et al. 2004; Tromp-van Meerveld & McDonnell 2006).

Why are stable isotopes needed?

Given the importance of overland and subsurface flow pathways to ecological processes (e.g. flushing of labile nutrients, etc), spatial and temporal resolution of these myriad pathways and processes is important. As Beven (1989) notes:

> . . . there is a continuum of surface and subsurface processes by which a hillslope [or watershed] responds to a storm rainfall, depending on the antecedent conditions, rainfall intensities, and physical characteristics of the slope and soil. . . . Individual storm responses may involve all of these processes [that we discuss above in this chapter] occurring in different parts of the same catchment, or different mechanisms occurring in the same part in different storms, or different times within the same storm.

It has been difficult to discern these processes using physical data alone. This is because the fluctuations in physical parameters, for instance ground-water levels, can arise from a variety of processes that can result in similar response patterns. In addition, many physical measures are point measurements and do not integrate hydrologic behavior to a scale that we are interested in such as a watershed or hillslope. Thus, other information is needed to help explain the movement and occurrence of water at more integrative scales.

Stable isotope tracers have been among the most useful tools employed to sort through Beven's surface–subsurface continuum to define the dominant runoff producing processes, geographic source of water comprising the storm hydrology, the time source separation of the flow response, and residence time of water in the subsurface. The next section of this chapter presents water stable isotope fundamentals as a starting point for how one might employ these techniques to resolve the age, origin, and pathway of runoff at the watershed scale.

General concepts in isotope hydrology

Isotope hydrology is based on the notion of tracing a water molecule through the hydrological cycle. Devine & McDonnell (2004) note that non-natural constituents have been widely used for centuries to characterize flowpaths and estimate ground water velocities. The Jewish historian Flavius Josephus recorded in approximately 10 CE that chaff was used as a tracer to link the spring source of the Jordan River to a nearby pond. More quantitative tracer tests using chloride, fluorescein and bacteria were employed in the large karst regions of Europe in the late 1800s and early 1900s (Devine & McDonnell 2004).

Stable isotopes of water (hydrogen (^{2}H or D for deuterium) and oxygen (^{18}O)) have been used since the pioneering work of Craig (1961). Unlike applied tracers, stable isotopes are added naturally at the watershed scale by rain and snowmelt events. These environmental isotopes (applied through meteoric processes) can be used to trace and identify different air and water masses contributing precipitation to a watershed since the stable isotope composition of water changes only through mixing and well-known fractionation processes that occur during evaporation and condensation. Once in the subsurface, and away from evaporative effects, the stable isotopes of water are conservative in their mixing relationships. This means that isotopic composition of the mixture of two water sources will fall on a straight line and its position is dependent only on the proportions of the two sources. Also, ^{2}H and ^{18}O, the elemental basis for H_2O molecules, are ideal tracers because they behave exactly as water would as it undergoes transport through a watershed. Water entering a watershed will have a characteristic fingerprint of its origin and therefore can help identify where the water in the stream comes from.

The isotopic composition of water is expressed as the ratio of the heavy to light isotopes (e.g., ^{18}O/^{16}O) relative to a standard of known composition:

$$\delta \text{ (in ‰ or per mil)} = (R_x/R_s - 1) \times 1000 \tag{11.2}$$

where R_x and R_s are the isotopic ratios of the sample and standard, respectively. The agreed upon standard issued by the International Atomic Energy Agency (IAEA) is Vienna-Standard Mean Ocean Water or VSMOW (Coplen 1996). The isotopic composition of water is determined by mass spectrometry (Kendall & Caldwell 1998).

Isotopic fractionation

Oxygen-18 and deuterium occur in water at abundances of 0.204% of all oxygen atoms and 0.015% of all hydrogen atoms, respectively (Clark & Fritz 1997). These relative abundances change slightly as a result of thermodynamic reactions that fractionate or partition atoms of different mass (isotopes vary in mass since they are defined as an element with the same number of protons, but different number of neutrons), which provides the unique isotopic composition indicative of the water source and process of formation. The isotopic fractionation in water occurs through diffusion during physical phase changes such as evaporation, condensation, and melt. Fractionation is strongly temperature dependent such that it is greater at low temperature (Majoube 1971). During phase changes, diffusion rates differ due to the differences in bond strength between lighter and heavier isotopes of a given element. Molecular bonds between lighter isotopes ($H_2^{16}O$) are more easily broken than molecular bonds between heavier isotopes ($HD^{16}O$ and $H_2^{18}O$).

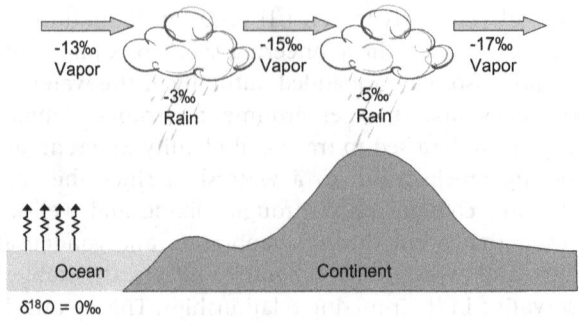

Figure 11.2 The diagram of isotopic composition of atmospheric water vapor over an ocean showing the processes of evaporation and rainout as the air mass proceeds over a continent. (Modified from Siegenthaler 1979.)

Heavy isotopic forms of water (i.e., with ^{18}O or ^2H) will require greater energy to break hydrogen bonds than water containing lighter isotopes and consequently, will react more slowly. For example, water vapor over large water bodies tends to be depleted in heavier isotopes (or enriched in lighter isotopes) relative to the evaporating water body (Figure 11.2). Stronger bonds indicate that heavy isotopic forms have lower saturation vapor pressures (i.e., the evaporation driving force) and thus, lower evaporation rates (i.e., diffusion across the water–atmosphere boundary layer). As the water vapor condenses from clouds to form precipitation, heavy isotopic forms will preferentially move into the liquid phase, which will be enriched in the heavy isotope compared with the residual water vapor. Under equilibrium conditions, the heavy isotopes are always enriched in the more condensed phases by an amount known as the fractionation factor, α. Further details of isotope fractionation can be found in Gat (1996), Kendall & Caldwell (1998), and Mook (2000).

Meteoric water line

The meteoric (or meteorological) water line (MWL) was first published by Craig (1961) and is a convenient reference for understanding and tracing water origin. It is a linear relation in the form of:

$$\delta D = 8\delta^{18}O + d \tag{11.3}$$

where d, the y-intercept, is the deuterium-excess (or d-excess) parameter when the slope $= 8$ (Dansgaard 1964). Craig's MWL, referred to as the Global MWL, with $d = 10$ and a slope of 8, was based on approximately 400 samples representing precipitation, rivers, and lakes from various countries (Figure 11.3). Cold regions are associated with waters depleted in heavy isotopes and

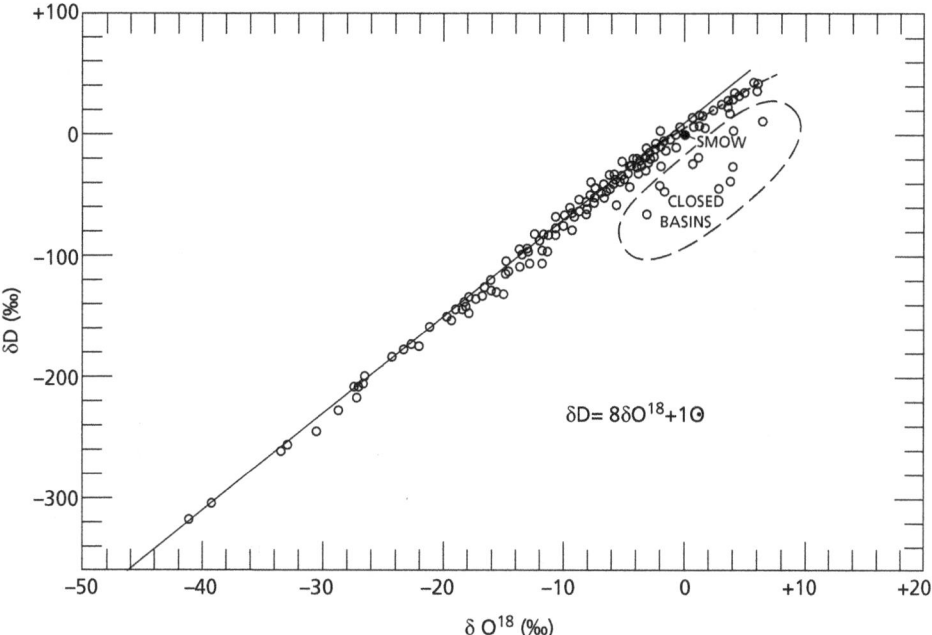

Figure 11.3 The global meteoric relationship between δD and $\delta^{18}O$ in water collected from rivers, lakes, rain, and snow by Craig (1961). Closed basins indicate areas where evaporation is significant and thus do not plot along the linear relation. Also, the dashed fit through the upper end of the data show enrichment of the heavy isotopes in samples collected from lakes in East Africa that experience evaporation effects. (Reprinted with permission from Craig, H. (1961) Isotopic variations in meteoric waters. *Science*, **133** (3465), 1702–1703. Copyright 1961 AAAS.)

warm regions tend to contain waters enriched in heavy isotopes (see Figure 11.4). The GMWL has been updated subsequently by (Rozanski et al. 1993) (δD = 8.17 (±0.07) $\delta^{18}O$ + 11.27 (±0.65) ‰) using weighted mean annual precipitation data from stations in the IAEA/World Meteorological Organization Global Network of Isotopes in Precipitation (GNIP). Local MWLs (linear δD – $\delta^{18}O$ relationships based on local precipitation measurements of at least a 1-year period) have been very useful for many water resource applications such as surface-water–groundwater interactions and evaporation effects. Local MWLs reflect variations in climate, rainfall seasonality, and geography by the deviations of the slope and *d*-excess value (see Figure 11.4). In most watershed studies, a LMWL would be constructed and used. Figure 11.4 shows that deviations from the GMWL can occur from humidity differences of the vapor source and from evaporation (as discussed later).

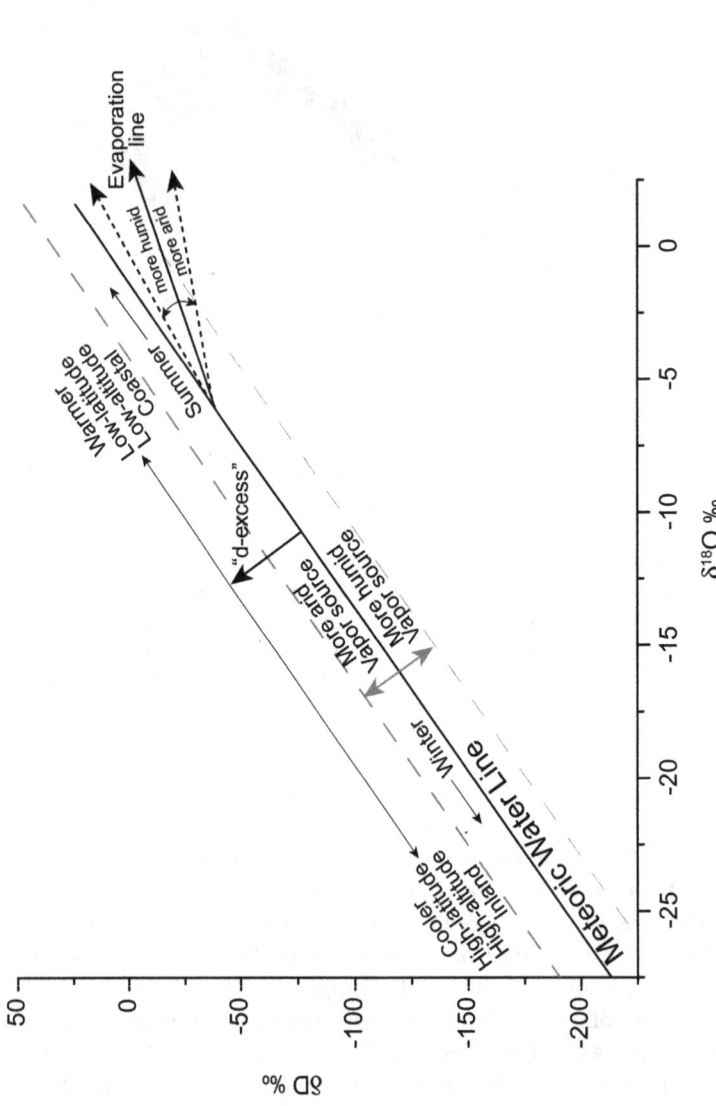

Figure 11.4 Schematic meteoric water line showing factors leading to deviations from the $\delta^{18}O$–δD relationship. (Modified after SAHRA 2005.)

Precipitation isotopic variation

An understanding of the processes that control the spatial and temporal distributions of precipitation isotopic composition is necessary since it is the ultimate source of water for all applications discussed in this chapter. Regional and global spatial distributions have been developed using interpolation schemes (Bowen & Revenaugh 2003) and atmospheric circulation models (Sturm et al. 2005), which help predict the isotopic input to watersheds. Temporal variations in the isotopic composition of precipitation have been used to evaluate climate change (Rozanski et al. 1992), recharge patterns (Winograd et al. 1998; Abbott et al. 2000), and residence time (Maloszewski et al. 1983; Pearce et al. 1986). The GNIP database contains stable isotope records for many sites around the world including spatial maps and animations of seasonal changes in the data for visualizing how precipitation stable isotope composition vary in time and space at the global scale (GNIP can be accessed from http://isohis.iaea.org/).

Precipitation has several so-called isotopic effects (or rules) that have been described and developed over the years and are useful to know for many of the watershed isotope tracing applications discussed in the following sections. The isotopic composition of precipitation is dependent upon several factors including the isotopic composition of its vapor source (typically from oceanic regions), fractionation that occurs as water evaporates into the air mass (sea-surface-temperature controlled), precipitation formation processes, and air mass trajectory (i.e., the influence of vapor source and rainout processes along the pathway of the air mass). Most of these factors are related to isotopic fractionation caused by phase changes.

As vapor masses form over ocean water, vapor pressure differences in water containing heavy isotopes impart disproportionate enrichments in the water phase during evaporation, which is dependent on sea surface temperatures (vapor pressure is higher for warmer regions such as equatorial regions), wind speed, salinity, and most importantly, humidity (Clark & Fritz 1997) (Figure 11.2). Rain will form from the vapor mass only through cooling that occurs from adiabatic expansion (no heat loss or gain) as warm air rises to lower pressures, or by heat loss through convection. Once the air cools to the dew point temperature, condensation and subsequent precipitation will occur and proceed to remove water vapor from the air mass. As the condensation temperature decreases, the δD and $\delta^{18}O$ values of precipitation also decrease. Then, as the system moves over continents, a rainout process causes the continual fractionation of heavy isotopes into the precipitation (i.e., according to a Rayleigh-like distillation, i.e., a slow process with immediate removal of the condensate) such that the residual vapor becomes progressively more depleted in heavy isotopes (Figure 11.2). Subsequent precipitation, while enriched with respect to the remaining vapor, will be depleted in heavy isotopes compared with previous precipitation from the same vapor

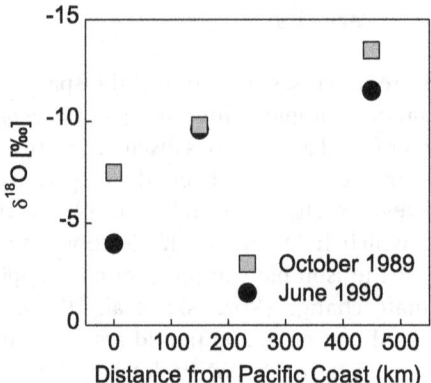

Figure 11.5 The $\delta^{18}O$ composition of precipitation (‰) collected during the first week of October in 1989 and during the second week of June in 1990 at three sites in Oregon: Alsea Guard Range Station, Andrews Experimental Forest, and at the Starkey Experimental Forest, which differ in their proximity to the Pacific Coast. (Modified after Welker 2000.)

mass (Clark & Fritz 1997). Of course, weather systems are not this simple, and are complicated by re-evaporation processes and atmospheric mixing with other vapor masses. Nevertheless, there are two major factors that control the isotopic composition of precipitation:

1 temperature (which controls the fractionation process);
2 the proportion of the original vapor that remains after the precipitation has begun.

Geographic and temporal variations associated with these factors are discussed below. They include the apparent effects of continental, elevation, amount, and latitude variations, which are due to temperature-dependent, continuous isotopic fractionation.

Continental effects

The process described above as rainout reduces the heavy isotopic composition of an air mass as it travels inland is known as the continental effect. Precipitation samples collected along a west to east transect in Oregon, USA (Figure 11.5) show a strong isotopic depletion in ^{18}O of approximately -1.5‰ per 100 km (Welker 2000), which is characteristic of the continental effect. Precipitation over inland temperate areas tends to be characterized by strong temperature variations (i.e., removed from moderating marine influences) and isotopically depleted precipitation with strong seasonal differences due to those temperature variations. Alternately, the isotopic composition of

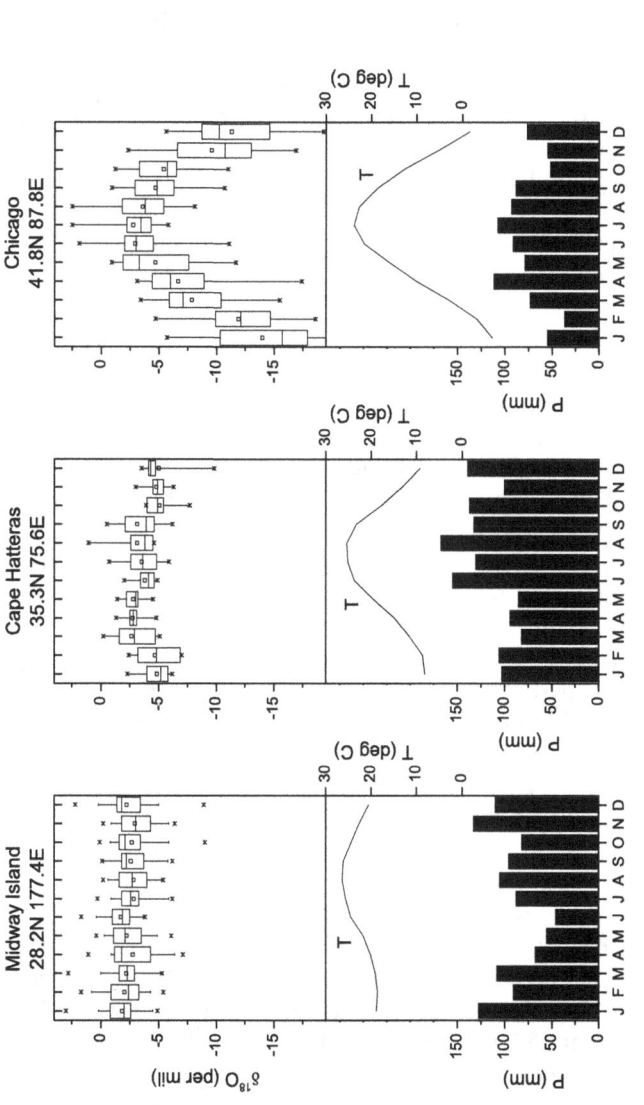

Figure 11.6 The seasonal variations in $\delta^{18}O$, temperature, and precipitation amount for Midway Island (a Pacific Ocean station), Cape Hatteras, North Carolina (an Atlantic coastal station), and Chicago, Illinois (A) (a continental station). The seasonal effect is more pronounced for continental sites with strong temperature variations. (Data are taken from the IAEA/WMO (2001) Global Network of Isotopes in Precipitation database.)

coastal precipitation tends to be less seasonally varied and isotopically enriched (Figure 11.6).

Pionke & DeWalle (1992) sampled 33 storms in central Pennsylvania and found that continental air masses originating from the Gulf of Mexico were generally less depleted than storms from oceanic air masses derived from the Atlantic. In addition, local storms that were not associated with frontal systems were the most depleted in ^{18}O. Celle-Jeanton et al. (2001) examined the typology of storms that affect the western Mediterranean region and found that based on 118 events, air masses originating from northern Atlantic and Mediterranean regions had very different isotopic compositions and rainfall amounts. Northern Atlantic storms had a strong continental effect since the air masses pass over Spain prior to reaching the monitoring station on the western Mediterranean coastline.

Elevation effects

Orographic precipitation caused by the cooling of air masses as they are lifted over higher elevation landforms, generally produces disproportionately higher rainfall with increased elevation on the windward side. This process forces rainout of heavier isotopic water; consequently, higher elevation regions receive more depleted precipitation. At higher elevations, isotopic depletion is further augmented by cooler average temperatures that cause increased fractionation. Gat (1980) suggests that secondary enrichment of raindrops resulting from partial evaporation during descent can contribute to the elevation effect. This process is dependent on the time raindrops are associated with unsaturated air, which is reduced in mountainous areas compared with valleys because raindrops fall shorter distances. Therefore, less enrichment of raindrops would be expected to occur in mountainous areas.

Observed elevation effects in the isotopic composition of precipitation have been reported in many studies around the world and generally vary from approximately −0.15 to −0.5‰ per 100 m increase in elevation and −1 to −4‰ per 100 m increase in elevation for ^{18}O and D, respectively (Clark & Fritz 1997). Detailed measurements in the western Cascades of Oregon showed that $\delta^{18}O$ from individual rainfall events were strongly elevation dependent (−0.22 to −0.32‰ per 100 m increase in elevation) and that elevation explained between 63 and 89% of the variance (Figure 11.7) (McGuire et al. 2005). The spatial pattern of these data suggests that elevation alone does not explain the isotopic variation in precipitation, but that other factors such as vapor source, air mass direction, and intensity (or amount) may affect the precipitation isotopic composition (Figure 11.7).

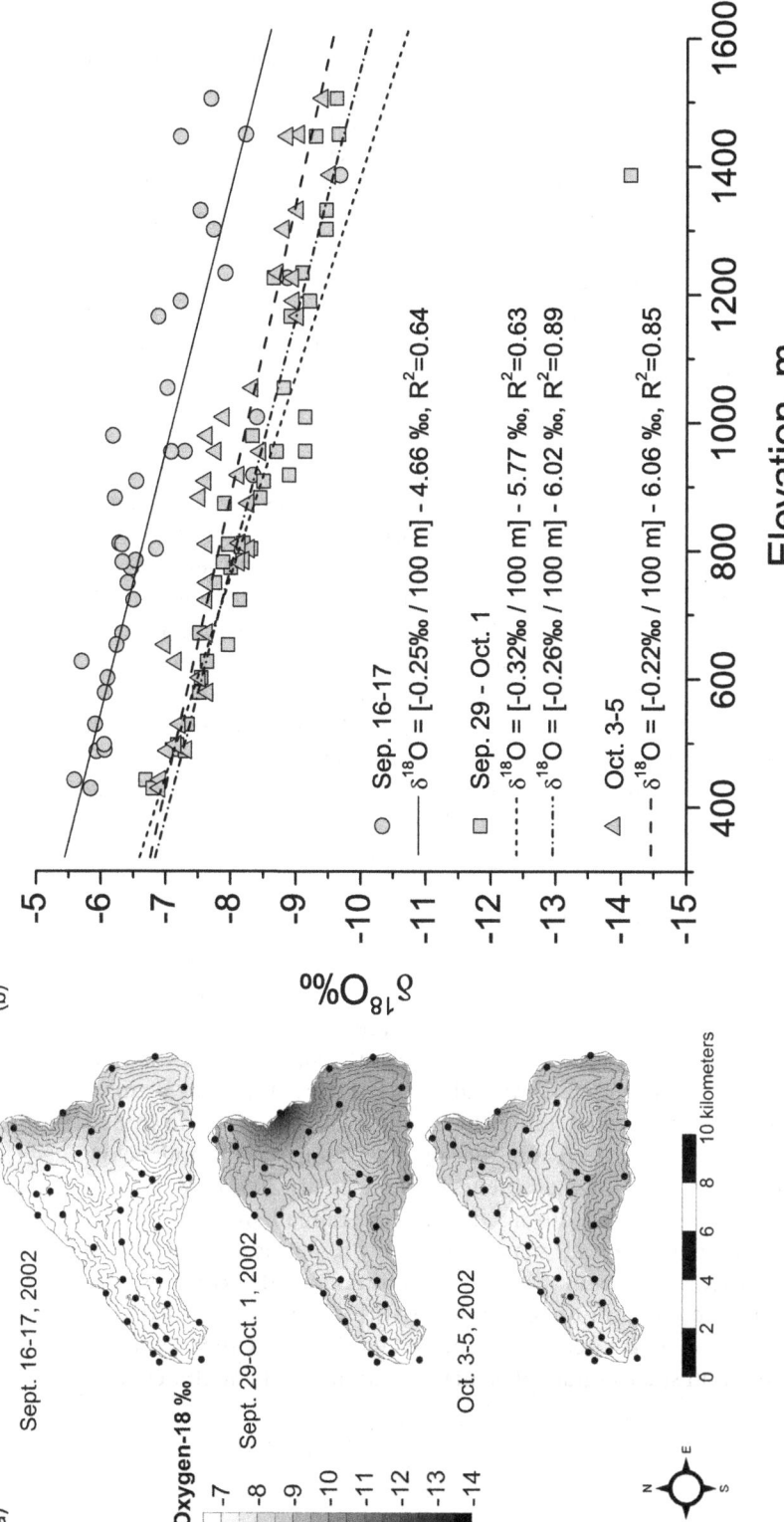

Figure 11.7 The variation of the ^{18}O composition in precipitation collected from three consecutive rain storms in the H.J. Andrews Experimental Forest, Oregon, USA (area = 64 km²). Both total rainfall amount (not shown) and $\delta^{18}O$ were highly correlated with elevation reflecting the amount and elevation effects, in addition to air mass trajectory (predominantly west to east). $\delta^{18}O$ and sampler elevation are plotted on the right with regression models fit to the data points. Two models were fit to the September 29 – October 1 storm (squares): one with the high-elevation data point of –14.14‰ (slope = –0.32‰ 100 m⁻¹) and one without that point (slope = –0.26‰ 100 m⁻¹).

Amount effects

Small rain storms are frequently observed to have more isotopically enriched water than larger storms. During brief rain showers, the amount effect has been attributed to evaporation and isotopic exchange of descending raindrops with atmospheric moisture, which more strongly affects storms of low rainfall intensity and low total rainfall amount (Dansgaard 1964; Gat 1980). As the storm proceeds, humidity beneath the cloud base increases through time, reducing the evaporation loss of the raindrops. The condensation of heavy isotopic forms early during larger rain events leaves subsequent rainfall with fewer heavy isotopes to acquire (Gedzelman & Arnold 1994). Thus, during longer duration rain storms, enrichment is less overall since evaporation is reduced in the later portion of the storm.

Latitude effects

Latitude effects are responsible for isotopic variations caused by cooler temperatures that air masses encounter as they proceed from equatorial regions, where 60% of the atmospheric vapor originates, to higher latitudes (Figure 11.8) (Yurtsever & Gat 1981). Condensation temperatures decrease which result in precipitation over higher latitudes having more negative isotopic composition. Rainout processes intensify this effect since polar regions (high latitudes) are situated at the end of the air mass trajectory where the isotopic-latitude gradient increases (Clark & Fritz 1997). The gradient over North America and Europe is approximately −0.5‰ per degree latitude for $\delta^{18}O$ (Yurtsever & Gat 1981). Once again, the animations of the GNIP data provide a very good visualization of the latitude effect (http://isohis.iaea.org/).

Intra-storm isotopic variations and throughfall

The initial isotopic composition of a rain event is heavy due to its formation by low altitude clouds and is typically followed by a gradual depletion in heavy isotopes, based on the amount effect, where the evaporation of raindrops below the cloud base is reduced over time as the air approaches saturation (Stewart 1975; Gedzelman & Arnold 1994) (Figure 11.9). As a storm progresses, the altitude at which rain is formed increases (i.e., due to frontal or convective rise), which decreases the air mass temperature and heavy isotopic composition of rainwater (Celle-Jeanton et al. 2004). Maximum depletion is usually achieved during the highest rainfall intensity which coincides with the maximum air mass cooling or lift, and is sometimes followed by an increase in heavy isotopic composition as the condensation altitude decreases, or by atmospheric mixing with a new air mass (Celle-Jeanton et al. 2004). This effect is shown in Figure 11.9. Other temporal patterns are common such as gradual depletion with no final enrichment and

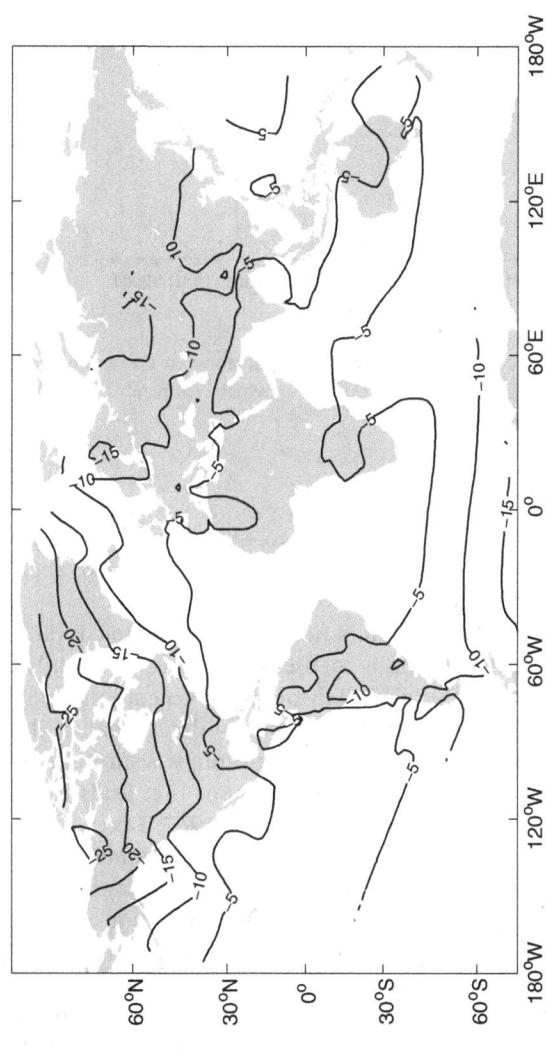

Figure 11.8 The global distribution of amount-weighted annual ^{18}O in precipitation based on 389 IAEA/WMO (2001) GNIP stations between 1961 and 1999 (created from data provided by Birks et al. 2002). These data illustrate the "latitude effect" where the isotopic composition of precipitation is lighter at higher latitudes. A simple cubic triangular interpolation was used for this map; however, more recent distribution maps have been created from the GNIP database using a Cressman objective analysis in a Grid Analysis and Display System (GrADS) (Birks et al. 2002), which includes monthly animations (see http://isohis.iaea.org) and using the Bowen–Wilkinson method (Bowen & Wilkinson 2001).

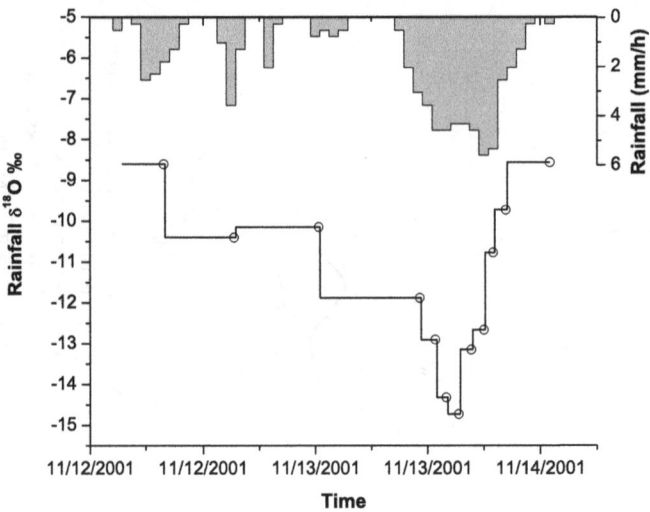

Figure 11.9 $\delta^{18}O$ temporal rainfall variations during a fall storm in the western Cascades of Oregon. The horizontal bars represent the time over which 5 mm rainfall increments were composited for each sample.

variations from mixing air masses related to successive frontal systems (McDonnell et al. 1990; Pionke & DeWalle 1992; Kubota & Tsuboyama 2003; Celle-Jeanton et al. 2004).

In many regions, gross precipitation is not the main isotopic input to the watershed; instead precipitation input to the soil surface is modulated by vegetation. In these instances, throughfall through the vegetation canopy and stemflow vertically down the stem (typically <5% of the annual rainfall) are the main inputs. Precipitation intercepted by the canopy is subject to evaporation and isotopic exchange with atmospheric vapor that leads to changes in the isotopic composition of rainwater (Gat & Tzur 1967; Saxena 1986) and snow (Claassen & Downey 1995; Cooper 1998). Throughfall is generally enriched in heavy isotopic forms (by approximately 0.5‰ and 3‰ for ^{18}O and D, respectively) through fractionation during evaporation and selective canopy storage for events with time-variable precipitation isotopic composition (Saxena 1986; Kendall & McDonnell 1993; DeWalle & Swistock 1994; Brodersen et al. 2000). Claasen & Dowy (1995) showed that intercepted snow enrichment can be much greater. They found that intercepted snow enrichments were about 2.1‰ and 13‰ for ^{18}O and D, respectively, according to results of a physically based model. The enrichment of intercepted snow was controlled primarily by the size of the snowfall and interception time.

As intercepted water evaporates from the canopy, fractionation (see above) processes usually lead to enrichment; however, molecular exchange with atmospheric water vapor may result in depletion (Brodersen et al. 2000). DeWalle & Swistock (1994) showed that selective canopy storage was more important than fractionation in governing the throughfall isotopic composition. The process of selection is related to the time-variable nature of the isotopic composition of precipitation. It has been suggested that canopy storage (i.e., interception) of the rainfall from the end of a storm event, a time when rainfall is typically depleted in heavy isotopes, would be lost to evaporation and produce higher isotope contents for throughfall compared with rainfall. However, if rain is lighter at the beginning of the event, then throughfall would be depleted. Less intense and intermittent rain showers would exacerbate the selection process, since the opportunity for interception loss is greater and because these portions of the event comprise the most isotopically enriched rainfall (see 'Amount effect' above). Therefore, the understanding of canopy storage behavior is imperative in controlling the isotopic composition of throughfall – the main input to forested watersheds (Saxena 1986; DeWalle & Swistock 1994; and also see Keim & Skaugset 2004).

Snowmelt

The isotopic composition of the snowpack profile generally represents the distinct isotopic composition of individual precipitation events. In spite of this, isotopic exchange, combined with snowpack metamorphism and surface sublimation, attenuates the signal in the snow layers provided by the individual events (Cooper 1998). Snowmelt isotopic composition that develops from these snowpacks results from two major processes:

1 sublimation and molecular exchange between vapor and the snowpack;
2 meltwater infiltration and exchange with snow and vapor in the snowpack (Taylor et al. 2001).

The isotopic fractionation associated with sublimation of snow surfaces was shown by Moser & Stichler (1975) to behave similarly to that of evaporating water, except that the well-mixed conditions of a water body are not present in snowpacks (Cooper 1998). During initial snowmelt, meltwater is depleted in heavy isotopes relative to the snowpack; however, it progressively enriches throughout the snowmelt season as the snowpack isotopic composition becomes homogeneous due to the preferential melt of lighter isotopes (Cooper et al. 1993; Taylor et al. 2001; Unnikrishna et al. 2002). Figure 11.10 shows a time series of snowmelt input to a forest clearing at the Central Sierra Snow Laboratory near the crest of the Sierra Nevada from the onset of spring melt on March 1 to the disappearance of snow on May 1 (Unnikrishna et al. 2002). Snowmelt $\delta^{18}O$ inputs rapidly decreased from −9.15‰ to approximately −15‰ on April 9 illustrating the ^{18}O depleted snowmelt caused by the preferential concentration of heavy isotopes in the solid snow phase. During

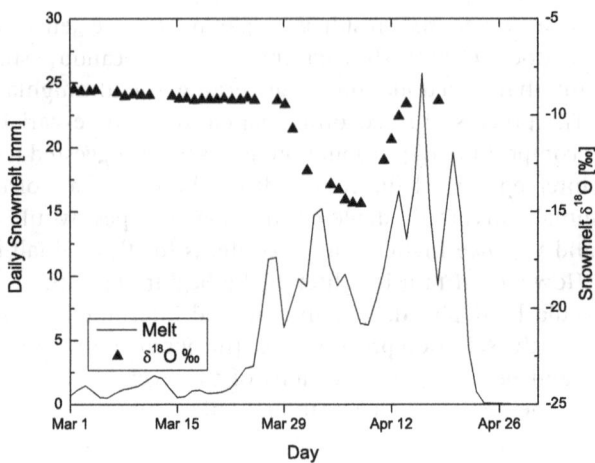

Figure 11.10 Daily snowmelt and snowmelt $\delta^{18}O$ from the Central Sierra Snow Laboratory, California. (Data from Unnikrishna et al. 2002.)

the final period of snowmelt, melt waters were progressively enriched in ^{18}O as the snowpack isotopic composition homogenized and resulting melt water increased to −9.20‰ on April 27 (Unnikrishna et al. 2002).

Applications of isotope hydrology in watershed and ecosystem studies

Evaporation rates

One direct application of stable isotopes in ecologically oriented studies is the calculation of evaporation. Since water isotopes fractionate upon phase changes, one can use isotopic enrichment during evaporation to estimate evaporation rates. This is a very simple and effective use of isotopes, used successfully by Gibson et al. (2002) for example, for computing lake surface evaporation in remote areas of northern Canada. Ambient humidity is the most important control on how evaporation from an open-water surface (or from a leaf surface or ponded water in a watershed) fractionates the isotopes of hydrogen and oxygen (Kendall & Caldwell 1998). Figure 11.11 shows that the higher the humidity during the evaporation process, the smaller the deflection from the meteoric water line in terms of change in $\delta^{18}O$ and δD during evaporation. For example, Figure 11.11 shows that at 95% humidity, the isotopic composition is constant for evaporation of the last 85% of the water. Evaporation results in lines with slopes <8 on a $\delta^{18}O$ vs. δD plot (i.e., the data plot on lines below the MWL that intersect the

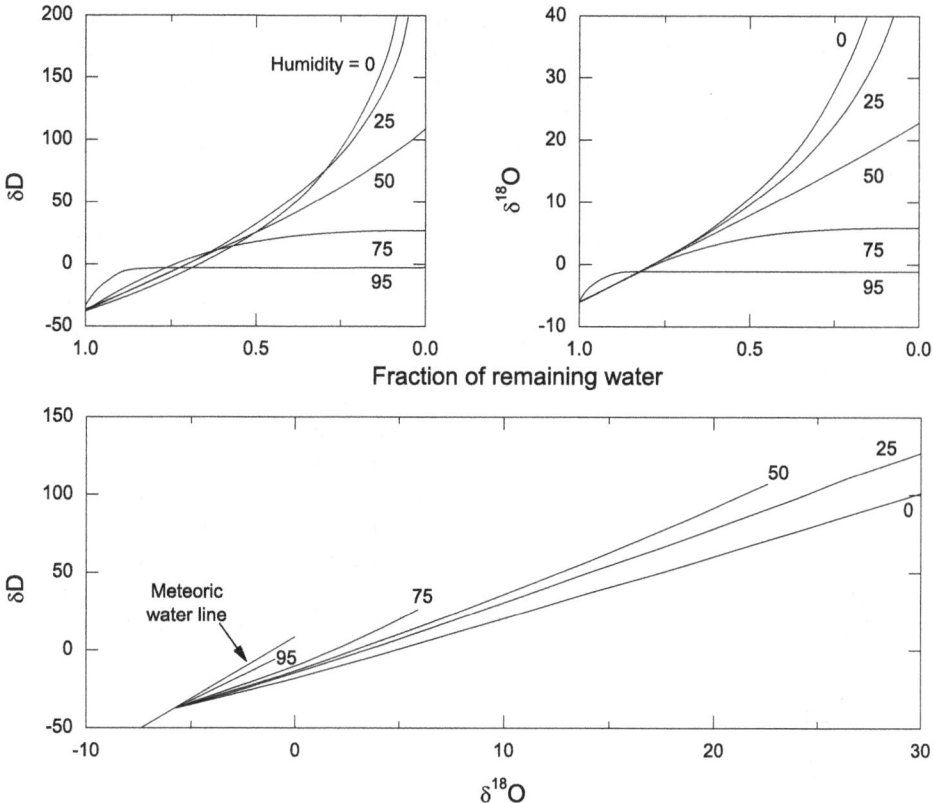

Figure 11.11 Humidity effects on $\delta^{18}O$ and δD of the residual water fraction during evaporation. Evaporation results in less fractionation at higher humidities and approaches a steady-state value for humidities >50% as the fraction of remaining water decreases. (Modified from Kendall & Caldwell 1998.)

MWL at the composition of the original water, see Figure 11.4) (Kendall & Caldwell 1998).

Lake evaporation as a fraction of precipitation can be calculated directly using the isotopic composition of a well-mixed lake that maintains a long-term constant volume (Gibson et al. 1993):

$$\frac{E}{P} = \frac{\delta_P - \delta_L}{\delta_E - \delta_L} \tag{11.4}$$

where E is lake evaporation, P is precipitation, δ_P is the weighted mean isotopic composition of local precipitation, δ_L is the isotopic composition of lake water, and δ_E is the isotopic composition of the evaporative flux. Values of δ_P and δ_L can be obtained readily by sampling precipitation and lake water; however, δ_E cannot be directly sampled. Estimates of δ_E are possible through calibration with a nearby lake of known water balance (Dinçer 1968), from pan evaporation experiments (Welhan & Fritz 1977), or from theoretical models (Craig & Gordon 1965) that require estimates of the isotopic composition of atmospheric vapor (δ_A). The Craig and Gordon model of δ_E is:

$$\delta_E = \frac{\delta_L - h\delta_A - \varepsilon}{1 - h + \varepsilon_K} \tag{11.5}$$

where h is the relative humidity normalized to the saturation vapor pressure at the temperature of the lake surface water–air interface and ε is the total isotopic enrichment factor, which accounts for both equilibrium ε^* and kinetic ε_K enrichment. Relative humidity and δ_L can be directly measured and ε^* is well-known for ^{18}O and ^{2}H as a function of temperature and ε_K is understood from theoretical and experiment studies (e.g., Merlivat 1978). The value of δ_A has been estimated by assuming that atmospheric vapor is in isotopic equilibrium with local precipitation (i.e., $\delta_A = \delta_P - \varepsilon^*$, and ε^* is approximated using mean air temperature), which generally holds if the slope of a local evaporation trend (see Figure 11.4) can be shown to be independent of h (Gibson et al. 1999). Equation 11.4 and 11.5 can then be combined as:

$$\frac{E}{P} = \frac{1 - h}{h} \frac{\delta_L - \delta_P}{\delta^* - \delta_L} \tag{11.6}$$

where δ^* is:

$$\delta^* = \frac{h\delta_A + \varepsilon}{h - \varepsilon} \tag{11.7}$$

Estimates of lake evaporation using equations 11.6 & 11.7 are best suited for longer-term studies involving complete annual cycles (Gibson et al. 1993, 1996).

Hydrograph separation

Hydrologists have traditionally separated the streamflow response (i.e., the discharge hydrograph) to rainfall and snowmelt inputs into its component parts using graphical techniques. Beyond the simple and rather arbitrary graphical measures used in engineering hydrology for channel routing and storm water drainage, other methods such as those introduced by Hewlett & Hibbert (1967), separate streamflow into "runoff components" (i) quickflow and (ii) delayed flow. Quickflow has been used frequently as a measure to describe the responsiveness of the watershed to a storm event. The term delayed flow is synonymous with baseflow (i.e., the flow in the stream between events) and conceptually represents the sum of the delayed shallow subsurface flow through the soil mantle and deep subsurface flow of groundwater (Ward & Robinson 2000). However, neither quickflow nor baseflow can be equated directly to precipitation–runoff conversion processes. While used extensively in watershed studies (Bonell, 1998), the quickflow separation method is still rather arbitrary for defining the relative rates of flow and neither it nor the engineering based approaches allow for the calculation of the geographic or time source of water contributing to streamflow.

Hydrograph separation has been perhaps the main use of environmental isotopes to date in small watershed hydrology (see reviews in Genereux & Hooper 1998; Rodhe 1998; Buttle & McDonnell 2004). Early isotopic hydrograph separations (IHS) used tritium (^{3}H) (Martinec 1975), but most studies in the past 25 years or so have used oxygen-18 (^{18}O) and deuterium (^{2}H) (Genereux & Hooper 1998; Burns 2002). Unlike ^{3}H, ^{18}O and ^{2}H are stable and do not undergo radioactive decay. Unlike the engineering approaches and the Hewlett & Hibbert approach, IHS can aid in quantifying the time source of water components of the storm hydrograph. When combined with additional tracers, IHS can also help to quantify the geographic source of water contributing to the hydrograph (Ogunkoya & Jenkins 1993).

Isotope tracers have a number of very useful attributes as water tracers for hydrograph separation (Buttle 1994).

1 They are applied naturally over entire catchments (unlike artificial tracers where application rates and extent are limited).
2 They do not undergo chemical reactions during contact with soil/regolith at temperatures encountered in the subsurface of watersheds.
3 New water is often different to old water. Numerous studies have shown that variations in the isotopic signature of precipitation are dampened as water transits the unsaturated zone to the water table (Clark & Fritz 1997). Groundwater isotopic composition may approach that of the mean annual precipitation isotopic values. In areas where seasonal isotopic variations in precipitation exist (e.g. middle and northern latitudes), there is frequently a difference between the isotopic composition of water input to the catchment's surface and water stored in the catchment before the event.

This difference between the isotopic signature of incoming water (event or "new" water) and water stored in the catchment before the event (pre-event or "old" water) often permits the separation of a stormflow hydrograph into a two-component mixing model: event and pre-event:

$$Q_t = Q_p + Q_e \tag{11.8}$$

$$\delta Q_t = \delta_p Q_p + \delta_e Q_e \tag{11.9}$$

$$X = (\delta_t - \delta_p)/(\delta_p - \delta_e) \tag{11.10}$$

where Q_t is streamflow; Q_p and Q_e are contributions from pre-event and event water; δ_t, δ_p and δ_e are isotopic compositions of streamflow, pre-event and event waters, respectively; and X is the pre-event fraction of streamflow. Event water is typically sampled in bulk or incrementally during storms and weighted by volume in the mixing model (equation 11.10). McDonnell et al. (1990) evaluated three weighting methods to determine the event water composition and found that incremental averaging methods were best for handling the temporal variability of the rainfall isotopic composition. Many of these methods assume an instantaneous mixing with pre-event water to produce the stream isotopic content at any time during the storm event; however, some recent studies have included delays or travel time distributions for the event water term in the mixing relationship shown in equation 11.10 (Joerin et al. 2002; Weiler et al. 2003).

The IHS results generally show that over half (more typically about 75%) the runoff and/or peakflow associated with rain storms is composed of pre-event water (Genereux & Hooper 1998). However, as Burns (2002) notes, most of our studies have focused on humid, temperate forested watersheds and little information is available for semi-arid and urban watersheds (see Buttle et al. 1995; Newman et al. 1998; Gremillion et al. 2000).

New techniques that combine simple rainfall–runoff models and IHS have made it possible to learn more about runoff generation processes than the use of the mixing model alone (Weiler et al. 2003; Iorgulescu et al. 2005). For example, Weiler et al. (2003) combined a transfer function model of the hydrology with an isotopic mixing model (equation 11.10) and were able examine the response time distributions of new water inputs (i.e., new water residence time) to a watershed for different storms and explore possible runoff generation processes.

Recharge rates and source

Given the clear and unambiguous signal of waters that have undergone evaporation (see above), quantifying recharge sources can be done quickly, clearly, and effectively with stable isotope tracers (Burns & McDonnell 1998). A simple and still relevant example was presented by Payne (1970) who illustrated how one could define sources of water recharge to springs around Lake Chala for basic water resources development questions. Here, villagers wanted to know if they could use water from Lake Chala for irrigation. They were mindful of the fact that using lake water could have a negative impact on flow from nearby springs used for drinking water supply if in fact the lake water recharged the springs. Payne (1970) shows a very clear example where plotting precipitation, lake water, and spring water on a meteoric water line can help reject possible recharge sources and connections (Figure 11.12). In this case, spring water plots on the meteoric water line and the lake waters

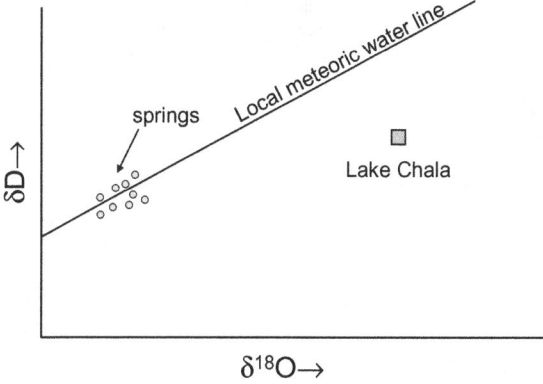

Figure 11.12 Lake Chala (square) and samples from a near-by spring (circles) plotted on the local meteoric water line. The figure is redrafted from a diagram of C. Kendall, as interpreted from the Payne (1970).

plot below the line indicating that lake water did not recharge the springs. If it did, the spring water would also plot off the MLW. The clear and unambiguous evaporation signal of the lake was very helpful in determining the lack of a groundwater and surface water connection.

Water travel time

Stable isotopes can be used as tracers to estimate how long it takes for water to travel through a watershed (Maloszewski & Zuber 1996; McGuire 2005; McGuire & McDonnell 2006). The travel time or residence time of water has important implications for water quality and the persistence of contaminants in the environment. Longer residence times indicate greater contact time and subsurface storage implying more time for biogeochemical reactions to occur (Scanlon et al. 2001; Burns et al. 2003).

Strong correlations between seasonally varying tropospheric temperature variations and the stable isotope composition in meteoric water provide an input isotopic signal that can be used in conjunction with the isotopic signal in a stream to estimate travel times. Estimating the distribution of travel times in a watershed using stable isotopes requires a well-monitored precipitation isotopic signal that should exceed at least 1 year; however, longer signals (>5 years) provide more reliable results (McGuire & McDonnell 2006). In addition, an input function is required to correct the precipitation isotopic composition to represent the recharge isotopic flux (Maloszewski et al. 1992; Vitvar & Balderer 1997; McGuire et al. 2002).

Water travel time distributions for catchments are typically inferred using lumped parameter models that describe integrated transport of

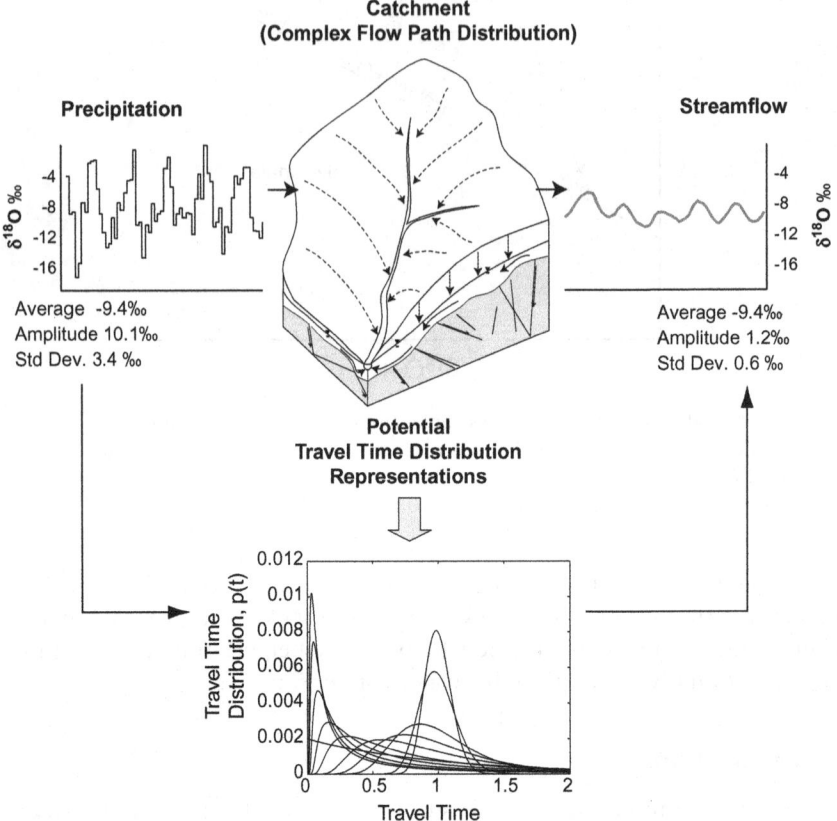

Figure 11.13 Conceptual diagram of the lumped parameter travel time modeling approach. Catchments receive $\delta^{18}O$ inputs that are transported along diverse flow paths in the unsaturated and saturated zones as the isotopes migrate through the subsurface toward the stream network. The result of differential transport within the catchment is an output streamflow $\delta^{18}O$ signal that is damped (i.e., decrease in standard deviation and amplitude) and lagged compared twith the input signal. The complex distribution of catchment flow paths is represented by a distribution of travel times, $p(t)$, that describe the integrated behavior of tracer transport through the catchment. (Modified from McGuire & McDonnell, 2006.)

the isotopic tracer through a catchment's subsurface via system response functions. Figure 11.13 illustrates the lumped parameter modeling approach for estimating the travel time distribution of water draining a catchment. The isotopic composition of precipitation that falls over the entire watershed area is transported to the stream network along diverse flow paths within the subsurface environment (see discussion of streamflow generation at the beginning of the chapter). The transport process along these diverse subsurface flow paths causes time delays (due to advection and dispersion) of precipitation isotopes as they arrive at the stream network, which is a direct manifestation of the catchment's flow path distribution, runoff processes, and subsurface hydrologic characteristics. The integrated response of isotopic arrival at the catchment outlet from all locations in the catchment is described by the travel time distribution (i.e., a probability density function of travel times). This process can be mathematically expressed by the convolution integral, which states that the stream isotopic composition at any time, $\delta_s(t)$, consists of precipitation with a unique isotopic signal, $\delta_p(t - \tau)$, that fell uniformly on the catchment in the past, which becomes lagged according to its travel time distribution, TTD(τ) (Maloszewski & Zuber 1982; Barnes & Bonell 1996; Kirchner et al. 2000):

$$\delta_s(t) = \int_0^\infty TTD(\tau)\delta_p(t - \tau)d\tau \qquad (11.11)$$

where τ are the lag times between precipitation and streamflow isotopic composition. A catchment's TTD could have various shapes depending on the exact nature of its flow path distribution and flow system assumptions. Equation 11.11 is only valid for the steady-state and when the mean subsurface flow pattern does not change significantly in time; however, it may be suitable for catchments where flow parameters (e.g., velocity) do not deviate significantly from the long-term mean values and when the water table fluctuations are small compared with the total subsurface volume (Zuber 1986).

The TTDs in equation 11.11 are typically composed of simple (i.e., one to three parameters) response functions that conceptually represent the dominant pathways, storages, and flow conditions of the real system. These TTDs can take the form of exponential flow reservoirs, piston-flow systems, exponential flow systems in parallel or series, and dispersive flow systems (Maloszewski & Zuber 1996; Turner & Barnes 1998). There has also been some evidence that the catchment geometry and topographic organization may exert some control on the shape of the catchment-scale TTD (Kirchner et al. 2001; McGuire et al. 2005). In current practice, TTDs are selected based either on an assumed flow system (Maloszewski & Zuber 1982) or through a fitting exercise resulting from numerous model simulations. This can be problematic since parameters are often not identifiable and different model types can yield non-unique results (Maloszewski & Zuber 1993). A full discussion on TTD models is beyond the scope of this chapter; however, there are many examples of the use of these models for catchment and groundwater systems (Stewart & McDonnell 1991; Vitvar & Balderer 1997; Buttle et al. 2001; McGuire et al. 2002).

Diagnostic tools in models

While this chapter has shown the usefulness of stable isotope approaches for understanding watershed function, very few studies have yet incorporated tracer data, interpretations, and concepts into current catchment-scale hydrologic models. The view that one's model captures the real-world processes correctly if one "fits" the hydrograph correctly still persists, but Hooper (2001, p. 2040) notes that "agreement between observations and predictions is only a necessary, not a sufficient, condition for the hypothesis to be correct." Seibert & McDonnell (2002) have argued that the experimentalist often has a highly detailed, yet highly qualitative, understanding of dominant runoff processes – and thus there is often much more information on the catchment than we use for calibration of a model. While modelers often appreciate the need for "hard data" for the model calibration process, there has been little thought given to how modelers might access this "soft data" or process knowledge, especially that derived from isotope tracer studies. Seibert & McDonnell (2002) presented a new method where soft data (i.e., qualitative knowledge from the experimentalist that cannot be used directly as exact numbers) are made useful through fuzzy measures of model-simulation and parameter-value acceptability. They developed a three-box lumped conceptual model for the Maimai catchment in New Zealand, where the boxes represent the key hydrological reservoirs that are known to have distinct groundwater dynamics, isotopic composition, and solute chemistry. The model was calibrated against hard data (runoff and groundwater-levels) as well as a number of criteria derived from the soft data (e.g. percent new water from isotope hydrograph separations). They achieved very good fits for the three-box model when optimizing the parameter values with only runoff ($E = 0.93$; E is the Nash & Sutcliffe (1970) efficiency, where 1 is a perfect fit). However, parameter sets obtained in this way in general showed a poor goodness-of-fit for other criteria such as the simulated new water contributions to peak runoff. Inclusion of soft-data criteria in the model calibration process resulted in lower E values (around 0.84 when including all criteria) but led to better overall performance, as interpreted by the experimentalist's view of catchment runoff dynamics. The model performance, with respect to the new water percentage, increased significantly and parameter uncertainty was reduced by 60% on average with the introduction of the soft data multi-criteria calibration. This work suggests that hydrograph separation information may have new applications in model calibration, where accepting lower model efficiencies for runoff is 'worth it' if one can develop a more 'real' model of catchment behavior based on the information content of the isotope approach. More recent work has suggested that these approaches can be useful for other model structures and other model applications (Vaché et al. 2004).

Conclusions

Watershed hydrology is a field of study very much related to ecology. Questions of where water goes when it rains, what flowpaths the water takes to the stream, and how long water resides in the watershed underpin many questions of plant water availability, biogeochemical cycling, microbial production, and other water-mediated ecological processes. Stable isotope tracing and analysis forms an important link between hydrologic and ecological processes at the watershed scale where knowledge of flow path, water source and age inform many water mediated ecological processes. We have tried to illustrate in this chapter how an understanding of watershed hydrology can be used to better understand water quality, sustainability, land-use change effects, nutrient cycling, and general terrestrial and aquatic system interactions via isotope-based techniques. The potential for future studies to explore the interface between hydrology and ecology using isotopic techniques is very positive.

Acknowledgments

We would like to express our appreciation to Carol Kendall for many useful discussions over the years. Her collaboration and insight on many projects has been an encouraging influence to us. We continue to draw upon her wide expertise in isotope biogeochemistry and hydrology and make use of some examples she features in her U.S. Geological Survey short course in this chapter. We also thank the editors for their comments and patience, and for inviting us to contribute to this second edition.

References

Abbott, M.D., Lini, A. & Bierman, P.R. (2000) $\delta^{18}O$, δD and ^3H measurements constrain groundwater recharge patterns in an upland fractured bedrock aquifer, Vermont, USA. *Journal of Hydrology*, **228**(1–2), 101–112.

Abdul, A.S. & Gillham, R.W. (1984) Laboratory studies of the effects of the capillary fringe on streamflow generation. *Water Resources Research*, **20**, 691–698.

Ambroise, B. (2004) Variable active versus contributing areas or periods: a necessary distinction. *Hydrological Processes*, **18**(6), 1149–1155.

Anderson, M.G. & Burt, T.P. (1990) Process studies in hillslope hydrology. John Wiley and Sons, Chichester.

Anderson, S.P., Dietrich, W.E., Montgomery, D.R., Torres, R., Conrad, M.E. & Loague, K. (1997) Subsurface flow paths in a steep, unchanneled catchment. *Water Resources Research*, **33**(12), 2637–2653.

Barnes, C.J. & Bonell, M. (1996) Application of unit hydrograph techniques to solute transport in catchments. *Hydrological Processes*, **10**(6), 793–802.

Bates, P.D., Stewart, M.D., Desitter, A., Anderson, M.G., Renaud, J.P. & Smith, J.A. (2000) Numerical simulation of floodplain hydrology. *Water Resources Research*, **36**(9), 2517–2529.

Bazemore, D.E., Eshleman, K.N. & Hollenbeck, K.J. (1994) The role of soil water in stormflow generation in a forested headwater catchment: synthesis of natural tracer and hydrometric evidence. *Journal of Hydrology*, **162**(1–2), 47–75.

Beven, K.J. (1978) The hydrological response of headwater and sideslope areas. *Hydrological Sciences Bulletin*, **23**, 419–437.

Beven, K. (1989) Interflow. In: *Unsaturated Flow in Hydrologic Modeling Theory and Practice* (Ed. H.J. Morel-Seyoux), pp. 191–219. Kluwer Academic Publishers, Boston, MA.

Beven, K. (2004) Robert E. Horton's perceptual model of infiltration processes. *Hydrological Processes*, **18**(17), 3447–3460.

Beven, K. & Germann, P. (1982) Macropores and water flow in soils. *Water Resources Research*, **18**(5), 1311–1325.

Birks, S.J., Gibson, J.J., Gourcy, L., Aggarwal, P.K. & Edwards, T.W.D. (2002) Maps and animations offer new opportunities for studying the global water cycle. *Eos (Transactions of the American Geophysical Union)*, **83**(37) (Available at http://www.agu.org/eos_elec/020082e.html).

Bishop, K., Seibert, J., Köhler, S. & Laudon, H. (2004) Resolving the double paradox of rapidly mobilized old water with highly variable responses in runoff chemistry. *Hydrological Processes*, **18**, 185–189.

Bonell, M. (1998) Selected challenges in runoff generation research in forests from the hillslope to headwater drainage basin scale. *Journal of the American Water Resources Association*, **34**(4), 765–786.

Bowen, G.J. & Revenaugh, J. (2003) Interpolating the isotopic composition of modern meteoric precipitation. *Water Resources Research*, **39**(10), doi:10.1029/2003WR002086.

Bowen, G.J. & Wilkinson, B. (2001) Spatial distribution of $\delta^{18}O$ in meteoric precipitation. *Geology*, **30**(4), 315–318.

Brodersen, C., Pohl, S., Lindenlaub, M., Leibundgut, C. & Wilpert, K.V. (2000) Influence of vegetation structure on isotope content of throughfall and soil water. *Hydrological Processes*, **14**(8), 1439–1448.

Burns, D.A. (2002) Stormflow-hydrograph separation based on isotopes: the thrill is gone – what's next? *Hydrological Processes*, **16**(7), 1515–1517.

Burns, D.A. & McDonnell, J.J. (1998) Effects of a beaver pond on runoff processes: Comparison of two headwater catchments. *Journal of Hydrology*, **205**(3–4), 248–264.

Burns, D.A., Plummer, L.N., McDonnell, J.J., et al. (2003) The geochemical evolution of riparian ground water in a forested piedmont catchment. *Ground Water*, **41**(7), 913–925.

Burt, T.P. & Butcher, D.P. (1985) Topographic controls of soil moisture distributions. *Journal of Soil Science*, **36**(3), 469–486.

Burt, T.P., Bates, P.D., Stewart, M.D., Claxton, A.J., Anderson, M.G. & Price, D.A. (2002) Water table fluctuations within the floodplain of the River Severn, England. *Journal of Hydrology*, **262**(1–4), 1–20.

Buttle, J.M. (1994) Isotope hydrograph separations and rapid delivery of pre-event water from drainage basins. *Progress in Physical Geography*, **18**(1), 16–41.

Buttle, J.M. (1998) Fundamentals of small catchment hydrology. In: *Isotope Tracers in Catchment Hydrology* (Eds C. Kendall & J.J. McDonnell), pp. 1–43. Elsevier, Amsterdam.

Buttle, J.M. & McDonnell, J.J. (2004) Isotope tracers in catchment hydrology in the humid tropics. In: *Forests, Water and People in the Humid Tropics Past, Present and Future Hydrological Research for Integrated Land and Water Management* (Eds M. Bonell & L.A.Bruijnzeel), pp. 770–789. Cambridge University Press, Cambridge, 994 pp.

Buttle, J.M. & Turcotte, D.S. (1999) Runoff processes on a forested slope on the Canadian shield. *Nordic Hydrology*, **30**, 1–20.

Buttle, J.M., Vonk, A.M. & Taylor, C.H. (1995) Applicability of isotopic hydrograph separation in a suburban basin during snowmelt. *Hydrological Processes*, **9**, 197–211.

Buttle, J.M., Hazlett, P.W., Murray, C.D., Creed, I.F., Jeffries, D.S. & Semkin, R. (2001) Prediction of groundwater characteristics in forested and harvested basins during spring snowmelt using a topographic index. *Hydrological Processes*, **15**, 3389–3407.

Buttle, J.M., Dillon, P.J. & Eerkes, G.R. (2004) Hydrologic coupling of slopes, riparian zones and streams: an example from the Canadian Shield. *Journal of Hydrology*, **287**(1–4), 161–177.

Cappus, P. (1960) Bassin expérimental d'Alrance – Étude des lois de l'écoulement – Application au calcul et à la prévision des débits. *La Houille Blanche* **A**, 493–520.

Celle-Jeanton, H., Travi, Y. & Blavoux, B. (2001) Isotopic typology of the precipitation in the Western Mediterranean region at three different time scales. *Geophysical Research Letters*, **28**(7), 1215–1218.

Celle-Jeanton, H., Gonfiantini, R., Travi, Y. & Sol, B. (2004) Oxygen-18 variations of rainwater during precipitation: application of the Rayleigh model to selected rainfalls in Southern France. *Journal of Hydrology*, **289**(1–4), 165–177.

Claassen, H.C. & Downey, J.S. (1995) A model for deuterium and oxygen 18 isotope changes during evergreen interception of snowfall. *Water Resources Research*, **31**(3), 601–618.

Clark, I.D. & Fritz, P. (1997) *Environmental Isotopes in Hydrogeology*. Lewis Publishers, Boca Raton, FL.

Cloke, H., Anderson, M.G., McDonnell, J.J. & Renaud, J.-P. (2006) Using numerical modelling to evaluate the capillary fringe groundwater ridging hypothesis of streamflow generation. *Journal of Hydrology*, **316**(1–4), 141–162.

Cooper, L.W. (1998) Isotopic fractionation in snow cover In: *Isotope Tracers in Catchment Hydrology* (Eds C. Kendall & J.J. McDonnell), pp. 119–136. Elsevier, Amsterdam.

Cooper, L.W., Solis, C., Kane, D.L. & Hinzman, L.D. (1993) Application of O-18 tracer techniques to arctic hydrological processes. *Arctic and Alpine Research*, **25**(3), 247–255.

Coplen, T. (1993) Uses of environmental isotopes. In: *Regional Ground-Water Quality* (Ed. W.M. Alley), pp. 227–254. Van Nostrand Reinhold, New York.

Coplen, T.B. (1996) New guidelines for reporting stable hydrogen, carbon, and oxygen isotope-ratio data. *Geochimica et Cosmochimica Acta*, **60**(17), 3359–3360.

Coplen, T., Herczeg, A. & Barnes, C. (2000) Isotope engineering – using stable isotopes of the water molecule to solve practical problems. In: *Environmental Tracers in Subsurface Hydrology* (Eds P. Cook & A. Herczeg), pp. 79–110. Kluwer Academic Publishers, Boston, MA.

Craig, H. (1961) Isotopic variations in meteoric waters. *Science*, **133**(3465), 1702–1703.

Craig, H. & Gordon, L.I. (1965) Deuterium and oxygen 18 variations in the ocean and marine atmosphere. In: *Stable Isotopes in Oceanographic Studies and Paleotemperatures* (Ed. E. Tongiorgi), pp. 9–130. Laboratorio di Geologia Nucleare, Pisa.

Crayosky, T.W., DeWalle, D.R., Seybert, T.A. & Johnson, T.E. (1999) Channel precipitation dynamics in a forested Pennsylvania headwater catchment (USA). *Hydrological Processes*, **13**(9), 1303–1314.

Dansgaard, W. (1964) Stable isotopes in precipitation. *Tellus*, **16**, 436–438.

Devine, C. & McDonnell, J.J. (2004) The future of applied tracers in hydrogeology. *Hydrogeology Journal*, **13**, 255–258.

DeWalle, D.R. & Swistock, B.R. (1994) Differences in oxygen-18 content of throughfall and rainfall in hardwood and coniferous forests. *Hydrological Processes*, **8**(1), 75–82.

Dinçer, T. (1968) The use of oxygen-18 and deuterium concentrations in the water balance of lakes. *Water Resources Research*, **4**, 1289–1306.

Dinçer, T., Payne, B.R., Florkowski, T., Martinec, J. & Tongiorgi, E.G.E.I. (1970) Snowmelt runoff from measurements of tritium and oxygen-18. *Water Resources Research*, **6**, 110–124.

Dunne, T. & Black, R.D. (1970) Partial area contributions to storm runoff in a small New England watershed. *Water Resources Research*, **6**(5), 1296–1311.

Dunne, T. & Leopold, L.B. (1978) *Water in Environmental Planning*. W.H. Freeman, San Francisco, CA.

Epstein, S. & Mayeda, T. (1953) Variation of ^{18}O content of water from natural sources. *Geochimica et Cosmochimica Acta*, **4**, 213–224.

Eshleman, K.N., Pollard, J.S. & O'Brien, A.K. (1993) Determination of contributing areas for saturation overland flow from chemical hydrograph separations. *Water Resources Research*, **29**(10), 3577–3587.

Freer, J., McDonnell, J.J., Beven, K.J., et al. (2002) The role of bedrock topography on subsurface storm flow. *Water Resources Research*, **38**(12), 1269, doi:1210.1029/2001WR000872.

Freeze, R.A. & Cherry, J.A. (1979) *Groundwater*. Prentice-Hall, Englewood Cliffs, NJ.

Gat, J.R. (1980) The isotopes of hydrogen and oxygen in precipitation. In: *Handbook of Environmental Isotope Geochemistry* (Eds P. Fritz & J.C. Fontes), pp. 21–47. Elsevier, Amsterdam.

Gat, J.R. (1996) Oxygen and hydrogen isotopes in the hydrologic cycle. *Annual Review of Earth and Planetary Sciences*, **24**, 225–262.

Gat, J.R. & Gonfiantini, R. (1981) *Stable Isotope Hydrology, Deuterium and Oxygen-18 in the Water Cycle*. International Atomic Energy Agency, Vienna, 337 pp.

Gat, J. & Tzur, Y. (1967) Modification of the isotopic composition of rainwater by processes which occur before groundwater recharge. *Proceedings of the Symposium on Isotopes in Hydrology*, International Atomic Energy Agency, Vienna, 14–18 November 1966, pp. 49–60.

Gedzelman, S.D. & Arnold, R. (1994) Modeling the isotopic composition of precipitation. *Journal of Geophysical Research*, **99**(D5), 10,455–410,471.

Genereux, D.P. & Hooper, R.P. (1998) Oxygen and hydrogen isotopes in rainfall-runoff studies. In: *Isotope Tracers in Catchment Hydrology* (Eds C. Kendall & J.J. McDonnell), pp. 319–346. Elsevier, Amsterdam.

Gibson, J.J., Edwards, T.W.D., Bursey, G.G. & Prowse, T.D. (1993) Estimating evaporation using stable isotopes: quantitative results and sensitivity analysis for two catchment in nrothern Canada. *Nordic Hydrology*, **24**, 79–94.

Gibson, J.J., Edwards, T.W.D. & Prowse, T.D. (1996) Development and validation of an isotopic method for estimating lake evaporation. *Hydrological Processes*, **10**, 1369–1382.

Gibson, J.J., Edwards, T.W.D. & Prowse, T.D. (1999) Pan-derived isotopic composition of atmospheric water vapour and its variability in northern Canada. *Journal of Hydrology*, **217**, 55–74.

Gibson, J.J., Prepas, E.E. & McEachern, P. (2002) Quantitative comparison of lake throughflow, residency, and catchment runoff using stable isotopes: modelling and results from a regional survey of Boreal lakes. *Journal of Hydrology*, **262**(1–4), 128–144.

Gillham, R.W. (1984) The capillary fringe and its effect on water-table response. *Journal of Hydrology*, **67**, 307–324.

Grayson, R.B., Western, A.W., Chiew, F.H.S. & Blöschl, G. (1997) Preferred states in spatial soil moisture patterns: Local and nonlocal controls. *Water Resources Research*, **33**(12), 2897–2908.

Gremillion, P., Gonyeau, A. & Wanielista, M. (2000) Application of alternative hydrograph separation models to detect changes in flow paths in a watershed undergoing urban development. *Hydrological Processes*, **14**(8), 1485–1501.

Harr, R.D. (1977) Water flux in soil and subsoil on a steep forested slope. *Journal of Hydrology*, **33**, 37–58.

Hewlett, J.D. (1961) *Watershed Management. Annual Report 1961.* USDA Forest Service, Southeastern Forest Experiment Station, Asheville, NC, pp. 61–66.

Hewlett, J.D. & Hibbert, A.R. (1967) Factors affecting the response of small watersheds to precipitation in humid areas. In: *Forest Hydrology* (Eds W.E. Sopper & H.W. Lull), pp. 275–291. Pergamon Press, New York.

Hill, D.E. & Parlange, J.-Y. (1972) Wetting front instability in layer soils. *Soil Science Society of America Journal*, **36**, 367–702.

Hillel, D. (1998) *Environmental Soil Physics.* Academic Press, San Diego.

Hooper, R.P. (2001) Applying the scientific method to small catchment studies: a review of the Panola Mountain experience. *Hydrological Processes*, **15**, 2039–2050.

Horton, J.H. & Hawkins, R.H. (1965) Flow path of rain from soil surface to water table. *Soil Science* **100**(6), 377–383.

Horton, R.E. (1933) The role of infiltration in the hydrologic cycle. *Transactions of the American Geophysical Union*, **14**, 446–460.

Hursh, C.R. (1936) Storm-water and adsorption. *Transactions of the American Geophysical Union*, **17**, 863–870.

Hursh, C.R. (1944) Subsurface flow. *Transactions of the American Geophysical Union*, **25**, 743–746.

IAEA/WMO (2001) *Global Network of Isotopes in Precipitation.* The GNIP Database. Accessible at: http://isohis.iaea.org. International Atomic Energy Agency/World Meteorological Organization.

Iorgulescu, I., Beven, K.J. & Musy, A. (2005) Data-based modelling of runoff and chemical tracer concentrations in the Haute-Mentue research catchment (Switzerland). *Hydrological Processes*, **19**(13), 2557–2573.

Joerin, C., Beven, K.J., Iorgulescu, I. & Musy, A. (2002) Uncertainty in hydrograph separations based on geochemical mixing models. *Journal of Hydrology*, **255**, 90–106.

Jones, J.A.A. (1971) Soil piping and stream channel initiation. *Water Resources Research*, **7**(3), 602–610.

Keim, R.F. & Skaugset, A.E. (2004) A linear system model of dynamic throughfall rates beneath forest canopies. *Water Resources Research*, **40**, W05208, doi:05210.01029/02003WR002875.

Kendall, C. & Caldwell, E.A. (1998) Fundamentals of isotope geochemistry. In: *Isotope Tracers in Catchment Hydrology* (Eds C. Kendall & J.J. McDonnell), pp. 51–86. Elsevier, Amsterdam.

Kendall, C. & McDonnell, J.J. (1993) Effect of intrastorm isotopic heterogeneities of rainfall, soil water, and groundwater in runoff modeling. In: *Tracers in Hydrology, Proceedings of the Yokohama Symposium*, pp. 41–48. Publication 215, International Association of Hydrological Sciences, Wallingford.

Kendall, C. & McDonnell, J.J. (1998) *Isotope Tracers in Catchment Hydrology.* Elsevier, Amsterdam.

Kendall, K.A., Shanley, J.B. & McDonnell, J.J. (1999) A hydrometric and geochemical approach to test the transmissivity feedback hypothesis during snowmelt. *Journal of Hydrology,* **219**(3–4), 188–205.

Kirchner, J.W., Feng, X. & Neal, C. (2000) Fractal stream chemistry and its implications for contaminant transport in catchments. *Nature,* **403**(6769), 524–527.

Kirchner, J.W., Feng, X. & Neal, C. (2001) Catchment-scale advection and dispersion as a mechanism for fractal scaling in stream tracer concentrations. *Journal of Hydrology,* **254**, 82–101.

Klaassen, W. (2001) Evaporation from rain-wetted forest in relation to canopy wetness, canopy cover, and net radiation. *Water Resources Research,* **37**(12), doi:10.1029/2001WR000480.

Kubota, T. & Tsuboyama, Y. (2003) Intra- and inter-storm oxygen-18 and deuterium variation of rain, throughfall, and stemflow, and two-component hydrograph separation in a small forested catchment in Japan. *Journal of Forest Research,* **8**, 179–190.

Laudon, H., Seibert, J., Köhler, S. & Bishop, K. (2004) Hydrological flow paths during snowmelt: congruence between hydrometric measurements and oxygen 18 in meltwater, soil water, and runoff. *Water Resources Research,* **40**, W03102, doi:03110.01029/02003WR002455.

Majoube, M. (1971) Fractionnement en oxygène-18 et en deutérium entre l'eau et sa vapeur. *Journal of Chemical Physics,* **197**, 1423–1436.

Maloszewski, P. & Zuber, A. (1982) Determining the turnover time of groundwater systems with the aid of environmental tracers. 1. models and their applicability. *Journal of Hydrology,* **57**, 207–231.

Maloszewski, P. & Zuber, A. (1993) Principles and practice of calibration and validation of mathematical models for the interpretation of environmental tracer data. *Advances in Water Resources,* **16**, 173–190.

Maloszewski, P. & Zuber, A. (1996) Lumped parameter models for the interpretation of environmental tracer data. In: *Manual on Mathematical Models in Isotope Hydrogeology,* pp. 9–58. TECDOC–910, International Atomic Energy Agency, Vienna.

Maloszewski, P., Rauert, W., Stichler, W. & Herrmann, A. (1983) Application of flow models in an alpine catchment area using tritium and deuterium data. *Journal of Hydrology,* **66**, 319–330.

Maloszewski, P., Rauert, W., Trimborn, P., Herrmann, A. & Rau, R. (1992) Isotope hydrological study of mean transit times in an alpine basin (Wimbachtal, Germany). *Journal of Hydrology,* **140**(1–4), 343–360.

Martinec, J. (1975) Subsurface flow from snowmelt traced by tritium. *Water Resources Research,* **11**(3), 496–498.

McDonnell, J.J. (1990) A rationale for old water discharge through macropores in a steep, humid catchment. *Water Resources Research,* **26**(11), 2821–2832.

McDonnell, J.J. & Buttle, J.M. (1998) Comment on "A deterministic-empirical model of the effect of the capillary-fringe on near stream area runoff. 1. description of the model" by Jayatilaka, C.J. and Gillham, R.W. (*Journal of Hydrology,* Vol. 184 (1996) 299–315). *Journal of Hydrology,* **207**, 280–285.

McDonnell, J.J., Bonell, M., Stewart, M.K. & Pearce, A.J. (1990) Deuterium variations in storm rainfall: implications for stream hydrograph separation. *Water Resources Research,* **26**, 455–458.

McDonnell, J.J., McGlynn, B.L., Kendall, K., Shanley, J. & Kendall, C. (1998) The role of near-stream riparian zones in the hydrology of steep upland catchments. In:

Conference HeadWater '98, 20–23 April, Meran, Italy (Eds K. Kovar, U. Tappeiner, N. Peters & R. Craig). Publication 248, International Association of Hydrological Sciences, Wallingford.

McGlynn, B.L. & McDonnell, J.J. (2003) Role of discrete landscape units in controlling catchment dissolved organic carbon dynamics. *Water Resources Research*, **39**(4), 1090, doi:1010.1029/2002WR001525.

McGuire, K.J. (2005) Water residence time, SW-1136. In: *The Encyclopedia of Water: Surface Water Hydrology* (Eds J.H. Lehr & J. Keeley). J. Wiley & Sons, New York.

McGuire, K.J. & McDonnell, J.J. (2006) A review and evaluation of catchment transit time modeling. *Journal of Hydrology*, **330**(3–4), 543–563.

McGuire, K.J., DeWalle, D.R. & Gburek, W.J. (2002) Evaluation of mean residence time in subsurface waters using oxygen-18 fluctuations during drought conditions in the mid-Appalachians. *Journal of Hydrology*, **261**(1–4), 132–149.

McGuire, K.J., McDonnell, J.J., Weiler, M., et al. (2005) The role of topography on catchment-scale water residence time. *Water Resources Research*, **41**(5), W05002, doi:05010.01029/02004WR003657.

Merlivat, L. (1978) Molecular diffusivities of $H_2^{16}O$, $HD^{16}O$, and $H_2^{18}O$ in gases. *Journal of Chemical Physics*, **69**, 2864–2871.

Mook, W.G. (2000) *Environmental Isotopes in the Hydrological Cycle: Principles and Applications.* International Atomic Energy Agency, Vienna.

Moser, H. & Stichler, W. (1975) Use of environmental isotope methods as a reconnaissance tool in groundwater exploration near San Antonio de Pichincha, Ecuador. *Water Resources Research*, **11**(3), 501–505.

Mosley, M.P. (1979) Streamflow generation in a forested watershed. *Water Resources Research*, **15**, 795–806.

Nash, J.E. & Sutcliffe, J.V. (1970) River flow forecasting through conceptual models, I, A discussion of principles. *Journal of Hydrology*, **10**, 282–290.

Newman, B.D., Campbell, A.R. & Wilcox, B.P. (1998) Lateral subsurface flow pathways in a semiarid ponderosa pine hillslope. *Water Resources Research*, **34**(12), 3485–3496.

Ogunkoya, O.O. & Jenkins, A. (1993) Analysis of storm hydrograph and flow pathways using a three-component hydrograph separation model. *Journal of Hydrology*, **142**(1–4), 71–88.

Payne, B.R. (1970) Water balance of Lake Chala and its relation to groundwater from tritium and stable isotope isotope data. *Journal of Hydrology*, **11**, 47–58.

Pearce, A.J., Stewart, M.K. & Sklash, M.G. (1986) Storm runoff generation in humid headwater catchments: 1. Where does the water come from? *Water Resources Research*, **22**, 1263–1272.

Penman, H.L. (1963) *Vegetation and Hydrology.* Commonwealth Bureau of Soils, Harpenden, 124 pp.

Pionke, H.B. & DeWalle, D.R. (1992) Intra- and inter-storm ^{18}O trends for selected rainstorms in Pennsylvania. *Journal of Hydrology*, **138**(1/2), 131–143.

Ragan, R.M. (1968) An experimental investiagation of partial area contributions. In: *General Assembly of Bern*, 25 September–7 October, pp. 241–251. Publication 76, International Association of Scientific Hydrology, Wallingford.

Rasmussen, T.C., Baldwin, R.H., Dowd, J.F. & Williams, A.G. (2000) Tracer vs. pressure wave velocities through unsaturated saprolite. *Soil Science Society of America Journal*, **64**, 75–85.

Rodhe, A. (1998) Snowmelt-dominated systems. In: *Isotope Tracers in Catchment Hydrology* (Eds C. Kendall & J.J. McDonnell), pp. 391–433. Elsevier, Amsterdam.

Rozanski, K., Araguás-Araguás, L. & Gonfiantini, R. (1992) Relationship between long-term trends of oxygen-18 isotope composition of precipitation and climate. *Science*, **258**, 981–984.

Rozanski, K., Araguas-Araguas, L. & Gonfiantini, R. (1993) Isotopic patterns in modern global precipitation. In: *Climate Change in Continental Isotopic Records* (Eds P.K. Swart, K.C. Lohmann, J. McKenzie & S. Savin), pp. 1–36. American Geophysical Union, Washington, DC.

SAHRA (2005) *Isotopes and Hydrology*. Sustainability of semi-Arid Hydrology and Riparian Areas, available: http://www.sahra.arizona.edu/programs/isotopes/oxygen.html (accessed Oct. 2005).

Saxena, R.K. (1986) Estimation of canopy reservoir capacity and oxygen-18 fractionation in throughfall in a pine forest. *Nordic Hydrology*, **17**(4/5), 251–260.

Scanlon, T.M., Raffensperger, J.P. & Hornberger, G.M. (2001) Modeling transport of dissolved silica in a forested headwater catchment: implications for defining the hydrochemical response of observed flow pathways. *Water Resources Research*, **37**(4), 1071–1082.

Seibert, J. & McDonnell, J.J. (2002) On the dialog between experimentalist and modeler in catchment hydrology: Use of soft data for multicriteria model calibration. *Water Resources Research*, **38**(11), 1241, doi:1210.1029/2001WR000978.

Sidle, R.C., Tsuboyama, Y., Noguchi, S., Hosoda, I., Fujieda, M. & Shimizu, T. (2000) Stormflow generation in steep forested headwaters: a linked hydrogeomorphic paradigm. *Hydrological Processes*, **14**(3), 369–385.

Sidle, R.C., Noguchi, S., Tsuboyama, Y. & Laursen, K. (2001) A conceptual model of preferential flow systems in forested hillslopes: evidence of self-organization. *Hydrological Processes*, **15**, 1675–1692.

Siegenthaler, U. (1979) Stable hydrogen and oxygen isotopes in the water cycle. In: *Lectures in Isotope Geology* (Eds E. Jäger & J.C. Hunziker), pp. 264–273. Springer-Verlag, Berlin.

Sklash, M.G. (1990) Environmental isotope studies of storm and snowmelt runoff generation. In: *Processes in Hillslope Hydrology* (Eds M.G. Anderson & T.P. Burt), pp. 401–435. J. Wiley & Sons, Chichester.

Sklash, M.G. & Farvolden, R.N. (1979) The role of groundwater in storm runoff. *Journal of Hydrology*, **43**, 45–65.

Smith, R.E. & Goodrich, D.C. (2005) Rainfall excess overland flow. In: *Encyclopedia of Hydrological Sciences* (Ed. M.G. Anderson), pp. 1707–1718. J. Wiley & Sons, New York.

Stewart, J.B. (1977) Evaporation from the wet canopy of a pine forest. *Water Resources Research*, **13**(6), 915–921.

Stewart, M.K. (1975) Stable isotope fractionation due to evaporation and isotopic exchange of falling waterdrops; applications to atmospheric processes and evaporation of lakes. *Journal of Geophysical Research*, **80**(9), 1133–1146.

Stewart, M.K. & McDonnell, J.J. (1991) Modeling base flow soil water residence times from deuterium concentrations. *Water Resources Research*, **27**(10), 2681–2693.

Stieglitz, M., Shaman, J., McNamara, J., Engel, V., Shanley, J. & Kling, G.W. (2003) An approach to understanding hydrologic connectivity on the hillslope and the implications for nutrient transport. *Global Biogeochemical Cycles*, **17**(4), 1105, doi:1110.1029/2003GB002041.

Sturm, K., Hoffmann, G., Langmann, B. & Stichler, W. (2005) Simulation of $\delta^{18}O$ in precipitation by the regional circulation model REMOiso. *Hydrological Processes*, **19**(17), 3425–3444.

Taha, A., Gresillion, J.M. & Clothier, B.E. (1997) Modelling the link between hillslope water movement and stream flow: application to a small Mediterranean forest watershed. *Journal of Hydrology*, **203**, 11–20.

Taylor, S., Feng, X., Kirchner, J.W., Osterhuber, R., Klaue, B. & Renshaw, C.E. (2001) Isotopic evolution of a seasonal snowpack and its melt. *Water Resources Research*, **37**, 759–770.

Torres, R., Dietrich, W.E., Montgomery, D.R., Anderson, S.P. & Loague, K. (1998) Unsaturated zone processes and the hydrologic response of a steep, unchanneled catchment. *Water Resources Research*, **34**(8), 1865–1879.

Tromp-van Meerveld, H.J. & McDonnell, J.J. (2006) Threshold relations in subsurface stormflow 2. The fill and spill hypothesis. *Water Resources Research*, **42**, WO2411, doi: 10.1020/2004 WRO03800.

Tsuboyama, Y., Sidle, R.C., Noguchi, S., Murakami, S. & Shimizu, T. (2000) A zero-order basin – its contribution to catchment hydrology and internal hydrological processes. *Hydrological Processes*, **14**(3), 387–401.

Tsukamoto, Y. (1961) An experiment on sub-surface flow. *Journal of the Japanese Forestry Society*, **43**, 62–67.

Turner, J.V. & Barnes, C.J. (1998) Modeling of isotopes and hydrochemical responses in catchment hydrology. In: *Isotope Tracers in Catchment Hydrology* (Eds C. Kendall & J.J. McDonnell), pp. 723–760. Elsevier, Amsterdam.

Uchida, T., Kosugi, K.I. & Mizuyama, T. (2001) Effects of pipeflow on hydrological process and its relation to landslide: a review of pipeflow studies in forested headwater catchments. *Hydrological Processes*, **15**, 2151–2174.

Uchida, T., Asano, Y., Ohte, N. & Mizuyama, T. (2003) Seepage area and rate of bedrock groundwater discharge at a granitic unchanneled hillslope. *Water Resources Research*, **39**(1), doi:10.1029/2002WR001298.

Unnikrishna, P.V., McDonnell, J.J. & Kendall, C. (2002) Isotope variations in a Sierra Nevada snowpack and their relation to meltwater. *Journal of Hydrology*, **260**, 38–57.

Vaché, K.B., McDonnell, J.J. & Bolte, J. (2004) On the use of multiple criteria for a posteriori model rejection: Soft data to characterize model performance. *Geophysical Research Letters*, **31**, L21504, doi:21510.21029/22004GL021577.

Vitvar, T. & Balderer, W. (1997) Estimation of mean water residence times and runoff generation by ^{18}O measurements in a pre-Alpine catchment (Rietholzbach, eastern Switzerland). *Applied Geochemistry*, **12**(6), 787–796.

Ward, R.C. & Robinson, M. (2000) *Principles of Hydrology*, 4th edn. McGraw-Hill, New York.

Weiler, M., McGlynn, B.L., McGuire, K.J. & McDonnell, J.J. (2003) How does rainfall become runoff? A combined tracer and runoff transfer function approach. *Water Resources Research*, **39**(11), 1315, doi:1310.1029/2003WR002331.

Welhan, J.A. & Fritz, P. (1977) Evaporation pan isotopic behaviour as an index of isotopic evaporation conditions. *Geochimica et Cosmochimica Acta*, **41**, 682–686.

Welker, J.M. (2000) Isotopic (δ^{18}O) characteristics of weekly precipitation collected across the USA: an initial analysis with application to water source studies. *Hydrological Processes*, **14**, 1449–1464.

Weyman, D.R. (1973) Measurements of the downslope flow of water in a soil. *Journal of Hydrology*, **20**, 267–288.

Williams, A.G., Dowd, J.F. & Meyles, E.W. (2002) A new interpretation of kinematic stormflow generation. *Hydrological Processes*, **16**(14), 2791–2803.

Winograd, I.J., Riggs, A.C. & Coplen, T.B. (1998) The relative contributions of summer and cool-season precipitation to groundwater recharge, Spring Mountains, Nevada, USA. *Hydrogeology Journal*, **6**(1), 77–93.

Yurtsever, Y. & Gat, J. (1981) Atmospheric water. In: *Stable Isotope Hydrology, Deuterium and Oxygen-18 in the Water Cycle* (Eds J.R. Gat & R. Gonfiantini), pp. 103–142. International Atomic Energy Agency, Vienna.

Zimmermann, U., Munnich, K.O., Roether, W., Kreutz, W., Schubach, K. & Siegel, O. (1966) Tracers determine movement of soil moisture and evapotranspiration. Science, **152**(3720), 346–347.

Zuber, A. (1986) On the interpretation of tracer data in variable flow systems. *Journal of Hydrology*, **86**, 45–57.

CHAPTER 12

Tracing anthropogenic inputs of nitrogen to ecosystems

CAROL KENDALL, EMILY M. ELLIOTT, AND SCOTT D. WANKEL

Introduction

Nitrate (NO_3^-) concentrations in public water supplies have risen above acceptable levels in many areas of the world, largely as a result of overuse of fertilizers and contamination by human and animal waste. The World Health Organization and the U.S. Environmental Protection Agency have set a limit of $10\,mg\,L^{-1}$ nitrate (as N) for drinking water because nitrate poses a health risk, especially for children, who can contract methemoglobinemia (blue-baby syndrome). Nitrate in lower concentrations is non-toxic, but the risks from long-term exposure are unknown, although nitrate is a suspected carcinogen. High concentrations of nitrate in rivers, lakes, and coastal areas can cause eutrophication, often followed by fish-kills, due to oxygen depletion. Increased atmospheric loads of anthropogenic nitric and sulfuric acids have caused many sensitive, low-alkalinity streams in North America and Europe to become acidified. Still more streams that are not yet chronically acidic could undergo acidic episodes in response to large rain storms and/or spring snowmelt, seriously damaging sensitive local ecosystems. Future climate changes may exacerbate the situation by affecting biogeochemical controls on the transport of water, nutrients, and other materials from land to freshwater ecosystems.

The development of effective management practices to preserve water quality, and remediation plans for sites that are already polluted, requires the identification of actual N sources and an understanding of the processes affecting local nitrate concentrations. In particular, a better understanding of hydrologic flowpaths and solute sources is required to determine the potential impact of contaminants on water supplies. Determination of the relation between nitrate concentrations in groundwater and surface water and the quantity of nitrate introduced from a particular source is complicated by:

1 the occurrence of multiple possible sources of nitrate in many areas;
2 the presence of overlapping point and non-point sources;
3 the co-existence of several biogeochemical processes that alter nitrate and other chemical concentrations.

In many circumstances, isotopes offer a direct means of source identification because different sources of nitrate often have distinct isotopic compositions. In addition, biological cycling of nitrogen often changes isotopic ratios in predictable and recognizable directions that can be reconstructed from the isotopic compositions. Nitrogen isotopes ($\delta^{15}N$) have been used to identify N sources and processes in hundreds of studies over the past several decades (Heaton 1986). Since the early 1990s, nitrate isotope studies have often included analysis of the oxygen isotopes of nitrate ($\delta^{18}O$), especially in studies of the role of atmospheric deposition in watersheds (Kendall 1998). Analysis of nitrate $\delta^{17}O$ is a promising new tool for determining nitrate sources and reactions, and complements conventional uses of $\delta^{15}N$ and $\delta^{18}O$. But the most promising forensic isotopic approaches combine nitrate isotopes with multi-isotope and multi-tracer approaches that track trace elements and organics specific to different sources of nitrate.

The primary goal of this chapter is to examine recent progress in the use of natural abundance isotopes of nitrate and other N-bearing species for identifying and quantifying the relative contributions of N from different anthropogenic sources (including fertilizer, sewage and animal waste, and atmospheric deposition) to various ecosystems, with an emphasis on applications to watersheds. This chapter contains sections on (i) the isotopic compositions of major N reservoirs, (ii) major processes affecting the isotopic composition of these reservoirs, (iii) how to distinguish the effects of mixing of sources from the effects of processes, (iv) applications to major ecosystem settings, and (v) a summary of the status of various "isotope tools".

Isotope techniques are a subset of tools available for hydrologists and biogeochemists studying nitrogen cycling in ecosystems; they are not a panacea. Most of the uses of nitrate isotopes are for source identification and qualitative estimations of source contributions, not quantitative determinations. The greatest problems for isotope studies are:

1 that different sources can have partially overlapping isotopic compositions;
2 sources can have considerable spatial and temporal variation in isotopic composition;
3 isotope fractionations can blur initially distinctive isotopic compositions.

These problems can often be minimized or eliminated by a multi-isotope, multi-tracer approach which also takes advantage of hydrologic and chemical data.

Why are stable isotope techniques underused in surface water studies in large agricultural basins?

While stable isotopes have become common tools for tracing sources of waters and solutes in small watersheds (e.g., Kendall & McDonnell 1998) and groundwater systems (e.g., Cook & Herczeg 2000), they are currently underutilized in larger basins, especially in agricultural rivers. This is probably because one of the first attempts to use natural abundance $\delta^{15}N$ to understand the causes of the increases in nitrate concentrations in surface waters in many agricultural areas (Kohl et al. 1971) elicited a very critical response by 10 prominent soil scientists and agronomists (Hauck et al. 1972), which concluded that use of natural abundance $\delta^{15}N$ was a "questionable approach."

The abundant publications over the next two decades that used natural abundance ^{15}N to determine the source of nitrate in groundwater evoked little such controversy. However, it appears that the critical response to the Kohl et al. study effectively inhibited similar investigations in rivers in agricultural areas in the USA until the late 1990s when some technological advances resulted in renewed interest in attempting to use stable isotopes to quantify nitrate sources in agricultural basins.

Since the study by Kohl et al. (1971) appears to have had such a dramatic and continuing impact, it is useful to briefly examine the nature of the original controversy. Kohl et al. (1971) investigated sources of nitrate in drainage waters of the Sangamon River (Illinois, USA). As part of their study, several dozen nitrate samples were collected from drain tile effluent, plus samples from a nearby lake, the Sangamon River, and a drainage ditch. The $\delta^{15}N$ of the two potential end-members, soils and fertilizer, was determined. A linear regression through the data on a plot of $\delta^{15}N$ vs. concentration intersected the values measured for fertilizer and incubated soils. The trends were attributed to mixing of NO_3^- from nitrification of soil N and fertilizer N, and they concluded that about half the nitrate was derived from soil sources and half from unfractionated fertilizer nitrate.

This conclusion was strongly criticized by Hauck et al. (1972) and others because they contended that fractionation effects and natural variability in soil systems would make it impossible to apply simple mixing models to the $\delta^{15}N$ values. Specifically, the response by Hauck et al. (1972) made five main points:

1 analytical precision of natural abundance measurements is insufficient for quantifying sources over the small range of differences in ^{15}N between fertilizer and soil end-members (0.004 atom % or 10‰);

2 fertilizer NH_4^+ mixes with soil N before it is oxidized to NO_3^-, thus losing its isotopic signature;

3 insufficient soil samples were analyzed to assess the true variability in the $\delta^{15}N$ of soils within the >900 square mile basin;

4 it is difficult to correct for the biological fractionation effects that cause great variability in $\delta^{15}N$ in soils;

5 the $\delta^{15}N$ of NO_3^- produced by nitrification of soil organic N is best determined by short-term incubations of soil, not the long-term incubations performed by Kohl et al.

The response by Kohl et al. (1972) to Hauck et al. (1972) carefully considered but ultimately dismissed most of the criticisms, responding that despite all the possible confounding complications, the surface water sample data themselves strongly supported their interpretation of mixing of soil and fertilizer NO_3^-, and that their evaluation method probably underestimated the true proportion of fertilizer-derived NO_3^-. The Sangamon River data, along with data from a somewhat similar study on the small Yerres River (France) described in Mariotti & Létolle (1977) and Létolle (1980), were reanalyzed by Hübner (1986). He noted that a logarithmic relation could be fitted to both data sets, with apparent enrichment factors of ca. 5‰, suggesting that denitrification or assimilatory fractionation effects could also be a factor. Because of the complications noted in both of these studies, Shearer & Kohl (1993) suggested that perhaps the best use of natural abundance ^{15}N research is not to try to estimate contributions from different sources, but to study N transformations. A recent review of agricultural $\delta^{15}N$ studies concluded that the most appropriate applications were for semi-quantitative to qualitative estimates of source proportions, pattern analysis, and generating hypotheses (Bedard-Haughn et al. 2003). However, this may be an overly conservative conclusion.

In the past few decades, numerous studies have shown that stable isotopic techniques are a powerful tool for determining sources and sinks of nutrients and organic matter in relatively small watersheds. Recently, these and newer isotope techniques have been successfully applied to tracing sources and sinks in large river basins, including the Mississippi River (Battaglin et al. 2001a,b; Kendall et al. 2001; Chang et al. 2002; Panno et al. 2006), large rivers in the northeastern USA (Mayer et al. 2002), the San Joaquin River in California (Kratzer et al. 2004), the Oldman River in Alberta (Canada; Rock & Mayer 2004), and the Seine River (France; Sebilo et al. 2006).

In reality, the isotopic compositions of nitrate are often the result of both mixing and cycling, and thus, a multi-tracer approach is usually the best approach. In the late 1980s and 1990s, a number of new approaches for studying the impact of agricultural sources of N on groundwater and surface water were developed. Most are based on using a multi-isotope and/or multi-tracer approach to resolve N source vs. cycling questions. The result has been scores of studies tracing sources of N and investigating N transformations in agricultural, urban, and forested watersheds, ranging from small to very large basins.

Perhaps the five most successful new isotopic approaches are:

1 Analysis of the N_2 gas produced by denitrification as a means for "correcting" for the fractionating effects of denitrification so that the initial $\delta^{15}N$ of the NO_3^- (and hence its source) can be determined. Examples: Vogel et al. (1981), Wilson et al. (1990), Böhlke & Denver (1995), McMahon & Böhlke (1996).

2 Development of methods for age-dating groundwater recharged in the past ca. 50 years with precisions of 1–3 years using chlorofluorocarbons (CFCs), $T/^3He$, etc., and applying this to understanding the history of agricultural N

contamination. Examples: Böhlke & Denver (1995), Böhlke (2002), McMahon & Böhlke (2006).

3 The analysis of nitrate for $\delta^{18}O$ as well as $\delta^{15}N$. Examples: Böttcher et al. (1990), Aravena & Robertson (1998), Campbell et al. (2002); Mayer et al. (2002), Wankel et al. (2006), Wassenaar et al. (2006).

4 Using the $\delta^{15}N$ of algae and fish as "proxies" for (or integrators of) the $\delta^{15}N$ of NO_3 contributed by different land uses. Examples: Harrington et al. (1998), Koerner et al. (1999); Hebert & Wassenaar (2001), Anderson & Cabana (2005, 2006).

5 The analysis of nitrate for $\Delta^{17}O$. While this approach is in its infancy, with most of the publications dealing with atmospheric processes, $\Delta^{17}O$ promises to be valuable in ecosystem studies because it is an unambiguous tracer of atmospheric NO_3^-. Examples: Michalski et al. (2003, 2004, 2005).

Perhaps one reason that isotope techniques have not yet become a mainstream tool in agricultural basins is that the extent of temporal and spatial variability in the biogeochemistry and isotopic composition of various soil components makes it seem improbable that isotopic compositions could meaningfully integrate the myriad of environmental variability inherent to natural systems. To the contrary, many researchers have found that isotopes indeed have a unique ability to integrate environmental variability such that major natural patterns emerge and can be meaningfully interpreted. In the following section we consider how methodological advances have allowed us to answer increasingly complex questions regarding N isotopes in environmental systems.

Methodological advances in analyzing nitrogen isotopes

Until recently, almost all NO_3^- for both $\delta^{15}N$ and $\delta^{18}O$ were analyzed using modifications of the silver nitrate method (Silva et al. 2000), where samples are concentrated on anion exchange resins, eluted, purified to produce silver nitrate, and then analyzed. The $\delta^{15}N$ of the silver nitrate can be measured using EA-IRMS or by pyrolysis. The original method used sealed-tube combustion to generate CO_2 for $\delta^{18}O$ measurement. However, automated pyrolysis systems that generate CO are now more commonly used. A number of modifications aimed at improved removal of dissolved organics have been described, including Chang et al. (1999), Hwang et al. (1999), and Heaton et al. (2004).

However, many laboratories are now analyzing nitrate, using the denitrifier method (Sigman et al. 2001; Casciotti et al. 2002), where samples are inoculated with a pure culture of denitrifying bacteria lacking the enzyme to reduce nitrate beyond N_2O. The resulting N_2O is stripped from the samples using an automated headspace analyzer, purified, and then analyzed for $\delta^{15}N$ and $\delta^{18}O$. This method is a significant improvement over the previous silver nitrate method because samples are about three orders of magnitude smaller and high-salinity seawater samples are easily analyzed.

With the ongoing refinement of our understanding of $\delta^{18}O_{NO_3}$ and limitations of previous methods, it is evident that earlier $\delta^{18}O_{NO_3}$ data generated using sealed-tube combustions were potentially biased because of exchange of O with the glass and/or contamination by O-bearing contaminants in the silver nitrate (Revesz & Böhlke 2002), especially for samples that produced less than the recommended minimum of 100–200 μmol CO_2 (unpublished U.S. Geological Survey data). There is some speculation that even data produced using pyrolysis may be affected by O contamination from organic material. If so, the earlier $\delta^{18}O_{NO_3}$ data may have been subject to a "permil-scale contraction". This topic will be discussed in more detail later in the section on **denitrification**.

Other recent methodological advances include methods for analyzing nitrite (NO_2^-) for $\delta^{18}O$ and $\delta^{15}N$ (McIlvin & Altabet 2003; Casciotti et al. 2007), nitrate for $\Delta^{17}O$ (Michalski et al. 2002; Kaiser et al. 2007), and marine dissolved organic N (DON) for $\delta^{15}N$ (Knapp et al. 2005).

Isotopic compositions of major N sources to ecosystems

Different sources of N to ecosystems have a wide range of $\delta^{15}N_{NO_3}$ and $\delta^{18}O_{NO_3}$ values (Figure 12.1). There is a vastly greater amount of $\delta^{15}N_{NO_3}$ and $\delta^{15}N_{NH_4}$ data available than $\delta^{18}O_{NO_3}$ data. Recent compilations of $\delta^{15}N$ data include Kendall (1998, which includes a compilation of $\delta^{18}O_{NO_3}$ data), Fogg et al. (1998), and Bedard-Haughn et al. (2003). The sections below provide brief discussions of the major sources.

Atmospheric N

Since the tightening regulation of SO_2 emissions in the USA and in Europe, nitrate has become an increasingly important component of acidic deposition. For example, sulfate concentrations in precipitation have decreased throughout most of the USA (Butler et al. 2001; Lehmann et al. 2005). As a result, NO_3^- has become a more significant contributor to soil acidification, stream acidification, and forest degradation, particularly in eastern USA. Moreover, NO_3^- concentrations have increased in many western states, in some cases by up to 20–50% (Nilles & Conley 2001; Lehmann et al. 2005).

Advances in analytical methods have had a tremendous influence on our understanding of atmospheric nitrate isotopes. Up until the early 1990s, only $\delta^{15}N$ data were available for precipitation. It had been generally assumed that the $\delta^{18}O$ of atmospheric nitrate would be similar to the isotopic composition of atmospheric O_2 (ca. +23‰) because the $\delta^{15}N$ of atmospheric nitrate was similar to the composition of N_2 (ca. 0‰). For this reason, it was not thought

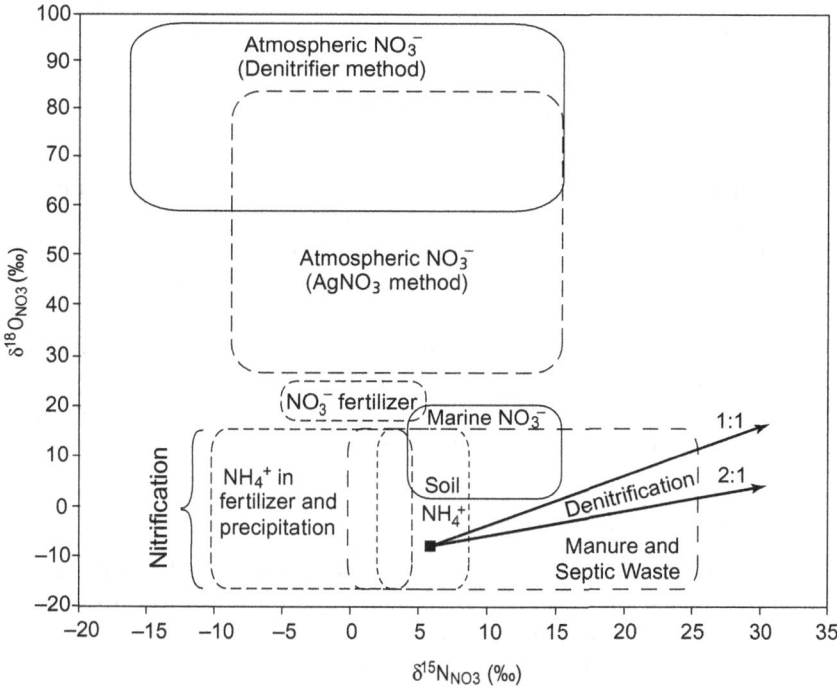

Figure 12.1 Typical values of $\delta^{15}N$ and $\delta^{18}O$ of nitrate derived or nitrified from various N sources. Atmospheric $\delta^{18}O_{NO_3}$ data are divided into the ranges observed for samples analyzed using the denitrifier and AgNO$_3$ (non-denitrifier) methods. The two arrows indicate typical expected slopes for data resulting from denitrification of nitrate with initial $\delta^{15}N = +6‰$ and $\delta^{18}O = -9‰$. The typical ranges of $\delta^{18}O_{NO_3}$ values produced by nitrification of ammonium and organic matter are denoted by "nitrification".

that analysis of $\delta^{18}O$ would provide much additional information. Since then, our understanding of the isotopic composition of atmospheric nitrate has experienced several major revisions, first in the mid-1990s with the development of silver nitrate methods for the $\delta^{18}O$ (first used by Kendall et al. (1995a,b) and Wassenaar (1995), but not published until Silva et al. (2000) and Chang et al. (1999)); then in the mid-2000s with the development of a denitrifier method for $\delta^{18}O$ (Sigman et al. 2001; Casciotti et al. 2002), and again in the early 2000s with the development of the methods for analyzing nitrate for $\delta^{17}O$ (Michalski et al. 2002; Kaiser et al. 2007).

The section below contains a brief discussion of whether different anthropogenic sources of atmospheric nitrate (e.g., power plant, vehicle, and agricultural emissions) may have distinguishable isotopic signatures and what is known about the $\delta^{15}N$, $\delta^{18}O$, and $\delta^{17}O$ of wet and dry precipitation. Later in the chapter, we discuss our current understanding of the atmospheric processes and **causes of variability** in the isotopic compositions of atmospheric nitrate.

Isotopic composition of sources contributing to atmospheric nitrate

NO_x is released to the atmosphere from human activities (e.g., fossil fuel combustion) and natural processes (e.g., biogenic soil emissions, lightning, biomass burning). Fossil fuel combustion from mobile (e.g., vehicles) and stationary sources (e.g., electricity generation, industrial processes) constitute the largest global NO_x input. The major sink for NO_x in the atmosphere is the oxidation to nitric acid (HNO_3), which readily dissociates to nitrate (NO_3^-) where it can be deposited as wet deposition. Dry deposition can contribute significant loads of atmospherically derived N to ecosystems as dry gases (HNO_3 vapor, NH_3, NO_2, HONO, NO, peroxyacetyl nitrate (PAN)), dry aerosols (particulate NO_3^-, particulate NH_4^+), or in fogwater.

Isotopic composition of NO_x sources

For the past several decades, $\delta^{15}N$ of nitrogen oxides (NO_x) has been identified as a potential means for distinguishing air pollution sources. Anthropogenic NO_x sources generally have positive $\delta^{15}N$ values. In the 1970s, Moore (1977) characterized the isotopic composition of vehicle exhaust (average = +3.7‰, $n = 3$). A subsequent study (Heaton 1990) reported negative $\delta^{15}N$ in vehicle NO_x emissions (−13‰ to −2‰, $n = 8$). More recently, roadside denuders (average = +5.7‰, $n = 9$; Ammann et al. 1999) and roadside vegetation (average = +3.8‰, $n = 10$) (Pearson et al. 2000) have been used to illustrate characteristic $\delta^{15}N$ values associated with vehicle NO_x emissions. The isotopic composition of NO_x from stationary source fossil fuel combustion has also been characterized in several studies. Heaton (1990) reported that NO_x emissions from coal-fired power plants have $\delta^{15}N$ values ranging from +6‰ to +13‰ ($n = 5$; Heaton 1990). In a more recent study, Kiga et al. (2000) reported $\delta^{15}N$ values in NO_x produced from coal combustion ranging from +4.8‰ to +9.6‰ ($n = 6$). In both vehicle and stationary source fossil fuel combustion, the isotopic value of the resulting NO_x is suggested to be a function of the N present in the original fuel (negligible in the case of gasoline), the N_2 pumped through the engine, and the fractionations associated with thermal NO_x production (oxidation of atmospheric N_2 at high temperatures). Because thermally produced NO_x is assumed to have lower $\delta^{15}N$ values than fuel-derived NO_x, it is generally assumed that vehicle NO_x emissions have lower $\delta^{15}N$ values compared with stationary source NO_x emissions.

Natural sources of NO_x to the atmosphere, including lightning, biogenic soil emissions, and biomass burning are not as well characterized as anthropogenic sources. However, it has been documented that relatively pristine sites generally have lower $\delta^{15}N_{NO_2}$ values than highly polluted or heavily traveled sites (Moore 1977; Ammann et al. 1999). Although $\delta^{15}N$ values of NO_x produced from biogenic soil emissions and biomass burning have not been directly characterized, as a volatile by-product of nitrification (and/or denitrification) and combustion respectively, $\delta^{15}N$ values are expected to be <0‰ due to the preferential volatilization of ^{14}N. The high temperature associated with lightning generates NO_x from the thermal oxidation of atmospheric N_2 and constitutes the other major natural NO_x source. The $\delta^{15}N_{NO_x}$ generated from laboratory discharges of lightning ranges from −0.5 to +1.4‰ (Hoering 1957). Although more extensive isotopic analyses are needed to more thoroughly characterize $\delta^{15}N_{NO_x}$ from various N sources, existing studies generally suggest that natural NO_x sources, including lightning and soil NO_x emissions, have lower $\delta^{15}N$ values than anthropogenically derived NO_x from fossil fuel combustion.

Variations in $\delta^{15}N$

Wet deposition

Wet deposition refers to all processes that transfer atmospheric N to the Earth's surface in aqueous form including rain, snow, and fog (Seinfeld & Pandis 1998). Complex chemical reactions in the atmosphere result in a large range of $\delta^{15}N$ values of N-bearing compounds depending on the reactants involved, the season, meteorological conditions, ratio of NH_4^+ to NO_3^- in the precipitation, types of anthropogenic inputs, proximity to pollution sources, distance from ocean, etc. (Hübner 1986; Heaton et al. 1997, 2004). Natural atmospheric sources of N-bearing gases (e.g., N_2O, HNO_3, NH_3, NO, NO_2, etc.) include volatilization of ammonia from soils and animal waste (with fractionations as large as −40‰), nitrification and denitrification in soils and surface waters, biomass burning, and lightning.

The $\delta^{15}N$ values of atmospheric NO_3^- and NH_4^+ are usually in the range of −15 to +15‰ (Figure 12.1), relative to atmospheric N_2 (0‰), however lower nitrate $\delta^{15}N$ values in polar regions have been observed in snow (Heaton et al. 2004) and minerals (Michalski et al. 2005). In general, NO_3^- in rain appears to have a higher $\delta^{15}N$ value than the co-existing NH_4^+, with the lower values for NH_4^+ attributed to washout of atmospheric NH_3 (Freyer 1978, 1991; Garten 1992). There is considerable literature on the $\delta^{15}N$ of N-bearing compounds in the atmosphere (see a review by Heaton et al. 1997). However, there have been few comprehensive studies of $\delta^{15}N$ of precipitation until recently, in part because of the difficulty of analyzing such dilute waters, prior to the development of the denitrifier method (Sigman et al. 2001). Below is a brief summary of the major findings of earlier studies.

Studies in Germany (Freyer 1978, 1991; Freyer et al. 1993), the USA (Russell et al. 1998), and South Africa (Heaton 1986, 1987) document that $\delta^{15}N_{NO_3}$ values show a seasonal cycle of low $\delta^{15}N$ values in spring and summer rain, and higher values in the winter. Russell et al. (1998) also showed seasonal shifts in $\delta^{15}N_{NH_4}$ that were attributed to increased springtime agricultural emissions. The $\delta^{15}N_{NO_3}$ value of throughfall (rain that intercepts the tree canopy before falling "through") was found to be higher than in open-air rain, whereas the $\delta^{15}N_{NH_4}$ in throughfall had a variable composition relative to rain in studies in Tennessee (Garten 1992) and Yorkshire (UK; Heaton et al. 1997). Although precipitation often contains unequal quantities of ammonium and nitrate, because ammonium is preferentially retained by the canopy relative to atmospheric nitrate (Garten & Hanson 1990), most of the atmospheric nitrogen that reaches the soil surface is in the form of nitrate.

More recent studies of nitrate isotopes in precipitation and snowpack using the denitrifier method also report seasonal $\delta^{15}N_{NO_3}$ variability, however, direction of the seasonal shifts vary. Hastings et al. (2004) report that $\delta^{15}N$ values are higher in spring and summer snowpack than in snowpack from fall and winter months in Greenland. Elliott et al. (2004; in preparation) observed strong seasonal and inter-event variability in $\delta^{15}N$ in a study of over 100 precipitation events at Connecticut Hill, New York (USA). Average $\delta^{15}N$ values in precipitation from this site were approximately 5‰ higher in the winter than in the summer. Back trajectory models, coupled with cluster analyses, indicate that inter-event variability in $\delta^{15}N$ at the site can be partially attributed to source areas of individual air masses. In comparison, in a series of 65 precipitation samples from Bermuda, Hastings et al. (2003) report opposite seasonal trends in $\delta^{15}N_{NO_3}$, with lower $\delta^{15}N$ values during the cool season (−5.9‰) than the warm season (−2.1‰). This pattern was explained by seasonal shifts in NO_x source and source region, which is dominated by lightning inputs during the warm season.

At a larger spatial scale, a recent investigation of temporal and spatial variations in the $\delta^{15}N$ of nitrate in wet deposition at ca. 150 precipitation monitoring sites across the USA showed values ranging from −11‰ to +3.5‰, with a mean value of −3.1‰ ($n = 883$; Elliott et al. 2006; in preparation). This range in $\delta^{15}N$ values is similar to those recently reported for rain in Bermuda (Hastings et al. 2003) and snow in Greenland (Hastings et al. 2004). Even at this large scale, seasonality was pronounced, with mean $\delta^{15}N$ values 3‰ higher during January-February than during May-June. For sites where bimonthly samples are available throughout the year, average annual $\delta^{15}N$ is calculated and the spatial distribution of these values is shown in Figure 12.2. The most prominent features of the data are several "hotspot" areas where $\delta^{15}N$ is consistently higher than the surrounding region, including areas in the Midwest, south of the Great Lakes, the central Front Range, and near Seattle. The lowest $\delta^{15}N$ values are generally located west of the Mississippi River and include sites in the Dakotas, Minnesota, Nebraska, and Texas.

At a subset of these sites spanning the nitrate and sulfate deposition gradient spanning the midwestern to northeastern USA, Elliott et al. (in press)

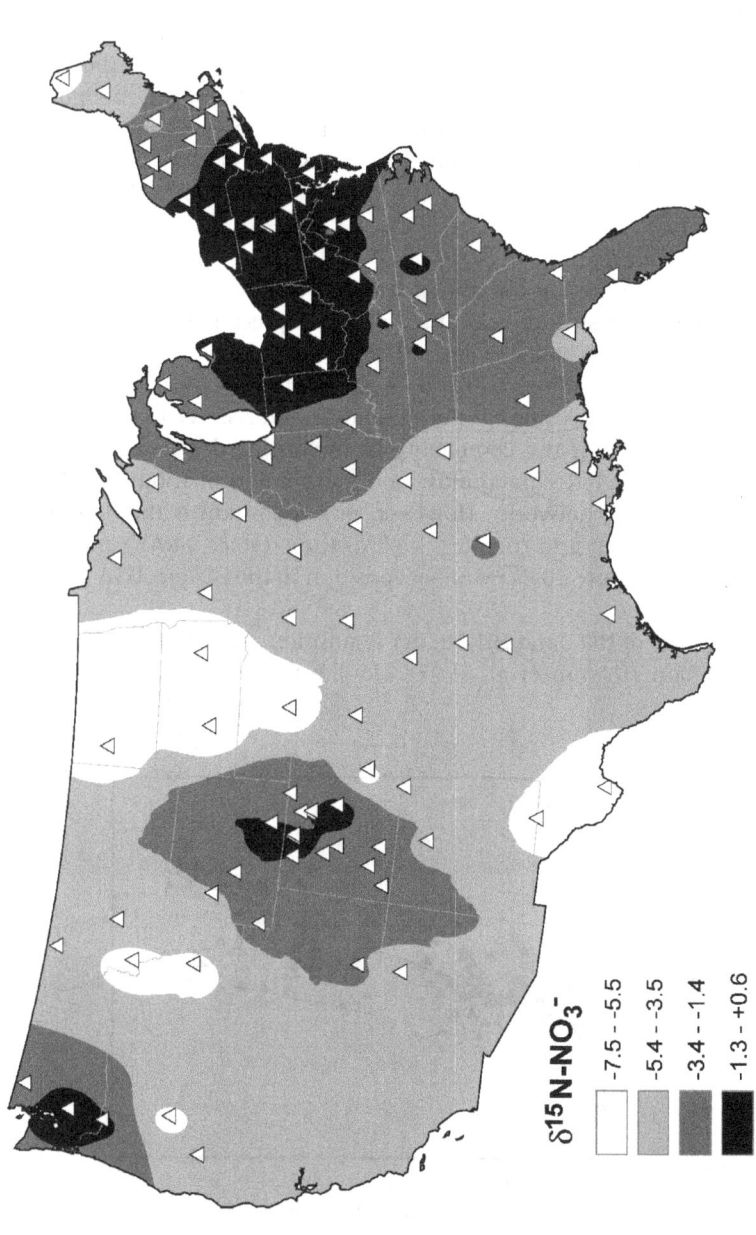

Figure 12.2 Spatial variability in the average $\delta^{15}N_{NO_3}$ of precipitation collected at National Atmospheric Deposition Program (NADP) sites in the USA in 2000. (Elliott & Kendall, unpublished U.S. Geological Survey data; in preparation.)

$\delta^{15}N\text{-}NO_3^-$

- -7.5 – -5.5
- -5.4 – -3.5
- -3.4 – -1.4
- -1.3 – +0.6

demonstrated that spatial variations in $\delta^{15}N$ are strongly correlated with amounts of NO_x emissions from surrounding electric generating units (Figure 12.3). Although vehicles comprise the single largest NO_x emission source in the eastern USA, $\delta^{15}N$ was not correlated with county-level vehicle NO_x emissions, suggesting that vehicle NO_x may not be as regionally distributed as stationary source NO_x. The results from this regional study suggest that nitrate isotopes in precipitation may be a "sharper tool" than concentration measurements and atmospheric transport models for assessing relative magnitude of various NO_x sources to landscapes, at any spatial or temporal scale.

Dry deposition and aerosols

Dry deposition is defined as the direct transfer of gaseous and particulate species to the Earth's surface without the aid of precipitation (Seinfeld & Pandis 1998) and can include dry gases (HNO_3 vapor, NH_3, NO_2, HONO, NO, peroxyacetyl nitrate (PAN)) or dry aerosols (particulate NO_3^-, particulate NH_4^+). In general, dry deposition is poorly understood relative to wet deposition, partially because of the complexity of measuring dry deposition, the array of dry deposited N compounds, and the limited distribution of dry deposition monitoring networks. However, dry deposition is the dominant form of N deposition in arid climates, such as the western USA (Fenn et al. 2003) and can contribute 20–50% of N deposition in the eastern USA (Butler et al. 2005).

The $\delta^{15}N$ values of NO_3^- and NH_4^+ in dry deposition are usually higher than in wet deposition (Heaton et al. 1997). Equilibrium exchange reaction of

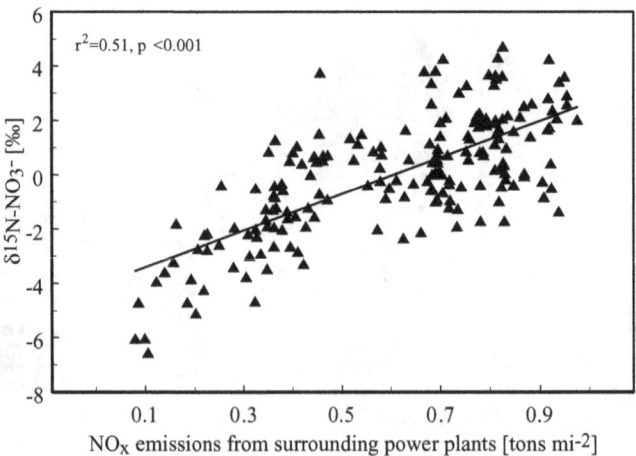

Figure 12.3 Correlation between precipitation $\delta^{15}N_{NO_3}$ from NADP sites in the northeastern and midwestern USA and NO_x emissions from power plants within 400 miles of individual NADP sites. (Modified from Elliott et al. in press.)

gaseous NO or NO_2 with dissolved NO_3 would likely result in ^{15}N enrichment of the NO_3. However, other studies have illustrated various complicated relations (Moore 1977; Heaton 1987), and considerable interstorm and seasonal variability in $\delta^{15}N$. In a recent study in France, Widory (2007) analyzed $\delta^{15}N$ of bulk N ($NO_3^- + NH_4^+$) of particulate matter less than $10\,\mu M$ in diameter (PM10) and determined that $\delta^{15}N$ was higher for particulates generated from unleaded and diesel fuels (+4.6‰, $n = 4$), coal (+5.3‰, $n = 1$), natural gas (+7.7‰, $n = 5$), and waste incineration (+6.7, $n = 3$), than in fuel oil (−7.8‰, $n = 8$).

Freyer (1991) examined $\delta^{15}N$ in particulate NO_3^- and HNO_3 vapor collected using both low volume and high volume samplers. Similar seasonal patterns (higher $\delta^{15}N$ in winter) were observed between fine and coarse particulate NO_3^- and accompanying wet nitrate. Seasonality in $\delta^{15}N$ of wet and particulate NO_3^- was strongly correlated with both temperature and solar radiation. $\delta^{15}N$ of particulate NO_3^- was always higher than wet deposition, and $\delta^{15}N$ of the coarse particulate fraction was generally lower than that of the fine particulate fraction. No seasonality was observed in HNO_3. In comparison, consistent seasonal variations in $\delta^{15}N$ of particulate NO_3^- and HNO_3 were observed at dry deposition sampling sites in New York, Ohio, and Pennsylvania (Elliott & Kendall, unpublished U.S. Geological Survey data).

Yeatman et al. (2001) measured $\delta^{15}N$ of aerosol NO_3^- and NH_4^+ using high volume samplers at two coastal sites in the UK and Ireland, and also near livestock sources, roadways, and in remote areas. Similar $\delta^{15}N_{NH_4}$ and $\delta^{15}N_{NO_3}$ values were observed near chicken, cow, and pig livestock ($\delta^{15}N_{NH_4} = +13.5$, $n = 7$ and $\delta^{15}N_{NO_3} = +10.6$‰, $n = 7$), whereas samplers deployed near three roadways resulted in lower $\delta^{15}N$ values ($\delta^{15}N_{NH_4} = +3.6$‰, $n = 3$; and $\delta^{15}N_{NO_3} = +11$‰, $n = 2$). Interestingly, samplers deployed in rural "remote" areas, not adjacent to immediate sources, have relatively high $\delta^{15}N$ for both NH_4^+ (+4.5‰) and NO_3^- (+11‰). Both $\delta^{15}N_{NH_4}$ and $\delta^{15}N_{NO_3}$ were higher at the coastal UK site (+6‰ , $n = 37$ and +7‰, $n = 25$, respectively) than at the coastal site in Ireland (−9‰, $n = 36$ and −1‰, $n = 21$, respectively). The proximity of the UK site to anthropogenic sources, coupled with differences in the relative influence of marine-derived N, are suggested as potential causes for the observed spatial differences.

Variations in $\delta^{18}O$

Wet deposition
There is much less known about the $\delta^{18}O$ of atmospheric NO_3^-, mainly because there were no methods for analyzing it until the late 1980s (Amberger & Schmidt 1987; Silva et al. 2000). There have been multiple investigations of the $\delta^{15}N$ and $\delta^{18}O$ of precipitation NO_3^- in localized areas (Durka et al. 1994; Russell et al. 1998; Burns & Kendall 2002; Campbell et al. 2002; Xiao & Liu 2002; Pardo et al. 2004), with many focusing on snowpack samples

during winter months. Several of these studies (e.g., Williard 1999; Hastings et al. 2003; Pardo et al. 2004) observed seasonal differences in the $\delta^{18}O_{NO_3}$ in precipitation, with higher values in the winter than in the summer.

A survey of existing nitrate $\delta^{18}O$ values of precipitation in the late 1990s observed values ranging from +14 to +75‰, with a highly non-normal distribution of values (Kendall 1998). More recently, in a spatially extensive survey of nitrate isotopes in precipitation across the USA, Elliott et al. (2006; in preparation) document $\delta^{18}O_{NO_3}$ values ranging from +63‰ to +94‰, with a mean value of +76.3‰ ($n = 883$) across ca. 150 precipitation monitoring sites. Similar to the case with $\delta^{15}N$, Elliott et al. determined that $\delta^{18}O$ is seasonally variable, with mean $\delta^{18}O$ values 9.5‰ higher during January–February than during May–June. This range in $\delta^{18}O$ values reported by Elliott et al. is similar to those recently reported for rain in Bermuda (Hastings et al. 2003), snow in Greenland (Hastings et al. 2004), and snow in the Arctic (Heaton et al. 2004).

To date, it appears that all $\delta^{18}O$ values for atmospheric nitrate samples produced thus far using the denitrifier method are >60‰ (Figure 12.1). Hence, the $\delta^{18}O_{NO_3}$ values observed using the denitrifier method are higher than those analyzed using either the closed-tube or pyrolysis methods for converting silver nitrate to gases. It is possible that some of the lower precipitation $\delta^{18}O$ values observed using earlier methods are a result of reaction with glass during combustion, exchange with O in the glass, or contamination by other O-bearing materials in the silver oxide (e.g., organic compounds, sulfate, carbonate), all of which would probably lower the observed range in $\delta^{18}O$ values (Revesz & Böhlke 2002). Samples that produced the recommended minimum of 100–200 µmol CO_2 during sealed-tube combustions usually show minimal offset (unpublished U.S. Geological Survey data).

Variations in $\delta^{17}O$ (or $\Delta^{17}O$) in wet and dry deposition

In all oxygen bearing terrestrial materials, there is a consistent relationship between $\delta^{18}O$ and $\delta^{17}O$ values because kinetic and equilibrium isotope fractionations depend on the relative differences in atomic mass. However, ozone (O_3) formation exhibits a unique kinetic isotope effect, producing $\delta^{17}O$ values higher than statistically expected (Mauersberger et al. 2003). This "mass independent fractionation" (MIF) results in ozone having anomalous or excess [17]O (beyond that expected from the abundance of ^{18}O). Further, because ozone is a photochemically reactive species, this isotopic anomaly is transferred to several other oxygen-bearing atmospheric compounds (Thiemens 1999, 2006; Lyons 2001).

Figure 12.4, a triple oxygen isotope plot, illustrates the concept of this mass independent anomaly. On this plot, mass **dependent** fractionations (MDF) for nitrate result in values approximated by the relation: $\delta^{17}O = 0.52$

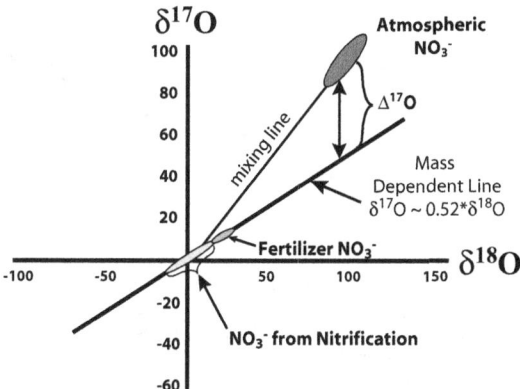

Figure 12.4 Schematic of relationship between $\delta^{18}O$ and $\delta^{17}O$ values. (Modified from Michalski et al. 2002.)

$\times\ \delta^{18}O$ (Michalski et al. 2002). This line is essentially fixed by the interactions of all conventional mass dependent kinetic and equilibrium isotopic fractionations involving oxygen. In contrast, MIFs cause values that deviate from this relation, and thus lie above the mass dependent line. Hence, MDF results in $\Delta^{17}O = 0$, whereas MIF results in $\Delta^{17}O \neq 0$ and $\Delta^{17}O$ values >0‰ are a useful tracer of O derived from atmospheric processes.

Current theory is that nitrate obtains its high $\delta^{17}O$ (and $\delta^{18}O$) due to chemical reactions with tropospheric ozone which has a $\Delta^{17}O$ of ca. +35‰ (Johnston & Thiemens 1997). Atmospheric nitrate $\Delta^{17}O$ values as high as ca. +30‰ have been observed (Michalski et al. 2003; Wankel 2006). Seasonal variation in the $\Delta^{17}O$ of aerosol nitrate (from +20 to +30‰) observed in Southern California (USA) was explained by a shift from nitric acid production by the OH* + NO_2 reaction, which is predominant in the spring and summer, to N_2O_5 hydrolysis reactions, which are more important in the winter (Michalski et al. 2003). Figure 12.5 illustrates seasonal patterns in $\Delta^{17}O$ of bimonthly volume-weighted precipitation samples across New England (USA) and the strong correlation of $\Delta^{17}O$ with $\delta^{18}O$ (Wankel 2006), presumably due to a seasonality in the relative proportions of NO_x oxidation by OH* or O_3 (Hastings et al. 2003; Michalski et al. 2003, 2004).

As discussed previously, many watershed studies have interpreted the seasonally high $\delta^{18}O$ of stream NO_3^- as an indicator of significant contributions of atmospheric nitrate. However, because of the strong isotopic discrimination involved in many processes that consume NO_3^-, there can also be increases in $\delta^{18}O_{NO3}$ values which cannot be attributed to inputs by atmospheric nitrate (e.g., denitrification). Besides providing an unequivocal quantification of atmospheric inputs, the non-zero $\Delta^{17}O$ of atmospheric NO_3^- offers a unique tracer of N cycling as well. The $\Delta^{17}O$ value of nitrate derived

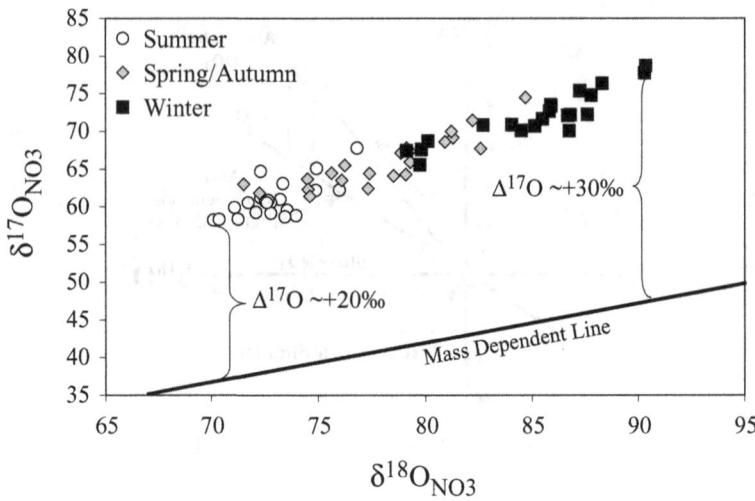

Figure 12.5 Seasonal variation of $\Delta^{17}O$ values in rain from NADP sites in northeastern USA. (Modified from Wankel & Kendall, unpublished U.S. Geological Survey data; in preparation.)

from atmospheric sources will remain unchanged regardless of the effects of the fractionating process (e.g., denitrification and assimilation), because all of these processes are strictly mass dependent and follow the slope of the mass dependent line (Figure 12.4). Thus, the triple oxygen isotopic composition of NO_3^- can be used to both calculate the proportion of atmospheric NO_3^- present and to estimate the relative amount of NO_3^- consumed (by back calculating the original $\delta^{18}O$ value). Production of new nitrate (nitrification) can dilute the $\Delta^{17}O$ signal of the original atmospheric NO_3^- to the point of being undetectable, and the recycling of atmospheric NO_3^- to organic matter and then back to NO_3^- will overprint the original atmospheric NO_3^- with the terrestrial $\Delta^{17}O$ signal (i.e., 0‰).

Fertilizers

Many kinds of fertilizers are added to soils. It is important to distinguish among "natural" nitrate fertilizers (e.g., guano, desert nitrate salts), "synthetic" nitrate produced by oxidation of ammonia produced via the Haber-Bosch process, and "microbial" nitrate derived from ammonium fertilizers. Inorganic fertilizers have $\delta^{15}N$ values that are uniformly low reflecting an origin from atmospheric N_2 (Figure 12.1), generally in the range of −4 to +4‰; however, some fertilizer samples have shown a total range of −8 to +7‰ (see compilations by Hübner 1986; Macko & Ostrom 1994; Vitoria et al. 2004). Nitrate fertilizers often have slightly higher $\delta^{15}N$ values than NH_4^+ fertilizers. Organic fertilizers, including cover crops and plant composts, and liquid and solid animal waste, generally have higher $\delta^{15}N$ values and a much wider range of compositions (generally +2 to +30‰) than inorganic fertilizers, reflecting their more diverse origins. Note that the $\delta^{15}N$ of NO_3^- in soils fertilized with NH_4^+ may not be the same as the fertilizer. During nitrification of applied ammonium, residual NH_4^+ and the NO_3^- formed in the soil can show $\delta^{15}N$ values that change by 20‰ during the process (Feigin et al. 1974), although ultimately the $\delta^{15}N$ of the total resulting nitrate may only be a few permil higher than the original NH_4^+.

Amberger & Schmidt (1987) determined that nitrate fertilizers have distinctive $\delta^{18}O_{NO_3}$ values. Synthetic fertilizers where the O is chiefly derived from atmospheric O_2 (ca. +23.5‰), have $\delta^{18}O$ values ranging from +17 to +25‰ whereas natural fertilizers derived from Chilean deposits have $\delta^{18}O$ values of +46 to +58‰ (Böhlke et al. 2003; Vitoria et al. 2004). Nitrate derived from nitrification of ammonium fertilizers has lower $\delta^{18}O$ values, usually in the range of −5 to +15‰. This range of $\delta^{18}O$ values (shown in the area labeled "NH_4^+ in fertilizer or precipitation" on Figure 12.1) reflects the normally observed range of $\delta^{18}O$ values for microbially produced nitrate in well-oxygenated soils. See the section on nitrification for more information on the controls on the $\delta^{18}O$ of microbial nitrate.

Animal and human waste

It has often been observed that consumers (microbes to invertebrates) are 2–3‰ enriched in ^{15}N relative to their diet. The increase in $\delta^{15}N$ in animal tissue and solid waste relative to diet is due mainly to the excretion of low $\delta^{15}N$ organics in urine or its equivalent (Wolterink et al. 1979). Animal waste products may be further enriched in ^{15}N because of volatilization of ^{15}N-depleted ammonia, and subsequent oxidation of much of the residual waste material may result in nitrate with a high $\delta^{15}N$. By this process, when animal waste with a typical $\delta^{15}N$ value of about +5‰ is converted to nitrate, the $\delta^{15}N$ values are generally in the range of +10 to +20‰ (Kreitler 1975, 1979). Nitrate derived from human and other animal waste becomes isotopically indistinguishable using $\delta^{15}N$ under most circumstances (an exception is Fogg et al. 1998). However, with a multi-isotope approach, some recent studies show that it is possible to distinguish waste from different animal sources (Spruill et al. 2002; Curt et al. 2004; Widory et al. 2004, 2005), and this is discussed further below.

Soils (organic N and dissolved inorganic nitrogen)

The $\delta^{15}N$ of total soil N ranges from about −10 to +15‰. Cultivated soils have slightly lower $\delta^{15}N$ values (+0.65 ± 2.6‰) than uncultivated soils (+2.73 ± 3.4‰), according to a major soil survey by Broadbent et al. (1980). Most of the N in soils is bound in organic forms and not readily available to plants;

hence, the $\delta^{15}N$ of total soil N is generally not a good approximation of the $\delta^{15}N$ of N available for plant growth.

Soluble dissolved inorganic nitrogen (DIN; mainly NO_3^-) constitutes about 1% of the N in soils, and represents a very small pool which can be much more sensitive to change than the larger organic pool. The $\delta^{15}N$ of soil nitrate ranges from about -10 to $+15$‰, with most soils having $\delta^{15}N_{NO_3}$ values in the range of $+2$ to $+5$‰ (Kendall 1998). The $\delta^{15}N$ of soil nitrate and organic matter is strongly affected by drainage, topographic position, vegetation, plant litter, land use, temperature, and rain amount (Shearer & Kohl 1988; Amundson et al. 2003). Depending on land use and proximity to possible atmospheric sources of anthropogenic contaminants, the soil nitrate $\delta^{15}N$ and $\delta^{18}O$ may reflect "natural conditions" or the effects of various anthropogenic activities. The discussion below is intended to provide some background on this important source of N to ecosystems. For a more detailed discussion of the $\delta^{15}N$ of soil N and DIN, see the literature review in Kendall & Aravena (2000).

There have been several investigations of the $\delta^{15}N_{NO_3}$ values of soils from different environments (i.e., "natural" soils (tilled and untilled), soils fertilized with synthetic fertilizers or manure, soils contaminated with septic waste, etc). For example, in a study of variability in the $\delta^{15}N$ of soil water in lysimeters, Ostrom et al. (1998) found low $\delta^{15}N_{NO_3}$ in the spring and fall related to mineralization of soil organic matter (OM), and high $\delta^{15}N_{NO_3}$ values in late summer because of denitrification. The $\delta^{15}N_{NO_3}$ of soil water in non-tilled land was ca. 1.5‰ higher than in tilled land, and both were >4‰ lower than soil OM (Ostrom et al. 1998).

In general, the soil NO_3^- produced from fertilizer (average $\delta^{15}N$ value = $+4.7 \pm 5.4$‰) and animal waste (average $\delta^{15}N = +14.0 \pm 8.8$‰) are isotopically distinguishable but they both overlap with the $\delta^{15}N_{NO_3}$ of precipitation and natural soils. However, given the large range of $\delta^{15}N$ values of the NO_3^- sources, the average values of sources from one site cannot be automatically applied to another. This is vividly illustrated by a compilation of nitrate $\delta^{15}N$ data (Fogg et al. 1998).

The higher $\delta^{15}N_{NO_3}$ values in soils on lower slopes or valley bottoms are usually attributed to either greater denitrification in more boggy areas (Karamanos et al. 1981), or higher relative rates of immobilization and nitrification in these bottom soils (Shearer et al. 1974). Soil nitrate is preferentially assimilated by tree roots relative to soil ammonium (Nadelhoffer & Fry. 1994). Surface soils beneath bushes and trees often have lower $\delta^{15}N$ values than those in open areas, presumably as the result of litter deposition (Nadelhoffer & Fry 1988; Shearer & Kohl 1988); the $\delta^{15}N$ of soil N and DIN generally increases with depth.

Plants

Plants, a major reservoir of organic N, can utilize a variety of materials from purely inorganic compounds (NH_4^+, NO_3^-, NO_2^-, N_2) to more complex forms of organic N, and can have a wide range in $\delta^{15}N$ values depending on environmental conditions. Plants fixing N_2 from the atmosphere have $\delta^{15}N$ values of ca. 0 to +2‰, close to the $\delta^{15}N$ value of atmospheric N_2 (= 0‰). Most terrestrial plants have $\delta^{15}N$ in the range of −5 to +2‰ (Fry 1991). Algae and other aquatic plants have a much larger range of $\delta^{15}N$ values (−15 to +20‰), with values typically in the range of −1 to +7‰ (Kendall, unpublished data). For more information about the isotopic compositions of plants, see (for terrestrial plants) Garten et al. (this volume, pp. 61–82) and Evans (this volume, pp. 83–98), and (for aquatic plants) Finlay & Kendall (this volume, pp. 283–333).

The agronomy literature is full of studies showing that plants grown on anthropogenic ammonium or nitrate fertilizers usually have lower $\delta^{15}N$ values than plants grown on natural soils or "green manure". Some of the spatial variability in foliar $\delta^{15}N$ that is commonly observed within forested catchments may be due to anthropogenic effects. For example, the lower foliar $\delta^{15}N$ values on ridgetops compared with valley bottom foliage in Tennessee (USA) may reflect the greater uptake of low-$\delta^{15}N$ atmospheric ammonium on ridges where soil dissolved inorganic N (DIN) is more limited, and the greater uptake of high-$\delta^{15}N$ soil NO_3^-, resulting from denitrification, by plants in the valleys (Garten 1993). Plants near busy roads have $\delta^{15}N$ values that are ca. 10‰ higher than in rural areas (Pearson et al. 2000).

There is considerable literature on using the $\delta^{15}N$ of aquatic plants and consumers to trace anthropogenic sources of N; these are reviewed by Finlay & Kendall (this volume, pp. 283–333) and briefly described in a later section on isotope **biomonitoring**. Perhaps the most generally useful observation is that in nutrient-rich environments, the $\delta^{15}N$ of the algae can closely track the $\delta^{15}N$ of the nitrate; studies in the Mississippi Basin (Battaglin et al. 2001a, 2001b) and San Joaquin Rivers (Kratzer et al. 2004) have shown an isotope fractionation of 4–5‰, offering considerable support for the usefulness of "isotope biomonitoring" of anthropogenic inputs.

Processes affecting the isotopic composition of DIN

In order for isotopes to be useful as tracers of various N sources to aquatic systems, an understanding of how biogeochemical cycling affects the isotopic composition of various inorganic forms of N is critical. It is impossible to do justice to this complex topic in the space allowed. However, here we briefly present the state of our understanding regarding major factors that can affect isotopic composition of N species in environmental systems, with an emphasis on processes with relevance to studying nitrate in watersheds and, to a lesser degree, coastal areas.

The basic fundamentals of isotope fractionation are discussed by Sulzman (this volume, pp. 1–21). Irreversible (unidirectional) kinetic fractionation effects involving metabolic nitrogen transformations are generally more important than equilibrium fractionation effects in low temperature environments. Many biological processes consist of a number of steps (e.g., nitrification: $NH_4^+ \rightarrow NO_2^- \rightarrow NO_3^-$). Each step has the potential for fractionation, and the overall fractionation for the reaction is often dependent on environmental conditions, including the number and type of intermediate steps, sizes of reservoirs (pools) of various compounds involved in the reactions (e.g., O_2, NH_4^+), soil pH, species of the organism, etc. Hence, estimation of fractionations in natural systems can be very complex.

Generally, most of the fractionation is caused by the rate-determining or slowest step. This step commonly involves a large pool of substrate where the amount reacting is small compared with the size of the reservoir. In contrast, a step that is not rate-determining generally involves a small pool of a compound that is rapidly converted from reactant to product. When the compound is converted to product as soon as it appears, there is little net fractionation at this step. The isotopic compositions of reactant and product pools during a multi-step reaction where the net fractionation is controlled by a single rate-determining step can be successfully modeled either by a Rayleigh closed-system approach or by a "steady-state" open-system approach. For more details, see reviews by Létolle (1980), Hübner (1986), Kendall (1998), Kendall & Aravena (2000), and Böhlke (2002).

The main biologically mediated reactions that control nitrogen dynamics in ecosystems are fixation, assimilation, mineralization, nitrification, and denitrification. These reactions commonly result in increases in the $\delta^{15}N$ of the substrate and decreases in the $\delta^{15}N$ of the product, unless the reactions go to completion. Physical processes, specifically ammonia volatilization, also significantly influence the $\delta^{15}N$ of the released ammonia, residual NH_4^+, and any subsequently formed NO_3^-. Processes that consume NO_3^- (primarily denitrification and assimilation by phytoplankton and/or prokaryotes) generally cause the $\delta^{15}N$ and $\delta^{18}O$ in the remaining pool of NO_3^- to increase in a relatively predictable pattern. These processes and their impact on isotopic compositions of selected N-bearing compounds are discussed below.

Fixation

The term **N-fixation** refers to processes that convert unreactive atmospheric N_2 into other forms of nitrogen (Cleveland et al. 1999). Although the term is usually used to mean fixation by bacteria, it has also been used to include fixation by lightning and, more importantly, by human activities (energy production, fertilizer production, and crop cultivation) that produce reactive N (NO_x, NH_y, and organic N). Nitrogen fixation by human activity (industrial and agricultural) was estimated to be ca. $160\,Tg\,N\,yr^{-1}$ in 1995, which is ca. 45% of all the nitrogen fixed on land and in the oceans (Galloway et al. 1995, 2004). These authors predict that the anthropogenic fixation rate will increase by 60% by the year 2020, mainly due to increased fossil-fuel combustion and fertilizer use, especially in the developing countries of India and Asia. This increase in N loading and N saturation is causing critical ecosystem changes on both the local and global scale (Galloway et al. 1995). Concern about the impact of these changes on human activities is the main reason for the increased interest in uses of nitrogen isotopes in environmental studies.

Bacterial fixation of atmospheric N_2 by the enzyme nitrogenase commonly produces organic materials with $\delta^{15}N$ values slightly less than 0‰. A compilation by Fogel & Cifuentes (1993) indicated measured fractionations ranging from −3 to +1‰. Because these values are generally lower than the values for organic materials produced by other mechanisms, low $\delta^{15}N$ values in organic matter are often cited as evidence for N_2 fixation (though other process can also give rise to low values). The isotopic compositions of N-bearing materials produced by anthropogenic fixation (atmospheric gases produced during fossil fuel combustion, and artificial fertilizers produced from atmospheric gases) are discussed in detail below.

Assimilation

Assimilation refers to the transformation of inorganic N-bearing compounds into an organic form during biosynthesis by living organisms. Generally, oxidized forms of N are initially reduced to NH_4^+ and then assimilated into organic matter. Assimilation, like other biological reactions, discriminates between isotopes and generally favors the incorporation of the isotope with the lower mass. A large range of N fractionations (−30 to 0‰) has been measured in field studies (Cifuentes et al. 1989; Montoya et al. 1991), and in laboratory experiments for nitrate and ammonium assimilation by algae (Pennock et al. 1996; Waser et al. 1998; Altabet et al.1999; Granger et al. 2004), and bacteria (Hoch et al. 1992) in aquatic environments. While there have been few studies of the effects of O fractionations during assimilation on the $\delta^{18}O$ of the residual NO_3^-, nitrate assimilation by marine phytoplankton seems to cause ca. 1:1 changes in the $\delta^{15}N$ and $\delta^{18}O$ of nitrate, regardless of species or the magnitude of the isotope effect; hence, these fractionations can be viewed as strongly "coupled" (Granger et al. 2004).

Fogel & Cifuentes (1993) present an elegant model for ammonium assimilation in aquatic algae that predicts total fractionations of −4, −14, or −27‰ depending on whether algae cells are nitrogen limited, enzyme limited, or diffusion limited, respectively. However, for the low pH values and low NH_4^+ concentrations common to soils and many aquatic environments, the model predicts that availability of N is the limiting condition and the transport of ammonium across cell walls is probably rapid, resulting in a small (<−4‰) overall fractionation.

More recently, Needoba et al. (2004) demonstrated that the isotopic effects or "apparent fractionation" imparted by phytoplankton during NO_3^-

assimilation on external NO_3^- pool results from several physiological factors. Phytoplankton cells transport and store high concentrations of internal NO_3^- in vacuoles, from which an internal enzyme (nitrate reductase) reduces NO_3^- for assimilation. While the transport step does not fractionate, the enzymatic reduction exhibits a large fractionation (−13 to −6‰) depending on growth rate and conditions. Interestingly, the only means by which the external nitrate pool is affected by this fractionation is through the efflux of internal, highly fractionated (i.e., ^{15}N-enriched) nitrate.

Mineralization

Mineralization is usually defined as the production of ammonium from organic matter and is also referred to as **remineralization** or **ammonification**. Mineralization usually causes only a small fractionation (±1‰) between soil organic matter and soil NH_4^+. Many other workers use the term **mineralization** for the overall production of nitrate from organic matter by several reaction steps. This usage results in literature that gives fractionations for mineralization that can range from −35 to 0‰, depending on which step is rate limiting (Delwiche & Steyn 1970; Feigin et al. 1974; Létolle 1980; Macko & Estep 1984). The large fractionations are caused by the nitrification of NH_4^+, not the conversion of organic N to NH_4^+. In general, the $\delta^{15}N$ of soil NH_4^+ is usually within a few permil of the $\delta^{15}N$ of total organic N in the soil.

Volatilization

Volatilization, the loss of ammonia gas to the atmosphere, is a highly fractionating process in which the ammonia gas produced has a lower $\delta^{15}N$ value than the residual NH_4^+. Volatilization involves several steps that can cause fractionation, including (i) the equilibrium fractionations between ammonium and ammonia in solution, and between aqueous and gaseous ammonia, and (ii) the kinetic fractionation caused by the diffusive loss of ^{15}N-depleted ammonia. The overall process causes a fractionation of ca. 25‰, but the actual fractionation depends on the pH, temperature, humidity and other factors (Hübner 1986).

Volatilization in farmlands results from applications of urea and manure to fields, and occurs within piles of manure. Ammonium produced from this organic N may have $\delta^{15}N$ values >20‰ due to ammonia losses. While there is little information about the $\delta^{15}N$ of the volatilized ammonia, it may reach values as low as −20‰. The downwind transport of this ^{15}N-depleted N may be a significant source of atmospheric ammonium and (when oxidized) nitrate to adjacent areas (W. Showers, pers. comm.). In a survey of fertilized soils in Texas, Kreitler (1975) attributed a 2–3‰ increase in $\delta^{15}N_{NO_3}$ in underlying groundwater relative to the applied fertilizer to volatilization, and noted that losses of ammonia in alkaline soils can be very large and cause dramatic shifts in $\delta^{15}N$ of the resulting nitrate.

Nitrification

Nitrification is the two-step process of NH_4^+ oxidation to NO_3^- mediated by several different autotrophic bacteria or archaea for the purpose of deriving metabolic energy. Various byproducts or intermediates can also be produced during nitrification and released into the environment including aqueous compounds (e.g., NH_2OH and NO_2^-) as well as gaseous compounds (e.g., NO and N_2O). In contrast to the "coupled" nature of isotope effects for nitrate-consuming processes such as assimilation and denitrification, nitrification can be considered "decoupled" because the sources of N and O atoms are unrelated. During nitrification, N atoms originate from NH_4^+ and/or NO_2^- molecules, while O atoms originate from O_2 and/or H_2O. Hence, the processes that control the $\delta^{15}N$ and $\delta^{18}O$ values during nitrification are discussed separately below.

Controls on $\delta^{15}N$

The total fractionation associated with nitrification depends on which step is rate determining. Because the oxidation of NO_2^- to NO_3^- is generally rapid in natural systems, this is generally not the rate-determining step, and most of the N fractionation is caused by the slower oxidation of NH_4^+ to NO_2^-. This first step of nitrification has been well studied in cultures of ammonium-oxidizing bacteria and been shown to have a large N isotope effect ranging from −38 to −14‰ (Mariotti et al. 1981; Casciotti et al. 2003). Similar to the effect of substrate concentration on NH_4^+ fractionation during assimilation, nitrification is expected to be diffusion limited at low NH_4^+ concentrations, and thus the isotope effect smaller (Casciotti et al. 2003). In diffusion-limited environments where nitrification is closely coupled with denitrification (e.g., benthic sediments), almost all the microbial nitrate may be rapidly denitrified, resulting in minimal efflux of nitrate with low $\delta^{15}N$ values to the water column (Lehmann et al. 2004).

The recent discovery of ammonium-oxidizing archaea may raise new questions about the role of archaea in N cycling (Francis et al. 2005; Könneke et al. 2005; Schleper et al. 2005). While it is generally believed that N isotope effects will be similar to those found in cultures of nitrifying bacteria, ammonium-oxidizing archaea have only recently been isolated in pure culture (Könneke et al. 2005) and the isotope effects are still unknown.

In general, the extent of fractionation during nitrification is dependent on the fraction of the substrate pool (reservoir) that is consumed. In N-limited systems, the fractionations are minimal. Hence, in soils where NH_4^+ is rapidly converted to NO_3^-, the $\delta^{15}N$ of soil NO_3^- is usually within a few permil of the

$\delta^{15}N$ of total organic N in the soil. If there is a large amount of NH_4^+ available (e.g., artificial fertilizer recently applied), nitrification is stimulated, and the oxidation of fertilizer NH_4^+ becomes the rate-determining step; this would result in a large fractionation. The $\delta^{15}N$ value of the first-formed NO_3^- would be quite low, but as the NH_4^+ pool is consumed, nitrification rate decreases, oxidation of NH_4^+ is no longer the rate-determining step, and the $\delta^{15}N$ of the total NO_3^- increases towards pre-fertilization values (Feigin et al. 1974).

As a result of the fractionations during transformation from NH_4^+ fertilizer to soil NO_3^-, one cannot accurately estimate the $\delta^{15}N$ value of NO_3^- being leaked to surface water or groundwater from an agricultural field from simple measurement of the average $\delta^{15}N$ of the NH_4^+ fertilizers. Even if the fertilizer applied were 100% synthetic KNO_3, there would still be a possibility of post-depositional increases in $\delta^{15}N$ caused by denitrification as the nitrate was slowly transported to the sampling point. Increases in $\delta^{15}N$ (and $\delta^{18}O$) of NO_3^- caused by denitrification are less likely in coarse-grained soils where waters percolate rapidly (and have higher concentrations of dissolved oxygen) than in finer-grained soils (Gormly & Spalding 1979). Hence, the best way to assess the "effective" $\delta^{15}N$ and $\delta^{18}O$ value of the fertilizer or manure **end-member** is to collect samples from beneath the field where the materials are applied, avoiding sample collection soon after application since the fractionations are greatest then.

Controls on $\delta^{18}O$

While the fractionation of N during NH_4^+ oxidation is relatively well understood (Mariotti et al. 1981; Casciotti et al. 2003), the source of O atoms ultimately incorporated into the NO_3^- molecule during nitrification remains somewhat unresolved. The $\delta^{18}O$ resulting from nitrification is controlled by the composition of the oxidant sources (i.e., H_2O and/or O_2). During the first step, NH_4^+ is oxidized to hydroxylamine (NH_2OH) via an ammonium mono-oxygenase; this reaction has been shown to incorporate O atoms from dissolved O_2 (Hollocher et al. 1981). It is currently unknown whether this process causes a kinetic fractionation and consequently preferential incorporation of $^{16}O_2$. Andersson & Hooper (1983) demonstrated that the resulting NO_2^- contains one O atom from dissolved O_2 and one from H_2O. However, they also revealed that during NH_4^+ oxidation to NO_2^- there can be considerable isotopic exchange between the O in H_2O and NO_2^-. The oxidation of NO_2^- to NO_3^- has been shown to incorporate O atoms from H_2O only (Aleem et al. 1965; Kumar et al. 1983; Hollocher 1984; Dispirito & Hooper 1986). Thus, $\delta^{18}O_{NO_3}$ largely has been interpreted as a mixture of two oxygen atoms from H_2O and one from O_2. Hence,

$$\delta^{18}O_{NO_3} = 2/3(\delta^{18}O_{H_2O}) + 1/3(\delta^{18}O_{O_2}) \tag{12.1}$$

where the $\delta^{18}O_{H_2O}$ is assumed to be that of ambient H_2O, and the $\delta^{18}O_{O_2}$ is assumed to be that of ambient O_2. For waters with $\delta^{18}O$ values in the normal range of −25 to +4‰, and soil O_2 with the $\delta^{18}O$ of atmospheric O_2 (ca. +23.5‰), soil NO_3^- formed from *in situ* nitrification, should be in the range of −10 to +10‰, respectively (Figure 12.1).

The simple equation above for calculation of the $\delta^{18}O_{NO_3}$ makes four critical assumptions:

1 the proportions of O from water and O_2 are the same in soils as observed in laboratory cultures;
2 there are no fractionations resulting from the incorporation of oxygen from water or O_2 during nitrification;
3 the $\delta^{18}O$ of water used by the microbes is equal to that of the bulk soil water;
4 the $\delta^{18}O$ of the O_2 used by the microbes is equal to that of atmospheric O_2.

However, the $\delta^{18}O$ of dissolved O_2 in aquatic systems reflects the effects of three primary processes:

1 diffusion of atmospheric O_2 (ca. +23.5‰) in the subsurface;
2 photosynthesis – resulting in the addition of O_2 with a low $\delta^{18}O$ similar to that of water;
3 respiration by microbes – resulting in isotopic fractionation and higher $\delta^{18}O$ values for the residual O_2.

Many studies have used measurement of $\delta^{18}O_{NO_3}$ in freshwater systems for assessing sources and cycling (see below). Often it has been found that the $\delta^{18}O$ of microbial NO_3^- is a few permil higher than expected for the equation and the assumptions above (e.g., Kendall 1998). A variety of explanations have been offered for these high $\delta^{18}O_{NO_3}$ values including:

1 nitrification in soil waters with higher than expected $\delta^{18}O$ values because of evaporation (Böhlke et al. 1997) or seasonal changes in rain $\delta^{18}O$ (Wassenaar 1995);
2 changes in the proportion of O from H_2O and O_2 sources (i.e., >1/3 from O_2) (Aravena et al. 1993);
3 nitrification using O_2 that has a high $\delta^{18}O$ due to respiration (Kendall 1998);
4 nitrification that occurs simultaneously via both heterotrophic and autotrophic pathways (Mayer et al. 2001).
At this time, it is still unresolved how each of these mechanisms affects $\delta^{18}O_{NO_3}$ during nitrification reactions.

Open ocean settings, because of the broader nature of the chemical gradients and relative isolation from interfering sources of contamination, may provide a simpler conceptual background in which to understand $\delta^{18}O_{NO_3}$.

Recent data from oceanic settings (Casciotti et al. 2002; Lehmann et al. 2004; Sigman et al. 2005; Wankel et al. 2007) indicate that NO_3^- formed in the deep ocean ultimately assumes a $\delta^{18}O_{NO_3}$ only slightly higher (ca. +3‰) than that of the $\delta^{18}O$ of seawater (ca. 0‰). It has been postulated that this low $\delta^{18}O_{NO_3}$ value is caused by water–nitrite O isotopic exchange which may be catalyzed by nitrifying bacteria (Andersson & Hooper 1983; Casciotti et al. 2002). Additionally, in a study of $\delta^{18}O_{NO_3}$ along an estuarine gradient (where $\delta^{18}O_{H_2O}$ ranged from −10 to ca. 0‰), it was postulated that rapid cycling of NO_3^- (assimilation → decomposition → nitrification) at low nitrate concentrations (~15 µM) increased the degree to which O from H_2O was incorporated into the NO_3^- molecule, potentially making $\delta^{18}O_{NO_3}$ a useful indicator of N recycling in such environments (Wankel et al. 2006).

Denitrification

Denitrification refers to the dissimilatory reduction of NO_3^- to gaseous products (N_2, N_2O, or NO) and usually occurs only where O_2 concentrations are less than 20 µM. Although denitrification does not generally occur in the presence of significant amounts of oxygen, it has been hypothesized that it can occur in anaerobic pockets within an otherwise oxygenated sediment or water body (Brandes & Devol 1997; Koba et al. 1997). Denitrification causes the $\delta^{15}N$ of the residual nitrate to increase exponentially as nitrate concentrations decrease; values >100‰ are not unusual. For example, denitrification of fertilizer NO_3^- with a $\delta^{15}N$ value of +0‰ can yield residual nitrate with much higher $\delta^{15}N$ values (e.g., +15 to +30‰) that are within the range of compositions expected for NO_3^- derived from a manure or septic-tank source (Figure 12.1). Additionally, denitrification causes the $\delta^{18}O$ values to increase in the residual NO_3^- pool. Thus, the effects of denitrification on the dual isotopic composition of NO_3^- are considered coupled since both the N and O atoms originate in the same molecule.

Nitrate reduction by heterotrophic microbes and the simultaneous respiration of CO_2 from the oxidation of organic matter is the generalized pathway of heterotrophic denitrification:

$$4NO_3^- + 5CH_2O + 4H^+ \rightarrow 2N_2 + 5CO_2 + 7H_2O \tag{12.2}$$

Most denitrifying heterotrophic microorganisms are actually facultatively anaerobic, switching from oxygen to nitrate respiration at O_2 levels of less than about 0.5 mg L^{-1} (Hübner 1986). However, chemo-autotrophic denitrification by bacteria such as *Thiobacillus denitrificans*, which oxidizes sulfur, can also be important (Batchelor & Lawrence 1978). The stoichiometry of the denitrification reaction mediated by *Thiobacillus denitrificans* is:

$$14NO_3^- + 5FeS_2 + 4H^+ \rightarrow 7N_2 + 10SO_4^{2-} + 5Fe^{2+} + 2H_2O \tag{12.3}$$

Measured enrichment factors for (apparent fractionation, or ε) associated with denitrification ($\varepsilon_{N_2\text{-}NO_3}$) range from -40 to $-5‰$ (Mariotti et al. 1981, 1982; Böttcher et al. 1990; Aravena & Robertson, 1998; Granger 2006) with the $\delta^{15}N$ of the N_2 lower than that of the NO_3^-. The N_2 produced by denitrification results in excess N_2 dissolved in groundwater; the $\delta^{15}N$ of this N_2 can provide useful information about sources and processes (Böhlke & Denver 1995).

The extent of fractionation is highly dependent on environmental conditions. Authors have distinguished between "benthic" denitrification in which NO_3^- diffuses into the anaerobic groundwater from a surficial aerobic environment before denitrification can occur (Brandes & Devol 1997), and "riparian" denitrification where there is partial conversion of the nitrate in the anaerobic groundwater (Sebilo et al. 2003). The apparent fractionation associated with benthic denitrification is small (ranging from -1.5 to $-3.6‰$), because NO_3^- diffusion through the water–sediment interface, which causes minimal fractionation, is the rate-determining step (Sebilo et al. 2003; Lehman et al. 2004). In contrast, riparian denitrification causes a much larger apparent fractionation (about $-18‰$; Sebilo et al. 2003). Similarly, one can distinguish between benthic and "pelagic" (or "water column") denitrification (Brandes & Devol 2002; Lehmann et al. 2004; Sigman et al. 2005), where again diffusion limits the effects of fractionations in the sediments on the $\delta^{15}N_{NO_3}$ in the overlying water column.

It is important to remember that even though benthic (sedimentary) denitrification has a minimal "isotope effect" on the overlying water column (i.e., does not cause a significant increase in the $\delta^{15}N$ or $\delta^{18}O$ of the NO_3^- in the water column), pore-water NO_3^- probably shows about the same isotopic fractionations observed in open-water environments (Sigman et al. 2001; Lehmann et al. 2004). The lack of isotope effect observed in the overlying water is simply the result of a lack of "communication" of fractionated (i.e., ^{15}N-enriched) pore-water with overlying water across the sediment–water interface. Hence, the potential large fractionation for denitrification is not "expressed" in the water column.

There are several methods for determining the presence, extent and/or rate of denitrification, including various enzyme-block methods (e.g., the acetylene block method) and ^{15}N tracer methods (Nielsen 1992; Mosier & Schimel 1993). Natural abundance isotope methods include comparison of decreasing NO_3^- concentrations with increases in (i) $\delta^{15}N_{NO_3}$, (ii) concentration and $\delta^{15}N$ of total N_2, or (iii) relative $\delta^{15}N$ and $\delta^{18}O$ of residual nitrate (see the section on "Fractionation due to denitrification" below).

It has been recognized for several decades that denitrification causes the $\delta^{15}N$ and $\delta^{18}O$ in the remaining pool of NO_3^- to increase in a relatively predictable pattern. (e.g., Olleros 1983; Amberger & Schmidt 1987; Voerkelius & Schmidt 1990). With the development of larger datasets, primarily from groundwater studies, relatively consistent patterns emerged that suggested denitrification causes $\delta^{15}N$ and $\delta^{18}O$ to increase in roughly a $2:1$ ratio; thus, $^{15}\varepsilon$ was approximately twice as large as $^{18}\varepsilon$ (e.g., Böttcher et al. 1990; Aravena & Robertson 1998; Mengis et al. 1999; Cey et al. 1999; Panno et al. 2006). While no clear mechanism was suggested for this phenomenon, it generally has been accepted that when nitrate isotope data from natural freshwater settings plot along a slope of ca. 0.5 ($\delta^{18}O/\delta^{15}N$), the pattern was consistent with an interpretation of denitrification.

There is currently some uncertainty about the expected fractionation caused by nondiffusion-limited denitrification. If earlier $\delta^{18}O_{NO_3}$ data generated using sealed-tube combustions (Revesz & Böhlke 2002) or pyrolysis were potentially biased because of exchange of O with the glass and/or contamination by O-bearing contaminants in the silver nitrate, then earlier $\delta^{18}O_{NO_3}$ data may have been subject to a "permil-scale contraction". Hence, if we expand the $\delta^{18}O_{NO_3}$ scale, a trend that previously had a slope of ca. 0.5 on plots such as Figure 12.1, would now have a higher slope. However, it is interesting to note that data from many studies, using a variety of different versions of the sealed tube and pyrolysis methods (e.g., Olleros 1983; Amberger & Schmidt 1987; Böttcher et al. 1990; Voerkelius & Schmidt 1990; Aravena & Robertson 1998; Mengis et al. 1999; Cey et al. 1999; Panno et al. 2006), show nitrate $\delta^{18}O$ and $\delta^{15}N$ values in areas where denitrification is likely plotting along slopes of 0.5 to 0.7. Evidently, there is more to be learned about controls on NO_3^- isotopic composition by denitrification in the environment.

Recent work with pure cultures of denitrifying bacteria indicate a slope equal to 1 for the respiratory process of denitrification (Sigman et al. 2005; Granger 2006), similar to the effects seen in NO_3^- assimilation by marine phytoplankton (Granger et al. 2004). Nevertheless, even with newer methods (e.g., the denitrifier method) which avoid the potential interferences of O isotope exchange with the glass (e.g., sealed tube combustion) and/or contamination by other oxygen-bearing compounds (e.g., pyrolysis), nitrate isotopic compositions in groundwater where denitrification occurs still give rise to slopes <1 (Wankel & Kendall, unpublished data) and thus may require additional explanation. Such alternate explanations might include the co-occurrence of respiratory denitrification with: (i) other nitrate consuming pathways such as bacterial nitrate assimilation (for which fractionation systematics are poorly characterized); (ii) anaerobic nitrification involving oxidants other than O_2; (iii) enzymatically catalyzed O exchange between NO_2^- and water (Andersson & Hooper 1983) and re-oxidation of NO_2^- to NO_3^-; or (iv) the process of "aerobic denitrification" or "auxiliary denitrification" (Granger 2006).

Interestingly, the results of Granger (2006) also indicate that while the truly respiratory process of denitrification gives rise to a slope of 1 (through isotope effects imparted by the NO_3^- reductase enzyme NAR, an additional nitrate reducing enzyme used by some bacteria for maintaining cellular redox balance), denitrification by the enzyme NAP (used for "auxiliary denitrification", which is sometimes referred to as "aerobic denitrification") results in a consistent slope of ca. 0.6. Thus, slopes <1 for nitrate samples from anaerobic environments may arise from a larger proportion of this NAP or "auxiliary denitrification" pathway. This might include conditions in which electron donor concentrations (i.e., organic C) are abundant relative to NO_3^- (Granger 2006). The ranges of observed denitrification slopes are bracketed by the two different lines on Figure 12.1. Clearly, further work is needed.

The coupled nature of N and O isotope effects during denitrification (or assimilation) offers a means for constraining other biogeochemical processes. In environments where the mixing of multiple sources of nitrate (sewage, fertilizer, precipitation, etc.) can be disregarded or well-constrained by other tracers, deviations from a coupled 1:1 pattern of $\delta^{15}N$ and $\delta^{18}O$ (by nitrate consuming processes) suggest the influence of additional processes such as nitrification. Sigman et al. (2005) used this approach in the oxygen minimum zone in the eastern tropical North Pacific Ocean where denitrification results in a net loss of NO_3^-. While denitrification was expected to cause both the $\delta^{15}N$ and $\delta^{18}O$ to increase equally, there were deviations from this pattern (denoted by the term "$\Delta(15,18)$"), which arose from the combined effects of denitrification and nitrification of organic N from N-fixation. Use of a steady-state dual isotope model suggested that up to 65% of the geochemical evidence for inputs by N fixation had been "erased" by denitrification (Sigman et al. 2005). Hence, the combined use of $\delta^{15}N$ and $\delta^{18}O$ provided some constraint on N cycling processes in this region.

Other dissimilatory N transformations

In strongly reducing environments with high sulfide concentrations, such as coastal marshes (e.g., Tobias et al. 2001) and estuarine sediments, the dissimilatory reduction of NO_3^- to ammonium (DNRA) has been shown to be as important, if not more so, than denitrification (Jorgensen 1989; Trimmer et al. 1998; An & Gardener 2002; Ma & Aelion 2005), which is inhibited by the high sulfide concentrations (Joye & Hollibaugh 1995). While direct reports on the N isotope fractionation occurring during DNRA are lacking, McCready et al. (1983) demonstrated that NH_4^+ produced from DNRA has a much lower $\delta^{15}N$ than the NO_3^-, which is consistent with a kinetic fractionation.

Similarly, there are no existing data for the fractionation occurring during the anaerobic oxidation of NH_4^+ to N_2 (e.g., anammox; Dalsgaard et al. 2003; Kuypers et al. 2003). This process has been recently attributed to a wide range of environments including the Black Sea (Kuypers et al. 2003), upwelling oceanic regions (Kuypers et al. 2005), Arctic marine sediments (Rysgaard et al. 2004), and even Arctic sea ice (Rysgaard & Glud 2004). While the extent of anammox research has focused mostly on estuarine and marine environments, very little is known regarding its existence or importance in terrestrial groundwater or surface water biogeochemistry.

Atmospheric reactions involving oxidized N species

Atmospheric chemistry of oxidized nitrogen species is complex, and closely coupled with ozone chemistry, hydroxyl radical chemistry, sulfate chemistry, and aerosol dynamics. Because of this complexity, and because we are in the early stages of being able to characterize the isotopic composition of small concentrations of NO_3^- and other N compounds, our understanding of the isotopic fractionations associated with atmospheric chemical reactions is limited. The section "Atmospheric N" (above) contained a brief discussion of whether different anthropogenic sources of atmospheric NO_3^- (e.g., power plant, vehicle, and agricultural emissions) may have distinguishable isotopic signatures. The section below discusses the current understanding of the atmospheric processes and **causes of variability** in the isotopic compositions of atmospheric NO_3^-. Specifically, we present the current understanding of the oxidation chemistry of NO_x in the atmosphere, as well as what is known about how these various pathways influence $\delta^{15}N$, $\delta^{18}O$, and $\Delta^{17}O$ compositions in atmospheric deposition.

NO_x is released to the atmosphere from human activities (e.g., fossil fuel combustion) and natural process (e.g., biogenic soil emissions, lightning, biomass burning). Once in the atmosphere, NO_x generally has a short lifetime, 1–3 days (Seinfeld & Pandis 1998, and references therein). During the daytime, O atoms are rapidly exchanged between ozone (O_3) and atmospheric NO_x (reactions R1 and R2). The major sink for NO_x in the atmosphere is the oxidation to nitric acid (HNO_3) which occurs via both daytime:

$$NO + O_3 \rightarrow NO_2 + O_2 \tag{R1}$$

$$NO_2 + hv \rightarrow NO + \tfrac{1}{3}O_3 \tag{R2}$$

$$NO_2 + OH^* \rightarrow HNO_3 \tag{R3}$$

and nighttime reactions:

$$NO_2 + O_3 \rightarrow NO_3 + O_2 \tag{R4}$$

$$NO_3 + NO_2 \rightarrow N_2O_5 \tag{R5}$$

$$N_2O_5 + H_2O \rightarrow 2HNO_3 \tag{R6}$$

A highly soluble strong acid, HNO_3 readily dissociates to NO_3^- where it can be deposited as wet deposition. In addition, nitrogenous species can be deposited as dry gases (HNO_3 vapor, NH_3, NO_2, HONO, NO, peroxyacetyl nitrate

(PAN)), dry aerosols (particulate NO_3^-, particulate NH_4^+), or in fogwater. Each of these forms can contribute significant loads of atmospherically derived N to ecosystems, and the relative importance of wet/dry deposition is spatially variable.

Atmospheric processes causing variations in $\delta^{15}N$

The question of whether the seasonal and spatial variability in the $\delta^{15}N$ of atmospheric NO_3^- is controlled mainly by mixing of sources with different $\delta^{15}N$ values or by variability in atmospheric processes – or some combination of the two – is a topic of active debate. As discussed previously, there is ample evidence that different NO_x sources have different isotopic signatures; however, it is not yet clear whether these signatures can be substantially overprinted by the effects of chemical reactions during long-range transport and deposition as nitrate. Therefore, it is important to consider several mechanisms that have been proposed in the literature as significant controls on the $\delta^{15}N$ of wet and dry precipitation. Freyer (1991) provides a thorough review of potential fractionating factors. Briefly, these potential factors include the following:

1 Isotope shifts of several permil can occur between and within storms because of selective washout of N-bearing materials (Heaton 1986). Long-term transport and progressive rainout of ^{15}N has been suggested as a potential factor in low $\delta^{15}N$ values (<−20‰) observed in polar regions (Wada et al., 1981; Heaton et al. 2004). However, Michalski et al. (2005) point out that because of advances in our understanding of HNO_3 production pathways, that NO_2 equilibrium with water droplets is not a viable explanation for low $\delta^{15}N$ values observed in polar regions. Further, Michalski et al. (2005) suggest that if kinetic or equilibrium fractionations associated with transport are responsible for low $\delta^{15}N$ values, that these fractionations are mass dependent and would therefore be accompanied by corresponding low $\delta^{18}O$ values. Due to the fact that very low $\delta^{18}O$ values have not been observed in high latitudes, Michalski et al. (2005) suggest that stratospheric NO_3^- characterized by low $\delta^{15}N$ values is the source of low $\delta^{15}N$ values observed in polar regions. Hastings et al. (2004) also suggest that interactions between NO_x and PAN (peroxyacetyl nitrate) over long distances may alter the isotopic composition of NO_x.

2 Freyer et al. (1993) reported that at an urban, polluted site in Germany, equilibrium reactions between NO and NO_2 can result in higher $\delta^{15}N$ values in NO_2 (and resulting HNO_3). High NO_2 concentrations at this site result in the incomplete oxidation of NO, particularly during the winter when O_3 concentrations are lowest. This equilibrium fractionation during the winter is suggested to be the cause of the seasonal patterns observed in $\delta^{15}N$ (higher in winter, lower in summer). However, subsequent studies have been able to rule out this potential effect as a factor in $\delta^{15}N$ seasonality by considering mixing ratios of NO_2, NO, and O_3. In particular, Hastings et al. (2003, 2004) determined that given abundant concentrations of O_3 relative to NO_2, equilibrium reactions between oxidized N species cannot fully account for variability observed in $\delta^{15}N_{NO_3}$ in Bermuda and Greenland (Hastings et al. 2003, 2004).

3 The preferential evaporation of ^{14}N from the dissociation of NH_4NO_3 has been suggested as a mechanism for causing higher $\delta^{15}N$ in residual nitrate (Freyer 1991).

4 Heaton et al. (1997) attributed seasonal variations in $\delta^{15}N$ to humidity, which can affect the $\delta^{15}N$ of NO_3^- and NH_4^+ by equilibrium exchange of N with gaseous HNO_3 and NH_3, respectively, and produce higher $\delta^{15}N$ values for NO_3^- and NH_4^+ when humidities are low.

Atmospheric processes causing variations in $\delta^{18}O$

A survey of the nitrate isotope literature in the late 1990s observed that there appeared to be a bimodal distribution of $\delta^{18}O_{NO_3}$ values in precipitation in North America, with mode values of +25‰ and +60‰ (Kendall 1998). Furthermore, precipitation in Europe showed a somewhat similar pattern, with $\delta^{18}O_{NO_3}$ >50‰ in Bavaria (Germany), which has high concentrations of NO_3^- in precipitation, many acid-rain damaged forests, and is downwind of the highly industrialized parts of central Europe (Voerkelius & Schmidt 1990; Durka et al. 1994). However, $\delta^{18}O_{NO_3}$ values were lower in Muensterland, which is farther from the pollution sources in central Europe (Mayer et al. 2001). Because of the seasonality in $\delta^{18}O$, the bimodal $\delta^{18}O_{NO_3}$ distribution in North America, and the trends in Europe (Kendall 1998) speculated that these patterns might reflect the presence of at least two sources and/or processes affecting $\delta^{18}O_{NO_3}$; specifically, that the higher $\delta^{18}O_{NO_3}$ values (i.e., >50‰) may be associated with anthropogenic NO_x pollution.

However, most new investigations using the microbial denitrifier method to analyze nitrate samples have concluded that the temporal variations in $\delta^{18}O_{NO_3}$ of precipitation are likely due to seasonal changes in atmospheric oxidation chemistry rather than source contributions. In particular, $\delta^{18}O_{NO_3}$ values are thought to vary according to the relative contributions of O_3 to HNO_3 molecule formation caused by seasonal shifts in HNO_3 production pathways. In a study of nitrate aerosols collected in southern California (USA), Michalski et al. (2004) report that seasonal variations in $\delta^{18}O$ and mass-independent $\Delta^{17}O$ result from seasonal shifts in temperature, hours of sunlight, and oxidant concentrations. Using $\Delta^{17}O$ coupled with a isotopic-photochemical box model, Michalski et al. estimate that during the spring, HNO_3 production is dominated (ca. 50%) by the homogeneous reactions R3 and R4; however, during the winter months, HNO_3 production (>90%) is driven by heterogeneous reactions R5 and R6.

Several studies in Bermuda (Hastings et al. 2003) and Greenland (Hastings et al. 2004, 2005) report that $\delta^{18}O$ seasonality depends on the reaction pathway for HNO_3 formation. In particular, during the daytime and summer months, reactions R1–R3 result in HNO_3 molecules where 2/3 of oxygen molecules are derived from O_3. In comparison, during the nighttime and winter months, reactions R4–R6 produce HNO_3 with up to 5/6 of the oxygen molecules from O_3 (Hastings et al. 2004). Because the $\delta^{18}O$ of O_3 in the troposphere generally is very high (>90‰; Johnston & Thiemens 1997) relative to OH and H_2O vapor (generally <0‰; Dubey et al. 1997), the resulting wintertime HNO_3 has higher $\delta^{18}O$ values.

A study of pre-industrial nitrate isotopes in ice cores from Summit, Greenland shows that $\delta^{18}O$ values ranged from +52‰ to +87‰ throughout the Holocene interglacial and preceding glacial period of the past 25,000 years (Hastings et al. 2005). This range in values is similar to what is reported in contemporary studies (e.g., Hastings et al. 2003, 2004; Elliott et al. 2006; in preparation) and further suggests that $\delta^{18}O$ values are reflective of HNO_3 production pathways.

Atmospheric processes causing variations in $\Delta^{17}O$

Seasonal variations in atmospheric $\Delta^{17}O$ (from +20 to +30‰) observed in Southern California (USA) were explained by a shift from nitric acid production by the $OH^* + NO_2$ reaction, which is predominant in the spring and summer, to N_2O_5 hydrolysis reactions that dominate in the winter (Michalski et al. 2004). Figure 12.5 illustrates seasonal patterns in $\Delta^{17}O$ of bimonthly volume-weighted precipitation samples across New England (USA) and the strong correlation of $\Delta^{17}O$ with $\delta^{18}O$ (Wankel 2006), presumably due to a seasonality in the relative proportions of NO_x oxidation by OH^* or O_3 (Michalski et al. 2003; Hastings et al. 2003). Ice-core samples from a glacier in Greenland show higher $\Delta^{17}O$ values during 1880s due to the effects of large biomass burning events in North America on NO_x (Alexander et al. 2004).

Separating mixing of sources from the effects of cycling

Under ideal circumstances, nitrate isotopes offer a direct means of source identification because the two major sources of NO_3^- in many agricultural areas, fertilizer and manure, generally have isotopically distinct $\delta^{15}N_{NO_3}$ values. In contrast, the two major sources of NO_3^- to more pristine watersheds, atmospheric NO_3^- and microbial NO_3^-, have isotopically distinct $\delta^{18}O_{NO_3}$ values (Figure 12.1). Hence, the relative contributions of these two sources to groundwater or surface water can be estimated by simple mass balance.

Figure 12.1 shows the normal range of $\delta^{18}O$ and $\delta^{15}N$ values for the dominant sources of nitrate. Nitrate derived from ammonium fertilizer, soil organic matter, and animal manure has overlapping $\delta^{18}O$ values; for these sources,

$\delta^{15}N$ is a better discriminator. In contrast, NO_3^- derived from nitrate fertilizer or atmospheric sources is readily separable from microbial NO_3^- using $\delta^{18}O$, even though the $\delta^{15}N$ values are overlapping. While these general ranges of isotopic values are useful starting points for distinguishing among various sources, because nitrogen is a major nutrient and thus undergoes significant amounts of cycling in most ecosystems, the actual isotopic values of the sources can be outside these ranges. Nitrogen cycling, as discussed above, imparts a wide variety of isotopic fractionations which tend to obscure the original source signal, whether there is a single source or a mixture of two or more sources. The following section discusses how isotopes can be used to determine the relative contributions of different sources to a mixed pool, as well as methods for recognizing and accounting for the impact of cycling (i.e., fractionation due to denitrification, assimilation, nitrification, etc.) on isotopic composition and water chemistry. Applications to different environmental settings are briefly discussed in a later section.

Mixing

If nitrate in groundwater or surface water derives from the mixing of two different sources that are known to have distinctive $\delta^{15}N_{NO_3}$ values, in the absence of any subsequent fractionations, the relative contributions of each can readily be calculated. Many studies have illustrated this point using $\delta^{15}N$ versus NO_3^- concentration plots, showing that mixtures must plot on a line between the two "end-member" compositions. However, such mixing lines are straight lines only when the nitrate concentrations of the two end-members are identical; otherwise, mixing lines are hyperbolic on such plots. Hence, a good test of whether $\delta^{15}N$ or $\delta^{18}O$ data can be explained by simple mixing is to plot the δ values vs. $1/NO_3^-$. An example of this is given in Figure 12.6 (modified from Mariotti et al. 1988), where two waters with NO_3^- concentrations of 200 and $1\,\mu M$ mix together. Note that the curvature of the mixing line is very slight for some concentrations (e.g., $50–200\,\mu M$) where NO_3^- concentrations of the end-members are very different.

Unfortunately, real-life studies are rarely this simple. The multiple potential sources of nitrate in various ecosystems rarely have constant isotopic compositions, and the initial compositions may be altered by various fractionating processes before, during, or after mixing. Hence, estimates of relative contributions will often be only qualitative (see Bedard-Haughn et al. 2003). In particular, denitrification can greatly complicate the interpretation of $\delta^{15}N$ values because the exponential increase in $\delta^{15}N$ of residual nitrate with decreasing NO_3^- content caused by denitrification can sometimes be confused with mixing of NO_3^- sources. For example, on Figure 12.6a, all three curves are almost linear for nitrate concentrations $100–200\,\mu M$. Thus, an incautious worker could try to interpret all three as mixing lines. However, as shown on Figure 12.6b, two of these curves are exponential relations resulting from denitrification, not mixing lines. Figure 12.6c illustrates that

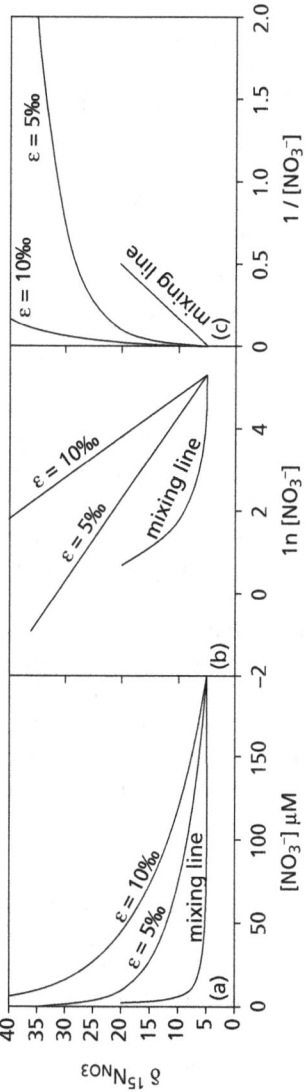

Figure 12.6 The curves on a plot of $\delta^{15}N$ vs. NO_3 (a) resulting from mixing of two sources of nitrate with different concentrations, can be distinguished from the curves resulting from denitrification with two different fractionations by plotting $\delta^{15}N$ vs. NO_3 (b), where different denitrification fractionations yield straight lines whereas mixing yields a curve, and by plotting $\delta^{15}N$ vs. $1/NO_3$ (c), where different denitrification fractionations yield curves whereas mixing yields a straight line.

true mixing will only be straight when $\delta^{15}N$ is plotted against the inverse of concentration.

Mixing of NO_3^- sources can sometimes be resolved by analysis of both $\delta^{15}N$ and $\delta^{18}O$ (or of other semi-conservative isotopic and/or chemical tracers). This dual-isotope approach has three main potential benefits:

1 $\delta^{18}O_{NO_3}$ separation of some sources is greater than for $\delta^{15}N_{NO_3}$, allowing better source resolution by having two tracers;
2 some nitrate sources that are usually indistinguishable with $\delta^{15}N$ alone (e.g., fertilizer vs. soil nitrate, or atmospheric vs. soil nitrate) may be identified only when the $\delta^{18}O$ is analyzed;
3 $\delta^{18}O$ values vary systematically with $\delta^{15}N$ during denitrification (as illustrated in Figure 12.1) and assimilation.

Thus, in systems where the dominant sources of nitrate are isotopically distinctive, source contributions can – in theory – be determined despite significant fractionation.

The greatest problems for using isotopes to determine mixing proportions are (i) that different sources can have partially overlapping isotopic compositions, (ii) sources can have considerable spatial and temporal variation in isotopic composition, and (iii) isotope fractionations can blur initially distinctive isotopic compositions. These problems can often be minimized or eliminated by a multi-isotope, multi-tracer approach – with a lot of hydrologic and chemical data. It is important to keep in mind that the successful solution of the mixing algebra does not ensure that the source determinations are accurate. Improvements in statistical analyses of mixing models offer increasingly sophisticated means to separate sources in situations of variable or partially overlapping isotope ratios (Phillips & Gregg 2001, 2003; Phillips & Koch 2002), and to make better use of multiple tracers. Furthermore, sometimes the isotopic variability may prove to be a useful natural "signal" that may actually enhance the power of isotope methods in these situations.

Fractionation due to denitrification

Denitrification is the process that poses most difficulties for simple applications of nitrate isotopes, because of both the large fractionations and its ubiquity in many landscape types. Hence, for successful applications of nitrate isotopes for tracing sources, it is critical to (i) determine if denitrification has occurred, and, if so (ii) determine the initial isotopic composition of the nitrate (which is a necessary prerequisite for later attempts to define sources).

There are several geochemical approaches for identifying and quantifying denitrification, and distinguishing it from mixing.

1 First, unless there is evidence of reducing conditions (e.g., low dissolved O_2, high H_2S or CH_4, etc) denitrification is unlikely. However, denitrification may also take place in anoxic microsites in an otherwise oxygenated sediment (Brandes & Devol 1997).

2 Geochemical modeling using chemical data and perhaps $\delta^{13}C_{DIC}$, $\delta^{34}S_{SO_4}$, $^{87}Sr/^{86}Sr$, and other isotopes may also be useful (Böhlke & Denver 1995) for evaluating whether mixing or denitrification best explains chemical and isotopic compositions.

3 Analysis of dissolved N_2 produced by denitrification for $\delta^{15}N$ can indicate whether there are systematic increases in $\delta^{15}N_{N_2}$ and $\delta^{15}N_{NO_3}$ with decreases in NO_3^- concentration (e.g., a "Rayleigh equation" relationship) that are consistent with denitrification. This method also requires estimation of recharge temperature (usually accomplished by analysis of dissolved inert gases such as Ar and Ne), verification that the groundwater samples analyzed represent closed N systems, calculation of the amount of excess N_2 produced by denitrification, and correction of $\delta^{15}N_{NO_3}$ for the amount of fractionation produced by denitrification (Vogel et al. 1981; Böhlke & Denver 1995).

4 Analysis of the NO_3^- for $\delta^{18}O$, as well as $\delta^{15}N$, can also be useful since systematic increases in $\delta^{18}O$ due to denitrification (or assimilation) will accompany increases in $\delta^{15}N$; however, $\delta^{18}O_{NO_3}$ usually is not as useful for determining extent of denitrification as $\delta^{15}N_{N_2}$.

5 Plotting $\delta^{18}O_{NO_3}$ vs. $\delta^{18}O_{H_2O}$ can also be useful because nitrification in contact with the ambient water is likely to result in $\delta^{18}O_{NO_3}$ values that show a strong correlation with $\delta^{18}O_{H_2O}$ (Wankel et al. 2006; McMahon & Bohlke 2006). These data can then be compared with the theoretical nitrification line defined in equation 12.1, or the $\delta^{18}O_{NO_3} = \delta^{18}O_{H_2O}$ line observed in recent marine studies (Casciotti et al. 2002). However, if denitrification is the main process affecting the $\delta^{18}O_{NO_3}$, there will be no correlation with $\delta^{18}O_{H_2O}$.

6 In special cases with significant atmospheric nitrate contributions, $\Delta^{17}O > 0$ and processes leading to increased $\delta^{18}O_{NO_3}$ (and $\delta^{15}N_{NO_3}$) will have no effect on the $\Delta^{17}O$ value (Figure 12.4), allowing separation of mixing of sources from fractionation.

7 Perhaps most importantly, mixing of sources will follow a hyperbolic relationship while fractionation is an exponential process. Hence, if mixing of two sources is responsible for the curvilinear relationship on a plot of $\delta^{15}N_{NO_3}$ (or $\delta^{18}O_{NO_3}$) and NO_3^-, plotting $\delta^{15}N_{NO_3}$ (or $\delta^{18}O_{NO_3}$) vs. $1/NO_3^-$ will result in a straight line (Figure 12.6c). In contrast, if denitrification (or assimilation) is responsible for the relationship, plotting $\delta^{15}N_{NO_3}$ (or $\delta^{18}O_{NO_3}$) vs. ln NO_3^- will produce a straight line.

Seasonal or storm-related cycles of denitrification and nitrification pose a considerable challenge to the use of isotope techniques for identifying nitrate sources and mixing proportions (Koba et al. 1997). A multi-isotope approach using triple isotopes of nitrate, combined with analysis of the concentrations and isotopic compositions of gases produced and consumed during denitrification and nitrification (e.g., $\delta^{15}N$ of N_2, $\delta^{18}O$ of O_2, $\delta^{13}C$ of CO_2), may allow determination of temporal changes in these processes if waters and soil gases can be sampled with sufficient temporal resolution.

Cycling of DIN through organic matter

The other main process that complicates NO_3^- source identification and apportionment is partial recycling of the NO_3^- through an organic matter pool. There are several geochemical approaches for identifying and quantifying assimilation in soil water or streams, and distinguishing it from mixing. For example, in aquatic systems chlorophyll is a sensitive indicator of productivity and can be used to approximate algal growth and N assimilation rates. Measurement of the concentration and $\delta^{13}C$ of DIC and/or concentration and $\delta^{18}O$ of O_2 can also be used to constrain photosynthesis (Parker et al. 2005). As with denitrification, analysis of the NO_3^- for $\delta^{18}O$ as well as $\delta^{15}N$, analysis of $\Delta^{17}O$, comparisons with $\delta^{18}O_{H_2O}$, and/or plotting $\delta^{15}N$ vs. ln NO_3^- can sometimes help distinguish fractionation vs. mixing.

Source identification and quantification becomes more complicated when NO_3^- assimilation co-occurs with nitrification. In this case, not only are the $\delta^{15}N$ and $\delta^{18}O$ values of the residual NO_3^- pool fractionated because of assimilation, but an additional source of NO_3^- has now been added (see the section on "Nitrification"). This kind of complicated mixture of processes is to be expected in biogeochemically and/or hydrologically active "hotspots" (McClain et al. 2003), and is probably quite common. Two good examples of this kind of complicated environmental setting are agricultural fields after fertilizer application (e.g., Feigin et al. 1974) and small forested catchments where atmospheric NO_3^- is a major source of N to the ecosystem (e.g., Burns & Kendall 2002). In both cases, the microbial cycling of the newly applied DIN is complicated by flushing of the soil by rain events and/or snowmelt. Ultimately, the isotopic composition of the fertilizer or atmospheric DIN is partially or totally overprinted in the soil zone. The $\delta^{15}N$ and $\delta^{18}O$ of the new NO_3^- are probably best assessed by analyzing waters that leach past the soil zone, or by leaching soil samples. Laboratory incubations may not be representative because disturbing the soil might cause changes in soil respiration and the resulting $\delta^{18}O$ of the ambient O_2. Hence, the NO_3^- produced during incubations might have a different $\delta^{18}O$ than the natural microbial NO_3^-.

Other good recent examples of using dual isotope approaches to "deconvolute" coupled biogeochemical processes include coupled denitrification–nitrification studies in lacustrine (Lehmann et al. 2003) and marine (Lehmann et al. 2004) environments, and coupled nitrification–assimilation studies in estuarine and marine environments (Wankel et al. 2007). This latter study investigated N cycling in surface waters of Monterey Bay (California, USA) to constrain nitrification occurring in the euphotic zone. While the NO_3^- in surface waters showed the expected pattern of increasing $\delta^{15}N$ and $\delta^{18}O$ due to phytoplankton assimilation, deviations from the $1:1$ pattern were used to estimate the degree of rapid organic matter remineralization and nitrification occurring in the euphotic zone. The authors suggested that, on average, ca. 30% of the NO_3^- in surface waters had been cycled through organic matter and regenerated via nitrification.

Applications to different environmental settings

Small forested catchment studies

One of the main applications of $\delta^{18}O_{NO_3}$ has been for determination of the relative contributions of atmospheric and soil-derived sources of NO_3^- to shallow groundwater and small streams. This problem is intractable using just $\delta^{15}N_{NO_3}$ because of overlapping compositions of soil and atmospherically derived NO_3^-, whereas these sources have very distinctive $\delta^{18}O_{NO_3}$ values (Figure 12.1). The $\delta^{18}O_{NO_3}$ values are such a sensitive indicator of NO_3^- sources that even diel changes in snowmelt, and consequently contributions of snowmelt-derived nitrate to streams, can be detected as diel oscillations in $\delta^{18}O_{NO_3}$ in streamwater (Ohte et al. 2004).

A number of studies have found that much of the NO_3^- in runoff from small catchments is microbial (i.e. from nitrification) instead of atmospheric (e.g., Burns & Kendall 2002; Campbell et al. 2002; Sickman et al. 2003; Pardo et al. 2004; Ohte et al. 2004; Piatek et al. 2005). For example, a multi-year investigation at the Loch Vale watershed in Colorado (USA), showed that half or more of the NO_3^- in the stream during the snowmelt period was microbial in origin (Kendall et al. 1995a,b), and probably originated from shallow groundwater in talus deposits (Campbell et al. 2002). Therefore, the NO_3^- eluted from the snowpack appears to go into storage, and most of the NO_3^- in streamflow during the period of potential acidification is apparently derived from pre-melt sources. Much of this NO_3^- was probably originally of atmospheric origin but had lost its atmospheric signature during microbial recycling in the talus (Campbell et al. 2002).

We expect that analysis of $\Delta^{17}O_{NO_3}$ will also be valuable in such studies because it is an even less ambiguous tracer of atmospheric NO_3^- than $\delta^{18}O_{NO_3}$ (e.g., Michalski et al. 2004). In theory, because all non-atmospheric sources have $\Delta^{17}O = 0$ and biogeochemical processes do not affect $\Delta^{17}O$ values, in the absence of any recycling of atmospheric nitrate in the watershed (admittedly a large caveat), the 0.1‰ analytical resolution of $\Delta^{17}O$ leads to a detection limit for atmospheric NO_3^- of 0.5% of total nitrate (Michalski et al. 2004). The few studies that have compared $\delta^{18}O$ and $\Delta^{17}O$ of NO_3^- in streamwater during storm events indicate that $\Delta^{17}O$ and $\delta^{18}O$ have different responses to discharge changes (Michalski et al. 2004; Showers & DeMasters 2005). Furthermore, both studies found that $\delta^{18}O$ significantly underestimated the contributions of atmospheric NO_3^- to runoff (Figure 12.7). Michalski

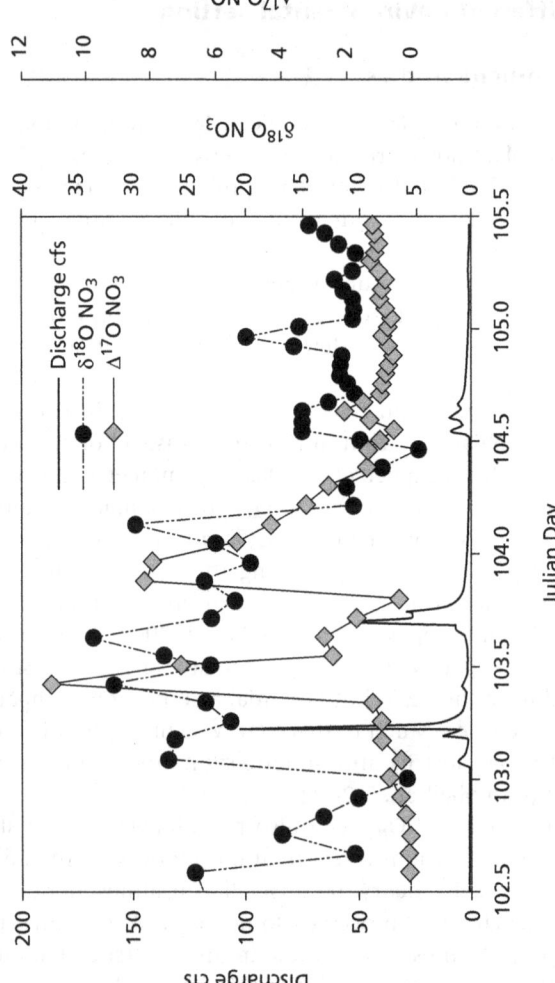

Figure 12.7 Temporal changes in $\Delta^{17}O$ and $\delta^{18}O$ of nitrate and discharge in an urban watershed in the Neuse River Basin, NC, USA. Note that the $\delta^{18}O$ shows a broad peak over the discharge event while the $\Delta^{17}O$ nitrate has distinct sharp peaks on the falling hydrograph portion of the storm event. (Plot courtesy of W.J. Showers, modified from Showers & DeMasters 2005.)

et al. (2004) noted that the larger amounts of atmospheric NO_3^- in runoff determined using $\Delta^{17}O$ implies that previous estimates for streams made using $\delta^{18}O$ may be too conservative. The temporal variability in stream $\Delta^{17}O$ at a small urban watershed in the Neuse River Basin (USA; Figure 12.7) suggests that calculations of atmospheric N flux in streams from $\Delta^{17}O$ measurements must be integrated over an entire event and not from discrete measurements (Showers & DeMasters 2005).

Urban stream studies

Several studies suggest that atmospheric NO_3^- may be a major contributor to streamflow in urban catchments. For example, a pilot study of NO_3^- sources in storm runoff in suburban watersheds in Austin, Texas (USA) found high $\delta^{15}N$ and low $\delta^{18}O$ values during baseflow (when Cl^- was high), and low $\delta^{15}N$ and high $\delta^{18}O$ values during storms (when Cl^- was low; Ging et al. 1996; Silva et al. 2002). The strong correspondence of $\delta^{15}N$ and $\delta^{18}O$ values during changing flow conditions, and the positive correlation of the percentage of impervious land-cover and the $\delta^{18}O_{NO_3}$, suggests that the stream composition can be explained by varying proportions of two end-member compositions (Figure 12.8), one dominated by atmospheric NO_3^- or nitrate fertilizer, that is the major source of water during storms, and the other a well-mixed combination of sewage and other NO_3^- sources, that contributes to baseflow (Silva et al. 2002). Analysis of stream NO_3^- samples for $\Delta^{17}O$ would provide more

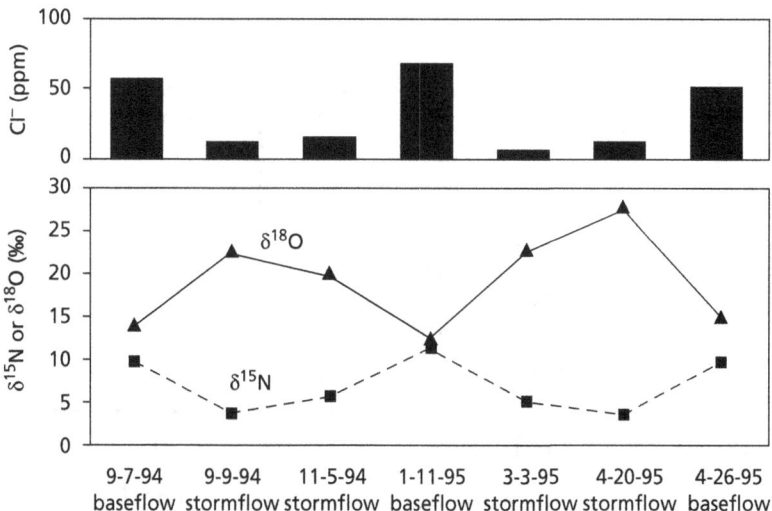

Figure 12.8 Average $\delta^{18}O$ and $\delta^{15}N$ values of NO_3^- during stormflow (low Cl^-) and baseflow (high Cl^-) conditions in urban streams in Austin, Texas (USA). (Modified from Silva et al. 2002.)

definitive evidence of an atmospheric source, as is shown in another urban watershed (Figure 12.7).

Several studies have shown that the $\delta^{15}N$ of nitrate, or "proxies" for nitrate (e.g., particulate organic matter (POM), plants, algae), are correlated with the percentage of wastewater inputs from urban areas. In a large-scale study of 16 large watersheds in the northeastern and mid-Atlantic, USA, Mayer et al. (2002) demonstrated that riverine $\delta^{15}N_{NO_3}$ values were positively correlated with wastewater inputs. Elliott & Brush (2006) report similar correlations at lower population densities by comparing historical reconstructions of watershed wastewater N loads and stratigraphic organic nitrogen in wetland sediments (Figure 12.9). Similarly, in a study of groundwater nitrate isotopes on Cape Cod, Massachusetts (USA), Cole et al. (2006) observed positive correlations between wastewater N loads and $\delta^{15}N$ in groundwater. The positive correlations between $\delta^{15}N$ and wastewater inputs in these and other studies suggest that the subsurface delivery of wastewater inputs make $\delta^{15}N$ a particularly effective indicator of wastewater contamination source; boron isotopes are also a useful tracer of wastewater (Widory et al. 2004), as will be described in a later section.

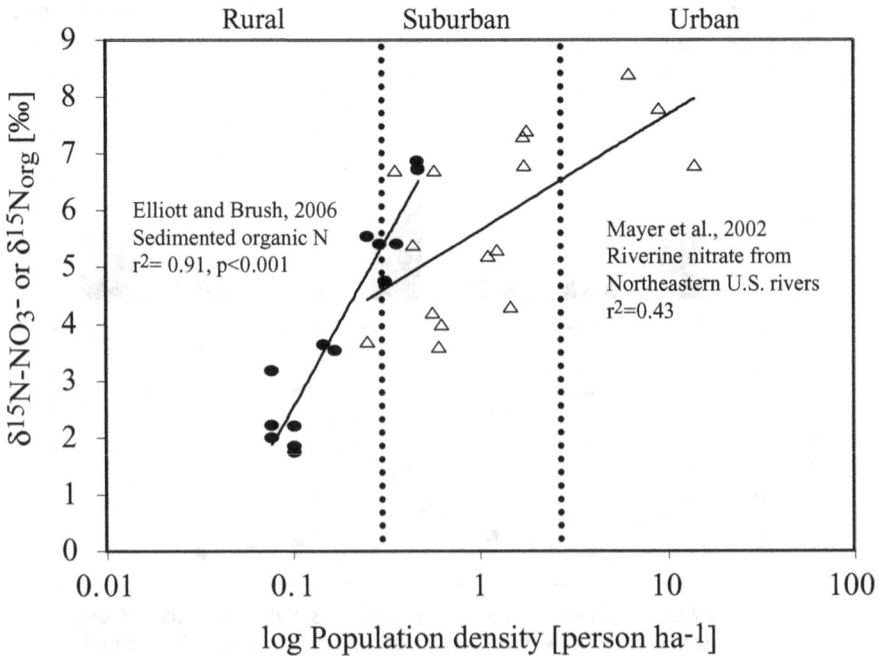

Figure 12.9 $\delta^{15}N$ vs. population density for nitrate (modified from Mayer et al. 2002) and organic matter (modified from Elliott & Brush 2006). Land use classifications from Theobald (2004).

Small agricultural rivers

There have been fewer nitrate isotope studies in small agricultural rivers than in small forested watersheds, perhaps because of the critical response elicited by the first such study by Kohl et al. (1971). However, several recent studies using combined tracer techniques have provided much insight on the dynamics of NO_3^- in agricultural systems. Surface-water NO_3^- studies by Böhlke & Denver (1995) in an agricultural watershed in Maryland (USA) and McMahon & Böhlke (1996) in the Platte River in Colorado (USA) used $\delta^{15}N$ data mainly to quantify the effects of denitrification and mixing between the river and aquifer, not to assess contributions from specific NO_3^- sources (although the $\delta^{15}N$ values did suggest the NO_3^- derived from animal waste and fertilizer). Denitrification was quantified by measurement of the excess N_2 and the $\delta^{15}N$ of dissolved N_2, and mixing relations and flowpaths were established using chlorofluorocarbon (CFC) and hydraulic data. Substantial denitrification was found in reducing zones within the aquifer in Maryland. However, there was probably limited denitrification in the wetlands and shallow organic soils adjacent to the streams because the deeper groundwater flowpaths avoided these buffer strips and converged directly beneath the streambeds and rapidly discharged upwards (Böhlke & Denver 1995). The hyporheic zone was found to be a major site for denitrification for the Platte River (McMahon & Böhlke 1996).

Böhlke et al. (2004) conducted an in-stream tracer experiment using Br^- and ^{15}N-enriched NO_3^- to determine the rates of denitrification and other processes in a high-nitrate gaining stream (Sugar Creek, Indiana) in the upper Mississippi Basin. The systematic downstream increase in $\delta^{15}N_{N_2}$ indicated high rates of in-stream denitrification. However, while N losses by processes other than denitrification were probably less than the denitrification rate, the overall mass fluxes of N_2 were dominated by discharge of denitrified groundwater and air–water gas exchange in response to changing temperature (Böhlke et al. 2004). The study concluded that the in-stream isotope tracer experiment provided a sensitive measurement of denitrification and related processes where other mass-balance methods were not suitable.

Large river basin studies

Several recent studies in North America evaluated whether the combination of nitrate $\delta^{18}O$ and $\delta^{15}N$ would allow discrimination of watershed sources of N and provide evidence for denitrification. A pilot study in the Mississippi Basin (Battaglin et al. 2001a,b; Chang et al. 2002) showed that large watersheds with different land uses (crops, animals, urban, and undeveloped) had overlapping but moderately distinguishable differences in nutrient isotopic compositions (e.g., $\delta^{18}O$ and $\delta^{15}N$ of NO_3^-, and $\delta^{15}N$ and $\delta^{13}C$ of POM). Atmospheric

NO_3^- was found to be a significant source of NO_3^- to large undeveloped and urban watersheds (Battaglin et al. 2001a, 2001b; Chang et al. 2002). A study in 16 large rivers in the northeastern USA found strong positive correlations between $\delta^{15}N$ and the calculated wastewater N contributions (Mayer et al. 2002); their $\delta^{15}N$ data are plotted relative to population density, a surrogate for wastewater, in Figure 12.9. Groundwater NO_3^-, probably derived from near-stream dairies, was found to be a significant source of NO_3^- to the San Joaquin River, the major agricultural basin in the Central Valley of California (USA; Kratzer et al. 2004). A study in the Oldman River Basin (Alberta, Canada) showed that the main source of nitrate in western tributaries draining a relatively pristine forested part of the basin was soil nitrate, whereas the main source in eastern tributaries draining agricultural and urban land uses was manure and/or sewage (Rock & Mayer 2004).

Prior N mass balance studies in the Mississippi Basin and in the northeastern USA rivers suggested appreciable losses of N via denitrification, especially in the headwaters. However, these studies did not find isotopic evidence for denitrification, perhaps because of continuous mixing with new nitrate, the small extent of denitrification, or the low fractionations resulting from diffusion-controlled (i.e., benthic) denitrification (Brandes & Devol 1997; Sebilo et al. 2003). A recent study in the Mississippi River in Illinois (Panno et al. 2006) concluded that the 1:2 relationship between $\delta^{18}O_{NO_3}$ and $\delta^{15}N_{NO_3}$ observed in both river and tile drain samples suggests that most of the denitrification probably occurred before discharge into the Mississippi River. They found most of the NO_3^- in the river is primarily derived from synthetic fertilizers and soil organic N, consistent with published estimates of N inputs to the Mississippi River. Depending on sample location and season, NO_3^- in the river and tile drains has undergone significant denitrification, ranging from about 0 to 55% (Panno et al. 2006).

Investigations of N sources and sinks in the Seine River Basin (France) have also shown denitrification to be a major sink for NO_3^-, especially in the summer (Sebilo et al. 2003, 2006). An investigation of riparian denitrification in various stream orders in the Seine River system during summer low-flow conditions concluded that riparian denitrification removed up to 50% of the N exported from agricultural soils; however, the extent of denitrification determined by shifts in the $\delta^{15}N$ of residual nitrate provided only a minimum estimate of denitrification (Sebilo et al. 2003). A subsequent study used $\delta^{15}N_{NH_4}$, $\delta^{15}N_{NO_3}$, and $\delta^{18}O_{NO_3}$ to assess the extent of nitrification and denitrification in the water column resulting from export of waste water treatment plant (WWTP) effluent from Paris into the Seine River. During summer low-flow conditions it was found that most of the NH_4^+ released from the WWTP was nitrified in the lower Seine River and its upper estuary, but there was no evidence for water-column denitrification (Sebilo et al. 2006).

A useful adjunct to tracing N sources and sinks in aquatic systems with nitrate isotopes is the analysis of POM for $\delta^{15}N$, $\delta^{13}C$, and $\delta^{34}S$. In many river systems, much of the POM is derived from *in situ* production of algae. Even if an appreciable percent of the POM is terrestrial detritus, the C:N value of the POM and the $\delta^{15}N$ and $\delta^{13}C$ can, under favorable conditions, be used to estimate the percent of POM that is algae, and its isotopic composition (Kendall et al. 2001). The $\delta^{15}N$ and $\delta^{13}C$ (and $\delta^{34}S$) of the POM reflect the isotopic compositions of dissolved inorganic N, C, and S in the water column. In turn, these compositions reflect the sources of N, C, and S to the system, and the biogeochemical processes (e.g., photosynthesis, respiration, denitri-fication, sulfate reduction) that alter the isotopic compositions of the dis-solved species; see Finlay & Kendall (this volume, pp. 283–333) for more details. Hence, the changes in the isotopic composition can be used to evalu-ate a variety of in-stream processes that might affect the interpretation of nitrate $\delta^{15}N$ and $\delta^{18}O$ (Kendall et al. 2001).

The $\delta^{15}N$ of POM may even serve as an integrator for $\delta^{15}N_{NO_3}$ when the POM is dominated by *in situ* production of algae; the $\delta^{15}N$ of POM appears to be ca. 4‰ lower than the associated nitrate in both the Mississippi Basin (Battaglin et al. 2001) and in the San Joaquin River (Kratzer et al. 2004). In the latter river, downstream changes in the $\delta^{15}N_{NO_3}$ in response to changes in NO_3^- sources are reflected, albeit shifted about 4‰ because of assimilatory fractionation, in the downstream changes in $\delta^{15}N_{POM}$ (see Figure 10.8).

Wetlands studies

The accumulation of organic matter in wetlands affected by human activities provides a historical archive of information about temporal changes in land uses and biogeochemical processes. In addition, due to the large fraction-ations associated with denitrification and sulfate reduction, the coupled use of N and S isotopes can be useful for examining redox reactions in wetland environments. For example, a study in Florida (USA) showed that export of high-sulfate waters from the agricultural areas around Lake Okechobee into canals and marshes in the Everglades caused increases in the $\delta^{34}S_{SO_4}$ as the sulfate (originally about +16‰) is progressively reduced to H_2S (Bates et al. 2002). These high concentrations of sulfate have apparently stimulated the proliferation of sulfate reducers in the marshes, causing the algae and fish in hypoxic marshes to be "labeled" with $\delta^{34}S$ values as high as +33‰; the biota in these "hot spots" of anoxic conditions are also labeled with low $\delta^{13}C$ and high $\delta^{15}N$ values (Kendall et al. 2000). The high $\delta^{34}S$ values of organic matter in hypoxic zones in the Everglades are unusual. Sulfate reduction in anoxic marshes and sediments usually results in low $\delta^{34}S$ of biota because of assimi-lation of a portion of ^{34}S-depleted H_2S (Trust & Fry 1992).

Wetland environments can also be used to extend the temporal span of our environmental observations. Elliott & Brush (2006) demonstrated that sediment accumulation in wetland environments provides a rich archive of historical information about redox conditions (inferred from coupled $\delta^{34}S$ and

$\delta^{15}N$ of organic matter), inorganic N sources to wetland plants ($\delta^{15}N$ of organic N), plant distributions (inferred from palynological records), and land-use changes (inferred from changes in sedimentation rates and palynology). Using this approach, they concluded that stratigraphic changes in $\delta^{15}N$ were most likely due to changes in N sources to wetland plants, rather than changes in redox status or diagenesis. In particular, as population density increased in the watershed over 300 years, increasing wastewater N contributions resulted in higher $\delta^{15}N$ values in sedimented organic N (Figure 12.9).

Groundwater studies

Most natural abundance nitrate $\delta^{15}N$ studies focus on attempts to trace relative contributions of fertilizer and animal waste to groundwater. This topic was discussed in detail in Kendall & Aravena (2000). Applications of $\delta^{15}N$ to trace relative contributions of fertilizer and animal waste to groundwater are complicated by a number of biogeochemical reactions, especially ammonia volatilization, nitrification, and denitrification. These processes can modify the $\delta^{15}N$ values of sources before and/or after mixing, causing estimations of the relative contributions of the sources of nitrate to be inaccurate. The combined use of $\delta^{18}O$ and $\delta^{15}N$ allows better resolution of these issues (e.g. Böttcher et al. 1990; Aravena & Robertson 1998; McMahon & Böhlke 2006). Analysis of the $\delta^{15}N$ of both nitrate and N_2 (e.g., Böhlke & Denver 1995) provides an effective means for investigating denitrification. Since it is likely that more studies in the future will utilize a multi-isotope or multi-tracer approach, the discussion below will concentrate on multi-isotope studies.

The first dual isotope investigation of groundwater investigated N cycling in municipal wells downgradient from heavily fertilized agricultural areas near Hannover (Germany). This study found that low concentrations of nitrate in groundwater were associated with high $\delta^{18}O$ and $\delta^{15}N$ values, and concluded that the decreases in nitrate away from the fields was caused by microbial denitrification, not mixing with more dilute waters from nearby forests (Böttcher et al. 1990). Furthermore, changes in $\delta^{18}O$ and $\delta^{15}N$ values along the flowpath were linearly related, with a slope of ca. 0.5. The linear relation between the isotope values and the logarithm of the fraction of residual nitrate (Figure 12.6) indicated that denitrification with constant enrichment factors was responsible for the increases in $\delta^{18}O$ and $\delta^{15}N$. Many subsequent studies of denitrification in freshwater systems showed slopes ranging from 0.5 to ca. 0.7 (e.g., Aravena & Robertson 1998; Mengis et al. 1999; Cey et al. 1999; Lehmann et al. 2004; Panno et al. 2006; Wassenaar et al. 2006).

One important use of nitrate isotopes is to assess the impact of temporal changes in agricultural practices on groundwater NO_3^- concentrations. For example, an early dual isotope study in British Columbia (Canada) attributed the high NO_3^- in groundwater to nitrification of poultry manure, with lesser amounts of ammonium fertilizers (Wassenaar 1995). A recent reappraisal of decadal trends in NO_3^- sources at this site, after implementation of best management practices (BMPs) aimed at reducing aquifer-scale NO_3^- contamination, showed increasing nitrate concentrations in young groundwater (<ca. 5 years), suggesting that voluntary BMPs were not having a positive impact in achieving groundwater nutrient reduction goals (Wassenaar et al. 2006). While the stable isotope data showed that animal manure is still the main source of nitrate in the aquifer, a recent decrease in $\delta^{15}N_{NO_3}$ suggests a BMP-driven shift away from animal wastes towards inorganic fertilizers. This study and others showed that when the extent of denitrification can be quantified, analyses of dated, denitrification-corrected groundwaters can provide a valuable record of past fertilizer loads in the recharge zone (e.g., Böhlke & Denver 1995). This type of record can be valuable in localities where data on long-term fertilizer-use are missing or unreliable.

The applicability of $\delta^{18}O$ and $\delta^{15}N$ of nitrate and other tracers to delineate contaminant plumes derived from domestic septic systems was evaluated by Aravena et al. (1993), in a study within an unconfined aquifer beneath an agricultural area in Ontario (Canada). They found that $\delta^{15}N_{NO_3}$, $\delta^{18}O_{H_2O}$, and water chemistry (especially Na^+) were effective for differentiating between the plume and native groundwater. The lack of a significant difference between the $\delta^{18}O_{NO_3}$ in the plume and in local groundwater suggests that nitrification of ammonium, from either human waste or agricultural sources, is the source of the NO_3^-. Another study of a septic plume determined that use of a multi-isotope approach (using $\delta^{13}C_{DIC}$, and $\delta^{18}O$ and $\delta^{34}S$ of sulfate in addition to nitrate $\delta^{18}O$ and $\delta^{15}N$) provided valuable insight into the details of the processes affecting nitrate attenuation in groundwater (Aravena & Robertson 1998).

Coastal and estuarine studies

Coastal and estuarine systems are extremely dynamic, highly productive, and abound with biogeochemical gradients (salinity, temperature, primary productivity, etc.). Over the past few decades, many studies have used the $\delta^{15}N$ of NO_3^- and NH_4^+ in estuarine systems to investigate sources of N pollution (McClelland & Valiela 1998; Kroeger et al. 2006).

With the advent of the denitrifier method, $\delta^{18}O_{NO_3}$ may now be used in estuarine systems to investigate the influence of both cycling and mixing of multiple sources. In their study in the San Francisco Bay estuary, a branched two-arm estuary, Wankel et al. (2006) showed that mixing of sources could explain most of the isotopic variability in the estuary. In particular, $\delta^{15}N_{NO_3}$ served as a useful tracer of sewage effluent from the southern arm, while $\delta^{18}O_{NO_3}$ was used to separate marine vs. riverine sources along the northern arm. Furthermore, the range of $\delta^{18}O_{NO_3}$ was more than twice that of $\delta^{15}N_{NO_3}$ and correlated strongly with the gradient in $\delta^{18}O_{H_2O}$. It was hypothesized that where concentrations of NO_3^- were relatively low, and thus turnover of the pool was relatively rapid, cycling of NO_3^- resulted in the incorporation of water O into the $\delta^{18}O_{NO_3}$ along this gradient of $\delta^{18}O_{H_2O}$. While the effects of cycling could not be independently verified, this study nonetheless demonstrated that $\delta^{18}O_{NO_3}$ may serve as a more useful tracer of cycling than $\delta^{15}N_{NO_3}$, especially in estuaries where large gradients in $\delta^{18}O_{H_2O}$ occur (Wankel et al. 2006). A recent estuarine study in the Seine River estuary (France) found that, despite the apparent conservative behavior of nitrate concentrations, coupled nitrification and denitrification in the water column was probably responsible for the significant shift of $\delta^{15}N_{NO_3}$ values from the trend expected for simple mixing of marine and riverine sources (Sebilo et al. 2006).

What sources of agricultural and urban sources of nitrate can be distinguished using isotopes?

Most of the interest in uses of nitrate isotopes centers on how to differentiate:

1 fertilizer versus animal waste contributions to groundwater or surface water;
2 septic tank leakages (or WWTPs) versus animal waste;
3 natural soil N from fertilizer and/or wastes.

Below is a brief discussion of what isotope tools seem to "work" or "not work", written in a format that we hope will be useful for readers skimming through the chapter and looking for specific answers.

For the purposes of this chapter, we have defined what "works" as follows. Given the usual case where there is a question of which of two main sources of NO_3^- is the dominant source of NO_3^- to groundwater or a stream, can the measurement of $\delta^{15}N_{NO_3}$ values (or $\delta^{15}N$ combined with other isotope and chemical tracers) along with the NO_3^- concentrations of the mixed source allow confident determination of the dominant NO_3^- source? Can the relative contributions be estimated within approximately ±20%? If so, we interpret this as a successful "quantitative" tool. A first step in such studies should be analysis of $\delta^{15}N_{NO_3}$ (and other isotope and chemical tracers) from near the two potential sources, to insure that the sources have distinctive compositions. Attention should be given to possible temporal and spatial variability in the end-member compositions. Despite the emphasis on the usefulness of isotopes, it is expected that the isotope data are accompanied by appropriate information on hydrology and water chemistry.

Fertilizer vs. animal waste source of nitrate

These sources usually can be differentiated successfully using $\delta^{15}N$ alone (Figure 12.1), especially in groundwater studies in sandy soils (where the effects of denitrification should be minimal). In contrast, studies in rivers, poorly drained soils, and poorly oxygenated groundwater are often more complicated, requiring the use of more tracers.

Groundwater and soil water studies

Quantification of the relative contributions of NO_3^- in groundwater or soil water (derived from fertilizer vs. animal waste, which includes human waste from septic systems and WWTPs) using only $\delta^{15}N_{NO_3}$ is usually successful if:

1 the groundwater is well-oxygenated and soils are sandy instead of clayey, so that denitrification can be (more-or-less) ruled out (Gormly & Spalding, 1979);
2 the NO_3^- from soil organic matter is insignificant (or its $\delta^{15}N$ is similar to fertilizer);
3 the fertilizer is nitrate, ammonium, or urea (not manure or green manure, which often has a high $\delta^{15}N$).

Quantification is much more difficult if the soils are clayey or the groundwater is not well-oxygenated, because the possibility of denitrification would have to be considered (see "Denitrification" section).

$\delta^{18}O_{NO_3}$ data are not essential for such determinations because of the large $\delta^{15}N$ differences between these sources (Figure 12.1). Furthermore, because most applied fertilizer is dominated by ammonium, analysis of $\delta^{18}O_{NO_3}$ may not add much additional information; however, this is not always true (see discussion below of Ging et al. 1996). $\delta^{18}O_{NO_3}$ could help improve the quantification if there is some difference in the $\delta^{18}O_{H_2O}$ of soil water vs. groundwater or lagoon water, or in the $\delta^{18}O_{O_2}$ in the soil zone or groundwater (these instances are not uncommon) where nitrification occurs. In these cases, measuring $\delta^{18}O_{NO_3}$ is recommended.

Quantification can be improved if other isotope (B, Sr, S, C, Li, U) or chemical tracers (caffeine, pharmaceuticals, rare earth elements (REEs), heavy metals, etc) specific to the different sources are used (Curt et al. 2004; Vitoria 2004; Vitoria et al. 2004; Widory et al. 2004). Improved quantification of waste sources of nitrate using a multi-isotope approach will be discussed in more detail below.

Surface water studies

Quantification of NO_3^- contributions in drains and streams using only $\delta^{15}N_{NO_3}$ (no $\delta^{18}O_{NO_3}$) can provide semi-quantitative to quantitative estimates of contributions if:

1 denitrification in the riparian zone and/or hyporheic zone can be ruled out or quantified (as above);
2 NO_3^- from soil organic matter is insignificant (or its $\delta^{15}N$ is similar to fertilizer);

3 uptake of nitrate by phytoplankton is minimal (or data are adjusted for this fractionation);
4 nitrification in the stream is minimal (or data are adjusted for this fractionation) (McMahon & Böhlke 1996; Sebilo et al. 2003, 2006).

Quantification can be improved if other chemical and/or isotopic tracers (e.g., $\delta^{18}O_{NO_3}$ or $\delta^{15}N$, $\delta^{13}C$, and/or $\delta^{34}S$ of POM) are used in addition to $\delta^{15}N_{NO_3}$ (e.g., Ging et al. 1996; Battaglin et al. 2001a,b; Silva et al. 2002; Kratzer et al. 2004; Panno et al. 2006; Sebilo et al. 2006).

Soil vs. animal waste source of nitrate

These sources can often be differentiated successfully using $^{15}N_{NO_3}$ alone. However, quantification may not be possible because soil nitrates often have variable $\delta^{15}N$ values, and often these values overlap the $\delta^{15}N$ values of nitrate derived from animal manure and human waste (see Fogg et al. 1998; Bedard-Haughn et al. 2003).

Quantification of the relative contributions of NO_3^- in soil or groundwater (derived from soil organic matter vs. animal waste) using only $\delta^{15}N_{NO_3}$ can be successful if:

1 the groundwater is well-oxygenated and the soils are sandy instead of clayey, so denitrification in the mixing zone can be ruled out;
2 NO_3^- from fertilizer is insignificant (or its $\delta^{15}N$ is similar to NO_3^- from soil organic matter).

There are many application studies (most not very quantitative), including Kreitler (1975), Kreitler & Browning (1983), Aravena et al. (1993), Komor & Anderson (1993), and Karr et al. (2001, 2003).

$\delta^{18}O_{NO_3}$ data are not essential. However, they could help improve the quantification if there is some difference in the $\delta^{18}O_{H_2O}$ of soil water vs. groundwater or lagoon water, or in the $\delta^{18}O_{O_2}$ in the soil zone or groundwater (these instances are not uncommon). Therefore, obtaining $\delta^{18}O_{NO_3}$ is recommended. For example, NO_3^- from manure piles or lagoons may have a high $\delta^{18}O$ due to water evaporation or high $\delta^{18}O_{O_2}$ due to respiratory fractionation. The $\delta^{18}O$ and δ^2H of the lagoon water may even be affected, with higher $\delta^{18}O$ and δ^2H values if water has been reduced via methanogenesis, and lower $\delta^{18}O$ values if there has been O exchange with CO_2. Analysis of the water for chemical constituents (animal waste has high chloride and other solutes), $\delta^{13}C$ of DIC, $\delta^{34}S$ of sulfate, or DOM for $\delta^{13}C$ and $\delta^{15}N$, might also help distinguish a waste lagoon source of nitrate because these other parameters are often sensitive to biogeochemical reactions in the highly reducing lagoon.

Soil vs. fertilizer source of nitrate

These sources usually cannot be differentiated using $\delta^{15}N$ alone because soil nitrates are often variable in $\delta^{15}N$ and usually overlap the $\delta^{15}N$ values of fertilizer-derived NO_3^-. Quantification of the relative contributions of NO_3^- derived from soil vs. fertilizer can be successful if:

1 it can be demonstrated that the $\delta^{15}N$ and/or $\delta^{18}O$ of different sources are sufficiently distinctive;
2 the system is well-oxygenated and the soils are sandy instead of clayey, so denitrification in the mixing zone can be ruled out;
3 other tracers that are specific for the fertilizer, including different isotopes (B, Sr, S, C, Li, U) or REEs are used.

Good examples include: Kohl et al. (1971, 1972), Hauck et al. (1972), Spruill et al. (2002), and Vitoria et al. (2004).

Septic waste vs. animal manure source of nitrate

These sources almost never can be differentiated using $\delta^{15}N_{NO_3}$ alone because the sources have overlapping $\delta^{15}N_{NO_3}$ values. However, see Fogg et al. (1998) for an example where the $\delta^{15}N$ values of the sources are significantly different.

Using a multi-isotope and/or multi-tracer approach, these types of waste can sometimes be distinguished if: (i) the average diets of the humans and animals are at different trophic levels (i.e., one group is herbivorous), causing the $\delta^{15}N$ of the diets and resulting waste, and hence the NO_3^- resulting from oxidation of the organic matter, to be several permil different; (ii) $\delta^{18}O_{NO_3}$ values are different (which could occur if there is some difference in the $\delta^{18}O_{H_2O}$ of public supply water vs. soil water or lagoon water, or in the $\delta^{18}O_{O_2}$ in the two environments); (iii) other isotope tracers (B, Sr, S, C, Li, U) or chemical tracers (caffeine, pharmaceuticals, hormones, DNA, K, REEs, heavy metals, etc.) that are specific for the different sources are used (see section on "Other tools") ; or (iv) the average diets of the humans and animals have different $\delta^{15}N$, $\delta^{13}C$, and $\delta^{34}S$ values (e.g., humans eat a marine-fish-rich diet, or the animals are fed mainly C4 plants, etc.), resulting in waste and resulting dissolved organic matter (DOM) with isotope signatures that are distinctive for the different waste sources. The combination of nitrate $\delta^{15}N$ and borate $\delta^{11}B$ can be particularly useful for distinguishing human vs. animal waste (see discussion of **boron isotopes** below). Good multi-tracer examples include: Aravena et al. (1993), Spruill et al. (2002), Curt et al. (2004), Vitoria (2004), and Widory et al. (2004, 2005). Sometimes a more sophisticated statistical or modeling approach can substitute for a lot of additional tracers (e.g., Spruill et al. 2002; Phillips & Gregg 2003; Otero 2004). Now that several studies have shown a multi-isotope approach can be successful, there is need for many more investigations of which kinds of tracers are most useful for which kinds of animals, land uses, and types of human waste.

Nitrate produced from waste from different kinds of farm animals

These sources almost never can be differentiated using $\delta^{15}N_{NO_3}$ alone because the sources have overlapping $\delta^{15}N_{NO_3}$ values. However, some recent studies suggest that a multi-isotope and/or multi-tracer approach may be successful in distinguishing between nitrates derived from different sources of manure.

Semi-quantification of contributions of NO_3^- derived from different kinds of farm animals is possible if: (i) $\delta^{18}O_{NO_3}$ values are different (which could occur if there is some difference in the $\delta^{18}O_{H_2O}$ of soil or lagoon water, or in the $\delta^{18}O_{O_2}$ in the two environments); (ii) other isotope tracers (B, Sr, S, C, Li) or chemical tracers (caffeine, pharmaceuticals, hormones, REEs, K, heavy metals, etc.) that are specific for the different sources are used; or (iii) the average diets of the different animals have different $\delta^{15}N$, $\delta^{13}C$, and $\delta^{34}S$ values (e.g., one animal group is fed a diet rich in marine fish or C4 plants), resulting in waste and resulting DOM with isotope signatures that are distinctive for the different animal types. Good examples include: Spruill et al. (2002), Vitoria (2004), and Widory et al. (2004). Sometimes a more sophisticated statistical or modeling approach can substitute for a lot of additional tracers (eg, Spruill et al. 2002; Phillips & Gregg 2003; Otero 2004). Now that several studies have shown a multi-isotope approach can be successful, there is need for many more investigations of which kinds of tracers are most useful for which kinds of animals and land uses.

Organic matter from animal waste vs. human waste vs. natural organic matter

These sources rarely can be differentiated using $\delta^{15}N_{NO_3}$ alone because the $\delta^{15}N_{NO_3}$ values overlap. However, the organic matter component of animal waste is often distinguishable from human waste or natural organic matter using a multi-isotope approach (e.g., the organic matter is also analyzed for $\delta^{13}C$ and $\delta^{34}S$). Quantification of waste from humans vs. animals can be successful if: (i) the average diets of the humans and animals are at different trophic levels (i.e., one group is herbivorous), causing the $\delta^{15}N$ of the diets and resulting waste, and hence the NO_3^- resulting from oxidation of the organic matter, to be several permil different; and/or (ii) one group eats a diet primarily composed of C3 plants (typical human diet in Asia, but not in Brazil; diets in the USA are intermediate), and the other eats a diet primarily composed of C4 plants (corn and sugar cane are the main C4 crops). $\delta^{34}S$ can also be useful for differentiating sources, especially when one source is marine (e.g., contamination of coastal waters with WWTP or animal farming operations (AFO) waste). Good examples for differentiating waste from different animals include: DeNiro & Epstein (1981), Minagawa (1992), and Bol & Pflieger (2002). Good examples for differentiating human/animal waste from natural organic matter include: Van Dover et al. (1992), McClelland et al. (1997), McClelland & Valiela (1998), Tucker et al. (1999), Hebert & Wassenaar (2001), and DeBruyn & Rasmussen (2002). For good discussions of the general topic of tracing sources of organic matter with isotopes see

Peterson & Fry (1987), Kendall et al. (2001), and Finlay & Kendall (this volume, pp. 283–333).

Nitrate derived from animal waste ammonium vs. other sources

There has been very little research on the question of whether different sources of NH_4^+ can be distinguished using the isotopic compositions of the resulting NO_3^-. Quantification of ammonium from AFOs vs. other sources (car exhaust, power plant exhaust, fertilizer volatilization) might be successful if:

1 there is a lot of volatilization of NH_3 from the waste (like hog lagoons), resulting in $\delta^{15}N_{NH_4}$ values that are very low;
2 the other sources of NH_4^+ have $\delta^{15}N$ values that are significantly higher;
3 when the NH_4^+ is nitrified, the two sources still have distinctive $\delta^{15}N$ values;
4 the sources show relatively little temporal and spatial variability relative to the difference between the mean $\delta^{15}N$ values.

Good examples include: Heaton (1986) and Karr et al. (2003).

Atmospheric nitrate derived from anthropogenic vs. natural sources

Although more extensive characterization of $\delta^{15}N_{NO_3}$ from individual NO_x sources is required, existing data suggest that $\delta^{15}N_{NO_3}$ of wet and dry atmospheric deposition can be used to help distinguish NO_x sources to deposition (see "Isotopic composition of NO_x sources" above). At the regional scale, $\delta^{15}N$ in wet deposition is strongly correlated with NO_x emissions from stationary sources surrounding precipitation monitoring sites (Figure 12.2; Elliott et al., in press). Several road gradient studies (e.g., Ammann et al. 1999; Pearson et al. 2000; Saurer et al. 2004) have illustrated how $\delta^{15}N$ isotopes in vegetation and NO_2 can be used to help assess the relative influence of vehicle emission in near-road environments. $\delta^{15}N$ can also be used to help distinguish sources of bulk particulate matter (Widory 2007), as well as aerosol NO_3^- and NH_4^+ (Yeatman et al. 2001).

Other tools for tracing anthropogenic contaminants

There are a variety of other isotope and geochemical "tools" that can help identify and (possibly) quantify anthropogenic sources of N (or contaminants related to N) to ecosystems. Some of these are described below.

Isotope biomonitoring

There is considerable literature on using the $\delta^{15}N$ of algae, terrestrial plants, and animals as "proxies" for the isotopic compositions of nitrate and/or ammonium – and hence for sources and the land uses that are specific to the N source. Some good examples include: Harrington et al. (1998), Hebert & Wassenaar (2001), Sauer et al. (2004), Anderson & Cabana (2005; 2006), Elliott & Brush (2006). See Finlay & Kendall (this volume, pp. 283–333) for more details.

Sulfur isotopes

The general terrestrial range of stable S isotope ($\delta^{34}S$) values is −50 to +50‰, with rare values much higher or lower. Analysis of sulfate for $\delta^{34}S$ and/or $\delta^{18}O$, or organic matter for $\delta^{34}S$, can provide information about:

1 atmospheric sources of acidic rain (Wadleigh et al. 1996; Wadleigh & Blake 1999);
2 fertilizer sources (Bates et al. 2002; Vitoria et al. 2004);
3 sources of animal/human waste (Bol & Pflieger 2002);
4 use of detergents in WWTP or animal waste caused by animal washing (Vitoria et al. 2004; Widory et al. 2004);
5 biogeochemical reactions occurring in the sediments or water column. Boron isotopes can be useful for these purposes as well (Widory et al. 2004).

Much of the literature on using $\delta^{34}S$ in ecosystems has been driven by the need to understand the effects of atmospheric deposition on S cycling in the natural environment, particularly in forest ecosystems. This is in response to increased S loadings to terrestrial ecosystems from anthropogenic S emissions, as sulfur is a dominant component of acid rain. Since nitrate is another major component of acid rain, it is obvious that linked studies of sulfate and nitrate isotopes are likely to be useful in tracing anthropogenic N sources.

Atmospheric $\delta^{34}S$ values are typically in the range of −5 to +25‰ (Krouse & Mayer 2000). Sulfate in precipitation derived from sea spray has a $\delta^{34}S$ value of +21‰ and $\delta^{18}O$ of +9.5‰, whereas rainwater sulfate derived from long-range transport of continental sources has a $\delta^{34}S$ of approximately +4‰ and $\delta^{18}O$ of approximately +11‰ (Wadleigh et al. 1996). Nriagu & Coker (1978) report that the $\delta^{34}S$ of precipitation in central Canada varies seasonally from about +2 to +9‰, with the low values caused mainly by biological S whereas the high values reflect S from fossil fuels.

The spatial distribution of $\delta^{34}S$ and the relative contributions from marine versus continental (including anthropogenic combustion) sources in Newfoundland (Canada) have been monitored by analyzing the $\delta^{34}S$ of rainfall and of epiphytic lichens that obtain their entire S content from the atmosphere (Wadleigh et al. 1996; Wadleigh & Blake 1999; Wadleigh 2003).

The lichen study (Wadleigh & Blake 1999) yielded a wonderful "bullseye" $\delta^{34}S$ contour plot showing low values in the interior of the island that are probably related to anthropogenic point sources, and progressively higher (more marine) values towards the coasts. These studies suggest that the study area is influenced by both marine (high $\delta^{34}S$ values) and continental sources (lower $\delta^{34}S$ values), with the possibility of anthropogenic influence from fossil-fuel powered plants.

Atmospheric sulfate (aerosol and rainfall) has recently been found to have a mass independent isotopic composition, with excess ^{17}O over what would have been expected based on the $\delta^{18}O$ of sulfate (Lee et al. 2002). For sulfate, the mass independent fractionation is described by: $\Delta^{17}O = \delta^{17}O - 0.52 \times \delta^{18}O$ (Lee et al. 2002). The $\Delta^{17}O$ values of wet and dry precipitation are generally <+2‰ (Lee et al. 2002; Johnson et al. 2001). Hence, $\Delta^{17}O$ can be used to identify the relative contributions of atmospheric sulfate versus terrestrial biological or geologic sources of sulfate to streams.

Sulfur-35 (^{35}S) is a naturally produced radioactive tracer (half-life = 87 days) that can be used to trace the movement of atmospherically derived sulfate in the environment. It is formed in the atmosphere from cosmic ray spallation of argon-40, and deposits on the Earth's surface in precipitation or as dryfall. It can be used both to trace the time scales for movement of atmospheric sulfate through the hydrosphere and, in ideal cases, to trace the movement of young (<1 year) water. It is an especially useful tracer in regions away from the ocean where sulfate concentrations are relatively low. Watershed application studies include: Cooper et al. (1991), (Michel et al. (2000), Novak et al. (2003), and Shanley et al. (2005).

Water isotopes

Analysis of water for $\delta^{18}O$ and δ^2H can provide extremely useful information about the sources of the nitrate and other solutes in water (Aravena & Robertson 1998; McMahon & Böhlke 2006). Very often, different sources of nitrate in rivers and groundwater are associated with different water $\delta^{18}O$ and δ^2H values because of evaporation or because the waters are derived from very different geographic areas. See Kendall & McDonnell (1998) for a review of the general topic, Kendall & Coplen (2001) for data on surface waters in the USA, and Dutton et al. (2005) for data on precipitation in the USA.

Boron isotopes

Boron (B) has two naturally occurring stable isotopes, ^{10}B and ^{11}B. The large relative mass difference between the boron isotopes leads to a wide (ca. 90‰) natural range of $\delta^{11}B$ values (Barth 1993). Because B is widely used in industrial, agricultural, cosmetics, and household products, $\delta^{11}B$ is a useful tool for determining sources of pollutants including nitrate (Eisenhut et al. 1996; Barth 2000). The main industrial source of B to waters is sodium perborate (NaBO$_3$), which is used in laundry detergents (primarily as a bleaching agent) and in household cleaners; consequently, B is commonly found in household sewage. Purification of waters in sewage treatment plants generally removes little or no B (Barth 2000); hence, δ^{11}B is a conservative tracer of a waste-water source. While B isotopes are not affected by denitrification, they are fractionated through processes such as adsorption on clays.

Boron isotopes have been shown to be useful for identifying anthropo-genic B sources in surface water and shallow groundwater systems:

1 municipal wastewater and sewage (Bassett 1990; Vengosh et al. 1994, 1999; Basset et al. 1995; Eisenhut et al. 1996; Barth 1998; Vengosh 1998; Seiler 2005);
2 irrigation return flows (Bassett et al. 1995);
3 fertilizer-affected irrigation waters from various agricultural settings (Komor 1997);
4 domestic solid waste deposit leachates from landfills (Eisenhut & Heumann 1997; Barth 2000);
5 mixed agricultural sources dominated by animal waste (Widory et al. 2004);
6 fly ash deposit leachates from a coal-fired power plant (Davidson & Bassett 1993).

Use of δ^{11}B coupled with δ^{15}N has proved to be an effective means for tracing agricultural nitrate sources (e.g., hog manure, cattle feedlot runoff, synthetic fertilizers) in surface and groundwaters (Basset et al. 1995; Komor 1997; Widory et al. 2004). In a recent study (Widory et al. 2004), δ^{11}B was used to distinguish between two types of sewage that were indistinguishable using δ^{15}N alone: a high-B/low-NO$_3$/low-δ^{11}B type that is derived from washing powders, and a moderate-B/moderate-NO$_3$ type with δ^{11}B values close to animal (probably human) manure. Some separation of different animal sources of B (e.g., sewage, cattle, hogs, poultry) is seen on plots of δ^{11}B vs. 1/B (Widory et al. 2004).

Strontium isotopes

Strontium (Sr) isotopes (^{87}Sr and ^{86}Sr) are another potentially useful tracer of anthropogenic sources of contaminants related to nitrate. Strontium iso-topes can be used to distinguish between a phosphorite and carbonatite origin of phosphate fertilizers (Vitoria et al. 2004). Several studies have shown sig-nificant differences between the Sr isotopic composition (denoted as δ^{87}Sr or as ^{87}Sr/^{86}Sr) of natural groundwater and human inputs (e.g., Négrel & Deschamps 1996; Négrel 1999; Böhlke & Horan 2000). In contrast to N and B isotopes, Sr isotopes will not fractionate through natural processes because of the low mass contrast between the ^{87}Sr and ^{86}Sr. Therefore Sr can, in theory, be used to identify mixing processes and water–rock interaction within the aquifer. Unfortunately, the flux coming from water–rock interaction is often large compared with the anthropogenic flux, and thus totally overprints the isotope signal (Widory et al. 2004).

Widory et al. (2004) tested whether the combined use of geochemical and isotopic tracers (N, B, and Sr) would provide a sensitive method for tracing sources of NO_3^- in contaminated groundwater. The basic idea was that N isotopes, as an intrinsic tracer of the NO_3^- molecule, will reflect both the sources and the fate (i.e. denitrification) of NO_3^- in groundwater. In contrast, B isotopes, because they are not affected by denitrification, will be isotopically labeled with the signature of the sources. Nitrogen and B isotopes proved extremely useful for distinguishing among agricultural sources. However, their study showed that differentiation between the different animals (hog, poultry, cattle manure; and sewage) was not possible using Sr alone due to the similarity of the isotope signatures; however, fertilizer is distinctive.

Lithium isotopes

Lithium (Li) has two naturally occurring stable isotopes, ^6Li and ^7Li. Considerable variability of δ^7Li has been reported for natural materials, with marine-derived waters and minerals having much higher δ^7Li values than minerals derived from igneous and metamorphic rocks and their associated waters (Bullen & Kendall 1998). Additionally, as a consequence of the processing of lithium to produce highly ^6Li-enriched materials for nuclear power plants, highly purified Li with a high δ^7Li value (often in the range of +200 to +400‰) is readily available for industrial, agricultural, and pharmaceutical use (Qi et al. 1997). Hence, δ^7Li is potentially a very valuable tracer of an anthropogenic source of wastewater.

A study in Pennsylvania, USA (Bullen & Senior 1992) showed that δ^7Li values in streams that are affected by discharges from lithium-processing plants and in groundwater downgradient from the processing plants are significantly higher than natural background δ^7Li values. As an example of how pharmaceuticals might have distinctive δ^7Li values, groundwater downgradient from a mental health facility in Pennsylvania (USA) was also found to have a substantially greater δ^7Li value than those found in natural materials. Because lithium is commonly used to treat manic depressive behavior, the elevated δ^7Li value was attributed to the transfer of the Li from pharmaceuticals into wastewater in this area of unsewered residential development (Thomas D. Bullen, pers. comm.).

Phosphate isotopes

Phosphorous (P) is an essential macronutrient in aquatic ecosystems. Excess anthropogenic inputs of P in the form of orthophosphate (PO_4^{3-}) can cause eutrophic conditions. Although phosphate has only one stable isotope, recent advances in analytical methods have made it possible to use the $\delta^{18}O$ of inorganic phosphate (McLaughlin et al. 2004) or organic phosphate (McLaughlin et al. 2006a) as a tracer of the phosphate source. Because the $\delta^{18}O$ of phosphate rapidly equilibrates with the O in water, PO_4-$\delta^{18}O$ usually is a usable tracer of phosphate sources only in waters that are not P-limiting and where the waters associated with the two sources (waste vs. natural) have water-$\delta^{18}O$ values that are different by several permil. Such conditions are likely in estuaries or near the coast (McLaughlin et al. 2006b). Other places where the water-$\delta^{18}O$ values might be sufficiently different to "label" the PO_4-$\delta^{18}O$ include WWTPs or waste lagoons where there is evaporative enrichment of water-$\delta^{18}O$, or locations where the public supply water is from a much different geographic location or elevation than the local soil water or groundwater (McLaughlin et al. 2006b). Under such conditions, it might be possible to distinguish phosphate from fertilizer vs. animal waste/septic waste vs. soil organic matter (Gruau et al. 2005; McLaughlin et al. 2006b).

A recent survey of the $\delta^{18}O$ values generated using the silver phosphate method of McLaughlin et al. (2004) for phosphate from many anthropogenic sources, reports ranges of values for sewage (+7 to +12‰), detergents (+13 to +19‰), and fertilizers (+16 to +23‰; Young et al. 2006). These sewage $\delta^{18}O$ values are considerably lower than the values (+16 to +19‰) previously reported by Gruau et al. (2005) generated by a different method, suggesting that in certain geographic locations $\delta^{18}O$ might be able to distinguish between waste and detergent or fertilizer sources of phosphate (Young et al. 2006), while in other areas the values may be indistinguishable.

Other isotopes can also be useful for tracing phosphate sources. For example, uranium is a trace constituent of geologic sources of phosphate. Hence, analysis of ^{234}U and ^{238}U can distinguish between "natural" and "geologic" sources (Zielinski et al. 1997, 2000). In addition, strontium is a trace constituent of geologic sources of phosphate. Different sources of geologic phosphate (phosphorites vs. carbonatites) appear to have different concentrations of trace metals and REEs; hence, these can be used as tracers of phosphate and/or fertilizer source and the different geologic units that different source waters traveled through (Böhlke & Horan 2000; Vitoria et al. 2004; Widory et al. 2004).

Age-dating nitrate contamination of groundwater

One powerful potential application of technological advances in the age-dating of young groundwater (Plummer et al. 1993; Dunkle et al. 1993) is to evaluate the impact of changes in agricultural management practices on water quality (e.g., Böhlke & Denver 1995; McMahon & Böhlke 1996). By combining nitrate isotope analyses with ground-water dating, it is possible to estimate the timing of nitrate-related events including:

1 the rate of natural denitrification;
2 when contaminated or remediated groundwaters will reach the streams (Böhlke & Denver 1995);
3 changes in contamination loads in the recharge zone over time (Böhlke & Denver 1995; Böhlke et al. 2006; McMahon & Böhlke 2006);
4 success (or otherwise) of implementation of BMPs (Wassenaar et al. 2006).

Böhlke & Denver (1995) showed that in places where the extent of denitrification can be quantified, analyses of dated, denitrification-corrected groundwaters can provide a valuable record of past contamination loads in the recharge zone. This type of record can be valuable in localities where long-term fertilizer-use data are missing or unreliable.

Statistical, geochemical, and hydrologic modeling

New interpretative techniques beyond simple bivariate plots and linear regressions can be used for isotopic data interpretation. One approach is End Member Mixing Analysis (EMMA: see Christophersen et al. 1990, 1993; Hooper & Christopherson 1992). Other approaches using isotopes and chemical data include: geochemical reaction path modeling (Plummer et al. 1983, 1991; Böhlke & Denver 1995); principal components analysis (Otero 2004); classification trees (Spruill et al. 2002); and uncertainty analysis (Phillips & Gregg 2001).

Pharmaceuticals, hormones, DNA, and other chemical constituents

Additional information for tracing N sources can be gained from: various chemicals specific to different animal types (including humans); trace and rare earth elements for tracing fertilizers; DNA and other molecular markers in the dissolved and particulate organic matter associated with the nitrate; and basic chemistry. For example, simple Cl^- concentrations can be used to identify animal waste contamination of rivers and groundwaters (Aravena & Robertson 1998; Silva et al. 2002; Karr et al. 2003; Seiler 2005).

Conclusions

Despite the initial controversy over 30 years ago regarding the use of nitrate isotopes in agricultural systems (e.g., Hauck et al. 1972; Kohl et al. 1972), there have been many very dramatic examples where isotope data have been critical in advancing our understanding of ecosystems. Although source estimates using isotope data are usually only qualitative to semi-quantitative, isotope hydrologists and biogeochemists have found that stable isotope data, collected at the appropriate temporal and spatial scale, often usefully integrate the natural patterns in complex systems. With the increasing automation of isotope techniques, it is becoming ever easier to acquire the large sets of isotope data required to distinguish the environmental patterns within the "noise".

This trend is expected to continue into the future, as new analytical advances allow isotope techniques to address an even wider range of important questions regarding both N biogeochemistry and human effects on N sources and cycling. For example, here we briefly list a few "cutting edge" questions:

1 What are the spatial and temporal patterns in $\Delta^{17}O$ (and $\delta^{18}O$ and $\delta^{15}N$) of nitrate in wet and dry deposition – and what causes them?
2 Will $\Delta^{17}O$ of nitrate verify or refute the estimates of atmospheric nitrate in small watersheds determined with $\delta^{18}O$?
3 Can different types of atmospheric N sources (e.g., vehicle exhaust vs. power plants vs. agricultural emissions) be distinguished isotopically and quantified?
4 What are the controls on the $\delta^{18}O$ of nitrate during nitrification, denitrification, and assimilation in different environments?
5 Do the $\delta^{18}O$ and $\delta^{15}N$ of nitrate get partially reset in the unsaturated zone (and shallow saturated zone) due to rapid oscillations in nitrification and denitrification as wetting fronts move through the profile?
6 What happens to atmospheric ammonium in watersheds?
7 What causes the $\delta^{15}N$ of biota in streams to increase with increasing watershed scale and human utilization of the land (regardless of the specific land use)?
8 To what extent does the nitrate in the stream only reflect molecules of nitrate from near-stream environments?

The challenges facing those who use natural abundance isotopic approaches to examine issues involving N biogeochemistry are:

1 that different sources of N can have partially overlapping isotopic compositions;
2 sources can have considerable spatial and temporal variation in isotopic composition;
3 isotope fractionations can overprint or blur initially distinctive isotopic compositions.

However, these problems can often be minimized or eliminated by using a multi-isotope, multi-tracer approach – with a lot of hydrologic and chemical data. In addition, it is hoped that as N isotope biogeochemistry moves forward, we will be able to improve our ability to quantify, rather than qualitatively estimate, contributions from multiple sources.

References

Aleem, M.I.H., Hoch, G.E. & Varner, J.E. (1965) Water as the source of oxidant and reductant in bacterial chemosynthesis. *Biochemistry*, **54**, 869–873.

Alexander, B., Savarino, J., Kreutz, K.J. & Thiemens, M.H. (2004) Impact of preindustrial biomass-burning emissions on the oxidation pathways of tropospheric sulfur and nitrogen. *Journal of Geophysical Research*, **109**, D08303, doi:10.1029/2003JD004218.

Altabet, M.A., Pilskaln, C., Thunell, R.C., et al. (1999) The nitrogen isotope biogeochemistry of sinking particles from the margin of the Eastern North Pacific. *Deep Sea Research, Part I*, **46**, 655–679.

Amberger, A. & Schmidt, H.L. (1987) Natürliche Isotopengehalte von nitrat als Indikatoren für dessen Herkunft. *Geochimica et Cosmochimica Acta*, **51**, 2699–2705.

Ammann, M., Siegwolf, R., Pichlmayer, F., et al. (1999) Estimating the uptake of traffic-derived NO_2 from N-15 abundance in Norway spruce needles. *Oecologia*, **118**(2), 124–131.

Amundson, A., Austin, A.T., Schuur, E.A.G., at el. (2003) Global patterns of the isotopic composition of soil and plant nitrogen. *Global Biogeochemical Cycles*, **17**, 31-1–31-10.

An, S. & Gardner, W.S. (2002) Dissimilatory nitrate reduction to ammonium (DNRA) as a nitrogen link, versus denitrification as a sink in a shallow estuary (Laguna Madre/Baffin Bay, Texas). *Marine Ecology Progress Series*, **237**, 41–50.

Anderson, C. & Cabana, G. (2005) delta N-15 in riverine food webs: effects of N inputs from agricultural watersheds. *Canadian Journal of Fisheries and Aquatic Sciences*, **62**, 333–340.

Anderson, C. & Cabana, G. (2006) Does $\delta^{15}N$ in river food webs reflect the intensity and origin of N loads from the watershed? *Science of the Total Environment*, **367**, 968–978.

Andersson, K.K. & Hooper, R.A.B. (1983) O_2 and H_2O are each the source of one O in NO_2 produced from NH_3 by Nitrosomonas: ^{15}N-NMR evidence. *FEBS Letters*, **164**, 236–240.

Aravena, R. & Robertson, W.D. (1998) Use of multiple isotope tracers to evaluate denitrification in ground water: study of nitrate from a large-flux septic system plume. *Ground Water*, **36**, 975–982.

Aravena, R., Evans, M.L. & Cherry, J.A. (1993) Stable isotopes of oxygen and nitrogen in sources identification of nitrate from septic systems. *Ground Water*, **31**, 180–186.

Barth, S. (1993) Boron isotope variations in nature: a synthesis. *Geologische Rundschau*, **82**, 640–651.

Barth, S. (1998) Application of boron isotopes for tracing sources of anthropogenic contamination in groundwater. *Water Research*, **32**, 685–690.

Barth, S.R. (2000) Boron isotopic compositions of near-surface fluids: A tracer for identification of natural and anthropogenic contaminant sources. *Water, Air and Soil Pollution*, **127**, 49–60.

Bassett, R.L. (1990) A critical evaluation of the available measurements for the stable isotopes of boron. *Applied Geochemistry*, **5**, 541–554.

Basset, R.L., Buszka, P.M., Davidson, G.R. & Chong-Diaz, D. (1995) Identification of groundwater solute sources using boron isotopic composition. *Environmental Science and Technology*, **29**, 2915–2922.

Batchelor, B. & Lawrence, A.W. (1978) A kinetic model for autotrophic denitrification using elemental sulfur. *Water Research*, **12**, 1075–1084.

Bates, A.L., Orem, W.H., Harvey, J.W. & Spiker, E.C. (2002) Tracing sources of sulfur in the Florida Everglades. *Journal of Environmental Quality*, **31**, 287–299.

Battaglin, W.A., Kendall, C., Chang, C.C.Y., et al. (2001a) *Chemical and Isotopic Composition of Organic and Inorganic Samples from the Mississippi River and its Tributaries, 1997–98*. Water Resources Investigation Report 01-4095, U.S. Geological Survey, 57 pp.

Battaglin, W.A., Kendall, C., Chang, C.C.Y., et al. (2001b) Chemical and isotopic evidence of nitrogen transformation in the Mississippi River, 1997–98. *Hydrological Processes*, **15**, 1285–1300.

Bedard-Haughn, A.K., Van Groenigen, J.-W. & Van Kessel, C. (2003) Tracing ^{15}N through landscapes: potential uses and precautions. *Journal of Hydrology*, **272**, 175–190.

Böhlke, J.K. (2002) Groundwater recharge and agricultural contamination, *Hydrogeology Journal*, **10**, 153–179.

Böhlke, J.K. & Denver, J.M. (1995) Combined use of ground-water dating, chemical, and isotopic analyses to resolve the history and fate of nitrate contamination in two agricultural watersheds, Atlantic coastal plain, Maryland. *Water Resources Research*, **31**, 2319–2339.

Böhlke, J.K. & Horan, M. (2000) Strontium isotope geochemistry of ground waters and streams affected by agriculture, Locust Grove, Maryland. *Applied Geochemistry*, **15**, 599–609.

Böhlke, J.K., Eriksen, G.E. & Revesz, K. (1997) Stable isotope evidence for an atmospheric origin of desert nitrate deposits in northern Chile and southern California, U.S.A. *Chemical Geology*, **136**, 135–152.

Böhlke, J.K., Mroczkowski, S.J. & Coplen, T.B. (2003) Oxygen isotopes in nitrate: New reference materials for ^{18}O:^{17}O:^{16}O measurements and observations on nitrate-water equilibration. *Rapid Communications in Mass Spectrometry*, **17**, 1835–1846.

Böhlke, J.K., Harvey, J.W. & Voytek, M.A. (2004) Reach-scale isotope tracer experiment to quantify denitrification and related processes in a nitrate-rich stream, midcontinent United States. *Limnology and Oceanography*, **49**, 821–838.

Böhlke, J.K., Smith, R.L. & Miller, D.N. (2006) Ammonium transport and reaction in contaminated groundwater: Application of isotope tracers and isotope fractionation studies. *Water Resources Research*, **42**, W05411, doi:10.1029/2005WR004349.

Bol, R. & Pflieger, C. (2002) Stable isotope (C-13, N-15 and S-34) analysis of the hair of modern humans and their domestic animals. *Rapid Communications in Mass Spectrometry*, **16**, 2195–2200.

Böttcher, J., Strebel, O., Voerkelius, S. & Schmidt, H.L. (1990) Using stable isotope fractionation of nitrate-nitrogen and nitrate-oxygen for evaluation of microbial denitrification in a sandy aquifer. *Journal of Hydrology*, **114**, 413–424.

Brandes, J.A. & Devol, A.H. (1997) Isotopic fractionation of oxygen and nitrogen in coastal marine sediments. *Geochimica et Cosmochimica Acta*, **61**, 1793–1801.

Broadbent, F.E., Rauschkolb, R.S., Lewis, K.A. & Chang, G.Y. (1980) Spatial variability in nitrogen-15 and total nitrogen in some virgin and cultivated soils. *Soil Science Society of America Journal*, **44**, 524–527.

Bullen, T.D. & Kendall, C. (1998) Tracing of weathering reactions and water flowpaths: a multi-isotope approach. In: *Isotope Tracers in Catchment Hydrology* (Eds C. Kendall & J.J. McDonnell), pp. 611–646. Elsevier, Amsterdam.

Bullen, T.D. & Senior, L.A. (1992) Lithogenic Sr and anthropogenic Li in an urbanized stream basin: isotopic tracers of surface water-groundwater interaction. *Eos (Transactions of the American Geophysical Union)*, **73**, F130.

Burns, D.A. & Kendall, C. (2002) Analysis of delta N-15 and delta O-18 to differentiate NO3- sources in runoff at two watersheds in the Catskill Mountains of New York. *Water Resources Research*, **38**(5), 1051.

Butler, T.J., Likens, G.E. & Stunder, B.J.B. (2001) Regional-scale impacts of Phase I of the Clean Air Act Amendments in the USA: the relation between emissions and concentrations both wet and dry. *Atmospheric Environment*, **35**(6), 1015–1028.

Butler, T.J., Likens, G.E., Vermeylen, F.M. & Stunder, B.J.B. (2005) The impact of changing nitrogen oxide emissions on wet and dry nitrogen deposition in the northeastern USA. *Atmospheric Environment*, **39**(27), 4851–4862.

Campbell, D.H., Kendall, C., Chang, C.C.Y., et al. (2002) Pathways for nitrate release from an alpine watershed: Determination using delta N-15 and delta O-18. *Water Resources Research*, **38**(5), 1052.

Casciotti, K.L., Sigman, D.M., Hastings, M.G., et al. (2002) Measurement of the oxygen isotopic composition of nitrate in seawater and freshwater using the denitrifier method. *Analytical Chemistry*, **74**, 4905–4912.

Casciotti, K.L., Sigman, D.M. & Ward, B.B. (2003) Linking diversity and stable isotope fractionation in ammonia-oxidizing bacteria. *Geomicrobiology Journal*, **20**, 335–353.

Casciotti, K.L., Böhlke, J.K., McIlvin, M.R., et al. (2007) Oxygen isotopes in nitrite: analysis, calibration, and equilibration. *Analytical Chemistry*, **79**, 2429–2436.

Cey, E.E., Rudolph, D.L., Aravena, R. & Parkin, R. (1999) Role of the riparian zone in controlling the distribution and fate of agricultural nitrogen near a small stream in southern Ontario. *Journal of Contaminant Hydrology*, **37**, 45–67.

Chang, C.C.Y., Langston, J., Riggs, M., et al. (1999) A method for nitrate collection for $\delta^{15}N$ and $\delta^{18}O$ analysis from waters with low nitrate concentrations. *Canadian Journal of Fisheries and Aquatic Science*, **56**, 1856–1864.

Chang, C.C.Y., Kendall, C., Silva, S.R., et al. (2002) Nitrate stable isotopes: tools for determining nitrate sources among different land uses in the Mississippi River Basin. *Canadian Journal of Fisheries and Aquatic Science*, **59**, 1874–1885.

Christophersen, N., Neal, C. & Hooper, R.P. (1990) Modeling streamwater chemistry as a mixture of soil-water end members, A Step towards second generation acidification models. *Journal of Hydrology*, **116**, 307–320.

Christophersen, N., Neal, C. & Hooper, R.P. (1993) Modeling environmental impacts: A challenge for the scientific method. *Journal of Hydrology*, **152**, 1–12.

Cifuentes, L.A., Fogel, M.F., Pennock, J.R. & Sharp, J.H. (1989) Biogeochemical factors that influence the stable nitrogen isotope ratio of dissolved ammonium in the Delaware Estuary. *Geochimica et Cosmochimica Acta*, **53**, 2713–2721.

Cleveland, C.C., Townsend, A.R., Schimel, D.S., et al. (1999) Global patterns of terrestrial biological nitrogen (N_2) fixation in natural systems. *Global Biogeochemical Cycles*, **13**, 623–646.

Cole, M.L., Kroeger, K.D., McClelland, J.W. & Valiela, I. (2006). Effects of watershed land use on nitrogen concentrations and $\delta^{15}N$ in groundwater. *Biogeochemistry*, **77**(2), 199–215.

Cook, P. & Herczeg, A.L. (Eds) (2000) *Environmental Tracers in Subsurface Hydrology*. Kluwer Academic Publishers, Boston, MA.

Cooper, L.W., Olsen, C.R., Solomon, D.K., et al. (1991) Stable isotopes of oxygen and natural fallout radionuclides used for tracing runoff during snowmelt in an arctic watershed. *Water Resources Research*, **27**, 2171–2179.

Curt, M.D., Aguado, P., Sanchez, G., Bigeriego, M. et al. (2004) Nitrogen isotope ratios of synthetic and organic sources of nitrate Water contamination in Spain. *Water, Air and Soil Pollution*, **151**, 135–142.

Dalsgaard, T., Canfield, D.E., Petersen, J., Thamdrup, B. & Acuna-Gonzalez, J. (2003) N_2 production by the anammox reaction in the anoxic water column of Golfo Dulce, Costa Rica. *Nature*, **6932**, 606–607.

Davidson, G.R. & Basset, R.L. (1993) Application of boron isotopes for identifying contaminants such as fly ash leachate in groundwater. *Environmental Science and Technology*, **27**, 172–176.

DeBruyn, A.M.H. & Rasmussen, J.B. (2002) Quantifying assimilation of sewage-derived organic matter by riverine benthos. *Ecological Applications*, **12**(2), 511–520.

Delwiche, C.C. & Steyn, P.L. (1970) Nitrogen isotope fractionation in soils and microbial reactions. *Environmental Science and Technology*, **4**, 929–935.

DeNiro, M.J. & Epstein, S. (1981) Influence of diet on the distribution of nitrogen isotopes in animals. *Geochimica et Cosmochimica Acta*, **45**, 341–351.

Dispirito, A.A. & Hooper, A.B. (1986) Oxygen exchange between nitrate molecules during nitrite oxidation by Nitrobacter. *Journal of Biological Chemistry*, **261**(23), 10,534–10,537.

Dubey, M., Mohrschladt, R., Donahue, N.M. & Anderson, J.G. (1997) Isotope specific kinetics of hydroxyl radical (OH) with water (H_2O): Testing models of reactivity and atmospheric fractionation. *Journal of Physical Chemistry*, **101**, 1494–1500.

Dunkle, S.A., Plummer, L.N., Busenberg, E., et al. (1993) Chlorofluorocarbons (CCl3F and CCl2F2) as dating tools and hydrologic tracers in shallow ground water of the Delmarva Peninsula, Atlantic Coastal Plain, United States. *Water Resources*, **29**, 3837–3860.

Durka, W., Schulze, E.D., Gebauer, G. & Voerkelius, S. (1994) Effects of forest decline on uptake and leaching of deposited nitrate determined from ^{15}N and [18]O measurements. *Nature*, **372**, 765–767.

Dutton, A., Wilkinson, B.H., Welker, J.M., et al. (2005) Spatial distribution and seasonal variation in $^{18}O/^{16}O$ of modern precipitation and river water across the conterminous USA. *Hydrological Processes*, **19**(20), 1421–4146.

Eisenhut, S. & Heumann, K.G. (1997) Identification of ground water contaminations by landfills using precise boron isotope ratio measurements with negative thermal ionization mass spectrometry. *Fresenius Journal of Analytical Chemistry*, **359**, 375–377.

Eisenhut, S. Heumann, K. & Vengosh, A. (1996) Determination of boron isotopic variations in aquatic systems with negative thermal ionization mass spectrometry as a tracer for anthropogenic influences. *Fresenius Journal of Analytical Chemistry*, **345**, 903–909.

Elliott, E.M., Kendall, C., Harlin, K., et al. (2004). Mapping the spatial and temporal distribution of N and O isotopes in precipitation nitrate across the northeastern and mid-Atlantic United States. *Eos (Transactions of American Geophysical Union)*, **85**, Abstract H52B-02.

Elliott, E.M. & Brush, G.S. (2006) Sedimented organic nitrogen isotopes in fresh-water wetlands record long-term changes in watershed nitrogen source and land use. *Environmental Science and Technology*, **26**(40), 2910–2916.

Elliott, E.M., Kendall, C., Burns, D.A., et al. (2006). Nitrate isotopes in precipitation to distinguish NO_x sources, atmospheric processes, and source areas in the United States. *Eos (Transactions of the American Geophysical Union)*, **87**(36).

Elliott, E.M., Kendall, C., Wankel, S.D., et al. (In press). Nitrogen isotopes as indicators of NO_x source contributions to atmospheric deposition across the Midwestern and Northeastern United States. *Environmental Science and Technology*.

Feigin, A., Shearer, G., Kohl, D.H. & Commoner, B. (1974) The amount and nitrogen-15 content of nitrate in soil profiles from two central Illinois fields in a corn–soybean rotation. *Soil Science Society of America Proceedings*, **38**, 465–471.

Fenn, M.E., Haeuber, R., Tonnesen, G.S., et al. (2003) Nitrogen emissions, deposition, and monitoring in the western United States. *Bioscience*, **53**(4), 391–403.

Fogel, M.L. & Cifuentes, L.A. (1993) Isotope fractionation during primary production. In: *Organic Geochemistry* (Eds M.H. Engel & S.A. Macko), pp. 73–98. Plenum Press, New York.

Fogg, G.E., Rolston, D.E., Decker, D.L., et al. (1998) Spatial variation in nitrogen isotope values beneath nitrate contamination sources. *Ground Water*, **36**, 418–426.

Francis, C.A., Roberts, K.J., Beman, J.M., et al. (2005) Ubiquity and diversity of ammonia-oxidizing archaea in water columns and sediments of the ocean. *Proceedings of the National Academy of Sciences of the United States of America*, **102**(41), 14,683–14,688.

Freyer, H.D. (1978) Seasonal trends of NH_4^+ and NO_3^- nitrogen isotope composition in rain collected at Julich, Germany. *Tellus*, **30**, 83–92.

Freyer, H.D. (1991) Seasonal variation of $^{15}N/^{14}N$ ratios in atmospheric nitrate species. *Tellus*, **43B**, 30–44.

Freyer, H.D., Kley, D., Volz-Thomas, A. & Kobel, K. (1993) On the interactions of isotopic exchange processes with photochemical reactions in atmospheric oxides of nitrogen. *Journal of Geophysical Research*, **98**(D8), 14791–14796.

Fry, B. (1991) Stable isotope diagrams of freshwater foodwebs. *Ecology*, **72**, 2293–2297.

Galloway, J.N., Schlesinger, W.H., Levy, H. II, Michaels, A. & Schnoor, J.L. (1995) Nitrogen fixation: anthropogenic enhancement–environmental response. *Global Biogeochemistry Cycles*, **9**, 235–252.

Galloway, J.N., Dentener, F.J., Capone, D.G., Boyer, E.W. et al. (2004) Nitrogen cycles: past, present, and future. *Biogeochemistry*, **70**(2), 153–226.

Garten, C.T., Jr. (1992) Nitrogen isotope composition of ammonium and nitrate in bulk precipitation and forest throughfall. Intern. J. Environ. *Analytical Chemistry*, **47**, 33–45.

Garten, C.T., Jr. (1993) Variation in foliar ^{15}N abundance and the availability of soil nitrogen on Walker Branch watershed. *Ecology*, **74**, 2098–2113.

Garten, C.T., Jr. & Hanson, P.J. (1990) Foliar retention of ^{15}N-nitrate and ^{15}N-ammonium by red maple (*Acer rubrum*) and white oak (*Quercus alba*) leaves from simulated rain. *Environmental and Experimental Botany*, **30**, 333–342.

Ging, P.B., Lee, R.W. & Silva, S.R. (1996) Water chemistry of Shoal Creek and Waller Creek, Austin Texas, and potential sources of nitrate. *U.S. Geological Survey Water Resources Investigations Report*, **96–4167**.

Gormly, J.R. & Spalding, R.J. (1979) Sources and Concentrations of Nitrate Nitrogen in Groundwater of the Central Platte Region, Nebraska. *Ground Water*, **17**, 291–301.

Granger J., Sigman D.M., Needoba J.A. & Harrison P.J. (2004) Coupled nitrogen and oxygen isotope fractionation of nitrate during assimilation by cultures of marine phytoplankton. *Limnology and Oceanography*, **49**(5), 1763–1773.

Granger, J. (2006) *Coupled nitrogen and oxygen isotope of nitrate imparted during its assimilation and dissimilarity reduction by unicellular plankton.* Unpublished PhD dissertation, University of British Columbia.

Gruau, G., Legeas, M., Riou, C., et al. (2005) The oxygen isotopic composition of dissolved anthropogenic phosphates: a new tool for eutrophication research?, *Water Research*, **39**, 232–238.

Harrington, R.R., Kennedy, B.P., Chamberlain, C.P., et al. (1998) ^{15}N enrichment in agricultural catchments: Field patterns and applications to tracking Atlantic salmon (*Salmo salar*). *Chemical Geology*, **147**, 281–294.

Hastings, M.G., Sigman, D.M. & Lipschultz, F. (2003) Isotopic evidence for source changes of nitrate in rain at Bermuda. *Journal of Geophysical Research – Atmospheres*, **108**(D24), 4790.

Hastings, M.G., Steig, E.J. & Sigman, D.M. (2004) Seasonal variations in N and O isotopes of nitrate in snow at Summit, Greenland: Implications for the study of nitrate in snow and ice cores. Journal Of Geophysical Research-Atmospheres, 109(D20).

Hastings, M.G., Sigman, D.M. & Steig, E.J. (2005) Glacial/interglacial changes in the isotopes of nitrate from the Greenland Ice Sheet Project 2 (GISP2) ice core. *Global Biogeochemical Cycles*, **19**(4), GB4024, doi:10.1029/2005GB002502.

Hauck, R.D., Bartholomew, W.V., Bremner, J.M., et al. (1972) Use of variations in natural nitrogen isotope abundance for environmental studies: a questionable approach. *Science*, **177**, 453–454.

Heaton, T.H.E. (1986) Isotopic studies of nitrogen pollution in the hydrosphere and atmosphere: a review. *Chemical Geology*, **59**, 87–102.

Heaton, T.H.E. (1987) $^{15}N/^{14}N$ ratios of nitrate and ammonium in rain at Pretoria, South Africa. *Atmospheric Environment*, **21**, 843–852.

Heaton, T H.E. (1990) 15N/14N ratios of NOx from vehicle engines and coal-fired power stations. *Tellus*, **42B**, 304–307.

Heaton, T.H.E., Spiro, B., Madeline, S. & Robertson, C. (1997) Potential canopy influences on the isotopic composition of nitrogen and sulphur in atmospheric deposition. *Oecologia*, **109**, 600–607.

Heaton, T.H.E., Wynn, P. & Tye, A.M. (2004). Low N-15/N-14 ratios for nitrate in snow in the High Arctic (79 degrees N). *Atmospheric Environment*, **38**(33), 5611–5621.

Hebert, C.E. & Wassenaar, L.I. (2001) Stable nitrogen isotopes in waterfowl feathers reflect agricultural land use in western Canada. *Environmental Science and Technology*, **35**, 3482–3487.

Hoch M.P., Fogel M.F. & Kirchman, D.L. (1992) Isotope fractionation associated with ammonium uptake by a marine bacterium. *Limnology and Oceanography*, **37**(7), 1447–1459.

Hoering, T. (1957) The isotopic composition of ammonia and the nitrate ion in rain. *Geochimica et Cosmochimica Acta*, **12**, 97–102.

Hollocher, T.C. (1984) Source of the oxygen atoms of nitrate in the oxidation of nitrite by Nitrocacter agilis and evidence against a P-O-N anhydride mechanism in oxidative phosphorylation. *Archives of Biochemistry and Biophysics*, **233**, 721–727.

Hollocher, T.C., Tate, M.E. & Nicholas, D.J.D. (1981) Oxidation of ammonia by Nitrosomonas europaea: Definitive ^{18}O-tracer evidence that hydroxylamine formation involves a monooxygenase. *Journal of Biology and Chemistry*, **256**, 10,834–10,836.

Hooper, R.P. & Christophersen, N. (1992) Predicting episodic acidification in the Southeastern United States: Combining a long-term acidification model and the end-member mixing concept. *Water Resources*, **28**, 1983–1990.

Hübner, H. (1986) Isotope effects of nitrogen in the soil and biosphere. In: *Handbook of Environmental Isotope Geochemistry*, Vol. 2b, *The Terrestrial Environment* (Eds P. Fritz & J.C. Fontes), pp. 361–425. Elsevier, Amsterdam.

Hwang, H.-H., Liu, J. & Hackley, J.C. (1999) Method improvement for oxygen isotope analysis in nitrates. *Geological Society of America, Abstracts with Programs*, **31**, A-23.

Johnson, C.A., Mast, M.A. & Kester, C.L. (2001) Use of $^{17}O/^{16}O$ to trace atmospherically-deposited sulfate in surface waters: a case study in alpine watersheds in the Rocky Mountains. *Geophysical Research Letters*, **28**(23), 4483–4486.

Johnston, J.C. & Thiemens, M.H. (1997) The isotopic composition of tropospheric ozone in three environments, *Journal of Geophysical Research*, **102**, 25,395–25,404.

Jorgensen, K.S. (1989) Annual pattern of denitrification and nitrate ammonification in estuarine sediment. *Applied and Environmental Microbiology*, **55**(7), 1841–1847.

Joye, S.B. & Hollibaugh, J.T. (1995) Influence of sulfide inhibition of nitrification on nitrogen regeneration in sediments. *Science*, **270**, 623–625.

Kaiser, J., Hastings, M.G., Houlton, B.Z., et al. (2007) Triple oxygen isotope analysis of nitrate using the denitrifier method and thermal decomposition of N₂O. *Analytical Chemistry*, **79**, 599–607.

Karamanos, E.E., Voroney, R.P. & Rennie, D.A. (1981) Variation in natural ^{15}N abundance of central Saskatchewan soils. *Soil Science Society of America Journal*, **45**, 826–828.

Karr, J.D., Showers, W.J., Jr., Gilliam, J.W. & Andres, A.S. (2001) Tracing nitrate transport and environmental impact from intensive swine farming using delta nitrogen-15. *Journal of Environmental Quality*, **30**, 1163–1175.

Karr, J.D., Showers, W.J. & Jennings, G.D. (2003) Low-level nitrate export from confined dairy farming detected in North Carolina streams using $\delta^{15}N$. *Agriculture Ecosystems and Environment*, **95**, 103–110.

Kendall, C. (1998) Tracing nitrogen sources and cycling in catchments. In: *Isotope Tracers in Catchment Hydrology* (Eds C. Kendall & J.J. McDonnell), pp. 519–576. Elsevier, Amsterdam.

Kendall, C. & Aravena, R. (2000) Nitrate isotopes in groundwater systems. *Environmental Tracers in Subsurface Hydrology* (Eds P. Cook & A. Herczeg), pp. 261–297. Kluwer Academic Publishers, Boston, MA.

Kendall, C. & Coplen, T.B. (2001) Distribution of oxygen-18 and deuterium in river waters across the United States. *Hydrological Processes*, **15**, 1363–1393.

Kendall, C. & McDonnell, J.J. (Eds) (1998) *Isotope Tracers in Catchment Hydrology*. Elsevier, Amsterdam, 839 pp.

Kendall, C., Campbell, D.H., Burns, D.A., et al. (1995a) Tracing sources of nitrate in snowmelt runoff using the oxygen and nitrogen isotopic compositions of nitrate: In: *Biogeochemistry of Seasonally Snow-covered Catchments* (Eds K. Tonnessen, M.W. Williams & M. Trantor), p. 339–347. Publication 228, International Association of Hydrological Sciences, Wallingford.

Kendall, C., Silva, S.R., Chang, C.C.Y., et al. (1995b) Use of the delta 18-O and delta 15-N of nitrate to determine sources of nitrate in early spring runoff in forested catchments. In: *Isotopes in Water Resources Management 1995*, Vol. 1, pp. 167–176. International Atomic Energy Agency, Vienna.

Kendall, C., Silva, S.R., Chang, C.Y., et al. (2000) Spatial changes in redox conditions and foodweb relations at low and high nutrient sites in the Everglades. *U.S. Geological Society Open File Report*, **00–449**, 61–63.

Kendall, C., Silva, S.R. & Kelly, V.J. (2001) Carbon and nitrogen isotopic compositions of particulate organic matter in four large river systems across the United States. *Hydrological Processes*, **15**, 1301–1346.

Kiga, T., Watanabe, S., Yoshikawa, K., et al. (2000) Evaluation of NOx Formation in Pulverized Coal Firing by Use of Nitrogen Isotope Ratios. *ASME 2000 International Joint Power Generation Conference*, Miami Beach, Florida. American Society of mechanical Engineers.

Koba, K., Tokuchi, N., Wada, E., et al. (1997) Intermittent denitrification: the application of a 15N natural abundance method to a forested ecosystem. *Geochimica et Cosmochimica Acta*, **61**, 5043–5050.

Koerner, W., Dambrine, E., Dupouey, J.L. & Benoit, M. (1999) Delta N-15 of forest soil and understorey vegetation reflect the former agricultural land use, *Oecologia*, **121**, 421–425.

Kohl, D.H., Shearer, G.B. & Commoner, B. (1971) Fertilizer nitrogen: contribution to nitrate in surface water in a corn belt watershed. *Science*, **174**, 1331–1334.

Kohl, D.H., Commoner, B. & Shearer, G.B. (1972) Use of variations in natural nitrogen isotope abundance for environmental studies – a questionable approach. *Science*, **177**, 453–456.

Komor, S.C. (1997) Boron contents and isotopic compositions of hog manure, selected fertilizers, and water in Minnesota. *Journal of Environmental Quality*, **26**, 1212–1222.

Komor, S.C. & Anderson, H.W. (1993) Nitrogen isotopes as indicators of nitrate sources in Minnesota sand-plain aquifers. *Ground Water*, **31**(2), 260–270.

Könneke, M., Bernhard, A.E., De la Torre, J., et al. (2005) Isolation of an autotrophic ammonia-oxidizing marine archaeon. *Nature*, **437**, 453–456.

Knapp, A.N., Sigman, D.M. & Lipschultz, F. (2005) N isotopic composition of dissolved organic nitrogen and nitrate at the Bermuda Atlantic Time-series Study site. *Global Biogeochemistry Cycles*, **19**, GB1018, doi:10.1029/2004GB002320.

Kratzer, C.R., Dileanis, P.D., Zamora, C., et al. (2004) *Sources and Transport of Nutrients, Organic Carbon, and Chlorophyll-a in the San Joaquin River upstream of Vernalis, California, during Summer and Fall, 2000 and 2001.* WRI 03-4127, U.S. Geological Survey. Online: http://water.usgs.gov/pubs/wri/wri034127/.

Kreitler, C.W. (1975) *Determining the Source of Nitrate in Groundwater by Nitrogen Isotope Studies.* Report of Investigation 83, Bureau of Economic Geology, University of Texas at Austin, Austin, TX, pp. 57.

Kreitler, C.W. (1979) Nitrogen-isotope ratio studies of soils and groundwater nitrate from alluvial fan aquifers in Texas. *Journal of Hydrology*, **42**, 147–170.

Kreitler, C.W. & Browning, L.A. (1983) Nitrogen-isotope analysis of groundwater nitrate in carbonate aquifers: natural sources versus human pollution. *Journal of Hydrology*, **61**, 285–301.

Kroeger, K., Cole, M.L., York, J. & Valiela, I. (2006) Nitrogen loads to estuaries from waste water plumes: Modeling and isotopic approaches. *Ground Water*, **44**, 188–200.

Krouse, H.R. & Mayer, B. (2000) Sulphur and oxygen isotopes in sulphate. In: *Environmental Tracers in Subsurface Hydrology* (Eds P. Cook & A. Herczeg), pp. 195–231. Kluwer Academic Publishers, Boston, MA.

Kumar, S., Nicholas, D.J.D. & Williams, E.H. (1983) Definitive ^{15}N NMR evidence that water serves as a source of 'O' during nitrite oxidation by Nitrobacter agilis. *FEBS Letters*, **152**, 71–74.

Kuypers, M.M.M., Sliekers A.O., Lavik G., et al. (2003) Anaerobic ammonium oxidation by anammox bacteria in the Black Sea. *Nature*, **422**, 608–610.

Kuypers, M.M.M., Lavik G., Woebken D., et al. (2005) Massive nitrogen loss from the Benguela upwelling system through anaerobic ammonium oxidation. *Proceedings of the National Academy of Sciences of the United States of America*, **102**(18), 6478–6483.

Lee, C.C.W., Savarino, J., Cachier, H. & Thiemens, M.H. (2002) Sulfur (^{32}S, ^{33}S, ^{34}S, ^{36}S) and oxygen (^{16}O, ^{17}O, ^{18}O) isotopic ratios of primary sulfate produced from combustion processes, *Tellus, Series B*, **54**, 193.

Lehmann, M.F., Reichert, P., Bernasconi, S.M., et al. (2003) Modeling nitrogen and oxygen isotope fractionation during denitrification in a lacustrine redox-transition zone. *Geochimica et Cosmochimica Acta*, **67**, 2529–2542.

Lehmann, M.F., Sigman, D.M. & Berelson, W.M. (2004) Coupling the 15N/14N and ^{18}O/^{16}O of nitrate as a constraint on benthic nitrogen cycling. *Marine Chemistry*, **88**, 1–20.

Lehmann, C.M.B., Bowersox, V.C. & Larson, S.M. (2005) Spatial and temporal trends of precipitation chemistry in the United States, 1985–2002. *Environmental Pollution*, **135**(3), 347–361.

Létolle, R. (1980) Nitrogen-15 in the natural environment. In: *Handbook of Environmental Isotope Geochemistry*, Vol. 1 (Eds P. Fritz & J.C. Fontes), pp. 407–433. Elsevier, Amsterdam.

Lyons, J.R. (2001) Transfer of mass-independent fractionation on ozone to other oxygen-containing molecules in the atmosphere. *Geophysical Research Letters*, **28**, 3231–3234.

Ma, H. & Aelion, C.M. (2005) Ammonium production during microbial nitrate removal in soil microcosms from a developing marsh estuary. *Soil Biology and Biochemistry*, **37**, 1869–1878.

Macko, S.A. & Estep, M.L.F. (1984) Microbial alteration of stable nitrogen and carbon isotopic compositions of organic matter. *Organic Chemistry*, **6**, 787–790.

Macko, S.A. & Ostrom, N.E. (1994) Pollution studies using nitrogen isotopes. In: *Stable Isotopes in Ecology and Environmental Science* (Eds K. Lajtha & R.M. Michener), pp. 45–62. Blackwell Scientific Publishers, Oxford.

Mariotti, A. & Letolle, R. (1977) Application de l'etude isotopique de l'azote en hydrologie et en hydrogeologie–analyse des resultats obtenus sur un exemple precis: Le Bassin de Melarchez (Seine-et-Marne, France). *Journal of Hydrology*, **33**, 157–172.

Mariotti, A., Pierre, D., Vedy, J.C., et al. (1980) The abundance of natural nitrogen 15 in the organic matter of soils along an altitudinal gradient (Chablais, Haute Savoie, France). *Catena*, **7**, 293–300.

Mariotti, A., Germon, J.C., Hubert, P., et al. (1981) Experimental determination of nitrogen kinetic isotope fractionation: some principles; illustration for the denitrification and nitrification processes. *Plant Soil*, **62**, 413–430.

Mariotti, A., Germon, J.C. & Leclerc, A. (1982) Nitrogen isotope fractionation associated with the NO_2^- → N_2O step of denitrification in soils. *Canadian Journal of Soil Science*, **62**, 227–241.

Mariotti, A., Landreau, A. & Simon, B. (1988) 15N isotope biogeochemistry and natural denitrification process in groundwater: application to the chalk aquifer of northern France. *Geochimica et Cosmochimica Acta*, **52**, 1869–1878.

Mauersberger, K., Krankowsky, D. & Janssen, C. (2003) Oxygen isotope processes and transfer reactions. *Space Science Reviews*, **106**, 265–279.

Mayer, B., Bollwerk, S.M., Mansfeldt, T., et al. (2001) The oxygen isotopic composition of nitrate generated by nitrification in acid forest floors. *Geochimica et Cosmochimica Acta*, **65**(16), 2743–2756.

Mayer, B., Boyer, E.W., Goodale, C., et al. (2002) Sources of nitrate in rivers draining sixteen watersheds in the northeastern US: Isotopic constraints. *Biogeochemistry*, **57**, 171–197.

McClain, M.E., Boyer, E.W., Dent, C.L., et al. (2003) Biogeochemical hot spots and hot moments at the interface of terrestrial and aquatic ecosystems. *Ecosystems*, **6**, 301–312.

McClelland, J.W. & Valiela, I. (1998) Linking nitrogen in estuarine producers to land-derived sources. *Limnology and Oceanography*, **43**, 577–585.

McClelland, J.W., Valiela, I. & Michener, R.H. (1997) Nitrogen-stable isotope signatures in estuarine food webs : A record of increasing urbanisation in coastal watershed. *Limnology and Oceanography*, **42**, 930–937.

McCready R.G.L., Gould W.D. & Barendregt R.W. (1983) Nitrogen isotope fractionation during the reduction of NO_3^- to NH_4^+ by *Desulfovibrio* sp. *Canadian Journal of Microbiology*, **29**, 231–234.

McIlvin, M. & Altabet, M.A. (2003) Nitrogen and oxygen isotopic detection of nitrate in seawater by chemical conversion of nitrate to nitrous oxide. *Eos (Transactions of the American Geophysical Union)*, **84**, Abstract B31D-0339.

McLaughlin, K., Silva, S., Kendall, C., et al. (2004) A precise method for the analysis of $\delta^{18}O$ of dissolved inorganic phosphate in seawater. *Limnology and Oceanography, Methods*, **2**, 202–212.

McLaughlin, K., Paytan, A., Kendall, C. & Silva, S. (2006a) Oxygen isotopes of phosphatic compounds – Application for marine particulate matter, sediments and soils. *Marine Chemistry*, **98**, 148–155.

McLaughlin, K., Kendall, C., Silva, S. & Paytan, A. (2006b) Phosphate oxygen isotope ratios as a tracer for sources and cycling of phosphate in North San Francisco Bay. *Journal of Geophysical Research*, **111**, G03003, doi: 10.1029/2005JG000079.

McMahon, P.B. & Böhlke, J.K. (1996) Denitrification and mixing in a stream-aquifer system: effects on nitrate loading to surface water. *Journal of Hydrology*, **186**, 105–128.

McMahon, P.B. & Böhlke, J.K. (2006) Regional patterns in the isotopic composition of natural and anthropogenic nitrate in groundwater, High Plains, USA. *Environmental Science and Technology*, **40**(9), 2965–2970.

Mengis, M., Schiff, S.L., Harris, M., et al. (1999) Multiple geochemical and isotopic approaches for assessing ground water NO3- elimination in a riparian zone. *Ground Water*, **37**, 448–457.

Michalski, G., Savarino, J., Bohlke, J.K. & Thiemans, M. (2002) Determination of the total oxygen isotopic composition of nitrate and the calibration of a $\Delta^{17}O$ nitrate reference material. *Analytical Chemistry*, **74**, 4989–4993.

Michalski, G., Scott, Z., Kabiling, M. & Thiemens, M.H. (2003) First measurements and modeling of $\Delta^{17}O$ in atmospheric nitrate. *Geophysical Research Letters*, **30**(16), doi:10.1029GL017015.

Michalski, G., Meixner, T., Fenn, M., et al. (2004) Tracing atmospheric nitrate deposition in a complex semiarid ecosystem using $\delta^{17}O$. *Environmental Science and Technology*, **38**(7), 2175–2181.

Michalski, G., Bockheim, J.G., Kendall, C. & Thiemens, M. (2005) Isotopic composition of Antarctic Dry Valley nitrate: Implications for NOx sources and cycling in Antarctica. *Journal of Geophysical Research Letters*, **32**, L13817, 4 p. doi:10.1029/2004GL022121.

Michel, R.L., Campbell, D., Clow, D. & Turk, J.T. (2000) Timescales for migration of atmospherically derived sulphate through an alpine/subalpine watershed, Loch Vale, Colorado. *Water Resources Research*, **36**(1), 27–36.

Minagawa, M. (1992) Reconstruction of human diet from $\delta^{13}C$ and $\delta^{15}N$ in contemporary Japanese hair: a stochastic method for estimating multi-source contribution by double isotopic tracers. *Applied Geochemistry*, **7**, 145–158.

Montoya, J.P., Korrigan, S.G. & McCarthy J.J. (1991) Rapid, storm-induced changes in the natural abundance of ^{15}N in a planktonic ecosystem, Chesepeake Bay, USA. *Geochimica et Cosmochimica Acta*, **55**(12), 3627–3638.

Moore, H. (1977). The isotopic composition of ammonia, nitrogen dioxide, and nitrate in the atmosphere. *Atmospheric Environment*, **11**, 1239–1243.

Mosier, A.R. & Schimel, D.S. (1993) Nitrification and denitrification. In: *Nitrogen Isotope Techniques* (Eds R. Knowles & T.H. Blackburn), pp. 181–208. Academic Press, New York.

Nadelhoffer, K.J. & Fry, B. (1988) Controls on natural nitrogen-15 and carbon-13 abundances in forest soil organic matter. *Soil Science Society of America Journal*, **52**, 1633–1640.

Nadelhoffer, K.J. & Fry, B. (1994) Nitrogen isotope studies in forest ecosystems. In: *Stable Isotopes in Ecology and Environmental Science* (Eds K. Lajtha & R.M. Michener), pp. 22–44. Blackwell Scientific Publishers, Oxford.

Needoba, J.A., Sigman, D.M. & Harrison, P.J. (2004) The mechanism of isotope fractionation during algal nitrate assimilation as illuminated by the 15N/14N of intracellular nitrate. *Journal of Phycology*, **40**, 517–522.

Négrel, P. (1999) Geochemical study of a granitic area – The Margeride Mountains, France: chemical element behaviour and $^{87}Sr/^{86}Sr$ constraints. *Aquatic Geochemistry*, **5**, 125–165.

Négrel, P. & Deschamps, P. (1996) Natural and anthropogenic budgets of a small watershed in the Massif Central (France): chemical and strontium isotopic characterization of water and sediments. *Aquatic Geochemistry*, **2**, 1–27.

Nielsen, L.P. (1992) Denitrification in sediment determined from nitrogen isotope pairing. *FEMS Microbiology Ecology*, **86**, 357–362.

Nilles, M.A. & Conley, B.E. (2001) Changes in the chemistry of precipitation in the United States, 1981–1998. *Water, Air and Soil Pollution*, **130**(1–4), 409–414.

Novak, M., Buzek, F., Harrison, A.F., et al. (2003) Similarity between C,N and S stable isotope profiles in European spruce forest soils: implications for the use of $\delta^{34}S$ as tracer. *Applied Geochemistry*, **18**(5), 765–779.

Nriagu, J.O. & Coker, R.D. (1978) Isotopic composition of sulphur in atmospheric precipitation around Sudbury Ontario, *Nature*, **274**, 883–885.

Ohte, N., Sebestyen, S.D., Shanley, J.B., et al. (2004) Tracing sources of nitrate in snowmelt runoff using a high-resolution isotopic technique, *Geophysical Research Letters*, **31**, L21506.

Olleros, T. (1983) *Kinetic isotope effects of the enzymatic splitting of arginine and nitrate; a contribution to the explanation of the reaction mechanisms.* Dissertation, Technical University Munchen-Weihenstephan, 158 pp.

Ostrom, N.E., Knoke, K.E., Hedin, L.O., et al. (1998) Temporal trends in nitrogen isotope values of nitrate leaching from an agricultural soil. *Chemical Geology*, **146**, 219–227.

Otero, N. (2004) *Dades isotopiques ($\delta^{34}S$, $\delta^{18}O$) i analisi estadistica aplicades a l'estudi contaminacio a les aigues superficials: el cas del Riu Llobregat.* PhD dissertation, University of Barcelona, 283 pp. (mostly English, part Catalonian).

Panno, S.V., Hackley, K.C., Kelly, W.R. & Hwang, H.H. (2006) Isotopic evidence of nitrate sources and denitrification in the Mississippi River, Illinois. *Journal of Environmental Quality*, **35**(2), 495–504.

Pardo, L.H., Kendall, C., PettRidge, J. & Chang, C.C.Y. (2004) Evaluating the source of streamwater nitrate using d15N and d18O in nitrate in two watersheds in New Hampshire, USA. *Hydrological Processes*, **18**, 2699–2712.

Parker, S.R., Poulson, S.R., Gammons, C.H. & DeGrandpre, M.D. (2005) Biogeochemical controls on diel cycling of stable isotopes of dissolved O2 and dissolved inorganic carbon in the Big Hole River, Montana. *Environmental Science and Technology*, **39**, 7134–7140.

Pearson, J., Wells, D.M., Seller, K.J., et al. (2000) Traffic exposure increases natural 15N and heavy metal concentrations in mosses. *New Phytologist*, **147**(2), 317–326.

Pennock, J., Velinsky, D., Ludlam, J., et al. (1996) Isotopic fractionation of ammonium and nitrate during uptake by *Skeletonema costatum*: Implications for $\delta^{15}N$ dynamics under bloom conditions. *Limnology and Oceanography*, **41**(3), 451–459.

Peterson, B.J. & Fry, B. (1987) Stable isotopes in ecosystem studies. *Annual Review of Ecology and Systematics*, **18**, 293–320.

Phillips, D.L. & Gregg, J.W. (2001) Uncertainty in source partitioning using stable isotopes. *Oecologia*, **127**, 171–179.

Phillips, D.L. & Gregg, J.W. (2003) Source partitioning using stable isotopes: coping with too many sources. *Oecologia*, **136**, 261–269.

Phillips, D.L. & Koch, P.L. (2002) Incorporating concentration dependence in stable isotope mixing models. *Oecologia*, **130**, 114–125.

Piatek, K.B., Mitchell, M.J., Silva, S.R. & Kendall, C. (2005) Sources of nitrate in Adirondack surface waters during dissimilar snowmelt events. *Water, Air and Soil Pollution*, 165, 13–35.

Plummer, L.N., Parkhurst, D.L. & Thorstenson, D.C. (1983) Development of reaction models for ground-water systems. *Geochimica et Cosmochimica Acta*, **47**, 665–686.

Plummer, L.N., Prestemon, E.C. & Parkhurst, D.L. (1991) An interactive code (NETPATH) for modelling net geochemical reactions along a flow path. *U.S. Geological Survey Water-Research Investigation Report*, **91–4078**, 227 pp.

Plummer, L.N., Michel, R.L., Thurman, E.M. & Glynn, P.D. (1993) Environmental tracers for age dating young ground water. In: *Regional Ground-Water Quality* (Ed. W.M. Alley), pp. 255–294. Van Nostrand Reinhold, New York.

Qi, H.P., Coplen T.B., Wang, Q.Z. & Wang, Y.H. (1997) Unnatural isotopic composition of lithium reagents. *Analytical Chemistry*, **69**, 4076–4078.

Revesz, K. & Böhlke, J.K. (2002) Comparison of $\delta^{18}O$ measurements in nitrate by different combustion techniques. *Analytical Chemistry*, **74**, 5410–5413.

Rock, L. & Mayer, B. (2004) Isotopic Assessment of Sources of Surface Water Nitrate within the Oldman River Basin, Southern Alberta, Canada. *Water, Air and Soil Pollution*, **4**, 545–562.

Russell, K.M., Galloway, J.N., Macko, S.A., et al. (1998) Sources of nitrogen in wet deposition to the Chesapeake Bay region. *Atmospheric Environment*, **32**(14–15), 2453–2465.

Rysgaard, S. & Glud R. (2004) Anaerobic N2 production in Arctic sea ice. *Limnology and Oceanography*, **49**(1), 86–94.

Rysgaard, S., Glud, R., Risgaard-Petersen, N. & Dalsgaard T. (2004) Denitrification and anammox activity in Arctic marine sediments. *Limnology and Oceanography*, **49**(5), 1493–1502.

Saurer, M., Cherubini, P., Ammann, M., et al. (2004) First detection of nitrogen from NO$_x$ in tree rings: a $^{15}N/^{14}N$ study near a motorway. *Atmospheric Environment*, **38**, 2779–2787.

Schleper, C, Jurgens, G. & Jonuscheit, M. (2005) Genomic studies of uncultivated archaea. *Nature Reviews of Microbiology*, **3**, 479–488.

Sebilo, M., Billen, G., Grably, M. & Mariotti, A. (2003) Isotopic composition of nitrate-nitrogen as a marker of riparian and benthic denitrification at the scale of the whole Seine River system. *Biogeochemistry*, **63**, 35–51.

Sebilo, M., Billen, G., Mayer, B., et al. (2006) Assessing nitrification and denitrification in the Seine River and Estuary using chemical and isotopic techniques. *Ecosystems*, **9**, 564–577.

Seiler, R.L. (2005) Combined use of 15N and 18O of nitrate and 11B to evaluate nitrate contamination in ground water. *Applied Geochemistry*, **20**(9), 1626–1636.

Seinfeld, J.H. & Pandis, S.N. (1998). *Atmospheric Chemistry and Physics: From Air Pollution to Climate Change*. Wiley-Interscience, Indianapolis.

Shanley, J.B., Mayer, B., Mitchell, M.J., et al. (2005) Tracing sources of streamwater sulfate during snowmelt using S and O isotope ratios of sulfate and 35S activity, *Biogeochemistry*, **76**, 161–185.

Shearer, G. & Kohl, D.H. (1988) ^{15}N method of estimating N$_2$ fixation. In: *Stable Isotopes in Ecological Research* (Eds P.W. Rundel, J.R. Ehleringer & K.A. Nagy), pp. 342–374. Springer-Verlag, New York.

Shearer, G. & Kohl, D. (1993) Natural abundance of N: fractional contribution of two sources to a common sink and use of isotope discrimination. In: *Nitrogen Isotope Techniques* (Eds R. Knowles & T.H. Blackburn), pp. 89–125. Academic Press, New York.

Shearer, G.B., Kohl, D.H. & Commoner, B. (1974) The precision of determination of the natural abundance of nitrogen-15 in soils, fertilizers, and shelf chemicals. *Soil Science*, **118**, 308–316.

Showers, W.J. & DeMaster, D. (2005) Nitrogen transport from atmospheric deposition and contaminated groundwater to surface waters on a watershed scale. *Abstracts of the American Geophysical Union, Fall Meeting*, **2005**, B54A-06.

Sickman, J.Q., Leydecker, A., Chang, C.C.Y., et al. (2003) Mechanisms underlying export of N from high-elevation catchments during seasonal transitions, *Biogeochemistry*, **64**, 1–24.

Sigman, D.M., Casciotti, K.L., Andreani, M., et al. (2001) A bacterial method for the nitrogen isotopic analysis of nitrate in seawater and freshwater. *Analytical Chemistry*, **73**, 4145–4153.

Sigman, D.M., Granger, J., DiFiore, P.J., et al. (2005) Coupled nitrogen and oxygen isotope measurements of nitrate along the eastern North Pacific margin. *Global Biogeochemical Cycles*, **19**, GB4022.

Silva, S.R., Kendall, C., Wilkison, D.H., et al. (2000) A new method for collection of nitrate from fresh water and the analysis of nitrogen and oxygen isotope ratios. *Journal of Hydrology*, **228**, 22–36.

Silva, S.R., Lee, P.B., Ebbert, R.W., et al. (2002) Forensic applications of nitrogen and oxygen isotopes of nitrate in an urban environment. *Environmental Forensics*, **3**, 125–130.

Spruill, T.B., Showers, W.J. & Howe, S.S. (2002) Application of classification-tree methods to identify nitrate sources in ground water. *Journal of Environmental Quality*, **31**, 1538–1549.

Theobald, D.M. (2004) Placing exurban land-use change in a human modification framework. *Frontiers in Ecology and the Environment*, **2**, 139–144.

Thiemens, M.H. (1999) Mass-independent isotope effects in planetary atmospheres and the early Solar system. *Science*, **283**, 341–345.

Thiemens, M.H. (2006) History and applications of mass-independent isotope effects. *Annual Reviews in Earth and Planetary Science*, **34**, 217–262.

Tobias, C.R., Anderson, I.C., Canuel, E.A. & Macko, S.A. (2001) Nitrogen cycling through a fringing marsh-aquifer ecotone. *Marine Ecology Progress Series*, **210**, 25–39.

Trimmer, M., Nedwell, D.B., Sivyer, D.B. & Malcolm, S.J. (1998) Nitrogen fluxes through the lower estuary of the river Great Ouse, England: The role of the bottome sediments. *Marine Ecology Progress Series*, **163**, 109–124.

Trust, B.A. & Fry, B. (1992) Stable sulphur isotopes in plants: a review. *Plant, Cell and Environment*, **15**, 1105–1110.

Tucker, J., Sheats, N., Giblin, A.E., et al. (1999) Using stable isotopes to trace sewage-derived material through Boston Harbor and Massachusetts bay. *Marine Environmental Research*, **48**, 353–375.

Van Dover, C.L., Grassle, J.F., Fry, B., et al. (1992) Stable isotope evidence for entry of sewage-derived organic material into a deep-sea food web. *Nature*, **360**, 153–155.

Vengosh, A. (1998) Boron isotopes and groundwater pollution. *Water Environment News*, **3**, 15–16.

Vengosh, A., Heumann, K.G., Juraske, S. & Kasher, R. (1994) Boron isotope application for tracing sources of contamination in groundwater. *Environmental Science and Technology*, **28**, 1968–1974.

Vengosh, A., Barth, S., Heumann, K.G. & Eisenhut, S. (1999) Boron isotopic composition of freshwater lakes from and possible contamination sources. *Acta Hydrochimica et Hydrobiologica*, **27**, 416–421.

Vitoria, L. (2004) *Multi-isotope approach ($\delta^{15}N, \delta^{34}S, \delta^{13}C, \delta^{18}O, \delta D, and {}^{87}Sr/{}^{86}Sr$) of nitrate contaminated groundwaters by agricultural and stockbreeder activities*. PhD thesis, University of Barcelona, 188 pp. (Part English, part Spanish, part Catalonian.)

Vitoria, L., Otero, N., Soler, A. & Canals, A. (2004) Fertilizer characterization: Isotopic data (N, S, O, C, and Sr). *Environmental Science and Technology*, **38** (12), 3254–3262.

Voerkelius, S. & Schmidt, H.L. (1990) Natural oxygen and nitrogen isotope abundance of compounds involved in denitrification. *Mitteilungen der Deutschen Bodenkundlichen Gesselschaft*, **60**, 364–366.

Vogel, J.C., Talma, A.S. & Heaton, T.H.E. (1981) Gaseous nitrogen as evidence for denitrification in groundwater. *Journal of Hydrology*, **50**, 191–200.

Wada, E., Shibata, R. & Torii, T. (1981) Nitrogen-15 abundance in Antartica: Origin of soil nitrogen and ecological implications. *Nature*, **292**, 327–329.

Wadleigh, M.A. (2003) Lichens and atmospheric sulphur: what stable isotopes reveal. *Environmental Pollution*, **126**, 345–351.

Wadleigh, M.A. & Blake, D.M. (1999) Tracing sources of atmospheric sulphur using epiphytic lichens. *Environmental Pollution*, **106**, 265–271.

Wadleigh, M.A., Schwarcz, H.P. & Kramer, J.R. (1996) Isotopic evidence for the origin of sulphate in coastal rain. *Tellus*, **48B**, 44–59.

Wankel, S.D. (2006) *Nitrogen sources and cycling in coastal ecosystems: insights from a nitrogen and oxygen stable isotope approach*. Unpublished PhD dissertation, Stanford University, 251 pp.

Wankel, S.D., Kendall, C., Francis, C.A. & Paytan, A. (2006) Nitrogen sources and cycling in the San Francisco Bay Estuary: A nitrate dual isotope approach. *Limnology and Oceanography*, **51**, 1654–1664.

Wankel, S.D., Kendall, C., Pennington, J.T., et al. (2007) Nitrification in the euphotic zone as evidenced by nitrate dual isotopic composition: Observations from Monterey Bay. *Global Biogeochemical Cycles*, **21**, xxxxxx, doi: 10.1029/2006GB002723.

Waser, N.A.D., Harrison, P.J., Nielsen, B. & Calvert, H.E. (1998) Nitrogen isotope fractionation during the uptake and assimilation of nitrate, nitrite, ammonium and urea by a marine diatom. *Limnology and Oceanography*, **43**(2), 215–224.

Wassenaar, L.I. (1995) Evaluation of the Origin and Fate of Nitrate in the Abbotsford Aquifer Using the Isotopes of ^{15}N and ^{18}O in NO_3^-. *Applied Geochemistry*, **10**, 391–405.

Wassenaar, L.I., Hendry, M.J. & Harrington, N. (2006) Decadal geochemical and isotopic trends for nitrate in a transboundary aquifer and implications for agricultural beneficial management practices. *Environmental Science and Technology*, **40**, 4626–4632.

Widory, D., Kloppmann, W., Chery, L., et al. (2004) Nitrate in groundwater: an isotopic multi-tracer approach. *Journal of Contaminant Hydrology*, **72**, 165–188.

Widory, D., Petelet-Giraud, E., Negrel, P. & Ladouche, B. (2005) Tracking the sources of nitrate in groundwater using coupled nitrogen and boron isotopes: A synthesis. *Environmental Science and Technology*, **39**(2), 539–548.

Widory, D. (2007) Nitrogen isotopes: tracers of origin and processes affecting PM_{10} in the atmosphere of Paris. *Atmospheric Environment*, **41**, 2382–2390.

Williard, K.W.J. (1999) *Factors affecting stream nitrogen concentrations from mid-Appalachian forested watersheds*. PhD dissertation, Pennsylvania State University.

Wilson, G.B., Andrews, J.N. & Bath, A.H. (1990) Dissolved gas evidence for denitrification in the Lincolnshire Limestone groundwaters, eastern England. *Journal of Hydrology*, **113**, 51–60.

Wolterink, T.J. (1979) *Identifying Sources of Subsurface Nitrate Pollution with Stable Nitrogen Isotopes*. EPA-600/4-79-050, U.S. Environmental Protection Agency, Washington, DC, pp 150.

Xiao, H.Y. & Liu, C.Q. (2002) Sources of nitrogen and sulfur in wet deposition at Guiyang, southwest China. *Atmospheric Environment*, **36**(33), 5121–5130.

Yeatman, S.G., Spokes, L.J., Dennis, P.F. & Jickells, T.D. (2001) Comparisons of aerosol nitrogen isotopic composition at two polluted coastal sites. *Atmospheric Environment*, **35**(7), 1307–1320.

Young, M., McLaughlin, K., Donald, E., et al. (2006) The oxygen isotopic composition of phosphate of various anthropogenic sources: fertilizers, manure, detergent, and sewage treatment products. *Eos (Transactions of the American Geophysical Union)*, **87**(36), Abstract OS45J-03.

Zielinski, R.A., Asher-Bolinder., S., Meier, A.L., et al. (1997) Natural or fertilizer-derived uranium in irrigation drainage: a case study in southeastern Colorado, U.S.A., *Applied Geochemistry*, **12**(1), 9–21.

Zielinski, R.A., Simmons, K.R. & Orem, W.H. (2000) Use of ^{234}U and ^{238}U isotopes to identify fertilizer-derived uranium in the Florida Everglades. *Applied Geochemistry*, **15**(3), 369–383.

CHAPTER 13

Modeling the dynamics of stable-isotope ratios for ecosystem biogeochemistry

WILLIAM S. CURRIE

Introduction

Models incorporating stable isotopes have been used for a number of purposes in ecosystem science and biogeochemistry. The primary motivation for their development and application has been to interpret the results of field and laboratory studies. In order to interpret the temporal or spatial patterns observed in isotopic ratios in field samples, investigators require mathematical representations of fluxes among or mixing among conceptual pools. Additional uses of these models include the planning of isotope experiments and the generation of hypotheses (Schimel 1993; Currie & Nadelhoffer 1999). A less common but promising application is the use of isotopically tested models to simulate other quantities of interest, such as energetic fluxes, interactions among element cycles, and ecosystem responses to global change (Tietema & Wessel 1992; van Dam & van Breemen 1995; Currie 2003; Currie et al. 2004).

Isotopically explicit models cover a broad range of complexity and a broad spectrum of mathematical approaches. At the simplest end of the spectrum lie mixing models and isotope dilution or "pool dilution" models having one or two conceptual pools (Davidson et al. 1991). These models can be expressed as algebraic equations and are typically solved analytically. The fundamental principles behind the pool-dilution equations were articulated over 50 years ago (Kirkham & Bartholomew 1954). More complex models developed in the past two decades rely on many of the same assumptions of simple pool dilution models, such as that each conceptual pool is well-mixed and isotopically homogeneous and that inputs to a pool can alter its isotopic ratio but outputs from a pool cannot. It has been conceptually straightforward to incorporate these principles into complex process models with numerous pools and fluxes, nonlinear interactions, tracer recycling, and nonsteady-state dynamics. However, developing the mathematics and computer programs to do so has been an intellectual and computational challenge. One aspect of the challenge is that enhanced physical detail, not originally needed in the conceptual model to describe bulk elemental flows, may need to be added to a model to provide any possibility of simulating isotopic mixing. When added to the algorithms needed to simulate bulk elemental flows, the computer coding algorithms required to carry out isotopic mixing computations can be substantially more elaborate. They also require more sophisticated testing and need to work properly to a higher number of significant digits (e.g. 6 to 8) with small numbers close to zero. Another challenge arises in the choice of a model time step. A relatively coarse time step (1 month, for example) may be suitable for modeling bulk elemental movements and transformations. A much finer time step may be needed to simulate isotopic mixing in pools that turn over rapidly or that attain values close to zero repeatedly on finer time scales.

Analytical solutions, unfortunately, are obtainable in only a minority of the interesting and important problems in the study of nutrient and carbon cycling (e.g. Riha et al. 1986). This limitation is inherent in applied mathematics more broadly. An illustrative example is the so-called "three-body problem" in physics. In this well-known problem, gravitational interactions can be written as a set of equations for any number of interacting celestial bodies (planets, moons, satellites, or stars), but analytical solutions can be found only in cases where no more than two bodies interact – one planet and the Sun, for example. When a third body is introduced (a moon, satellite, or another planet), the equations of motion cannot be solved analytically. Numerical methods must be used and computers have made such problems tractable. In an analogous way, when isotopes are allowed to recycle among ecosystem pools and when there are more than two pools in the conceptual model, we essentially have the three-body problem and we require numerical or algorithmic approaches.

Intermediate in the spectrum of complexity among stable-isotope models are those expressed as sets of differential equations. Under highly restricted assumptions (such as steady-state and no tracer recycling), the simplest of these can be solved analytically. In other cases, numerical integration techniques are used. At the most complicated end of the spectrum lie dynamic biogeochemical models expressed as computer algorithms. Driven by the increasingly widespread laboratory analysis of stable-isotope ratios in field studies, several complex process models have now been developed that include stable-isotopic ratios in ecosystem pools as state variables.

This chapter reviews the present state and recent advances in the mathematical modeling of stable isotopes for research in ecosystem biogeochemistry. I draw primarily on models that explicitly include the isotopes of nitrogen (N) and carbon (C). However, many of the principles and techniques described here can potentially be extended to the modeling of stable isotopes generally. By "model" I include all manner of mathematical formulations from the simplest algebraic equations to elaborate computer models. Virtually all interpretations of stable-isotope data require equations that are derived from a conceptual framework of pools or end-members together with fluxes or mixing. I begin with a review of principles important in designing model–data comparisons, then describe modeling frameworks across the spectrum of mathematical approaches, and close with a discussion of qualitative behavior and uncertainty in complex models.

Conceptual and mathematical models

The conceptual model of an environmental system or subsystem stands between the natural system of interest and the mathematical model used to interpret observations (Figure 13.1). In planning an experiment or in interpreting data, an investigator first establishes a conceptual model that simplifies nature by introducing some assumptions and approximations. In interpreting stable-isotope data, investigators translate their conceptual model into sets of equations that can be analyzed analytically or into computer algorithms that can be run and studied numerically.

Inconsistencies are likely to exist between both (i) the natural system and the conceptual model and (ii) the conceptual and mathematical models. A key part of the modeling activity is to understand and to minimize such inconsistencies and to understand the trade-offs to be made. While important consistencies can be achieved, some amount of inconsistency seems unavoidable. Investigators judiciously choose which aspects of a natural system to represent best in a conceptual model while allowing known fictions to exist in other aspects. Hobbie et al. (1999) have pointed out that many conceptual fluxes in biogeochemistry are, in reality, combinations of several chemical and physical processes. For example the uptake of NH_4 by plants involves the animation of glutamate to form glutamine, which is then used to animate other compounds in plants. The question to ask in assessing a conceptual model is not whether it is completely realistic (meaning it faithfully represents the natural system in every detail), but rather whether a model is reasonably faithful to the present understanding of biogeochemical processes while being **useful** in a particular context for advancing the goals of the research. To be useful, a model should have enough realism and complexity to capture the key interactions to be addressed in a research project, but should eschew gratuitous detail.

In choosing trade-offs to make in model construction, one should consider how the model will be used in conjunction with field or laboratory experiments. Models can make strong contributions to empirical research when they are tailored to the design of field experiments. For example, imagine an experiment in which a ^{15}N tracer was to be added to a soil surface, then at a later time mineral soil samples were taken and $^{15}N/^{14}N$ ratios measured in sorted fractions. The conceptual model of the pools and fluxes of N should guide the construction of both the field experiment and the model (Figure 13.1). Close consistency in the definitions of compartments, or pools, is key (Schimel 1993). If mineral soil samples from the field study were to have living fine roots separated out for analysis but dead fine roots composited with mineral soil organic matter (SOM) for analysis, the model should be

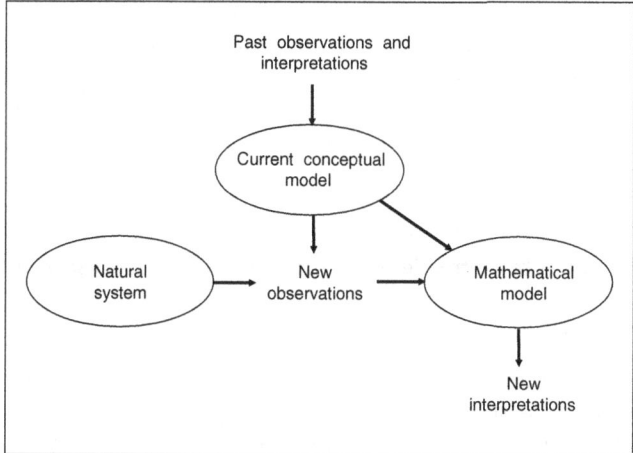

Figure 13.1 A conceptual model is used both to guide observations and to guide the construction of a mathematical model, which is then used to interpret observations.

designed to allow a direct comparison between model results and field data. If, in this example, the model pooled living and dead fine roots together, then a direct comparison of $^{15}N/^{14}N$ ratios between the modeled and the field-sampled pools would be impossible. In this example the mismatch is obvious; in other cases they are more subtle. To take this example further, investigators studying the forest soil would need to consider whether microbial biomass will be implicitly included in soil pools or explicitly separated; whether detrital layers in the forest floor will be sorted in the field study according to physical layering or to a presumed degree of decomposition, and so on. Careful consideration should be given to designing a conceptual model that facilitates both (i) addressing the objectives of the field study and (ii) guidance in the construction of a mathematical model that will be useful and testable using the field data to be obtained.

Some assumptions are common to all biogeochemical models because they are part of the systems paradigm that is fundamental to the field. One such approximation is that each pool in a conceptual model represents a homogeneous, well-mixed pool with a single overall rate of export or turnover. In most cases this approximation is probably false. Biogeochemical fluxes and transformations are dynamic at multiple scales and "hot spots" of activity exist potentially even at fine scales. Within organisms and tissues, isotopes may not penetrate evenly and an organism or a tissue may not behave as a single well-mixed pool (Conover & Francis 1973; Schimel 1993). Similarly, the material we define as mineral soil organic matter, whether we conceive of it as one pool with a single turnover rate (Aber & Federer 1992), two pools with distinct turnover rates (Clark 1977; Currie et al. 2004), or three pools with distinct turnover rates (Parton et al. 1987), may in reality lie closer to a continuum of materials with a continuum of turnover rates (Agren & Bosatta 1996). Future modeling efforts could address means to incorporate a continuum of pool turnover rates into an isotope mixing model, but to date, models have incorporated discrete pools with discrete rates of turnover.

Designing consistent model–data linkages and comparisons

An additional dimension of observation

A ratio of stable isotopes such as $^{15}N/^{14}N$, which can be measured with high precision using mass spectroscopy, offers intrinsic benefits as a model state variable. Isotopic ratios are sensitive to gross fluxes of the element into or out of an ecosystem pool, even where net fluxes may be small, may be zero, or may flow in the opposite direction to a gross flux of interest (Schimel 1993). A simple model example using steady-state N fluxes and pool sizes in two conceptual pools (shown in Figures 13.2 & 13.3) illustrates this principle.

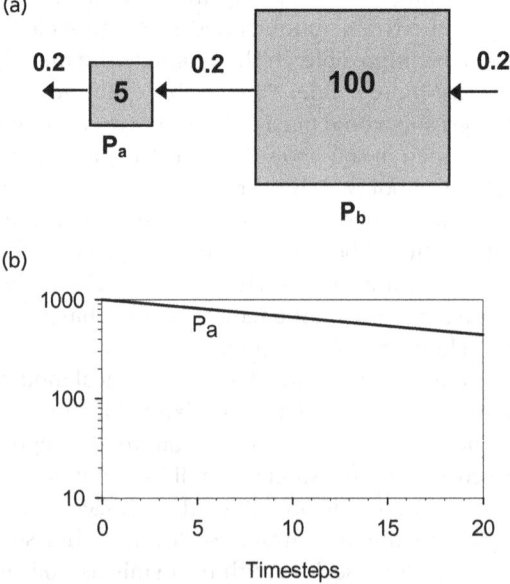

Figure 13.2 Results of a simple model illustration: (a) pools and fluxes of N in g m^{-2} and g m^{-2} time step^{-1}, respectively, for steady-state N fluxes and pool sizes shown; (b) the value of $\delta^{15}N$ in pool P_a over time, given a 1000‰ ^{15}N tracer introduced into pool P_a at time zero. Here, $\delta^{15}N$ begins at zero and remains at zero in pool P_b. (Contrast with Figure 13.3.) Note logarithmic scale on the $\delta^{15}N$ axis.

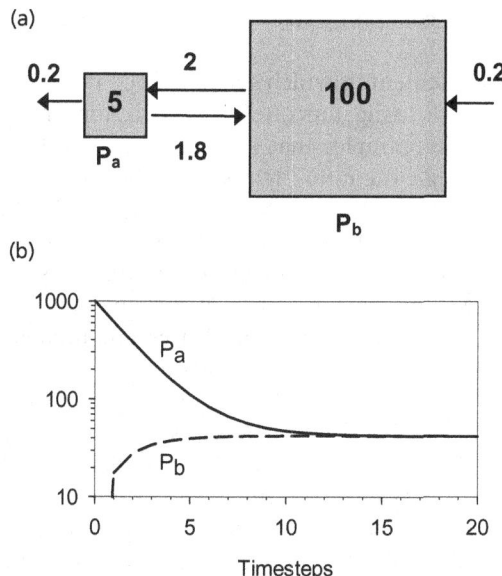

Figure 13.3 Results of a simple model illustration: (a) shows pools and fluxes of N in g m⁻² and g m⁻² time step⁻¹, respectively, for steady-state N fluxes and pool sizes shown; (b) the values of $\delta^{15}N$ in pool P_a and P_b over time, given a 1000‰ ^{15}N tracer introduced into pool P_a at time zero. Here, $\delta^{15}N$ begins at zero in pool P_b and rises to 40‰ within five time steps. Note that the net flux of N from P_b to P_a is identical to that in Figure 13.2, while the gross flux from P_b to P_a is ten times greater. Axes and scales are as in Figure 13.2.

In the model results shown in both of these figures, the net flux of N from pool P_b to P_a is 0.2 g N m⁻² timestep⁻¹, and a ^{15}N tracer with $\delta^{15}N = 1000‰$ is introduced to pool P_a at time zero. Note that a field measurement of net N fluxes, lacking isotopic information, would be unable to distinguish between the two cases depicted (Figures 13.2 & 13.3). The ^{15}N tracer, in contrast, when introduced to pool P_a allows us to distinguish between these two cases because in the model results we see the ^{15}N appear in pool P_b quite rapidly, even though the **net** flux of N is from pool P_b to P_a, or opposite the direction of flow of the ^{15}N.

Investigators do not analyze or model an isotopic ratio in isolation. Quite the contrary: the ratio $^{15}N/^{14}N$, for example, is viewed as an additional state variable that is measured and modeled together with an existing array of state variables that includes measures of molecular forms of N or total N. It is of limited value to compare isotope ratios between model and data if modeled pool sizes or fluxes of N, C, or organic matter are widely unrealistic, or if the degree of realism in these is inadequately known. Furthermore, such comparisons would miss the value of using isotopic ratios because they would replace one state variable (elemental pool size) with another (isotopic ratio in a pool), rather than combining state variables to gain an increased dimension of comparison.

Many pools of an element in which stable-isotope ratios are measured are operationally defined. Investigators often analyze isotopic ratios and elemental concentrations in subsamples that have undergone other extractions or separations. For example, the ratio $^{15}N/^{14}N$ might be measured in extractable NH_4 from soils (together with NH_4–N concentrations), in extractable NO_3 from soils (together with NO_3–N concentrations), and in the inextractable soil residue (together with total N concentrations in the residue). As another example, the $^{15}N/^{14}N$ ratio (together with N concentrations) might be measured in subsamples of fine-root litter, foliar litter, and unidentifiable detritus each sorted by hand from a soil organic horizon.

When a ^{15}N tracer is added and followed over time, this can be viewed as labeling a temporal cohort of N inputs to the system. Investigators can then follow that cohort of N inputs as it is partitioned initially and then as it redistributes within the ecosystem over time. For example, Nadelhoffer et al. (2004) applied $^{15}NH_4$ and $^{15}NO_3$ tracers to forest floors at the Harvard Forest, Massachusetts over a 16-month period in 1990–1991 and sampled numerous vegetation and soil pools over the subsequent 8-year period. Over this time period, differences in values of $\delta^{15}N$ from natural-abundance values in vegetation and soil pools reflected the location of the "1990–1991-Nitrogen" as it moved through the ecosystem. To enhance this interpretation of the labeling experiment, the label should be added in a molecular form and in a manner that mimics natural (or ambient) N inputs to the system or to a particular pool. It is also important to consider that a labelled cohort of N will move through the ecosystem in a way that reflects the state of the system at the time of the labeling. For example, suppose a hypothetical field manipulation to increase the downward transport of N through the soil, perhaps through irrigation, began in the year 2000. The initial partitioning of a label applied to the forest floor in 1999 is likely to be very different from that of a similar label applied in 2001 once the irrigation treatment were under way. The redistributions of these hypothetical labels over 10-year periods would reflect the differences in this initial partitioning.

Model–data comparisons

Because the logical basis of process-based or causal models is deductive, model realism must be checked against empirical observations. This may involve formal or informal approaches to model testing, verification, or validation. These terms are used with wide ranges in meaning (Oreskes et al. 1994; Gardner & Urban 2003); I simply point out that any such efforts require direct model–data comparisons. Additionally, as noted above, the primary motivation in developing stable-isotope models has been to aid in the interpretation of field and laboratory data, which necessarily requires an assumed correspondence between measured and modeled quantities.

Although we ultimately seek to understand systems in nature, it is through observational methods and techniques (with their inherent limitations) that we use empirical data to assess models or use models to interpret observational data (Oreskes et al. 1994). In this sense, **observables** stand between nature and our mathematical models (Figure 13.1). Whether for model testing, model parameterization, model calibration, or model use to interpret observations, it is important to keep this in mind when designing or finding key points of model–data comparison. Observables do not necessarily reflect our current conceptualizations of fundamental processes in the ecosystems we study; often they instead reflect operational definitions. Comparisons between observed and modeled quantities may be mismatched or ill-defined. This consideration enters any model–data comparison (Magid et al. 1997), but it is amplified in the case of isotope models. Both an elemental pool size and its isotopic signature enter into a calculation of tracer recoveries (explained below) and both may be mismatched between a modeled quantity and the corresponding field measurement. Given these considerations, model–data comparisons should be constructed not necessarily to represent fundamental processes. Instead, a close correspondence is needed between a quantity as modeled and the same quantity **as measured by observational techniques**.

Ecosystem biogeochemistry is rich in operationally defined quantities, but these may or may not match the pools and fluxes in our conceptual models. For example, during decomposition of plant litter, C is partly respired to CO_2 and partly transformed to a quasi-stable material, humus. Humification is often conceived, quantitatively, as occurring to 15–20% of the initial litter mass for foliar and fine-root litter (Aber et al. 1990). This conceptual model works well in computer models: when a certain portion of initial mass remains, the material is passed to a more stable pool with a lower turnover rate. But a problem arises when we try to compare the modeled size of the humus pool, or its isotopic signature, with field observations. Field definitions of components of the soil organic horizon are based on rubbed fiber content (an operational definition), or on whether the initial source of the material is identifiable as to leaves, needles, roots, bark, and so on, or not so identifiable (an operational definition), or by horizontal layering in the horizon (also an operational definition). None of these definitions used in the field matches the conceptualization upon which the model operates in this example, i.e. the definition of humus as what remains after 80 to 85% of the initial litter mass has been lost. Rather than being a clear point of model–data comparison, the size of the humus pool or its isotopic signature is in this case a vague comparison and involves the entry of yet another conceptual model (the correspondence of "percent of initial mass remaining," in the model, to "identifiable subhorizon," in the field). This corollary conceptual model is not explicit and not falsifiable in this context. Mismatches like this occur to varying degrees in many, if not most, points of model–data comparison in isotope models. Examples abound, including operational definitions of plant-available nutrient pools (e.g. through buried-bag field incubations; Hart et al. 1994), operational definitions of leaching fluxes (material passing through a filter of a certain pore size; Lajtha et al. 1999), and the list goes on. Together these seriously hamper our ability to view model–data comparisons as firm quantitative tests or interpretations.

In some cases models can have internal pools, parameters, or fluxes designed to match precisely to operationally defined quantities from a particular field study. This, however, can result in a loss of generality or a substantial increase in modeling effort to include multiple operationally defined options for the model user. Currie et al. (2004) addressed this problem in the TRACE model (Tracer Redistributions Among Compartments in Ecosystems; Currie et al. 1999) by allowing isotopic tracers in dead fine-root tissue to be accounted for under to three different definitions used in field studies: either together with live roots, or composited together with soil organic matter, or as its own pool separate from live roots and from soil organic matter. To interpret isotopic tracer recoveries in a field study, the user of the TRACE model chooses the set of pool formulations that best matches the field-study design. This solution, however, has drawbacks. It requires a large amount of computer code dedicated to carrying out "bookkeeping" calculations in three different ways, each needing to be thoroughly tested and thoroughly integrated with the rest of the model.

Metrics: stable isotope ratios, mass balances, tracer percent recoveries

Several metrics are available to compare modeled ratios of isotopes against observations. In modeling natural-abundance isotopic ratios, investigators typically use the familiar δ notation. This is a widely recognized quantity (Lajtha & Michener 1994) that is based on the mass ratio of stable isotopes in a sample, or, when averaged across replicate samples, in an organism tissue or ecosystem pool at a point in time. It is the metric typically reported in mass-spectroscopic results and can be easily calculated in a computer model. Note that δ is an intensive variable; the size of the elemental pool does not enter into its calculation. Other intensive variables that are related solely to the isotopic ratio include the ratio itself (e.g. $^{13}C/^{12}C$) or the atom percent of the less common isotope (e.g. atom % ^{13}C).

In tracer studies, investigators are interested in the relative apportionment of the tracer into various compartments in an organism, subsystem, or ecosystem. This information is not conveyed by the δ notation or atom percent notation alone. A small pool receiving a small amount of enriched ^{15}N tracer could show a large change in $\delta^{15}N$ or atom % ^{15}N, whereas a large pool receiving a large amount of the tracer could show a small change in $\delta^{15}N$ or atom % ^{15}N. To interpret tracer studies, in any ecosystem pool in which the isotopic ratio, δ value (e.g. $\delta^{15}N$), or atom percent (e.g. atom % ^{15}N) is reported, the pool size of elemental mass (e.g. total N mass in g N m^{-2}) must also be scaled up and reported (e.g. Nadelhoffer et al. 2004).

In a mass-balance approach to interpret recoveries of enriched tracers, investigators combine an extensive variable describing an elemental pool size with an intensive variable related to its average isotopic character. Extensive variables (e.g. mass or volume) are those that scale with the size of a pool, whereas intensive variables (e.g. pH or temperature) are independent of pool size. For example, one could combine bulk N mass in a pool (an extensive variable) with either its $^{15}N/^{14}N$ ratio or atom % ^{15}N (each intensive variables) to calculate ^{15}N mass in each pool before and after the introduction of a ^{15}N tracer (Zak et al. 2004.) Such mass-balance calculations do scale directly with the elemental pool size and thus do convey the relative apportionment of enriched tracers among compartments in a soil, plant, or ecosystem. Many investigators divide a mass-balance expression of tracer recovery in each pool by the mass of isotopic tracer (above background) applied in the study to express recoveries as "percent recovery" of the tracer. Several different formulations of percent recovery have been used (Nadelhoffer & Fry 1994; Hart et al. 1994; Hauck et al. 1994). Currie et al. (1999) defined $PR^{15}N$, the percent of the ^{15}N tracer mass (above background) that is recovered (above background) in a particular ecosystem pool, or compartment (C_i), at a particular point in time (t) as follows:

$$PR^{15}N(C_i, t) = \frac{N_{C_i}(t)\left(\text{atom}\%^{15}N_{C_i}(t) - \text{atom}\%^{15}N_b\right)}{A(t - t_0)\left(\text{atom}\%^{15}N_a - \text{atom}\%^{15}N_b\right)} \tag{13.1}$$

where $N_{C_i}(t)$ is the amount of N [g m^{-2}] in C_i at time t, $A(t - t_0)$ is the sum of N amendments (g m^{-2}) to time t, C_i is an ecosystem compartment, and the subscripts "a" and "b" denote amendment and background.

This quantity, the percent recovery of enriched tracers, has proven useful both in expressing field-study results alone and for direct comparison against model results. Because it includes the amount of tracer added to time (t) and recovered at time (t), it can be used in the midst of a study in which tracers are added over several applications or added continually. It has been useful in following temporal redistributions of isotopic tracers over long time periods (Currie et al. 2004). Because it corrects for pre-existing differences in natural abundance and corrects for different strengths or amounts of tracer added, it is useful in making succinct and intuitive comparisons among different field treatments (Currie & Nadelhoffer 1999). For example, in a large-scale field study at the Harvard Forest, Massachusetts (USA), Nadelhoffer et al. (1999, 2004) used ^{15}N inputs to label both ambient (untreated) and fertilized (N-amended) plots with tracers differing in values of atom % ^{15}N. Comparisons of $\delta^{15}N$ values reveal instrument sensitivity and standard errors of groups of samples. Comparisons of percent recovery of the tracers allowed differing patterns of tracer movement among forest types and treatments to be clearly compared and conveyed.

Other benefits of $PR^{15}N$ for model–data comparisons arise from its inclusion of extensive pool sizes and its ability to be summed directly. If an observational dataset and a model used slightly different sizes for a pool of N – for example, the stock of N in fine roots expressed in g N m^{-2} – this mismatch could be missed in a model–data comparison of $\delta^{15}N$ values. But a comparison of $PR^{15}N$ values would be sensitive to model–data differences in both $^{14}N/^{15}N$ ratios and sizes of the pools being compared. Additionally, quantities of $PR^{15}N$ can be summed. For example, values of $PR^{15}N$ in current-year needles and in prior-year needles can be summed to express $PR^{15}N$ in all needles; similarly, $PR^{15}N$ values in separate soil horizons can be summed to express $PR^{15}N$ in the solum. Percent recovery of tracers is thus a convenient quantity for constructing budgets in a tracer study.

The expression for $PR^{15}N$ (equation 13.1) includes subtraction of natural-abundance values of atom % ^{15}N in each pool. This introduces some uncertainty because one needs to measure background values of atom % ^{15}N prior to the label application, or in a separate unlabelled plot and assume that these represent background values of atom % ^{15}N in the labeled plot over the period of sampling.

Model results sum and balance, but field results may not

A source of disparity in model–data comparisons is often evident when tracer recoveries are summed over the components of a system. Model results are constrained by a mathematical mass balance of the isotope. If a mathematical model is logically self-consistent, its values of percent recovery of tracers (e. g. $PR^{15}N$) will sum to 100% after accounting for system losses. Field observations, in contrast, are typically scaled up from fine-scale samples and may not quantitatively account for all of the isotope in the system under study. Nutrient and carbon budgets based on field observations need not exhibit mass balance (Lajtha 2000) and field-measured values of percent recoveries of tracers typically do not sum to 100%. In the field, different components of a nutrient budget may be measured by different teams of investigators at different locations or at different scales. In some cases, experimental uncertainty is large; in other cases, some pools may be incompletely sampled or omitted from the field study altogether. For example, mineral soils may be sampled to a limited depth (10–20 cm), limiting the model–data comparison of percent recovery of a tracer in the solum.

Different pool sizes and fluxes of N in a field study may be measured at different points in time, introducing additional uncertainties into mass-balance calculations based on field data. For example, foliar N content and

$^{15}N/^{14}N$ ratio might have been measured in July 2000, while the forest floor N content and $^{15}N/^{14}N$ ratio might have been measured in October 2001. Under such circumstances, model–data comparisons are a strength in helping to interpret field data. Each of the quantities the model simulates are quantified in each time step; a process model can simulate values for N content and $^{15}N/^{14}N$ ratios in foliage in July 2000 and the soil O horizon in October 2001 that are consistent with each other, consistent with the conceptual model of system dynamics, and able to be compared directly against field observations that are separated in time.

Another consideration in making model–data comparisons is that in some sense one is comparing one model against another. Scaling up observations from individual samples to ecosystem pools requires a conceptual and mathematical model; this may be expressed as sets of equations in a spreadsheet. Additionally, it is important to realize that some ostensibly "measured" numbers are better considered as modeled quantities. Nutrient uptake by vegetation provides an example. Field investigations do not measure fluxes of nutrient uptake; instead, tissue concentrations of nutrients are measured, and a conceptual model and set of equations is employed to calculate fluxes of nutrient uptake. In such cases, consistency in model–data comparisons will be aided if investigators use the same set of scaling assumptions in the process model and the spreadsheet equations used to scale up observations. Close collaboration between modelers and field investigators at each stage of the research is advisable because it furthers such consistency.

Principles and techniques of stable isotope modeling

Algebraic mixing models

I outline two types of models commonly formulated as algebraic equations: mixing models and isotope dilution or "pool dilution" models. In the simplest form of mixing model, the isotopic ratio in a measured pool R_M is considered to result from a mixture of two "end-member" ratios R_A and R_B (Figure 13.4). (R refers to an isotopic ratio such as $^{13}C/^{12}C$.) In equation 13.2, a and b are parameters representing the weighted contributions of end-member ratios R_A and R_B, respectfully:

$$R_M = aR_A + bR_B \tag{13.2}$$

If one stipulates that the parameters a and b are proportions such that $0 \leq a \leq 1$, $0 \leq b \leq 1$, and $a + b = 1$, then there is a unique (a, b) pair for a given R_A, R_B, and R_M (the measured sample). Simple substitution of $(b = 1 - a)$ into equation 13.2 and solving for a yields the following:

$$a = \frac{R_B - R_M}{R_B - R_A} \tag{13.3}$$

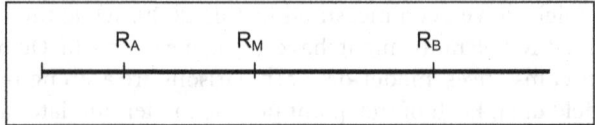

Figure 13.4 A two end-member mixing-model conceptualization in which R_M, the measured isotopic ratio in a pool, is conceived as a mixture of the two end-member isotopic ratios R_A and R_B.

This two-component mixing model has been widely used in several types of applications. These have included isotopic analysis of food webs, analysis of strontium isotopes to determine relative inputs from weathering versus atmospheric deposition (Vitousek et al. 2003), analysis of the contributions of C3 versus C4 plant sources to soil organic carbon (Balesdent et al. 1987, 1988), and analysis of the contribution of biologically-fixed N versus soil N to plants. For the latter case, Shearer & Kohl (1986) developed the following equation (Amundson & Baisden 2000):

$$a = \left(\frac{\delta^{15}N_S - \delta^{15}N_P}{\delta^{15}N_S - \delta^{15}N_F} \right) \tag{13.4}$$

where a is the proportion of plant N deriving from N fixation and $\delta^{15}N_S$, $\delta^{15}N_P$, and $\delta^{15}N_F$ refer to the $\delta^{15}N$ values of soil N, plant N, and fixed N, respectively. Equation 13.4 is analogous to equation 13.3 when the isotopic signature of plant N is viewed as the mixture of the isotopic signatures of the two end-members soil N and biotically fixed N. Schimel (1993) provided an alternate equation for this mixing model and indicated that it may be preferable to use isotopic ratios (as in equation 13.3) rather than δ values. An analogous equation to calculate the relative inputs of C3 versus C4 vegetation to soil organic matter, based on a two-component isotopic mixing model for natural-abundance differences in C isotopes in C3 versus C4 plants, was developed by Balesdent et al. (1987, 1988) and reproduced in Schimel (1993), Wolf et al. (1994), and Amundson & Baisden (2000).

When isotopic signatures of two elements are available simultaneously (e.g., $\delta^{15}N$ and $\delta^{13}C$, or $\delta^{15}N$ and $\delta^{18}O$), then a three-component mixing model can be used; a unique solution can be found for the contributions of each of three end-members to the two measured isotopic ratios in the mixed pool or sample. In this case, where $R(N)$ denotes the isotopic ratio $^{15}N/^{14}N$, $R(C)$ denotes the ratio $^{13}C/^{12}C$, subscripts A, B, C refer to the three end-members and subscript M to the measured sample, and a, b, c refer to the mixing proportions, the three equations and three unknowns a, b, c are written as in equation 13.2, which can be reduced algebraically for a, b, and c through substitution:

$$R(N)_M = aR(N)_A + bR(N)_B + cR(N)_C \tag{13.5}$$
$$R(C)_M = aR(C)_A + bR(C)_B + cR(C)_C$$
$$a + b + c = 1$$

Uncertainty enters into these two or three-component mixing models from several sources. One source is the variabilities, which could be expressed as standard errors of measurement, of the isotopic ratios within each of the end members and within the mixed pool or sample. Uncertainty also enters in the structure of the model, meaning a two-component model may approximate a situation in which three or more isotopically distinct sources exist in nature. Another source of uncertainty arises from the possibility of isotopic fractionation occurring as part of the mixing process or subsequent to the mixing process. Examples of the former include during diffusion or advection through a soil or during uptake by vegetation. An example of fractionation subsequent to the mixing process occurs when the mixed quantity is the isotopic signature in an organism. There are typically differences in both $\delta^{13}C$ and $\delta^{15}N$ values between organisms and their food sources, due to fractionation during metabolism and excretion. Among types of organisms in a trophic level, the variabilities in the net effects of such fractionation can be on the same order of magnitude as the effect itself, limiting the precision with which a mixing model can be used for food sources (Ponsard & Arditi 2000).

Algebraic models of pool dilution

Isotope dilution or "pool dilution" models are so named because any flux of non-enriched element into an artificially enriched pool or compartment will dilute the enrichment of the pool. Relatively simple pool-dilution models can be drawn with one to three elemental pools or compartments together with associated fluxes. These models can be expressed as simple algebraic or differential equations. Typically, unique algebraic solutions can be found only when several assumptions are made. These include the assumption that total elemental fluxes among pools are in steady state, or that the enriched tracer does not recycle among pools over the time of the study. Solutions describe values of elemental fluxes that are consistent with the assumptions together with the observed changes in isotopic enrichment of the pools at points in time following the addition of an enriched tracer (thus the mathematical models are used to interpret the observations in terms of conceptual pools and fluxes). This analytical modeling approach is widely used in N-cycling research to calculate gross N transformations in soil such as gross NH_4 mineralization, gross nitrification, or gross NH_4 and NO_3 uptake by soil micro-organisms. Methods for pool dilution and mass-balance techniques for fine-scale soil cores and for plot-level studies are standard procedures that are widely covered in the literature (Hart et al. 1994; Robertson et al. 1999; Schimel 1993).

Tietema & Wessel (1992) assessed uncertainties and errors introduced by the assumption that enriched tracers do not recycle. They tested two simultaneous single-pool models of pool dilution of NH_4 and NO_3 by comparing the algebraic solutions with the results of a computer simulation. Using standard techniques, $^{15}NH_4$ and $^{15}NO_3$ were added separately in the laboratory to forest soil cores from three forest sites on each of four dates. Standard equations of production and consumption of NH_4 and NO_3 (Kirkham & Bartholomew 1954) were first used to calculate gross rates of NH_4 mineralization, NH_4 immobilization, nitrification, and NO_3 immobilization. According to these authors, the equations contained the following five assumptions:

1 that fractionation was negligible;
2 that NH_4 and NO_3 pools after enrichment with ^{15}N were each homogeneous;
3 that the natural-abundance ratios of $^{15}N/^{14}N$ did not vary among pools at the start of the experiment;
4 that gross N transformation rates were constant for the duration of the experiment (i.e. that nutrient fluxes were in steady-state);
5 that there was no recycling of enriched ^{15}N tracers (Tietema & Wessel 1992). (Note that the absence of tracer recycling is a necessary approximation made by all algebraic models as well as most differential-equation models.)

Tietema & Wessel (1992) then constructed a numerical (computer) simulation model that included each of the first four assumptions but allowed ^{15}N tracer recycling. The simulation model calculated higher gross rates of N transformation, on average, than the algebraic model. In most cases the simulation model estimated rates higher than the algebraic equations by 20% or less; in a few cases (particularly where gross fluxes were small), differences between the models were as high as 100%. When this source of uncertainty was combined with statistical measures of confidence in observations of $^{15}N/^{14}N$ (such as standard errors), overall uncertainties were even greater. The authors concluded that, because of its limiting assumptions, the standard algebraic model could reveal semi-quantitative insights into differences in gross N transformation among soils and dates, as opposed to quantitative rate calculations.

Differential equation models

Formulations relying on differential equations have the benefit of making use of widely understood conceptual archetypes but, at the same time, introduce conceptual limitations. Such models are often designed for use in a specific study, for example to infer rates of fractionation from a particular or generalized pattern in natural-abundance observations. Schimel (1993) described a differential-equation approach used by Alperin et al. (1988) to calculate a discrimination coefficient for C during methane oxidation based on vertical patterns of carbon isotopes and hydrogen-deuterium in marine sediments. Key assumptions were that diffusion dominated methane movement and that observed vertical gradients in methane were in steady state. Somewhat similarly, Amundson & Baisden (2000) developed a differential-equation model to explain vertical patterns of C isotopes in soil organic matter. This example is informative because it illustrates both the power and limitation of the differential analytical approach.

Amundson & Baisden (2000) began with a partial differential equation describing changes in the concentration of each isotope of soil organic C over time (t) and soil depth (z):

$$\frac{\partial C}{\partial t} = D\frac{\partial^2 C}{\partial z^2} - v\frac{\partial C}{\partial z} - kC + f_d \tag{13.6}$$

$$\frac{\partial C^*}{\partial t} = D\frac{\partial^2 C^*}{\partial z^2} - v\frac{\partial C^*}{\partial z} - \alpha kC^* + f_d R_I \tag{13.7}$$

where C represents ^{12}C and C* represents ^{13}C, the first term on the right-hand side of equations 13.6 & 13.7 represents vertical diffusion of ^{12}C and ^{13}C (where D is the diffusion constant), the second term on the right-hand side represents vertical advection of ^{12}C and ^{13}C (and v the advection constant), kC and kC^* represent first-order turnover (heterotrophic respiration) of soil C, and f_d represents the rate of C inputs from plant roots. The isotopic fractionation constant is denoted by α and the ^{13}C/^{12}C ratio in root inputs is denoted R_I. After specifying boundary conditions (including plant-litter inputs of C at the soil surface) and introducing other assumptions including the approximation that root-litter inputs are zero and that the system is in steady state, the authors found an analytical solution. The solution amounted to an equation describing changes in R, the ratio of ^{13}C/^{12}C, with depth (z):

$$R(z) = \frac{R_p}{\sqrt{\alpha}} e^{(\sqrt{k/d})(z\sqrt{\alpha}-1)} \tag{13.8}$$

where R_p is the ratio of ^{13}C/^{12}C in plant litter inputs at the surface. This solution was used to calculate modeled changes in δ^{13}C with depth. The δ values were then compared directly against field observations, both to assess the model and to provide a theoretical explanation of the field observations in terms of the conceptual processes represented by the equations: diffusion, advection, and turnover. (The authors pointed out that "diffusion" is intended in an abstract sense and probably includes biotically driven mixing across soil depths.) This highlights a key shortcoming in the use of differential equation models: it is the nature of such analysis to introduce concepts such as diffusion and steady state into the model and the analysis because analytical solutions can then be obtained.

An exception to the typical use of differential equations in isotope modeling is the NESIS model (Non-Equilibrium Stable Isotope Simulator) developed by Rastetter et al. (2005). A goal in developing NESIS was to help "overcome the problems associated with the (often implicit) assumption of steady-state used in many isotope studies" (Rastetter et al. 2005). One of the most promising aspects of the NESIS model is that it can essentially add isotopically explicit calculations for any element or elements to an existing nonlinear, dynamic model of ecosystem biogeochemistry (referred to as the "parent model"). NESIS runs independently of the parent model. For inputs, NESIS requires values of all of the pools and fluxes of an element at the start and end of each time step in the parent model, isotopic ratios in each pool at the start of the simulation, isotopic fractionation constants associated with each flux, and any external inputs to any pool over time together with their isotopic signatures (which can include enriched tracers). NESIS then uses a linear, donor-controlled set of differential equations to describe isotopic ratios in each pool of the parent model in each time step and to calculate separate fluxes of the heavy and light isotope associated with each flux in the parent model. The set of simultaneous equations describing heavy and light isotope fluxes are solved, in each time step, through a numerical integration technique. Nonlinear, non-steady-state dynamics can be accommodated because NESIS self-corrects by re-reading all pool sizes and flux values from the parent model in each time step and re-estimates the parameters for the linear, donor-controlled model. This is a promising direction in isotope modeling. Although most biogeochemical models do not print out enough information in each time step for NESIS, they could be altered to do so much more easily than they could be rewritten to include isotopically explicit internal equations.

Algorithmic process models

In causal-dynamic models of ecosystem biogeochemistry, equations or algorithms that represent individual ecological or biogeochemical processes such as photosynthesis, nitrification, or evapotranspiration are linked together in chains of cause and effect. Within this category, models vary significantly in the degree of empiricism or mechanism in their formulations of individual processes as well as in model complexity. By including ecosystem-level feedbacks in element cycles and other nonlinear processes these models can be used to investigate interactions across levels of organization. Investigators are able to simulate interactions among individual causal-mechanistic processes that give rise to emergent properties at the ecosystem level or at the level of key subsystems. This approach typically involves construction of a mechanistic set of cause-and-effect relationships first written as algebraic or differential equations. These are translated into difference equations with a discrete time step and translated further into a set of computer instructions (an algorithm) to simulate the system on a discrete time interval (Currie et al. 1999). This produces a dynamic, nonlinear evolution of state variables. Because analytical solutions are not sought, these models can include element (including tracer) recycling and non-steady-state dynamics, requiring fewer such assumptions relative to analytical models (Figure 13.5). The use of algorithmic computer models to simulate ecosystem biogeochemistry in this manner has been popular for decades (e.g. Pastor & Post 1986; Parton et al. 1988). The use of such an approach to simulate redistributions of stable isotopic tracers for ecosystem biogeochemistry has grown over the past decade (Tietema & Wessel 1992; van Dam & van Breemen 1995; Tietema & van Dam 1996; Koopmans & van Dam 1998; Currie et al. 1999, 2004; Hobbie et al. 1999). An important reason to consider using such a model is that well-known interactions between element cycles can be included as well as the effects of well-understood forcing functions such as temperature and moisture. An isotopically explicit formulation of a model, when combined with isotope measurements, makes it possible to test model mechanisms more fully and thus increase confidence in the model for non-isotope applications.

As an isotopically explicit model runs, it requires four pieces of information for each elemental pool (e.g. for each pool of N): the total elemental pool size (e.g. $^{14}N + {}^{15}N$ in g N m^{-2}), the mass of ^{14}N (in g ^{14}N m^{-2}), the mass of ^{15}N (in g ^{15}N m^{-2}), and the ratio of $^{14}N/^{15}N$ in the pool. For a number of reasons, including minimizing computer round-off errors, it is best to store and continually update only two pieces of information and calculate the other information as needed in the model. The model could store the ^{14}N mass and ^{15}N mass in each pool and calculate the total N and the $^{15}N/^{14}N$ ratio when needed; alternately, the model could store the total N and the isotopic ratio in each pool and calculate the ^{14}N mass and ^{15}N mass when needed. In the TRACE model (Currie et al. 1999), written in the MicroSoft VisualBasic language, ^{14}N and ^{15}N masses for each pool are stored in a one-dimensional vector variable. For example, ^{14}N and ^{15}N masses (in g m^{-2}) in a pool of N named *NPoolA* would be stored as *NpoolA(1)* and *NpoolA(2)*, respectively, where the indices ($i = 1$) and ($i = 2$) allow two different values to be stored in the vector variable *NpoolA(i)*. The proportions of ^{14}N and ^{15}N in this pool (here denoted by the variable names *f14NpoolA* and *f15NpoolA*, respectively), and the total N in the pool (*TotNpoolA*) would be recalculated whenever needed, using the following code:

```
TotNpoolA = NpoolA(1) + NpoolA(2)
f14NpoolA = NpoolA(1) / (NpoolA(1) + NpoolA(2))      (13.9)
f15NpoolA = NpoolA(2) / (NpoolA(1) + NpoolA(2))
```

Note that these are 'assignment' statements, in which the quantity on the right-hand side of the equation is calculated and then stored in the variable named on the left side of the equation. A gross flux of total N between *NPoolA* and *NPoolB*, named *NfluxAB* and defined as positive in the direction from A to B (thus negative if from B to A), would be carried out as follows, debiting the source or donor pool and crediting the sink, or target pool in an isotopically explicit manner:

```
If NfluxAB > 0 then                                          (13.10)
   NpoolB(1) = NpoolB(1) + NfluxAB * f14NpoolA
   NpoolB(2) = NpoolB(2) + NfluxAB * f15NpoolA
   NpoolA(1) = NpoolA(1) - NfluxAB * f14NpoolA
   NpoolA(2) = NpoolA(2) - NfluxAB * f15NpoolA
Else
   NpoolA(1) = NpoolA(1) + NfluxAB * f14NpoolB
   NpoolA(2) = NpoolA(2) + NfluxAB * f15NpoolB
   NpoolB(1) = NpoolB(1) - NfluxAB * f14NpoolB
   NpoolB(2) = NpoolB(2) - NfluxAB * f15NpoolB
End if
```

Note that according to the principle of pool dilution, the isotopic ratio of the gross flux between *NPoolA* and *NPoolB* is equal to the isotopic ratio of the source pool, which varies according to the direction of the gross flux. This coding approach lends itself to the process of augmenting an existing model to add isotopically explicit calculations. For example, if an existing model had the pool *NpoolA* as a variable, this could simply be replaced by a vector variable of the same name, initialized with a background isotopic ratio, and all model algorithms amended to include the principles of pool dilution and mass balance in each isotope. For a simple model this could be straightforward, but for a complex model this substantially increases the amount of code and the level of model complexity. Note that in a non-isotope model of N cycling, the 11 lines of the code in equation 13.10 could be accomplished with two, much simpler, lines of code. Process models implement the principle of pool dilution by recalculating the proportions such as $^{14}N/(^{14}N + ^{15}N)$ and $^{15}N/(^{14}N + ^{15}N)$ in each model pool after each time step (van Dam & van Breemen 1995) or, alternatively, in a source pool prior to any gross flux of N and the target pool after any gross flux of N (Currie et al. 1999). Isotopically explicit models provide the ability to simulate gross (as well as net) elemental fluxes; at the same time, they require the simulation of gross elemental fluxes if they are to incorporate the principle of pool dilution.

Isotopic fractionation has been either included or excluded from particular algorithmic process models. In using the NICCCE model (Nitrogen Isotopes and Carbon Cycling in Coniferous Ecosystems; van Dam & van Breemen 1995), which included both isotopic fractionation and enriched-tracer redistributions for N, Koopmans & van Dam (1998) concluded that fractionation was insignificant to model simulations when enriched tracers were well above natural-abundance levels. Currie et al. (1999) used this assumption in the development of the TRACE model, which omitted fractionation in order to simplify the modeling of the redistributions of highly enriched tracers.

When fractionation is included, the mathematics is often simplified by approximating the behavior of the more common isotope of an element (i.e. ^{12}C or ^{14}N) as equal to that parameterized for the total element (i.e. C or N). For example, in the NICCCE model, Koopmans & van Dam (1998) described generalized flux calculations for ^{14}N and ^{15}N (denoted J_{14N} and J_{15N} respectively) as:

$$J_{14N} = K(^{14}N + {}^{15}N)(1 - R) \qquad (13.11)$$

$$J_{15N} = K(^{14}N + {}^{15}N)R(1 - \alpha) \qquad (13.12)$$

where R is the ratio ($^{15}N/(^{14}N + {}^{15}N)$), α is the fractionation constant, and K is a reaction rate constant that is multiplied by the mass of N in the donor pool; K is parameterized or defined for whole-element (i.e. total N) fluxes. Substituting the value for R in equations 13.11 & 13.12) and canceling terms, they simplify to:

$$J_{14N} = K^{14}N \qquad (13.13)$$

$$J_{15N} = K^{15}N(1 - \alpha) \qquad (13.14)$$

This illustrates that the reaction-rate constant K is applied to ^{14}N mass alone, but the ratio of fluxes J_{14N} and J_{15N} is correctly preserved. This approximation introduces a small error in whole-element cycling rates (Amundson & Baisden 2000), but this error is well beyond the precision with which reaction-rate constants are typically specified.

The decision whether or not to include fractionation impacts model uncertainty either way. Incorporating fractionation processes in a complex ecosystem model adds an additional set of parameters, i.e. fractionation rates, each with an additional uncertainty (van Dam and van Breemen 1995; Hobbie et al. 1999). Another consideration is the metric that will be used in model–data comparisons. Depending on the level of tracer enrichment, it may not be advisable to omit fractionation processes in the model if values of $\delta^{15}N$ will be compared directly against observations, whereas it may be appropriate if percent recoveries of ^{15}N tracers are compared. Omitting fractionation rates could impact model results more significantly over longer time periods or as tracers become more dilute in ecosystem pools. Finally, an additional and related source of uncertainty is the problem of kinetic fractionation. Isotope models typically employ thermodynamic fractionation constants, i.e. differences in reaction rates when both isotopes are plentiful in the source or donor pool. In nature, an important effect is the kinetic change in fractionation, in which an isotopic difference in reaction rates decreases as the size of the source pool decreases (Schimel 1993). Representing such effects has not received much attention in the modeling of stable isotopes.

Complex model code should be tested for errors prior to use in planning or interpreting field experiments. Subtle errors can arise in logic, in the translation of equations to algorithms, and in code writing. Computer round-off errors can also arise and propagate because some computations are repeated numerous times. Error-checking can involve operational tests performed by the user and can also involve model enhancements to include automated error-checking code. A common operational test is to run a model with all isotope calculations active, but all pools initialized to the isotopic ratio of the standard, and all fractionation constants set to zero (Hobbie et al. 1999; Currie et al. 2004). Values of δ should then remain zero throughout the simulation. In this test, computer round-off errors can lead to a nonzero value in the second decimal place of δ, e.g. $\delta^{15}N = 0.02‰$ when it should be 0.00‰. This is insignificant in model–data comparisons because it is an order of magnitude below the precision with which $\delta^{15}N$ is typically specified.

An example of a semi-automated test is an ecosystem mass-balance test for each element. In a simple approach to test mass balance for N, initial total ecosystem N is calculated by summing all N pools at the start of the model run. A cumulative ecosystem-level input–output balance of N is saved as a book-keeping variable, initialized to zero at the start of the model run and incremented any time there is an N influx or outflux (positive values for influxes). Total ecosystem N can be calculated at the end of each time step by summing all N pools in the model; when the cumulative input–output flux of N to that time step is subtracted, the difference should equal the initial value of total ecosystem N. This test can be incorporated into model code easily for each element and printed and checked by the user prior to or during each model run (Currie et al. 2004). Another simple semi-automated test is to check that values of percent recovery of tracers sum to 100% in the model, when corrected for ecosystem losses of the tracer. Introducing model code for fully automated tests is more involved but potentially important. In the code of the TRACE model, for example, automated tests are performed to raise an error flag if any pool of ^{14}N or ^{15}N mass becomes less than zero at any point in time (with the exception of computer round-off errors). Under normal operation this never occurs, but in Monte-Carlo simulations in which numerous model parameters are varied stochastically and independently (Currie & Nadelhoffer 1999), infrequently a random combination of parameter values causes a slight violation of mass balance to develop, which is then flagged by the error-checking code and the model run is automatically discarded.

Sources and analysis of uncertainty in interpreting field study results

While the sensitivity of stable-isotope ratios to gross elemental fluxes and to fine-scale processes provides an additional dimension of observation, it can also create difficulties in interpreting natural-abundance patterns or tracer movements. The dynamics of an isotopic ratio in an ecosystem pool provides only a signature of the net effects of multiple processes that may be interwoven. This can hinder interpretation of elemental movements and transformations where multiple pathways are possible, where pool sizes are inadequately known, or where detailed mechanisms are poorly understood. While this limitation is true of both empirical and modeling studies (Nadelhoffer & Fry 1994), it may seem more vivid in a modeling analysis. In this section I consider uncertainty and limitations inherent in stable-isotope modeling as well as some approaches, such as inverse modeling or iterative model–data comparisons, that have been used to overcome these limitations.

In complex isotopically explicit models, uncertainties arise from a somewhat different set of sources and approximations than in simple models. In model–data comparisons using either type of model, uncertainties are present in standard errors of measurement (SEM) of background isotopic ratios, in SEM of observed isotopic ratios through time, in SEM of elemental pool sizes, and in the degree of success in labeling a particular pool with an enriched tracer (if used). Algebraic and some differential-equation models require a greater number of assumptions such as steady-state, no tracer recycling, and linear interactions or near-equilibrium conditions. These assumptions contribute to uncertainty in model–data comparisons (Tietema & Wessel 1992). Many of these assumptions can be dropped in algorithmic simulation models, but at the same time, other types of assumptions and approximations increase

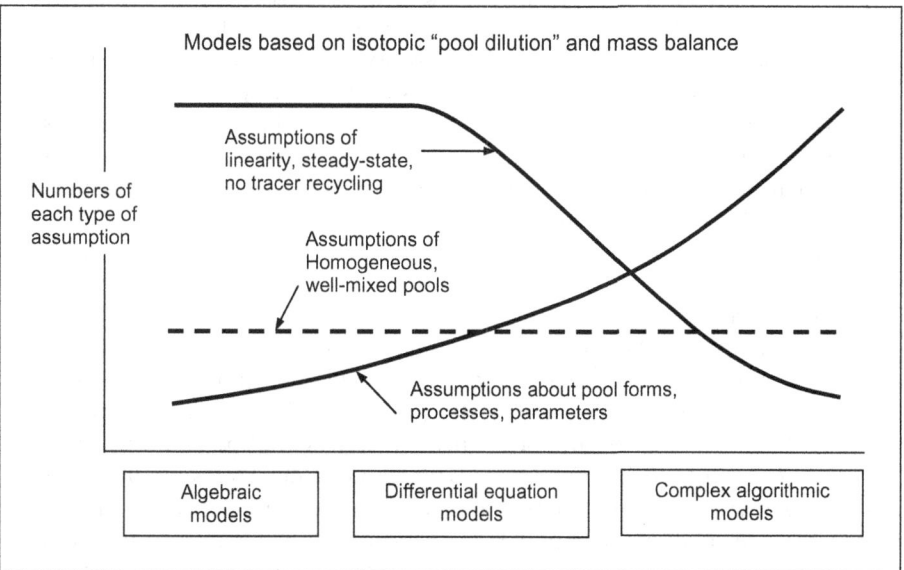

Figure 13.5 Relative relationships among numbers of assumptions of various types that are needed by models incorporating principles of isotopic "pool dilution" and mass balance.

(Figure 13.5). These include assumptions about conceptual model structure (which increase necessarily as the number of pools and fluxes in the model increase) and assumptions made in specifying values for parameters (which increase in number rapidly as the process-based complexity of the model increases). An additional source of uncertainty in algorithmic models that is not present in analytical models (where the latter employ steady-state) lies in temporal scaling of dynamic processes. If the dynamics of a gross flux in nature is much more rapid than the model time step (e.g. minutes or hours for microbially mediated processes, versus a monthly time step in a model), the model must endeavor to make an empirical approximation of the isotopically explicit effects of the gross flux on a monthly basis – which may be a coarse approximation.

Another key source of model uncertainty lies in sensitivity to parameters that are difficult to quantify in the field or to historical initial conditions that are difficult to estimate. For example, C and N cycling processes in the TRACE model are sensitive to turnover rates of humified matter in the organic and mineral soil horizons (Currie et al. 1999). Unfortunately, turnover rates of humified matter *in situ* are very difficult to quantify meaningfully in a process-based field study and are typically estimated in a coarse manner from other data (e.g. Currie & Aber 1997). Specifying historical initial conditions with confidence is equally problematic (Yanai et al. 2003). Even at locations where site management and disturbance histories are well known, the biogeochemical and ecological history is typically not known precisely enough to initialize past pool sizes of C and N, or to set the historical rates of NH_4 and NO_3 deposition, or to test the modeled vegetation history against historical rates of key ecosystem processes such as photosynthesis or nutrient uptake (Currie et al. 2004). Investigators infer or extrapolate such historical pool sizes and fluxes from present-day observations and from other historical data; in an environment that has seen dramatic change in the past 100 years, such inferences are made with limited confidence.

One approach to confront such uncertainty is to adopt precise best-estimates of initial or past conditions and conduct several comparative model runs in a formal or informal analysis of model uncertainty. For a complex algorithmic model, Monte-Carlo techniques can be used. Van Dam & van Breemen (1995) assessed the range of model predictions that resulted when 52 parameters were varied stochastically by ±20%; similarly, Currie & Nadelhoffer (1999) assessed model uncertainty when 63 parameters varied stochastically in normal distributions with standard deviations of 10%. This formal analysis, while important because it includes nonlinearities in the model, addresses only one aspect of model uncertainty, i.e. that arising from confidence limits on parameters. An overall assessment of uncertainty in model–data comparisons would need to combine the results of a parameter-based analysis with uncertainties in observations such as pool sizes, together with an analysis of uncertainties in model structure (Rastetter 2003) and an analysis of model results under 'bracketed' historical conditions (e.g. Goodale et al. 2002). Even with such a thorough analysis of uncertainties, a modeling analysis would contain additional assumptions and approximations, for example the assumption that elemental pools are homogeneous and well-mixed and the assumption that kinetic isotopic fractionation (Schimel 1993) can be neglected.

Over-determinacy

When a model has a large number of parameters that are difficult to specify with confidence, relative to the number of model predictions that can be compared against observations, the model might be referred to as being mathematically "over-determined". In this situation, agreements between model results and field observations could be obtained with numerous different parameter sets that one might use. This is the case not only for isotopically explicit models but for virtually all complex models of ecosystem biogeochemistry. Parsimony in model construction can help to combat this problem (Hobbie et al. 1999). The trade-offs include omission of interactions with other element cycles or physical environmental variables (e.g., temperature, moisture), or simplification of pools and fluxes in the conceptual model at the expense of realism (e.g. combining microbial and detrital material into a single pool; Currie et al. 1999). Over-determinacy can be an ever-growing problem because as an already complex model is enhanced and applied to new situations there is a tendency to include additional processes and mechanisms (Bankes 1993).

Given a model structure and number of parameters, at least three techniques have been used to address over-determinacy. Each of these techniques has drawbacks. First, a multivariate statistical technique can be used to simultaneously calibrate numerous parameters while minimizing a statistic, such as the sum of squared deviations, in a direct model–data comparison (Rastetter et al. 1992). The second technique involves a Monte-Carlo randomization of parameter values, with rejection of a large number of model runs in which certain key results are outside of the confidence range of field-observed values (van Dam & van Breemen 1995). A third technique makes use of stepwise calibration of a small number of sensitive parameters that are difficult to quantify empirically, while comparing a larger number of key pools and fluxes against field-measured values (Currie et al. 2004). Drawbacks with these approaches are that a test statistic for optimization of model–data agreement or thresholds of confidence used for parameter-set rejection may seem arbitrary and the resulting parameter set may not be unique. Additionally, over-use of calibration techniques creates a circularity if all of the available data to test a model are used to calibrate the model.

Incorporating isotopes into a complex model can either exacerbate the problem of over-determinacy or help to mitigate the problem. The additional dimension of observational data (an isotopic ratio) should help to combat the problem, but the number of additional model parameters may grow faster than the number of additional observations. Keeping the number of conceptual pools and fluxes no larger than necessary, and close to the operationally defined observables, should help, but a trade-off is that an operational field or laboratory method may measure conceptually inconvenient quantities. One approach taken by two teams of investigators using complex isotopically explicit ecosystem models (van Dam & van Breemen 1995; Currie et al. 2004) has been to parameterize and calibrate overall pool sizes and fluxes of N-cycling first, without reference to N-isotope observations, and then to predict dynamics in isotopic ratios. This has the benefit of ensuring that the magnitude of nutrient cycling pools and fluxes are realistic for a particular field study, while allowing the modeled stable-isotope redistributions, which are sensitive to model mechanisms, to be falsifiable predictions (Currie et al. 2004).

Conclusions

The greatest present strength of isotopically explicit biogeochemical models lies in aiding the interpretation of field and laboratory observations – this reflects the main purpose for which such models have been developed. The use of mathematical models to elucidate the ramifications of mass balance or pool dilution in simple one-box or two-box models, or slightly more complex models that disallow tracer recycling, is well standardized and widely used. Complex ecosystem models that include stable isotopes, both at enriched-tracer and natural-abundance levels, have been applied to interpret observational data from large-scale field studies, although these models are still in relatively early stages of advancement. Some areas of model development and application could still be considered to be in experimental or exploratory stages. These include, for example, the combination of isotopic fractionation and enriched-tracer redistributions in the same model, or the combination of multiple stable isotopes of different elements in the same model.

A legitimate question to ask is the following: Why should one add more processes or more detail to a model if this exacerbates problems that complex models already face regarding parameterization? Oreskes (2003) has argued convincingly that when more processes are included in a model to increase its realism, our uncertainty increases regarding whether our tests of the model are adequate or meaningful. The inclusion of isotopes in a model may provide a counter-example because this provides an additional set of state variables to test against observational data – it provides a clear constraint on elemental flows in ecological systems of all kinds. Because the new state variables (ratios of stable isotopes) are highly sensitive to mechanistic processes, when used in combination with existing state variables their inclusion could increase our confidence in the model relative to an isotope-free formulation. A model with isotope ratios, even though it contains additional complexity and parameters, has the potential to be more fully testable. For example, we might have greater confidence in a model of C pools and fluxes if it simulated not only testable $^{13}C/^{12}C$ ratios in its pools of C, but also C/N interactions and testable $^{15}N/^{14}N$ ratios in its pools of N. Extending this idea, we might have increased confidence in a C model if it also included energy fluxes, H and O fluxes and isotopes in water vapor and biomass, and so on. Reiners advanced this idea as the principle of using "complementary" models in the study of ecosystems (Reiners 1986). In the practical aspects of coding and parameterizing models, there are trade-offs between the Reiners' principle of complementarity and Oreskes' argument from epistemology. It may not be worthwhile to add new dimensions to a model if, in each case, so many new parameters would need to be added that the problems of over-determinacy, sensitivity to unknown quantities, and unclear interactions among model processes were exacerbated (Bankes 1993).

Computational, conceptual, and philosophical advances are all needed in stable-isotope modeling. Continued advance is needed in the development of mathematics and algorithms. For example, concisely formulated computer algorithms are needed that will be able to simulate isotopic fractionation and enriched tracer redistributions in sets of interacting processes, potentially with isotopes of multiple elements at once. The complications in encoding algorithms grow rapidly with each new set of processes that are added to an already complicated model. Unfortunately, few new investigators in ecology and biogeochemistry are receiving training in the computational tools and techniques needed to make these advances. While the "democratization" of modeling, training more ecologists to understand and to use models (Cottingham et al. 2003), is highly desirable, it is likely that the development of concise algorithms for advanced stable-isotope modeling will be pursued by investigators who specialize to some degree in informatics and computation.

The second need for advance is conceptual. A continued merging of conceptual models and empirical (field and laboratory) methods is needed so that inconsistencies can be removed. Improvement is needed on both ends. Conceptual models are needed that more accurately represent operationally defined observations that are made in field studies. At the same time, operationally defined observations are required that come closer to our theoretical ideas about the causative mechanisms that we aim to incorporate into causative-mechanistic models.

Third, progress is needed on methodological questions related to modeling approaches. How will we overcome the problem of over-determinacy in complex models while maximizing our ability to gain increased insight and understanding into complex natural systems? Ecologists sometimes adopt the stance that the only informative modeling results are those that can be shown to agree significantly with observations in a traditional or "frequentist" (Reckhow 2003) statistical approach (e.g. Gardner & Urban 2003). While such a criterion is important to consider, it is not a necessary condition for complex modeling analyses to be useful. Field observations themselves are subject to uncertainty. Observational data at times reflect idiosyncrasies of field measurement methods rather than a clear correspondence with conceptual pools or fluxes (Yanai et al. 2003). Observational data may also reflect idiosyncrasies of a particular year in which measurements were made, not necessarily representative of longer time periods (McClain et al. 2003), with no means of assessing that fact from the limited data themselves. When observations of disparate pools and fluxes, collected by various investigators for various purposes, are assembled in an attempt to synthesize a view of elemental cycling in a particular system, the resulting dataset may lack internal self-consistency (e.g., field-observed percent recoveries of ^{15}N tracers typically do not sum to 100%). If we required synthetic process models to meet both the criteria that (i) all modeled pool and flux values must match field data within a traditional statistical uncertainty and (ii) the model must be self-consistent mathematically, these could be mutually exclusive requirements.

Clearly, methodological progress is needed in the manner in which we use models to gain increased insight into natural systems. The problem I refer to here as over-determinacy is a serious hurdle that will require creative solutions to overcome. The heuristic value of inverse modeling or exploratory modeling are key areas for future research (Bankes 1993; Luo et al. 2001; Currie et al. 2004). Future research should also explore the use of alternate statistical underpinnings, such as Bayesian inference, that allow the incorporation of multiple modes of evidence including indirect evidence supplied by complex causative models (Reckhow 2003). Progress is needed in developing computational approaches, in improving consistency between conceptual models and observational techniques, and in developing creative methodological approaches to combine evidence from field and modeling investigations. Since the field of isotope modeling in ecosystem biogeochemistry is young, much progress is possible in each of these areas.

Acknowledgments

This work was partially supported by grants from the National Science Foundation (DEB-0235380) and from the USDA Forest Service, Northeastern Research Cooperative (Grant 01-CA-11242343-051). I thank Dr Don Zak for comments and suggestions on the manuscript and Drs Wim Wessel and Knute Nadelhoffer for insightful conversations.

References

Aber, J.D. & Federer, C.A. (1992) A generalized, lumped-parameter model of photosynthesis, evapotranspiration and net primary production in temperate and boreal forest ecosystems. *Oecologia*, **92**, 463–474.

Aber, J.D., Melillo, J.M. & McClaugherty, C.A. (1990) Predicting long-term patterns of mass loss, nitrogen dynamics, and soil organic matter formation from initial fine litter chemistry in temperate forest ecosystems. *Canadian Journal of Botany*, **68**, 2201–2208.

Agren, G.I. & Bosatta, E. (1996) *Theoretical Ecosystem Ecology. Understanding Element Cycles*. Cambridge University Press, Cambridge.

Alperin, M.J., Reeburgh, W.S. & Whiticar, M.J. (1988) Carbon and hydrogen isotope fractionation resulting from anaerobic methane oxidation. *Global Biogeochemistry Cycles*, **2**, 297–288.

Amundson, R. & Baisden, W.T. (2000) Stable isotope tracers and mathematical models in soil organic matter studies. In: *Methods in Ecosystem Science* (Eds O. Sala, R.B. Jackson, H.A. Mooney & R.W. Howarth), pp. 117–137. Springer-Verlag, New York.

Balesdent, J., Mariotti, A. & Guillet, B. (1987) Natural C-13 abundance as a tracer for the studies of soil organic matter dynamics. *Soil Biology and Biochemistry*, **19**, 25–30.

Balesdent, J., Wagner, G.H. & Mariotti, A. (1988) Soil organic matter turnover in long-term field experiments as revealed by carbon-13 natural abundance. *Soil Science Society of America Journal*, **52**, 118–124.

Bankes, S. (1993) Exploratory modeling for policy analysis. *Operations Research*, **41**, 435–449.

Clark, F.E. (1977) Internal cycling of 15nitrogen in shortgrass prairie. *Ecology*, **58**, 1322–1333.

Conover, R.J. & Francis, B. (1973) The use of radioactive isotopes to measure the transfer of materials in aquatic food chains. *Marine Biology*, **18**, 272–283.

Cottingham, K.L., Bade, D.L., Cardon, Z.G., et al. (2003) Increasing modeling savvy: strategies to advance quantitative modeling skills for professionals within ecology. In: *Models in Ecosystem Science* (Eds C.D. Cahnham, J.J. Cole & W.K. Lauenroth), pp. 429–436. Princeton University Press, Princeton, NJ.

Currie, W.S. (2003) Relationships between carbon turnover and bioavailable energy fluxes in two temperate forest soils. *Global Change Biology*, **9**, 919–929.

Currie, W.S. & Aber, J.D. (1997) Modeling leaching as a decomposition process in humid, montane forests. *Ecology*, **78**, 1844–1860.

Currie, W.S. & Nadelhoffer, K.N. (1999) Dynamic redistribution of isotopically labelled cohorts of nitrogen inputs in two temperate forests. Ecosystems 2, 4–18.

Currie, W.S., Nadelhoffer, K.J. & Aber, J.D. (1999) Soil detrital processes controlling the movement of ^{15}N tracers to forest vegetation. *Ecological Applications*, **9**, 87–102.

Currie, W.S., Nadelhoffer, K.J. & Aber, J.D. (2004) Redistributions of ^{15}N highlight turnover and replenishment of mineral soil organic N as a long-term control on forest C balance. *Forest Ecology and Management*, **196**, 109–127.

Davidson, E.A., Hart, S.C. & Firestone, M.K. (1991) Measuring gross nitrogen mineralization, immobilization, and nitrification by ^{15}N isotope dilution in intact soil cores. *Journal of Soil Science*, **42**, 335–349.

Gardner, R.H. & Urban, D. (2003) Model validation and testing: Past lessons, present concerns, future prospects. In: *Models in Ecosystem Science* (Eds C.D. Cahnham, J.J. Cole & W.K. Lauenroth), pp. 184–203. Princeton University Press, Princeton, NJ.

Goodale, C.L., Lajtha, K. Nadelhoffer, K.J. Boyer E.W. & Jaworski N.A.(2002) Forest nitrogen sinks in large eastern U.S. watersheds: estimates from forest inventory and an ecosystem model. *Biogeochemistry*, **57/58**, 239–266.

Hart, S.C., Stark, J.M., Davidson, E.A. & Firestone, M.K. (1994) Nitrogen mineralization, immobilization, and nitrification. In: *Methods of Soil Analysis. Part 2. Microbiological and Biochemical Properties* (Eds R.W. Weaver, S. Angle & P. Bottomley), pp. 987–1018. Soil Science Society of America, Madison, WI.

Hauck, R.D., Meisinger, J.J. & Mulvaney, R.L. (1994) Practical considerations in the use of nitrogen tracers in agricultural and environmental research. In: *Methods of Soil Analysis. Part 2. Microbiological and Biochemical Properties* (Eds R.W. Weaver, S. Angle & P. Bottomley), pp. 907–950. Soil Science Society of America, Madison, WI.

Hobbie, E.A., Macko, S.A. & Shugart, H.H. (1999) Interpretation of nitrogen isotope signatures using the NIFTE model. *Oecologia*, **120**, 405–415.

Kirkham, D. & Bartholomew, W.V. (1954) Equations for following nutrient transformations in soil, utilizing tracer data. *Soil Science Society of America Proceedings*, **18**, 33–34.

Koopmans, C. & van Dam, D. (1998) Modelling the impact of lowered atmospheric nitrogen deposition on a nitrogen saturated ecosystem. *Water, Air and Soil Pollution*, **104**, 181–203.

Lajtha, K. (2000) Ecosystem nutrient balance and dynamics. In: *Methods in Ecosystem Science* (Eds O. Sala, R.B. Jackson, H. Mooney & R.W. Howarth), pp. 249–264. Springer-Verlag, New York.

Lajtha, K. & Michener, R.H. (1994) *Stable Isotopes in Ecology and Environmental Science*. Blackwell Scientific Publications, Oxford, 316 pp.

Lajtha, K., Jarrell, W., Johnson, D.W. & Sollins, P. (1999) Collection of soil solution. In: *Standard Soil Methods for Long-Term Ecological Research* (Eds P. Robertson, D. Coleman, C. Bledsoe & P. Sollins), pp. 115–142. Oxford University Press, Oxford.

Luo, Y.L., Wu, J.A., Andrews, L., et al. (2001) Elevated CO_2 differentiates ecosystem carbon processes: deconvolution analysis of Duke Forest FACE data. *Ecological Monographs*, **71**, 357–376.

Magid, J., Mueller, T., Jensen, L.S. & Nielsen, N.E. (1997) Modelling the measurable: interpretation of field-scale CO_2 and N-mineralization, soil microbial biomass and light fractions as indicators of oilseed rape, maize and barley straw decomposition. In: *Driven by Nature: Plant Litter Quality and Decomposition* (Eds G. Cadisch & K.E. Giller), pp. 349–362. CAB International, Wallingford.

McClain, M.E., Boyer, E.W., Dent, C.L., et al. (2003) Biogeochemical hot spots and hot moments at the interface of terrestrial and aquatic ecosystems. *Ecosystems*, **6**, 301–312.

Nadlehoffer, K.J. & Fry, B. (1994) Nitrogen isotopes in forest ecosystems. In: *Stable Isotopes in Ecology and Environmental Science* (Eds K. Lajtha & R.M. Michener), pp. 22–44. Blackwell Scientific Publishers, Oxford.

Nadelhoffer, K.J., Downs, M. & Fry, B. (1999) Sinks for ^{15}N-enriched additions to an oak forest and a red pine plantation. *Ecological Applications*, **9**, 72–86.

Nadelhoffer, K.J., Colman, B.P., Currie, W.S., Magill, A. & J.D. Aber (2004) Decadal scale fates of ^{15}N tracers in oak and pine stands under ambient and elevated N inputs at the Harvard Forest (USA). *Forest Ecology and Management*, **196**, 89–107.

Oreskes, N. (2003) The role of quantitative models in science. In: *Models in Ecosystem Science* (Eds C.D. Cahnham, J.J. Cole & W.K. Lauenroth), pp. 13–31. Princeton University Press, Princeton, NJ.

Oreskes, N., Shrader-Frechette, K. & Belitz, K. (1994) Verification, validation, and confirmation of numerical models in the earth sciences. *Science*, **263**, 641–646.

Parton, W.J., Schimel, D.S., Cole, C.V. & Ojima, D.S. (1987) Analysis of factors controlling soil organic matter levels in great plains grasslands. *Soil Science Society of America Journal*, **51**, 1173–1179.

Parton, W.J., Stewart, J.W.B. & Cole, C.V. (1988) Dynamics of C, N, P and S in grassland soils: a model. *Biogeochemistry*, **5**, 109–131.

Pastor, J. & Post, W.M. (1986) Influence of climate, soil moisture, and succession on forest carbon and nitrogen cycles. *Biogeochemistry*, **2**, 3–27.

Ponsard, S. & Arditi, R. (2000) What can stable isotopes (δ-^{15}N and δ-^{13}C) tell about the food web of soil macro-invertebrates? *Ecology*, **81**, 852–864.

Rastetter, E.B. (2003) The collision of hypotheses: What can be learned from comparisons of ecosystem models? In: *Models in Ecosystem Science* (Eds C.D. Cahnham, J.J. Cole & W.K. Lauenroth), pp. 211–224. Princeton University Press, Princeton, NJ.

Rastetter, E.B., McKane, R.B., Shaver, G.R. & Melillo, J.M. (1992) Changes in C-storage by terrestrial ecosystems – How C–N interactions restrict responses to CO_2 and temperature. *Water, Air and Soil Pollution*, **64**, 327–344.

Rastetter, E.B., Kwaitkowski, B.L. & McKane, R.B. (2005) A stable isotope simulator that can be coupled to existing mass-balance models. *Ecological Applications*, **15**, 1772–1782.

Reckhow, K.H. (2003) Bayesian approaches in ecological analysis and modeling. In: *Models in Ecosystem Science* (Eds C.D. Cahnham, J.J. Cole & W.K. Lauenroth), pp. 168–183. Princeton University Press, Princeton, NJ.

Reiners, W.A. (1986) Complementary models for ecosystems. *American Naturalist*, **127**, 59–73.

Riha, S.J., Campbell, G.S. & Wolfe, J. (1986) A model of competition for ammonium among heterotrophs, nitrifiers, and roots. *Soil Science Society of America Journal*, **50**, 1463–1466.

Robertson, G.P., Wedin, D., Groffman, P., et al.(1999) Soil carbon and nitrogen availability: Nitrogen mineralization, nitrification, and soil respiration potentials. In: *Standard Methods for Long-Term Ecological Research* (Eds G.P. Robertson, D.C. Coleman, C.S. Bledsoe & P. Sollins), pp. 258–271. Oxford University Press, New York.

Schimel, D.S. (1993) *Theory and Application of Tracers*. Academic Press, San Diego. 119 pp.

Shearer, G. & Kole, D.H. (1986) N_2-fixation in field settings: Estimations based on natural ^{15}N abundance. *Australian Journal of Plant Physiology*, **13**, 699–756.

Tietema, A. & van Dam, D. (1996) Calculating microbial carbon and nitrogen transformations in acid forest litter with 15-N enrichment and dynamic simulation modeling. *Soil Biology and Biochemistry*, **28**, 953–965.

Tietema, A. & Wessel, W.W. (1992) Gross nitrogen transformations in the organic layer of acid forest ecosystems subjected to increased atmospheric nitrogen input. *Soil Biology and Biochemistry*, **24**, 943–950.

Van Dam, D. & van Breemen, N. (1995) NICCCE: A model for cycling of nitrogen and carbon isotopes in coniferous forest ecosystems. *Ecological Modelling*, **79**, 255–275.

Vitousek, P., Chadwick, O.A. Matson, P., et al. (2003) Erosion and the rejuvenation of weathering-derived nutrient supply in an old tropical landscape. Ecosystems 6, 762–772.

Wolf, D.C., Legg, J.O. & Boutton, T.W. (1994) Isotopic methods for the study of soil organic matter dynamics. In: *Methods of Soil Analysis. Part 2. Microbiological and Biochemical Properties* (Eds R.W. Weaver, S. Angle & P. Bottomley), pp. 865–906. Soil Science Society of America, Madison, WI.

Yanai, R.D., Currie, W.S. & Goodale, C.L. (2003) Soil carbon dynamics following forest harvest: an ecosystem paradigm reconsidered. *Ecosystems*, **6**, 197–212.

Zak, D.R., Pregitzer, K.S., Holmes, W.E., Burton, A.J. & Zogg, G.P. (2004) Anthropogenic N deposition and the fate of ^{15}NO$_3$- in a northern hardwood ecosystem. *Biogeochemistry*, **69**, 143–157.

CHAPTER 14

Compound-specific stable isotope analysis in ecology and paleoecology

RICHARD P. EVERSHED, IAN D. BULL, LORNA T. CORR,
ZOE M. CROSSMAN, BART E. VAN DONGEN,
CLAIRE J. EVANS, SUSAN JIM, HAZEL R. MOTTRAM,
ANNA J. MUKHERJEE, AND RICHARD D. PANCOST

Introduction

Compound-specific stable isotope ratio mass spectrometry enables the molecular specificity and isotopic signature of compounds to be exploited concomitantly, to provide a powerful tool for tracing the origin and fate of organic matter in both extant and fossil ecosystems. The scope of applications is enormous, with the pollution, paleoenvironmental and archeological fields being examples of the increasing number of subject areas benefiting from the use of compound-specific approaches. Indeed, as will be shown below such approaches can be used to investigate chemical and biological processes and metabolism in both whole ecosystems (marine, terrestrial, freshwater) and individual organisms of all classes, i.e. microbes to mammals.

Even before the advent of new on-line technologies, the stable isotopic compositions of individual compounds had begun to be rigorously investigated, e.g. Abelson & Hoering (1961). Such studies relied upon classic purification methods and required relatively large sample sizes. The results of such investigations underpin much of our biochemical understanding of interpretations of bulk stable isotope determinations and certainly provided the impetus for the development of improved instrumental approaches (Galimov 1981). The introduction of high performance liquid chromatography provided an effective means of isolating milligram quantities of highly purified compounds of widely varying polarity for off-line combustion and stable isotope analysis. The major breakthrough in compound-specific stable isotope analysis came with the commercial production of the gas chromatograph/combustion/isotope ratio mass spectrometer (GC/C/IRMS) in the early 1990s. The technical aspects of such instruments are discussed in detail below. Their availability resulted in an explosion of compound-specific stable isotope analyses, particularly amongst the organic geochemical community, for the purposes of paleoenvironmental reconstruction (Merritt et al. 1994). However, opportunities rapidly began to be recognized for the application of GC/C/IRMS by members of other scientific communities, including ecologists (Meier-Augenstein 1999; Boschker & Middelburg 2002).

Why use compound-specific stable isotopes?

There are a number of compelling reasons for employing compound-specific stable isotope approaches (rather than more traditional bulk measurements), which derive from the fact that:

1 Different biochemical components, even those within a single organism, can possess different stable isotope values, e.g. $\delta^{13}C$ values of carbohydrates and lipids in plants, and isoprenoids vs. aliphatic lipids.

2 Structurally similar biochemical components of ecological materials can derive from a range of sources potentially exhibiting different stable isotopic signatures, e.g. palmitic acid in soil can derive from plants, invertebrates, or microbes.

3 Biogenic organic matter, either living or dead, is chemically complex such that changes in the bulk stable isotope value may occur as a result of changes in chemical composition, i.e. losses or gains of components with stable isotope values that differ from the bulk (mean) stable isotope value may be erroneously interpreted as fractionation phenomena, e.g. preferential degradation of isotopically heavy components would lead to an overall lowering of bulk stable isotope values.

4 Stable isotopic analysis of individual components in chemically complex materials, e.g. sedimentary biomarkers, can reveal characteristic stable isotopic signatures of contributors mediating processes that would otherwise be masked in bulk stable isotope values.

5 The complementary use of structurally diagnostic biomarkers together with their compound-specific stable isotope values provides biological process or biochemical pathway-based information inaccessible through bulk stable isotope analyses.

6 Different biochemical components, even identical components within different pools, of a living organism can possess significantly different turnover times that would be undetectable by bulk analysis and may even lead to erroneous interpretations of trends seen in recorded stable isotope values, especially in temporal studies, e.g. essential and nonessential components or structural vs. storage substances.

7 Where genuine kinetic isotope-based fraction effects exist their source can only be determined through stable isotopic assessments at the level of the biochemical component and specific pathway.

It is therefore essential that in order to use light stable isotopes effectively in ecological research a mechanistic understanding of the biochemical factors that underpin stable isotope signals be developed. This principle applies no matter what stable isotope or combination of isotopes is being employed or what ecosystem is under investigation. Unless we strive for a level of understanding that links biochemistry to stable isotope composition it is unlikely that we can ever rigorously interpret the results of bulk stable isotopes or exploit the use of light stable isotopes in ecological research to their fullest potential.

Compound-specific stable isotope analyses are more complex to undertake than bulk stable isotope analyses and require that careful consideration be given to sample preparation protocols. Such factors that need to be considered are:

1 What are the target compound(s)?
2 Whether there is a possibility of compounds existing in more than one physical or chemical state within a complex environmental or biological matrix?
3 Whether or not the target compounds are amenable to gas chromatography (GC) analysis directly or will require derivatization?
4 Which is the most suitable derivatizing agent for GC/C/IRMS analysis; this is not necessarily the most appropriate derivative if only GC analysis was intended (the specific considerations to GC derivative selection are given later)?
5 What is the most suitable GC column (stationary phase and column dimensions)?
6 Finally, and probably most important of all, are the target compounds fully separable from other eluting compounds, i.e. can **baseline resolution** be achieved?

Each of these factors will be dealt with in the following sections.

Analytical considerations in compound-specific stable isotope analysis

The overall aim of any compound-specific stable isotope analysis is to provide accurate stable isotope value(s) for a specific component(s) of what is likely to be a biochemically complex matrix containing many tens, hundreds or even thousands of components of widely varying chemical and physical states. Thus, the aim will be to purify that compound avoiding, or accounting for, protocol-induced isotope effects such that the recorded stable isotope value will be comparable, within reasonable analytical error, to that of the compound in its natural state. Given that the primary means of obtaining compound-specific stable isotope values will be via GC/C/IRMS, then sample preparation schemes will comprise the following steps: (i) extraction, (ii) separation, and (iii) derivatization (in the case of functionalized compounds). General and specific features of such schemes are presented in turn below.

Sample preparation

The majority of the practical aspects of sample preparation have been covered by Teece & Fogel (2004), hence, only a brief summary will be

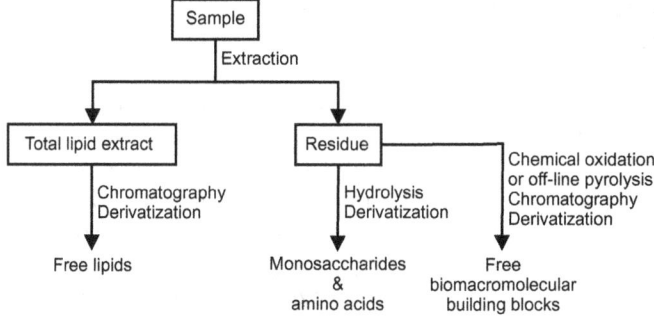

Figure 14.1 A schematic overview of an isolation procedure.

given here. Figure 14.1 shows a typical protocol for the isolation of the most common compounds targeted for compound-specific stable isotope analysis. Before extraction, samples are dried and crushed to ensure homogeneity and increase the effectiveness of solvent penetrating the sample matrix. Methods available for lipid extraction include: Soxhlet (large samples; tens to hundreds of grams; heat stable), ultrasonication (small samples, <10 g) and liquid/liquid extraction. The Bligh-Dyer method is specifically designed for the extraction of fresh biological tissues (Bligh & Dyer 1959; Smedes & Askland 1999; Manirakiza et al. 2001). No matter which extraction method is employed the resulting total lipid extract (TLE) is then further separated into different compound classes using various chromatographic methods (Kim & Salem 1990; Touchstone 1991, 1993, 1995; Alvarez & Touchstone 1992; Myher & Kuksis 1995; Abidi, 2001). Occasionally, a single chromatographic technique does not yield base-line-resolved peaks in the GC/C/IRMS analysis and a combination of techniques is required. Whichever technique(s) is selected it is important to verify whether or not aspects of the analytical protocol will introduce an isotopic fractionation effect by analysis of reference compounds of known stable isotopic composition.

Many important components of ecological materials cannot be analyzed directly by GC. For example, complex lipids require chemical cleavage to yield GC amenable components, i.e. commonly occurring phospholipids are saponified and methylated to generate fatty acid methyl esters (FAMEs; Crossman et al. 2004). Likewise, solvent unextractable biopolymers such as carbohydrates, proteins, lignin, or aliphatic biomacromolecules, e.g. suberin or cutin, must be chemically, enzymatically, or pyrolytically cleaved (Hedges & Ertel 1982; Poole & van Bergen 2002; Anwar et al. 2004; Poole et al. 2004; Stefanova et al. 2004). Care must be taken to ensure that the cleavage reactions are complete, since incomplete reactions may result in kinetic isotope effects (Jim et al. 2003a).

Derivatizations for compound-specific stable isotope analysis

GC/C/IRMS instrumentation enables the compound-specific isotope analysis of individual organic compounds, for example, n-alkanes, fatty acids, sterols, amino acids, and monosaccharides, extracted and purified from bulk organic materials. The principle caveat of compound-specific work is the requirement for chemical modification, or derivatization, of compounds containing polar functional groups primarily to enhance their volatility prior to introduction to the GC/C/IRMS. Table 14.1 summarizes the most commonly employed procedures for derivatization of polar, nonvolatile compounds for compound-specific stable isotope analysis using GC/C/IRMS.

The addition of a derivative group (or several derivative groups in the case of polyfunctionalized compounds, such as amino acids and monosaccharides) introduces exogenous carbon which ultimately alters the $\delta^{13}C$ value of the compound of interest. Derivatization of fatty acids and sterols, to FAMEs and sterol trimethylsilyl ethers, respectively, is relatively straightforward since the resulting sample-to-derivative carbon molar ratio is high, resulting in minimal analytical error. Additionally, derivatization reactions, such as esterification and silylation are typically rapid and quantitative thereby precluding kinetic isotope effects (Rieley 1994).

In contrast, derivatization of small, polyfunctional compounds, such as amino acids and monosaccharides presents a more complex analytical challenge. The most prevalent procedures for derivatization of amino acids for GC analysis involve either silylation, to *tert*-butyldimethylsilyl (tBDMS) or trimethylsilyl (TMS) derivatives or a multi-step procedure, involving esterification of the carboxylic acid group with an acidified alcohol and acylation of amino, hydroxyl, and thiol groups with an anhydride, to form N(O,S)-acyl alkyl esters (Demmelmair & Schmidt 1993; Metges et al. 1996; Macko et al. 1997; Metges & Daenzer 2000; Docherty et al. 2001; O'Brien et al. 2002; Jim et al. 2003a; Corr et al. 2005). While trialkylsilylation represents a rapid, one-step reaction, trialkylsilylated derivatives possess relatively poor stability and the addition of several carbon atoms in the derivative alters the $\delta^{13}C$ value of the compound of interest, which results in considerable analytical error, especially in small molecules (Derrien et al. 2003). Also, the requirement of baseline resolution, a prerequisite for obtaining reliable isotope measurements, is rarely achieved using silylated derivatives in these cases due to their apolar character. Derivatization of amino acids and monosaccharides with TMS groups can involve the formation of several products of one compound (i.e., mono-, di-, and tri- TMS, and isomeric derivatives; Evershed 1993; Meier-Augenstein 1997). Finally, it has been postulated that silicon in *t*BDMS groups forms silicon carbide (SiC_x) in the oxidation reactor of the GC/C/IRMS, resulting in incomplete analyte combustion to CO_2 (Shinebarger et al. 2002; Derrien et al. 2003). However, after nearly 10 years

Table 14.1 Summary of the most commonly employed procedures for derivatization of polar, nonvolatile compounds for compound-specific stable isotope analysis using GC/C/IRMS.

Procedure	Functional group	Mechanism	Reagent	Product	References
Silylation (sterols, alcohols, amino acids, monosaccharides, lignin phenols)	$-OH$ $-CO_2H$ $-NH_2$ $-NHR$ $-SH$	*tert*-butyldimethyl-silylation (tBDMS) Trimethylsilylation (TMS)	*N-tert*-butyldimethylsilyl-*N*-methyl-trifluoroacetamide (MTBSTFA) *N,O*-bis(trimetylsilyl) trifluoro-acetamide (BSTFA)	O/N/S-TMS O/N/S-tBDMS	Lockheart et al. (1997) Stott & Evershed (1996) Derrien et al. (2003)
Esterification (fatty acids, amino acids)	$-CO_2H$	Methylation $n/2$-propylation	BF_3/methanol Acetyl chloride/methanol Acetyl chloride/ $n/2$-propanol	CO_2CH_3 $CO_2(CH_2)_2CH_3$ $CO_2CH(CH_3)_2$	Howland et al. (2003) Docherty et al. (2001)
Acylation (amino acids, monosaccharides)	$-OH$ $-NH_2$ $-NHR$ $-SH$	Acetylation Trifluoroacetylation Pivaloylation	Acetic anhydride Trifluoroacetic anhydride Pivaloyl chloride	$O/N/S-C=OCH_3$ $O/N/S-C=OCF_3$ $O/N/S-C=OC(CH_3)_3$	Demmelmair & Schmidt (1993) Docherty et al. (2001) Metges & Daenzer (2000)
Methylboronatrion (MBA) (monosaccharides)	$-OH$	Cyclization of two adjacent $-OH$ groups	Methane boronic acid (MBA)		Van Dongen et al. (2001)

of undertaking GC/C/IRMS analyses of cholesterol as its TMS ether, no undue deterioration of instrument performance has ever been observed (Stott & Evershed 1996; Jim 2000).

The alternative approach, acylation, represents a widely employed method of derivatizing hydroxyl groups of amino acids and monosaccharides, despite the kinetic isotope effect (KIE) associated with the procedure, which precludes the direct calculation of compound-specific $\delta^{13}C$ values via a simple mass balance equation. However, because such reactions have been shown to be reproducible for each amino acid and monosaccharide, the KIE can be accounted for by employing empirical correction factors ($\delta^{13}C_{corr}$) to calculate the "effective" $\delta^{13}C$ value of the reagent where fractionation occurs (Demmelmair & Schmidt 1993; Rieley 1994; Docherty et al. 2001). Hence, three acylation procedures, in combination with esterification, are now routinely employed for amino acid $\delta^{13}C$ measurements: acetylation (+2C; Demmelmair & Schmidt 1993; Metges et al. 1996), trifluoroacetylation (+2C; Macko et al. 1997; Docherty et al. 2001; O'Brien et al. 2002; Howland et al. 2003; Jim et al. 2003a, 2006) and pivaloylation (+5C; Metges et al. 1996; Metges & Petzke 1997; Metges & Daenzer 2000), each advantageous over TMS (+3C) and *t*BDMS (+6C) esters because of their overall lower carbon addition to each amino acid. *N*-trifluoroacetyl iso-propyl (*N*TFA-IP) esters are extensively employed in the stable carbon isotope analysis of amino acids due to their excellent chromatographic properties compared with *N*-acetyl iso-propyl (*N*AIP) esters. However, their compatibility with GC/C/IRMS is now uncertain because combustion of fluorinated derivatives irreversibly poisons the oxidation reactor, forming extremely stable CuF_2 and NiF_2 products with associated reduction in oxidation efficiency (Meier-Augenstein 1997). This analytical challenge has, nevertheless, been resolved by analysing amino acid *N*AIP esters on a high-polarity stationary phase to prevent the previously reported poor chromatography and co-elution problems identified for these derivatives on the traditionally employed non-/low-polarity phases.

Existing acetylation techniques for monosaccharides have also been adapted for determining $\delta^{13}C$ values of these compounds, of which the most common is the formation of alditol acetates via a two step-reaction, comprising their reduction with sodium borohydride, followed by acetylation of all alcohol groups with acetic anhydride using either pyridine (Moers et al. 1993; Macko et al. 1998) or *N*-methylimidazole (Blakeney et al. 1983; Docherty et al. 2001) as a catalyst. A rigorous approach to error propagation is essential in the use of these derivatives to combine both the imprecision associated with the use of correction factors to account for the KIE and also the high derivative-to-sample carbon molar ratio of alditol acetates (Rieley 1994; Docherty et al. 2001).

Van Dongen et al. (2001) avoided the KIE associated with acetylation reactions in their derivatization of monosaccharides for carbon isotope work, instead reacting analytes containing two or more adjacent hydroxyl groups with methylboronic acid, followed by trimethylsilylation of the remaining hydroxyl groups using N,O-bis(trimethylsiyl)trifluoroacetamide (BSTFA). While this method also involves lower carbon addition (+2/3C) compared with alditol acetate derivatization (+10/12C), reduced yields have been observed for several monosaccharides indicating an incomplete reaction with associated isotopic fractionations (van Dongen et al. 2001). It was subsequently postulated that TMS groups could potentially substitute methylboronic acid on hydroxyl groups, generating by-products (Derrien et al. 2003; Gross & Glaser 2004).

Instrumentation related aspects of compound-specific stable isotope determinations

Determining the $\delta^{13}C$ and $\delta^{15}N$ values of individual compounds

GC/C/IRMS was first demonstrated by Matthews & Hayes (1978). However, it was somewhat later that Barrie et al. (1984) coupled a GC, via a combustion interface, to a dual collector mass spectrometer to produce the forerunner of today's GC/C/IRMS instruments. Even so, true determinations of $\delta^{15}N$ values of individual compounds by GC/C/IRMS remained elusive until finally demonstrated by Merritt et al. (1991). More recently the precision of GC/C/IRMS instruments has been improved further with uncertainties in $\delta^{13}C$ values as low as ±0.5‰ for samples containing 5 pmol C and ±0.1‰ for 100 pmol analytes having been demonstrated (Merritt & Hayes 1994). Instruments available commercially today all conform to the same general principals of design.

Figure 14.2a depicts a generalized schematic of a GC/C/IRMS instrument configured for determinations of $\delta^{13}C$ (or $\delta^{15}N$) values of individual compounds. Briefly, mixtures of compounds are separated by high resolution capillary GC then individually combusted online over a catalyst (CuO/Pt, 850°C or CuO/NiO/Pt, 940°C; Merritt et al. 1995) generating CO_2 and H_2O. For the determination of $\delta^{15}N$ values or $\delta^{13}C$ values of compounds containing nitrogen (N_2O is generated and causes isobaric interference with m/z 44 and 45; Metges & Daenzer 2000) the effluent is then passed through a second reactor where nitrogen oxides are catalytically (Cu, 600°C) reduced to N_2 (Brand et al. 1994). H_2O is removed by a water separator, typically comprising a length of water permeable Nafion™ tubing, thereby avoiding the formation of HCO_2^+ ions that would otherwise result in isobaric interference with $^{13}CO_2$ (Leckrone & Hayes 1998) it should be noted that the efficiency of this process is temperature dependent (Leckrone & Hayes 1997). For determination of $\delta^{13}C$ values the remaining CO_2 is introduced into a MS equipped with a triple collector comprising three Faraday cups monitoring simultaneously m/z 44, 45 and 46 corresponding to the ions of the three isotopomers $^{12}C^{16}O_2$, $^{13}C^{16}O_2$ and $^{12}C^{18}O^{16}O$, respectively. For the determination of $\delta^{15}N$ values, the eluting CO_2 is retained in a cryogenic trap to avoid the production of isobaric CO^+ resulting from the unimolecular decomposition of CO_2^+ ions (Brand et al. 1994). Remaining N_2 is then introduced in an identical manner but this time monitoring simultaneously m/z 28 and 29 using the first two Faraday cups corresponding to the isotopomers $^{14}N^{14}N$ and $^{15}N^{14}N$. The resultant output currents are then amplified, digitized, and recorded by computer, which then integrates each signal and calculates the corresponding stable carbon or nitrogen isotope ratio, represented by the signal, relative to either co-injected standards or a gas standard, returning it as a δ value (Merrit et al. 1994; Ricci et al. 1994). As with any analytical technique a number of methods have been developed, ostensibly by the user base, to maximise the accuracy and precision of determinations.

The single most important requirement in performing a valid and robust determination of the $\delta^{13}C$ or $\delta^{15}N$ value of an individual compound by GC/C/IRMS is good chromatographic separation of the target compound(s) as shown in Figure 14.3, achieved by optimization of the GC operating conditions and judicious column selection. Peaks that are not fully resolved (up to 25% co-elution) can still be integrated separately as long as a minimum estimate of analytical error is gained by running the sample at different concentrations (Ricci et al. 1994). For any larger overlap, co-eluting components must be integrated as one peak using the integration software, although the errors associated with such determinations may be substantial. Another consideration when analysing isotopically enriched compounds is the possibility of carryover effects between analytical runs, i.e residual amounts of an isotopically enriched component from a previous analysis can affect the δ value subsequently determined for the same compound from a different experiment (Mottram & Evershed 2003). In resolving this problem points in the flowpath of the GC prone to the accumulation of such residues, e.g. injector liners and metal connectors, should be adequately heated and, when necessary, changed to avoid the build-up of isotopically enriched contaminants; checks with standards to assess such effects should be made at regular intervals.

In addition to the reference gas calibrant it is important that the performance of the GC/C/IRMS instrument be constantly monitored using a suite of compounds of known relative stable isotopic composition. Such references should ideally belong to the same compound class as the target compounds since the performance of the instrument for one particular class of compounds will not necessarily replicate that of a different class of compounds (Meier-Augenstein et al. 1996). In addition, a range of homologues should be utilized as a standard mixture thereby assessing the performance of the instrument across the entire temperature range utilized by the GC. By doing this factors such as leaks and/or blockages at varying temperatures can be quickly identified and resolved.

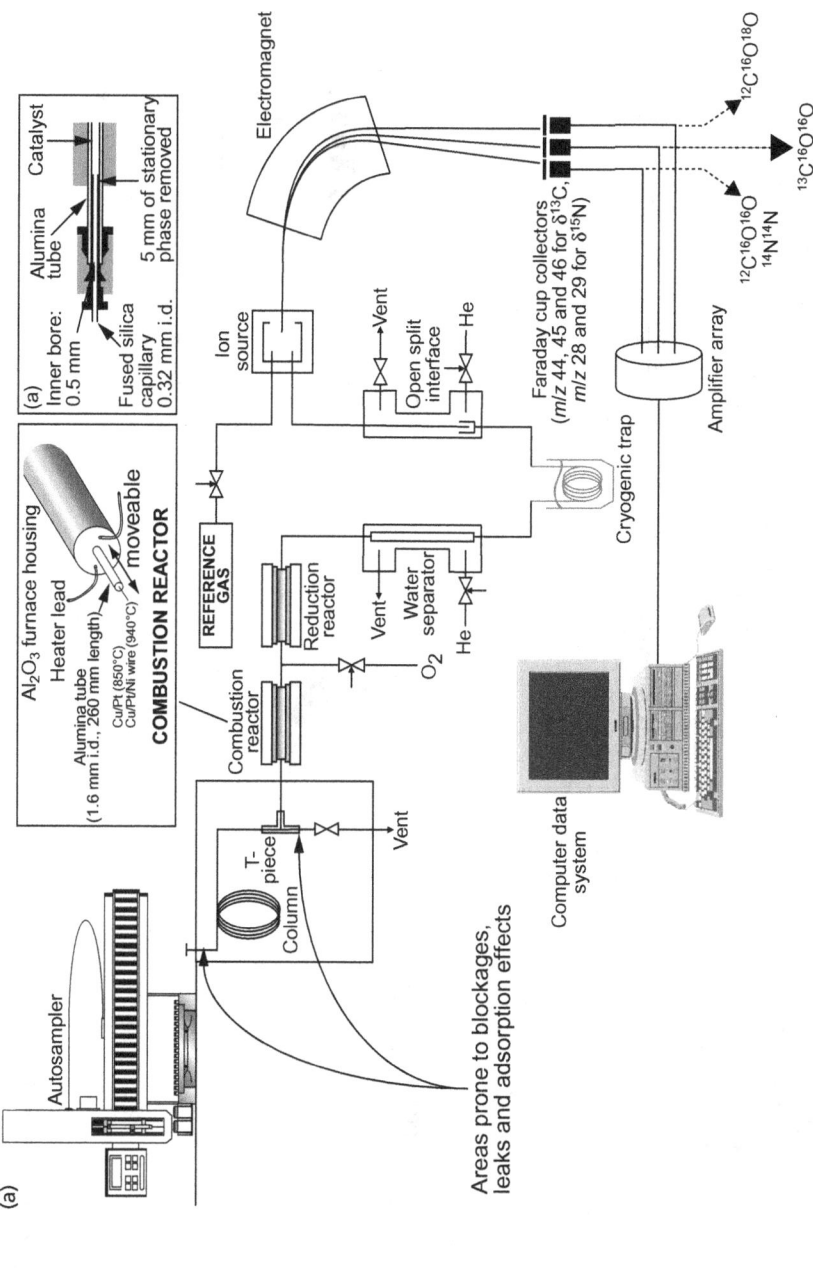

Figure 14.2 A generalized schematics of a GC/C/IRMS configured for (a) δ^{13}C or δ^{15}N analysis. (b) δD analysis. (c) δ^{18}O analysis.

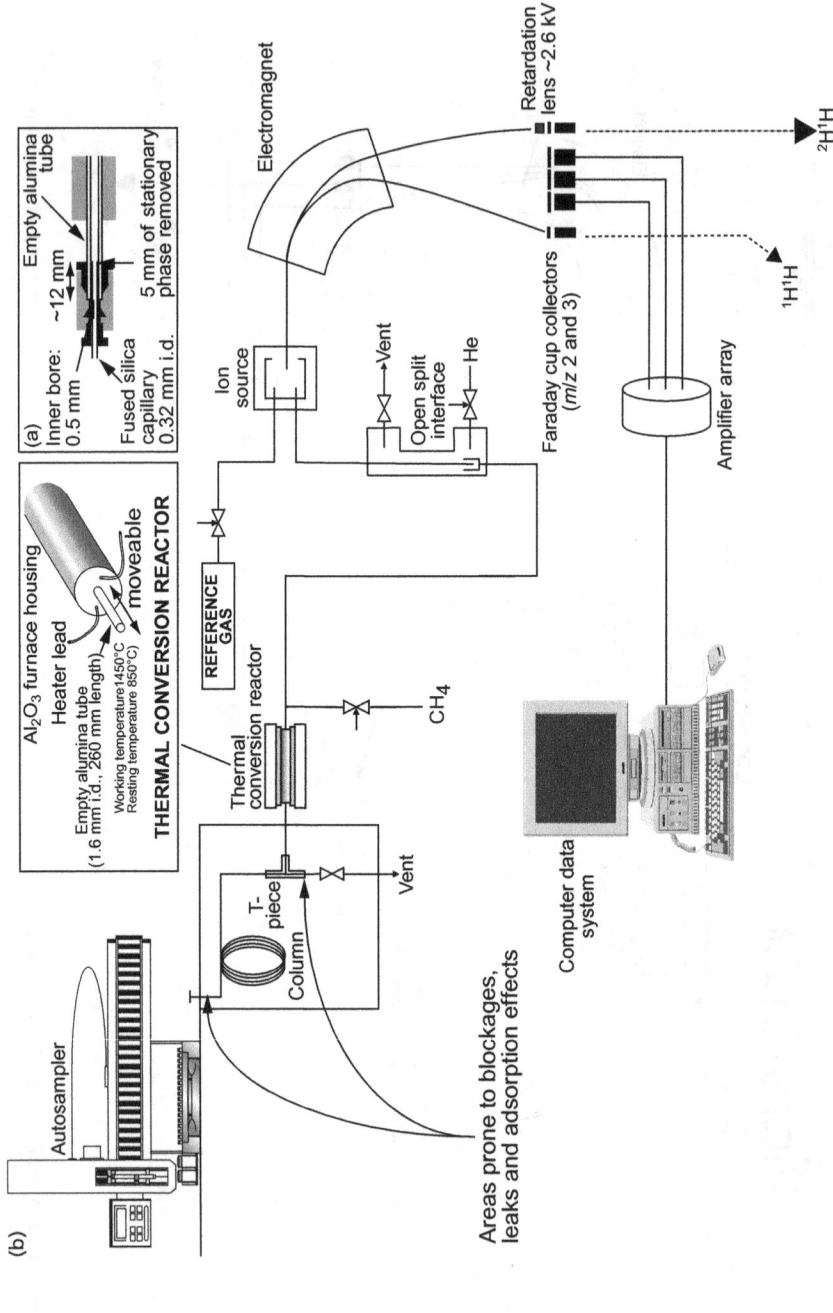

Figure 14.2 (*Continued*)

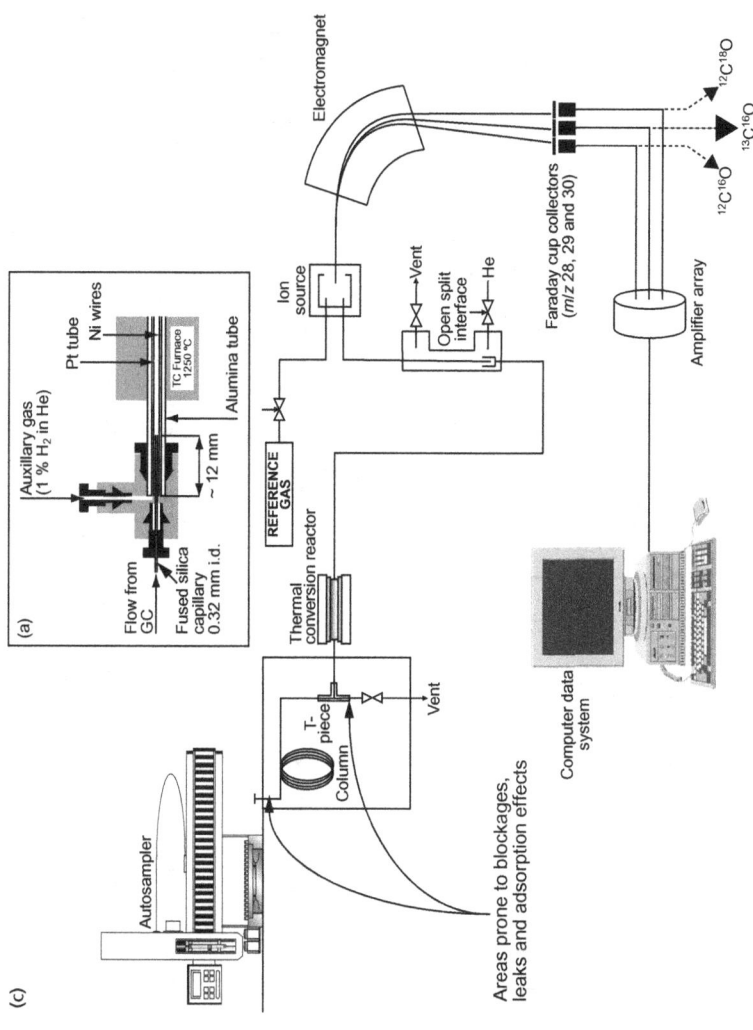

Figure 14.2 (*Continued*)

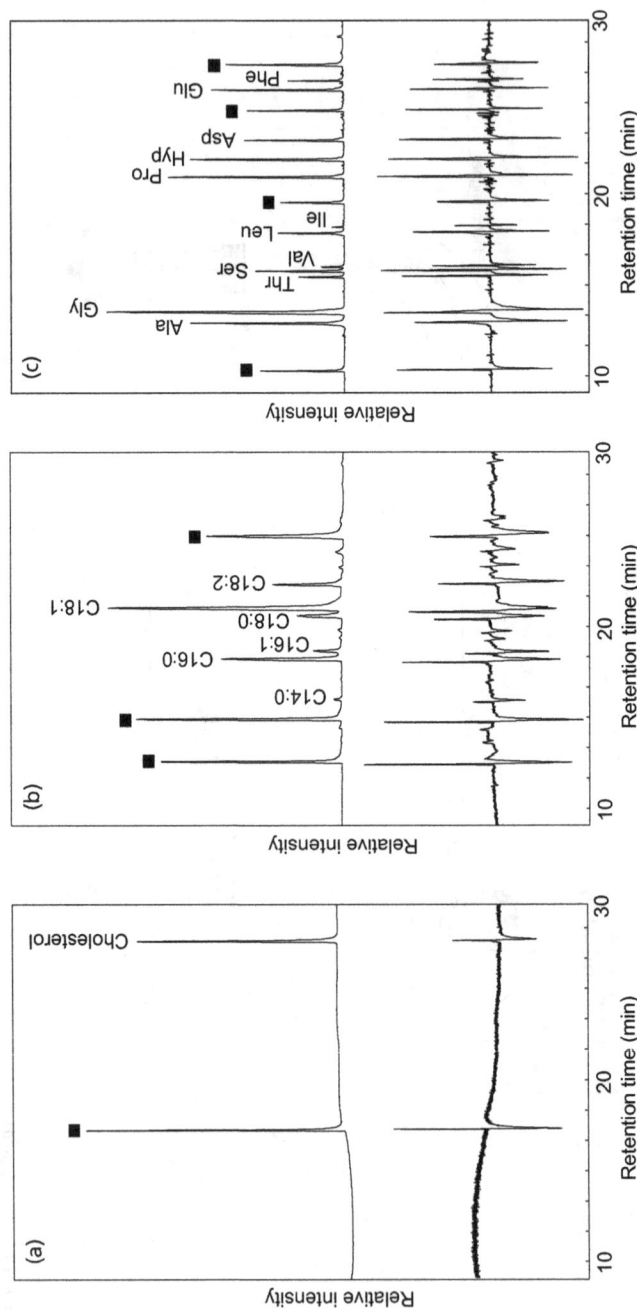

Figure 14.3 GC/C/IRMS chromatograms for biochemical components of pig bone from an isotopically controlled feeding experiment: (a) cholesterol (as its TMS ether), (b) fatty acids (as their methyl esters), and (c) amino acids (as their TFA-IPA derivatives). Peaks annotated with black squares are internal standards. (From Howland et al. 2003.)

One factor of particular importance is the position of the combustion reactor within the heated zone of the furnace. Altering this position may have a significant effect on determined δ values. This is probably due to the partial decomposition and/or pyrolysis of compounds eluting into the alumina reactor prior to the zone of optimal quantitative combustion. This problem may be reduced or resolved by moving the position of the reactor tube within the furnace. In addition, the fused silica capillary connecting the flow path from the GC to the reactor should be adjusted so that it extends beyond the metal connector to line the initial part of the reactor devoid of catalyst and insufficiently hot to enable quantitative combustion of eluting compounds. When making this modification ca. 5mm of the polyimide coating the glass capillary should be removed, use of a gas burner is usually sufficient, to avoid deleterious effects caused by pyrolysis and/or combustion of the stationary phase.

One final important consideration is the degree by which compounds of interest have become enriched in the heavier isotope during the course of a labeling experiment. Analysis of highly isotopically enriched samples, as well as requiring a highly enriched reference, will require an adjustment to the amplification range of the middle Faraday cup corresponding to m/z 45 (or m/z 29) since high levels of ^{13}C or ^{15}N incorporation will quickly saturate the detector when set to a range suitable for natural abundance work. Commercially available GC/C/IRMS instruments now incorporate switchable resistors to enable rapid reconfiguration between amplification ranges. Since the linear range of the amplifiers of different instruments is variable, determining the point of saturation of the middle Faraday cup is of particular importance in determining accurate δ values of compounds from isotopically enriched labeling experiments.

Specific considerations for $\delta^{15}N$ analyses

In addition to the potential problem of isobaric interference from CO^+ ions (m/z 28) there are a number of additional factors that should be considered when attempting compound-specific $\delta^{15}N$ analysis. First, the actual abundance of N in an organic compound is typically <10% (cf. C > 60%), in a gas sample 1 molecule of N_2 contains two N atoms (only 1 in CO_2), the natural abundance of ^{15}N is 0.732% (cf. 1.08% ^{13}C) and N_2 has an ionization efficiency of 70% relative to that of CO_2. All of these factors result in the average N containing organic compound requiring, for $\delta^{15}N$ analysis, a minor isotope abundance around 50 times that required for the corresponding $\delta^{13}C$ analysis (Brand et al. 1994). This problem may be partially overcome by using shorter GC columns (e.g. 30m) with both a smaller internal diameter (e.g. 0.25mm) and thicker stationary phase (e.g. 0.4μm) thereby generating a higher signal to noise ratio and enabling a higher sample loading. An increase in the sensitivity of the MS (e.g. a higher accelerating voltage, optimized tuning) will also help alleviate this problem. The second major drawback relating to compound-specific $\delta^{15}N$ analyses results from the fact that the atmosphere comprises >78% N_2. The risk of introducing atmospheric N_2 into the ion source either by inadequate sample handling or leaks in the flow path of the GC/C/IRMS and thereby compromising the accuracy of determined $\delta^{15}N$ values is particularly high. Those parts of the GC/C/IRMS exposed to repetitive changes in temperature, i.e. those parts associated with the GC oven, are particularly prone to leaking and should be regularly checked and made gas tight. There can be no doubt that, despite its undeniable value in a wide range of applications, $\delta^{15}N$ analysis of individual compounds still remains a nontrivial methodology, which probably accounts for the current paucity of communications reporting the use of the technique.

Determining the δD and $\delta^{18}O$ values of individual compounds

The main difference between compound-specific δD and $\delta^{18}O$ analysis and the analytical techniques outlined above lies in the method by which the analyte gas is generated. For both δD and $\delta^{18}O$ analyses organic compounds are converted online to produce $^1H^1H$, $^2H^1H$, $^{12}C^{16}O$, $^{13}C^{16}O$, and $^{12}C^{18}O$. Several methods have been adopted by different manufacturers and researchers to achieve this chemical conversion, including reduction using a chromium metal catalyst at 1050°C (e.g. Gehre et al. 1996) and pyrolysis in a graphitized alumina reactor at 1450°C (Begley & Scrimgeour 1997; Tobias & Brenna 1997); only the latter technique shall be considered here.

Figure 14.2b is a schematic of a typical GC/C/IRMS instrument configured for compound-specific δD analysis. The combustion reactor has been replaced with a thermal conversion reactor (the term "thermal conversion" is used since there is currently no consensus on how to refer to this reaction and terms such as quantitative pyrolysis or calcination are too specific), which comprises an empty alumina tube maintained at a working temperature of 1450°C. At this temperature compounds eluting from the GC into the reactor are quantitatively converted to H_2, CO, and their corresponding isotopomers (after Burgoyne & Hayes 1998). Since the reaction produces no water the Nafion™ water trap can be omitted. The analyte gases are introduced to the mass spectrometer via an open split and the ion beams corresponding to m/z 2 and 3 are monitored by two Faraday cups for a simultaneous determination of H_2 and HD, respectively. An electrostatic filter is placed inline with the Faraday cup measuring the HD ion beam to remove the effect of the disperse $^4He^+$ ion beam that would otherwise superimpose and corrupt the measurement of the small HD beam (Hilkert et al. 1999). The resulting ion currents are recorded and converted to δD values in a process similar to that detailed above.

One crucial difference is the need to determine the effect that protonation reactions occurring in the ion source have on the determination of δD values.

Such reactions result in the formation of H_3^+ ions. Since the abundance of H_3^+ is proportional to the square of the hydrogen pressure a correction factor, termed an "H_3-Factor", may be determined based on the peak area information obtained from a series of reference gas pulses of different magnitude. The calculations of H_3-Factors and associated corrections have been reviewed previously in detail (Sessions et al. 2001a, 2001b).

Another vital factor for the determination of accurate δD values is the need to ensure that the alumina reactor in the thermal conversion interface has been suitably graphitized prior to performing any analyses. Burgoyne & Hayes (1998) reported that such treatment resulted in better peak shapes and reproducibility. This "conditioning" may be achieved by the direct injection of *n*-hexane or alternatively Bilke & Mosandl (2002) reported that conditioning the reactor by backflushing CH_4 resulted in greater stability over the ensuing sampling period. In addition, the accuracy of δD values may also be improved by the use of a concentrated "sacrificial" compound that elutes prior to the compound(s) of interest; this is a particularly useful method by which the accuracy of δD values determined for early eluting components may be improved. Finally, the flow through the reactor should ideally be fairly low, e.g. ca. $0.8\,mL\,min^{-1}$ and certainly no greater than $1.2\,mL\,min^{-1}$, since an overly short residence time in the central ca. 10 cm zone of optimum performance can result in nonquantitative thermal conversion.

Determinations of compound-specific $\delta^{18}O$ values utilize an instrument configuration very similar to that used for δD analyses. Figure 14.2c depicts a schematic of a typical instrument configuration. The main point of divergence between the two techniques is in the design of the thermal conversion reactor. The empty alumina reactor is replaced by an alumina reactor containing a Pt tube that also contains Ni wire. The function of the Pt tube is to prevent any exchange of oxygen between the CO formed during thermal conversion and the alumina (Al_2O_3) reactor, the Ni is required as a support for graphitized carbon since carbon is not retained by the surface of the Pt tube. A sheath gas of He + 1% H_2 is applied as a counter flow between the alumina reactor and the platinum tube to prevent effluent flow returning towards the beginning of the reactor. Addition of 1% H_2 helps maintain reductive conditions in the reactor and improve overall stability. Eluting CO is introduced into a mass spectrometer equipped with a triple collector comprising three Faraday cups monitoring simultaneously *m/z* 28, 29, and 30 corresponding to the ions of the three isotopomers $^{12}C^{16}O$, $^{13}C^{16}O$, and $^{12}C^{18}O$, respectively (Hener et al. 1998; ThermoElectron 2003). $\delta^{18}O$ values are calculated in an analogous fashion to the methods outlined above. Of the four isotope techniques detailed above compound-specific $\delta^{18}O$ analysis is by far the most nascent and is likely to undergo many further developmental changes as the methodology is slowly adopted by the wider user community.

For an in depth review of this area see Meier-Augenstein (1999). In addition, readers are directed towards the ISOGEOCHEM list server (http://list.uvm.edu/archives/isogeochem.html) as an invaluable resource for troubleshooting and the reporting of recent developments in GC/C/IRMS.

Data treatment: kinetic isotope effects (KIEs) and error analysis

Derivatization: general considerations

As discussed above, account must be taken of exogenous atoms added to the analyte molecule during analysis. In addition, any potential derivatization reaction must be evaluated for fractionation resulting from KIEs. Bonds involving heavier isotopes have a higher potential energy which can result in different reaction rates in reactions involving different isotopic species (Melander & Saunders 1980). For derivatization the most significant isotopic fractionation is the primary KIE that occurs at a specific molecular position causing an alteration in the δ value at that position. Hence, if the rate-determining step of a derivatization reaction involves the making or breaking of a bond to the atom under consideration, then the reaction involving that atom may be nonquantitative and any KIE must be quantified. If the KIE is reproducible, a correction can be made.

Correcting for derivative groups

If no KIE is present, the contribution of the derivative atom to the measured δ value of the derivatized compound can be calculated using a simple mass balance equation (14.1), where n is number of moles of the isotope of interest, F is the fractional abundance of the isotope of interest, c refers to compound of interest, d refers to the derivative group, and cd refers to the derivatized compound (Rieley 1994). For compounds at natural abundance, F can be replaced with the corresponding δ value (equation 14.2):

$$n_{cd}F_{cd} = n_cF_c + n_dF_d \tag{14.1}$$

$$n_{cd}\delta_{cd} = n_c\delta_c + n_d\delta_d \tag{14.2}$$

The application of these equations requires the isotope value of the derivatizing molecule to be established. If all the atoms of interest in the derivatizing reagent are transferred to the analyte, the contribution can be measured directly offline. However, if this is not the case or if the reagent is purchased in numerous small batches thereby precluding off-line analysis, the contribution of the derivatizing group can be measured indirectly by derivatizing a compound of known isotope value. However, this will have implications for the errors associated with the measurement, as discussed below.

Estimating KIEs

If the observed δ values for derivatized analytes do not equal those predicted by equation 14.1, then a kinetic isotope effect is present. Correction for the KIE can be made through the use of correction factors, as long as the KIE is proved to be reproducible across a range of analyte concentrations. These correction factors can be defined as the "effective" stable isotope composition of the derivative carbon introduced during derivatization taking into account the isotopic fractionation associated with the reaction. Correction factors are determined indirectly by measuring the δ value of an underivatized standard of the molecule of interest (by IRMS), the value of the standard after derivatization (by GC/C/IRMS), and using a rearranged equation 14.2 to determine δ_d. The δ_d term can then be replaced with δ_{corr} to represent the correction factor for the analyte of interest (Silfer et al. 1991; Macko et al. 1998).

Analysis of errors

Where no KIE is present, the measurement of the δ value is a result of two measurements, each with their own associated precision. During correction for the added derivative carbon, where σ is the standard deviation associated with a given δ determination, the errors propagate according to equation 14.3.

$$\sigma_c^2 = \sigma_{cd}^2 \left(\frac{n_c + n_d}{n_c} \right)^2 + \sigma_d^2 \left(\frac{n_d}{n_c} \right)^2 \tag{14.3}$$

In derivatization reactions with a KIE, correction factors are first calculated, this calculation introduces another step where errors propagate. The propagation of errors under these circumstances is calculated using equation 14.4, where subscript s stands for the standard used in correction factor determination and sd stands for the derivatized standard. The magnitude of the errors associated with the correction factors themselves can be calculated using equation 14.3, along with the precisions for each determination (Docherty et al. 2001).

$$\sigma_c^2 = \sigma_d^2 \left(\frac{n_s}{n_c} \right)^2 + \sigma_{ad}^2 \left(\frac{n_a + n_d}{n_c} \right)^2 + \sigma_{cd}^2 \left(\frac{n_c + n_d}{n_c} \right)^2 \tag{14.4}$$

The precision associated with the determination of δ values is derived from the errors associated with the correction factors and is also dependent on the molar ratio of the element of interest between the sample and derivative. Therefore in order to minimize the final error it is preferable to employ a derivatization reaction with no associated KIE and to minimize the amount of derivative atoms added.

Specific considerations when working at high stable isotopic enrichments

In recent years, the use of highly enriched stable isotope tracers in "pulse-chase" type experiments has become far more widespread since they circumvent the considerable loss of sensitivity that can be experienced through depletion of natural abundance tracer levels as a result of respiration, low incorporation rates, and dilution effects. Whilst highly enriched tracers have many benefits, there are also potential pitfalls associated with interpretation of the resulting data. Carryover within a run has been demonstrated to have a significant effect on the $\delta^{13}C$ values reported for compounds eluting immediately after a highly enriched component (Mottram & Evershed 2003). Whilst this effect does not represent a serious drawback for qualitative studies, careful consideration is needed when quantitatively interpreting results from analysis of complex mixtures containing enriched compounds. The recommendation here is that experiments resulting in excessively enriched compounds are probably of questionable value except for rather specific investigations where dilution effects are likely to be very substantial.

Applications of compound-specific stable isotope approaches in ecology and paleoecology

The application of GC/C/IRMS in organic geochemical studies

The major impetus for the development of the GC/C/IRMS technique came from the work of Hayes and co-workers during the 1980s and 1990s, such that the vast majority of early instruments were acquired by organic geochemistry laboratories. The range of applications of GC/C/IRMS in this area has been truly spectacular, and such a volume of published works makes a comprehensive review beyond the scope of a contribution of this nature. Thus, what follows is a brief overview of the applications of GC/C/IRMS in this area, with specific papers being used to highlight areas of novel application.

A variety of processes, including environmental conditions during organic matter production (Bidigare et al. 1997; O'Leary 1981), diagenesis (e.g. Spiker et al. 1985), catagenesis (e.g. Hayes et al. 1992), and heterotrophic reworking (Hayes et al. 1990; Ostrom & Fry 1993; Logan et al. 1997) all exert significant controls on the carbon isotopic composition of sedimentary organic matter. Although this presents the opportunity for the application of stable carbon isotope methodologies to diverse investigations, it also makes it difficult to interpret the bulk organic carbon isotope record; consequently, the use of GC/C/IRMS to determine the carbon isotopic compositions of specific compounds has become widely used in organic geochemistry. Using this approach, primary vs. secondary and allochthonous vs. autochthonous organic materials can be distinguished based on their isotopic compositions.

Moreover, the information recovered from a single sample is much more diverse: insights into algal photosynthesis, higher plant community structure, and bacterial recycling of organic matter can be elucidated from a few analyses (Freeman et al. 1990). However, the compound-specific approach also introduces additional complications related to carbon isotope fractionation during the biosynthesis of specific compound classes (e.g. Lockheart et al. 1997; Schouten et al. 1998).

The utility of compound-specific isotope analysis is directly related to the specificity of the compounds being analyzed. Compounds with very specific sources, such as alkenones derived from certain species of haptophyte algae, provide more precise data, allowing the most constrained interpretation (e.g. Pagani et al. 1999), but less diagnostic compounds can still provide useful information in the proper context. The lipids of a range of organisms (including marine and lacustrine algae, bacteria, archaea, and higher plants), their degradation pathways and representative biomarkers in the geologic record have been studied for decades, and this literature is only briefly summarized here, with particular emphasis on compounds used in carbon isotope studies.

Alkenones, long-chained (C_{37}–C_{39}) unsaturated ethyl and methyl ketones produced by only a few species of Haptophyte algae in the modern ocean (Volkman et al. 1980; Marlowe et al. 1984) are the most commonly used algal biomarkers in environmental investigations due to their relative ease of preparation and isotopic analysis, source specificity, diagenetic robustness, and the use of their distributions as a sea-surface temperature proxy. One of the most striking applications of alkenone $\delta^{13}C$ values has been to the development of atmospheric pCO_2 records from the Eocene to the Miocene (Pagani et al. 1999, 2005). Pagani and colleagues investigated sediments from ocean gyres, where growth rate variations were thought to be minor. Thus, changes in alkenone (and carbonate) derived ε_p values could be interpreted as changes in pCO_2.

After alkenones, steroids are the algal lipids most commonly investigated using isotopic approaches. Because of their structural diversity (Volkman 1986; Volkman et al. 1998), certain sterols are relatively diagnostic for specific compound classes. For example, 24-methylcholesta-5,22E-dien-3β-ol and especially 24-methylcholesta-5,24(28)-dien-3β-ol have both been invoked as diatom biomarkers, although these sterols are also present in other algae (Volkman et al. 1998). More diagnostic are the 4-methylsterols, especially 4α,23,24-trimethyl-5α-cholest-22E-en-3β-ol (dinosterol), as biomarkers for dinoflagellates (Withers et al. 1979). Several workers have determined sterol $\delta^{13}C$ values in modern surface-waters and shallow marine sediments (Pancost et al. 1997, 1999; Popp et al. 1989; Eek et al. 1999). Sterol $\delta^{13}C$ values have been used to either clarify the sterol sources (e.g. that 24-ethylcholesterols in Peru surface-waters derive from diatoms and not higher plants; Pancost et al. 1999), or evaluate controls on algal growth rates (Pancost et al. 1997).

In ancient sediments, steranes are among the most abundant preserved hydrocarbons; however, thermal isomerization typically results in a complex distribution of steranes and determination of specific compounds' $\delta^{13}C$ values is difficult (e.g. Schouten et al. 2000).

Chlorophylls and their degradation products can also be used as tracers for the isotopic composition of primary photosynthate. Since they are not GC-amenable, it is difficult to directly determine chlorophyll (and porphyrin) $\delta^{13}C$ values, and most efforts have focused on their degradation products. Phytol, the esterified side chain of most chlorophylls, and its inferred hydrocarbon degradation products (pristane, phytane) have been the subject of compound-specific carbon isotopic analysis since the advent of the technique (Freeman et al. 1990; Hayes et al.1990). Maleimides, 1-H-pyrrole-2,5-diones, are direct degradation products of the chlorophyll and bacteriochlorophyll tetrapyrrole structure, and are common in extracts of ancient sediments and are GC-amenable (Grice et al. 1997). Thus, maleimide $\delta^{13}C$ values are relatively easy to determine and have been used to gain insight into carbon cycling in ancient settings (e.g. Permian Kupferschiefer; Grice et al. 1997).

Long-chain n-alkyl compounds are major components of epicuticular waxes from the leaves of vascular plants (Eglinton et al. 1962). These compounds, especially n-alkanes, are relatively resistant to degradation, which makes them suitable for use as higher plant biomarkers (Cranwell 1981). It is relatively easy to determine carbon isotopic compositions of higher plant n-alkyl compounds using GC/C/IRMS, because they commonly occur in high abundances and relatively simple adduction procedures can be used to obtain pure fractions; thus, the carbon isotopic compositions of these compounds in modern plants, soils, and lacustrine and marine sediments have been extensively studied. Another higher plant derived component that is commonly investigated is lignin, a relatively stable and microbially resistant heteropolymeric structure comprising phenyl-propanoid subunits (Sarkanen & Ludwig 1971) and a significant component of wood. Moreover, lignin monomers have different sources (Hedges & Parker 1976) such that the isotopic compositions of syringyl, vanillyl and cinnamyl phenols can be used to distinguish isotopic signals of different plant types (e.g. Huang et al. 1999). Although preparation of lignin monomers for isotopic analysis requires careful chemical work-up, typically involving CuO oxidation of the lignin macromolecule, generated fractions are readily analyzable by GC-C-IRMS (Goñi & Eglinton 1996).

Sedimentary higher plant biomarker $\delta^{13}C$ values are commonly used to identify shifts in the relative proportion of C3 vs. C4 plants in the adjacent terrestrial ecosystem (e.g. Bird et al. 1995; Yamada Ishiwatari 1999; Huang et al, 2000). Often subtle changes can be identified, but also dramatic isotopic shifts in n-alkane $\delta^{13}C$ values have been observed and used to infer significant changes in past ecosystems. Freeman & Colarusso (2001) revisited Bengal Fan sediments that had previously been investigated using bulk sedimentary $\delta^{13}C$

values (France-Lanord & Derry 1994); observing a shift in *n*-alkane $\delta^{13}C$ values from −30 to −22‰, they confirmed the shift from C3-dominated to C4-dominated vegetation on the Himalayan foreland. The same approach was utilized in Mesozoic sediments from the southern proto North Atlantic (Kuypers et al. 1999), but was feasible only because sulfurization of algal organic matter had prevented dilution of terrestrial *n*-alkanes with marine-derived counterparts. In DSDP Site 367, *n*-alkane $\delta^{13}C$ values exhibited a dramatic shift of ca. 14‰ at the Cenomanian=Turonian boundary (ca. 90 Ma), suggesting a nearly complete vegetation shift in parts of North Africa at this time (Kuypers et al. 1999). This was attributed to a pCO_2 decrease associated with the C=T oceanic anoxic event but was nonetheless surprising as the C4 photosynthetic pathway was not thought to have evolved by this time.

The most common bacterial biomarkers in marine sediments are free and bound (phospholipid) fatty acids, of which the latter comprise the membranes of bacteria; however, eukaryotes also contain membranes comprised of phospholipid fatty acids and these compounds are not diagnostic as a class. The most common, such as saturated C_{16} and C_{18} fatty acids, are particularly widespread and appear to have little utility as tracers of explicit prokaryotic processes. However, some fatty acids, characterized by site-specific methyl groups, double bonds, or cyclic moieties are less common. Other bacterial membrane lipids are the hopanoids, pentacyclic triterpenoids common in a range of primarily aerobic bacteria (Ourisson et al. 1987; Farrimond et al. 2000). Many hopanoids are well resolved during GC analysis of marine sediments and their $\delta^{13}C$ values can be readily determined.

In addition to the above widespread bacterial biomarkers, a variety of compounds are biomarkers for photosynthetic bacteria. These include diverse pigments, including isorenieratene (Liaaen-Jensen 1978), chlorobactene, and bacteriochlorophylls *d* and *e*, which are diagnostic for green sulphur bacteria and thus, photic zone euxinia in the depositional environment. Of these, it is relatively easy to determine $\delta^{13}C$ values for isorenieratene derivatives (especially isorenieratane) due to their high molecular weight and long retention time eliminating most co-elution problems. Because green sulphur bacteria utilize the reverse tricarboxylic acid (TCA) cycle during carbon assimilation, their $\delta^{13}C$ values are less depleted relative to other organic matter (Quandt et al. 1977; Sirevag et al. 1977). Combined with the structural specificity of isorenieratene, this provides confirmation of the presence of green sulphur bacteria and, critically, photic zone euxinia, in paleoceanographic investigations spanning Earth history (e.g. Koopmans et al. 1996).

Compound-specific isotope analysis of bacterial and archaeal lipids has been particularly useful when applied to the investigation of anaerobic oxidation of methane. Determination of extremely low $\delta^{13}C$ values (−80 to −120‰) for archaeal (archeol, hydroxyarcheol and GDGTs) and sulphate reducing bacterial membrane lipids (including fatty acids) in cold seep sediments has been used to confirm the role of these organisms in the anaerobic oxidation of methane at cold seeps (Hinrichs et al. 1999; Thiel et al. 1999; Pancost et al. 2001; Elvert et al. 2001). The low $\delta^{13}C$ values offer direct evidence that the Archaea and sulfate-reducing bacteria (SRB) are incorporating methane. A wide range of settings and compounds have now been investigated, with almost all revealing a predominance of archaeal lipids (e.g. Figure 14.4), confirming the important role of these organisms in methane cycling.

The application of GC/C/IRMS in archeology

Organic residues in archeological pottery

Early work in the use of stable isotopes in archeological pottery involved bulk isotopic analysis (e.g. Hastdorf & DeNiro 1985; Morton & Schwarcz 1988; Sherriff et al. 1995). Compound-specific stable isotope analysis was first applied to archeological potsherds in 1994 (Evershed et al. 1994), confirming that an extract with lipid distribution consistent with wild type *Brassica* species was of a C3 plant origin. The application of compound-specific stable isotope analysis via GC/C/IRMS to lipid residues in archeological pottery allows greater specificity to be achieved than is possible with bulk analyses, since the structures of diagnostic (biomarker) components of complex mixtures can be unambiguously linked to their stable isotope values. Thus, compound-specific stable isotope analysis avoids ambiguities arising from contamination (e.g. plasticizers originating from plastic bags in which sherds are often stored), which cannot be resolved from endogenous components in bulk isotope analyses. Compound-specific $\delta^{13}C$ values also afford insights into the biochemical sources of carbon even when chemical structures are identical. $\delta^{13}C$ values of fatty acids provide the basis for distinguishing between ruminant (e.g. sheep/goat and cattle) and porcine (pig) adipose fats. The potential of GC/C/IRMS in this area (Table 14.2) was realized by Evershed and co-workers in 1997; $\delta^{13}C$ values of the $C_{16:0}$ and $C_{18:0}$ fatty acids in medieval lamps and dripping dishes, when compared with modern reference fats, revealed that ruminant fat had been used as fuel in the lamps, whereas porcine fat was collected in the dripping dishes, disproving the theory that fat was collected from spit-roasting and recycled as lamp fuel (Evershed et al. 1997; Mottram et al. 1999). The $\delta^{13}C$ values exhibited by these animals reflect their different diets and variations in their metabolisms and physiologies (Evershed et al. 1999)

Ruminant adipose and dairy fats can also be distinguished by the $\delta^{13}C$ values of their fatty acids (Dudd & Evershed 1998). The $C_{18:0}$ fatty acid in dairy fat is significantly more depleted in ^{13}C (ca. 2.1‰; Copley et al. 2003). Fatty acids in ruminant adipose are mainly synthesized from acetate (as acetyl CoA), originating predominately from the fermentation of dietary carbohydrate in the rumen. The mammary gland is incapable of biosynthesizing the $C_{18:0}$ fatty acid; instead, it is obtained via the remobilization of adipose

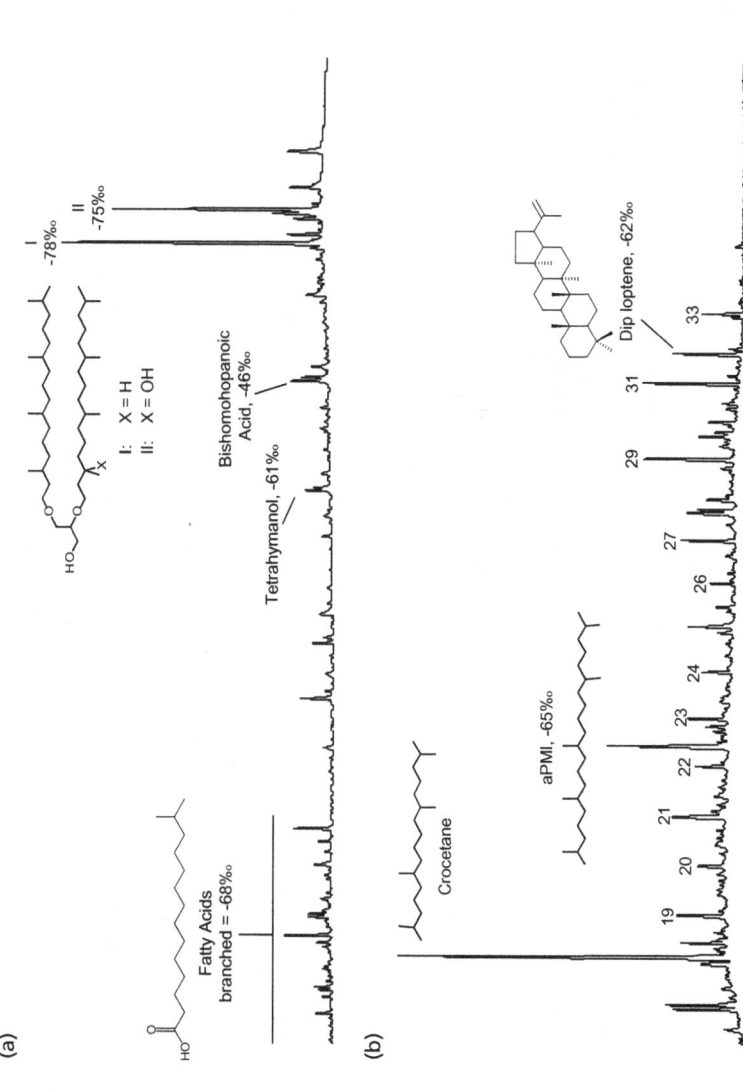

Figure 14.4 Partial gas chromatograms showing the relative abundances of archaeal and bacterial biomarkers (and their carbon isotopic compositions) in the polar fraction (a) and apolar fraction (b) of the total lipid extract of a Napoli Mediterranean mud volcano cold seep. (Adapted from Pancost et al. 2000; Pancost & Sinninghe Damsté 2003.)

Table 14.2 Summary of compound-specific stable isotope analyses applied to archeology.

Object or find type	Biomarker	Uses	References
Potsherd	$C_{16:0}$ and $C_{18:0}$ fatty acids	Distinguishes between remnant animal fats of ruminant adipose, ruminant dairy or porcine adipose origin	Evershed et al. 1997, 2003; Mottram et al. 1999; Dudd & Evershed 1998; Copley et al. 2003, 2005a,b,c; Berstan et al. 2004; Craig et al. 2005
	n-Dotriacontanol	Biomarker for maize processing	Reber et al. 2004 Reber & Evershed 2004a,b
Bone lipid	Cholesterol	Indicator of $\delta^{13}C$ value of whole diet. Depleted in ^{13}C relative to whole diet. Neosynthesis > assimilation of dietary cholesterol in governing $\delta^{13}C$ value of bone cholesterol. Indicator of short-term diet.	Stott & Evershed, 1996; Stott *et al.*, 1997b; Jim *et al.*, 2001, 2004
	Fatty acids	Good indicators of general short-term trends in diet.	Howland et al. 2003; Copley et al. 2004
	Nonessential fatty acids	$\delta^{13}C$ values correlates well with whole diet.	
	Linoleic acid	$\delta^{13}C$ value correlates well with dietary linoleic acid.	
Bone protein	Amino acids: Alanine, glycine, threonine, serine, valine, leucine, isoleucine, proline, hydroxyproline, aspartate, glutamate, phenylalanine	Mass balance calculations using the $\delta^{13}C$ values of single amino acids accurately predicts $\delta^{13}C$ value of whole collagen	Howland et al. 2003
	$\Delta^{13}C_{\text{Glycine-Phenylalanine}}$	Marine dietary indicator	Corr et al. 2005
	Alanine and glutamate	$\delta^{13}C$ values from bone collagen correlate well with whole diet	Howland et al. 2003
	Leucine and phenylalanine	Little isotopic fractionation between diet and bone collagen	Howland et al. 2003
	Leucine	Long-term indicator of the protein component of diet	Copley et al. 2004
	Glutamate	Long-term whole diet indicator providing an internal check of apatite	Copley et al. 2004

fatty acids and directly from the dietary C_{18} fatty acids, after biohydrogenation in the rumen (Moore & Christie 1981). The difference between the $C_{18:0}$ fatty acids from ruminant adipose and dairy fat can be explained by the fact that lipids are more depleted in ^{13}C than carbohydrates (DeNiro & Epstein 1977) and approximately 60% of the $C_{18:0}$ fatty acid in dairy fat is derived via biohydrogenation of dietary unsaturated C_{18} fatty acids (i.e. $C_{18:1}$, $C_{18:2}$ and $C_{18:3}$) in the rumen. The $\delta^{13}C$ values of the contributors to the $C_{18:0}$ fatty acid in dairy fat are summarized in Figure 14.5. Compound-specific $\delta^{13}C$ values are readily determined by GC/C/IRMS for fatty acids derivatized to FAMEs.

GC/C/IRMS analysis of remnant animal fats of archeological origin has now been extensively used to address some key questions concerning animal husbandry in prehistory, for example the earliest evidence for dairying in prehistoric Britain (Dudd & Evershed 1998; Copley et al. 2003, 2005a, 2005b, 2005c), and the exploitation of pigs in the late Neolithic (Mukherjee, 2004; Mukherjee et al. in press). $\delta^{13}C$ values of fatty acids extracted from potsherds have also been used to identify prehistoric dairying activities in the Western Isles of Scotland (Craig et al. 2005).

Clearly many archeological vessels will have been used to process commodities from more than one type of animal. In order to account for this a mixing model is used to calculate theoretical $\delta^{13}C$ values. This mathematical model has been used elsewhere for the detection of the mixing of vegetable oils of differing stable carbon isotope composition (Woodbury et al. 1995; Mottram et al. 2003) and sedimentary lipids (Bull et al. 1999) and utilises the percentage abundance of each specific fatty acid and its associated $\delta^{13}C$ value:

$$\delta^{13}C_{mix} = \delta^{13}C_{(A)}\left(\frac{(X \times A)}{(X \times A) + (Y \times B)}\right) + \delta^{13}C_{(B)}\left(\frac{(Y \times B)}{(X \times A)(Y \times B)}\right) \qquad (14.5)$$

where $\delta^{13}C_{mix}$ is the predicted $\delta^{13}C$ value of the fatty acid with contributions from fats A and B, $\delta^{13}C_{(A)}$ is the $\delta^{13}C$ value of the individual fatty acid in fat A, $\delta^{13}C_{(B)}$ is the $\delta^{13}C$ value of the individual fatty acid in fat B, X is the percentage of fat A present, Y is the percentage of fat B present, A is the percentage of the individual fatty acid in fat A, and B is the percentage of the individual fatty acid in fat B.

Compound-specific stable carbon isotope analysis of bone biochemical components

The use of bone in paleodietary reconstruction was motivated by the variations in collagen $\delta^{13}C$ values observed during ^{14}C dating, and light stable isotopes were first used in archeology in the late 1970s (Vogel & Van der Merwe 1977). Since then dietary analysis through both carbon and nitrogen isotopes

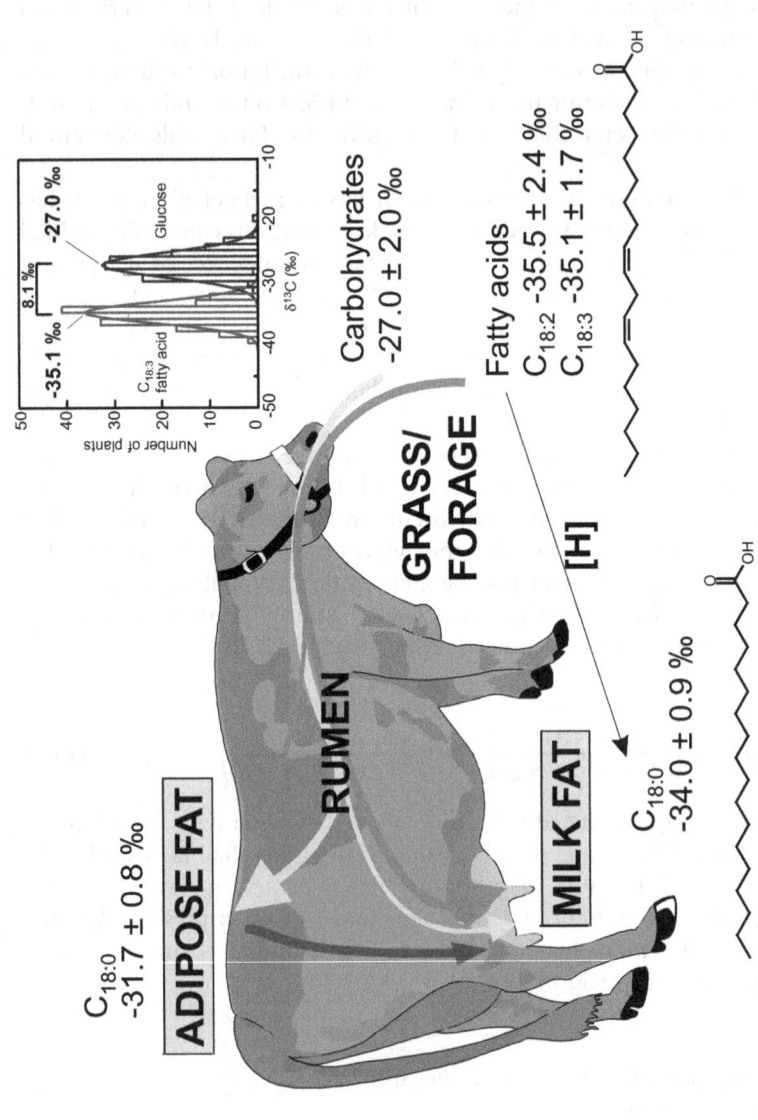

Figure 14.5 Diagram showing the routing of dietary fatty acids and carbohydrates in the rumen, adipose tissue, and mammary gland of the ruminant animal and histogram of the δ¹³C values of C$_{18:3}$ fatty acid and glucose extracted from plants demonstrating the 8.1‰ mean difference between them. (Adapted from Copley et al. 2003; Docherty 2002.)

has been applied extensively to archeological studies. The reconstruction of ancient diets is possible because $\delta^{13}C$ and $\delta^{15}N$ values of fossil bone reflect the isotopic signatures of the local environment, specifically the plants that lie at the base of the food chain (Gannes et al. 1998).

Very extensive investigations of human diet in the past have been undertaken using the stable isotopic signals recorded in collagen and hydroxyapatite in bones, with important evidence also coming from teeth. Although isotopic analysis may not always enable the precise reconstruction of an animal's diet, it does allow discrimination of animals belonging to particular dietary niches (Gannes et al. 1998). Compound-specific stable carbon isotope analysis of bone components such as cholesterol, fatty acids, and individual collagen amino acids is finding increasing application to paleodietary reconstruction. An application of compound-specific isotopic analyses is in improving our understanding of the relationship between dietary macronutrient composition and the $\delta^{13}C$ values of bone components (Hare et al. 1991; Ambrose 1993; Jim et al. 2001, 2003b, 2004, 2006).

Cholesterol in archeological bone

The use of cholesterol as a paleodietary indicator is a relatively recent development. Cholesterol was demonstrated to survive in archeological bone in 1995 (Evershed et al. 1995). Archeological human and animal bones, including a 75,000 year old whale bone from a permafrost deposit, were found to contain free cholesterol and cholesteryl fatty acyl esters, and diagenetic products (5α- and 5β-cholestan-3-one, 5α- and 5β-cholestanol and cholest-5-en-7-one-3β-ol; Evershed et al. 1995; Stott et al. 1997a). The cholesterol found in bone may derive from either the remnants of the original bloodborne lipid (in the case of vascular bones), the fat component of bone marrow that would have been present at the time of death of the organism, or a component of cellular lipids present in bone-forming cells (Stott et al. 1997a). The use of cholesterol as a paleodietary indicator has been extensively investigated (Stott & Evershed 1996; Stott et al. 1997b, 1999; Jim et al. 2004). The $\delta^{13}C$ values of cholesterol are readily recorded by GC/C/IRMS as its TMS ether derivative (Figure 14.3) and were constant across different skeletal members for a given individual (Stott & Evershed 1996). Assessment of $\delta^{13}C$ values for cholesterol from animals raised on isotopically distinct diets (Stott et al. 1997b; Jim, 2000; Jim et al. 2001, 2003b, 2004; Corr 2003), indicate that: (i) cholesterol is a good indicator of whole diet, (ii) neosynthesis of cholesterol is more significant than assimilation in determining the $\delta^{13}C$ value of cholesterol, and (iii) bone cholesterol $\delta^{13}C$ values respond to changes in the isotopic composition of whole diet more rapidly than collagen and apatite such that cholesterol is an indicator of short-term diet (Stott et al. 1997a; Jim, 2000). These results have been applied, alongside collagen and apatite analysis, to address archeological questions relating to the diets of a range of ancient populations (Stott et al. 1999; Jim, 2000; Jones, 2002: Corr 2003; Howland, 2003; Copley et al. 2004).

Fatty acids in archeological bone and those of experimental animals

As discussed above $C_{16:0}$ and $C_{18:0}$ fatty acids have been used as biomarkers for animal fats in the study of lipid residues in archeological pottery sherds. However, the analysis of fatty acids extracted from archeological bone has been much less explored (Evershed et al. 1995). This is mainly due to the low survival of bone fatty acids in the archeological record; fatty acids only seem to be preserved in significant abundances under exceptional burial environments, for example arid and waterlogged sites (Evershed & Connolly 1987; Copley et al. 2004). Fatty acids present in bone mainly derive from bone marrow fat (Evershed et al. 1995). Studies on rats and pigs raised on isotopically controlled diets have shown that bone fatty acid $\delta^{13}C$ values are ^{13}C-depleted by up to 3.4‰ with respect to whole diet (Jim, 2000; Jim et al. 2001, 2003b; Howland et al. 2003). This phenomenon results from a kinetic isotope effect occurring during the oxidation of pyruvate by pyruvate dehydrogenase to acetyl CoA, the common precursor in lipid biosynthesis (DeNiro & Epstein 1977; Hayes 1993). The $\delta^{13}C$ values of bone fatty acids have recently been used together with those of individual amino acids and apatite as indicators of trends in the management of domesticated animals in Egypt (Copley et al. 2004).

Amino acids in archeological bones and those of experimental animals

Compound-specific stable isotope analyses of the building blocks of complex biopolymers, such as collagen, are essential to unraveling the stable isotope signals expressed in bulk protein signals. The exploitation of individual collagenous amino acids has great potential in paleodietary reconstruction, however, surprisingly only a handful of studies have determined the $\delta^{13}C$ values of amino acids from ancient bone collagen (Hare & Estep 1983; Tuross et al. 1988; Hare et al. 1991; Fogel & Tuross, 2003; Jim et al. 2003b; Copley et al. 2004; Corr et al. 2005). Amino acids are difficult to isolate for isotopic analysis. Ion exchange liquid chromatography has been used but is slow and the reproducibility of retention times is poor (Hirs et al. 1954, Gaebler et al. 1966, Hare & Estep 1983, Macko et al. 1983, 1987, Tuross et al. 1988, Hare et al. 1991). Moreover, van Klinken (1991) has reported that isotopic fractionation is a major disadvantage with this technique.

GC/C/IRMS enables amino acids to be separated by GC and combusted on-line, for stable isotopic analysis, thereby avoiding manual preparative steps (see Figure 14.3). The $\delta^{13}C$ values of individual collagen amino acids are highly robust and by use of mass balance calculations can even be used to reconstruct the bulk $\delta^{13}C$ value of whole collagen (Jim et al. 2003a). A recent study of pigs raised on six isotopically controlled diets investigated the routing of dietary macronutrients to bone biochemical components (Howland et al. 2003), showing that: (i) the $\delta^{13}C$ values of single amino acids accurately predicted the $\delta^{13}C$ value of whole collagen; (ii) the $\delta^{13}C$ values of nonessential amino acids alanine and glutamate from bone collagen correlated well with whole diet; and (iii) the essential amino acids leucine and phenylalanine showed little isotopic fractionation between diet and bone collagen. Still more recently a feeding experiment involved rats fed on diets where the $\delta^{13}C$ values of the major dietary macronutrients were switched between C3 and C4 enabling quantitative assessment to be made of the carbon sources used in the *de novo* synthesis of nonessential amino acids (Jim et al. 2006).

Applying these techniques to archeological populations enabled Evershed and co-workers to demonstrate the wider utility of compound-specific stable isotope analysis of individual bone collagen amino acids to distinguish between high marine protein and terrestrial consumers (Corr et al. 2005). This was previously shown to be problematic in extremely arid environments using bulk collagen $\delta^{15}N$ values alone (Heaton 1987; Heaton et al. 1986; Schwarcz et al. 1999; Sealy 1997) since herbivore bone collagen $\delta^{15}N$ values overlap with the range for marine species (Heaton et al. 1986; Sealy et al. 1987). Due to their contrasting metabolic pathways, $\delta^{13}C$ values of the essential amino acid phenylalanine and the nonessential amino acid glycine in bone collagen preserve different paleodietary signals and this difference ($\Delta^{13}C_{Glycine-Phenylalanine}$) can be exploited to distinguish between high marine protein and terrestrial consumers (Figure 14.6). Compound-specific stable isotope analysis of single collagenous amino acids along with fatty acids, collagen, and apatite have been used to investigate foddering and foraging strategies of domesticated animals from Qasr Ibrim, Egypt (Copley et al. 2004). Essential and nonessential collagenous amino acids provided long-term indicators of the diet of cattle and sheep/goat. The essential amino acid leucine, incorporated directly from dietary protein (Beynon 1998), has a $\delta^{13}C$ value that directly reflects the $\delta^{13}C$ values of the leucine in the protein component of the plants that are incorporated into animals' diets.

The recent investigations discussed above now mean that a range of bone-based biochemical proxies exists for investigating: (i) whole diet, (ii) specific elements of the diet, e.g., protein and energy components, and (iii) long- and short-term dietary variation within such proxies, and as such are highly applicable to the investigation of ancient diet.

^{13}C-labeling of lipids to investigate environmental microbes

The complexity of the microbial populations in sedimentary environments is universally acknowledged, with major challenges to their study arising from the unculturable nature of the major proportion of such populations (Amann

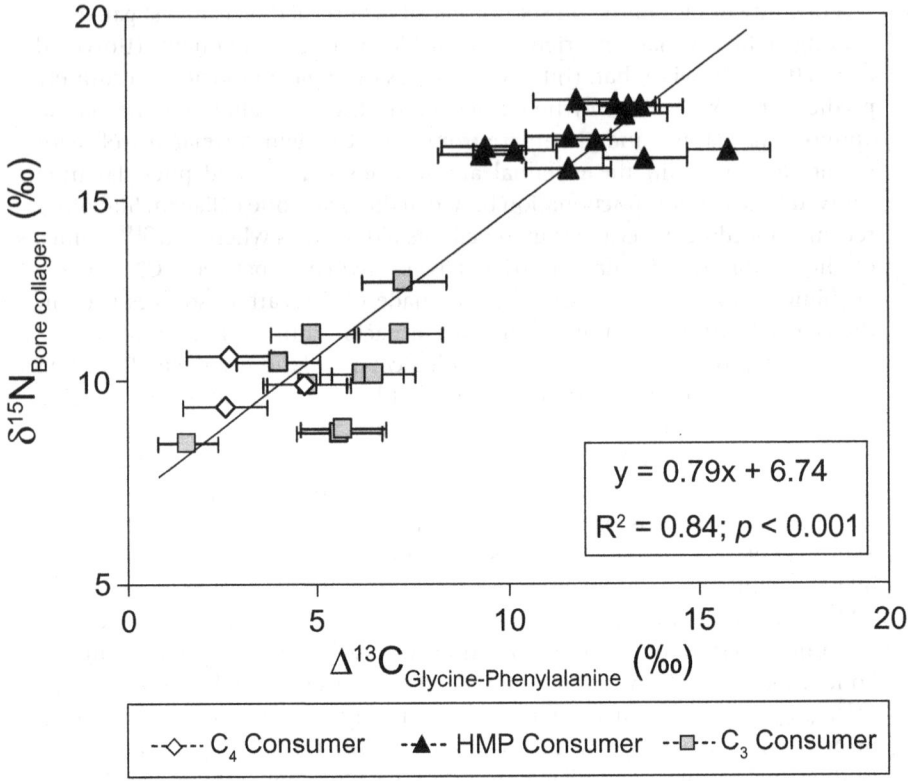

Figure 14.6 Plot showing the correspondence between $\Delta^{13}C_{\text{gly-phe}}$ and $\delta^{15}N$ values of collagen from hunter/gatherers from the South western Cape, indicating the potential of compound-specific glycine carbon isotope values as new marine dietary proxy. (Adapted from Carr et al. 2005.)

et al. 1995). The use of compound-specific stable isotope approaches to study living microbial populations rests on matching specific biomarkers, to a particular group(s) of organism(s). The biomarkers chosen are required to indicate both the presence and the activity of a given microbial group, therefore, on cell death, that biomarker must be rapidly removed from the system. For this reason the majority of studies focus on microbial membrane lipids, which have been shown to be rapidly degraded on cell death. Combining biomarker analysis with isotopic labeling studies provides the opportunity of linking specific processes, e.g. methane-oxidation, with the microbial taxa responsible. Labeling approaches involve adding a ^{13}C-labeled substrate to a microcosm, sediment, soil, or culture medium, then following its fate temporally and/or spatially. Table 14.3 lists examples of the application of compound-specific stable isotopes in this area.

A major area of utility of this methodology is the use of $^{13}CH_4$ to investigate methanotrophic bacteria in both aerobic and anaerobic environments, including sediments (Boschker et al. 1998; Nold et al. 1999), soils (Bull et al. 2000; Crossman et al. 2001, 2004, 2006; Knief et al. 2003), microbial mat (Blumenberg et al. 2005), and peat bogs (Raghoebarsing et al. 2005). A major advantage of using methane arises from its ease of addition to microcosms providing the opportunity to target such an important group of microorganisms. This approach was especially effective in investigating unculturable high-affinity methanotrophs in soils revealing novel type II methanotrophs producing a br17:0 PLFA (Bull et al. 2000). In addition to the classification of bacteria, Crossman et al. (2004) investigated how communities adapted to their environment. They used $^{13}CH_4$ incubations in laboratory microcosms to demonstrate variations in methanotrophic bacterial populations with depth through a landfill cover soil (Figure 14.7); type I methanotrophs were found to be more active in the surface layers, where concentrations of oxygen were highest and methane concentration low, while type II methanotrophs dominated in the deepest layers of the cap where methane concentrations were high and oxygen low.

Other research has also exploited the fact that the $\delta^{13}C$ value of biogenic methane is highly depleted (-50 to $-100‰$) as a result of isotopic fractionation (Whiticar et al. 1986) and there have been a variety of reports of the presence of ^{13}C-depleted archaeal ether lipids and sulphate-reducing bacterial biomarkers in marine sediments near methane seeps (Pancost et al. 2001; Teske et al. 2002), mud volcanoes (Pancost et al. 2000), carbonates (Thiel et al. 1999, 2001) and in the water column (Schouten et al. 2001). These have been interpreted as communities of methane-consuming archaea, possibly methanogenic archaea operating in reverse, and sulphate-reducing bacteria performing anaerobic methane oxidation and are reviewed in greater detail by Pancost & Sinnighe-Damsté (2003).

Microbial communities utilizing products of organic matter mineralization have also been investigated (Boschker et al. 2001; Pombo et al. 2002, 2005). For example, Boschker et al. (2001) have shown through laboratory incubation of small anoxic/brackish sediment cores that ^{13}C-acetate and ^{13}C-propionate were utilized by different members of the microbial community. ^{13}C-label from acetate was recovered mainly from even carbon numbered PLFAs (16:1ω7c, 16:0, 18:1ω7c) while primarily odd carbon numbered fatty acids (a15:0, 15:0, 17:1ω6, 17:0) became labeled upon incubation with propionate. These findings clearly indicate that the two substrates were predominantly consumed by different specialized groups of sulfate-reducing bacteria. The PLFA labeling pattern for the acetate consumers was similar to *Desulfotomaculum acetoxidans* and *Desulfofrigus* spp., two acetate-consuming sulfate-reducing

Table 14.3 Summary of compound-specific stable isotope based investigations of environmental microbes.

Environment	^{13}C-labeled substrate	Biomarker	Detection method	Conclusion	Reference
Laboratory microbial cultures	Natural abundance glycerol, glucose, mannose, lactose, complex medium	PLFA	GC/C/IRMS	Complex fractionation patterns varying with substrate and organism	Abraham et al. (1998)
Woodland and grassland soils	Universally labeled starch, xylose, vanillin and litter	PLFA	GC/C/IRMS	Similar microbial groups responsible for degrading simple substrates in woodland and grassland soils but different communities degraded complex substrates	Waldrop & Firestone (2004)
Estuarine sediments	[U-^{13}C]acetate	PLFA	GC/C/IRMS	Acetate consumed by sulphate reducing bacteria similar to Gram +ve *Desulfotomaculum acetoxidans* and not by a population of Gram -ve *Desulfobacter* spp.	Boschker et al. (1998)
Experimental soils	[1-^{13}C]sodium acetate	PLFA, neutral lipids, glycolipids	GC/MS SIM	Incorporation greatest into PLFA; bacterial growth limited at low pH but occurred at pH 7 and 8	Arao (1999)
Rhizosphere rice paddy soil	$^{13}CO_2$	PLFA	GC/C/IRMS	Microbial populations in rice soil differ in their response to plant photosynthate input	Lu et al. (2004)
Rhizosphere grassland soil	$^{13}CO_2$	PLFA	GC/C/IRMS	^{13}C-labeling showed fundamental differences in the way rhizodeposition was cycled through microbial community during different stages of plant development	Butler et al. (2003)

Sample	Substrate	Biomarker	Analysis	Comments	Reference
Rhizosphere grassland soil	$^{13}CO_2$	PLFA	GC/C/IRMS	Fungal and Gram -ve bacterial PLFAs showed most ^{13}C enrichment. Liming did not affect assimilation or turnover rates of ^{13}C-labeled C	Treonis et al. (2004)
Sediments from petroleum contaminated aquifer	[methyl-^{13}C]toluene	PLFA	GC/C/IRMS	PLFAs resemble those of PHC-degrading *Azoarcus* spp.	Pelz et al. (2001)
Petroleum contaminated groundwater	[2-^{13}C]acetate	PLFA (FISH)	GC/C/IRMS	Field-scale application of acetate to investigate carbon assimilation and mineralization	Pombo et al. (2002)
Petroleum contaminated aquifer water and sediment	[2-^{13}C]acetate	PLFA (FISH)	GC/C/IRMS	Main sulfate reducing bacteria degrading acetate in water *Desulfotomaculum acetoxidans* and *Desulfobacter* in sediment	Pombo et al. (2005)
Antarctic soil bacteria	^{13}C-labeled grass	Ergosterol, PLFA, NLFA	GC/MS SIM	Incorporation of ^{13}C increased over incubation period which was not seen in PLFA and NLFA fractions	Malosso et al. (2004)
Soil	Ring labeled [^{13}C]toluene and [U-^{13}C]glucose	PLFA	GC/C/IRMS	Specific labeling patterns for microbial PLFAs from ^{13}C-toluene incubation contrasting universal labeling of PLFAs from incubating with ^{13}C-glucose	Hanson et al. (1999)
Batch culture	[U-^{13}C]toluene	PLFA	GC/C/IRMS	Quantified carbon flow along substrate–bacteria–protist food chain	Mauclaire et al. (2003)
Soil	$^{13}CH_4$	Hopanoids	GC/C/IRMS	Specific bacteriohopanoids labeled	Crossman et al. (2001)
Upland soils	$^{13}CH_4$	PLFA, DGGE	GC/C/IRMS	Different methanotrophs are present in different soils that oxidize atmospheric methane	Knief et al. (2003)
Landfill cover soils	$^{13}CH_4$	PLFA	GC/C/IRMS	Changes in methanotrophic community from type I to type II with depth	Crossman et al. (2004)

(Continued)

Table 14.3 *Continued*

Environment	¹³C-labeled substrate	Biomarker	Detection method	Conclusion	Reference
Soil	$^{13}CH_4$	PLFA, hopanoids	GC/C/IRMS	Novel population of methane-oxidizing bacteria related to type II methanotrophs, *Methylocapsa* and *Methylocella*	Crossman et al. (2005)
Sediment/soil	$^{13}CH_4$	PLFA	GC/C/IRMS	Shift in the composition of the methane oxidizing bacterial community in the sediments/soils treated with ammonium	Nold et al. (1999); Crossman et al. (2006)
Peat bog	$^{13}CH_4$	Hopanoids, sterols (FISH)	GC/C/IRMS	Methanotrophic bacteria associated with *Sphagnum* mosses providing CO_2 for photosynthesis	Raghoebarsing et al. (2005)
Sediment	Biogenic CH_4	Hopanoids	GC/C/IRMS	Hopanoids exhibit depleted $\delta^{13}C$ values as a result of their production by methanotrophic bacteria	Freeman et al. (1990)
Mussel gill tissue	Biogenic CH_4	Methanotroph biomarkers	GC/C/IRMS	Detection of methanotrophic symbionts associated with the mussel	Jahnke et al. (1995)
Sediment	Biogenic CH_4	Archaeal lipids	GC/C/IRMS	Archaeal lipids are depleted in ^{13}C indicating that archaea are involved in methane consumption. Abundances differ between sites suggesting multiple archaeal species	Pancost et al. (2001)
Anaerobic oxidation of methane (AOM)	$^{13}CH_4$	PLFA, archaeal lipids	GC/C/IRMS	^{13}C uptake into specific lipids indicate phylogenetically distinct microbes participate in AOM	Blumenberg et al. (2005)

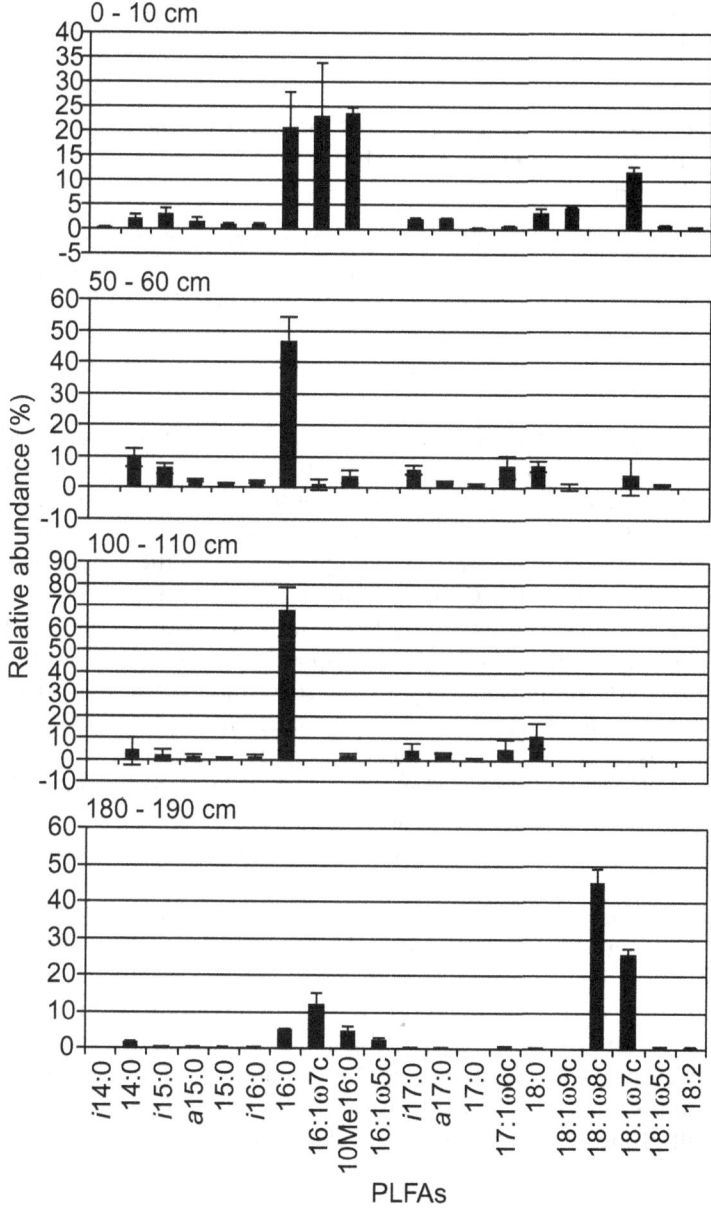

Figure 14.7 Relative abundances of ^{13}C-labeled PLFAs extracted from four sections of the profile of a landfill cap following incubation with 10,000 ppm methane containing 1% ^{13}CH$_4$. (Adapted from Crossman et al. 2004.)

bacteria, while those of the propionate consumers did not resemble any known strain.

Use of compound-specific stable isotopes to investigate soil organic matter cycling, pollution, and biodegradation

Table 14.4 summarizes a number of key environmental applications that have successfully utilized a compound-specific stable isotope approach. The three main environmental areas are considered in turn below.

Soil organic matter cycling

The majority of investigations into soil organic matter preservation and dia-genesis have focused on the ^{13}C content of bulk fractions with surprisingly few adopting molecular approaches (Lichtfouse 1995, 1998; Lichtfouse et al. 1994, 1995; Huang et al. 1996). Lichtfouse (1995) used the difference in carbon isotope composition of C3 and C4 plants to trace the increasing incor-poration, with time, of the C_{29} long chain *n*-alkane from maize epicuticular wax into a soil that had previously been cultivated with C3 plants. Huang et al. (1996) determined the isotopic composition of *n*-alkanes and triterpe-noids through the profile of three different soils, finding that, with increasing depth, the *n*-alkanes became progressively more enriched in ^{13}C. They inter-preted this as an addition of enriched alkanes from microbial sources. Addi-tionally, they also reported hopanoids enriched by 4–5‰ compared with plant *n*-alkanes, leading to the conclusion that these lipids were biosynthe-sized from carbohydrates or proteins from the plant. Lichtfouse (1998) observed that long-chain fatty acids, *n*-alkanes, and alcohols in soils exhibit carbon isotope values consistent with their derivation from C3 plant waxes, however, shorter chain fatty acids (C_{16}–C_{18}) have higher δ^{13}C values indicating that they are most likely biosynthesized by the microbial biomass in the soil. This is consistent with the results from a previous study by Licht-fouse et al. (1995) where ^{13}C-labeled glucose was applied to a soil and the extractable fatty acids, *n*-alkanes, and *n*-alkanols were analyzed by GC/C/IRMS and the short-chain fatty acids exhibited significant enrichment in ^{13}C as a result of their biosynthesis by microbes from the labeled glucose.

Pollution studies involving compound-specific stable isotope analyses

Source apportionment

Different manufacturing pathways and differing raw materials result in the production of materials with distinct isotopic signatures for their site of origin.

Table 14.4 Environmental studies utilizing compound-specific stable isotope approaches.

Environment	^{13}C-labeled substrate	Biomarker	Detection method	Conclusion	Reference
Groundwater	MTBE fractionation	MTBE	GC/C/IRMS	Biodegradation of MTBE causes enrichment in ^{13}C more than physical processes	Hunkeler et al. (2001)
Coastal Lagoon	Biogenic CH_4	PAHs	GC/C/IRMS	PAHs derive from industrial plant utilizing methane as a carbon source rather than a petrogenic source	McRae et al. (2000)
River	PCBs of known isotopic composition	PCBs in duck tissue, duck liver, duck egg and grass carp	GC/C/IRMS	Isotopic patterns provide information on the relative metabolic stability and degradation processes of various PCB congeners in different animals	Yanik et al. (2003)
Soil	Coal derived PAHs	PAHs	GC/C/IRMS	Coal derived PAHs can be resolved from those derived from petroleum in contaminated land	Sun et al. (2003)
Sediments	Fuel n-alkanes	n-alkanes	GC/C/IRMS	Isotopic values of n-alkanes from contaminated sediment correlate with those shipping bunker fuel n-alkanes	Rogers & Savard (1999)
Soil	Fuel derived PAHs	PAHs	GC/C/IRMS	The resolution of PAHs derived from automobile exhaust, tyre abrasion and tailpipe soot is attempted, however differences were too small to be conclusive	Glaser et al. (2005)
Groundwater – microcosms and field site	MTBE isotopic fractionation	MTBE	GC/C/IRMS	Anaerobic biodegradation of MTBE causes extensive fractionation of hydrogen and carbon isotopes	Kuder et al. (2005)
Tropical and temperate soils and termite nests	PAHs	PAHs	GC/C/IRMS	Tropical soil PAHs exhibit significant inputs of biogenic PAHs compared with temperate soil PAHs that are mainly from pyrolytic sources	Wilcke et al. (2002)
Laboratory microcosm and aquifer	BTEX and PAH fractionation	BTEX and PAHs	GC/C/IRMS	Significant biodegradation of toluene and o-xylene occurs along the groundwater flow path. Certain aromatic hydrocarbons do not exhibit isotope fractionation with microbial degradation, e.g., indane and indene	Richnow et al. (2003a)

Table 14.4 *Continued*

Environment	^{13}C-labeled substrate	Biomarker	Detection method	Conclusion	Reference
Landfill leachate plume	BTEX fractionation	BTEX	GC/C/IRMS	Isotope fractionation similar to previous laboratory based study indicated that biodegradation of ethyl benzene and m/p-xylene had occurred. Other compounds were at such low concentrations that isotope values could not be recorded accurately	Richnow et al. (2003b)
Oil spill, tar balls and oil-coated feathers	*n*-alkane and PAHs from Erika oil spill	*n*-alkanes and PAHs	GC/C/IRMS	Oil on bird feathers is correlated by biomarkers and isotope ratios with the oil spilled by Erika tanker. Tar balls derive from elsewhere	Mazeas & Budzinski (2002)
Soil	Fatty acids, alcohols and *n*-alkanes	Fatty acids, alcohols and alkanes	GC/C/IRMS	Long chain fatty acids, *n*-alkanes and *n*-alcohols are derived from C_3 plant sources, however, C_{14}-C_{18} fatty acids have a higher content of ^{13}C consistent with biosynthesis by microbial or fungal biomass	Lichtfouse et al. (1995)
Soil	Maize C_{31} *n*-alkane	C_{31} *n*-alkane	GC/C/IRMS	Enrichment in ^{13}C of soil *n*-alkane over time as maize epicuticular wax *n*-alkane is incorporated into the soil	Lichtfouse (1995)
Soil	Alkanes and triterpenoids	*n*-alkanes and triterpenoids	GC/C/IRMS	More enriched δ^{13}C values interpreted as inputs from microbes utilizing plant protein or carbohydrate	Huang et al. (1996)
Soil	^{13}C-labeled glucose	Fatty acids	GC/C/IRMS	C_{14}, C_{16} and C_{18} fatty acids became enriched in ^{13}C as a result of their biosynthesis by microbes, however, long-chain fatty acids, *n*-alkanes and alcohols derived from plants and therefore were not labeled	Lichtfouse (1998)
Soil	C_4 Plant	*n*-alkanes	GC/C/IRMS	Isotope values of soil *n*-alkanes increased with time of maize cultivation as a result of maize carbon integration into soil organic matter	Lichtfouse et al. (1994)

These distinct signatures can be used in conjunction with molecular analysis in order to establish the source of particular compounds. This information is useful for determining how the substance entered the environment, in particular, when identifying responsible parties for further examination of practices. Furthermore it is possible to assess whether these compounds are a result of contamination, produced naturally, or are degradation products of a contaminating agent. Prior to the use of isotopic analysis the majority of this type of work rested upon biomarker analysis, whereby particular sources are matched with residues by their molecular compositions: the robustness of this technique is improved considerably by combining it with GC/C/IRMS analysis.

The isotopic composition of certain biomarker components of pollutants can be used to determine the source of these pollutants in various environmental samples. The majority of these investigations have focused on the source of n-alkanes or polyaromatic hydrocarbons (PAHs) deriving from oil pollution. Rogers & Savard (1999) correlated the isotopic composition of n-alkanes deriving from sediment samples with those of a shipping bunker fuel illustrating that a combination of isotopic and biomarker analyses can be used to determine the source of the sedimentary pollution. This technique was also employed by Mazeas & Budzinski (2002) who sampled oil residues and oiled bird feathers from the coast of France following an oil spillage. By combining PAH and n-alkane compositional and isotopic analyses they were able to determine that the bird feathers were contaminated with oil from the slick, however, many of the oil residues found likely derived from other oil tankers cleaning their tanks. McRae et al. (2000) observed isotopically light PAHs in lagoon sediments which they concluded were emitted by a nearby plant that used biogenic methane as a carbon source rather than petrogenic sources. In addition to pollution of marine environments, studies have also focused on soils (Wilcke et al. 2002; Sun et al. 2003; Glaser et al. 2005). Sun et al. (2003) were able to resolve coal-derived PAHs from those derived from petroleum sources in contaminated land as a result of their differing isotopic compositions. Wilcke et al. (2002) used the isotopic signature of perylene to show that in tropical environments recent biological sources of PAHs from termites were present, whereas, temperate soil PAHs were primarily formed from pyrolytic sources.

Biodegradation

A further use of the isotopic compositions of pollutant residues is in the investigation of the extent and processes of their biodegradation. A more in depth overview of the majority of studies of biodegradation of organic contaminants investigated by compound-specific isotope analysis is given by Schmidt et al. (2004). In general, as compounds are biologically degraded their isotopic compositions are changed as a result of the organism preferentially removing the lighter isotope to leave the residual compound isotopically enriched compared with its initial composition. This is illustrated by Hunkeler et al. (2001) who showed that as methyl *tert*-butyl ether (MTBE) was biodegraded an isotope fractionation occurred leaving the residual MTBE isotopically heavier, however, when phase partitioning occurred the fractionation was much smaller. This was confirmed by studies of Kuder et al. (2005) who determined $\delta^{13}C$ and δD values of MTBE in an anaerobic enrichment culture. They reported extensive fractionation of both isotopes as a result of anaerobic biodegradation and were able to detect biodegradation in groundwater samples at nine gasoline spill-sites, in some cases exceeding 90% of the original mass. In addition to MTBE biodegradation, Richnow et al. (2003a, 2003b) have used the isotopic composition of benzene, toluene, ethylbenzene, and xylenes (BTEX) to investigate their biodegradation. They performed a laboratory based study of biodegradation to obtain the fractionation factors for a variety of BTEX and applied this factor in the field to observe biodegradation of BTEX along a groundwater flow path and a landfill leachate plume. This was found to correlate with the change in concentration along the gradient. However, they found that not all BTEX tested produced an observable isotope fractionation with biodegradation.

Application of compound-specific stable isotope analysis for trophic studies of invertebrates

Compound-specific stable isotopic techniques have proven to be valuable in studies involving C flow and trophic studies in aquatic and terrestrial systems, especially when direct observations of feeding are difficult (Rieley et al. 1999; Ziegler & Fogel, 2003; Chamberlain et al. 2004; McCarthy et al. 2004). In particular, fatty acid compositions and their associated $\delta^{13}C$ values have been utilized effectively in confirming trophic interactions, since certain fatty acids are conserved during trophic transfer, thus can be used as biomarkers. Studies using this approach have found evidence of chemosynthetic mussels and tube worms incorporating the fatty acids of symbiotic bacteria (MacAvoy et al. 2002). The fatty acid $\delta^{13}C$ values determined for the mussels (−45.4‰ to −39.6‰ or −78.8‰ to −68.4‰), reflected the methane source utilized by their symbionts, whereas the tube worm fatty acid $\delta^{13}C$ values (−29.3‰ to −18.0‰) reflected the local dissolved inorganic carbon (DIC) pools due to the tube worm symbionts utilizing the DIC as their C source. The results demonstrated that chemosynthetically produced essential and precursor fatty acids isolated from host tissue retain the $\delta^{13}C$ values of the synthesizing bacteria. Consequently, the C source used by the symbionts can be identified and the essential fatty acid $\delta^{13}C$ values can be used to trace the fatty acids transferred between organisms. Similarly, Pond et al. (1998) determined the stable carbon isotope composition of fatty acids in chemosynthetic mussels and confirmed that thio- and methanotrophic bacterial endosymbionts were equally important in the nutrition of the mussel.

The trophic behavior of shrimps has been studied by adopting similar compound-specific methods (Pond et al. 1997, 2000; Rieley et al. 1999; Pakhomov et al. 2004). Rieley et al. (1999) investigated the trophic ecology of the alvinocaridid shrimp *Rimicaris exoculata* by determining the molecular composition and $\delta^{13}C$ values of its PLFAs and sterols. The values were compared with the three main sources of dietary C available, and the results suggested that the predominant source of dietary C for the shrimp was from the epibiotic bacteria living within their carapaces, and that the shrimp cholesterol was likely to be derived from the oceanic photic zone.

Nitrogen stable isotope signatures are also important in trophic studies, and can be used as an indicator of trophic level (DeNiro & Epstein 1981). The $\delta^{15}N$ values of some amino acids becomes heavily enriched from food source to consumer, while others exhibit little or no change and thus can provide information regarding the source of N (McClelland & Montoya, 2002). Pakhomov et al. (2004) determined the $\delta^{15}N$ values of amino acids of the bottom-dwelling caridean shrimp *Nauticaris marionis* to study its spatial and temporal variability. The amino acid $\delta^{15}N$ values suggested that the *N. marionis* from the inter-island realm were second-order carnivores, whereas the shrimps at the nearshore realm were first-order carnivores. This was not revealed in bulk $\delta^{15}N$ values, which suggested that the trophic levels did not differ, thus highlighting the advantage of compound-specific stable isotope analyses.

Carbon isotope analysis of PLFAs can be used to delineate algal and bacterial material from other organic matter in estuaries, and to trace the differential dynamics and isotope fractionation of green algae and diatoms along the estuarine gradient. Boschker et al. (2005) investigated the planktonic community structure along the Scheldt estuary using $\delta^{13}C$ values of PLFAs. This study revealed that the $\delta^{13}C$ values of the diatom fatty acid biomarker, 20:5ω3, were primarily related to DIC $\delta^{13}C$ values, and that green algae and diatom markers use a different inorganic C source or carbon-fixation metabolism, due to the difference in the $\delta^{13}C$ values of the green algal fatty acid biomarker, 18:3ω3, and the diatom fatty acid biomarker, 20:5ω3. The isotopic composition of the three bacterial PLFAs, i15:0, a15:0, and 18:1ω7c, closely followed particulate organic carbon (POC) $\delta^{13}C$ values, indicating that POC was the main source for bacterial growth. At the marine end of the estuary, algal PLFAs were similar in $\delta^{13}C$ values to bacterial PLFAs and POC, suggesting that local production by phytoplankton may be an important source for bacterial growth in that region of the estuary. However, in the upper estuary algal PLFAs were more depleted than bacterial PLFAs, and the isotope signatures suggest that the bacterial C substrate was primarily of terrestrial or anthropogenic sewage origin.

Combining compound-specific stable isotope analysis with ^{13}C-labeling provides further opportunities for monitoring the presence and growth of individual populations in planktonic and benthic communities, and to track C and fatty acid transfer along the food chain to primary and secondary consumers (Pel et al. 2004, Middelburg et al. 2000). Such studies have revealed rapid and significant transfer of C from benthic algae to bacteria and nematodes, with maximum labeling after 1 day (Middelburg et al. 2000), and that zooplankton, previously thought to be predaceous, grazed mainly on algae (Pel et al. 2004).

Compound-specific stable isotope analysis can also provide new insights into food web analyses; $\delta^{13}C$ values of individual compounds offer the advantage of providing additional isotopic information regarding the assimilation and biosynthesis of specific compounds from an organism's diet, which also assist in tracing the origins of nutrients necessary for biomass production. However, stable isotope composition can change between diet and consumer due to differential digestion or fractionation during assimilation and metabolic processes. Metabolic fractionation can also result in $\delta^{13}C$ values of different tissues varying substantially within a consumer (Hobson & Clark 1992). Furthermore, there are differences in the isotopic composition of the major biochemical classes, e.g. lipids usually possess more depleted $\delta^{13}C$ values than other major compound classes and the whole organism, due to enzymatic discrimination against ^{13}C during lipid biosynthesis (DeNiro & Epstein 1977). Hammer et al. (1998) determined the $\delta^{13}C$ values of individual fatty acids extracted from subcutaneous fat tissues of redhead ducks, *Aythya americana*, and the roots and rhizomes of the seagrass, *Haloduke wrightii*, assumed to be the main dietary source for at least a month. The $\delta^{13}C$ values of the fatty acids in the ducks were more positive than the identical fatty acids in the seagrass. The $\delta^{13}C$ values of the duck fatty acids were generally closer to the $\delta^{13}C$ value of the whole tissue of the seagrass roots and rhizomes. This strongly indicates that the ducks do not directly incorporate the fatty acids of their diet into their fat tissues, at least not to a large extent, and that the fatty acids are biosynthesized from precursors with more positive $\delta^{13}C$ values, such as carbohydrates and/or proteins. However, there may also be isotopic fractionations associated with the biosynthesis and catabolism of these compounds.

Carbon isotope compositions of individual amino acids can also be used to define isotope fractionations related to specific metabolic processes. Fantle et al. (1999) demonstrated that the experimentally cultured juvenile blue crab, *Callinectes sapidus*, fractionates the C isotopes of nonessential amino acids to a greater extent than essential amino acids, implying different metabolic pathways. Depletions in $\delta^{13}C$ values in the detritus fed crab also revealed that crabs feeding on detritus selectively assimilated an isotopically distinct pool of C, rather than the bulk detritus. In addition, the $\delta^{13}C$ values of the individual amino acids from crabs collected from the bay and marsh of a natural estuarine system, indicated that crabs in the bay and marsh utilize different sources of C, but shared a protein-rich dietary component, most likely plankton, which contributed significantly to their essential amino acid requirement. The $\delta^{13}C$ values of amino acids have further been employed to study the dietary sources of egg amino acids in nectar-feeding insects (O'Brien et al. 2002, 2005). The insects were raised on isotopically contrasting diets (C3 and C4) and the egg amino acid $\delta^{13}C$ values determined. The results showed that the adult diet had a highly significant effect on nonessential amino acid $\delta^{13}C$ values but no adult dietary C was incorporated into any of the essential amino acids, therefore essential egg amino acids originate entirely from the larval diet.

The potential of compound-specific stable isotope analysis as a powerful tool for the study of C dynamics and trophic interactions in soil ecosystems has only recently been explored. Chamberlain et al. (2004) examined the implications for the use of compound-specific carbon isotope analysis in animal dietary studies by investigating the rate of incorporation of dietary C into collembolan lipids by switching Collembola (*Folsomia candida* and *Proisotoma minuta*) from C3 to C4 diets, and then determined the $\delta^{13}C$ values of fatty acids over a period of 39 days (Chamberlain et al. 2006a). The $\delta^{13}C$ values of the collembolan lipids were often different to those of the same components in the diet, suggesting that fractionation or partitioning occurs during digestion, assimilation, and biosynthesis within the Collembola. The rates of change of $\delta^{13}C$ values differed among compounds, and half-lives ranged from 29 min to 14 days, some of which were related to the abundance of the component in the diet. This highlights the importance of determining potential sources of isotopic fractionation so that meaningful interpretations can be achieved.

Trophic preferences of soil invertebrates are mainly conducted by gut content analysis, or through visual observations, which can be labor intensive and often unreliable. Chamberlain et al. (2006b) have recently applied compound-specific stable carbon isotope analysis to determine the trophic preference of two species of Collembola, *F. candida* and *P. minuta*. The diets were isotopically distinct to facilitate the determination of collembolan food sources. Furthermore, any isotopic fractionation or partitioning occurring would be small relative to the stable C isotope signature of C derived from the C3 and C4 diets. The results showed that the collembolan fatty acid $\delta^{13}C$ values increased over 21 days, reflecting those of the nematode diet (C3 signature), suggesting a strong feeding preference for the nematodes over the fungi, even though the species of Collembola are classically considered to be mycophagous. Figure 14.8 shows the $\delta^{13}C$ values of fatty acids 16:0 vs. 18:0 and 16:1ω7 vs. 18:2ω6, for each of the Collembola and the offered diets. The $\delta^{13}C$ values of 16:0 and 18:0 in *F. candida* (Figure 14.8a) lie close to those of the nematode, but for *P. minuta* they appear to be intermediate between both diets, implying that *P. minuta* consumed both fungi and nematodes. However, the $\delta^{13}C$ values of 16:1ω7 and 18:2ω6 (Figure 14.8b) for both *F. candida* and *P. minuta* plot close to the $\delta^{13}C$ values of *P. redivivus*, consistent with this diet being the only

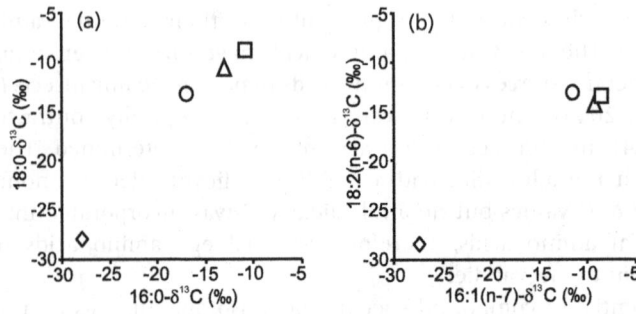

Figure 14.8 δ^{13}C values of selected fatty acids derived from diets (*C. cladosporioides* and *P. redivivus*), and consumers (*F. candida* and *P. minuta*) after 21 days of the preference test. (a) 16:0 vs. 18:0; (b) 16:1ω7) vs. 18:2ω6. Errors associated with the δ^{13}C values are smaller than the symbols. ◇ *C. cladosporioides*; □ *P. redivivus*; △ *F. candida*; ○ *P. minuta*. (Adapted from Chamberlain et al. 2006b).

significant C source consumed by the Collembola. This study demonstrates compound-specific stable isotope analysis as a powerful tool for determining trophic preferences of Collembola, and potentially many other classes of soil invertebrates, offering significant improvements over traditional methods.

Pulse labeling of above-ground vegetation with ^{13}CO$_2$ *in situ* has recently been combined with compound-specific analysis to determine C assimilation by individual enchytraeid worms (Black et al., unpublished results). Cholesterol was extracted from the individual enchytraeids and δ^{13}C values determined by GC/C/IRMS. Rhizosphere C assimilation was demonstrated by significant changes in the cholesterol δ^{13}C values. Results showed that no significant differences were observed in the δ^{13}C values of cholesterol in the two dominant genera, but variation in the intra-population indicate that the C may be assimilated in different ways, and large variations in the δ^{13}C values suggest that individual worms may have a range of food sources. In addition, Derrien et al. (2003, 2004) applied ^{13}C-labeling and compound-specific stable isotope techniques to investigate the nature and dynamics of individual carbohydrates in the rhizosphere. They demonstrated that in wheat, neutral sugars were the dominant components released into the soil by living roots, and that polymerized glucose was the most abundant compound of the rhizodeposits. The combined use of ^{13}C-labeling and compound-specific stable isotope analysis clearly shows potential for providing critical information regarding C dynamics and trophic interactions in soil environments.

Conclusions

Major impetus was given to the field of compound-specific stable isotope analysis through the development of the GC/C/IRMS technique and the past decade has seen an explosion in the number and range of applications of the approach.

Application of compound-specific stable isotope approaches in the field of organic geochemistry has become routine and has led to many advances in our understanding of past environments. The major advantage in using compound-specific approaches in paleoecological reconstructions stems from the ability to match structures to isotope values to give greater specificity to biomarker studies, and this provides enhanced insights into sedimentary processes and biogeochemical cycles. Adoption of the approach into other research areas has been surprisingly slow, possibly due to unfamiliarity of workers with the molecular level analytical approaches that are implicit in organic geochemistry. Familiarity with the separation and characterization methods underpins all applications of compound-specific stable isotope analyses and is essential to the success of the approach.

The potential for the further application of compound-specific approaches to improve our understanding of biogeochemical cycles in contemporary environments is considerable indeed. The accumulating examples of studies involving stable isotopes as tracers, essentially replacing radioactive tracers in some investigations, are testimony to the possibilities offered by the approach. As laboratories acquire the necessary instrumentation the next few years is set to see a major expansion in the application of compound-specific approaches to studies of microbes in the environment, combined with genetic approaches. Such approaches are proving to be especially powerful in providing new insights, especially into unculturable microbes, since ^{13}C-labeled tracers can be introduced with minimal disturbance into natural systems. The major challenges lie with the design of imaginative and robust experiments.

Likewise, such approaches have the potential to probe the activities of micro-invertebrates in environments where possibilities of visual observations are limited or impossible. The possibilities in this area are enhanced by the development of techniques for the stable carbon isotope analysis of individual organisms (e.g. Evans et al. 2003), which have the capacity to reveal the behavior of a species at the level of the individual rather than the population. Unlike radiolabeled tracers no ethical problems exist in the application of stable isotope tracers in natural systems, thereby allowing biochemical pathways and ecosystem processes to be studied in a diverse range of natural and anthropogenically altered environments. The introduction of novel stable isotope tracer delivery systems is helping to increase the range of applications still further (Ostle et al. 2000).

Considerable promise is offered in the field of archaeology for the further application of compound-specific stable isotope approaches due to the survival of biochemical components in association with archaeological materials, such as skeletal remains and pottery. This has proved to be an area where stable isotope approaches focus mainly on mammals, including both humans and domesticated species. In the field of paleodietary reconstruction the compound-specific approaches have clear potential and complement the bulk stable isotope determinations that have been extensively used for around three decades. Recent work has shown that otherwise inaccessible information becomes available via compound-specific studies of the amino acids that comprise collagen (Corr et al. 2005). Likewise, new possibilities exist for using largely untapped sources of light stable isotope information contained in such compounds as cholesterol, which reflect different aspects of diet and environment (Jim et al. 2004). A surprisingly neglected area appears to be the application of compound-specific stable isotope approaches in our understanding of avian behavior (Hammer et al. 1998) despite the intense interest in the use of stable isotopes in this field.

While instrumentation for the stable isotope analysis of other light isotopes is commercially available surprisingly few publications have emerged for compound-specific studies of nitrogen (e.g. Petzke et al. 2005), deuterium (e.g. Xie et al. 2000) and oxygen (Hener et al. 1998). The next few years will undoubtedly see an expansion in the range of applications of these techniques in a variety of fields.

A recent major development has been the first reported applications of an operational high-performance liquid chromatograph (HPLC)–IRMS. The new HPLC–IRMS method appears to offer significant advantage over the GC/C/IRMS for certain applications. First, it extends the range of compounds that can be investigated. For example, in the case of proteins or peptides they can be studied intact, or if $\delta^{13}C$ values are required for their component amino acids then only purification and hydrolysis is required, rather than derivatization. Analytes separated by aqueous based HPLC are quantitatively converted to CO_2 in a reactor containing sodium peroxodisulphate and phosphoric acid. Preliminary analyses have shown that amino acids can be analyzed directly on such HPLC-IRMS systems to provide accurate $\delta^{13}C$ values (Krummen et al. 2004). Analytical precisions reported for a range of amino acids are ca. ±0.2–0.3‰ for 100–300 ng amino acid injected. Clearly, there is great potential for the application of this technique in a wide range of fields adding yet further scope to the application of compound-specific stable isotope analyses in ecological and paleoecological studies.

References

Abelson, P.H. & Hoering, T.C. (1961) Carbon isotope fractionation in the formation of amino acids by photosynthetic organisms. *Procedings of the National Academy of Sciences USA*, **47**(5), 623–632.

Abidi, S.L. (2001) Chromatographic analysis of plant sterols in foods and vegetable oils. *Journal of Chromatography A*, **935**(1–2), 173–201.

Abraham, W.-R., Hesse, C. & Pelz, O. (1998) Ratios of carbon isotopes in microbial lipids as an indicator of substrate usage. *Applied and Environmental Microbiology*, **64**, 4202–4209.

Alvarez, J.G. & Touchstone, J.C. (1992) Separation of acidic and neutral lipids by aminopropyl-bonded silica-gel column chromatography. *Journal of Chromatography – Biomedical Applications*, **577**(1), 142–145.

Amann, R.I., Ludwig, W. & Schleifer, K.H. (1995) Phylogenetic identification and *in situ* detection of individual microbial cells without cultivation. *Microbiological Reviews*, **59**, 143–169.

Ambrose, S.H. (1993) Isotopic analysis of palaeodiets: methodological and interpretive considerations. In: *Investigations of Ancient Human Tissue* (Ed. M.K. Sandford), pp. 59–130. Gordon & Breach Science Publishers, Langhorne, PA.

Anwar, S., Kosaki, T. & Yonebayashi, K. (2004) Cupric oxide oxidation products of tropical peat soils. *Soil Science and Plant Nutrition*, **50**(1), 35–43.

Arao, T. (1999) *In situ* detection of changes in soil bacterial and fungal activities by measuring ^{13}C incorporation into soil phospholipid fatty acids from ^{13}C acetate. *Soil Biology and Biochemistry*, **31**, 1015–1020.

Beynon, S. (1998) *Metabolism and Nutrition*. Mosby International, London.

Barrie, A., Bricout, J. & Koziet, J. (1984) Gas-chromatography – stable isotope ratio analysis at natural abundance levels. *Biomedical Mass Spectrometry*, **11**, 439–447.

Begley, I.S. & Scrimgeour, C.M. (1997) High-precision δ^2H and $\delta^{18}O$ measurement for water and volatile organic compounds by continuous-flow pyrolysis isotope ratio mass spectrometry. *Analytical Chemistry* **69**, 1530–1535.

Berstan, R., Dudd, S.N., Copley, M.S., Morgan, E.D., Quyec, A. & Evershed, R.P. (2004) Characterisation of "bog butter" using a combination of molecular and isotopic techniques. *Analyst*, **129**, 270–275.

Bidigare, R.R., Fluegge, A., Freeman, K.H., et al. (1997) Consistent fractionation of C-13 in nature and in the laboratory: Growth-rate effects in some haptophyte algae. *Global Biogeochemical Cycles*, **11**(2), 279–292.

Bird, M.I., Summons, R.E., Gagan, M.K., et al. (1995) Terrestrial vegetation change inferred from *n*-alkane $\delta^{13}C$ analysis in the marine-environment. *Geochimica et Cosmochimica Acta*, **59**, 2853–2857.

Bilke, S. & Mosandl, A. (2002) Measurements by gas chromatography/pyrolysis/mass spectrometry: fundamental conditions in $^2H/^1H$ isotope ratio analysis. *Rapid Communications in Mass Spectrometry*, **16**, 468–472.

Blakeney, A.B., Harris, P.J., Henry, R.J. & Stone, B.A. (1983) A simple and rapid preparation of alditol acetates for monosaccharide analysis. *Carbohydrate Research*, **113**(2), 291–299.

Bligh, E.G. & Dyer, W.J. (1959) A rapid method of total lipid extraction and purification. *Canadian Journal of Biochemistry and Physiology*, **37**, 911–918.

Blumenberg, M., Seifert, R., Nauhaus, K., Pape, T. & Michaelis, W. (2005) *In vitro* study of lipid biosynthesis in an anaerobically methane-oxidising microbial mat. *Applied and Environmental Microbiology*, **71**, 4345–4351.

Boschker, H.T.S. & Middelburg, J.J. (2002) Stable isotopes and biomarkers in microbial ecology. *FEMS Microbiology Ecology*, **40**, 85–95.

Boschker, H.T.S., Nold, S.C., Wellsbury, P., et al. (1998) Direct linking of microbial populations to specific biogeochemical processes by ^{13}C-labelling of biomarkers. *Nature*, **392**, 801–805.

Boschker, H.T., de Graaf, W., Koster, M., Meyer-Reil, L.A. & Cappenberg, T.E. (2001) Bacterial populations and processes involved in acetate and propionate consumption in anoxic brackish sediment. *FEMS Microbiology Ecology*, **35**, 97–103.

Boschker, H.T.S., Kromkamp, J.C. & Middelburg, J.J. (2005) Biomarker and carbon isotopic constraints on bacterial and algal community structure and functioning in a turbid, tidal estuary. *Limnology and Oceanography*, **50**, 70–80.

Brand, W.A., Tegtmeyer, A.R. & Hilkert, A. (1994) Compound-specific isotope analysis: extending toward $^{15}N/^{14}N$ and $^{18}O/^{16}O$. *Organic Geochemistry*, **21**, 585–594.

Bull, I.D., van Bergen, P.F., Bol, R., et al. (1999) Estimating the contribution of *Spartina anglica* biomass to salt-marsh sediments using compound specific stable carbon isotope measurements *Organic Geochemistry*, **30**, 477–483.

Bull, I.D., Parekh, N.R., Hall, G.H., Ineson, P. & Evershed, R.P. (2000) Detection and classification of atmospheric methane oxidizing bacteria in soil. *Nature*, **405**(6783), 175–178.

Burgoyne, T.W. & Hayes, J.M. (1998) Quantitative production of H_2 by pyrolysis of gas chromatographic effluents. *Analytical Chemistry*, **70**, 5136–5141.

Butler, J.L., Williams, M.A., Bottomley, P.J. & Myrold, D.D. (2003) Microbial community dynamics associated with rhizosphere carbon flow. *Applied and Environmental Microbiology*, **69**, 6793–6800.

Chamberlain, P.M., Bull, I.D., Black, H.I.J., Ineson, P. & Evershed, R.P. (2004) Lipid content and carbon assimilation in Collembola: implications for the use of compound-specific carbon isotope analysis in animal dietary studies. *Oecologia*, **139**, 325–335.

Chamberlain, P.M., Bull, I.D., Black, H.I.J., Ineson, P. & Evershed, R.P. (2006a) The effect of diet on isotopic turnover in Collembola examined using the stable carbon isotopic compositions of lipids. *Soil Biology and Biochemistry*, **38**, 1063–1076.

Chamberlain, P.M., Bull, I.D., Black, H.I.J., Ineson, P. & Evershed, R.P. (2006b) Collembolam trophic preferences determined using fatty acid distributions and compound-specific stable carbon isotope analysis. *Soil Biology and Biochemistry*, **38**, 1275–1281.

Copley, M.S., Berstan, R., Dudd, S.N., et al. (2003) Direct chemical evidence for widespread dairying in prehistoric Britain *Proceedings of the National Academy of the United States of America*, **100**, 1524–1529.

Copley, M.S., Jim, S., Jones, V., et al. (2004) Short- and long-term foraging and foddering strategies of domesticated animals from Qasr Ibrim, Egypt. *Journal of Archaeological Science*, **31**, 1273–1286.

Copley, M.S., Berstan, R., Dudd, S.N., Straker, V., Payne, S. & Evershed, R.P. (2005a) Dairying in antiquity. I. Evidence from absorbed lipid residues dating to the British Iron Age. *Journal of Archaeological Science*, **32**, 485–503.

Copley, M.S., Berstan, R., Dudd, S.N., Straker, V., Payne, S. & Evershed, R.P. (2005b) Dairying in antiquity. II. Evidence from absorbed lipid residues ating to the British Bronze Age. *Journal of Archaeological Science*, **32**, 505–521.

Copley, M.S., Berstan, R., Mukherjee, A.J., et al. (2005c) Dairying in antiquity. III. Evidence from absorbed lipid residues dating to the British Neolithic *Journal of Archaeological Science*, **32**, 523–546.

Corr, L.T. (2003) *The evaluation of a multi-proxy stable isotope approach to palaeodietary reconstruction*, PhD thesis, University of Bristol.

Corr, L.T., Sealy, J.L., Horton, M.C. & Evershed, R.P. (2005) A novel marine dietary indicator utilising compound-specific collagen amino acid $\delta^{13}C$ values of ancient humans *Journal of Archaeological Science*, **32**, 321–330.

Craig, O.E., Taylor, G., Collins, M.J. & Parker Pearson, M. (2005) The identification of prehistoric dairying activities in the Western Isles of Scotland: an integrated biomolecular approach. *Journal of Archaeological Science*, **32**, 91–103.

Cranwell, P.A. (1981) Diagenesis of free and bound lipids in terrestrial detritus deposited in a lacustrine sediment. *Organic Geochemistry*, **3**, 79–89.

Crossman, Z.M., McNamara, N., Parekh, N., Ineson, P. & Evershed, R.P. (2001) A new method for identifying the origins of simple and complex hopanoids in sedimentary materials using stable isotope labelling with $^{13}CH_4$ and compound specific stable isotope analyses. *Organic Geochemistry*, **32**, 359–364.

Crossman, Z.M., Abraham, F. & Evershed, R.P. (2004) Stable isotope pulse-chasing and compound specific stable carbon isotope analysis of phospholipid fatty acids to assess methane oxidizing bacterial populations in landfill cover soils. *Environmental Science and Technology*, **38**(5), 1359–1367.

Crossman, Z.M., Ineson, P. & Evershed, R.P. (2005) The use of ^{13}C labelling of bacterial lipids in the characterisation of ambient methane-oxidising bacteria soils. *Organic Geochemistry*, **36**, 769–778.

Crossman, Z.M., Wang, Z.P., Ineson, P. & Evershed, R.P. (2006) Investigation of the effect of ammonium sulfate on populations of ambient methane oxidising bacteria by ^{13}C-labelling and GC/C/IRMS analysis of phospholipid fatty acids. *Soil Biology and Biochemistry*, **38**, 983–990.

Demmelmair, H. & Schmidt, H.-L. (1993) Precise δ^{13}C-determination in the range of natural abundance on amino acids from protein hydrolysates by gas chromatography – isotope ratio mass spectrometry. *Isotopenpraxis*, **29**, 237–250.

DeNiro, M.J. & Epstein, S. (1977) Mechanism of carbon isotope fractionation associated with lipid synthesis. *Science*, **197**, 261–263.

DeNiro, M.J. & Epstein, S. (1981) Influence of diet on the distribution of nitrogen isotopes in animals. *Geochimica et Cosmochimica Acta*, **45**, 341–351.

Derrien, D., Balesdent, J., Marol, C. & Santaella, C. (2003) Measurement of the ^{13}C/^{12}C ratio of soil–plant individual sugars by gas chromatography/combustion/isotope-ratio mass spectrometry of silylated derivatives. *Rapid Communications in Mass Spectrometry*, **17**(23), 2626–2631.

Derrien, D., Marol, C. & Balesdent, J. (2004) The dynamics of neutral sugars in the rhizosphere of wheat. An approach by ^{13}C pulse-labelling and GC/C/IRMS. *Plant and Soil*, **267**, 243–253.

Docherty, G. (2002) *Natural variations in plant tissue individual fatty acid and monosaccharide δ^{13}C values: Implications for defining the origins of animal fats in archaeological pottery.* PhD thesis, University of Bristol.

Docherty, G., Jones, V. & Evershed, R.P. (2001) Practical and theoretical considerations in the gas chromatography/combustion/isotope ratio mass spectrometry δ^{13}C analysis of small polyfunctional compounds. *Rapid Communications in Mass Spectrometry*, **15**, 730–738.

Dudd, S.N. & Evershed, R.P. (1998) Direct demonstration of milk as an element of archaeological economies. *Science*, **282**, 1478–1481.

Eek, M.K., Whiticar, M.J., Bishop, J.K.B. & Wong, C.S. (1999) Influence of nutrients on carbon isotope fractionation by natural populations of Prymnesiophyte algae in NE Pacific. *Deep-Sea Research Part II – Topical Studies in Oceanography*, **46**, 2863–2876.

Eglinton, G., Raphael, R.A., Gonzalez, A.G., et al. (1962) Hydrocarbon constituents of the wax coatings of plant leaves: A taxonomic survey. *Phytochemistry*, **1**, 89–102.

Elvert, M., Suess, E., Greinert, J. & Whiticar, M.J. (2001) Archaea mediating anaerobic methane oxidation in deep-sea sediments at cold seeps of the eastern Aleutian subduction zone. *Organic Geochemistry*, **31**, 1175–1187.

Evans, C., Evershed, R.P., Black, H.I.J. & Ineson, P. (2003) Compound-specific stable isotope analysis of soil mesofauna using thermally assisted hydrolysis and methylation for ecological investigations. *Analytical Chemistry*, **75**, 6056–6062.

Evershed, R.P. (1993) Advances in silylation. In: *Handbook of Derivatives for Chromatography* (Eds K. Blau & J. McK. Halket), pp. 52–107. John Wiley & Sons, Chichester.

Evershed, R.P. & Connolly, R.C. (1987) Lipid preservation in Lindow Man. *Naturwissenschaften*, **75**, 143–145.

Evershed, R.P., Arnot, K.I., Collister, J., Eglington, G. & Charters, S. (1994) Application of isotope ratio monitoring gas chromatography-mass spectrometry to the analysis of organic residues of archaeological origin. *Analyst*, **119**, 909–914.

Evershed, R.P., Turner-Walker, G., Hedges, R.E.M., Tuross, N. & Leyden, A. (1995) preliminary results for the analysis of lipids in ancient bone. *Journal of Archaeological Science*, **22**, 277–290.

Evershed, R.P., Mottram, H.R., Dudd, S.N., et al. (1997) New criteria for the identification of animal fats preserved in archaeological pottery *Naturwissenschaften*, **84**, 402–406.

Evershed, R.P., Dudd, S.N., Charters, S., et al. (1999) Lipids as carriers of anthropogenic signals from prehistory. *Philosophical Transactions of the Royal Society of London Series B-Biological Sciences*, **354**, 19–31.

Evershed, R.P., Dudd, S.N., Copley, M.S. & Mukherjee, A.J. (2003) Identification of animal fats via compound specific $\delta^{13}C$ values of individual fatty acids: assessments of results for reference fats and lipid extracts of archaeological pottery vessels. In: *Documenta Praehistorica, XXIX; 9th Neolithic Studies* (Ed. M. Budja), pp. 73–96.

Fantle, M.S., Dittel, A.I., Schwalm, S.M., Epifanio, C.E. & Fogel, M.L. (1999) A food web analysis of the juvenile blue crab, Callinectes sapidus, using stable isotopes in whole animals and individual amino acids. *Oecologia*, **120**, 416–426.

Fogel, M.L. & Tuross, N. (2003) Extending the limits of palaeodietary studies of humans with compound specific carbon isotope analysis of amino acids. *Journal of Archaeological Science*, **30**, 535–545.

France-Lanord, C. & Derry, L.A. (1994) $\delta^{13}C$ of organic carbon in the Bengal Fan: Source evolution and transport of C_3 and C_4 plant carbon to marine sediments. *Geochimica et Cosmochimica Acta*, **58**, 4809–4814.

Farrimond, P., Head, I.M. & Innes, H.E. (2000) Environmental influence on the biohopanoid composition of recent sediments. *Geochimica et Cosmochimica Acta*, **64**, 2985–2992.

Freeman, K.H. & Colarusso, L.A. (2001) Molecular and isotopic records of C_4 grassland expansion in the late Miocene. *Geochimica et Cosmochimica Acta*, **69**, 1439–1454.

Freeman, K.H., Hayes, J.M., Trendel, J.-M. & Albrecht, P. (1990) Evidence from carbon isotope measurements for diverse origins of sedimentary hydrocarbons. *Nature*, **343**, 254–256.

Gaebler, O.H., Vitti, T.G. & Vukmirovich, R. (1966) Isotope effects in metabolism of ^{14}N and ^{15}N from unlabelled dietary proteins *Canadian Journal of Biochemistry*, **44**, 1249–1257.

Galimov, E.M. (1981) *The Biological Fractionation of Isotopes*. Academic Press, London.

Gannes, L.Z., del Rio, C.M. & Koch, P. (1998) Natural abundance variations in stable isotopes and their potential uses in animal physiological ecology. *Comparative Biochemistry and Physiology A Molecular and Integrative Physiology*, **119**, 725–737.

Gehre, M., Hoefling, R., Kowski, P. & Strauch, G. (1996) Sample preparation device for quantitative hydrogen isotope analysis using chromium metal. *Analytical Chemistry*, **68**, 4414–4417.

Glaser, B., Dreyer, A., Bock, M., Fiedler, S., Mehring, M. & Heitmann, T. (2005) Source apportionment of organic pollutants of a highway-traffic-influenced urban area in Bayreuth (Germany) using biomarker and stable carbon isotope signatures. *Environmental Science and Technology*, **39**, 3911–3917.

Goñi, M.A. & Eglinton, T.I. (1996) Stable carbon isotopic analyses of lignin-derived CuO oxidation products by isotope ratio monitoring-gas chromatography-mass spectrometry (irm-GC-MS). *Organic Geochemistry*, **24**, 601–615.

Grice, K., Schaeffer, P., Schwark, L. & Maxwell, J.R. (1997) Changes in palaeoenvironmental conditions during deposition of the Permian Kupferschiefer (Lower Rhine Basin, N.W. Germany) from variations in isotopic compositions of biomarker components. *Organic Geochemistry*, **26**, 677–690.

Gross, S. & Glaser, B. (2004) Minimization of carbon addition during derivatization of monosaccharides for compound-specific $\delta^{13}C$ analysis in environmental research. *Rapid Communications in Mass Spectrometry*, **18**(22), 2753–2764.

Hammer, B.T., Fogel, M.L. & Hoering, T.C. (1998) Stable carbon isotope ratios of fatty acids in seagrass and redhead ducks. *Chemical Geology*, **152**, 29–41.

Hanson, J.R., Macalady, J.L., Harris, D. & Scow, K.M. (1999) Linking toluene degradation with specific microbial populations in soil. *Applied and Environmental Microbiology*, **65**, 5403–5408.

Hare, P.E. & Estep, M.L.F. (1983) Carbon and nitrogen isotopic composition of amino acids in modern and fossil collagens. *Carnegie InstitutionWashington Yearbook*, **82**, 410–414.

Hare, P.E., Fogel., M.L., Stafford, T.W., Mitchell, A.D. & Hoering, T.C. (1991) The isotopic composition of carbon and nitrogen in individual amino acids isolated from modern and fossil proteins. *Journal of Archaeological Science*, **18**, 277–292.

Hastdorf, C.A. & DeNiro, M.J. (1985) Reconstruction of prehistoric plant production and cooking practice by a new isotopic method. *Nature*, **315**, 489–551.

Hayes, J.M. (1993) Factors controlling ^{13}C contents of sedimentary organic compounds – principles and evidence. *Marine Geology*, **113**, 111–125.

Hayes, J.M., Freeman, K.H., Popp, B.N. & Hoham, C. (1990) Compound-specific isotopic analyses: A novel tool for the reconstruction of ancient biogeochemical processes. In: *Advances in Organic Geochemistry* (Eds B. Durand & F. Behar), pp. 1115–1128. Pergamon Press, Oxford.

Hayes, J.M., Des Marais, D.J., Lambert, I.B., Strauss, H. & Summons, R.E. (1992) Proterozoic biogeochemistry. In: *The Proterozoic Biosphere: A Multidisciplinary Study* (Eds J.W. Schopf & C. Klein), pp. 81–134. Cambridge University Press, Cambridge.

Heaton, T.H.E. (1987) The $^{15}N/^{14}N$ ratio of plants in South Africa and Namibia: relationship to climate and coastal/saline environments. *Oecologia*, **74**, 236–246.

Heaton, T.H.E., Vogel., J.C., von la Chevallerie, G. & Collett, G. (1986) Climatic influence on the isotopic composition of bone collagen. *Nature*, **322**, 822–823.

Hedges, J.I. & Parker, P.L. (1976) Land-derived organic matter in surface sediments from the Gulf of Mexico. *Geochimica et Cosmochimica Acta*, **40**, 1019–1029.

Hedges, J.L. & Ertel, J.R. (1982) Characterization of lignin by gas capillary chromatography of cupric oxide oxidation products. *Analytical Chemistry*, **54**, 174–178.

Hener, U., Brand, W.A., Hilkert, A.W., Juchelka, D., Mosandl, A. & Podebrad, F. (1998) Simultaneous online analysis of $^{18}O/^{16}O$ and $^{13}C/^{12}C$ ratios of organic compounds using GC-pyrolysis-IRMS. *Zeitschrift fur Lebensmittel-Untersuchung und-Forschung*, **206**, 230–232.

Hilkert, A.W., Douthitt, C.B., Schlüter, H.J. & Brand, W.A. (1999) Isotope ratio monitoring gas chromatography/mass spectrometry of D/H by high temperature conversion isotope ratio mass spectrometry. *Rapid Communications in Mass Spectrometry*, **13**, 1226–1230.

Hinrichs, K.-U., Hayes, J.M., Sylva, S.P., Brewer, P.G. & DeLong, E.F. (1999) Methane-consuming archaebacteria in marine sediments. *Nature*, **398**, 802–805.

Hirs, C.H.W., Moore, S. & Stein, W.H. (1954) The chromatography of amino acids on ion exchange resins: use of volatile acids for elution. *Journal of the American Chemical Society*, **76**, 6063–6065.

Hobson, K.A. & Clark, R.G. (1992) Assessing avian diets using stable isotopes. 2. Factors influencing diet–tissue fractionation. *Condor*, **94**, 189–197.

Howland, M.R., Corr, L.T., Young, S.M.M., et al. (2003) Expression of the dietary isotope signal in the compound-specific $\delta^{13}C$ values of pig bone lipids and amino acids *International Journal of Osteoarchaeology*, **13**, 54–65.

Huang, Y., Bol, R., Harkness, D.D., Ineson, P. & Eglinton, G. (1996) Post-glacial variations in distributions, ^{13}C and ^{14}C contents of aliphatic hydrocarbons and bulk organic matter in three types of British acid upland soils. *Organic Geochemistry*, **24**, 273–287.

Huang, Y.S., Freeman, K.H., Eglinton, T.I. & Street-Perrott, F.A. (1999) $\delta^{13}C$ analyses of individual lignin phenols in Quaternary lake sediments: A novel proxy for deciphering past terrestrial vegetation changes. *Geology*, **27**, 471–474.

Huang, Y., Freeman, K.H., Wilkin, R.T., Arthur, M.A. & Jones, A.D. (2000) Black Sea chemocline oscillations during the Holocene: molecular and isotopic studies of marginal sediments. *Organic Geochemistry*, **31**, 1525–1531.

Hunkeler, D., Butler, B.J., Aravena, R. & Barker, J.F. (2001) Monitoring biodegradation of methyl tert-butyl ether (MTBE) using compound-specific carbon isotope analysis. *Environmental Science and Technology*, **35**, 676–681.

Jahnke, L.L., Summons, R.E., Dowling, L.M. & Zahiralis, K.D. (1995) Identification of methanotrophic lipid biomarkers in cold-seep mussel gills – chemical and isotopic analysis. *Applied and Environmental Microbiology*, **61**, 576–582.

Jim, S. (2000) *The development of bone cholesterol $\delta^{13}C$ values as a new source of palaeodietary information: models of its use in conjunction with bone collagen and apatite $\delta^{13}C$ values*. PhD thesis, University of Bristol.

Jim, S., Stott, A.W., Evershed, R.P., Rogers, J.M. & Ambrose, S.H. (2001) Animal feeding experiments in the development of cholesterol as a palaeodietary indicator. In: *Archaeological Sciences '97* (Ed. A. Millard), pp. 68–77. International Series, British Archaeological Research, Oxford.

Jim, S., Jones, V., Copley, M.S., Ambrose, S.H. & Evershed, R.P. (2003a) Effects of hydrolysis on the delta C-13 values of individual amino acids derived from polypeptides and proteins. *Rapid Communications in Mass Spectrometry*, **17**(20), 2283–2289.

Jim, S., Ambrose, S.H. & Evershed, R.P. (2003b) Natural abundance stable carbon isotope evidence for the routing and de novo synthesis of bone FA and cholesterol. *Lipids*, **38**, 179–186.

Jim, S., Ambrose, S.H. & Evershed, R.P. (2004) Stable carbon isotopic evidence for differences in the dietary origin of bone cholesterol, collagen and apatite: Implications for their use in palaeodietary reconstruction. *Geochimica et Cosmochimica Acta*, **68**, 61–72.

Jim, S., Jones, V., Ambrose, S.H. & Evershed, R.P. (2006) Quantifying dietary macronutrient sources of carbon for bone collagen using natural abundance stable carbon isotope analysis. *British Journal of Nutrition*, **95**, 1055–1062.

Jones, V. (2002) *Investigating the routing and synthesis of amino acids between diet and bone collagen via feeding experiments and application to palaeodietary reconstruction*. PhD thesis, University of Bristol.

Knief, C., Lipski, A. & Dunfield, P.F. (2003) Diversity and activity of methanotrophic bacteria in different upland soils. *Applied and Environmental Microbiology*, **69**, 6703–6714.

Kim, H. & Salem, N., Jr. (1990) Separation of lipid classes by solid phase extraction [published erratum appears in *Journal of Lipid Research* (1993), **34**, 166]. *Journal of Lipid Research*, **31**(12), 2285–2289.

Koopmans, M.P., Köster, J., van Kaam-Peters, et al. (1996) Dia- and catagenetic products of isorenieratene: Molecular indicators for photic zone anoxia. *Geochiica et Cosmochimica Acta*, **60**, 4467–4496.

Krummen, M., Hilkert, A.W., Juchelka, D., Duhr, A., Schluter, H.J. & Pesch, R. (2005) A new concept for isotope ratio monitoring liquid chromatography/mass spectrometry. *Rapid Communications in Mass Spectrometry*, **18**, 2260–2266.

Kuder, T., Wilson, J.T., Kaiser, P., Kolhatkar, R., Philp, P. & Allen, J. (2005) Enrichment of stable carbon and hydrogen isotopes during anaerobic biodegradation of MTBE:

microcosm and field evidence compound-specific stable isotope analysis is used to characterize microcosm and field anaerobic biodegradation of MTBE. *Environmental Science and Technology*, **39**, 213–220.

Kuypers, M.M.M., Pancost, R.D. & Sinninghe Damsté, J.S. (1999) A large and abrupt fall in atmospheric CO_2 concentration during Cretaceous times. *Nature*, **399**, 342–345.

Leckrone, K.J. & Hayes, J.M. (1997) Efficiency and temperature dependence of water removal by membrance dryers. *Analytical Chemistry*, **69**, 911–918.

Leckrone, K.J. & Hayes, J.M. (1998) Water-induced errors in continuous-flow carbon isotope ratio mass spectrometry. *Analytical Chemistry*, **70**, 2737–2744.

Liaaen-Jensen, S. (1978) Chemistry of carotenoid pigments. In: *Photosynthetic Bacteria* (Eds R.K. Clayton & W.R. Sistrom), pp. 233–247. Plenum Press, New York.

Lichtfouse, E. (1995) ^{13}C labelling of soil *n*-hentriacontane (C_{31}) by maize cultivation. *Tetrahedron Letters*, **36**, 529–530.

Lichtfouse, E. (1998) Isotope and biosynthetic evidence for the origin of long-chain aliphatic lipids in soils. *Naturwissenschaften*, **85**, 76–77.

Lichtfouse, E., Elbisser, B., Balesdent, J., Mariotti, A. & Bardoux, G. (1994) Isotope and molecular evidence for direct input of maize leaf wax *n*-alkanes into crop soils. *Organic Geochemistry*, **22**, 349–351.

Lichtfouse, E., Berthier, G., Houot, S., Barriuso, E., Bergheaud, V. & Vallaeys, T. (1995) Stable carbon isotope evidence for the microbial origin of C_{14}–C_{18} *n*-alkanoic acids in soils. *Organic Geochemistry*, **23**, 849–852.

Lockheart, M.J., van Bergen, P.F. & Evershed, R.P. (1997) Variations in the stable carbon isotope compositions of individual lipids from the leaves of modern angiosperms: implications for the study of higher land plant-derived sedimentary organic matter. *Organic Geochemistry*, **26**, 137–153.

Logan, G.A., Summons, R.E. & Hayes, J.M. (1997) An isotopic biogeochemical study of neoproterozoic and early Cambrian sediments from the Centralian Superbasin, Australian. *Geochimica et Cosmochimica Acta*, **61**, 5391–5409.

Lu, Y., Murase, J., Watanabe, A., Sugimoto, A. & Kimura, M. (2004) Linking microbial community dynamics to rhizosphere carbon flow in a wetland rice soil. *FEMS Microbiology Ecology*, **48**, 179–186.

MacAvoy, S.E., Macko, S.A. & Joye, S.B. (2002) Fatty acid carbon isotope signatures in chemosynthetic mussels and tube worms from gulf of Mexico hydrocarbon seep communities. *Chemical Geology*, **185**, 1–8.

Macko, S.A., Estep., M.L.F., Hare, P.E. & Hoering, T.C. (1983) Stable nitrogen and carbon isotopic composition of individual amino acids isolated from cultured microorganisms. *Carnegie Institution Washington Yearbook*, **82**, 404–410.

Macko, S.A., Fogel, M.L., Hare, P.E. & Hoering, T.C. (1987) Isotopic fractionation of nitrogen and carbon in the synthesis of amino acids by microorganisms. *Chemical Geology (Isotope Geoscience Section)*, **65**, 79–92.

Macko, S.A., Uhle, M.E., Engel, M.H. & Andrusevich, V. (1997) Stable nitrogen isotope analysis of amino acid enantiomers by gas chromatography combustion/isotope ratio mass spectrometry. *Analytical Chemistry*, **69**(5), 926–929.

Macko, S.A., Ryan, M. & Engel, M.H. (1998) Stable isotopic analysis of individual carbohydrates by gas chromatographic combustion/isotope ratio mass spectrometry. *Chemical Geology*, **152**, 205–210.

Malosso, E., English, L., Hopkins, D.W. & O'Donnell, A.G. (2004) Use of ^{13}C-labelled plant materials and ergosterol, PLFA and NLFA analyses to investigate organic matter decomposition in Antarctic soil. *Soil Biology and Biochemistry*, **36**, 165–175.

Manirakiza, P., Covaci, A. & Schepens, P. (2001) Comparative study on total lipid determination using Soxhlet, Roese-Gottlieb, Bligh & Dyer & modified Bligh & Dyer extraction methods. *Journal of Food Composition and Analysis*, **14**, 93–100.

Marlowe, I.T., Green, J.C., Neal, A.C., Brassell, S.C., Eglinton, G. & Course, P.A. (1984) Long-chain (N-C_{37}–C_{39}) alkenones in the Prymnesiophyceae – distribution of alkenones and other lipids and their taxonomic significance. *British Phycological Journal*, **19**, 203–216.

Matthews, D.E. & Hayes, J.M. (1978) Isotope-ratio-monitoring gas chromatography-mass spectrometry. *Analytical Chemistry*, **50**, 1465–1473.

Mauclaire, L., Pelz, O., Thullner, M., Abraham, W.-R. & Zeyer, J. (2003) Assimilation of toluene carbon along a bacteria-protist food chain determined by ^{13}C-enrichment of biomarker fatty acids. *Journal of Microbiological Methods*, **55**, 635–649.

Mazeas, L. & Budzinski, H. (2002) Molecular and stable carbon isotopic source identification of oil residues and oiled bird feathers sampled along the Atlantic coast of France after the Erika oil spill. *Environmental Science and Technology*, **36**, 130–137.

McCarthy, M.D., Benner, R., Lee, C., Hedges, J.I. & Fogel, M.L. (2004) Amino acid carbon isotopic fractionation patterns in oceanic dissolved organic matter: an unaltered photoautotrophic source for dissolved organic nitrogen in the ocean? *Marine Chemistry*, **92**, 123–134.

McClelland, J.W. & Montoya, J.P. (2002) Trophic relationships and the nitrogen isotopic composition of amino acids in plankton. *Ecology*, **83**, 2173–2180.

McRae, C., Snape, C.E., Sun, C.G., et al. (2000) Use of compound-specific stable isotope analysis to source anthropogenic natural gas-derived polycyclic aromatic hydrocarbons in a lagoon sediment. *Environmental Science and Technology*, **34**, 4684–4686.

Meier-Augenstein, W. (1997) The chromatographic side of isotope ratio mass spectrometry – Pitfalls and answers. *LC*GC*, **15**, 244.

Meier-Augenstein, W. (1999) Applied gas chromatography coupled to isotope ratio mass spectrometry. *Journal of Chromatography A*, **842**, 351–371.

Meier-Augenstein, W., Watt, P.W. & Langhans, C.-D. (1996) Influence of gas chromatographic parameters on the measurement of ^{13}C/^{12}C isotope ratios by gas-liquid chromatography-combustion isotope ratio mass spectrometry. I. *Journal of Chromatography A*, **752**, 233–241.

Melander, L. & Saunders, W.H. (1980) *Reaction Rates of Isotopic Molecules*. Wiley, New York.

Merritt, D.A. & Hayes, J.M. (1994) Factors controlling precision and accuracy in isotope-ratio-monitoring mass-spectrometry. *Analytical Chemistry*, **66**(14), 2336–2347.

Merritt, D.A., Ricci, M.P., Studley, S.A., Hayes, J.M. & Brand, W.A. (1991) Carbon and nitrogen isotopic compositions of gas chromatographic effluents. *Proceedings of the 39th Annuaul Conference on Mass Spectrometry and Related Matters*, 19–24 May, Nashville, TN.

Merritt, D.A., Brand, W.A. & Hayes, J.M. (1994) Isotope-ratio-monitoring gas chromatography-mass spectrometry: methods for isotopic calibration. *Organic Geochemistry*, **21**, 573–583.

Merritt, D.A., Freeman, K.H., Ricci, M.P., Studley, S.A. & Hayes, J.M. (1995) Performance and optimisation of a combustion interface for isotope ratio monitoring gas chromatography/mass spectrometry. *Analytical Chemistry*, **67**, 2461–2473.

Metges, C.C. & Daenzer, M. (2000) ^{13}C gas chromatography-combustion isotope ratio mass spectrometry analysis of N-pivaloyl amino acid esters of tissue and plasma samples. *Analytical Biochemistry*, **278**, 156–164.

Metges, C.C. & Petzke, K.J. (1997) Measurement of ^{15}N/^{14}N isotopic composition in individual plasma free amino acids of human adults at natural abundance by gas chromatography combustion isotope ratio mass spectrometry. *Analytical Biochemistry*, **247**, 158–164.

Metges, C.C., Petzke, K.J. & Hennig, U. (1996) Gas chromatography combustion isotope ratio mass spectrometric comparison of *N*-acetyl- and *N*-pivaloyl amino acid esters to measure ^{15}N isotopic abundances in physiological samples: A pilot study on amino acid synthesis in the upper gastro- intestinal tract of minipigs. *Journal of Mass Spectrometry*, **31**, 367–376.

Middelburg, J.J., Barranguet, C., Boschker, H.T.S., Herman, P.M.J., Moens, T. & Heip, C.H.R. (2000) The fate of intertidal microphytobenthos carbon: An *in situ* C-13-labeling study. *Limnology and Oceanography*, **45**, 1224–1234.

Moers, M.E.C., Jones, D.M., Eakin, P.A., Fallick, H., Griffiths, H. & Larter, S.R. (1993) Carbohydrate diagenesis in hypersaline environments: application of GC-IRMS to the stable isotope analysis of derivatised saccharides from surficial and buried sediments. *Organic Geochemistry*, **20**, 927–933.

Moore, J.H. & Christie, W.W. (1981) Lipid metabolism in the mammary gland of ruminant animals. In: *Progress in Lipid Research, Supplement 1* (Ed. W.W. Christie), pp. 227–276. Pergamon Press, London.

Morton, J.D. & Schwarcz, H.P. (1988) Stable isotope analysis of food residues from Ontario ceramics. In: *Proceedings of the 26th International Archaeometry Symposium* (Eds R. Farquarhar, R.G.V. Hancock & L. Pavlish), pp. 89–93. Toronto University Press, Toronto.

Mottram, H.R. & Evershed, R.P. (2003) Practical considerations in the gas chromatography/ combustion/isotope ratio monitoring mass spectrometry of ^{13}C-enriched compounds: detection limits and carryover effects. *Rapid Communications in Mass Spectrometry*, **17**, 2669–2674.

Mottram, H.R., Dudd, S.N., Lawrence, G.J., Stott, A.W. & Evershed, R.P. (1999) New chromatographic, mass spectrometric and stable isotope approaches to the classification of degraded animal fats preserved in archaeological pottery. *Journal of Chromatography A*, **833**, 209–221.

Mottram, H., Woodbury, S.E., Rossell, J.B. & Evershed, R.P. (2003) High-resolution detection of adulteration of maize oil using multi-component compound-specific δ^{13}C values of major and minor components and discriminant analysis. *Rapid Communication in Mass Spectrometry*, **17**, 706–712.

Mukherjee, A.J. (2004) *The importance of pigs in the later British Neolithic: Integrating stable isotope evidence from .lipid residues in archaeological potsherds, animal bone, and modern animal tissues.* PhD thesis, University of Bristol.

Mukherjee, A.J., Berstan, R., Copley, M.S., Gibson, A.M. & Evershed, R.P. (In press) Compound-specific stable carbon isotope detection of pig product processing in British late neolithic pottery. *Antiquity*.

Myher, J.J. & Kuksis, A. (1995) General strategies in chromatographic analysis of lipids. *Journal of Chromatography B – Biomedical Applications*, **671**(1–2), 3–33.

Nold, S.C., Boschker, H.T.S., Pel, R. & Laanbroek, H.J. (1999) Ammonium addition inhibits ^{13}C-methane incorporation into methanotroph membrane lipids in a freshwater sediment. *FEMS Microbiology Ecology*, **29**, 81–89.

O'Brien, D.M., Fogel, M.L. & Boggs, C.L. (2002) Renewable and nonrenewable resources: amino acid turnover and allocation to reproduction in Lepidoptera. *Proceedings of the National Academy of Sciences*, **99**(7), 4413–4418.

O'Brien, D.M., Boggs, C.L. & Fogel, M.L. (2005) The amino acids used in reproduction by butterflies: A comparative study of dietary sources using compound-specific stable isotope analysis. *Physiological and Biochemical Zoology*, **78**, 819–827.

O'Leary, M.H. (1981) Carbon isotope fractionation in plants. *Phytochemistry*, **20**, 552–567.

Ourisson, G., Rohmer, M. & Poralla, K. (1987) Prokaryotic hopanoids and other polyterpenoid sterol surrogates. *Annual Reviews in Microbiology*, **41**, 301–333.

Ostle, N., Ineson, P., Benham, D. & Sleep, D. (2000) Carbon assimilation and turnover in grassland vegetation using an *in situ* $^{13}CO_2$ pulse labelling system. *Rapid Communications in Mass Spectrometry*, **14**, 1345–1350.

Ostrom, P.H. & Fry, B. (1993) Sources and cycling of organic matter within modern and prehistoric food webs. In: *Organic Geochemistry: Principles and Applications* (Eds M.H. Engle & S.A. Macko), pp. 785–798. Plenum Press, New York.

Pagani, M., Arthur, M.A. & Freeman, K.H. (1999) Miocene evolution of atmospheric carbon dioxide. *Paleoceanography*, **14**(3), 273–292.

Pagani, M., Zachos, J.C., Freeman, K.H., Tipple, B. & Bohaty, S. (2005) Marked decline in atmospheric carbon dioxide concentrations during the Paleogene. *Science*, **309**, 600–603.

Pakhomov, E.A., McClelland, J.W., Bernard, K., Kaehler, S. & Montoya, J.P. (2004) Spatial and temporal shifts in stable isotope values of the bottom-dwelling shrimp *Nauticaris marionis* at the sub-Antarctic archipelago. *Marine Biology*, **144**, 317–325.

Pancost, R.D. & Sinninghe-Damsté, J.S. (2003) Carbon isotopic compositions of prokaryotic lipids as tracers of carbon cycling in diverse settings. *Chemical Geology*, **195**, 29–58.

Pancost, R.D., Freeman, K.H., Wakeham, S.G. & Robertson, C.Y. (1997) Controls on carbon isotope fractionation by diatoms in the Peru upwelling region. *Geochimica et Cosmochimica Acta*, **61**(23), 4983–4991.

Pancost, R.D., Freeman, K.H. & Wakeham, S.G. (1999) Controls on the carbon-isotope compositions of compounds in Peru surface waters. *Organic Geochemistry*, **30**(5), 319–340.

Pancost, R.D., Sinninghe-Damsté, J.S., de Lint, S., van der Maarel, M.J.E.C. & Gottschal, J.C. (2000) Biomarker evidence for widespread anaerobic methane oxidation in Mediterranean sediments by a consortium of methanogenic archaea and bacteria. *Applied and Environmental Microbiology*, **66**, 1126–1132.

Pancost, R.D., Hopmans, E.C. & Sinninghe-Damsté, J.S. (2001) Archaeal lipids in Mediterranean cold seeps: Molecular proxies for anaerobic methane oxidation. *Geochimica et Cosmochimica Acta*, **65**, 1611–1627.

Pel, R., Floris, V. & Hoogveld, H. (2004) Analysis of planktonic community structure and trophic interactions using refined isotopic signatures determined by combining fluorescence-activated cell sorting and isotope-ratio mass spectrometry. *Freshwater Biology*, **49**, 546–562.

Pelz, O., Chatzinotas, A., Andersen, N., et al. (2001) Use of isotopic and molecular techniques to link toluene degradation in denitrifying aquifer microcosms to specific microbial populations. *Archives of Microbiology*, **175**, 270–281.

Petzke, K.J., Boeing, H. & Metges, C.C. (2005) Choice of dietary protein of vegetarians and omnivores is reflected in their hair protein C-13 and N-15 abundance. *Rapid Communications in Mass Spectrometry*, **19**(11), 1392–1400.

Pombo, S.A., Pelz, O., Schroth, M.H. & Zeyer, J. (2002) Field scale ^{13}C-labeling of phospholipid fatty acids (PLFA) and dissolved inorganic carbon tracing acetate assimilation and mineralisation in a petroleum hydrocarbon-contaminated aquifer. *FEMS Microbiology Ecology*, **41**, 259–267.

Pombo, S.A., Kleikemper, J., Schroth, M.H. & Zeyer, J. (2005) Field-scale isotopic labeling of phospholipid fatty acids from acetate-degrading sulfate-reducing bacteria. *FEMS Microbiology Ecology*, **51**, 197–207.

Pond, D.W., Segonzac, M., Bell, M.V., Dixon, D.R., Fallick, A.E. & Sargent, J.R. (1997) Lipid and lipid carbon stable isotope composition of the hydrothermal vent shrimp Mirocaris fortunata: evidence for nutritional dependence on photosynthetically fixed carbon. *Marine Ecology – Progress Series*, **157**, 221–231.

Pond, D.W., Bell, M.V., Dixon, D.R., Fallick, A.E., Segonzac, M. & Sargent, J.R. (1998) Stable-carbon-isotope composition of fatty acids in hydrothermal vent mussels containing methanotrophic and thiotrophic bacterial endosymbionts. *Applied and Environmental Microbiology*, **64**, 370–375.

Pond, D.W., Sargent, J.R., Fallick, A.E., Allen, C., Bell, M.V. & Dixon, D.R. (2000) delta C-13 values of lipids from phototrophic zone microplankton and bathypelagic shrimps at the Azores sector of the Mid-Atlantic Ridge. *Deep-Sea Research Part I – Oceanographic Research*, **47**, 121–136.

Poole, I. & van Bergen, P.F. (2002) Carbon isotope ratio analysis of organic moieties from fossil mummified wood: establishing optimum conditions for off-line pyrolysis extraction using gas chromatography/mass spectrometry. *Rapid Communications in Mass Spectrometry*, **16**, 1976–1981.

Poole, I., van Bergen, P.F., Kool, J., Schouten, S. & Cantrill, D.J. (2004) Molecular isotopic heterogeneity of fossil organic matter: implications for delta(13) C-biomass and delta(13) C-palaeoatmosphere proxies. *Organic Geochemistry*, **35**, 1261–1274.

Popp, B.N., Takigiku, R., Hayes, J.M., Louda, J.W. & Baker, E.W. (1989) The post-Paleozoic chronology and mechanism of ^{13}C depletion in primary marine organic matter. *American Journal of Science*, **289**, 436–454.

Quandt, I., Gottshalk, G., Ziegler, H. & Stichler, W. (1977) Isotope discrimination by photosynthetic bacteria. *FEMS Microbiology Letters*, **1**, 125–128.

Raghoebarsing, A., Smolders, A.J.P., Schmid, M.C., et al. (2005) Methanotrophic symbionts provide carbon for photosynthesis in peat bogs. *Nature*, **436**, 1153–1156.

Reber, E.A. & Evershed, R.P. (2004a) How did Mississippians prepare maize? The application of compound-specific carbon isotope analysis to absorbed pottery residues from several Mississippi Valley sites. *Archaeometry*, **46**, 19–33.

Reber, E.A. & Evershed, R.P. (2004b) Identification of maize in absorbed organic residues: a cautionary tale. *Journal of Archaeological Science*, **31**, 399–410.

Reber, E.A., Dudd, S.N., van der Merwe, N.J. & Evershed, R.P. (2004) Direct detection of maize in pottery residues via compound specific stable carbon isotope analysis. *Antiquity*, **78**, 682–691.

Ricci, M.P., Merritt, D.A., Freeman, K.H. & Hayes, J.M. (1994) Acquisition and processing of data for isotope-ratio-monitoring mass spectrometry. *Organic Geochemistry*, **21**, 561–571.

Richnow, H.H., Annweiler, E., Michaelis, W. & Meckenstock, R.U. (2003a) Microbial in situ degradation of aromatic hydrocarbons in a contaminated aquifer monitored by carbon isotope fractionation. *Journal of Contaminant Hydrology*, **65**, 101–120.

Richnow, H.H., Meckenstock, R.U., Ask Reitzel, L., et al. (2003b) *In situ* biodegradation determined by carbon isotope fractionation of aromatic hydrocarbons in an anaerobic landfill leachate plume (Vejen, Denmark). *Journal of Contaminant Hydrology*, **64**, 59–72.

Rieley, G. (1994) Derivatization of organic-compounds prior to gas-chromatographic combustion-isotope ratio mass-spectrometric analysis – identification of isotope fractionation processes. *Analyst*, **119**, 915–919.

Rieley, G., Van Dover, C.L., Hedrick, D.B. & Eglinton, G. (1999) Trophic ecology of Rimicaris exoculata: a combined lipid abundance stable isotope approach. *Marine Biology*, **133**, 495–499.

Rogers, K.M. & Savard, M.M. (1999) Detection of petroleum contamination in river sediments from Quebec City region using GC-IRMS – interpreting molecular fossils in petroleum and ancient sediments. *Organic Geochemistry*, **30**, 1559–1569.

Sarkanen, K.V. & Ludwig, C.H. (1971) *Lignins: Occurrence, Formation, Structure and Reactions*. Wiley-Interscience, New-York.

Schmidt, T.C., Zwank, L., Elsner, M., Berg, M., Meckenstock, R.U. & Haderlein, S.B. (2004) Compound-specific stable isotope analysis of organic contaminants in natural environments: a critical review of the state of the art, prospects, and future challenges. *Analytical and BioAnalytical Chemistry*, **378**, 283–300.

Schouten, S., Breteler, W., Blokker, P., et al. (1998) Biosynthetic effects on the stable carbon isotopic compositions of algal lipids: Implications for deciphering the carbon isotopic biomarker record. *Geochimica et Cosmochimica Acta*, **62**, 1397–1406.

Schouten, S., Van Kaam-Peters, H.M.E., Rijpstra, W.I.C., Schoell, M. & Damste, J.S.S. (2000) Effects of an oceanic anoxic event on the stable carbon isotopic composition of Early Toarcian carbon. *American Journal of Science*, **300**, 1–22.

Schouten, S., Wakeham, S.G. & Damste, J.S.S. (2001) Evidence for anaerobic methane oxidation by archaea in euxinic waters of the Black Sea. *Organic Geochemistry*, **32**(10), 1277–1281.

Schwarcz, H.P., Dupras, T.L. & Fairgrieve, S.I. (1999) ^{15}N enrichment in the Sahara: in search of a global relationship. *Journal of Archaeological Science*, **26**, 629–636.

Sealy, J. (1997) Stable carbon and nitrogen isotope ratios and coastal diets in the Later Stone Age of South Africa: a comparison and critical analysis of rtwo data sets. *Ancient Biomolecules*, **1**, 131–147.

Sealy, J.C., van der Merwe, N.J., Lee-Thorp, J.A. & Lanham, J.L. (1987) Nitrogen isotope ecology in Southern Africa: implications for environmental and dietary tracing. *Geochimica et Cosmochimica Acta*, **51**, 2707–2717.

Sessions, A.L., Burgoyne, W.W. & Hayes, J.M. (2001a) Correction of H_3^+ contributions in hydrogen isotope ratio monitoring mass spectrometry. *Analytical Chemistry*, **73**, 192–199.

Sessions, A.L., Burgoyne, W.W. & Hayes, J.M. (2001b) Determination of the H_3 factor in hydrogen isotope ratio monitoring mass spectrometry. *Analytical Chemistry*, **73**, 200–207.

Sherriff, B.L., Tisdale, M.A., Sayer, B.G., Schwarcz, H.P. & Knyf, M. (1995) Nuclear magnetic resonance spectroscopic and isotopic analysis of carbonized residues from subartic Canadian prehistoric pottery. *Archaeometry*, **37**, 95–111.

Shinebarger, S.R., Haisch, M. & Matthews, D.E. (2002) Retention of carbon and alteration of expected ^{13}C-tracer enrichments by silylated derivatives using continuous-flow combustion-isotope ratio mass spectrometry. *Analyical Chemistry*, **74**, 6244–6251.

Silfer, J.A., Engel, M.H., Macko, S.A. & Jumeau, E.J. (1991) Stable carbon isotope analysis of amino-acid enantiomers by conventional isotope ratio mass-spectrometry and combined gas chromatography-isotope ratio mass spectrometry. *Analytical Chemistry*, **63**, 370–374.

Sirevag, R., Buchanan, B.B., Berry, J.A. & Throughton, J.H. (1977) Mechanisms of CO_2 fixation in bacterial photosynthesis studied by the carbon isotope technique. *Archives of Microbiology*, **112**, 35–38.

Smedes, F. & Askland, T.K. (1999) Revisiting the development of the Bligh and Dyer total lipid determination method. *Marine Pollution Bulletin*, **38**, 193–201.

Spiker, E.C., Hatcher, P.G. & Orem, W.H. (1985) Stable isotope effects of selective preservation during early diagenesis. *Estuaries*, **8**, A90–A90.

Stefanova, M., Maman, O., Guillet, B. & Disnar, J.-R. (2004) Preserved lignin structures in Miocene-aged lignite lithotypes, Bulgaria. *Fuel*, **83**, 123–128.

Stott, A.W. & Evershed, R.P. (1996) $\delta^{13}C$ analysis of cholesterol preserved in archaeological bones and teeth. *Analytical Chemistry*, **68**, 4402–4408.

Stott, A.W., Davies, E., Evershed, R.P. & Tuross, N. (1997a) Monitoring the routing of dietary and biosynthesised lipids through compound-specific stable isotope (δ^{13}C) measurements at natural. abundance *Naturwissenschaften*, **84**, 82–86.

Stott, A.W., Evershed, R.P. & Tuross, N. (1997b) compound-specific approach to the δ^{13}C analysis of cholesterol in fossil bones. *Organic Geochemistry*, **26**, 99–103.

Stott, A.W., Evershed, R.P., Jim, S., et al. (1999) Cholesterol as a new source of palaeodietary information: Experimental approaches and archaeological applications. *Journal of Archaeological Science*, **26**, 705–716.

Sun, C., Snape, C.E., McRae, C. & Fallick, A.E. (2003) Resolving coal and petroleum-derived polycyclic aromatic hydrocarbons (PAHs) in some contaminated land samples using compound-specific stable carbon isotope ratio measurements in conjunction with molecular fingerprints. *Fuel*, **82**, 2017–2023.

Teece, M.A. & Fogel, M.L. (2004) Preparation of ecological and biochemical samples for isotope analysis. In: *Handbook of Stable Isotope Analytical Techniques*, Vol. I (Ed. P.A. De Groot), pp. 177–202. Elsevier, Amsterdam.

Teske, A., Hinrichs, K.U., Edgcomb, V., et al. (2002) Microbial diversity of hydrothermal sediments in the Guayamas Basin: Evidence for anaerobic methanotrophic communities. *Applied and Environmental Microbiology*, **68**, 1994–2007.

ThermoElectron (2003) *Oxygen Analysis by GC/TC. Operating Manual.* ThermoElectron (Advanced Mass Spectrometry), Bremen, 10 pp.

Thiel, V., Peckmann, J., Seifert, R., Whrung, P., Reitner, J. & Michaelis, W. (1999) Highly isotopically depleted isoprenoids: molecular markers for ancient methane venting. *Geochimica et Cosmochimica Acta*, **63**, 3959–3966.

Thiel, V., Peckmann, J., Richnow, H.H., Luth, U., Reitner, J. & Michaelis, W. (2001) Molecular signals for anaerobic methane oxidation in Black Sea seep carbonates and a microbial mat. *Marine Chemistry*, **73**, 97–112.

Tobias, H.J. & Brenna, J.T. (1997) On-line pyrolysis as a limitless reduction source for high-precision isotopic analysis of organic-derived hydrogen. *Analytical Chemistry*, **69**, 3148–3152.

Touchstone, J.C. (1991) Thin-layer chromatography in food analysis. *Abstracts of Papers of the American Chemical Society*, **202**, 180-AGFD.

Touchstone, J.C. (1993) History of chromatography. *Journal of Liquid Chromatography*, **16**, 1647–1665.

Touchstone, J.C. (1995) Thin-layer chromatographic procedures for lipid separation. *Journal of Chromatography B – Biomedical Applications*, **671**, 169–195.

Treonis, A.M., Ostle, N.J., Stott, A.W., Primrose, R., Grayston, S.J. & Ineson, P. (2004) Identification of groups of metabolically-active rhizosphere microorganisms by stable isotope probing of PLFAs. *Soil Biology and Biochemistry*, **36**, 533–537.

Tuross, N., Fogel., M.L. & Hare, P.E. (1988) Variability in the preservation of the isotope composition of collagen from fossil bone. *Geochimica et Cosmochimica Acta*, **52**, 929–935.

Van Dongen, B.E., Schouten, S. & Damste, J.S.S. (2001) Gas chromatography /combustion/isotope-ratio-monitoring mass spectrometric analysis of methylboronic derivatives of monosaccharides: a new method for determining natural ^{13}C abundances of carbohydrates. *Rapid Communication in Mass Spectrometry*, **15**, 496–500.

Van Klinken, G.J. (1991) *Dating and dietary reconstruction by isotopic analysis of amino acids in fossil bone collagen – with special reference to the Caribbean.* PhD thesis, University of Groningen.

Vogel, J.C. & Van der Merwe, N.J. (1977) Isotopic evidence for early maize cultivation in New York State. *American Antiquity*, **42**, 238–242.

Volkman, J.K. (1986) A review of sterol markers for marine and terrigenous organic-matter. *Organic Geochemistry*, **9**, 83–99.

Volkman, J.K., Eglinton, G., Corner, E.D.S. & Forsberg, T.E.V. (1980) Long-chain alkenes and alkenones in the marine Coccolithophorid Emiliania-Huxleyi. *Phytochemistry*, **19**, 2619–2622.

Volkman, J.K., Barrett, S.M., Blackburn, S.I., Mansour, M.P., Sikes, E.L. & Gelin, F. (1998) Microalgal biomarkers: A review of recent research developments. *Organic Geochemistry*, **29**, 1163–1179.

Waldrop, M.P. & Firestone, M.K. (2004) Microbial community utilization of recalcitrant and simple carbon compounds: impact of oak-woodland plant communities. *Oecologia*, **138**, 275–284.

Whiticar, M.J., Faber, E. & Schoell, M. (1986) Biogenic methane formation in marine and freshwater environments – CO_2 reduction vs acetate fermentation isotope evidence. *Geochimica et Cosmochimica Acta*, **50**, 693–709.

Wilcke, W., Krauss, M. & Amelung, W. (2002) Carbon isotope signature of polycyclic aromatic hydrocarbons (PAHs): evidence for different sources in tropical and temperature environments? *Environmental Science and Technology*, **36**, 3530–3535.

Withers, N.W., Tuttle, R.C., Goad, L.J. & Goodwin, T.W. (1979) Dinosterol side-chain biosynthesis in a marine dinoflagellate, *Crypthecodinium-Cohnii. Phytochemistry*, **18**, 71–73.

Woodbury, S.E., Evershed, R.P., Rossell, J.B., Griffith, R.E. & Farnell, P. (1995) Detection of vegetable oil adulteration using gas- chromatography combustion isotope ratio mass-spectrometry. *Analytical Chemistry*, **67**, 2685–2690.

Xie, S., Nott, C.J., Avsejs, L., A., Volders, F., et al. (2000) Palaeoclimate records in compound-specific delta D values of a lipid biomarker in ombrotrophic peat. *Organic Geochemistry*, **31**, 1053–1057.

Yamada, K. & Ishiwatari, R. (1999) Carbon isotopic compositions of long-chain *n*-alkanes in the Japan Sea sediments: implications for paleoenvironmental changes over the past 85 kyr. *Organic Geochemistry*, **30**, 367–377.

Yanik, P.J., O'Donnell, T.H., Macko, S.A., Qian, Y. & Kennicutt Ii, M.C. (2003) Source apportionment of polychlorinated biphenyls using compound specific isotope analysis. *Organic Geochemistry*, **34**, 239–251.

Ziegler, S.E. & Fogel, M.L. (2003) Seasonal and diet relationships between the isotopic compositions of dissolved and particulate organic matter in freshwater ecosystems. *Biogeochemistry*, **64**, 25–52.

Index